BIRD STUDIES
AT OLD
CAPE MAY

BIRD STUDIES
AT OLD
CAPE MAY

An Ornithology of Coastal New Jersey

WITMER STONE

Introduction by Roger Tory Peterson
Author biography by James A. G. Rehn

STACKPOLE
BOOKS

To my wife

Published by
STACKPOLE BOOKS
5067 Ritter Road
Mechanicsburg, PA 17055
www.stackpolebooks.com

Printed in the United States of America
First edition
10 9 8 7 6 5 4 3 2 1

First published in 1937 by the Delaware Valley Ornithological Club. Published in 1965 by Dover Publications.

Stackpole Books' 2000 publication of *Bird Studies at Old Cape May* is a reprint of the 1965 edition of the book, published in two volumes by Dover Publications. The Dover edition added an introduction by Roger Tory Peterson, an author biography by James A. G. Rehn, and a listing of additional species compiled by Ernest A. Choate to the complete text and artwork of the original two-volume edition of the book, published in 1937. Both the 1937 and 1965 editions of *Bird Studies at Old Cape May* included black-and-white photographs, which are not reproduced here.

Cover and interior drawings by Earl L. Poole, Conrad Roland, J. Fletcher Street, and others. Cover design by Caroline Stover

Library of Congress Cataloging-in-Publication Data
Stone, Witmer, 1866–1939.
Bird studies at Old Cape May: an ornithology of coastal New Jersey/Witmer Stone; introduction by Roger Tory Peterson.—1st ed.
    p. cm.
Originally published as a two volume work: New York : Dover Publications, 1965.
Photographs from previous ed. not reproduced.
Includes bibliographical references (p.).
ISBN 0-811731510 (pbk.)
1. Birds—New Jersey—Cape May. 2. Birds—New Jersey. I. Title.

QL684.N5 S77 2000
598'.097498—dc21   00-026046

INTRODUCTION TO THE STACKPOLE EDITION

The Delaware Valley Ornithological Club, publisher of the two-volume classic *Bird Studies at Old Cape May* in 1937, is delighted to grant Stackpole Books the opportunity to produce this current edition and thus make an important work available again for new readers to learn of Cape May's unique capacity for attracting birds and birders.

That Witmer Stone chose the D.V.O.C. to be his publisher, not only for this work but also for his *Birds of Eastern Pennsylvania and New Jersey* in 1894, is a strong indication of the very close relationship which existed between Stone and the club, a relationship which commenced in 1890, when he and six other young men founded the club in Philadelphia, and continued throughout his lifetime. Stone served a term as president of the club the following year and became the first editor of its new journal, *Cassinia*, when it made its debut in 1901. His reputation as its editor led to his being chosen in 1912 to edit *The Auk*, the prestigious quarterly journal of the American Ornithologists' Union, a position he held for the next twenty-five years. Even though he was no longer the editor of *Cassinia*, he continued to serve on the publications committee, making sure the current editor's product was one of continued excellence.

Stone was elected vice president of the A.O.U. in 1914 and was its president from 1920 to 1923. He was a member of the important Committee on Nomenclature and became its chairman in 1919; he had the herculean task of supervising the production of the Third Edition of the A.O.U. Checklist of North American Birds. He was also the principal author of the Fourth Edition of the Checklist, published in 1931, and a member of the International Committee on Zoological Nomenclature.

In spite of these many national and international obligations, the D.V.O.C. always remained his favorite organization and its members his closest friends. George Spencer Morris, also a D.V.O.C. founder, best summarized this closeness in a paper read at the club's twentieth annual meeting. "And then there is Stone, with his hand ever on the tiller, quiet in manner but potent in influence. No matter who may be president, we all recognize him as the power behind the throne. With infinite tact he gives a push here and a pull there, as occasion requires, keeping us all in line. In our hearts we know that the guiding hand of Stone has made the D.V.O.C. what it is."

PHILLIPS B. STREET

March, 2000

INTRODUCTION TO THE DOVER EDITION

You must pardon a senior ornithologist if he reminisces. A historical sense grows with the years. To me, even John James Audubon now seems real, a man who actually lived—not merely a legend.

Going halfway back to Audubon we encounter Witmer Stone, who lovingly recorded what went on in southern New Jersey during the first third of our century and, toward the end of his life, published this two-volume work on the birds of Old Cape May.

I recall the summer, nearly forty years ago, when I first met Dr. Stone. Three of us, all teenagers—Clarence Beal, Ed Stearns and I—had thumbed our way southward across the state of New Jersey on our first great ornithological adventure. It turned out that Clarence Beal's uncle, Walker Hand, with whom we were to stay at Cape May, was a bird man. And it was at his modest home that we were introduced to Witmer Stone, one of the great field ornithologists of his time. Smiling benignly beneath his white moustache he seemed to enjoy thoroughly the three boys who plied him with so many questions. We heard of great rarities and red-letter days: the time four wood ibises were seen at Lighthouse Pond, and the time when he and Julian Potter and William Bailey spotted a Mississippi kite on the road to Higbee's Beach.

Since then I have driven to Cape May many times and have seen more than my own share of rarities. I have also witnessed changes in the avifauna of southern New Jersey that Stone could not have foreseen—some for the better, some definitely not. Some birds have benefited from a more enlightened public attitude, backed up by protective laws and even refuge areas. Other species are inexorably losing out because of the subtle pressures of "progress," which, in addition to the obvious competition for living space, involves perils and poisons undreamed of when I started watching birds. It seemed just a matter of protection in those days. That certainly was my view in 1935 when I was sent to Cape May Point by the National Audubon Society to observe the hawk shooting. In *Birds Over America* (1948), I wrote:

The pioneer exploits; later generations conserve. So the hunter has always preceded the aesthete and the ornithologist. At Cape May, as at Hawk Mountain, the hawks were slaughtered for years before anyone became concerned. A concrete highway runs from east to west through the groves of Spanish oaks just north of the town of Cape May Point. Here the local gunners formerly lined the road and waited for the hawks to come over. You could depend on the three boys who lived over the grocery store to be there, and the old Italian who ran the taxi from the railroad station, and "Pusey," the colored man who owned a boat over on the bay. Sometimes sportsmen came from as far away as Trenton and Camden. One September morning in 1935, I watched 800 sharp-shins try to cross the firing line. Each time a "sharpy" sailed over the treetops it was met by a pattern of lead. Some folded up silently; others, with head wounds, flopped to the ground, chattering shrilly. By noon 254 birds lay on the pavement.

That evening, in a Cape May home, I sat down to a meal of hawks—twenty sharp-shins, broiled like squabs, for a family of six. I tasted the birds, found them good, and wondered what my friends would say if they could see me. Like a spy breaking bread with the enemy, I felt uneasy. I could not tell my hosts I disapproved, for their con-sciences were clear—weren't they killing the hawks as edible game and at the same time saving all the little song birds? It would have done no good to explain predation, ecology and the natural balance to these folk. Having lived at Cape May all their lives, they had a distorted idea of the abundance of hawks. They did not realize that a single season's sport by the Cape May gunners could drain the sharp-shins from thousands of square miles of northern woodland.

Shooting has now been prohibited from the highways in New Jersey, and shooting from the sandhills in the woods at Cape May Point has been largely eliminated through the creation of the Witmer Stone Wildlife Sanctuary by the National Audubon Society. The sanctuary has now changed hands and is guarded by the New Jersey Audubon Society. As at Hawk Mountain, the binocular has replaced the gun and hawks now enjoy a relatively safe passage.

Since I wrote these paragraphs the "model hawk law" has been passed by the New Jersey Legislature, extending protection to all species of birds of prey except those individuals caught in the act of damaging property. Hawk-shooting is now illegal at Cape May. With the heat taken off it would seem that the future holds more promise for these maligned birds. But this may not be so.

The fumes from the tall stacks of the magnesium plant, erected where the highway reaches the bay, have turned the pines and the oaks of the sanctuary into a ghost forest. However, this is of far less consequence to the birds of prey than the effects of other chemicals now polluting the environment. When Rachel Carson told her story of *Silent Spring* it was easy to see how songbirds might be affected by eating poisoned insects, but few of us foresaw the chain reaction on the higher predators.

Since World War II, and particularly during the last ten years, the chlorinated hydrocarbons have been widely promoted, like wonder drugs, as a panacea for all conceivable insect ills and plant diseases. These hydrocarbons murder differentially, because they mostly have a long life without chemical breakdown and *accumulate* in soil and water and in the bodies of all members of the animal pyramid based upon the earthworms, plankton, insects and other vertebrates of the soil and water. Those birds at the top of the pyramid are particularly vulnerable, for they take poison biologically concentrated by their prey and their prey's prey.

Therefore, the thing that disturbs me most is not that a million song-birds should die with DDT tremors, upsetting though it may be. We still have a great many warblers and robins. Their reproductive potential is high, and they will probably survive until we get some sense and the hydrocarbon syndrome is a thing of the past.

I am more concerned about those species that are at the ends of long food chains—particularly bird-eating birds and certain of the fish-eating birds that eat large fish. A lifetime of observation in many parts of the world and in every state in the union has convinced me that those species are in the greatest danger and some may even face eventual extinction.

Let me tell you about the ospreys on the Connecticut River where I live. The Connecticut River is a beautiful river—and it is dying. I moved to Old Lyme, Connecticut, just ten years ago, largely because of the colony of ospreys around the mouth of the river. There was even a nest on our property. In 1954 there were about 150 nests in the general area, and they were a pleasure to everyone. We had lived there two or three years when I investigated the concentration of nests on Great Island at about the time there should have been full-grown young. Most of the nests were empty. The next year was a similar failure. Some of the birds sat on unhatched eggs for sixty and seventy days. (An osprey's egg should hatch in thirty-two days.)

At about that time Peter Ames, a graduate student in ornithology at Yale University, started his studies of our local ospreys, and we became deeply involved. There were about twenty nests in one concentration on Great Island. One season they produced six young, one year three, and one year only one. Normal success should have produced between one and two young per nest—perhaps thirty birds out of twenty active nests. Twenty-one raccoon-proof poles with platforms were erected to rule out predation by raccoons and other disturbances, but even after the birds adopted these sites the survival percentages remained the same, about ten per cent of the norm. Finally several eggs were analyzed. They contained significant amounts of DDT, DDE and other metabolites of DDT. Thirty samples of fish taken from the nests all contained these poisons. Only one conclusion could be drawn.

Without proper replacement by young birds the colony dwindled. Here are some of the statistics: In 1954 there were approximately 150 nests in the area. In 1960 there were 71; in 1963, 24; in 1964, 17. They were dropping out at the rate of more than thirty per cent yearly. Projecting this decline and assuming the trend is not reversed, we might well see our last nest in Connecticut in 1970 or shortly thereafter.

There is not much spraying at Old Lyme, but many towns upriver spray. There is undoubtedly air-drift and runoff. We cannot confine these poisons as long as wind blows, water flows or fishes swim. Where we are likely to find the magnification effect is in the estuarine waters at the mouths of rivers. Traces of poisons ingested by little fish upriver—either in the runoff or through poisoned insects—make them easier prey for larger fish. Numbers of affected fingerlings compound their poison in their predators, and it is the large fish that is wobbly, swimming near the surface, that is most likely to be caught by the osprey which transfers the accumulated poisons to its own tissues.

Ospreys in Massachusetts and Rhode Island and on the coast of New Jersey are also dropping fast, apparently for the same reasons. At Gardiner's Island, off the end of Long Island, where there were formerly two hundred nests there are now only a few and these produce very few young.

We had suspected that the residues of chemicals were affecting bald eagles ever since Charles Broley, "the eagle man," first suggested it (and was scoffed at) in the mid-fifties when reproduction in his west Florida eyries dropped to almost nil. We now know that our national bird is in grave trouble along the Atlantic Seaboard. The many nests in the Chesapeake Bay area have virtually stopped producing young. There still are adults about but they won't live forever. We know that pesticides are largely responsible; experimental work by Dr. DeWitt and other biologists of the U.S. Fish and Wildlife Service has proven this to our satisfaction.

To narrow things down to New Jersey: For many years there have been between seven and ten active eagle nests in the southern half of the state. In the year 1936 it was recorded that the four active nests near Cape May produced a total of nine young. This was prior to the era of DDT. But consider the picture today. During the seasons of 1956, 1957 and 1958 not a single young bird was raised in any of the known nests in Cape May, Cumberland, Gloucester and Salem counties. The following four years were little better: 1959, four young; 1960, 1961 and 1962, one young each year; 1963, two young. Some birds sat on their eggs for as long as eighty-four days (the normal incubation period is thirty-five days). Three such eggs were analyzed chemically and were found to contain significant quantities of DDT.

It came as a shock in the early sixties to discover that the osprey, always far more abundant than the eagle, was also in trouble in southern New Jersey. The handsome fish hawk that Stone used as a frontispiece for Volume I is no longer to be seen, wherever one looks in the sky. The big bulky nests are no longer a part of every coastal vista. True, there are local groups, such as the colony at Avalon which are still producing a fair number of young, but the bird is virtually gone at Cape May.

And what of the other birds of prey that funnel through Cape May on the northwest winds of autumn? What about the bird-eating sharp-shins and Cooper's hawks? Hawk watchers everywhere have commented on the sharp decline of the Cooper's hawk. When Robert Allen and William Rusling were Audubon wardens at Cape May in the mid-thirties, monitoring the hawk shoots, the autumnal passage of Cooper's hawks ran between 800 and 1200 birds. Today a yearly tally would not reveal one-tenth as many. Similarly, at Hawk Mountain in Pennsylvania, where close to 500 Cooper's hawks were counted yearly in the late thirties, only 73 were recorded in 1963.

Normally the Accipiters tone up the populations they prey upon by catching the marginal or sick birds. Poisoned birds, of course, are sick birds, easier to catch. But because of the poison syndrome, natural selection has now become unnatural selection. In the British Isles and on the continent of Europe the sparrow hawk (an Accipiter, unlike our sparrow hawk) has almost disappeared and nearly every other bird of prey is also down in numbers, for the same reasons.

The peregrine falcon, the finest and fastest bird that flies, has entirely disappeared as a breeding bird from the eastern states since 1950. I visited fourteen occupied eyries in the Hudson Valley and nearby valleys during a single weekend in 1946. Today they are all gone—and gone from the Susquehanna, the Delaware and all our big eastern river bluffs. A 1964 survey from Georgia to Canada revealed not one bird. We cannot say for certain that pesticides did it, but early in the fifties the birds went through the same suspicious pattern as the osprey. Many of them sat on eggs that did not hatch. With no replacements the existing pairs finally disappeared.

Peregrines on autumnal migration may still be seen at Cape May by the hawk watcher; these, however, are certainly migrants from the subarctic tundra, shore-bird followers whose passage to tropical America is swift. But the resident peregrines of the eastern river valleys are gone. Not all changes in the avifauna since Stone's day have been for the worse. On the credit side of the ledger is the return of the glamorous herons and egrets that had been eliminated by the plumers during the previous century. This return developed in spite of restricted habitat and ditching of the coastal marshes. When Dr. Stone published his great work in 1937, the American egret was just beginning to nest again in the state, and so was the little blue heron (in fact, I was the first to spot the colony south of Camden in 1936). Since then the snowy egret has returned to breed in considerable numbers and small numbers of Louisiana herons and glossy ibises have become established.

Most extraordinary of all has been the invasion or, should we say, explosion of the Old-World cattle egret, unknown in North America prior to 1952 when, in hardly more than a month's time, individuals were observed in Massachusetts, Cape May and Florida. The first Cape May bird was discovered on the farm of Michael McPherson on May 25, 1952. Later a second bird appeared. Bird watchers by the hundreds from Philadelphia, Washington, New York and even farther, poured in by car, by train and by bus to see these two distinguished visitors that remained for months.

Now there are hundreds of cattle egrets in the southern half of the state during the summer, and many not only roost in the Stone Harbor heronry but also nest there.

The story of the Stone Harbor heronry, a half hour's drive north of Cape May, is one of the heartening success stories of conservation. The herons started nesting there, within the town limits, about 1938 or 1939. I recall spending a hot July day there, twenty-odd years ago, filming little blue herons. Hot, sticky and exhausted, I struggled with my heavy camera gear through a clinging, clawing, cat-brier entanglement and across a rank poison-ivy thicket while nervous young herons disgorged half-digested fish from above. Later, when I returned through the streets of Stone Harbor, a half-dozen mongrel dogs followed me for several blocks. I was probably the most interesting-smelling visitor they had met in some time.

Nowadays visitors are not allowed to prowl through the rookery as I did. The sanctuary was officially established in October, 1947, and has now become a proud attraction of the seaside community of Stone Harbor—in fact, virtually the last bit of wildness left on that coastal strip. But it was touch-and-go for a while. Birds do not pay taxes and more than one real-estate operator was all for bull-dozing out the herons' grove to make room for more beach bungalows. Fortunately, good sense prevailed. It is no accident that the herons have chosen this isolated grove on the outer strip for their nursery instead of the extensive pine barrens across the salt meadows. Here there are no raccoons; on the mainland, nest predation by raccoons would make the herons' survival impossible.

When I last visited the colony, in 1964, it was having a great boom. Every heron in New Jersey (except the two bitterns) was nesting there—mostly snowies, but also cattle egrets (about seventy-five), American egrets, little blues, Louisianas, both species of night heron, green herons, even a pair of great blues and a fair-sized group, perhaps one hundred, of glossy ibises: ten species in all, totaling five-to six-thousand birds. Little wonder Stone Harbor has become a mecca for bird watchers from all over the East! Late in the day, toward sundown, they park their cars beside the highway to watch the evening flight. A roadside exhibit identifies the various herons for the tyro and for a coin he may peer through a large pair of binoculars on a stand.

A side trip to Stone Harbor is now an automatic feature of a Cape May weekend. The biggest event of the year is the convention of the New Jersey Audubon Society in October when more than one thousand *aficionados* converge on Cape May to listen to talks, see wildlife films and go on field trips. Everyone hopes for a northwest wind to make things interesting at Lily Pond and at the Witmer Stone Sanctuary.

The second most important event of the Cape May ornithological year is the Christmas Count. Every year since 1930, on a Sunday near Christmas, the Delaware Valley Ornithological Club has made a bird count which is published by the National Audubon Society. The cumulative number of species through 1963 was 212. At least one or two more are added each year. The highest number on a single count was 153 on December 27, 1959, an all-time high for any area north of the Carolinas.

The crowds, larger every year, have opened the eyes of the townspeople in this old summer resort who now realize a far greater economic asset in the horde of birders than they ever did when they catered to the handful of hawk-shooters. New highways such as the Garden State Parkway have cut the driving time from New York or Philadelphia.

The establishment of the Witmer Stone Sanctuary and the Stone Harbor heronry are not the only milestones in the wildlife conservation picture of the New Jersey coast. The Brigantine National Wildlife Refuge, established in 1939, embraces twenty-one square miles of marsh and water north of Atlantic City. Like so many other federal refuges, managed by expert technicians, it is now far richer in bird life than it was originally. Canada geese, probably descended from feral stock, nest in numbers; so do a few gadwall and shovellers which we once regarded unequivocally as birds of the Western prairies. Recently, on one of the sheltered backwaters, a group of us saw a small flock of fulvous tree ducks, the first occurrence of this tropical species in New Jersey. A complex of roads and dikes makes the refuge accessible to the birders. They now greatly outnumber the sportsmen who, since 1953, have been allowed to shoot on part of the marsh, although the original purpose of the refuge was to provide inviolate sanctuary for the birds.

Whole regiments of recruits to the brotherhood of the binocular and the balscope have sprung up since the original edition of *Bird Studies at Old Cape May* went out of print. They will welcome this Dover edition as a part of their ornithological library. Unlike so many other regional works, it is not just an annotated list, though all the detailed information is in its pages; it is also a book to be read for pleasure.

Few areas in the New World have been documented as consistently as Cape

May since 1900; perhaps only Concord, Massachusetts, and the New York City region can claim as complete a record. And no publication has set things down in more detail than Dr. Stone's two-volume work. But do not be misled by the title. The canvas on which the portraits and vignettes are drawn is not limited to Cape May, but includes the entire coast of New Jersey with emphasis on the southern half of the state. Everyone interested in the distribution and the conservation of birds, whether he lives in the Garden State or elsewhere, will find Dr. Stone's writings worth study and reflection.

Nearly thirty years have passed since Witmer Stone put the finishing touches on his intensive study of the birds at Cape May. During that interval many changes have taken place. To me the fascinating part of any regional work is this evidence of flux, the ebb and flow of populations, the disappearance of species, the invasion by others.

Man, of course, is the great disturber. The decline of the birds of prey due to shooting and poisons testifies to that. We find, if we study the record, that clapper rails, bitterns, marsh wrens, seaside and sharp-tailed sparrows, marsh hawks and short-eared owls have also declined. They certainly are not as numerous as during Stone's earlier years. Marsh drainage has deprived them of much of their habitat. Attrition goes on at nearly five per cent each year. Even the recent come-back of the southern herons may be reversed by continued ditching, spraying and disturbance. Certain other species—the barred owl, purple martin and parula warbler—have also become less common, but for reasons that are not as obvious.

These debit-birds are balanced by credit-species that have become noticeably commoner—the gnatcatcher, mockingbird, cardinal, prothonotary warbler, blue grosbeak and chuck-wills-widow. The increase and spread of these "southern" birds seem in line with the general climatic warming of the Northern Hemisphere.

In addition to the herons mentioned earlier, the oystercatcher and the Wilson's plover have returned to nest on the beaches and spoil banks of southern New Jersey. It was no surprise when the boat-tailed grackle, a new invader, established small colonies near Delaware Bay. Breeding willets, formerly confined to the Delaware Bay side, near Fortescue, now reside locally on the east coast as well, having nested since 1952. Extending southward from New England and Long Island shores, the herring gull now nests on the New Jersey coast and nesting black-backs must be anticipated.

New birds are being added almost yearly to the state list. Such western strays as the western grebe, eared grebe and white pelican, among others, were not on Stone's list; nor was the Greenland wheatear that delighted hundreds of watchers during its sojourn at Cape May in October, 1951, nor the European great snipe, filmed recently on the shores of Lily Lake.

Birds have wings. But the funneling effect of Delaware Bay and the Atlantic traps many of them in the peninsula where they are almost certain to be spotted by their human admirers. Anything is possible at Cape May.

ROGER TORY PETERSON

December, 1964

ADDITIONAL SPECIES RECORDED IN CAPE MAY COUNTY
SINCE THE ORIGINAL PUBLICATION OF THIS BOOK

In the twenty-six years after the first publication of *Bird Studies at Old Cape May*, fifty-one additional species were recorded in Cape May County. They vary from the Glossy Ibis and Cattle Egret, which are now well-established breeding birds, to accidentals from the Old World such as the European Woodcock, and possible escapes such as the Baikal Teal. Following is a list of these additional species prepared by Ernest A. Choate.

SPECIES	DATES	LOCATION	RECORDERS
Eared Grebe (*Colymbus caspicus*)	2/14/54 10/28/61	Stone Harbor Cape May Point	Ray J. Brooke, Jr. Herbert and Elizabeth Cutler, E. A. Choate, William Russell
Western Grebe (*Aechmophorus occidentalis*)	2/4/54 1/21/62	Stone Harbor Cape May Point	V. A. Debes *et al.* Mr. and Mrs. Malcolm B. Sheldrich
White Pelican (*Pelecanus erythrorhynchos*)	11/7–14/51	Stone Harbor	W. B. Wright
Common (European) Cormorant (*Phalacrocorax carbo*)	9/28/54 10/10/54	Cape May Point Cape May Point	Joseph M. Cadbury, E. A. Choate
Cattle Egret (*Bubulcus ibis*) (first record—2 birds) (now breeding)	5/25/52	Higbee's Beach *2 birds*	R. Smart, J. Baird, R. Bates and B. Murray
	7/6/52	Higbee's Beach *1 bird*	E. A. Choate, Joseph Jacobs
	7/4/58	Stingaree Point *breeding*	Russell S. Fowler, F. Russell Lyons, Henry Rodberg
Glossy Ibis (*Plegadis falcinellis*) (now breeding)	7/4/55	Stingaree Point Heronry *breeding*	Russell S. Fowler, F. Russell Lyons, Henry Rodberg
	8/5/53	South Cape May	E. A. Choate (*9 individuals*)
White Ibis (*Guara alba*)	8/17/51	House Roof Stone Harbor	Edward J. Reimann
	9/7–13/58	Stone Harbor Heronry	B. Kimple *et al.*, J. K. Wright
	10/3/58	Stone Harbor	Harry Brown, Norman McDonald, Chas. Wonderly
	10/28/62	Cape May Point	P. Livingston
Baikal Teal (*Anas formosa*)	Feb. 1961 left early March 1961	Lily Lake	E. A. Choate *et al.*

SPECIES	DATES	LOCATION	RECORDERS
European Widgeon (*Mareca penelope*)	11/8/59	Lily Lake, Cape May Point	William Bailey E. A. Choate (2 birds)
	Dec. 1961	Lily Lake, Cape May Point	E. A. Choate
	10/15/60 to Jan. 1961	Cape May Point	
Black Vulture (*Coragyps atratus*)	6/7/54	Cape May Point	B. Murray *et al.*
Swallow-tailed Kite (*Elanoides forficatus*)	9/1/46	Cape May Point	J. D'Arcy Northwood
Golden Eagle (*Aquila chrysaetus*)	9/2/46	Cape May Point	E. A. Choate
	10/7/54	Cape May Point	J. Kramer
	10/6/62	Cape May Point	Chandler Robbins, E. A. Choate
Sandhill Crane (*Grus canadensis*)	10/6/58	Lighthouse Pond, Cape May Point	Alfred Nicholson, Kenneth Wright
Black Oystercatcher (*Haematopus bachmani*)	8/29/51	Nummy's Island	E. L. Altemus, J. C. Seegers, E. A. Choate
European Woodcock (*Scolopax rusticola*)	1/2/56	Goshen	Dale Coman
Baird's Sandpiper (*Erolia bairdii*)	8/7/47	Nummy's Island	E. A. Choate
Curlew Sandpiper (*Erolia ferruginea*)	7/10/47	Anglesea Wildwood	E. A. Choate
Long-billed Dowitcher (*Limnodromus scolopaceus*)	12/26/54	Salt marshes, between Wildwood and Stone Harbor	Phillips Street
Ruff (*Philomachus pugnax*)	8/24/48	Marsh by Cape May Harbor	E. A. Choate, Earle Poole
	8/14/52	Marsh by Cape May Harbor	John Springer
Pomarine Jaeger (*Stercorarius pomarinus*)	10/8/55	Cape May Point	E. A. Choate, Dave Cutler
Long-tailed Jaeger (*Stercorarius longicaudus*)	10/18–19/47	Cape May Point	D. Fables, Jr., E. Kramer, J. D. Northwood
	10/9/54	Cape May Point	D. Cutler, E. A. Choate
Glaucous Gull (*Larus hyperboreus*)	4/1/61	Lily Lake, Cape May Point	Alan Brady
Iceland Gull (*Larus glaucoides*)	12/27/59	Schellenger's Landing	D. and H. Cutler, John Sawyer
	3/1–6/62	Lily Lake	E. A. Choate, Alan Brady
Black-headed Gull (*Larus ridibundus*)	1/2/56	Cape May Harbor	H. Cutler
	12/22/57	Cape May Harbor	D. and H. Cutler, H. Armistead

SPECIES	DATES	LOCATION	RECORDERS
Little Gull (*Larus minutus*)	12/24/54	Hereford Inlet	E. J. Reimann, D. Cutler
Kittiwake (*Rissa tridactyla*	12/27/57	Wildwood	Edward J. Reimann
	12/27/59	Inland waterway between Wildwood and Stone Harbor	Joseph Cadbury, Peter McCord, Robert Murtz
Sandwich (Cabot's) Tern (*Thalasseus sandvicensis*)	9/30/61	Stone Harbor	Herbert Miller
Noddy Tern (*Anous stolidus*)	Sept. 1960	Stone Harbor	Kenneth Wright
Snowy Owl (*Nyctia scandiaca*)	1937, 1945, 1954	Cape May Christmas Count	D.V.O.C.
	1/3/54	Stone Harbor Christmas Count	Julian K. Potter *et al.*
Saw-whet Owl (*Aegolius acadicus*)	1/3/54	Wildwood Area Christmas Count	M. Albert Linton *et al.*
Kiskadee Flycatcher (*Pitangus sulphuratus*)	12/26/62	Cape May Court House	St. Stauffer, H. Kreuger, S. Thomas
Scissor-tailed Flycatcher (*Muscivora forficata*)	9/2/41	Cold Spring	D. Fables, Jr.
Black-capped Chickadee (*Parus atricapillus*)	12/22/57 12/31/61	Cape May Point Cape May Christmas Census	A. Brady D.V.O.C.
Bewick's Wren (*Thryomanes bewickii*)	10/31/62	Cape May Point	Richard Benedict
Wheatear (*Oenanthe oenanthe*)	10/7/51	Cape May Point	E. A. Choate *et al.*
Orange-crowned Warbler (*Vermivora celata*)	12/28/52 12/26/56	*1 individual* *11 individuals*	William Middleton 3 parties J. Cadbury, J. K. Potter, M. A. Linton
	12/27/59 10/8/61	Stone Harbor Sanctuary	J. K. Potter Albert F. Matlach
Cerulean Warbler (*Dendroica cerulea*)	5/6–7/62	Cape May Point	E. A. Choate *et al.*
Black-throated Gray Warbler (*Dendroica nigrescens*)	9/30/61	Cape May Point	Harry Armisted, Russell Fowler
Bullock's Oriole (*Icterus bullockii*)	12/18/53	Cape May Court House	K. Kreuger, M. Stauffer, S. Thomas
	Dec. 1960	Cape May Court House	Minnie Stauffer

SPECIES	DATES	LOCATION	RECORDERS
Brewer's Blackbird (*Euphagus cyanocephalus*)	12/27/59	Higbee's Beach	Steven Harty, Robert Haines, Geo. Reynard
	12/26/60	Higbee's Beach	J. Kramer, Howell Penneston
Summer Tanager (*Piranga rubra*)	5/15/56	Cape May Point	E. A. Choate, William Bailey
Black-headed Grosbeak (*Pheucticus melanocephalus*)	12/26/60	Cape May Court House	Joseph Cadbury, John McIlvanie
Painted Bunting (*Passerina ciris*)	5/4/58	Cape May Point	Russell S. Fowler
Pine Grosbeak (*Pinicola enucleator*)	10/17/39	Cape May Point *flock of 5*	D. Fables, Jr., L. Holland
	Dec. 1961	Cape May Point	Alan Brady
Red Crossbill (*Loxia curvirostra*)	11/11/59	Cape May Point	Alfred Williams, Harry Steatsmeyer
Clay-colored sparrow (*Spizella pallida*)	9/11–14/57	Stone Harbor	Julian K. Potter *et al.*
	10/6–7/56	Cape May Point *2 birds*	E. A. Choate, D. Cutler, E. Reimann
	10/5/56	Cape May Point	Robert Grant
	9/28/62	Cape May Point	William Russell *et al.*
Golden-crowned sparrow (*Zonotrichia atricapilla*)	10/6/62	Cape May Point	Joseph M. Cadbury
Lark Bunting (*Calamospiza melanocorys*)	9/16/56	Cape May Point	R. Grant, J. Kramer, I. Black, D. Kunkle *et al.*
Black-backed Three-toed Woodpecker (*Picoides arcticus*)	10/4–11/63	Cape May Point	Goldie Lichten and 300–400 others
Great Snipe (*Gallinago media*)	9/8/63	Lily Lake, Cape May Point	Harold Price *et al.* (Substantiated by photographs and pronounced by A. Wetmore and R. T. Peterson to be apparently this species.)

A BIOGRAPHICAL NOTE ON
WITMER STONE*

(1866–1939)

By James A. G. Rehn

On May 23, 1939, the Delaware Valley Ornithological Club lost the most outstanding figure in its history in the passing of Dr. Witmer Stone. No other individual tied the Club, as we know it today, so closely to the limited group of enthusiasts who composed it in the nineties, and during the virtual fifty years of his association, no one exerted a more constructive, yet restrained and tactful, influence upon its natural development.

During these decades of its history, with their kaleidoscopic passing through the membership of the Club of many promising amateur ornithologists, often in their student years and some now distinguished college professors elsewhere, Witmer Stone stood in our ranks as the symbol of authoritative ornithology, the arbiter whose knowledge, wit, and kindliness made him beloved to the beginners and the seasoned "wheel horses" alike. Drawing upon his experience with birds, bird literature and bird men, and their history, he gave to the discussions at the meetings of our Club a breadth, value, and appreciation of ornithology in its broadest sense, which we shall miss increasingly with the years.

The progressive development of human knowledge evolves new intellectual associations, and destroys others. Witmer Stone belonged to one of these latter, now passing groups, that of the old school general naturalist, as his interest in animate nature was so broad and vivid that he had a more than casual acquaintance with the components of many groups of animals, and of plants as well. The increasing complexity of our knowledge of living things, the consequent need for specialization and the multiplication of all scientific literature have united in bringing to a close the era of the general naturalist, which was a necessary stage in the development of research in the natural sciences, and, of the enthusiastic group which characterized that era in America, Witmer Stone was one of the last. Equally at home with birds or mammals, he also possessed a good knowledge of reptiles, of mollusks, and particularly our local land and marine forms, of many of our insects and crustaceans, as well as critical ability in the systematics of our local flora surpassed only by that of Simon-pure botanists. This knowl-

* This article is reprinted with permission from the *Proceedings* of the Delaware Valley Ornithological Club, No. XXXI, 1938–41.

edge, accumulated in the mind of our old associate, was freely drawn upon in and out of season by members of our Club. Truly, as our late President, Wharton Huber, has said, "It is an inestimable privilege to have known Witmer Stone."

Witmer Stone was born in Philadelphia, September 22, 1866. He was the second son of Frederick D. and Anne E. Witmer Stone. His father, a distinguished Pennsylvania historian, was for years Librarian of the Historical Society of Pennsylvania. Much of the love and regard of Witmer Stone for good literature, graceful writing, and for books as creations of human minds and hands probably can be traced to the influence of a scholarly father. Young Witmer Stone early exhibited an interest in nature and the accumulation of boyhood collections proceded along the usual pattern of the times, aided and abetted by youthful acquaintance, soon life-long friendship, with members of the neighboring Brown family, one of whom, Amos, later became a distinguished professor of geology at the University of Pennsylvania, another, Stewardson, a botanist of attainments, Curator of Botany at the Academy and also a President of the D. V. O. C. Stewardson Brown's association with our Club extended from his election in 1891 until his death in 1921.[1] This juvenile circle, of the type which, in the latter part of the past century, inspired a number of our eminent scientists of today, was doubtless as important in shaping the future of Witmer Stone as it was in influencing those of his juvenile associates.

The historic Germantown Academy supplied Stone's early training, followed by the University of Pennsylvania, from which he graduated in 1887. Elected in under-graduate years to the Secretaryship of his college class, which has produced a number of men of exceptional ability, he held this post until his death. His A.B. degree of 1887 was followed by that of A.M. in 1891, and in 1913, the honorary one of Sc.D. was conferred upon him by his Alma Mater.   In 1937, the University of Pennsylvania further honored him with the Alumni Award of Merit.

Following his graduation from the University of Pennsylvania, Stone served for a period as assistant in the library of the Historical Society of Pennsylvania. It is not difficult to understand from this and his parental inheritance why the history of ornithology and of the lives of ornithologists occupied so large a niche in his ever active mind.   In March, 1888, he was appointed a Jessup Fund Student at the Academy of Natural Sciences of Philadelphia, and then began the association which continued through his life, productive of opportunities for ornithological association for the individual and constructive direction in many fields for the institution. Largely through Stone's energy and initiative, an Ornithological Section of the Academy was established in May, 1891, which supplemented within the legal structure of the institution the independent and less formal D. V. O. C. The Section was maintained for nearly thirty-five years, passing out of existence in 1925 when a change in Academy By-laws abolished all section organizations.

[1] See Cassinia, No. XXIV (1920-1921), pp. 1-7, portrait, (1923).

During his more than fifty years of association with the Academy, Witmer Stone served the institution in many capacities. From a Jessup Fund Student he passed in 1891 to the Conservatorship of the Ornithological Section, together with an assistantship to the Board of Curators, becoming a member of that body of four in 1908, and its executive in 1918. In 1925, he moved by an administrative reorganization to the newly created post of Director of the Museum, becoming Emeritus Director in 1928. With these broadly administrative posts he also held that of Curator of Vertebrate Zoology from 1918 to 1934, Curator of North American Birds from 1934 to 1938, and, due to failing health, in 1938, that of Honorary Curator of Birds. In 1927, he was elected one of the two Vice-Presidents of the Academy, an honor he held at the time of his death. The scope of his official Academy duties during these many years can hardly be summarized in a few sentences. With a supporting staff of truly skeleton proportions, his energy, initiative, and foresight unquestionably saved for posterity priceless collections in many fields of the natural sciences. As one who labored with him for years in this work, the writer can testify to what was accomplished with few hands, meager funds, and make-shift housing, the latter often built by the hands of men worth far more as scientific research workers. As Dr. Stone's long-time associate, Dr. J. Percy Moore, has so graphically said in a memorial minute of the Council of the Academy: "His life became so merged with that of the Academy that for many years it was difficult to think of them apart. With characteristic industry and thoroughness, and the help of other members, the then wasting biological collections were salvaged, renovated, and conserved. Many valuable specimens and records were saved and many reforms in labelling, cataloguing, and storing were established . . . Poverty greatly hampered the progress of the Academy. Only through the loyalty and enthusiasm of the staff was its place in the scientific world maintained. The personal sacrifices were little short of heroic and Stone was an acknowledged leader and an inspiration to the younger men."

In 1888, when Dr. Stone became associated with the Academy, its bird collections numbered 26,000 specimens. In 1939, at the time of his death, the same collections totalled 143,000. Much of this growth took place in years when a concentration of administrative and non-ornithological curatorial responsibility gave but exceedingly few hours for birds or bird problems. Yet between 1889 and 1908 thirty-eight papers, chiefly on birds, appeared from his pen in the "Proceedings" of the Academy, together with joint authorship in two others. Aside from these, numerous contributions by him appeared during the same years in the "Auk," "Cassinia," and other journals.

The field activities of Dr. Stone in behalf of the Academy were many and varied, although largely carried on in the eastern United States. Virtually hundreds of field trips were taken to the New Jersey pine barrens and the coastal section of the southern part of the state, while many others were made to the lower Susquehanna Valley and the Sullivan-Wyoming county portion of Pennsylvania and western Maryland. The Cumberland mountain section of Kentucky, and coastal plain of South Carolina, the Duluth region of Minnesota, and the Chiracahua Mountains of Arizona were also visited and studied by him, and from all of these extensive and important collections were brought back for the Academy. To gather material for others to study was second nature to him. In 1890, he was also a member of an Academy expedition, under the leadership of Professor Angelo Heilprin, to Yucatan and eastern Mexico, which early gave him a personal acquaintance with tropical American bird life. Several years before, at the beginning of his Academy association, in 1888, he had visited Bermuda on another Academy expedition, similarly under Heilprin's leadership.

The area, however, idelibly associated with the name of Witmer Stone is the Cape May peninsula, where some of his earliest observations and collections were made, and there in later years, in the afternoon of life, by months of residence and through many special visits, he built the background which enabled him to give us and to the world in his "Bird Studies at Old Cape May" the most lasting of his memorials. There, most fittingly, the National Association of Audubon Societies has established the "Witmer Stone Bird Sanctuary" to permanently link our memory of the man, his service in the protection of bird life, and the place he loved and has made ornithologically famous.

The first serious contribution to the science of ornithology from the pen of Witmer Stone was "The Turkey Buzzard Breeding in Pennsylvania," published in the "American Naturalist" for 1885, in which two Chester County nests of the species were reported. In the following six years, of the sixteen of his ornithological contributions, two dealt with a subject always of surpassing interest to him, that of bird migration and means for its graphic recording. Others of these papers evidence the work in which he was engaged at the Academy in the rejuvenation of its bird collections, such as the catalogues of the Old World Flycatchers, the Owls, the Crows, the Birds of Paradise and the Old World Orioles in the series there, as well as a report on the birds of the Mexican Expedition and those of the several Peary Greenland expeditions sponsored by the Academy. Stone's interest in spiders is also shown by the publication in 1890 of a catalogue of the Pennsylvania and New Jersey members of the arachnid family Lycosidae.

Of the ornithological contributions which appeared from his pen before the turn of the century, probably that of greatest permanent value in itself, and because it led the way for more comprehensive studies of the same problem by others, was "The Molting of Birds, with special reference to the Plumages of the Smaller Land Birds of Eastern North America." This relatively pioneer paper of 1896 dealt with one of the subjects which remained throughout his life a most absorbing one—the molts and plumage sequences of birds. His interest in this phase of ornithology was constantly reflected in his writings dealing with purely systematic or faunistic ornithology, and much original information on plumages will be found in those papers.

It was but shortly after Witmer Stone became a student at the Academy that he made the acquaintance of Spencer Trotter, who as a Jessup Student in 1876 had helped move the Academy bird collections from the Broad and Sansom

Streets building to that at Nineteenth and Race.   As Stone later wrote, this was his first acquaintance with a real ornithologist—one who knew more about birds than he did.   The friendship which developed between these men was of the type Witmer Stone possessed the faculty for creating, an abiding, life-long, mutual appreciation of respective worth.   Many of those who knew Spencer Trotter as a friend and fellow club-member can understand this mutual attraction of personalities which possessed in common brilliant minds, broad culture, ready wit, and a touch of whimsey.

Through Spencer Trotter's acquaintance with them, Witmer Stone met William Baily, George Spencer Morris, and Samuel N. Rhoads.   These five, with Charles A. Voelker and J. Harris Reed, formally organized the Delaware Valley Ornithological Club on February 3, 1890.   Thence the life of Witmer Stone had as an integral part the activities and progress of the D. V. O. C., and its members the value of an interest in their observations, their writings, and their many field excursions, which broadened their knowledge, helped them in their perplexities, and gave them the pleasure of an association more deeply valued with each passing year.   The history of the Delaware Valley Ornithological Club has been written by those who were its founders, and the part which Witmer Stone played in this history needs now but the stressing of high points.

Dr. Witmer Stone was elected the second President of the Club in 1891, and served a single year, as was customary at that time.   During the first twenty years of the Club's existence, he made numerous formal communications on a variety of topics, such as the birds of Fulton, Sullivan, Wyoming, and Susquehanna Counties of Pennsylvania, Cape May and Salem Counties, and the Pine Barrens of New Jersey, of Delaware, parts of Maryland, the Duluth region of Minnesota, Bermuda, Yucatan and the Eastern Cordillera of Mexico; molts and early plumages of North American birds; mutation; the forms of the genus *Ammodramus*, the relationship of the Rallidae; migration theories, recording migration; lost species of Wilson and Audubon, ornithology at Philadelphia in the past, Dr. Elliott Coues, the A. O. U. of 1840–45, and first describers of American birds.   His interest in the history of ornithology and the biography of ornithologists is evident from these communications, plus the numerous sketches or memorabilia of earlier or contemporary ornithologists, botanists, and other naturalists, which have appeared in the pages of "Cassinia," "Bartonia," the "Auk," and in certain volumes of biographical character.

Soon after the organization of the D. V. O. C., a committee of three was appointed to superintend the preparation and publication of a catalogue of the birds known to occur in the vicinity of Philadelphia and on the New Jersey coast.   This committee consisted of George Spencer Morris, Samuel N. Rhoads, and Witmer Stone.   The preparation and editing of this work was at once placed in the hands of Dr. Stone.   Based upon previously published literature, the very important notes and collections of the members of the Club, as well as information supplied by many correspondents in various parts of Pennsylvania and New

Jersey, "The Birds of Eastern Pennsylvania and New Jersey" appeared from his pen in 1894, as a publication of the D. V. O. C.

This important regional work has remained for nearly a half century the best single volume of its character and coverage.   Divided into two parts, the first of these is entitled, "Geographical Distribution of Birds and Bird Migration," and contains brief but excellent summaries of the "general laws of geographical distribution" with an analysis of the life-zones to be encountered in our territory, followed by descriptions of the faunal areas of eastern Pennsylvania and New Jersey with their characteristic resident or nesting birds.   The portion on bird migration presents a summarized exposition of the subject as reflected in the bird life of the area covered.   The second part of the volume is an "Annotated List of the Birds of Eastern Pennsylvania and New Jersey."   To this very carefully prepared treatment of the local occurrence of three hundred and fifty-two forms of birds, which also presents for most of them the summarized breeding range and winter distribution, is appended a bibliography of ornithological literature relating to eastern Pennsylvania and New Jersey published to the year 1894. It is safe to say this volume gave to the D. V. O. C. a permanent place as a serious ornithological organization, and it also helped establish Witmer Stone, early in his life as one of our leading American ornithologists.

While the D. V. O. C. during its first decade published four brochures containing abstracts of its proceedings, and also a few short papers by its members, the need for a more ambitious serial publication on local ornithology, also containing the Club's migration data, was constantly discussed.   This resulted, in 1901, in the establishment of "Cassinia," to the editorship of which Witmer Stone was at once appointed.   He had been one of the supporters of the journal idea, and its name, perpetuating the memory of Philadelphia's early master ornithologist, was his happy suggestion.   Most appropriately, the first article in the serial was a sketch of the life of John Cassin by Dr. Stone, the frontispiece a reproduction of an original Cassin photograph which hangs in the ornithological department of the Academy.   It is fair to say that the existence of "Cassinia" through nearly forty years has been due very largely to the interest and labors of Witmer Stone, who, serving first as its official Editor, continued through the years unofficially to aid and advise those who succeeded him in that post.   The exceedingly laborious analysis of the invaluable migration reports, which give "Cassinia" its greatest value, were until recent years the labor of love which Witmer Stone assumed for this one of his numerous brain children.   The continuity and exhaustive, fifty year coverage of the Club's migration reports made them, he felt, of the utmost ornithological importance—in his mind, as he often said, the most valuable of our local contributions to bird knowledge.

During the decade and a half from 1920 to 1935, Dr. Stone spent a number of his summers at Cape May, a locality which had appealed to him from his first visit many years before.   The notes and observations which he accumulated on the bird life of the Cape May Peninsula interested an increasing number of D. V. O. C. members, and soon Stone's Cape May observations, already extensive and important, were supplemented by those of a score or so of similar enthusiasts, and Cape May became our local ornithological mecca.   By its geographical position and importance as a migration fly-way concentration point, this selection of Cape May was fully justified.   The plan for a work on the ornithology of coastal New Jersey was the definite result of these studies made over several decades, and, in 1937, Dr. Stone produced the two volumes of "Bird Studies at Old Cape May," which are his greatest single ornithological contribution.   Illustrated by reproductions of paintings, drawings, and photographs, largely made by D. V. O. C. members, these volumes weave into a unit Stone's innumerable personal observations, those of many other Club members, and the relevant published record, the whole presented, in a background of sympathetic understanding, with the charm which characterizes so many of his writings, its verbal pictures broadened by the vision of one who in viewing the particular never neglected the broad scene.   As Dr. J. P. Moore has so truly said, this work "reflects Stone at his best, both as a naturalist and a writer."

While Philadelphia remained the center of all of Stone's activities, the country as a whole drew upon his knowledge and ability.   Becoming an Associate of the American Ornithologists' Union in 1885, he was elected a Fellow in 1892.   In 1912, he succeeded Dr. J. A. Allen as Editor of the "Auk,", probably one of the most outstanding recognitions of ability and scholarship which can be accorded an American ornithologist.   During his incumbency, Dr. Stone worked unceasingly for the diversification of the contents of the "Auk" to cover the many ramifying fields of ornithological research and observation.   His reviews of important literature and listing of the less outstanding contributions made this side of the journal of broader usefulness.   His interest in so many angles of ornithology gave to his critiques a particular value not found in mere contents abstracts. For twenty-five years, Witmer Stone gave to this work the best years of his life, relinquishing it only after the need for conserving his strength made this imperative.   To those who were his constant associates, the exactions of this editorship and what it demanded of his energy were clearly evident.

From 1914 to 1920, Dr. Stone was Vice-President of the American Ornithologists' Union, and from 1920 to 1923 he served as its President.   But a relatively few days after his death, the Brewster Medal of the Union was awarded posthumously for his outstanding ornithological contribution "Bird Studies at Old Cape May."   In early years, when a vigorous champion of the need for comprehensive bird protection, Dr. Stone was from 1897 for a number of years a member, and for part of the time Chairman, of the American Ornithologists' Union's Committee on Protection of North American Birds.   During this period, he worked actively for state and national legislation for the protection of bird life, and his continued but eminently practical interest in the subject was evident in his services as President of the Pennsylvania Audubon Society, and for some years as a member of the National Committee of Audubon Societies.   He was an outspoken opponent of poisoning campaigns by governmental or other agencies in the "control" or extermination of so-called noxious and predatory mammals.

In 1901, Witmer Stone was appointed a member, and from 1919, Chairman, of the Committee on Nomenclature and Classification of the A. O. U., from 1905 to 1908, taking an active part in the revision of the A. O. U. Code of Nomenclature, this being an angle of systematics to which he had given special study. In recognition of preeminence in this field of zoological science, he was elected one of the few American members of the International Commission on Zoological Nomenclature, a post he held at the time of his death.

The Third Edition of the A. O. U. Check-list of North American Birds issued in 1910 by the Union's Committee on Nomenclature and Classification states in its preface that "the preliminary revision of the geographical ranges of the species and subspecies was undertaken by Mr. Stone," and again that "the Union owes a lasting debt of gratitude to Mr. Stone and Dr. Merriam" for their work in the preparation of the check-list.

In 1924, the Union authorized the preparation of a fourth edition of the check-list, and Dr. Stone was appointed chairman of a special committee in charge of this work, power being given him to select the other members of this committee. His years between 1924 and 1931 were crowded with check-list labors, and when this catalogue—for such it was, and of a most unusually comprehensive type—appeared from the press in 1931, it was recorded in the preface that the work was largely written by Dr. Stone. Few but those associated with him in this project, or close enough to him officially to know its virtually day-to-day demands, appreciate what a great amount of his time and energy this task consumed. The verification of all references alone was a tedious and day-consuming responsibility, while the refurbishing of the distributional statements to include the enormous increase of recorded information since the previous check-list was, in itself, enough to deter any but the most serious student. To those of us who knew him well, Stone's work on the Fourth Edition of the A. O. U. Check-list seemed to afford him a deeper and more lasting personal satisfaction than any other purely technical undertaking with which he had been associated.

Of Witmer Stone's contributions to ornithology and other sciences not already mentioned, a number are of special importance. In 1902, jointly with W. E. Cram, he prepared a popular volume on American mammals entitled "American Animals," which remained until relatively recent years the standard work of its character. In 1908, his "Mammals of New Jersey" appeared, followed the subsequent year by "The Birds of New Jersey, their Nests and Eggs." The latter volume is of special value to local ornithologists as it contains a full bibliography of New Jersey ornithology up to the year of the preparation of his work.

In 1912, there appeared from Stone's pen one of his most outstanding contributions, one which showed very clearly his authoritative versatility. This was "The Plants of Southern New Jersey with Especial Reference to the Flora of the Pine Barrens and the Geographic Distribution of the Species." An exhaustive work of over eight hundred pages and one hundred and twenty-nine plates, this study will long remain the authoritative, systematic, and distributional guide to the flora of the Pine Barrens. It is not a mere compilation of previously published records and a tabulation of material in established herbaria. While including these necessary data, it had as its chief reason for existence the specially assembled collections, totalling over twelve thousand specimens, secured by its author on many scores of field trips to southern New Jersey localities, these extending over a period of approximately ten years. On many of these trips I was privileged to accompany him, collecting insects while he busied himself with the plant life. His personal plant records thus always had the support of preserved material, subsequently given without restriction to the herbarium of the Academy. Dr. Francis Pennell, Curator of Plants at the Academy, in paying tribute to Dr. Stone's work as a botanist, has said this volume "stands forth increasingly with time as the most careful geographic study of any comparable part of the flora of eastern North America."

From Dr. Stone's pen Dr. Pennell lists twenty botanical titles; Mr. Huber in his tribute to Stone's work as a mammalogist lists nineteen in that field. Three papers on reptiles and amphibians appeared in the Academy's "Proceedings" prior to 1911, and allusion has already been made to his important paper of 1890 in lycosid spiders. His purely ornithological contributions, which include a large number of reviews of ornithological works, total many hundreds. In local field entomology he found, in the later years of his life, escape from the limitations laid down by failing health, and, while at Cape May, accumulated an extensive, important, and representative series of the insects of that area, himself collecting, sorting, mounting, labelling, and determining much of the material, securing the determination of more by specialists in the various fields of entomology. This work gave marked satisfaction to the deeply implanted collector's instinct, and all of these collections were assembled for eventual addition to the Academy series. This desire, often expressed by him, has now been consummated, and in more than one order of insects the most extensive local representation possessed by the Academy is that accumulated by Witmer Stone.

Many recognitions of Witmer Stone's attainments, other than those already mentioned, were accorded him. These included the Presidency of the American Society of Mammalogists, of which he was a charter member, and membership in the American Philosophical Society and the Society of the Sigma Xi; fellowship in the American Association for the Advancement of Science; the Otto Hermann Medal of the Hungarian Ornithological Society (conferred in 1931); honorary membership in the Nuttall Ornithological Club, the Cooper Ornithological Club, the Linnean Society of New York, the Zoological Society of Philadelphia (of which he was also a Director), the Philadelphia Botanical Club, the British Ornithologists' Union, the Ornithological Society of France, the Ornithological Society of the Netherlands and the Hungarian Ornithological Society; and foreign membership in the German Ornithological Society and the Bavarian Ornithological Society.

The dominant traits of Witmer Stone's life were kindliness and helpfulness. As our D. V. O. C. fellow-member Dr. Cornelius Weygandt, writing of him before Dr. Stone's passing, has so well said, "He has a genius for friendship, and he regards it as part of a duty done cheerfully to help others working in any field that approaches his own." Again Dr. Weygandt says, "I find mention of a thrush of the high mountains of Northern Italy. I go to the Academy of Natural Sciences to ask my friend to identify it, and to show me a specimen. He digs out the skin of the rock thrush for me, giving me of his time so infinitely precious. He is of those years in which he begins to wonder, as I do of myself, whether he will be able to finish up all the work that is within his power to do. Yet such is his human fellowliness and unselfishness that he will spend an hour any time to help you in your little concerns."

The saving grace of humor gave Witmer Stone a corrective balance to a nature inclined to be retiring in consequence of deafness left by juvenile whooping-cough. The merry twinkle of the eye, the clever quips or even the touch of whimsey, and the healthy abandon of a hearty laugh, would show that beneath all the serious work, the grind of editorship, the often monotonous routine of curatorial duties, and the piles of sometimes pointless and often trivial correspondence requiring attention, there was a spirit fully in tune with its fellowmen, akin to their problems, their joys, and their paradoxes. As Dr. Weygandt observantly writes, "He accompanies a list made for you of the principle men in natural science of our state of Pennsylvania with a story of a drunk in Washington who stole an ocelot from the Zoo there." Those of us who knew Witmer Stone as a daily associate can recall many instances when his humor, as vivid as in boyhood, but always kindly, lightened our burdens, made the task shorter, the problem easier to solve, the long trudge shorter.

To the younger man privileged to work with Witmer Stone, this eternal buoyancy of spirit and the enthusiasm which he kept until the last years of his life created inspiration of the sort kindled by few. Many great scientists completely lack this ability to enthuse others, or to paint their discoveries in language most of us can read. Witmer Stone had that personal magnetism without pose, directive ability without presumption or arrogance, and kindliness without patronizing which inspired both young and old and placed them forever in his debt.

In a memorial of the late Stewardson Brown, Witmer Stone penned a few sentences which, while referring to a kindred organization, deserve to be read by members of the D. V. O. C., so expressive are they of the measure of the man. "While technical knowledge and activity in research are admittedly necessary for the successful program of a club such as ours, there would be no club were it not for further qualities exhibited in the membership. There must be personalities that hold us together—that inspire respect and affection and weld bonds of friendship that will not yield with the lapse of time." As our botanical colleague Dr. Pennell has so well said, "How many of us to the end of our days will feel a debt of gratitude to Witmer Stone for his encouragement to us during our formative years."

It is given to few of us to have our life's work, and the reflection of our span of years upon our associates, as beautifully expressed as in the memorial minute to Witmer Stone adopted on October 3, 1939, by the Council of the Academy of Natural Sciences. "Of Doctor Stone it may be said that the seed within him on fertile soil grew into a sturdy tree of many branches, bearing fragrant flowers and nourishing fruit. Naturalist, scientist, faithful custodian of collections, biographer and historian of scientists and their science, interpreter of the rules of zoological nomenclature, protector of birds, writer of exceptional beauty and vigor, sometimes poetical, lecturer and teacher, helpful adviser, delightful companion and valued friend, Witmer Stone gave the best of his life and labor to the historic Academy and of the riches of his personality to colleagues and associates. His works and our memories are a fitting memorial, and may his spirit long abide in the lives of those on whom he spent it."

PREFACE TO THE FIRST EDITION

The primary object in the preparation of the present work has been to furnish, for purposes of future comparison, as accurate a picture as possible of the bird life of Cape May during the decade, 1920–1930, with an account of the changes that have taken place in the years that followed. I have been so impressed by the failure of the ornithologists who were active in the half century subsequent to the time of Alexander Wilson—and for some years later—to leave us any adequate sketch of the bird life of the New Jersey coast of their day that it seems to me a duty that we owe to our successors to provide such a picture for our own time.

The "studies" were written originally in 1920 and 1921, as the result of an intensive investigation of the bird life immediately about Cape May and have been amplified in later years as opportunities occurred for the further observation of familiar birds under different conditions or the study of additional species.

As the work proceeded it seemed desirable to include data on local bird-life prior to the period above mentioned as well as to gather together such information as exists on the birds of the coast in still earlier times with pertinent quotations from the classic "American Ornithology" of Alexander Wilson, the pioneer ornithologist of the region.

It also seemed that observations on birds along the entire New Jersey coast should be included as this area really constitutes but a single ecologic unit, as well as on those of the shores of the Delaware River and of the Delaware coast just across the Bay. It has also developed that only a small number of birds have occurred in New Jersey that have not been recorded in Cape May County and brief mention of these has been added. However, while all of the birds of New Jersey have been listed, it should be kept in mind that this work deals primarily with the coastal fauna and does not pretend adequately to cover the bird life of the interior or of the northern counties. Only such data on this region has been included as was necessary to explain the nature of the birds' occurrence in the state.

The "studies" are intended not only to describe the character of occurrence of the various species but to picture, as well, their activities and their environment and I trust that certain of the observations here recorded may have value in connection with the study of bird behavior and migration. A detailed account of the birds of a single locality has a certain interest that is enhanced by intimate items connected with its history and physical features and such have been added. To this end, too, the old names of the beaches have been used in preference to those of the communities and "developments" that have sprung up upon them while the names of the original proprietors of the old estates—names by which they are better known than by those of the present day owners, have been retained, with no disrespect to the latter! So, too, I have used the word "town" instead of the more politically accurate "city" in referring to Cape May itself as a defence of the resort since to proclaim this delightful community of 2600 inhabitants as a "city" would tend to drive away prospective visitors attracted by its wealth of bird life and wild flowers and whose object, like that of many of the present summer residents, is to escape from the "cities."

The effort has been made to present a readible account of wild bird life, rather than a mere tabulation of scientific data and as systematics has no place in such a work all family and other headings have been omitted except the names of the birds which, together with the order of sequence, follow the "American Ornithologists' Union Check-List" except for the shortening of some of the English names where qualifying adjectives were unnecessary.

No descriptions of the birds are presented as everyone interested in bird study today possesses one or more manuals for identification. There are, however, many comments on the appearance of birds in the field, of the deceptive effect of light and shadow upon coloration and the relative value of field characters. Even these are perhaps unnecessary in view of the appearance, after most of the "studies" had been written, of Roger Peterson's admirable "Field Guide to the Birds," but so intimately are they associated with other parts of the text that it seemed best to allow them to stand.

In the preparation of the "studies" I have relied mainly upon my personal observations of the Cape May birds as recorded in my field notes covering in all some forty-eight years but mainly the period from 1920 to 1930. I have also had the valuable coöperation of the members of the Delaware Valley Ornithological Club who have visited Cape May or other parts of the coast and have generously placed their notes and observations at my disposal. Published matter relative to the birds of southern New Jersey has been consulted and quoted while many records have been taken from Julian K. Potter's painstaking compilation in "Bird-Lore" comprising the Philadelphia section of "The Season" which is our most important record of bird life on the New Jersey coast ever since 1917; other records have been taken from the annual reports on local ornithology in "Cassinia," the proceedings of the Delaware Valley Club (1890–1932) and the "Proceedings of the Linnaean Society of New York" (1888–1836). For most quotations full reference to place of publication is added but in the case of individual records taken from the above compilations the name of the observer only is given as the reference can easily be ascertained if desired. The reader is referred to the bibliography in my "The Birds of New Jersey" (1909) for a list of publications relating to New Jersey ornithology prior to that date and, for later publications, to the annual bibliographies in "Cassinia." (cf. also pp. 930–932, beyond.)

There has been no compilation of matter relating to habits and behavior from the published observations of bird students in other parts of the country; this work is not a monograph and its local character has always been kept in mind and only such extralimital information has been included as bears directly upon the occurrence of the birds at Cape May.

In my frequent mention of my colleagues of the Delaware Valley Ornithological Club I have omitted middle names, not only for brevity but because the informality tends to bring observers and readers into closer touch just as the use of local names of ponds, woods and meadows as well as of the birds themselves makes for a closer familiarity with the Cape May of which I write. The maps will indicate the location of the latter and the list of acknowledgments will furnish the full names of the former. In the use of syllabic representation of songs and call notes I do not intend to question the more technical and accurate character of the musical and diagrammatic methods, but I feel that these are not understandable to the average field student and that the syallabic representation does convey a pretty accurate idea of the variations in a bird's utterances, especially to one familiar in a general way with its song.

During the period of my field work at the Cape the greater perfection and more general use of the binocular glass has entirely changed the method of the field ornithologist and there are today no collectors of bird skins in southern New Jersey. The modern glass brings the bird so close to the observer that the experienced student can identify practically all of the birds that he sees at a reasonable distance but the less experienced and less conscientious observer certainly can not. Unfortunately he is inclined to make identifications in the nonchalant manner that he is led to think constitutes him an ornithologist while the Christmas and other censuses, into which the spirit of rivalry enters, put a further premium upon unintentionally careless identification as opposed to scientific accuracy. The latter would leave a certain proportion of the birds seen by an observer of limited experience, unidentified. An author is therefore confronted with a mass of "sight records" upon which he has to pass judgment. In all such cases I have considered the personal equation and the experience of others in the same region and acted accordingly.

The Delaware Valley Club is fortunate in having among its members a number of excellent artists and to them I am indebted for the drawings and paintings used in illustrating the present work. Earl L. Poole has contributed the water colors from which the frontispieces to the two volumes have been engraved as well as the greater part of the line drawings of birds used as headings for the several "studies." Conrad Roland has contributed the remainder of these bird drawings. Richard E. Bishop has loaned one of his aquatints of Black Ducks from which a halftone has been made. J. Fletcher Street has drawn the maps and he and Herbert Brown the line drawings from which the landscape headings and vignettes were engraved. I am under obligations to James Bond for the privilege of having cuts made from some of Mr. Poole's drawings that were originally prepared for his "Birds of the West Indies." (These unfortunately lack the artist's initials.)

In addition to artists the Club has developed many skillful photographers and from the numerous prints that they have generously placed at my disposal I have selected 240 representing birds, nests and landscapes almost all of them from photographs taken in southern New Jersey which appear on the 119 halftone plates presented herewith.

Of these eighty-eight were taken by Wharton Huber, forty-two by William L. Baily, twenty-five by Julian K. Potter, eleven each by John Bartram and J. Fletcher Street, ten by Normen J. McDonald, seven by Benjamin Hiatt with others by Dr. Alfred C. Redfield, Dr. Francis Harper, Dr. C. Brooke Worth, John Gillespie, Edward Woolman, Conrad Roland, Henry Troth, Harry Parker and the late Henry R. Carey. Additional photographs were contributed by Harry P. Baily, C. M. Beal, Al. Cummins, R. F. Engle, Chas. D. Kellogg, A. B. Miller, J. A. Moore, Montgomery Lightfoot, Roger T. Peterson, T. Gilbert Pearson, H. C. Scott, William Vogt and Smith Studio of Cape May. The use of several of these was made possible through the courtesy of "Bird-Lore" in which journal they originally appeared.

When I first visited Cape May in August 1890, the only line of railway traversed West Jersey by way of Vineland and Millville and the facilities were not such as to make possible a single day's trip from Philadelphia and return except on special occasions. The excursion steamer "Republic" ran between the city and Cape May Point but it required all day for the trip and it spent but an hour at the Cape. In winter, moreover, there were practically no accommodations at the resort for visitors. It was, therefore, some years before the Club was able to include Cape May among the convenient localities for its one day outings and such trips as were made were by individuals and usually in summer.

Meanwhile my studies of the birds of the Jersey shore were conducted on the sounds and meadows back of Atlantic City as the guest of Norris DeHaven on his catboat "The Widgeon" while William L. Baily, who spent several summers on Five Mile Beach, David McCadden and Dr. William E. Hughes on Peck's Beach and Stewardson and Herbert Brown at Pt. Pleasant gathered information and specimens which were later put at my disposal.

When I began seriously to study the birds of Cape May in 1916 as a summer resident, a trolley line still ran from the Point to the Harbor, the successor to an earlier steam road, but there was no way of reaching points farther north in the peninsula except by the regular morning and evening trains. The result was that my field work was carried on on foot and one thought nothing of covering ten or fifteen miles on a day's tramp. A trip to Seven Mile Beach was a difficult matter, from Cape May, as it necessitated a rail journey to and from the Court House and a walk of four miles each way over the meadows. With the advent of rather infrequent bus service, not only from Cape May to the Point but also to the other beaches, one could reach Seven Mile in less than an hour while the general availability of automobiles a little later made it possible to cover all parts of the County on a morning trip.

It is questionable, however, whether the ornithological results of a day's outing under present conditions are as important or as thorough as in the pedestrian days, although of course a far greater area can be covered.

I was able to spend a large part of the months of July and August 1891, as well as of the years 1916–1918, at the Cape and all of these months from 1920 to 1937, although for the past eight seasons, owing to impaired health, I was able to do only such ornithological work as could be accomplished from an automobile. In other years and other seasons I have visited Cape May for week end trips or sometimes for a week at a time, but every month of the year has been covered. The activities of my colleagues in the Club have been mainly confined to similar short trips, many to single days, but J. Fletcher Street, Conrad Roland, Ernest A. Choate and the late David G. Baird, have spent large parts of several summers at the Cape.

In my many years' study of the bird life of the New Jersey shore I am deeply indebted to the authorities of the Academy of Natural Sciences of Philadelphia who have coöperated in every way and have granted me leave of absence during July and August of many years. The local collection of bird skins at the Academy, comprising the personal collections of some sixteen members of the Delaware Valley Club has always been available for study and to it have been added all specimens secured during the progress of my studies. The Academy has for many years generously provided a meeting place for the Club so that it has become, as it were, the home of the latter organization.

To the Delaware Valley Ornithological Club, in addition to the assistance rendered by the members individually, I am deeply indebted for sponsoring the present publication on the occasion of the dinner tendered me on my seventieth birthday on September 22, 1936. While I am under obligations to every member of the Club who has carried on field work in South Jersey there are some to whom I am particularly indebted.

To the late Henry Walker Hand, of Cape May, a lifelong resident of Cape May County whose knowledge both of the history and natural history of the region was unequalled. Through an intimate companionship of over a quarter of a century I feel that I owe to him much of my knowledge of the water birds, and of the lore of the Cape, and it is a matter of the deepest regret that he could not have lived to see the completion of this work in the planning of which he took such a keen interest. His migration records published from year to year in "Cassinia" have been freely used as well as his verbal accounts of Cape May bird life. To Julian K. Potter, of Collingswood, N. J., companion on many trips about Cape May, I am indebted for the free use of his voluminous notes and of his admirable compilation on the Philadelphia section of "The Season" in "Bird-Lore." Mr. Potter also read the galley proofs of the book. To Charles A. Urner of Elizabeth, N. J., I am indebted for most of my information on the water birds of the northern part of the coast, not only through his published reports but from much unpublished information that he has been kind enough to submit. To J. Fletcher Street of Beverly, N. J., my companion on many a field trip through South Jersey in search of bird lore, I owe much assistance in collecting data on the birds of the coast especially on those of Ludlam's Beach. To Turner E.

McMullen of Camden, N. J., and Richard F. Miller of Philadelphia, for lists, with data, of all nests examined by them in the southern counties of New Jersey.

Other Club members to whom acknowledgments are made in the text or who have aided me in the preparation of the work are:

*Dr. William L. Abbott
William L. Baily
*David G. Baird
Richard Bender
*George B. Benners
James Bond
Beecher S. Bowdish
*J. Farnum Brown
*Stewardson Brown
Herbert Buckalew
Joseph M. Cadbury
Charles M. B. Cadwalader
*Henry R. Carey
Hampton L. Carson, Jr.
John D. Carter
Dr. Ernest A. Choate
Henry H. Collins, 3d
*William B. Crispin
W. Stuart Cramer
Delos E. Culver
Victor A. Debes
*I. Norris DeHaven
Charles B. Doak
Arthur C. Emlen
Dr. John T. Emlen, Jr.
Richard Erskine
William B. Evans
Henry W. Fowler
Dr. Henry Fox
Henry Gaede
John A. Gillespie
Ludlow Griscom
Dr. Francis Harper
Richard C. Harlow
Harry M. Harrison
John Hess
Benjamin C. Hiatt
Wharton Huber
*Chreswell J. Hunt
Dr. William E. Hughes
William W. Justice, Jr.
David P. Leas
M. Albert Linton

Philip A. Livingston
David McCadden
Horace D. McCann
Norman J. McDonald
Clifford Marburger
Bennett K. Matlack
F. Guy Meyers
*Gilbert H. Moore
Robert T. Moore
*George Spencer Morris
Edward Norris
Edward H. Parry
Dr. Max M. Peet
*Charles J. Pennock
William Pepper, Jr.
Earle L. Poole
Richard H. Pough
Dr. James F. Prendergast
Nelson DeW. Pumyea
J. Harris Reed
Edward J. Reimann
James A. G. Rehn
Samuel N. Rhoads
Charles H. Rogers
Conrad Roland
Horace W. Rolston
Frederick C. Schmid
Samuel Scoville
William A. Shryock
George H. Stuart, 3d
W. Gordon Smith
*Dr. Reynold A. Spaeth
*Joseph W. Tatum
*Dr. Spencer Trotter
Dr. Henry Tucker
*C. Eliot Underdown
Charles A. Voelker
Edward S. Weyl
Mark L. C. Wilde
Edward Woolman
Dr. C. Brooke Worth
William Yoder
Robert T. Young

* Deceased.

To many members of the Linnaean Society of New York and others I am indebted for records quoted from "The Season" in "Bird-Lore" or for direct assistance. Among these are:

T. Donald Carter	F. W. Loetscher
Allan D. Cruikshank	Dr. Ernst Mayr
*Warren F. Eaton	Charles K. Nichols
James L. Edwards	John T. Nichols
Richard Herbert	Graham Rebell
Joseph J. Hickey	William Vogt
J. M. Johnson	Lester L. Walsh
Irving Kassoy	J. A. Weber
John F. Kuerzi	William A. Weber
Daniel S. Lehrman	Laidlaw O. Williams

To John H. Baker for access to the reports of the Audubon Association wardens at the sanctuary at Cape May Point and to the wardens personally:

George B. Saunders	James T. Tanner
Robert P. Allen	Richard Kuerzi
William J. Rusling	

Also to Mrs. Cordelia H. Arnold, R. Dale Benson, Jr., F. W. Laughlin, Charles C. Page, Isaac G. Roberts and the late William H. Werner for valuable information.

To H. Raymond Otter and the late John W. Mecray I am indebted for important information on the wildfowl of Cape May County based upon their many years' experience in gunning on these waters. To Otway H. Brown, of Cold Spring, Cape May County, I am indebted for migration records covering recent years and for many records for years gone by.

And to other residents of Cape May for information of various kinds:

Frank Dickinson	J. R. Moon
Miss Mary Doak	Lewis Sayre
Dr. J. Smallwood Eldredge	S. Irwin Stevens
Mrs. Emlen H. Fisher	Lewis T. Stevens
Harry F. Graves	Wilfred M. Swain
*Henry Hazelhurst	Charles York

Also to Mr. and Mrs. H. Walker Hand, Mr. and Mrs. Otway H. Brown and Mr. and Mrs. Sydney R. Goff for generous hospitality when visiting Cape May during winter, spring and autumn, to Messrs. Stephen and Victor Simon for similar hospitality at Sea Isle City, to Franklin Cook on a visit to Little Beach Island, and to the U. S. Biological Survey during a several days' cruise on its patrol boat on Barnegat Bay.

As I look back upon the many years that are identified in mind with Cape May I realize that my greatest pleasure has been in the delightful association with men of kindred interests. Whether while watching the

* Deceased.

shore birds from a skiff on Jarvis Sound with Walker Hand or living in one of those little stilted cabins on the meadows with Frank Dickinson, photographing Wood Ibises under a broiling sun with Fletcher Street or counting Skimmers and Terns on wind-swept Gull Bar with Julian Potter, the personal contact has always meant as much or more than the ornithological association. And there have been those gatherings for dinner at the Court House at the close of the Christmas Census with one party after another coming in half frozen from boats on the sounds and the Bay or from stations out on the end of the jetty or on remote ponds and in dense woodlands, to prepare their combined report.

Then there have been the frequent visits of Club members during my summer residence at the Cape. But my visitors have not all been from near at hand and I can count among the ornithologists from elsewhere who have visited me at Cape May, Drs. Alexander Wetmore, Theodore S. Palmer, A. K. Fisher, and Paul Bartsch and Frederick C. Lincoln, all of Washington, John T. Nichols of New York and Ludlow Griscom then also of that city, Arthur C. Bent of Taunton, Mass. and Dr. Tracy I. Storer of Davis, Calif., all of whom are identified with trips along the shore when it was my privilege to introduce them to the bird life of Old Cape May.

On October 25, 1929, one hundred and sixty members of the American Ornithologists' Union spent the day at Cape May Point, following the meeting of the society at Philadelphia, and for several years the National Association of Audubon Societies has conducted an excursion to the Cape at the time of the autumn hawk flight which has made many other ornithologists acquainted with this historic spot.

WITMER STONE

September 22, 1937.

CONTENTS

VOLUME ONE

CONTENTS

PART I

INTRODUCTION

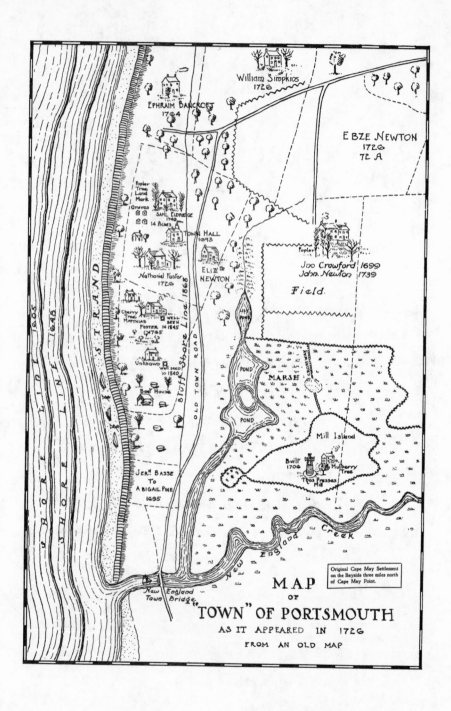

MAP
OF
"TOWN" OF PORTSMOUTH
AS IT APPEARED IN 1726
FROM AN OLD MAP

Original Cape May Settlement
on the Bayside three miles north
of Cape May Point.

William Simpkins
1726

Ephraim Bancroft
1754

EBZE NEWTON
1726
72 A

Poplar Tree
Land Mark
Graves
Saml. Eldredge
1740
14 Acres

TOWN HALL
1693

Poplar

Jno Crawford 1699
John. Newton 1739

Eliz th
NEWTON

Nathaniel Foster
1726

Field.

LILY
POND

Cherry
Tree
Matthias
Foster
1745
WELL
SEEN
IN 1845

POND

MARSH

POND

Unknown Well seen
in 1840

POND

Mill Island

Built
1706

Mulberry
Tree

Boat House

Thos Presses Mill

Jerm. Basse
To
Abigail Pine
1695

New England Creek

SHORE LINE 1605

SHORE LINE 1646

THE STRAND

Bluff Shore Line 1866

OLD TOWN ROAD

New England
Town Bridge

OLD CAPE MAY

When one speaks of Old Cape May it is not to be inferred that there is any definite barrier between the old and the new, but rather that, along with the accompaniments of the modern seaside resort which are common to the long line of summer towns that dot our New Jersey coast, there are, about Cape May, characteristics which are distinctive and which bear evidence of a history dating back for many generations, full of the rigorous life of fishermen, mariners and whalers. It is these old associations, these constant suggestions of residence long-established and of a long-maintained communion with the ocean, that constitute the charm of Old Cape May.

We delight in the heavily shaded streets of the town and in the farms and woodlands stretching down to the salt water—the linking as it were of upland and sea. The old fashioned weatherboarded houses, wind swept and bleached, with their steep gables and dormer windows, and their colonial doorways with sometimes a low-backed bench on either side of the stoop, recall days long passed and a type of architecture discarded, for the most part, today.

Towering above nearly every old house, be it in town or country, is a silver poplar, its smooth gray branches stretching up like the arms of some great supplicant giant and its deep green foliage ever quivering in the breeze, the white under surfaces of the leaves forming a silver frosting to the darker green. There are hedge pear trees, too, about the old houses, loaded with fruit until the branches break with their burden, and pink and white althea bushes with red tiger lilies and blue larkspurs in the gardens, the flower beds and paths all bordered with rows of conch shells. The old wagon road across the farm, with its fringe of bayberry and sweetbrier—the latter long established, leads down to a patch of salt hay which may not be cut till midwinter, or to a little landing at the head of some salt creek by which the oyster bed can be reached, while oyster tongs and seines hung about the farm buildings or a basket of duck decoys, bear further testimony to the

(3)

close association of land and sea which makes every farmer a fisherman or a bayman at the proper season.

These are the things that go to make up Old Cape May and create that distinctive atmosphere which lures one back season after season. There is also the charm of the seafaring dialect and its application to the mainland. We find ourselves getting "to louard" of a patch of woodland in our ramble across the country just as naturally as we do when we skirt the shore in a boat; and we knew a venerable bird dog who "came to anchor" when so ordered, just as other dogs "charge" or "stand." Then too the early isolation of the community and a subsequent conservatism have held the descendants of the early settlers to the Cape, where they comprise no small part of the population. And with them are preserved old traditions and stories of early adventure.

As one falls under the spell of Old Cape May and breathes in the atmosphere of the place, one can almost picture Captain Cornelius Jacobese Mey with his sturdy followers entering the Bay, in 1623, and bestowing his name upon the Cape which it has retained ever since even though the spelling has been slightly changed. Then we see David Pieterson de Vries landing in 1631 on the Bay shore and establishing his colony near Cape Henlopen, only to find it a year later annihilated by the Indians.

Probably as early as 1640 came the whalers from Long Island descended, in turn, from those of the New England coast. They seem to have established the first permanent colony, bringing with them many traits of the New England fisherfolk. They gave the name of Portsmouth to their little settlement and called the creek which there enters Delaware Bay, New England Creek. Many of the family names now most common in Cape May are of New England origin and two descendants of the Pilgrims of Plymouth, who came with the early voyagers on the *Mayflower*, have provided Pilgrim descent to many present day residents.

An ancient map (see inside cover)*presents the colony at Town Bank, apparently at the height of its prosperity, while the shore lines of different dates show how the battering waters of the Bay have worn away the bluff shore until no trace of the settlement remains, not even the graves of the pioneers. (cf. Maurice Beesley, Early History of the County of Cape May, 1857, and Lewis T. Stevens, History of Cape May County, New Jersey, 1897.)

And what of the bird life of the Cape? The proximity of the woodland and upland pastures to the salt marshes and the beach offers unusual conditions in that it brings into close association several very different groups of birds, making it possible to observe a very large number of species in a single day. On September 1, 1920, by way of example, I identified no less than eighty-six species on a morning's walk from Cape May to the Point while on their joint Christmas census the members of the Delaware Valley Ornithological Club have, on several occasions, recorded upwards of ninety species and on December 22, 1935, reached 111. On September 7 and 8, 1935,

---

* The maps which were printed on the inside book covers in the original edition of this book are reproduced in this Dover edition on page 2 and following page 5.

five men identified 113 species and on October 27 of the same year members of the Audubon Association, on their annual trip to the Point, reported 123. The southerly location of Cape May, lying as it does seventy-five miles south of Philadelphia in the same latitude as Washington, D. C., attracts a number of southern birds, either summer residents or casual visitors, which find, in the low pineland and swamps, conditions similar to those of their haunts in Virginia or the Carolinas. The Gnatcatcher and the Mockingbird breed here and possibly too the Chuck-will's-widow and Yellow-throated Warbler which are often present in summer. The Wood Ibis, Louisiana Heron, Purple Gallinule, Mississippi Kite, Red-bellied Woodpecker and Gray Kingbird have occurred as casual visitors. The Arkansas Flycatcher and Lark Sparrow, stragglers from the west, have, on several occasions, found their way to Cape May.

Then again there is its position at the southernmost tip of a long peninsula, which makes it a favorable spot for the study of migration. Autumn flights come, according to some opinions, down the coast and down the Delaware River Valley to join forces at Cape May Point, or if, as others think, the migration is on a much broader front, then with winds from the northwest the whole flight is gradually blown coastward until concentrated at the same spot. Then if strong northwest winds still prevail the migrants hesitate to cross the waters of the Bay and the whole vicinity will be alive with birds until the wind abates when they drift away again on their course. While the absence of such large bodies of water as Barnegat Bay, to the northward, is unfavorable for the occurrence of so great a variety and abundance of ducks and geese as prevail there, the shallow sounds of the Cape furnish wonderful feeding grounds for the shore birds.

The charm of the whole situation, due to these various conditions, is that one never knows what he may see. There is always the element of uncertainty. Some straggler from the south or north may come down that great seacoast way farther than any of his kind has come before; some waif may be driven in from his true home far out on the ocean; or we may awake some morning, in autumn, to find that with a change in the wind, the whole country is deluged with birds, where but a few individuals had been seen the day before. These are the things that lure the bird-lover to Cape May and make an intensive study of its bird life so interesting.

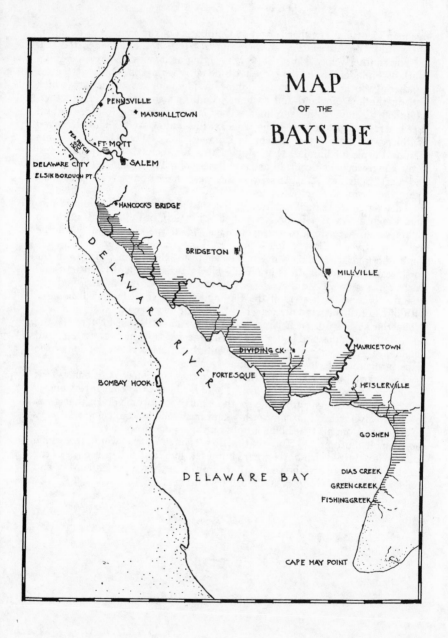

MAP

OF THE

BAYSIDE

PENNSVILLE
MARSHALLTOWN

PEA PATCH
FT. MOTT
DELAWARE CITY
ELSINBOROUGH PT.
SALEM

HANCOCKS BRIDGE

BRIDGETON

MILLVILLE

DELAWARE RIVER

DIVIDING CK.
FORTESQUE

MAURICETOWN

HEISLERVILLE

BOMBAY HOOK

GOSHEN

DIAS CREEK
GREEN CREEK
FISHING CREEK

DELAWARE BAY

CAPE MAY POINT

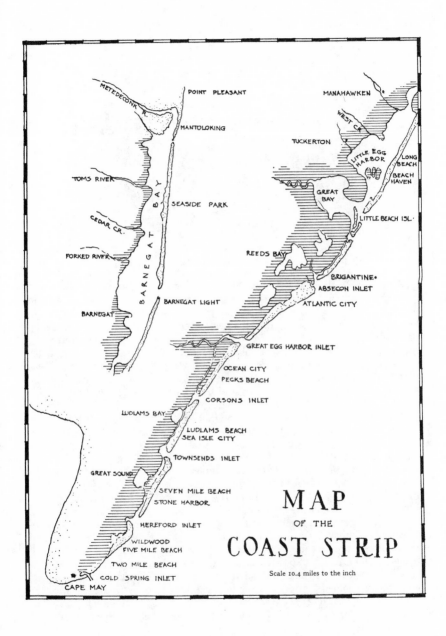

MAP

OF THE

COAST STRIP

Scale 10.4 miles to the inch

BIRDS OF THE SEA

One who strolls along the sea beach at Cape May during late June and early July will gain the idea that both strand and ocean are almost devoid of bird life. An occasional Osprey may go flapping out to sea in search of fish; a stray tern or gull, from some nesting colony not too far distant, may pass along the coast at this season; but most of the breeding birds are engaged in incubation or in patrolling the vicinity of their nests. Since the transients have not yet appeared, our bird list for the beach is likely to be a blank. We are wont to regard the winter months of December and January as the period of lowest ebb in our land bird fauna and the fact that, in this respect, June and July seem to constitute the "winter period" in our sea bird life, only illustrates once more the inherent difference both in origin and controlling conditions between the bird life of the land and the sea.

In July, however, long before the southward migration of land birds has started, we begin to see evidence of this movement along the beach front. Little bunches of Semipalmated Sandpipers and Ring-necked Plovers appear on the strand, the former running rapidly along the edge of the receding waves darting their bills right and left as they pick up various minute crustaceans and other marine animals; the latter standing like Robins on a lawn as if listening for their prey, and after darting forward a few steps to retrieve some choice morsel of food on the wet sand they once more come to rest. Their numbers increase during August until the shore is lined with flocks of them, which take wing on our approach and whirl out over the surf only to come back to the beach a little farther on. About the first of August, too, they are joined by flocks of the larger lighter-colored Sanderlings, which venture farther into the water, wading boldly in until their legs are completely submerged. Now and then several Turnstones with their mottled plumage and short coral red legs, will be found in a flock of Sanderlings, and once in a while some Willet or a straggling Black-bellied Plover will be seen on the beach, but the first three are the only common species of regular occurrence. The delicate sandy gray, snowy-breasted Piping Plover which formerly nested on the Cape May strand now appears regularly in small numbers, as a transient, piping its bell-like call in the spring flight but silent on the return journey, and always frequenting the dry white sand of the upper beach where its similar coloration renders it almost invisible, like the pale ghost crabs which inhabit the same spots.

In August, too, the sea birds begin to assemble in flocks to rest on the beach, especially just above Cape May Point, where they are within easy reach of the fishing grounds out at the mouth of the Bay. A flock of terns assembles there soon after the first of the month, increasing in size until it may number several hundred. It is composed mostly of Common Terns, but there is a good proportion of Black Terns, and not a few Least. One would suppose that these Common Terns represented the breeding colonies which are to be found a little farther north along the Jersey coast, but we know that the Black Terns do not breed nearer than the shores of Lake Erie, and if they come from such a distance there is no certainty that the Common Terns do not do so as well, and quite possibly the breeding terns of the New Jersey coast bring up the rear of the migration. The Laughing Gulls begin to assemble on the beach at the same spot as the terns, during August, and are later joined by straggling Herring Gulls; while occasional yellow-legged Ring-billed Gulls may be seen wading in the water here and there along the beach. From late July on there is a constantly increasing assemblage of sea birds busy fishing at the "ripps" where the waters of Bay and ocean meet—a whirling maze of wings, circling and diving or settling on the water. The glasses will disclose the dark-backed Laughing Gulls, the large gray and white adult Herring Gulls and the dusky birds of the year, the silvery white Common Terns and the smaller Black Terns in a variety of transitional plumages, and the Ospreys which appear perfectly black in the distance and which, in the opinion of some of my fisherman friends, are as much "sea birds" as "land birds"!

By November the terns and Laughing Gulls have usually moved off to the south and the delicate little Bonaparte's Gulls come from the north to take their place, they and the Herring Gulls, largely augmented by later migrants, now make up the feeding throngs. Late autumn and winter, however, bring other additions to the bird life of the sea. The black and white-winged Gannets which fly high above the water and make a peculiar diagonal dive, the Loons which float solitary and peacefully on the rough water and are past masters in the art of diving, and the streams of black Scoter Ducks, which drift along over the water far out to sea like long whisps of black smoke, or cover the surface of the ocean like great mats of floating seaweed. One can hardly realize that so many thousands of any species of large bird still persist. Then, there are the V-shaped flocks of honking

Canada Geese which come in November crossing overland to rest on the sea out beyond the breaker line. All of these as well as some other species of ducks will linger offshore so long as the temperature remains moderate and move south only when intense cold and ice in the Bay force them to seek more open water.

There are other sea birds, too, which rarely or never come within range of the beach. Offshore fishing parties during the summer regularly encounter the leaden-black, white-rumped petrels—Leach's and Wilson's, and several species of shearwaters, and in winter the Kittiwakes occur about the fishing banks, but to the shoreman knowledge of these is confined to stray specimens washed up on the beach, and in this category must be placed the Little Auk and the Razor-billed Auk, winter visitors from the coast of Greenland, and the Sooty Shearwater, a large petrel which breeds far south in the Antarctic.

As one gazes out from shore over the endless stretch of tossing waves, it seems as if there must always be a chance of sighting something unusual, so great is the waste of waters, with apparently no barriers to hinder visitors from remote seas. And so it is that a view that may seem commonplace to one who sees it day after day is always fraught with possibilities to the bird-lover and I never gaze upon the ocean without that feeling of expectancy. When one does catch a glimpse of a passing gull far out to sea or a bunch of ducks or flock of shore birds passing down the coast, beyond the surf, the time that they are in sight is so very brief that we realize that had we been a moment later we should have missed them, and we cannot but consider how many, many sea birds the casual observer actually does miss, and what a small period his observations cover.

BIRDS OF THE OPEN

The extent of sky as we near the sea always appears so vast, compared with inland spots where it is cut off by wooded hills and mountains, that it seems as if the great outdoors had been suddenly expanded. At Cape May, on the one hand it stretches far away to a point where it meets the ocean, either in a sharp line of tossing waves or in a misty haze that seems to blend them together, while on the other side it covers wide marshes and open fields flanked by an irregular barrier of low green woodland far in the distance.

One never looks out over either land or sea that there are not birds of one kind or another circling overhead or passing leisurely across the sky. Over the town, in spring or early summer, there are always Martins circling about mounting higher and higher until they are mere specks against the blue vault, or dropping low over the marshes where they skim at great speed above the top of the sedge. With them are the smaller Chimney Swifts, with narrow saber-like wings which flutter so rapidly that the eye cannot follow the pulsations or are set in bow-like curves as they cleave the air.

Up about the Harbor or back over the hayfields, where boat houses or farm buildings respectively furnish nesting sites, are the more graceful Barn Swallows, with their deep-forked tails, lilting this way and that low down over water and land in erratic flight. Later they line up with their young on the telegraph wires and in August go drifting in endless procession down the beach. In late summer and all through the autumn come the myriads of Tree Swallows, gathering in masses on the bayberry bushes like clusters of fluttering butterflies, and completely covering the telegraph wires, packed close together, shoulder to shoulder, until they look like heavy ropes festooned from pole to pole. Then in a moment they are away again in a cloud, twisting and turning in an intricate maze as they mount higher and higher till they seem like a swarm of bees and then drift with the wind and spread out over the marshes, gradually to reassemble at the starting point.

Of the larger birds of the air there are the Laughing Gulls which are constantly passing over the town, from the nesting grounds a little to the north, to the "ripps" at the mouth of the Bay, where they feed. It is a constant surprise and delight to glance up between the tops of the shade trees and see these graceful gray and white sea birds passing overhead. Sometimes in August, over the salt meadows between Cape May and the Point, they gather in large numbers and perform intricate aërial evolutions rivalling those of the Tree Swallows, for what purpose, except sheer joy of flying, I have been unable satisfactorily to determine.

The Fish Hawk is another large bird of the open, which is almost constantly in sight. They are usually seen singly and their customary line of flight is at right angles to the beach; to and fro they go from the ocean to their nests on the mainland, carrying fish to their young. Their heavy, labored flight is easily recognized though when sailing or circling on set pinions they somewhat resemble soaring gulls. Late in summer when the young are on the wing half a dozen sometimes sail in company winding higher and higher in the air until they appear no larger than swallows.

The third big bird of the air is the Turkey Vulture, past master in the art of soaring. Half a dozen or more may be seen far back on the horizon drifting lazily round and round under the blazing sun of midsummer, and occasionally, when the wind blows strong, they come low down over the trees and housetops or out over the green marshes. In autumn large numbers are to be seen in the air for hours at a time, now circling in close association, now passing on like airplanes in battle formation, directly in the teeth of the wind with never the movement of a feather.

In autumn, too, come great flights of hawks which circle about over Cape May Point, hundreds being in view at once, mostly the Sharp-shinned Hawk, but representatives of nearly all the hawk species of the East are present in the assemblage. Every few days one or two Bald Eagles come down the Bay sailing about among the lesser birds of prey with impressive, majestic flight, their pinions stretched abruptly straight away from body with none of the gentle curve of the Vultures' wings and with white head and tail glistening in the sunlight.

Such are the characteristic birds of the air in summer and autumn as one views them from the shore or the marshes. Other birds, of course, cross continually in passing from one spot to another, Green or Great Blue Herons, Black Ducks, Crows, Blackbirds and Meadowlarks, but they belong more properly in other categories and there are other sea birds, too, but they likewise form a group by themselves. In late autumn and winter Turkey Vultures are still present sailing proudly far overhead. The larger Herring Gulls have taken the place of the Black-heads but stay more closely to the water, while a solitary Marsh Hawk may be seen quartering back and forth over the brown meadows, its snow white rump gleaming in the sun as it wheels in the air and, so long as there is open water, the V-shaped lines of Canada Geese with broad, slow-beating wings may be seen passing overhead.

The smaller birds which may now be flushed from the ground and rise into the air are the Horned Larks and Pipits, Goldfinches and sometimes Siskins, all in drifting flocks, now present now gone, according to weather conditions, and the long straggling flights of Meadowlarks passing across the sky from marsh to upland in the yellow glow of the winter evening.

BIRDS OF THE TOWN AND FARMLAND

Besides the "overhead" species the bird life of the southern tip of the Cape May Peninsula may be divided into three groups: birds of the town and farmland, birds of the salt meadows and birds of the Point.

The shade trees and garden shrubbery have attracted to the town a certain number of species which normally belong to the thickets and hedgerows, while others, like the English Sparrow and Starling, are always associated as closely as possible with man and his habitations. Next to these ubiquitous species the Robin is today the most abundant bird of the shaded yards and gardens of Cape May, building its nests in the trees along the streets and watching for earthworms on the close-cropped lawns. Later in the summer, when the speckled-breasted young are on the wing, the smooth sward of the golf links just before sunset is thickly populated with Robin families in search of food and, as the dusk of evening settles down, they used to resort to the thick growth of trees on the Physick property to pass the night in company with Grackles, Red-wings, Martins and Starlings. The old gardens, however, shelter other birds as well. Catbirds are frequent, mewing their displeasure if we approach too close to the shrubbery which conceals their nest. Several pairs of yellow-eyed Thrashers also make their home in the town and usually a pair of Cardinals. Some years there is a pair of Wood Pewees or Kingbirds established in one of the groves of shade trees, in gardens along Washington Street, where early in the year Grackles have their nests and here, too, the *cow, cow, cow* of a Cuckoo may be heard well into midsummer and we sometimes catch a glimpse of its long sleek gray body and slender tail as it floats from one tree to another. Doves, a few Flickers and some Song Sparrows and Goldfinches frequent the larger gardens and occasionally a pair of Orchard Orioles, Crested Flycatchers or Red-eyed Vireos, while many Red-wings, and a few Yellow Warblers and Maryland Yellow-throats come in from the marshy areas close at hand. Cape Island Creek formerly wound its canal-like course through the western end of the town, bordered with picturesque hedges of tamarisk—gray green and pink, and furnished an avenue of approach for an occasional Belted Kingfisher or Green Heron from the marshes farther on.

While the Martin has already been referred to among the birds of the air, it is during early summer one of the most conspicuous birds of the town. The nesting boxes scattered here and there about the gardens are scenes of constant animation, and as the birds leave the holes to seek food and recreation, and return again to take their place on the eggs, the air fairly vibrates with the harsh grating calls of the males, as they change places with their mates. In July when the young emerge from the nests and throng the narrow porches, there is a continual pandemonium as the parents constantly arrive laden with food. And when the first effort at flight is attempted to a nearby wire or dead tree top, the whole colony gives voice with harsh gutturals of alarm or encouragement. In autumn and winter flocks of Myrtle Warblers and Juncos invade the town, drifting like flights of wind-blown leaves, and when the marshes are ice-bound the starving Herring Gulls gather in flocks in the streets in search of food.

Out in the open country immediately about the town one finds bird life of a different character. Here the Meadowlark is perhaps the most characteristic bird. As we tramp in early summer through the fields of ripening red-top grass and the prostrate tangles of early fruiting dewberries, the Meadowlarks are constantly taking wing and drifting away on wings now rapidly vibrating, now set rigidly for a protracted sail. Perhaps they will come to rest on the top of some bunch of bayberry bushes or on the telegraph wires along the road, where they give vent to their spluttering notes of protest or to their sweet familiar song. If we stop short as the bird first takes wing we are very likely to find at our feet the domed nest which sometimes holds eggs as late as the third week of July. The Meadowlarks are about as much birds of the marshes, especially smaller ones, as they are of the upland and when the young are on the wing they resort there in considerable numbers, forming into definite flocks of fifty to a hundred individuals which drift about the open country throughout the autumn and winter.

In pastures where the cattle have kept the grass well cropped, or in the open cultivated fields which alternate with the brambly uplands, the characteristic bird of spring and early summer is the Killdeer—the Lapwing of America. Ever on the alert, he sees the intruder while still far away and sounds his shrill cry of fear and warning as he skims over the ground or runs rapidly ahead or, apparently feigning a broken wing, flops about on the ground in pathetic efforts to lure the intruder away from the young hiding somewhere on the ground nearby. Later, in family parties, the Killdeers range more widely and we flush them from moist spots on the upland, from the edge of the marsh or the borders of the Harbor, or from inland ponds. Even in winter a few are to be found in the latter situations and the familiar cry occasionally breaks the silence though at this season they are more likely to be noiseless. We also start other birds from the old grass fields. The familiar Field Sparrow, whose plaintive song may be heard all summer through, flits away to some bramble thicket where he will perch until danger is past, while the Grasshopper Sparrow, springing from beneath our feet in short twisting flight, drops again into the sheltering grass. Early in the season it may be seen anxiously watching from the top of some bush or uttering its buzzing insect-like song from a wire or fence post, but once the young are awing they are distinctly birds of the ground. Then there is the Henslow's Sparrow a characteristic bird of the strip where marsh and upland meet, or of some damp hollow in the briery fields overgrown with reeds and sedges. The male is detected sitting motionless on some dead weed stalk that rises a little above the surrounding vegetation. At intervals he throws back his head and gives vent to a single explosive effort of two notes so closely welded that they may almost be called one; *chisek* he seems to say. As we approach he drops to the grass and is off like a mouse or with a laborious fluttering flight he gains another perch a little farther off. In autumn and winter the Savannah Sparrow inhabits the same areas and great mixed flocks of Red-wings, Cowbirds, Grackles and Starlings roll across the open country as far north as Cold Spring.

The old fields also harbor Bobwhites and, though we may see them but seldom, their clear ringing notes come to us from over the hill and far away borne on the breezes that carry with them also the breath of the sea and link up once more the land and water. As we tramp through the coarse grass our steps constantly impeded by the trailing dewberry vines, an explosion of bird life occurs immediately before us even more sudden than the flushing of a covey of Bobwhites and away go some English Pheasants, usually dull-colored, half-grown young, during the summer months, but brilliant coppery long-tailed adults in autumn. These birds are now thoroughly established but why they had to be added to our all sufficient natural fauna it is hard to see.

Thrashers and Catbirds, Chipping Sparrows and Yellow-throats, constitute the regular inhabitants of the old fencerows which are densely bordered with wild cherries and narrow-leaved crab apple—gorgeous in their springtime glory of white and pink bloom and flanked by masses of blackberries and greenbrier.

Here and there in the upland bramble pastures are low marshy spots or ponds, and of these my favorite was the old Race Track Pond (now completely drained). Its dark waters were studded with white and gold pond lilies and bordered by bayberry bushes and wild roses. At one end was a dense cattail growth over which there were always to be found anxious Red-wings, the males flaunting their brilliant epaulets as they hovered overhead voicing their protests, and the females flushing with a frightened chatter from the waving sea of narrow dark green leaves which sheltered nests and young. This pond too, was visited by Green Herons, Killdeers and Spotted Sandpipers and during late summer by various migrant shore birds. Some years, too, a pair of King Rails and a pair of Bitterns nested in the shelter of the cattails and a bunch of Black Duck would drop down in passing and rest among the lily pads. The destruction of this and other similar ponds is fatal to the bird life that once thronged them and a sorrow to all bird students. Other cattail thickets closer to the border of the salt marsh were the home of the Long-billed Marsh Wren. The spluttering song of the bird which, if it does not resemble "water flowing from a bung hole," as has been said, surely has something very liquid and bubbling about it, which seems to suit its environment. Following the sound in among the cattails we may see the tiny performer with tail cocked over his back fairly boiling over with song, as he clings, swaying up and down, to the stalks of the cattails and nearby is the globular or ovoid nest of plaited grass with the tiny entrance hole on the side. The Short-billed Marsh Wren, too, a bird unknown to both Wilson and Audubon, occurs on the border of the upland and marsh, usually in patches of fine marsh grass (*Spartina* or *Distichlis*) amongst which its nest is supported, smaller and rounder than that of the Long-bill and close to the ground.

It is hard to differentiate definitely between the birds of the uplands as contrasted with the open marsh on the one hand and the wood edges which border them on the north and west, and one species, at least, the Song Sparrow, seems equally common throughout, but with it our list of the birds of the open upland seems complete. Migration, of course, brings to this as to other areas many visitors from afar; Snipe and Woodcock in the damp spots and hosts of small birds to the hedgerows and thickets. They are mostly silent transients, however, and do not give character to these brambly, grassy uplands which are ever associated in my mind with the wild plaintive calls of the Killdeer, the spluttering flight song of the Meadowlark and the dreamy trill of the Field Sparrow—the three dominating summer species.

BIRDS OF THE SALT MEADOWS

It is hard to conceive of two regions so absolutely different in plant and animal life as the salt marshes, or "meadows" in the language of the natives, and the adjacent uplands. Lying side by side and within a stone's throw of one another they have scarcely a species in common. The difference in plant life is not so surprising as plants are subjected to the direct action of the salt, and so too, the lower forms of animal life, but we marvel to find the birds also so sharply divided. An occasional far flying land bird—swallow or bird of prey, may course over the meadows, or a flock of shore birds may whirl over the land but each returns at once to its own environment and only a few species such as herons, Ospreys, or Eagles, which nest on the upland and feed on the marsh, may be regarded as overstepping the imaginary line. When protracted rain storms flood the lower areas of the uplands and form temporary shallow pools, flocks of shore birds and gulls frequently gather there to feed but leave again as soon as the feast is over.

The meadows began originally just east of the Lighthouse and extended to the edge of the town but most of this area has now been filled in and is a dry waste of weeds or bayberry bushes. Back of the town in an ever widening stretch and following the course of Cape Island Creek, they reach the Harbor at Schellenger's Landing and thence in their full glory they sweep northward for more than ninety miles to the head of Barnegat Bay reaching a width of nearly four miles back of Seven Mile Beach and nine miles at Brigantine, just north of Atlantic City. (Plates 2 and 4.)

There is a charm about the meadows at all seasons whether they are clad in the green of summer, with still blue waterways interspersed, or wrapped in the dull brown of midwinter, with channels and ponds ice-bound. It is in spring, however, that they especially appeal to me, when the mainland is bright with opening buds and everywhere tinted with green and crimson but the meadows still apparently in the grip of winter with scarcely a sign of spring vegetation. To offset this, however, they are all astir with the hosts of migrating shore birds and with flocks of gulls and terns *en route* to their breeding grounds. Here on many occasions I have spent several days at a time in one of those little weather-beaten gunning shacks that stand on stilt-like legs far out on the edge of some sound or thoroughfare. To quote from my journal of one such trip:

May 21, 1921. It was just sunset as we pulled our boat out from Weeks' Landing under the two silver poplar trees and floated slowly down the little creek and on out into the wider thoroughfare. It was nearly high tide and the water reached almost to the top of the banks, so that we had an uninterrupted view over the wide stretches of salt meadows. Far away to the east across the open sound was the low-lying shrubbery of Two Mile Beach with a glint of sand dunes here and there amidst the green, and the roar of the surf beyond. To the west lay the long border of the mainland woods growing black in the shadows. Great masses of kelp with its countless floating bladders lie in the water just below the surface as we slip quietly along between the slender stakes that mark the location of the oyster beds.

There are many clouds on the western horizon but the fiery red orb of the sun comes into view for a moment between their dark masses, and lights up all the water and land to the eastward while across its disc, as if traversing a field of fire, a couple of Crows pass by. Almost immediately it disappears again and sinks below the horizon, but for fully ten minutes the fleecy clouds overhead and to the eastward are all edged with crimson and pink, the color constantly spreading and changing, and all its varied tints are clearly reflected in the smooth glassy waters of the channel. The two poplars at the landing and a little oyster shack supported on poles become black silhouettes against the fading light and the far off woods are now merged with the shore line in a uniform dusky band. The full moon already risen a little above the horizon, is growing brighter with every moment as the sunlight fades, and makes a long path of yellow light on the water, broken by the ripples of a passing breeze. In the gathering dusk the last wild flights of little sandpipers wheel over the meadows; we hear the cackle of the Mudhens among the taller grass; and solitary Night Herons come winging their way across from the woods on the shore to drop into some shallow pool for their night's feeding. Then to the south there flashes out, at regular intervals, the brilliant yellow star of the Cape May Light and night settles down over the meadows.

May 22.—At dawn all the light effects are reversed. At 4:30 it is already quite light, and the crimson tints on the fleecy clouds almost exactly duplicate the effect at sunset, the glow along the horizon increases and then gradually the great orange red ball comes up over the sea. In the gray light of early dawn a Barn Swallow comes twittering over the meadows and here and there a Night Heron, weary of his long vigil, is rising and beating his way back to shore. From far away in the mainland fields comes the clear whistle of the Bobwhite, and an early gull flying past us breaks into his harsh grating laugh. As the sun's rays lighten the landscape and warm the air flocks of sandpipers begin to take wing leaving the flats, which are rapidly being submerged by the rising tide, to seek the meadows and the higher sand stretches and beaches. Later on, before the tide begins to ebb, we have anchored our skiff in the middle of Jarvis Sound, where a few wisps of marsh grass showing above the water indicate where the shoal will first appear, and await the coming of the birds to their feeding ground on the mussel beds. It is very still, the gray water of the sound is scarcely rippled. Beyond it stretch the yellow green meadows, with brown splotches where the dead grass of last year is still in evidence and bands of deeper green where the tall grass, bordering the little creeks, has attained a ranker growth, while on the dry shell heaps left by the channel dredges there are some marsh elder bushes of a peculiarly vivid green. Farther still is the low gray green woodland of Two Mile Beach broken here and there by protruding crests of high sand dunes glistening silvery white in the sun and back of all the haze of the ocean and the far off boom of the surf.

Southward stretch the meadows again, with scattered shallow ponds here and there and in the distance a mass of billowy green foliage spreading over the town, against which white houses stand out here and there, with church spires and thin columns of smoke reaching skyward, and far behind it all the Lighthouse tower white in the bright sunshine. The whole western horizon is lined with dark woodland cut here and there by cultivated fields. Green Herons flap lazily past us as we lie there with the water lapping the sides of the skiff and Laughing Gulls wing their way across the sky, while little flocks of Semipalmated Sandpipers flash into sight as they turn their snowy breasts to the sun and then disappear instantly as their dark backs blend with the background. Far off on the Harbor a solitary Loon floats on the glassy water while nearby several belated Red-breasted Mergansers are playing, diving and swimming about under shelter of the muddy banks. One would not suppose that there were any shore birds on the meadows when suddenly a small flock of Dowitchers wheels about over the grass. They have located the spot exactly but are a little too early as the water is still several inches deep and away they go again. Other little bunches pass over and finally when the tops of the largest shells begin to show above the surface the first birds, four Semipalmated Sandpipers, drop down although they are forced to stand in water nearly up to their bodies. They gather close together and stand perfectly still, now a Dowitcher joins them, then ten more Semipalmateds, and four of the same, seven more Dowitchers, and three Common Terns, then a Yellow-legs far off by himself in deep water jerking up his body in characteristic manner, six Ring-necks, four Turnstones, two and four Semipalmateds, fifteen Black-bellied Plover, a Turnstone, two Curlew, six Turnstones and so on until in less than an hour some three thousand birds are scattered over the flat. The species already mentioned were the most numerous but there were a number of Red-backed Sandpipers, many Least, and a few Knot. At first most of the birds remain resting, the Dowitchers often on one leg with head bent backward over the shoulder. Soon, however, as the flock increases and the exposed flat widens, the smaller birds begin to feed actively and before long all are busily engaged probing among the beds of mussels or dabbing their bills in the soft black mud in pursuit of prey. (Plate 43.)

Every now and then the clams or razor shells, buried deep below, eject water suddenly from their siphons showering the birds with spray like miniature fountains and cause them to leap into the air in alarm. Stray terns and gulls alighting here often cause many of the smaller birds to take wing but they settle again almost at once. The scene is one of the greatest activity and a constant murmur of soft calls and trills, like a continuous conversation, comes up from the assemblage. The contrasts of color against the black mud is striking, the jet black and snowy white of the Bullheads (Black-bellied Plovers); the mottled black white and maroon chestnut of the Calicobacks (Turnstones) with their bright red feet; the rusty back and black bellies of the Red-backed Sandpipers; and the more modest studies in brown and gray presented by the other species. Gulls, both Laughing and Herring, gather on the larger flats to feed on small fresh fish or dead crabs and fishes, as the case may be, and Ospreys and an occasional Bald Eagle pass overhead.

There were one hundred and ten birds on our flat when it measured but ten by thirty feet and by the time it reached one hundred by seventy-five there were nearly one thousand. The feast is over and the tide begins to rise and away go the birds one small flock after another to seek roosting places about the shallow ponds, or perhaps back on the upper beaches, and once again the water, creeping up, spreads over the mussel beds and the white shell heaps until naught is left but the tips of marsh grass waving above the ripples. One wonders how these birds of passage whose stay is so brief know so accurately not only the place, but the time, for their feast, especially as the latter is not a function of daylight, which affects the activities of land birds, but of the wholly independent rise and fall of the water and this again brings up the question as to whether these birds of the meadows feed also at ebb tides which happen to fall in the night? There is still much of mystery about their lives!

When full moon and high tide coincide the water will rise until it covers the entire meadows from coast islands to mainland and the familiar expanse of green becomes essentially an arm of the sea. We awoke one night during our stay to find ourselves facing such a flood. Our skiff left stranded on the bar, is afloat once more, the steps of our stilted shack are completely under water and the waves lap against the floor. Nearer to town the waters creep silently up until the lower portion of the golf links are flooded as well as low lying gardens and pasture lands. Strong northeast winds produce similar conditions when at flood tide they blow the waters of the inlets back into the sounds and over the meadows. These floods are disastrous to the nesting birds of the meadows and many times the entire tern, gull and Skimmer colonies are washed out and eggs and young scattered far and wide to furnish food for the scavenger Herring Gulls. Sometimes several nestings are destroyed in a single season and there have been years when the Black

Skimmers and Laughing Gulls, in spite of numerous attempts, have failed to raise a single brood.

Scattered all along the coast between the mainland and the island beaches are numerous sounds and bays communicating with the inlets and thoroughfares and dividing the meadows into various sections quite separate from one another. Beginning on the south there was originally Cape Island Sound, now entirely filled up, then to the north of the Harbor come Jarvis Sound, back of Two Mile Beach; Richardson's and Grassy Sounds, back of Five Mile Beach; and Jenkins Sound, Great Sound and Stites Sound, back of Seven Mile Beach. Back of Ludlam's Beach are Townsend's Sound, Ludlam's Bay and Corson's Sound, the last really back of the inlet of the same name; and back of Peck's Beach, lie Peck's Bay and Great Egg Harbor, the largest of all. This brings us to the northern boundary of Cape May County.

Farther north come Scull's, Lake's and Absecon Bays, back of Absecon Beach; Reed's, Grassy and Little Bays, back of Brigantine Beach; Great Bay, back of Island Beach, and finally Little Egg Harbor, Manahawkin and Barnegat Bays, back of Long Beach. These are all resorts for ducks and geese throughout the winter as well as in time of migrations, while gulls and terns and other waterfowl course over them in search of food, in their appropriate season.

THE COASTAL ISLANDS

The eastern edge of the salt meadows is bordered by narrow sand islands
all the way from Cape May to upper Barnegat Bay where the most northern
of these barriers forms a long peninsula reaching from Bayhead to Barnegat Inlet. They form a check to the inroads of the sea except where numerous inlets break through. Considering these islands from Cape May northward we find that the shore line from the Point to the Harbor is now intact,
the former Cape Island, upon which most of the town is situated, having
long since been joined to the mainland on the west, while openings that at
one time or another existed near South Cape May and at the end of Madison
Avenue have closed up. These gave outlets to a branch of Cape Island
Creek and to Cape Island Sound, respectively, and the strip of beach lying
between Madison Avenue and Sewell's Point, at the mouth of the Harbor,
was known in former times as "Poverty Beach." To the north of the Harbor came Two Mile Beach, until about 1925 a wild beach with no inhabitants except the crew of the Coast Guard Station. At that time, however,
Turtle Gut Inlet, which marked its northern extremity, was filled in and Two
Mile became a part of Five Mile Beach while a railroad terminus and a fish
dock were established upon it. There were formerly several extensive
thickets of low trees and bayberry bushes on this island which formed a
breeding place for many Green Herons and Fish Crows, while Black Ducks,
Bitterns, Woodcock, and such passerine birds as Red-winged Blackbirds,
Tree Swallows, Maryland Yellow-throats, etc. nested there.

Five Mile Beach was originally covered with a splendid forest and boasted
of a few farms and many wild cattle but with the establishment of Wildwood,
Holly Beach, Anglesea and other summer resorts all has changed and the
entire island has been built over until not an original forest tree remains.

Lewis T. Stevens, writing of this beach (History of Cape May Co.)
says: "About fifty acres of Five Mile Beach are in woods, grand timber,
some of the trees being nearly one hundred feet high, white and black oak,
sassafras (six feet in circumference), red cedar, holly, magnolia, wild cherry,
persimmon, sweet gum, plum, etc. and from the branches of many of them

hang festoons of beautiful green moss (Usnea) three to six feet in length. Gigantic grapevines here flourish, one monster, nearly a yard in circumference ten feet from the ground, spreading away over the branches of the oaks a distance of two hundred feet. In the center of the forest is a charming little body of fresh water about three feet in depth, fed by a small stream that rises a mile or so away." This was the condition about 1890.

Of the "wild cattle" much has been said. Charles S. Westcott ("Homo") writing in "Forest and Stream," August 27, 1885, says, "They are descended from blooded stock escaped from a wrecked English brig that came ashore many years ago. They frequent the wooded meadows, wild as deer, and regularly hunted with dogs, twenty or thirty of which are kept for the purpose. I saw the herd last week on the surf line, headed by an old bull, endeavoring to rid themselves of the green-heads and mosquitoes." Previously (January 10, 1884) he had written that "William Kern of Forest Hill, Cumberland Co., had purchased all the wild cattle and will shortly begin the work of running them down. Bulls and cows with calf are dangerous to approach. A few years ago Gladding Brothers of Philadelphia bought the right to shoot them but gave it up as too expensive and dangerous." Judging from the note of 1885, Kern also gave up the attempt but the animals were eventually destroyed. While this explanation of the origin of the cattle is current legend, others with whom I have talked discount the story of the wreck and say that it was a common practice to take young stock over to both Two and Five Mile Beaches for the summer and they think that the herd originated from local cattle that were never recaptured.

Philip Laurent, who had a cabin at Anglesea, at the northern extremity of the island, has published an account of the bird life (Ornithologist and Oölogist, 1892), the earliest comprehensive account of Cape May County birds since the scattered records of the pioneers.

Seven Mile Beach, the next to the north, now the site of Stone Harbor, Peermont and Avalon, maintained its natural features for a number of years later. This island was a veritable paradise for birds back in the seventies and eighties of the last century. Here were the highest sand dunes on the coast with dense strips of forest wherein nested herons of several sorts, Fish Hawks, Fish Crows and various smaller birds, while on the broad beaches nested Common and Least Terns, Black Skimmers, Piping Plovers and doubtless other species of which we have no record. While a few terns and Skimmers nested at the extreme southern point of the island in 1936 and a few Fish Hawks, Crows, and Green and Night Herons still frequent what is left of the woodland, the ornithological glory of Seven Mile has departed. Uncalled for development and draining has left only a barren waste of sand while trees have been cut, many of the tall sand dunes destroyed, and what few ponds remain poisoned with oil. Had we but realized what was to happen the entire island could doubtless have been saved as a wildlife sanctuary and its abundant bird life preserved forever. Charles S. Schick began an account of the birds of this island (Auk, 1890) but it was never finished. (Plates 4, 5 and 8.)

Ludlam's Beach, the next in line, still presents spots where terns nest and where transient shore birds may be studied, as on Seven Mile, but the populous resorts—Townsend's Inlet, Sea Isle City and Strathmere are year by year encroaching upon them. (cf. J. Fletcher Street, Cassinia, 1931–32.) Peck's Beach, the former home of the Snowy Egrets and we know not of how many other interesting water birds, is now entirely occupied by Ocean City and the natural conditions almost destroyed. This was the beach visited by Wilson and Audubon and others who put up at Beesley's Point in the olden days. It brings us to the end of the Cape May County beaches. The next beach is Absecon, the site of Atlantic City, and above that Brigantine, the best bird beach of the coast today, which the residents in their wisdom have established as a bird sanctuary. Island Beach next in line is small in extent but still essentially in a state of nature. Finally come Long Beach and the Seaside Park Peninsula, very narrow and supporting but little more than low bushes, with few summer resident species of birds, although rich in transients along its salt marsh borders and on the broad waters of Barnegat Bay which lie behind it.

CAPE MAY POINT

Cape May Point seems like a bit of the Pine Barrens which characterize so much of southern New Jersey, isolated here at the southwestern extremity of the Cape May Peninsula. To the north of the "Point" lie the extensive Pond Creek meadows while to the east there is farm land and below it the head of the Cape Island Creek marsh. The ocean might be said to bound it on the south and Delaware Bay on the west, though just where Bay and ocean join may be a debatable question; shifting from day to day it is usually marked by a juncture of smooth and rough water known as "the ripps."

The Point is almost entirely wooded. As one approaches from the east along the old turnpike the green barrier of low woodland cuts straight across at right angles. First comes a dense scrub of bayberry bushes with low, round-topped pond pines interspersed, and along with these sassafras, persimmon and sumac, the last, as summer advances, showing its bright yellow clusters of pollen-laden blossoms and here and there a bright red leaf forerunner of the brilliant masses of yellow and red which turn the whole place into a dazzling wealth of color in the crisp days of October. There are wild cherry trees raising their heads higher than the rest and in late July heavily laden with clusters of fruit. Wild plum bushes and early goldenrod form a low outer fringe. At other points oaks creep in and chicken and fox grapes cover the entire growth with their broad leaves, while trumpet creepers hang their festoons of scarlet tubes from convenient limbs. About these Hummingbirds constantly dart, seeking food in the gaudy trumpets or chasing one another through the air. Greenbrier and poison ivy help to weave the vegetation into an almost impenetrable tangle, which forms a welcome retreat for numerous Catbirds, Maryland Yellow-throats and White-eyed Vireos. Farther back toward the Bay are taller, more open woods of pine and oak with a low scrub of huckleberries. Here all summer long may be heard the sharp *chewink* of the Towhee and until mid-July his more elaborate song. The drowsy notes of the Field Sparrow seem to be always in the air and the call of the Chickadee and Chippy-like trill of the Pine Warbler come to us now and then from the dense cone-covered tops of the pines. Brown Thrashers scurry over the sandy ground or along some old path through the brush, Cedar Waxwings sigh from the very tops of the tall trees, and from certain dense thickets where tall bush huckleberries grow, laden in late July with fruit in clusters as thick as chicken grapes, we may hear the elusive and varied notes of the Chat and the clear whistle of the Cardinal. About the scattered negro cabins or the bungalows of summer sojourners the cheering prattle of House Wrens greets the ear, the ringing notes of the larger Carolina Wren, and, less frequently of late years, the low warble of a Bluebird or the piping call of a Tufted Tit. It is here, too, in this bit of southern pineland, that we hear the wheezy song of the Prairie Warbler and at intervals have the rare privilege of meeting a Mockingbird, Gnatcatcher, or rarer still, a Yellow-throated Warbler, stragglers or frontier outposts of the bird life of the great pinelands of the South Atlantic States.

There is one stand of pines, taller and denser than usual and sheltered on the west by a ridge of oak-covered sand dunes, where one looks early in August for the first autumnal migrants. Wood Pewees and Great Crested Flycatchers are there all summer through, a pair or two of Fish Crows nest there, and there a Downy or Hairy Woodpecker often seeks shelter. With the first turn in the tide of migration, however, the flashing yellow and salmon of the Redstarts, the black and gold of Hooded Warblers, and the striped backs of the Black and White Warblers will be sure to catch our eye as we peer into the dark recesses of these pines.

Just beyond is Lake Lily a considerable body of fresh water, full of aquatic plants of various kinds from the large snowy stars of the water lilies, with their disc-like leaves turning up in the breeze until their maroon under-surfaces are exposed, down to the little yellow blooms of the bladder-worts, supported on slender stalks above the water. Here Green Herons skulk along under the banks and watchful Kingfishers perch on the dead limbs of overhanging pines. Spotted Sandpipers, and later Solitaries, run along the sandy bars and Laughing Gulls drop down in squads to bathe with much splashing and churning. Here, too, in late autumn and winter, Pied-billed Grebes may sometimes be found floating about or noiselessly diving beneath the surface of the water, while a duck or two or an occasional Canada Goose will come in to rest. A deep pond close to the Lighthouse, known as the Lighthouse Pond, into which flows much sewage and the edge of which has been converted into a public dump, still is attractive to migrant ducks,

Florida Gallinules, Coots, grebes and Green Herons but owing to the cutting off of its connection with Cape Island Creek it is rapidly filling up with cattails and other aquatic vegetation so that the area of open water has been sadly depleted and its attraction to waterfowl will soon disappear. A shallow brackish pond formerly existed just east of the Lighthouse which was a great resort for shore birds and resting gulls but it too has suffered from ditching and draining and is now mainly overgrown with cattails while a large part of it has completely dried up. (Plates 3, 6 and 7.)

Far over on the Bay side of the Point are rows of dunes of loose shifting white sand, broken at one point by the mouth of Pond Creek. Some of these rise to a height of twenty or thirty feet and are covered with a scattered growth of red cedar, pine, oak, holly, persimmon and wild cherry with thickets of beach plum and bayberry. At one spot the operations of a sand plant left a deep depression which has filled with water and constitutes "Davis Lake"; at others the vegetation is very dense and bound together with greenbrier and poison ivy, forming splendid shelter for winter birds. Much of this dune strip has been leased by the National Association of Audubon Societies and designated the "Witmer Stone Wild Life Sanctuary" where all birds, including the sadly persecuted migrant hawks, are offered a haven of refuge. It is here that the great autumnal congregations of hawks, Flickers, Woodcock and Kingbirds occur.

These dune thickets extend northward along the Bay to Higbee's Beach, and Town Bank, while taller woodland continues far to the north. Just above the point where New England Creek enters the Bay lies Price's Pond, formerly an open body of salt water famous for its crabs but, since the closing of its contact with the creek and the shutting off of tidewater, it has become wholly fresh and is largely covered with cattails and other aquatic plants forming an attractive resort for nesting Bitterns, Wood Duck and Black Duck and many summering herons. Up to 1936 this pond had escaped the disastrous draining operations but in 1937 the further check of the water flow of New England Creek has already seriously affected its water level and this beautiful pond seems headed for destruction, the same fate that has overtaken practically every body of fresh water in the county. It seems that wild life conservation and draining policies will not mix and when those in authority decide on the latter much of our wildlife, both animal and plant, must disappear and with it no small part of the glory of the countryside! (Plates 1 and 9.)

BIRDS OF THE WOODLAND

While Cape May Point brings the woodland into close proximity to the sea there is not much woodland elsewhere until one passes Cold Spring two miles above the Cape. There is a wooded border, it is true, along the eastern side of the Pond Creek Meadows and a narrow strip along the ocean side of the peninsula, but the large tracts of forest are located farther north. Some of these are of great extent and furnish retreats for nesting Bald Eagles and Great Horned Owls and, in years gone by, for the Pileated Woodpeckers. The woodland avifauna, however, is not extensive and in winter scarcely any birds are to be found except along the wood edges. Wood Thrushes, Ovenbirds, Red-eyed Vireos and Blue Jays are the most widely distributed species peculiar to the forest in the summer, but the low swampy areas abound in all the species that are common in similar situations at the Point. In the great swamps at Dennisville, and in other similar situations, the Hooded and Parula Warblers are common breeders, with occasional Black and White Warblers. (Plate 1.)

While most of the great forest areas lie to the west of the shore road one of the finest pieces of primeval woodland was located along the edge of the meadows east of Rio Grande; unfortunately most of it was sacrificed to the woodman's axe a few years ago, including many venerable holly trees with trunks over a foot in diameter; another tract lies east of Burleigh. Oaks of several species make up most of the Cape May woodlands but there are frequent stands or scattered trees of pine and many sweet and sour gums, hickories etc. (cf. Stone, Flora of Southern New Jersey, Report of the New Jersey State Museum for 1910.)

J. F. S

OLD MILESTONE FORMERLY STANDING ON TURNPIKE EAST OF CAPE MAY POINT.

THE CHANGING BIRD LIFE OF THE CAPE

Our earliest records show that the bird life of the Cape was always noteworthy. De Vries mentions incidentally that in April, 1633, when he visited the spot for the first time he saw there immense flocks of Pigeons which obscured the sky while Master Evelin who visited the Cape May Indians in 1648 states: "I saw there an infinite quantity of bustards, swans, geese and fowls, covering the shores as within a like multitude of Pigeons and stores of Turkeys, of which I tried one to weigh 46 pounds." It would seem from the latter statement to be as dangerous a thing to record the weight of game birds as to mention the length of a fish—but perhaps Master Evelin's scales were not of the present day standard. We have also the record of Swedish settlers on the lower Delaware visiting the Maurice River near the head of the Cape May Peninsula and killing great quantities of geese for their feathers leaving the carcasses behind them, and of the "great abundance of eggs which the swans and geese and ducks and other wild fowls" laid about Cape May and which were responsible for the names of Egg Island, Little and Great Egg Harbors, though from more recent knowledge the eggs probably belonged to the gulls, terns and Skimmers rather than to the Anatidae.

There is a certain ambiguity in the use of the name Cape May in the early records. The older writers referred to the entire peninsula, nearly coincident with the present county of Cape May, and extending all the way from Great Egg Harbor southward. Then it was used for the county town now referred to as Cape May Court House. The present Cape May, or Cape May City, at the tip of the peninsula some fifteen miles south was not established until 1850 and its site was known to early writers as Cape Island. Therefore many of the early bird records refer to the upper marshes rather than to the Cape May of today, but the same conditions must have prevailed all along the coast and what was said of one beach applies to the others as well.

Up until 1850 this Cape May coast was a delectable spot where gunners and ornithologists found an abundance of all sorts of water birds. It was the favorite resort of Alexander Wilson and George Ord, who went down by water, landing on the Delaware Bay shore, and doubtless may have put up at Higbee's Beach, where a curious old wooden hotel building still stands, close to the dense Bay side dune forest; or they may sometimes have traversed the Pine Barrens crossing with some fish or oyster peddler from Camden to Great Egg Harbor. Audubon made at least one such trip to this latter locality and his experiences are presented in his "Episode," entitled "Great Egg Harbor." Frank L. Burns, who has made a careful study of Wilson's field activities, tells me that he made six trips to the coast of Cape May County the earlier ones in May, late June and December, prior to 1811, and later, accompanied by George Ord, in July 1811, November 1812, and four weeks, May–June, 1813. Most of his time was spent in the vicinity of Great Egg Harbor but part, at least, of the last excursion was spent on Cape Island. Later the ornithologists, John K. Townsend and the Baird brothers —Spencer F. and William M.—visited Cape May; Townsend and William Baird marrying sisters of the Holmes family who lived near Cape May Court House in the old homestead which still remains in the family.

Among Prof. Baird's correspondence is a letter from his brother William, dated Schellenger's Landing, Cape Island, July 16, 1843, which gives some idea of conditions at that time. He writes:—"Gunners are so very numerous that almost all the birds are driven off except the two common species of terns (*hirundo* and *minuta*) the latter very abundant and easily got. I have seen a few Black Terns. I have seen the following: four species of Tern, two Plovers,—*melodus* and *wilsonius*, Red-breasted Snipe, *Tringa semipalmata: pusilla* and *alpina*, *Totanus maculatus, flavipes* and *vociferus* (not a single Willet) one Gull (*atricilla*) *Rhynchops nigra*, several Black-necked Stilts which I chase day after day but cannot get a shot at, Clapper Rail and the two seaside finches, two White Herons, Night Heron. There is no chance at the Herons however. I have put up nine birds, including two Havell's Tern and two Wilson's Plover. Perhaps a Black-necked Stilt (Lawyers they call them here) may give me a chance. Townsend's brother-in-law who lives at the Court House, fifteen miles from here came to see me on Sunday and invited me to come up and spend some time with him. The shooting in the neighborhood is very fine. Dr. Leib of Philadelphia has been there all summer and is making a collection. He knows all the localities and I have no doubt would go out with me. The shooting will probably be better before long as I see the birds are beginning to collect in flocks and pass over-head towards what is called Five Mile Beach, every evening. They spend the day somewhere along the Bay Shore."

Another letter dated June 16, 1845, when Wm. Baird was staying with the Holmes family at the Court House has the following:—"I saw a great many birds especially Hudsonian Curlews, large flocks of which flew over our heads in the evening, constantly, as we were fishing. They were breeding [an error]. I also went to visit a breeding place of the two species of White Herons and Night Herons but the nests were only commenced and no eggs laid. There was a great number of nests; one small tree must have had twenty upon it." In this letter he also gives Spencer Baird instructions as to how to get to Cape May. He writes:—"Go to Cape Island by steam boat and to Court House 13 miles by stage. Mrs. Hand keeps a tavern, board $3. per week; a man and boat can be got for 75c a day." The Cape Island landing was apparently at the end of the turnpike (present Sunset Boulevard) where the Philadelphia excursion boat *Republic* docked for many years.

What a pity that we have no comprehensive account of the bird life of the Cape May coast in still earlier times; but none of the men who could have written it have left us more than scattered fragments which only make us long for more. By piecing together the accounts of various species given by Wilson and Ord on the pages of the "American Ornithology" we can form some idea of conditions at the beginning of the nineteenth century.

Picture to yourself beach after beach, along the coast, with the surf rolling in just as it does today but in place of the lines of hotels and cottages, the boardwalks and electric lights, fishing piers and throngs of bathers, we see only an endless stretch of gleaming strand flanked by sand dunes capped with beach grass and behind these, sometimes partly buried by them, dense thickets of cedar, scrub oak, wild cherry, holly and sumac. There are great colonies of Common Terns with nests scattered about among the dunes and at other points groups of the delicate Least Terns, with their eggs in hollows scooped in the sand just above high water mark. There are colonies, too, on the lower sand bars and flats of the curious Black Skimmers with their black and white contrasting plumage and long red bills, flattened vertically like shears. There were Gull-billed Terns, too, and doubtless Roseate and Forster's or perhaps a few of the great Royal and Caspian Terns though our records are silent as to these latter species. As the intruder advances thousands of birds mount in the air and, circling about overhead in an intricate maze, give vent to a perfect bedlam of harsh cries, dashing about his head in frantic defense of their nests and young. Scurrying across the beaches go the beautiful little Piping Plovers, uttering their bell-like notes, and the larger Wilson's Plovers, first discovered by that famous ornithologist on Cape Island beach where the city of Cape May now stands, and here and there like giant plover a pair of Oyster-catchers run along, always wary and keeping at a good distance from their pursuers. They also nest among the low dunes above high water mark. Back in the cedar thickets are the heronries where nest vast numbers of the beautiful Snowy Egrets along with Black-crowned Night Herons, Little Blue, Green, and Great Blue Herons. The trees are not tall but contain three to four nests each, while all over the ground below are scattered great quantities of egg shells, the result of the depredations of the Fish Crows which are continually hovering over the place. The great White Egrets nest back in the Cedar Swamps of the mainland which border the larger streams flowing down into the sounds.

On the salt meadows, well concealed by the coarse grass, the Mudhens (Clapper Rails) swarm and are estimated to be "more than double in number of all the other marsh fowl." Here, too, in the wetter spots, are large colonies of Laughing Gulls which, as one approaches, rise in the air with their weird cries like maniacal laughter, fully equalling the volume of sound produced by the tern colonies on the beach. Back here, too, nest the Willets in scattered pairs, and they also mount in the air and follow the intruder, circling over his head with constant protests. Where open places occur on the marsh, bald patches as they are called, surrounding small shallow pools, groups of the grotesque long-legged Stilts are nesting and the still larger Avocets with their curious upturned bills. The whistle of the Osprey is heard on every side and there are always a few Bald Eagles in sight. Marsh Hawks, Black Ducks and Bitterns, Spotted Sandpipers and Killdeer are also present in numbers along the edge of the marshes. Such was the summer resident bird life of the shore but in May and June, and again in late July and August, came the hordes of shore birds. No man living today can appreciate their numbers, but from the scattered records extant, we know that they were tremendous, and in late fall and winter came the ducks and geese to throng the bays and sounds.

The casual shooting for their tables by the seaboard natives had little effect upon the host nor did their gathering of the eggs of gulls, terns and Mudhens but with the increasing advent of sportsmen from the interior the toll began to tell. In June, 1843, we learn from a letter of J. P. Giraud, Jr., the Long Island ornithologist, that his friend Brasher had been at Egg Harbor for a few days and had killed upwards of 1200 shore birds, and this, be it remembered, with an old muzzle-loading gun. Baird's letter, already quoted, and written a little later in the same year, shows how many gunners were engaged in the slaughter. The outcome is not surprising. With their numbers reduced the uncontrolled shooting and egg gathering of the natives had its effect on the birds, while the establishment of summer resorts drove them away to the wilder spots. Year by year the nesting birds became fewer and the transients—plover, snipe and ducks, were subject to ruthless bombardment, not only in Cape May but all along the Atlantic coast, on their journeys to and from their breeding grounds in the Far North, and naturally they shrank rapidly in numbers.

The herons and terns, not being of value as game, escaped serious persecution until about 1883 when millinery collecting reached its height. At that date apparently a few of all the breeding species that I have mentioned, except the Stilt and Avocet, still nested on the Cape May shore, but the final destruction was rapid and complete (see also pp. 364, 365).

Speaking of the Least Tern at a point on Long Beach, a little farther up the Jersey coast, the late George Spencer Morris has written me:—"In the summer of 1884 I could no longer find terns' eggs on the beach and natives told me that they no longer found them there, and in the next year I remember coming upon two professional millinery gunners who had two piles about knee-high of Least and Common Terns which they said they were sending to New York." The white herons were said still to nest on Seven Mile

Beach in 1886, but by 1888 they were exterminated, one gunner having shot seventy-three birds in a single day. While these were supposed to have been Snowy Egrets we have not, so far as I know, a single breeding specimen from New Jersey to show which white Herons really did nest there. Thus passed the ornithological glory of the Cape May coast.

In 1921, I wrote: "As breeding birds the white herons, Willet, Oyster-catcher, Gull-billed Tern, Avocet and Stilt are gone forever. The last four and the Wilson's Plover are never seen at any time; and the others only in small numbers, or very rarely, as transients. A few years ago we should have added the Least Tern, Piping Plover and Black Skimmer to the list of birds that breed no more on the shores of Cape May County, but lately they have been found nesting again, sparingly, at remote places where primitive conditions remain."

Since then important things have transpired; both the American Egret and the Little Blue Heron have been found nesting in the state, in small numbers, though not in the immediate vicinity of the Cape; Wilson's Plover has several times been seen on the beaches and we have several records of the Oyster-catcher, while a large number of Willet have been found nesting on the marshes near Fortesque, somewhat west of the Cape May border. The Least Terns, Black Skimmers and Piping Plovers have greatly increased in numbers and nest in several different localities.

The abolition of spring shooting and the adoption of the Migratory Bird Treaty with Canada, with the ultimate ban on the shooting of all shore birds, with the exception of the Woodcock and Snipe, have been mainly responsible for these recoveries, but much is also due to the general increase of interest in the protection of wildlife through the activities of the National Association of Audubon Societies and kindred organizations.

Following the limitation and abolition of shooting, however, and even prior to this action, other agencies have arisen fully as detrimental to coastal bird life as was uncontrolled shooting. These have had to do with the destruction of the haunts of the birds rather than of the birds themselves, but the result is the same. The steady increase of summer resorts along the coast and on Delaware Bay has reduced the "wild beaches" to a minimum. When these "developments" have been in response to definite needs no objection can be raised as we must needs realize that we cannot check the spread of civilization, and the resultant destruction of nature can only be met by the establishment of wildlife sanctuaries. But it is regrettable to see large areas, where natural conditions still prevailed, destroyed simply to permit of speculative "developments" which have never developed farther than the filling in of ponds and marshes and the cutting down of woods and thickets, and for which there was no legitimate call.

The construction of the Cape May Harbor and the accompanying development of the Fill were natural extensions of the town and its requirements but they have, none the less, had a marked effect upon the bird life and it is interesting to trace the changes that have taken place. Up through the late nineties Cape Island Sound with its borders of salt meadows stretched north-ward from Madison Avenue with a fringe of sand dunes separating it from the beach, while to the westward it extended almost to Washington Street for some distance south of Schellenger's Landing. The "Landing," then separated from the town, was located on Cape Island Creek which wound its way through the great meadows to broaden into the "Inlet" at Sewell's Point. All about the Sound there nested hundreds of Clapper Rails, Seaside and Sharp-tailed Sparrows, with Long-billed Marsh Wrens, Virginia Rails, and Red-winged Blackbirds thronging the adjoining cattail swamps. There, in autumn, came hosts of Yellow-legs and other shore birds to feed in the shallows and on the sand bars. Numerous stakes marked the location of oyster beds and the oyster men could be seen plying their tongs only a short distance behind the spot where the Admiral Hotel now stands. Then, in 1904, came the construction of the Harbor involving the cutting away of part of the meadows and the dredging of the whole area while the resultant mud and sand were used to fill in the sound and thus the present "Fill" was developed. Where once had been water and marsh there now stretched a vast waste of sand and broken shells. The breeding birds of the entire area disappeared at once. For a time pools, formed as a result of rain storms or from incomplete drainage, attracted migrant ducks or shore birds in season. Then, as grass covered the sand, Meadowlarks, Spotted Sandpipers, Kill-deers, Grasshopper and Henslow's Sparrows established themselves and Upland Plover and Snipe were to be seen during the migrations, while the Red-wings, loath to leave their old haunts, adapted themselves to the change and nested along the edge of the tract in tussocks or low bushes. As the years passed the changes continued and now we have a dense jungle of bayberry in some places reaching to a height of eight or ten feet. Doves, Thrashers, Goldfinches and Cuckoos now frequent these thickets and some of them nest there, while thousands of Purple Martins, Starlings, Grackles and Robins took possession of the place for several years as a summer roost. Meanwhile the original flora, a few salt marsh sedges and grasses, has been replaced by a collection of weeds or weedy wild flowers of great variety and Otway Brown has recorded four hundred and thirty-nine species growing there at some time or other since the development began! Thus we have seen three distinct "bird associations" succeed one another in the thirty years that have passed:—salt marsh, open grassland, and bayberry thicket. To the west of Schellenger's Landing a similar change took place when the Harbor branch of the railroad was constructed, also in 1904, connecting the fish wharves with the main line. This formed a virtual dam across the creek which with its salt meadows stretched north to Mill Lane and in spite of a sluice gate, where the creek itself comes through, practically all of the area has been converted into a fresh marsh grown high with reed grass, cattail and brambles. Here again the original bird population of shore birds, herons, etc., has disappeared and has been succeeded by a sparse assemblage of Song Sparrows, Maryland Yellow-throats, etc.

CAPE MAY
1890

SEVEN MILE BEACH
HEREFORD INLET

GRASSY SOUND
RICHARDSON'S SOUND

ANGLESEA
WILDWOOD
HOLLY BEACH
FIVE MILE BEACH

TURTLE GUT INLET

TWO MILE BEACH

JARVIS SOUND

COLD SPRING INLET

SEWELL'S POINT

POVERTY BEACH

RIO GRANDE
FISHING CREEK
COLD SPRING
BENNETT'S STA.
CAPE MAY CITY
POINT

SCALE
1 MILE

1 LAKE LILY, CAPE MAY PT.
2 LIGHTHOUSE POND
3 SHALLOW POND
4 LIGHTHOUSE
5 COAST GUARD STATION

6 OLD BOAT LANDING
7 POND CREEK MEADOWS
8 PONDS BACK OF S. CAPE MAY
9 HIGBEE'S BEACH
10 NEW ENGLAND MEADOW

11 PRICE'S SALT POND
12 TOWN BANK
13 POND NEAR BROADWAY
14 RACE TRACK POND

15 W. CAPE MAY PONDS
16 CAPE ISLAND CREEK
17 SCHELLENGER'S LANDING
18 GOLF LINKS

19 PHYSICK PLACE
20 BRIER ISLAND
21 MILL LANE
22 WEEKS' LANDING

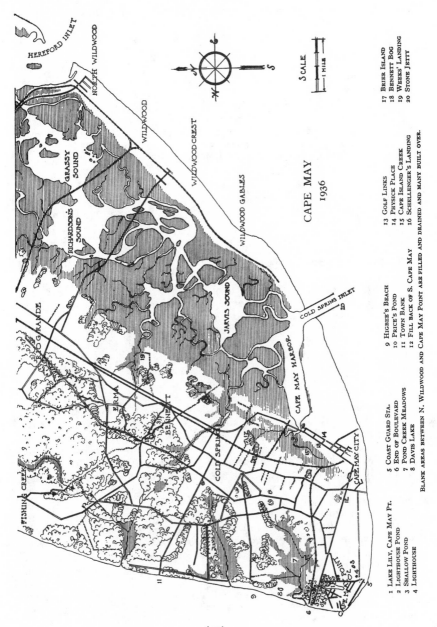

HEREFORD INLET

NORTH WILDWOOD

GRASSY SOUND

WILDWOOD

RICHARDSON'S SOUND

WILDWOOD CREST

WILDWOOD GABLES

JARVIS SOUND

FISHING CREEK

RIO GRANDE

ERMA

BENNETT

COLD SPRING

COLD SPRING INLET

CAPE MAY HARBOR

CAPE MAY CITY

UNION

CAPE MAY PT.

CAPE MAY
1936

SCALE

1 MILE

1 LAKE LILY, CAPE MAY PT.   5 COAST GUARD STA.   9 HIGBEE'S BEACH   13 GOLF LINKS   17 BRIER ISLAND
2 LIGHTHOUSE POND   6 END OF BOULEVARD   10 PRICE'S POND   14 PHYSICK PLACE   18 BENNETT BOG
3 SHALLOW POND   7 POND CREEK MEADOWS   11 TOWN BANK   15 CAPE ISLAND CREEK   19 WEEKS' LANDING
4 LIGHTHOUSE   8 DAVIS LAKE   12 FILL BACK OF S. CAPE MAY   16 SCHELLENGER'S LANDING   20 STONE JETTY

BLANK AREAS BETWEEN N. WILDWOOD AND CAPE MAY POINT ARE FILLED AND DRAINED AND MANY BUILT OVER.

(35)

To the west of the town, draining and developments in 1926 and later, have been responsible for quite a striking change in the wildlife, both animal and plant. The open stretch between Cape May and the Point was formerly covered with cattail and rose mallow swamps with salt meadows bordering on Cape Island Creek. Here the black grass was cut as an autumn or winter hay crop. The Creek wound its way back of South Cape May from the Lighthouse Pond east to the western boundary of the town. Here the Clapper and Virginia Rails and Least Bitterns bred regularly and numbers of Green Herons came to feed, while Red-wings and Long-billed Marsh Wrens thronged the cattails, and Seaside and Sharp-tailed Sparrows abounded on the open marsh. To the north of this area lay the Race Track Pond, where much of my studies were conducted in 1920 and 1921, and north of this the West Cape May Ponds, all of them resorts of nesting Bitterns, King and Virginia Rails and feeding grounds for many shore birds, Green Herons and families of Black Ducks. The shallow ponds east of the Lighthouse were still more notable as feeding grounds for shore birds and herons of various kinds—Egrets, Little and Great Blue, Green and Night Herons, as well as resting places for Laughing Gulls after the nesting season was over.

Today practically the entire area thus described ceases to exist as a resort for birds and the numerous population of ten to fifteen years ago has departed. The Mosquito Commission long since began the draining of the ponds and marshes and the distribution of poisonous oil which destroyed much of the vegetation and rendered the water unfit for animal life of any sort. One of the unfortunate and uncalled for developments, too, was started in the middle of this area. The meadows were filled in by transfer of surface soil from the adjacent farm land to the north, rendering the latter unfit for further agricultural uses, and both sections have grown up in bay-berry bushes as did the Fill, the "development" having failed to develop. Still later the Federal agencies seeking to find work for the unemployed have continued the ditching and draining, regardless of whether mosquitos were breeding there or not, and now only a vestige of water or marsh remains. The Race Track Pond and the ponds of West Cape May have disappeared, along with many rare plants which had attracted the attention of botanists throughout the East. The two larger ponds at the Lighthouse still remain but greatly reduced in size and the abundant bird life—the herons, egrets, rails, bitterns, marsh finches and wrens, is but a memory. Where ten years ago one could list forty kinds of summer resident birds in a few minutes walk from the town not twenty can be found today and these mostly common dry ground species. (For a detailed account of the effect of salt meadow ditching and draining cf. Charles Urner, Abstr. Proc. Linnaean Society, New York, No. 43–44, pp. 40–42.)

As the marsh bird habitats immediately about Cape May disappeared bird students have been forced to go farther afield; to the great meadows back of Five and Seven Mile Beaches and to the Bay shore, but the inevitable draining and ditching have now reached these spots, the last stronghold of the wading birds and the ducks. Pond Creek and New England Meadows, once constituting valuable salt hay tracts, have been largely drained and are now grown up in marsh elder bushes which have later died, as the water disappeared, and they are now for the most part stretches of dead brush and a waste of weeds. The ocean meadows, with their shallow ponds drained, are yearly becoming less attractive to the migrant shore birds while further despoiling of the Bay side marshes will reduce the already scanty winter duck population to a sad remnant, and drive away the nesting Willet from the Fortesque Meadows.

There is a little hope extended to the nature-lover by the establishment of the Chilcohook Wild Life Refuge at Fort Mott, on the lower Delaware River, and another, just authorized, at Bombay Hook, Delaware; both will prove excellent sanctuaries for ducks and certain other marsh birds while the Audubon Association's Sanctuary at Cape May Point (Plate 10) offers protection to the host of autumnal migrants that congregate there and serves to attract public attention to the importance of wildlife conservation. (Nat. Asso. Audubon Societies, 1775 Broadway, New York City.) The action of the Borough of Cape May Point in prohibiting gunning within its limits is most commendable and the establishment of the Atlantic City Bird Sanctuary covering the lower part of Brigantine Island, deserves the support of all nature lovers. (Irving W. Street, Sec'y, 723 Pacific Ave., Atlantic City, N. J.)

But in spite of all the changes that we have enumerated bird life at Cape May is still of abounding interest. Every year intensive study yields new facts in the life history of the supposedly well known species and as contrasted with conditions elsewhere in the eastern United States, and without comparison with the glorious days long past, Cape May still proves a worthy Mecca for the ornithologist.

It may still be possible for the summer resorts along the coast to establish themselves as wildlife sanctuaries. Following the lead of the Audubon Association the borough of Cape May Point has forbidden gunning within its limits and the residents of Brigantine have made a sanctuary of that island. It is nothing less than a calamity that a portion of Seven Mile Beach and the area between Cape May and the Point were not secured when their purchase was easily possible and dedicated to the preservation of Nature but there is still opportunity to bring common sense into the mosquito draining work and instill a proper appreciation of wildlife preservation into the popular mind. With the enormous increase in the ranks of nature-lovers, today, there is no better way to exploit our seaside resorts than by advertising their wealth of wild birds and wild flowers and if such a plan were followed here it would attract thousands of visitors of the most desirable character and redound to the reputation and financial benefit of Cape May.

BIRD MIGRATION AT THE CAPE

Cape May, or more particularly Cape May Point, is an outstanding spot at which to study bird migration and while this phenomenon is discussed in detail in connection with my studies of the hawks, Flickers, Woodcock, Kingbirds and Robins—species in which its peculiarities are most manifest, a brief general summary of the movements of the birds at the time of migration and my interpretation of them may not be out of place.

The spring and autumn movements, it should be understood at the outset, are totally different both in character and scope. In the spring the northward flight does not seem to differ materially from that with which we are familiar in the vicinity of Philadelphia or elsewhere in the interior. There is the same steady progress of the transients and while there may be more or less massed flights of Woodcock, Snipe, Fox Sparrows, etc., and occasional "waves" of warblers and other small passerine birds, the latter are never, so far as my experience goes, at all comparable to the great warbler waves that we see in the Delaware Valley and from all accounts in the Susquehanna Valley as well. Furthermore such massed flights do not occur in the Pine Barrens, which lie between the coast and the Delaware, the Barrens indeed, are noted as poor country for transient birds. Moreover it should be realized that there are, at the Point, no great congested spring flights of Flickers, Kingbirds, or Tree Swallows such as we see there in the autumn and, what is more remarkable, there is no return flight whatsoever of the countless hawks that pass south at the latter season.

Lighthouses have always contributed to our knowledge of bird migration as birds during stormy weather have been attracted and bewildered by the light and have flown against the glass or the tower to their death, and the Cape May Light, situated near to the beach a little west of the actual point of the peninsula, has been responsible for not a few dead birds. So far as I am aware nearly all of the birds that have struck the light have been in the autumn migration and the same is true of the light at Atlantic City, where

Major A. G. Wolf, the keeper, reported that four hundred and twenty-four birds representing twenty-six species struck that light during the autumn of 1889 (Warren's, Birds of Pennsylvania, second edition, pp. 400–401). While most of the birds that have been killed at the City Hall tower in Philadelphia also have been in the autumn (William L. Baily, Abstract Proceedings Delaware Valley Ornithological Club, III, pp. 15–19) one of the worst fatalities at this spot was on May 21, 1915 (D. E. Culver, Cassinia, 1915, pp. 33–37), when upwards of three hundred birds were killed on a single night.   At Barnegat Light, at the entrance to Barnegat Bay, another bird catastrophy took place on the night of October 28, 1925, and birds of all kinds and sizes, from Cormorants to warblers, were killed.   Charles Urner, who visited the spot a few days later, was still able to count no less than five hundred dead birds of thirty species strewn all about, twenty-five to the square yard.

From the data that I have been able to assemble, and the testimony of students of migration elsewhere, there are two major groups of migrant land birds in the East.   One of these consists mainly of species that winter in our South Atlantic States or in the West Indies or which travel to South America by way of the latter islands or the peninsula of Florida, such as the Kingbird, Bobolink, Savannah, White-throated, Chipping and Fox Sparrows, Black and White, Black-throated Blue and Myrtle Warblers, Maryland Yellow-throat, Redstart and Northern Water-Thrush.   They travel normally along the Atlantic coast and have been recorded regularly on the coast of South Carolina.   They also occur at Cape May in both spring and fall.   The second group comprises species which winter in South America or southern Central America and which cross the Gulf of Mexico and follow the Alleghany Mountains in both the northward and southward flight. Needless to say they are only casually recorded from the coast of the South Atlantic states if at all.   These include the Olive-sided, Least and Yellow-bellied Flycatchers, Rose-breasted Grosbeak, Scarlet Tanager, White-crowned Sparrow, Baltimore Oriole, and the following warblers, Tennessee, Nashville, Blue-winged, Golden-winged, Blackburnian, Magnolia, Chestnut-sided, Bay-breasted, Connecticut, Wilson's, Canada and Mourning.   Of six of these one or two specimens have been recorded in coastal South Carolina, but of the remaining thirteen, none.   Most of them must cross the mountains somewhere in Virginia in order to reach eastern Pennsylvania and New York, and New England, and it seems likely that many of them, as they progress in a northeasterly direction, turn north when they reach Chesapeake Bay and follow the Susquehanna Valley, while another large detachment follows Delaware Bay and the Delaware Valley, leaving but a very small number to reach the coast of New Jersey.   This is purely theoretical but it would satisfactorily explain the paucity of birds of this latter group at Cape May or elsewhere on the New Jersey coast, especially in spring.   Carolinian species which reach their northern breeding limit (in the East) in the Delaware Valley, central New Jersey north of the Pine Barrens, or in southern New England and New York, and which also are very rare at Cape May may follow the same line of flight.   At all events there seems to be nothing really abnormal in the northward migration at the Point except for the absence of hawks, the meager warbler waves, and the scarcity of certain warblers, vireos, etc., of group two.   The absence of hawks in spring is also characteristic of many points in the interior where these birds abound on their southward journeys.

When we come to study the autumnal migration at Cape May we face quite a different problem and this flight is frequently abnormal and difficult

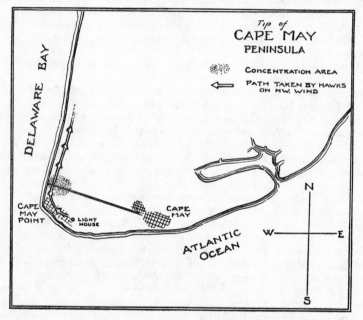

From Allen and Peterson, Auk, 1936.

to explain.   There is, in the first place, a steady night flight down the coast as is evidenced by the calls of the transients as they pass overhead, while during the daytime the Barn Swallows stream down the beach and compact bunches of Bobolinks, Siskins, Robins and Bluebirds pass regularly in their appropriate season.   These, as well as the species mentioned in the first group, above, probably continue down the coast to their winter quarters. Members of group two probably cross New Jersey above the Pine Barrens and follow the high ground bordering the Delaware Valley on the west to retrace their spring course along the western side of the Alleghanies. The prevailing winds at the Cape are south and, while they blow, migration proceeds without interruption, while some birds pass during easterly winds. At intervals, however, the wind will suddenly shift to the northwest, during the night, with a sharp fall in the temperature, and on such occasions the whole Cape May Point area will be deluged with birds; it may be with great flights of Kingbirds or with thousands of warblers and other small passerine birds, and later in the season Flickers will arrive in enormous numbers and Woodcock will throng the countryside, while at any time during the autumn northwest winds will bring the phenomenal hawk flights for which the Point is noted. These congested migratory movements always occur early in the morning or, at least, they are over by 8 o'clock except possibly in the case of the hawks which apparently continue to arrive for a somewhat longer period, or perhaps they, like the other species, are simply feeding and are the more conspicuous from having to fly about in the open in order to secure their prey.

The perplexing thing about these congested migrations, which seems impossible to those who have not witnessed the phenomenon, is that the birds are flying north along the Bay in exactly the opposite direction to that in which they are supposed to be progressing! (See fig. p. 40.) They come steadily in from "somewhere" and press on directly in the teeth of the wind until shortly after sunrise when the flight is over. On days of Flicker flights, when thousands of the birds were passing steadily northward along the Bay shore, there were none in evidence, even before dawn, about Cape May or on the ocean front to the north, and I saw no trace of the flight until I had crossed the Bay Shore Road less than a mile from the Bay. During the great Kingbird flight on the morning of August 30, 1926, the birds passed north by thousands where I was located, a mile above the turnpike and close to the Bay, but there was no trace of a flight passing Cape May, yet, from my position, I could see the flocks coming in from the southeast but at a much greater elevation than that at which they passed my station. The Woodcock flights are also restricted to the immediate vicinity of the Bay.

As to the source of these flights, I have seen Robins at South Cape May coming in large numbers from the open ocean as far offshore as the eye or the glass could distinguish them and after reaching land they continued their flight northwest until they reached the woodland north of the Point before attempting to rest. Woodcock have frequently been reported by members of the Coast Guard crew at the Point, flying in from the ocean at night and Walker Hand told me of a great flight of warblers that he saw on October 6, 1909, about 9 or 10 P. M., coming in from the sea directly over the bathing beach. They crossed the boardwalk where a number of them struck the railing or nearby wires and he gathered a hatful of dead birds next morning. Again on the beach at the Lighthouse, in October, Mrs. Stone and I saw apparently the last stragglers of a night flight come in from the sea as late as 10:00 a. m. and Red-breasted Nuthatches and Brown Creepers alighted on our shoulders and backs as we stood on the sand, apparently taking us for tree stumps. It is evident from these and other observations that a considerable number of birds are blown out to sea by northwest winds which have overtaken them while on their normal southward course. It would seem that in some way or other they realize that safety lies to windward and they laboriously beat their way in that direction and eventually reach land, after which they keep on for some time on the same course. It seems hardly credible, however, that all of the birds in these congested flights have been blown offshore.

Observations of hawks on days of south winds have shown them moving south at a great altitude, well above the limit of vision, and this suggests the possibility of most migrating birds passing at a greater height than we have supposed. My observation of the Kingbirds coming in at a great height would be explained in this way and also the arrival of Flickers and other birds at the Point which were not seen anywhere along the coast to the north. Birds flying at a great elevation might be alarmed when their course brought them in sight of the sea and if so they would be likely to seek shelter in the woodland that lay below them, while some that did not take heed would find themselves offshore and over the water. In connection with flight at a considerable height, might not this also explain our failure to see hawks on their northward journey in spring? The tendency of birds to fly into the wind which is so marked in their actions at the Point requires more study. Prof. Cooke believes (Auk, 1913, p. 205) that birds proceed on their migrations irrespective of the direction of the wind but I do not consider that his evidence covers all aspects of the problem by any means. The north flights at the Point cannot be dismissed as erratic and we have the flight of the Curlew over Cape May in the face of a south wind and its abatement during north winds, while many birds persistently head into a strong wind even if its force may be sufficient to carry them backwards in the other direction. Resting shore birds and gulls, moreover, all face the wind. If flight into the wind is a deep-rooted inherited tendency it would explain the action of birds blown offshore without involving any question of seeking safety by flying in that direction. Just why birds approaching the sea should wish to check their flight when a northwest wind is blowing brings us to the still more complicated problems of intelligence and behavior! What experience have they had in flying over water? Why should they fear a flight over a portion of the western Atlantic Ocean when they boldly launch forth over the Gulf of Mexico? Why should they hesitate to cross Delaware Bay in a northwest wind when they cross in a south wind without a moment's hesitation? Avoidance of flying with the wind may have more to do with it than fear of the water.

As to the destination of these north-flying south-bound migrants, and their immediate objective, opinions differ not a little. Walker Hand who was familiar with the autumnal bird flights all his life always maintained that at their conclusion at about 8:00 a. m. the birds scattered over the Bay side woods and thickets to feed and remained there over night or until the wind shifted when they resumed their southward flight in the usual scattered formation and crossed the Bay from Cape May Point to Cape Henlopen. My observations fully confirm this theory and I do not think, as some have

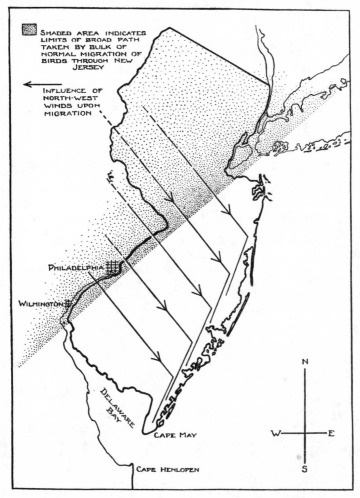

SHADED AREA INDICATES LIMITS OF BROAD PATH TAKEN BY BULK OF NORMAL MIGRATION OF BIRDS THROUGH NEW JERSEY

INFLUENCE OF NORTH-WEST WINDS UPON MIGRATION

PHILADELPHIA

WILMINGTON

DELAWARE BAY

CAPE MAY

CAPE HENLOPEN

N
W——E
S

From Allen and Peterson, Auk, 1936.

suggested, that the birds travel up the Bay to find a narrower stretch of water over which to fly. Such action would involve remarkable intelligence and judgement, moreover, I have been unable to secure any evidence that they cross the upper Bay or the lower Delaware River at any point. The flights do, however, extend on some occasions as far north as Dennisville and the extent of country occupied by the birds, on the morning following the flight, probably depends upon the length of time that the northwest wind has been blowing and the number of birds that have been affected by it. The number of birds present after the conclusion of a flight compared with those to be found on the day previous would seem to show that none of them had crossed the Bay while the paucity of bird life on the day following, if the wind has changed, as it usually does within twenty-four hours, is striking.

The warblers and other small birds seem to spread over a larger area for morning feeding than do the Flickers or Woodcock and I have found them on the early morning of a flight day, while the wind was still blowing half a gale, all the way from Broadway to the Bay, flushing at every step from fields, swamps and thickets, only to drop again into the grass as the wind swept them on. There are often quite a number of migrants to be found in the Physick garden in the heart of the town, on the morning of a warbler flight but they are probably stragglers that have concentrated in a place of shelter and I do not think that many transients coming down from a higher level or wind-blown birds carried out to sea, come to earth or to shore farther east than South Cape May. Farther up the Bay shore I have found the transient warblers, in the morning, spread all through the woods and thickets as far as Fishing Creek and east to the Bay Shore Road while sometimes they extend east of that line and Otway Brown tells me that on one occasion the flight covered his premises to the east of the road and that many of the warblers struck a chicken wire fence on his grounds and falling to the ground were promptly devoured by a flock of domestic ducks which had gathered there to partake of the unexpected feast.

On several mornings when evidences of these congested flights were abundantly present at the Point, I have visited, by automobile, various sections of the town as well as the Fill and the farming country east of the Bay Shore Road but found only scattered transients, if any, and nothing to indicate the remarkable phenomenon that had just taken place only a couple of miles to the west!

To summarize the autumn migration at Cape May Point as I interpret the evidence that has been presented above: I think that the birds that I have included in Group I on the spring flight proceed regularly southward along the coast under normal conditions of wind and weather, while those constituting Group II retrace their route across New Jersey above the Pine Barrens and thence down the Delaware Valley. When a north wind arises both groups are blown off their course in a southeasterly direction and more or less of them are carried out to sea while the others, battling against the gale, bank down at the Point when they find themselves confronted with water on three sides as well as being exhausted and in need of food, and there await a change of weather conditions (see fig. p. 43, from Allen and Peterson's discussion of hawk flights at the Point, Auk, 1936, pp. 393–404).

A sharp fall in temperature, which always accompanies a northwest wind, stimulates migration everywhere, a fact well known to all who have kept migration records in the interior, and this naturally increases the number of birds on the wing and helps to produce the congestion that we see at the Point. The source of the migrating host is either high in the air or off the ocean, which accounts for their not being seen until they reach the Point, and their destination on their northward progress along the Bay is simply such feeding grounds as may be reached by dawn or shortly after. As soon as the wind changes they reverse the direction of their mad early morning flight and proceed normally southward to the Point and across the Bay.

The movements of shore birds and other water birds are apparently not governed by the same conditions as determine the flights of the land birds—at least not entirely. The Woodcock is an exception since it really is a "land bird" so far as its life and activities are concerned. Such shore birds as we see *en route* fly into the wind as do land birds and therefore travel during south winds, but Night Herons and Great Blues that we have seen definitely on migration were moving during east winds. The bulk of the shore bird migration is earlier than that of the land birds, covering July and August rather than September and October, furthermore there seems to be a greater tendency for transients to settle temporarily on good feeding grounds and not to move on immediately as is customary with our smaller land birds. This, however, can only be proven by banding, if it shall ever be possible to band transient shore birds! The same problem is presented by roosting Purple Martins which we presume use the same roost for some days (or weeks?) before passing on but we cannot prove it. So too, with the roosts of Chimney Swifts. To return to the shore birds, the study of their migration is further complicated by the fact that a number of individuals never reach their nesting grounds and linger on all summer, far to the southward, so that we may have several species represented at Cape May from early spring until late autumn while the true transients are here for only short periods. Such "hang over" birds have usually failed to assume the full breeding plumage and will be found to be sexually immature which illustrates how closely the molt and the migratory instinct are dependent upon physiological development. Precisely the same thing occurs with the gulls and we see immature Herring Gulls present all summer and many Laughing Gulls which failed to develop the black heads and which take no part in the nesting activities. Some of the latter do reach the breeding grounds but most of them spend the summer farther south.

There are other phases of migration which, while not comparable to the regular periodic north and south movements, must be considered in our study of migration at the Point. The herons, here as elsewhere, travel north after the nesting is over and to this peculiar habit we owe the presence of so many Egrets and Little Blue Herons in late summer and early autumn, as well as the two visits of the Wood Ibis. The presence of straggling individuals of other distinctly southern birds such as the Purple Gallinule, Audubon's Shearwater, Man-o'-war-bird, Brown Pelican and Gray Kingbird is due to the violent hurricanes that occasionally sweep the South Atlantic coast and carry birds far to the northward of their normal range. The occurrence of occasional petrels, shearwaters, phalaropes, Kittiwakes or jaegers is usually due to storms off the coast, as all of these are birds of the open sea and do not normally come in to the beach. Then there are rare winter visitors such as the Murre, Razor-billed Auk and Guillemot which belong in the same group although they occur here, even on the ocean, only as severe cold drives them south. The Little Auk differs slightly in its manner of occurrence as its flights seem to be periodic like those of the Goshawk and Snowy Owl and it is not easy to account for them. The same is true of the irregular occurrences of the Stilt Sandpiper, Forster's Tern, and of the Red-breasted Nuthatch and Cedar Waxwing. A few birds like the Lesser Yellow-legs, Pectoral Sandpiper, Black Tern, Connecticut Warbler etc. are well known to have a different line of migration in spring and fall going up the Mississippi Valley on their northward journey and south along the Atlantic coast which explains their rarity here in spring. The Arkansas Flycatcher, Varied Thrush, Yellow-headed Blackbird, Lark Sparrow etc. are merely strays from farther west caught up by easterly storms or associating with flocks of other species normally traveling in a southeasterly direction.

While bird migration may not be quite the "mystery of mysteries" that it was considered to be in the past, its details, origin and method are still involved in obscurity and there is much to be learned about what is undoubtedly one of the most fascinating phenomena of Nature.

A MONTHLY SUMMARY OF THE BIRDS OF SOUTHERN CAPE MAY COUNTY

We have records of the occurrence of 318 species and subspecies of birds in Cape May County with fifty-six additional kinds farther north in the state or offshore. Of these twenty have probably occurred in the county. Six have become extinct within historic time, and others have disappeared from our region. In order to present a seasonal summary of bird life in the immediate vicinity of Cape May I have prepared from my records and those of my colleagues for the years 1920 to 1936, and a few for 1890 to 1919, a list of the birds seen during each month in the area extending from Cape May and Cape May Point north to Higbee's Beach and Cold Spring, which has been about the limit of our daily tramps. Records for the meadows and the coast, however, have been included from as far north as Townsend's Inlet since our intensive study of the bird life has extended to this point, while, as a matter of fact, the fauna of the meadows is almost uniform and to divide the portion that we have studied would be misleading. (See map p. 35.)

The species have been grouped under several headings:

Regular Residents. Species that are present throughout the year.

Irregular Residents. Species that may occur at any time during the year but which do not breed regularly within the area. Most of them do, however, breed a little farther north in the peninsula, or in adjoining counties to the north.

Regular Summer Residents. Birds that nest regularly in the area and usually depart southward in the autumn. A few individuals of some of them remain through the winter.

Irregular Summer Residents. Birds that have nested at least once in our area and are likely to occur at any time during the summer; as a rule they nest a little farther north in the county. They almost all go south for the winter and some of them are plentiful here as transients.

Transients. Birds that pass through our area in spring and autumn on their way to breeding grounds to the north and winter quarters to the south. None of this group has been found nesting in the Cape May area although a few of them breed only a short distance to the north. Some individuals of other species, notably shore birds, fail to reach full maturity and linger here during the summer without breeding.

Winter Visitants. Birds that come to us from the north and remain more or less regularly throughout the winter. Individuals of some of these, notably ducks and gulls, may remain through the summer without breeding.

Summer Visitants from the South. These are either herons, which have a characteristic habit of ranging north of their nesting area after the breeding season, or other species which seem to overshoot their normal range in the northward migration and either spend the summer without breeding or

(47)

return to the south; a few have been blown here by storms.   The Egret and Little Blue Heron, which are included here, have begun to nest again in New Jersey, as they did in former times, but not as yet within our area.

Accidental Visitants from the West.   Strays apparently blown off their course during migration.

Oceanic Birds Occasionally Blown Ashore by Storms.

Such an arrangement is, to a certain extent, arbitrary since, while most birds fall naturally into one group or another, some would seem with equal pertinence to belong to several groups.   A large number of our Catbirds, Towhees and Chipping Sparrows, for instance, are summer residents, many others are transients, while a small number remain with us throughout the winter; these species might therefore be included in each of the three groups mentioned.   Furthermore since some individuals are present in every month of the year, they might also be regarded as residents!   Whether the same individuals occur in both summer and winter can only be determined by banding, but it seems probable that they do not and, even in the case of recognized resident species, some individuals probably pass south at the approach of winter and their places are taken by others from farther north.

JUNE

140 Species Recorded

It is in June that the avifauna of Cape May is more nearly stable than at any other time of year.   We have present during the month all of the resident and summer resident species together with a few belated transients and a few individuals of winter visitors which for some reason or other failed to go north with their fellows in the spring.

The first two groups constitute the breeding birds of the Cape and while the nests of some species in the "irregular" section have not actually been found within the limits above set forth, they nest a little north of our boundary and may nest here at any time, since they occur more or less regularly throughout the summer.

I. REGULAR RESIDENTS—30

Common Black Duck	Flicker	English Sparrow
Turkey Vulture	Downy Wood-	Meadowlark
Marsh Hawk	pecker	Red-wing
Bob-white	Tree Swallow	Purple Grackle
Ring-necked Pheasant	Eastern Crow	Cowbird
Killdeer	Fish Crow	Cardinal
American Woodcock	Carolina Chickadee	Goldfinch
Rock Dove (Domestic	Carolina Wren	Field Sparrow
Pigeon)	Mockingbird	Eastern Song Sparrow
Mourning Dove	Robin	Atlantic Song Sparrow
Belted Kingfisher	Starling	

Ia. Irregular Residents—15

Great Blue Heron
(no nesting record)
Red-shouldered Hawk
(no nesting record)
Bald Eagle
(no nesting record)
Sparrow Hawk
(rare breeder)
Ruffed Grouse
(no nesting record)

Barn Owl (rare breeder)
Screech Owl (rare breeder)
Great Horned Owl
(no nesting record)
Barred Owl
(no nesting record)
Hairy Woodpecker
(rare breeder)

Blue Jay (rare breeder)
Tufted Titmouse
(rare breeder)
Bluebird (rare breeder)
Cedar Waxwing
(rare breeder)
Savannah Sparrow
(rare breeder)

II. Regular Summer Residents—46

Green Heron
Black-crowned Night
Heron
Yellow-crowned Night
Heron (rare)
American Bittern
Osprey
King Rail
Clapper Rail
Virginia Rail
Piping Plover
Spotted Sandpiper
Laughing Gull
Common Tern
Least Tern
Black Skimmer
Yellow-billed Cuckoo

Black-billed Cuckoo
Whip-poor-will
Chimney Swift
Ruby-throated Humming-
bird
Kingbird
Great Crested Flycatcher
Wood Pewee
Rough-winged Swallow
Barn Swallow
Purple Martin
House Wren
Long-billed Marsh Wren
Short-billed Marsh Wren
Catbird
Brown Thrasher
White-eyed Vireo

Red-eyed Vireo
Yellow Warbler
Pine Warbler
Prairie Warbler
Ovenbird
Maryland Yellow-throat
Yellow-breasted Chat
Orchard Oriole
Indigo Bunting
(rare breeder)
Towhee
Grasshopper Sparrow
Henslow's Sparrow
Sharp-tailed Sparrow
Seaside Sparrow
Chipping Sparrow

IIa. Irregular Summer Residents—16

For those marked with an * we have no actual nesting record

*Pied-billed Grebe
Least Bittern
Wood Duck
*Broad-winged Hawk
*Black Rail
*Florida Gallinule

Roseate Tern
*Chuck-will's-widow
Nighthawk
Wood Thrush
Gnatcatcher

Black and White Warbler
Parula Warbler
*Yellow-throated Warbler
Hooded Warbler
Bobolink

III. Belated Spring Transients—17

Double-crested Cormorant
Black-bellied Plover
Semipalmated Plover
Turnstone
Solitary Sandpiper
Willet

Greater Yellow-legs
Lesser Yellow-legs
Knot
White-rumped Sandpiper
Least Sandpiper
Semipalmated Sandpiper

Sanderling
Red-headed Woodpecker
Blackburnian Warbler
Wilson's Warbler
Blue-winged Warbler

IV. WINTER VISITORS NOT MIGRATING—10

Common Loon   American Scoter   Red-backed Sandpiper
Buffle-head   White-winged Scoter   Herring Gull
Old-squaw   Red-breasted Merganser   Ring-billed Gull
                     Sharp-shinned Hawk

V. SUMMER VISITORS FROM THE SOUTH—2

American Egret   Little Blue Heron

VI. OCEANIC BIRDS OFFSHORE—4

Sooty Shearwater   Greater Shearwater   Parasitic Jaeger
                     Cory's Shearwater

JULY

148 Species Recorded

During July the entire series of residents and summer residents is present as in June but certain species are of more frequent occurrence due to the completion of the nesting season and the ranging of the birds over a wider area. In some species, moreover, there is a distinct influx of individuals from farther north. This southbound migration is strongly marked in the case of the shore birds. Those species which occurred in early June as delayed northern transients are mostly present in July traveling in the opposite direction along with others which had not been seen since the beginning of the northward flight in May.

About the same number of winter visitors that failed to go north have been recorded in July as in June although the species are not quite the same while a greater number of stragglers from the south have been observed. We have made but few trips offshore and the number and character of the oceanic birds recorded for any month does not, therefore, indicate accurately the bird life of the high seas.

    I.   REGULAR RESIDENTS (as in June list)—30.

    Ia. IRREGULAR RESIDENTS (as in June list)—15.

    II. REGULAR SUMMER RESIDENTS (as in June list)—46.

    IIa. IRREGULAR SUMMER RESIDENTS (as in June list)—16.

Although we have no actual record of the Little Black Rail or the Florida Gallinule for July they must occur within our area in this month.

III. SOUTHBOUND TRANSIENTS—23

Semipalmated Plover   Lesser Yellow-legs   Western Sandpiper
Black-bellied Plover   Knot   Sanderling
Turnstone   Pectoral Sandpiper   Black Tern
Hudsonian Curlew   White-rumped Sandpiper   Bank Swallow
Upland Plover   Least Sandpiper   Blue-winged Warbler
Solitary Sandpiper   Dowitcher   Northern Water-Thrush
Willet.   Stilt Sandpiper   Redstart
Greater Yellow-legs   Semipalmated Sandpiper

IV. WINTER VISITANTS NOT MIGRATING—12

Common Loon	Surf Scoter	Red-backed Sandpiper
Canada Goose	American Scoter	Herring Gull
Old-squaw	Red-breasted Merganser	Ring-billed Gull
White-winged Scoter	Sharp-shinned Hawk	Red-breasted Nuthatch

V. SUMMER VISITANTS FROM THE SOUTH—4

American Egret	Wood Ibis
Little Blue Heron	Oyster-catcher

VI. OCEANIC BIRDS—2

Leach's Petrel	Wilson's Petrel

AUGUST

180 Species Recorded

August finds all of the resident and summer resident birds still present although some of the latter are noticeably reduced in numbers. Other species on the contrary are greatly augmented by the arrival of migrant individuals from the north. All of the transient shore birds recorded in July are present also in August, usually in much larger numbers, while toward the end of the month many new arrivals among the transient land birds are recorded, as well as a definite increase in the numbers of certain winter visitors in addition to those that failed to go north in the spring, especially of the Herring and Ring-billed Gulls.

Additional visitants from the south are often recorded in August as a result of tropical storms prevalent at this season.

Birds flock more noticeably in August and one sees the height of the roosting of the Purple Martins, and the massing of Grackles and Robins which began in July. A northwest wind at the end of the month will bring the first marked flight of hawks as well as of the smaller land birds, movements which become more pronounced in September.

I. REGULAR RESIDENTS (as in June list)—30.

Although we have no August record of the Yellow-crowned Night Heron it is undoubtedly present.

Ia. IRREGULAR RESIDENTS (as in June list)—15.

II. REGULAR SUMMER RESIDENTS (as in June list)—46.

IIa. IRREGULAR SUMMER RESIDENTS (as in June list)—16.

No actual record for the Little Black Rail, Roseate Tern and Chuck-will's-widow.

III. SOUTHBOUND TRANSIENTS—52

Cormorant
Mallard
Blue-winged Teal
Green-winged Teal
Sora Rail
Duck Hawk
Pigeon Hawk
Semipalmated Plover
Golden Plover
Black-bellied Plover
Turnstone
Hudsonian Curlew
Upland Plover
Solitary Sandpiper
Willet
Greater Yellow-legs
Lesser Yellow-legs
Knot

Least Sandpiper
Pectoral Sandpiper
Dowitcher
Semipalmated Sandpiper
Western Sandpiper
Stilt Sandpiper
Marbled Godwit
Sanderling
Caspian Tern
Black Tern
Red-headed Woodpecker
Least Flycatcher
Yellow-bellied Flycatcher
Olive-sided Flycatcher
Bank Swallow
Cliff Swallow
Veery
Olive-backed Thrush

Gray-cheeked Thrush
Migrant Shrike
Worm-eating Warbler
Golden-winged Warbler
Blue-winged Warbler
Tennessee Warbler
Magnolia Warbler
Cape May Warbler
Black-throated Blue
    Warbler
Chestnut-sided Warbler
Blackburnian Warbler
Northern Water-Thrush
Wilson's Warbler
Canada Warbler
Redstart
Baltimore Oriole

IV. WINTER RESIDENTS—10

American Scoter
Red-breasted Merganser
Sharp-shinned Hawk

Cooper's Hawk
Herring Gull
Ring-billed Gull
Bonaparte's Gull

Red-backed Sandpiper
Red-breasted Nuthatch
Myrtle Warbler

VI. OCEANIC BIRDS—2

Leach's Petrel   Wilson's Petrel

V. SUMMER VISITANTS FROM THE SOUTH—8

Audubon's Shearwater
Man-o'-war-bird
American Egret

Snowy Egret
Louisiana Heron
Little Blue Heron

Wood Ibis
Purple Gallinule

VI. STRAGGLER FROM THE WEST—1

Lark Sparrow

SEPTEMBER

213 Species Recorded

September is the time of the maximum migration of the land birds and, with a few exceptions, marks the passing of the last of the shore birds and of the terns, while at its end come the first flights of ducks from the north.

The summer residents begin to depart and we lack records for a few of them while others have been seen but once or twice. The southward migration, while it may progress regularly and continuously, seems to be very irregular due to the succession of great bird waves coincident with northwest

winds and falling temperature. At such times the countryside along the
Bay shore is flooded with migrants while between the flights we get some of
the poorest daily lists of the whole year. The individuals of many species
which have bred at the Cape seem to pass on to the south while the migrants
from farther north do not stop in abundance except during the great waves
so that for periods of several days at a time certain common species may be
all but lacking. On September 25, 1921, I recorded only five species of land
birds at the Point, and on October 1, following, only eighteen species all told.

It is during the latter part of this month that the great Flicker flights
occur, usually accompanied by a host of smaller passerine species, and the
first great hawk flights. So great is the number of transients occurring
during September that the total number of species recorded is larger than
or any other month.

> I.   REGULAR RESIDENTS (as in June list)—30.
> Ia.   IRREGULAR RESIDENTS (as in June list)—15.
> II.   REGULAR SUMMER RESIDENTS (as in June list)—46.
> IIa. IRREGULAR SUMMER RESIDENTS (as in June list)—16.

No actual records of Little Black Rail, Chuck-will's-widow or Yellow-
throated Warbler.

III. SOUTHBOUND TRANSIENTS—71

- Cormorant
- Mallard
- Baldpate
Pintail
Green-winged Teal
Blue-winged Teal
Sora Rail
Coot
Duck Hawk
Pigeon Hawk
Semipalmated Plover
Golden Plover
Black-bellied Plover
Turnstone
Wilson's Snipe
Hudsonian Curlew
Upland Plover
Solitary Sandpiper
Willet
Greater Yellow-legs
Lesser Yellow-legs
Knot
Pectoral Sandpiper
Least Sandpiper

White-rumped Sandpiper
Dowitcher
Stilt Sandpiper
Semipalmated Sandpiper
Western Sandpiper
Buff-breasted Sandpiper
Sanderling
Black Tern
Forster's Tern
Long-eared Owl
Red-headed Woodpecker
Yellow-bellied Sapsucker
Phoebe
Least Flycatcher
Yellow-bellied Flycatcher
Alder Flycatcher
Olive-sided Flycatcher
Bank Swallow
White-breasted Nuthatch
Veery
Olive-backed Thrush
Gray-cheeked Thrush
Ruby-crowned Kinglet
Migrant Shrike
Solitary Vireo

Worm-eating Warbler
Nashville Warbler
Tennessee Warbler
Magnolia Warbler
Cape May Warbler
Black-throated Blue
  Warbler
Black-throated Green
  Warbler
Blackburnian Warbler
Chestnut-sided Warbler
Bay-breasted Warbler
Black-poll Warbler
Western Palm Warbler
Yellow Palm Warbler
Northern Water-Thrush
Connecticut Warbler
Wilson's Warbler
Canada Warbler
Redstart
Baltimore Oriole
Scarlet Tanager
Rose-breasted Grosbeak
White-crowned Sparrow

IV. Winter Visitors—25

American Scoter
White-winged Scoter
Surf Scoter
Red-breasted Merganser
Sharp-shinned Hawk
Cooper's Hawk
Red-tailed Hawk
Red-backed Sandpiper

Herring Gull
Ring-billed Gull
Bonaparte's Gull
Black-backed Gull
Short-eared Owl
Red-breasted Nuthatch
Brown Creeper
Golden-crowned Kinglet
Winter Wren

Hermit Thrush
Pipit
Myrtle Warbler
Pine Siskin
Purple Finch
Junco
Swamp Sparrow
White-throated Sparrow

V. Summer Visitors from the South—5

Brown Pelican
American Egret

Snowy Egret
Little Blue Heron
Royal Tern

VI. Oceanic Birds—2

Leach's Petrel

Parasitic Jaeger

VII. Stragglers from the West—3

Arkansas Kingbird   Dickcissel   Lark Sparrow

OCTOBER

190 Species Recorded

About half of the summer resident birds take their departure during September so that the October list is much reduced and the same is true of the August and September transients for many of which we have no October records. Many new transients appear during the month and there are a number of additional winter visitants arriving from the north.

The month is notable for the great flocks of Tree Swallows and soaring Turkey Vultures, while the hawk migration reaches its height and the gathering of Laughing Gulls, Meadowlarks and miscellaneous assemblages of sparrows are conspicuous. Ducks, rails and owls are also of more frequent occurrence.

I. Regular Residents (as in June list)—30.

Ia. Irregular Residents (as in June list)—15.

II. Regular Summer Residents—36

Green Heron
Black-crowned Night
    Heron
American Bittern
Osprey
King Rail
Clapper Rail
Virginia Rail

Spotted Sandpiper
Laughing Gull
Common Tern
Black Skimmer
Yellow-billed Cuckoo
Black-billed Cuckoo
Whip-poor-will
Chimney Swift

Ruby-throated Humming-
    bird
Kingbird
Barn Swallow
House Wren
Long-billed Marsh Wren
Short-billed Marsh Wren
Catbird

Brown Thrasher
White-eyed Vireo
Red-eyed Vireo
Yellow Warbler
Pine Warbler

Prairie Warbler
Maryland Yellow-throat
Indigo Bunting
Towhee
Grasshopper Sparrow

Henslow's Sparrow
Sharp-tailed Sparrow
Seaside Sparrow
Chipping Sparrow

IIa. Irregular Summer Residents—9

Pied-billed Grebe
Wood Duck
Broad-winged Hawk

Florida Gallinule
Nighthawk
Wood Thrush

Black and White Warbler
Bobolink
Savannah Sparrow

III. Southbound Transients—60

Cormorant
Gannet
Snow Goose
Baldpate
Pintail
Shoveller
Green-winged Teal
Blue-winged Teal
Ruddy Duck
Ring-necked Duck
Redhead
Duck Hawk
Pigeon Hawk
Yellow Rail
Sora Rail
Coot
Semipalmated Plover
Golden Plover
Black-bellied Plover
Turnstone

Wilson's Snipe
Solitary Sandpiper
Greater Yellow-legs
Lesser Yellow-legs
Knot
Least Sandpiper
Pectoral Sandpiper
White-rumped Sandpiper
Least Sandpiper
Semipalmated Sandpiper
Dowitcher
Sanderling
Long-eared Owl
Red-headed Woodpecker
Yellow-bellied Sapsucker
Phoebe
White-breasted Nuthatch
Olive-backed Thrush
Gray-cheeked Thrush
Veery
Ruby-crowned Kinglet

Migrant Shrike
Solitary Vireo
Yellow-throated Vireo
Tennessee Warbler
Magnolia Warbler
Cape May Warbler
Black-throated Blue
  Warbler
Black-throated Green
  Warbler
Blackburnian Warbler
Western Palm Warbler
Yellow Palm Warbler
Northern Water-Thrush
Wilson's Warbler
Redstart
Baltimore Oriole
Scarlet Tanager
White-crowned Sparrow
Rose-breasted Grosbeak
Lincoln's Sparrow

IV. Winter Visitants—38

Common Loon
Red-throated Loon
Horned Grebe
Canada Goose
Scaup Ducks
Old-squaw
Buffle-head
King Eider
American Scoter
White-winged Scoter
Surf Scoter
Red-breasted Merganser
Sharp-shinned Hawk

Cooper's Hawk
Red-tailed Hawk
Rough-legged Hawk
Red-backed Sandpiper
Herring Gull
Ring-billed Gull
Bonaparte's Gull
Black-backed Gull
Short-eared Owl
Horned Lark
Red-breasted Nuthatch
Brown Creeper
Winter Wren

Hermit Thrush
Golden-crowned Kinglet
Pipit
Myrtle Warbler
Purple Finch
Pine Siskin
Ipswich Sparrow
Vesper Sparrow
Junco
White-throated Sparrow
Swamp Sparrow
Fox Sparrow

V. Summer Visitants from the South—2

American Egret   Little Blue Heron

NOVEMBER

147 Species Recorded

While there is considerable migration of waterfowl, notably of Canada Geese, during November, and great flights of Robins and Bluebirds passing south, the bird life of the Cape has largely settled down to winter conditions by the middle of the month. A few summer residents of a number of species are still present during the early days and also a few laggard transients but these as a rule belong to species which are represented by scattered individuals throughout the winter.

I. REGULAR RESIDENTS—30

Ia. IRREGULAR RESIDENTS—15

II. REGULAR SUMMER RESIDENTS—17

Black-crowned Night Heron	Woodcock	Catbird
American Bittern	Laughing Gull	Brown Thrasher
King Rail	Common Tern	Towhee
Clapper Rail	Barn Swallow	Sharp-tailed Sparrow
Virginia Rail	Long-billed Marsh Wren	Seaside Sparrow
	Short-billed Marsh Wren	Chipping Sparrow

IIa. IRREGULAR SUMMER RESIDENTS—3

Pied-billed Grebe	Florida Gallinule	Savannah Sparrow

III. SOUTHBOUND TRANSIENTS—39

Cormorant	Canvas-back	Pectoral Sandpiper
Gannet	American Merganser	Least Sandpiper
Whistling Swan	Hooded Merganser	White-rumped Sandpiper
Greater Snow Goose	Duck Hawk	Semipalmated Sandpiper
Blue Goose	Pigeon Hawk	Sanderling
Mallard	Yellow Rail	Long-eared Owl
Baldpate	Sora Rail	Phoebe
Pintail	Coot	White-breasted Nuthatch
Shoveller	Semipalmated Plover	Ruby-crowned Kinglet
Green-winged Teal	Black-bellied Plover	Migrant Shrike
Blue-winged Teal	Wilson's Snipe	Cape May Warbler
Ruddy Duck	Greater Yellow-legs	Western Palm Warbler
Redhead	Knot	Yellow Palm Warbler

IV. WINTER VISITANTS—41

Common Loon	Golden-eye	Red-backed Sandpiper
Red-throated Loon	American Scoter	Herring Gull
Horned Grebe	White-winged Scoter	Black-backed Gull
Holboell's Grebe	Surf Scoter	Ring-billed Gull
Canada Goose	Red-breasted Merganser	Bonaparte's Gull
Scaup Ducks	Sharp-shinned Hawk	Short-eared Owl
Buffle-head	Cooper's Hawk	Horned Lark
Old-squaw	Red-tailed Hawk	Red-breasted Nuthatch

Brown Creeper
Winter Wren
Hermit Thrush
Golden-crowned Kinglet
Pipit
Myrtle Warbler

Purple Finch
Pine Siskin
Ipswich Sparrow
Vesper Sparrow
Junco
Tree Sparrow

White-throated Sparrow
Fox Sparrow
Swamp Sparrow
Snow Bunting
Lapland Longspur

V. VISITANTS FROM THE SOUTH—2

American Egret

Blue Grosbeak

DECEMBER

139 Species Recorded

In December we find bird life at the Cape in the condition of winter stability. It consists mainly of the residents and winter visitants although a few individuals of several summer resident and transient species remain, especially in mild seasons. If very cold weather prevails many of these disappear but whether they perish or pass farther south we do not know. There is certainly a distinct southward movement of waterfowl when the waterways to the north are frozen over; and birds of the high seas more frequently come in to the coast.

I. REGULAR RESIDENTS (as in June list)—30

Ia. IRREGULAR RESIDENTS (as in June list)—15

II. REGULAR SUMMER RESIDENTS—13

Black-crowned Night
  Heron
American Bittern
Virginia Rail
Laughing Gull

Barn Swallow
Long-billed Marsh Wren
Short-billed Marsh Wren
Catbird
Brown Thrasher

Towhee
Sharp-tailed Sparrow
Seaside Sparrow
Chipping Sparrow

IIa. IRREGULAR SUMMER RESIDENTS—2

Pied-billed Grebe

Savannah Sparrow

III. SOUTHBOUND TRANSIENTS—28

Cormorant
Gannet
Whistling Swan
Snow Goose
Brant
Baldpate
Pintail
Green-winged Teal
Ruddy Duck

Canvas-back
American Merganser
Hooded Merganser
Duck Hawk
Coot
Black-bellied Plover
Wilson's Snipe
Greater Yellow-legs
Least Sandpiper
Western Sandpiper

Sanderling
Long-eared Owl
Red-headed Woodpecker
Yellow-bellied Sapsucker
Phoebe
White-breasted Nuthatch
Ruby-crowned Kinglet
Western Palm Warbler
Yellow Palm Warbler

IV. Winter Visitants—46

All the birds listed for November with the addition of

Purple Sandpiper	Rough-legged Hawk	Evening Grosbeak
Goshawk	Northern Shrike	

V. Oceanic Birds from Offshore—4

Dovekie	Razor-billed Auk	Puffin
	Black Guillemot	

VI. Summer Visitor that Failed to Return South

American Egret

JANUARY

105 Species Recorded

January bird life is the same as that of December except for the disappearance of a few species in years of severe cold. Our trips to the shore have been less frequent, however, and the absence of some birds from our records may be due to less careful observation. Such movements as occur in January are due entirely to temperature and the freezing of ponds and other bodies of water.

I. Regular Residents (as in June list)—30

Ia. Irregular Residents (as in June list)—15

II. Regular Summer Residents—8

Black-crowned Night	Long-billed Marsh Wren	Towhee
Heron	Short-billed Marsh Wren	Sharp-tailed Sparrow
American Bittern	Catbird	Seaside Sparrow

IIa. Irregular Summer Resident—1

Savannah Sparrow

III. Transients—9

Gannet	Ruddy Duck	Red-headed Woodpecker
Snow Goose	Wilson's Snipe	Phoebe
Baldpate	Sanderling	Western Palm Warbler

IV. Winter Visitants—42

All of the birds listed for November with the addition of the Northern Shrike.

FEBRUARY

99 Species Recorded

There is no change in the make-up of February bird life except for the appearance, during the latter half of the month, of flocks of migrant Redwings, Grackles, Robins and Fox Sparrows and increases in the numbers of other species which remain in varying numbers through the winter. There is also some movement of waterfowl to the north but in some recent years, when very cold weather prevailed all through February, migration was at a standstill.

I.   REGULAR RESIDENTS (as in June list)—30

Ia. IRREGULAR RESIDENTS (as in June list)—15

II. REGULAR SUMMER RESIDENTS—4

Catbird	Brown Thrasher	Chipping Sparrow
	Towhee	

IIa. IRREGULAR SUMMER RESIDENT—1

Savannah Sparrow

III. TRANSIENTS REMAINING THROUGH THE WINTER—8

Pintail	Wilson's Snipe	Red-headed Woodpecker
Baldpate	Black-bellied Plover	Migrant Shrike
Ruddy Duck	Sanderling	

IV. WINTER VISITANTS—41

All species on the November list. While a few of these have not actually been recorded in February this has been undoubtedly due to lack of observation as they were present both in January and March.

MARCH

111 Species Recorded

In March the northward migration becomes distinctly apparent. While we have records of most of the winter visitants there are a few which seem to have already retired northward while there is an evident increase in the numbers of others and of transients and summer residents which have been represented by stray individuals during the winter. In addition to these several transients and summer residents, which have not been seen since autumn, are observed on their return flight. There is also a conspicuous northward movement of waterfowl.

I.   REGULAR RESIDENTS (as in June list)—30

Ia. IRREGULAR RESIDENTS (as in June list)—15

II. REGULAR SUMMER RESIDENTS—10

Those marked with an * are migrants from the south, the others are individuals that have remained over the winter.

Black-crowned Night	*Laughing Gull	Towhee
Heron	Catbird	Sharp-tailed Sparrow
*Osprey	Brown Thrasher	*Chipping Sparrow
*Piping Plover	*Pine Warbler	

IIa. Irregular Summer Residents—3

*Pied-billed Grebe	Savannah Sparrow	*Yellow-throated Warbler

III. Transients—17

Gannet	Ruddy Duck	Semipalmated Sandpiper
Whistling Swan	Green-winged Teal	Sanderling
Brant	Blue-winged Teal	Phoebe
Mallard	American Merganser	Migrant Shrike
Baldpate	Hooded Merganser	Western Palm Warbler
Pintail	Wilson's Snipe	

IV. Winter Visitants—36

All the birds listed in November have been recorded in March with the exception of the Red-backed Sandpiper, Short-eared Owl, Pipit, Pine Siskin, Snow Bunting and Lapland Longspur, but individuals of these may well have been present. The Northern Shrike has also been recorded.

APRIL

139 Species Recorded

In April we see the return of the majority of our summer resident birds and the passing northward of the early transients and the great majority of the winter visitants, leaving only a few individuals of the latter to be recorded in May. The great spring migration continues throughout the month and some species are nesting before its close.

I. Regular Residents (as in June list)—30
Ia. Irregular Residents (as in June list)—15

II. Regular Summer Residents—30

Green Heron	Chimney Swift	Yellow Warbler
Black-crowned Night Heron	Hummingbird	Pine Warbler
	Kingbird	Prairie Warbler
American Bittern	Rough-winged Swallow	Ovenbird
Osprey	Barn Swallow	Maryland Yellow-throat
Virginia Rail	Purple Martin	Towhee
Piping Plover	House Wren	Grasshopper Sparrow
Spotted Sandpiper	Short-billed Marsh Wren	Henslow's Sparrow
Laughing Gull	Catbird	Sharp-tailed Sparrow
Whip-poor-will	Brown Thrasher	Chipping Sparrow
	White-eyed Vireo	

IIa. Irregular Summer Residents—4

Pied-billed Grebe	Black and White Warbler
Gnatcatcher	Yellow-throated Warbler

III. Northbound Transients—29

Gannet	Semipalmated Plover	Sanderling
Cormorant	Black-bellied Plover	Yellow-bellied Sapsucker
Whistling Swan	Turnstone	Bank Swallow
Brant	Wilson's Snipe	Cliff Swallow
Mallard	Hudsonian Curlew	Phoebe
Baldpate	Greater Yellow-legs	Ruby-crowned Kinglet
Pintail	Lesser Yellow-legs	Veery
Blue-winged Teal	Least Sandpiper	White-breasted Nuthatch
Green-winged Teal	Dowitcher	Yellow Palm Warbler
Coot	Semipalmated Sandpiper	

IV. Winter Visitants—31

Common Loon	Sharp-shinned Hawk	Myrtle Warbler
Red-throated Loon	Cooper's Hawk	Evening Grosbeak
Horned Grebe	Red-tailed Hawk	Pine Siskin
Canada Goose	Herring Gull	Ipswich Sparrow
Scaup Ducks	Ring-billed Gull	Vesper Sparrow
American Scoter	Bonaparte's Gull	Junco
White-winged Scoter	Red-breasted Nuthatch	Tree Sparrow
Surf Scoter	Brown Creeper	White-throated Sparrow
Old-squaw	Winter Wren	Swamp Sparrow
Red-breasted Merganser	Hermit Thrush	Fox Sparrow
	Golden-crowned Kinglet	

MAY

178 Species Recorded

By the middle of May the northward migration both of small land birds and shore birds reaches its height and by the last of the month all but a few straggling transients have passed, while all the summer residents are nesting. A few individuals of the winter visitant species linger on to the middle of the month before they take their departure while a few Herring Gulls and Scoter Ducks fail to go north with their fellows and remain all summer, probably non-breeding individuals.

I. Regular Residents (as in June list)—30

Ia. Irregular Residents (as in June list)—15

II. Regular Summer Residents (as in June list)—46

IIa. Irregular Summer Residents (as in June list)—16

III. Northbound Transients—49

Cormorant	Wilson's Snipe	White-rumped Sandpiper
Hooded Merganser	Solitary Sandpiper	Dowitcher
Semipalmated Plover	Willet	Hudsonian Godwit
Black-bellied Plover	Greater Yellow-legs	Semipalmated Sandpiper
Turnstone	Knot	Sanderling
Hudsonian Curlew	Least Sandpiper	Black Tern

Red-headed Woodpecker
Yellow-bellied Sapsucker
Bank Swallow
Cliff Swallow
Veery
Olive-backed Thrush
Gray-cheeked Thrush
Ruby-crowned Kinglet
Solitary Vireo
Yellow-throated Vireo
Blue-winged Warbler

Golden-winged Warbler
Tennessee Warbler
Cape May Warbler
Magnolia Warbler
Black-throated Blue
    Warbler
Black-throated Green
    Warbler
Chestnut-sided Warbler
Bay-breasted Warbler
Black-poll Warbler

Yellow Palm Warbler
Northern Water-Thrush
Blackburnian Warbler
Kentucky Warbler
Canada Warbler
Wilson's Warbler
Redstart
Baltimore Oriole
Scarlet Tanager
Rose-breasted Grosbeak
White-crowned Sparrow

IV. Winter Visitants—18

Common Loon
Red-throated Loon
Canada Goose
Scaup Ducks
American Scoter
Surf Scoter

American Merganser
Cooper's Hawk
Herring Gull
Ring-billed Gull
Bonaparte's Gull
Red-backed Sandpiper

Red-breasted Nuthatch
Myrtle Warbler
Evening Grosbeak
Vesper Sparrow
Junco
White-throated Sparrow

V. Summer Visitants from the South—4

American Egret

Little Blue Heron
Mississippi Kite

Gray Kingbird

J. F. S.

BREEDING AND WINTER RANGES OF
NEW JERSEY BIRDS

To properly understand the migratory movements of our birds it is necessary to know where they spend the summer and where they winter as between the two lie their semiannual migratory flights, sometimes by the same route, sometimes differing in spring and autumn. The grouping which follows is based primarily upon the American Ornithologists' Union "Check List" to which the reader is referred for more detailed statements as those here given are necessarily only approximate.

Under "Summer Homes" all New Jersey birds are included but under "Winter Homes" only those which range south of the United States. In stating "Breeding Ranges" only the eastern portion of the range is considered. Species inclosed in brackets are extinct in New Jersey.

SUMMER HOMES OF OUR BIRDS

BIRDS BREEDING IN NEW JERSEY

I. Species that nest through most of the state, where conditions are suitable, including Cape May County (the few that do not nest in the county are marked with an *).

The breeding range of these birds in the East extends north to the Canadian Provinces, sometimes just across the line, sometimes far to the northwest, even to Alaska, following the trend of the timber line. To the south some of them breed to Florida and the Gulf Coast; others (as indicated) not so far. Where a species is represented in Florida or elsewhere by a slightly different race this has been included, *i. e.*, the range of the species rather than the subspecies is indicated.

Pied-billed Grebe
Great Blue Heron
Green Heron
Black-crowned Night Heron
American Bittern (to N. J.)
Least Bittern
Mallard (to Va.)
Black Duck (to Del.)
Blue-winged Teal (to Md.)
Wood Duck
Sharp-shinned Hawk
Cooper's Hawk
Red-tailed Hawk
Red-shouldered Hawk
Broad-winged Hawk
Bald Eagle
Marsh Hawk (to Va.)
Osprey

Sparrow Hawk
Ruffed Grouse (to Va., mts. to Ga.)
Virginia Rail (to N. C.)
Piping Plover (to N. C.)
Killdeer
Woodcock
Spotted Sandpiper (to S. C.)
Common Tern (to N. C. and Fla. Keys)
Roseate Tern (to Va. and Tortugas)
Mourning Dove
[Passenger Pigeon (to Pa.)]
Yellow-billed Cuckoo
Black-billed Cuckoo (mts. to Ga.)
Screech Owl
Great Horned Owl
Barred Owl
Short-eared Owl (to N. J.)
Whip-poor-will (to Ga.)

Nighthawk
Chimney Swift
Ruby-throated Hummingbird
Belted Kingfisher
Flicker
Pileated Woodpecker
Red-headed Woodpecker
Hairy Woodpecker
Downy Woodpecker
Kingbird
Great Crested Flycatcher
Wood Pewee
Tree Swallow (to Va.)
Barn Swallow (to N. C.)
Purple Martin
Blue Jay
Crow
House Wren (to Va., mts. to S. C.)
Short-billed Marsh Wren (to Del.)
Catbird
Brown Thrasher
Robin
Bluebird

Cedar Waxwing (to N. C.)
Yellow-throated Vireo
Red-eyed Vireo
Black and White Warbler (to Ga.)
Parula Warbler
Yellow Warbler (to Ga.)
Pine Warbler
Ovenbird (to Ga.)
Maryland Yellow-throat
Bobolink (to W. Va.)
Meadowlark
Redwing
Cowbird (to Va.)
Scarlet Tanager (mts to Ga.)
Indigo Bunting (to Ga.)
Goldfinch (to n. Ga.)
Savannah Sparrow (to Del.)
Sharp-tailed Sparrow (to Va.)
Vesper Sparrow (to N. C.)
Chipping Sparrow (to Ga.)
Field Sparrow
Song Sparrow (mts. to Ga.)

II. Additional species that nest in northern or central New Jersey and thence north to the Canadian Provinces and south to parts of the Southern States but not in Cape May County. Either the southern boundary of their breeding range crosses the middle of the state in a northeast-southwest line, or conditions in the county are not suitable.

Sora (to Md.)
Upland Plover (to Va.)
Phoebe (mts. to Ga.)
Long-eared Owl (to Va.)
Least Flycatcher (mts. to N. C.)
Alder Flycatcher (mts. to N. C.)
Prairie Horned Lark (to Md.)
Bank Swallow (to Va.)
Raven (mts. to Ga.)

Black-capped Chickadee (mts. to N. C.)
White-breasted Nuthatch (to Fla.)
Veery (mts. to N. C.)
Warbling Vireo (to N. C.)
Redstart (to N. C.)
Baltimore Oriole (to Ga.)
Rose-breasted Grosbeak (mts. to Ga.)
Swamp Sparrow (mts. to W. Va.)

III. Species of northern affinities which breed north to the Canadian Provinces but which nest in New Jersey only in the mountains of Passaic, Sussex or Morris Counties. Most of them breed also in the Alleghany Mountains to Virginia and North Carolina. While all have been found in the New Jersey mountains in the breeding season the nests of some of them have not yet been actually found there.

Duck Hawk (in mts. to Tenn.)
Wilson's Snipe (in mts. to Pa.)
Coot (locally further south)
Olive-sided Flycatcher (mts. to N. C.)
Brown Creeper (mts. to N. C.)
Hermit Thrush (mts. to Va.)
Blue-headed Vireo (mts. to N. C.)
Nashville Warbler (mts. to Pa.)
Magnolia Warbler (mts. to Va.)

Black-throated Blue Warbler (mts. to N. C.)
Black-throated Green Warbler (mts. to Ga.)
Blackburnian Warbler (mts. to Ga.)
Chestnut-sided Warbler (mts. to Ga.)
Northern Water-Thrush (mts. to W. Va.)
Canada Warbler (mts. to Ga.)
Purple Finch (mts. to Md.)

IV. Species of southern affinities many of which nest only so far north as New Jersey while others breed, less plentifully, north to the lower Hudson and Connecticut Valleys or to southern New England.

a. Breeding north to New England often only casually:

Bobwhite
[Heath Hen]
[Wild Turkey]
Florida Gallinule
King Rail
Clapper Rail
Black Rail
Laughing Gull
Least Tern
Rough-winged Swallow
Barn Owl
Acadian Flycatcher
Long-billed Marsh Wren

Mockingbird
Wood Thrush
White-eyed Vireo
Blue-winged Warbler
Golden-winged Warbler
Prairie Warbler
Louisiana Water-Thrush
Orchard Oriole
Towhee
Grasshopper Sparrow
Henslow's Sparrow
Seaside Sparrow

b. Breeding north to the lower Hudson and Connecticut Valleys:

Turkey Vulture
Fish Crow
Carolina Wren
Worm-eating Warbler
Yellow-breasted Chat

Kentucky Warbler
Hooded Warbler
Purple Grackle
Cardinal
Atlantic Song Sparrow (to Long Island)

c. Breeding north to New Jersey. (Present in breeding season but nests not yet actually found*.)

American Egret
Little Blue Heron
Yellow-crowned Night Heron
Willet (also in N. S.)
Black Skimmer (once on L. I.)
*Chuck-will's-widow

Carolina Chickadee
Tufted Titmouse
Blue-gray Gnatcatcher
Prothonotary Warbler (once)
*Yellow-throated Warbler

V. Species of the Middle West which formerly or occasionally have nested in New Jersey.

Cliff Swallow   Bewick's Wren   Dickcissel

Birds Breeding Entirely North of New Jersey

I. Species breeding from the Canadian Provinces, Alaska, or Greenland south to northern New England and the mountains of New York; occasionally further. They occur in New Jersey as transients or winter visitants.

Common Loon
Horned Grebe
Leach's Petrel
Double-crested Cormorant
Golden-eye
American Eider
Hooded Merganser
American Merganser
Red-breasted Merganser
Goshawk (in mts. to Md.)
Pigeon Hawk
Yellow Rail
Solitary Sandpiper (in mts. to Pa.?)
Herring Gull
Ring-billed Gull
Arctic Tern
Black Guillemot
Puffin
Saw-whet Owl (in mts. to Pa.)
Yel.-bellied Sapsucker (in mts. to N. C.)
Arctic Three-toed Woodpecker
Yellow-bellied Flycatcher (mts. to Pa.)
Red-breasted Nuthatch (mts. to N. C.)
Acadian Chickadee
Winter Wren (mts. to Ga.)

Olive-backed Thrush (mts. to W. Va.)
Bicknell's Thrush
Golden-crowned Kinglet (mts. to N. C.)
Ruby-crowned Kinglet
Migrant Shrike (mts. to N. C.)
Philadelphia Vireo
Tennessee Warbler
Cape May Warbler
Myrtle Warbler
Bay-breasted Warbler
Black-poll Warbler
Yellow Palm Warbler
Mourning Warbler (mts. to W. Va.)
Wilson's Warbler
Rusty Blackbird
Bronzed Grackle
Pine Grosbeak
Pine Siskin (mts. to N. C.)
Red Crossbill (mts. to Ga.)
White-winged Crossbill
Ipswich Sparrow (Sable Island only)
Acadian Sparrow
Slate-colored Junco (mts. to Pa. and N.C.)
White-throated Sparrow (mts. to Pa.)
Lincoln's Sparrow

II. Species breeding from Alaska or Alberta to Hudson Bay and into the north-central United States. Transients or winter visitants in N. J.

White Pelican
[Whooping Crane]
Gadwall
Baldpate
Pintail
Green-winged Teal
Shoveller
Redhead
Canvas-back
Ring-necked Duck
Buffle-head
Ruddy Duck
Golden Eagle
Marbled Godwit
Long-billed Curlew

Western Willet
Avocet
Caspian Tern (also Gulf coast)
Black Tern
Arkansas Kingbird
Magpie
Willow Thrush
Western Palm Warbler
Grinnell's Water-Thrush
Connecticut Warbler
Yellow-headed Blackbird
Evening Grosbeak (once in Vt.)
Nelson's Sparrow
Clay-colored Sparrow

III. Species breeding wholly north of the United States.   Transients or winter visitants in New Jersey.

Red-throated Loon
Holboell's Grebe
Fulmar
Gannet
European Cormorant
Whistling Swan
White-fronted Goose
Blue Goose
Lesser Snow Goose
Canada Goose
Hutchins' Goose
American Brant
Black Brant
Greater Snow Goose
Red-legged Black Duck
Greater Scaup (N. Dak.?)
Lesser Scaup (N. Dak.?)
Oldsquaw
Harlequin
[Labrador Duck]
King Eider
White-winged Scoter (N. Dak.?)
Surf Scoter
American Scoter
Rough-legged Hawk
Semipalmated Plover
Black-bellied Plover
Golden Plover
Turnstone
Hudsonian Godwit
Hudsonian Curlew
[Eskimo Curlew]
Pectoral Sandpiper
White-rumped Sandpiper
Baird's Sandpiper
Least Sandpiper
Red-backed Sandpiper
Dowitcher
Long-billed Dowitcher
Stilt Sandpiper

Greater Yellow-legs
Lesser Yellow-legs
Knot
Purple Sandpiper
Semipalmated Sandpiper
Western Sandpiper
Buff-breasted Sandpiper
Sanderling
Red Phalarope
Northern Phalarope
Pomerine Jaeger
Parasitic Jaeger
Long-tailed Jaeger
Kittiwake
Glaucous Gull
Iceland Gull
Black-backed Gull
Thayer's Gull
Bonaparte's Gull
Razor-billed Auk
Brunnich's Murre
Dovekie
Snowy Owl
Great Gray Owl
Hawk Owl
Horned Lark
Gray-cheeked Thrush
Varied Thrush
Pipit
Bohemian Waxwing
Orange-crowned Warbler
Northern Shrike
Common Redpoll
Greater Redpoll
Tree Sparrow
White-crowned Sparrc
Fox Sparrow
Snow Bunting
Lapland Longspur

BIRDS BREEDING ENTIRELY SOUTH OF NEW JERSEY

I. Species breeding in the South Atlantic States as far north as indicated. They occur in New Jersey only as stragglers.

Brown Pelican (to S. C.)
Snowy Egret (to N. C.)
Louisiana Heron (to N. C.)

Glossy Ibis (Fla.)
White Ibis (to S. C.)
Wood Ibis (to S. C.)

Black Vulture (to Va.)
Swallow-tailed Kite (to S. C.)
Mississippi Kite (to S. C.)
Purple Gallinule (to S. C.)
Oyster-catcher (to Va.)
Wilson's Plover (to Va.)
Black-necked Stilt (Fla.)
Sooty Tern (Tortugas)
Cabot's Tern (to Va.)
Royal Tern (to Va.)

Gull-billed Tern (to Va.)
Ground Dove (to S. C.)
Red-bellied Woodpecker (to Del.)
Red-cockaded Woodpecker (to Va.)
Gray Kingbird (to Ga.)
Brown-headed Nuthatch (to Del.)
Boat-tailed Grackle (to Del.)
Summer Tanager (to Del.)
Blue Grosbeak (to Md.)

II. Species breeding wholly south of the U. S.; accidental in N. J.

Audubon's Shearwater (Bahamas)   Forked-tailed Flycatcher (S. A.)
Man-o'-war-bird (West Indies)

III. Species breeding in the Antarctic regions and spending the summer (their winter) period on the Atlantic off our eastern seaboard.

Greater Shearwater (Inaccessible Is-   Sooty Shearwater (New Zealand)
      land)   Wilson's Petrel (Antarctic Islands)

EUROPEAN BIRDS THAT HAVE OCCURRED IN NEW JERSEY

All but Cory's Shearwater are accidental stragglers.

Cory's Shearwater (Azores; winters off   Corn Crake
      our coast)   Curlew Sandpiper
European Teal   Ruff
European Woodcock   Lesser Black-backed Gull
European Widgeon   Little Gull

BIRDS INTRODUCED BY MAN

Mute Swan   Starling
Ring-necked Pheasant   English Sparrow
Domestic Pigeon   British Goldfinch
Skylark

WINTER HOMES OF OUR BIRDS

A large number of our familiar summer birds and some of those that pass through New Jersey on migration in spring and autumn, spend the winter in the Southern States sometimes extending their range over the line into northern Mexico. Many others do not go farther south than the Middle States to spend the winter while some individuals of species that normally go south are content to remain here until spring.

There are many other birds that winter wholly south of the United States. Some cross to the West Indies and the Bahamas while others proceed around the Gulf coast to southern Mexico or Central America or boldly cross the waters of the Gulf of Mexico to northern South America and of the latter some continue their flight down the Andes to Peru or Bolivia or travel across Brazil to the Pampas of the Argentine.

The groupings given below are only approximate as straggling individuals complicate any attempt to make a complete statement. They represent however, the winter range of the bulk of each species. Only species wintering south of the United States are listed.

I. Wintering from the southern border of the United States to South America:

Broad-winged Hawk
Osprey
Pigeon Hawk
Sora
Semipalmated Sandpiper
Turnstone
Wilson's Snipe
Spotted Sandpiper
Solitary Sandpiper
Willet

Greater Yellow-legs
Least Sandpiper
Dowitcher
Semipalmated Sandpiper
Western Sandpiper
Marbled Godwit
Laughing Gull
Common Tern
Black Skimmer

II. Wintering from the southern border of the United States to Central America or Panama:

Whip-poor-will
Ruby-throated Hummingbird
Catbird
Gnatcatcher

Ruby-crowned Kinglet
White-eyed Vireo
Blue-headed Vireo

III. Wintering in the West Indies and the Bahamas, often also in Florida.

Black and White Warbler
Parula Warbler
Cape May Warbler
Black-throated Blue Warbler
Maryland Yellow-throat
Yellow-throated Warbler

Prairie Warbler
Western Palm Warbler
Ovenbird
Northern Water-Thrush
Redstart

IV. Wintering from southern Mexico or northern Central America to the Andes of South America:

Kingbird
Crested Flycatcher
Yellow-bellied Flycatcher
Alder Flycatcher
Wood Pewee
Barn Swallow
Olive-backed Thrush
Philadelphia Vireo
Yellow-throated Vireo

Black and White Warbler
Prothonotary Warbler
Worm-eating Warbler
Golden-winged Warbler
Tennessee Warbler
Yellow Warbler
Magnolia Warbler
Lousiana Water-Thrush
Northern Water-Thrush

Kentucky Warbler
Mourning Warbler
Hooded Warbler
Redstart

Orchard Oriole
Baltimore Oriole
Rose-breasted Grosbeak

V. Wintering in the Andes of South America:

Yellow-billed Cuckoo
Black-billed Cuckoo
Acadian Flycatcher
Olive-sided Flycatcher
Gray-cheeked Thrush
Veery
Red-eyed Vireo

Cerulean Warbler
Blackburnian Warbler
Bay-breasted Warbler
Connecticut Warbler
Canada Warbler
Scarlet Tanager

VI. Wintering from southern Mexico to Panama, or wholly in Central America:

Rough-winged Swallow
Wood Thrush
Warbling Vireo
Blue-winged Warbler
Nashville Warbler
Parula Warbler
Black-throated Green Warbler

Chestnut-sided Warbler
Maryland Yellow-throat
Yellow-breasted Chat
Wilson's Warbler
Redstart
Indigo Bunting

VII. Wintering in central or southern Brazil:

Hudsonian Curlew
Nighthawk
Chimney Swift (probably)

Purple Martin
Bobolink
Black-poll Warbler

VIII. Wintering on the Pampas of Argentina etc., or in Patagonia:

Golden Plover
Upland Plover
Lesser Yellow-legs
Knot
Pectoral Sandpiper

White-rumped Sandpiper
Baird's Sandpiper
Stilt Sandpiper
Buff-breasted Sandpiper

PART II

BIRD STUDIES AT OLD CAPE MAY

WITH BRIEF MENTION OF
OTHER BIRDS THAT HAVE BEEN RECORDED
ELSEWHERE IN THE STATE OF
NEW JERSEY

In preparing the accounts of the birds found in the Cape May area it seemed desirable to include records of their occurrence elsewhere along the shore, as well as New Jersey records of other oceanic or coastal species not yet found in Cape May County.

As but few additional species have been recorded from elsewhere in the state, brief reference to these has also been included so that the following pages contain mention of all the birds of New Jersey.

Birds not yet recorded from Cape May County, fifty-six of a total of 374, are marked by an * before the name.

COMMON LOON

Gavia immer immer (Brünnich)

The Common Loon, though not the most conspicuous, is probably the most characteristic bird of the Harbor and ocean front during the winter. No matter how cold it may be, or what the weather, a careful survey of the inlet and the waters out beyond the surf line will disclose scattered individuals resting quietly on the surface or buoyantly riding the waves if the sea be rough.

Frequently we find them feeding, diving with an ease and grace that enlists our admiration, and bobbing up again so quietly that we often fail to see them emerge. The number to be seen on a winter's walk skirting the Harbor and along the beach to Cape May Point may vary from one to twenty, probably more in December, before the ice begins to collect about the mouth of the Bay, than later in the season. My records show Loons present at Cape May from mid-November to the end of April and most abundant during the first and last months of this period when the migration is in progress. Individuals, of course, occur earlier in the autumn and I have a record for October 18, while Irwin Stevens tells me that he has occasionally seen them during the last week of September. Stragglers likewise remain later in the spring after the main flight goes north. Even as late as May 21, 1922, I saw three on the Harbor and single birds up to May 25, 1922, and May 27, 1929, while Julian Potter saw one on June 18, 1923, off the end of the jetty swimming out beyond the surf line. Together we saw another on the flat sand bar at Anglesea, on July 8, 1922, which may have had its plumage covered with crude oil, as it is very unusual to see a Loon in this vicinity, out of the water; it, however, took to the sea apparently uninjured. Possibly it was the same bird that had been reported by Richard Miller off Two Mile Beach on June 17. On June 9, 1929, Walker Hand saw two Loons on the Harbor which acted as if mating. These late occurrences, however, were probably all non-breeding birds which may have remained in the vicinity all summer. While usually occurring singly in winter time we often find two or three Loons together on the Harbor, or on the thoroughfares or salt creeks of the meadows to which they frequently resort. In winter plumage Loons have the entire under surface of the body white, including the fore neck and throat, and the back gray or blackish, very different from the black-necked checker-backed adults of spring and summer.

On February 11, 1921, I approached quite closely to three birds floating on the surface of Cape Island Creek back of Schellenger's Landing. One, evidently an old adult, was larger and blacker than the others and purer white below, with a snowy patch on each side of the head, the others were grayer above with whitish edgings to the feathers. They rested with the body rather high in the water but seemed able to sink lower at will and one bird, coming to the surface after a dive and finding himself too close to a passing boat, swam for some time with only the head and neck above the water while later only the head was visible. While swimming they several times stretched the head and neck forward along the surface so that the water was level with the eyes. Their object was not apparent as they were not frightened or in any need of concealment, being entirely undisturbed at the time. Every now and then the old bird would spread his tail feathers like a fan, holding them vertical as he swam about. Several times, too, he reared up in the water and flapped his wings vigorously.

On March 24, 1923, six Loons were seen on the Harbor, two of which were preening their plumage. They would stretch the neck far back pecking at the feathers just above the rump and then would roll over on one side while they probed into the plumage of the breast and belly. During this operation the white breast was exposed to the sunlight and flashed out conspicuously against the water. As the bird lay on its side one foot would be thrown up in the air like a balancer and sometimes the body would revolve until the head pointed in exactly the opposite direction from that which it held at first. At other times the vigorous preening would cause the bird to turn almost over on its back so that it required a sudden effort to right itself. Another individual was going through precisely the same action on January 19, 1924, lying on its side for a considerable period, with one leg outstretched in the air, tail spread, and the other leg presumably paddling as the bird slowly revolved; presenting first the white belly, then the tail, then the black back until finally, completing the revolution, it faced me. Half an hour before it must have been doing the same thing as from far across the Harbor I saw its white breast flash into sight and disappear again at regular intervals, like the revolving light of the old Lighthouse. Two other Loons accompanied this bird which was larger and blacker than his companions. Three Loons observed swimming down the coast beyond the breakers (February 11, 1921) carried themselves high in the water and the rays of the setting sun lit up their snowy white breasts conspicuously. Frequently they turned their heads back over their shoulders.

One bird studied carefully (November 12, 1920) was swimming and diving in almost still water near the mouth of the Bay. The neck appeared relatively short for such a large bird, the bill large and distinctly light in color, while the back and head were glistening silvery gray in the bright sunlight. Indeed the effect of light and shadow on the apparent color of Loons is an important factor in determining their actual coloration. This bird carried its neck directed a little forward as it swam and the back was distinctly arched, the lowest point being at the juncture with the neck which was almost on the water line. The bill was slightly elevated as the bird looked now this way and now that for any possible danger, and all being satisfactory it bent the neck over and slid head foremost under the water, turning up the tail just as it disappeared. The whole action was exceedingly graceful and without apparent effort; just a slight heave of the shoulders and it had submerged. It remained under water for thirty to thirty-five seconds on a number of successive dives, a period of fifteen to twenty seconds elapsing between them. Another bird timed on the Harbor (March 24, 1923) in perfectly still water remained submerged sixty, fifty-five and fifty-seven seconds on successive dives. When frightened, Loons can remain under water for longer periods and the adult bird that I watched on the creek at Schellenger's Landing, upon being approached, dived and passed under the bridge and out into the Harbor for a couple of hundred yards without coming to the surface, so far as I could see, though it might have stuck its head out and dived again without being detected.

Occasionally one is fortunate enough to see a Loon rise in flight. A bird swimming on the sea on April 3, 1922, off Cape May Point, took to the air for some reason or other while I was watching it through the glass. It stretched its neck to the fullest extent, arching it slightly and flapped along vigorously on the surface, kicking lustily with its feet and finally drawing them up against the tail as it got clear of the water. The rather long, narrow, rapidly beating wings seemed inadequate to keep the body aloft. After gaining momentum it rather suddenly raised to an elevation of twelve or fifteen feet, traveling for a distance at that altitude, and then descending on a slope until with a splash it dropped again into the water. Another individual (May 23, 1922) started from shallow water on Jarvis Sound and flew close to the surface with rapid wing strokes and labored flight, but soon dropped with a considerable splash into deeper water. It held the body at an angle of about 45° with the surface, the head being elevated, and struck with the tail first, dragging the body, as it were, through the water. Single birds seen flying northward overhead, May 24 and 25, 1922, and several on May 10, 1925, were traveling at considerable speed at an altitude of fifty feet or more with constant rapid wingbeats. The body, neck and head of the flying Loon are in a perfectly straight line and the bird seems to taper gradually from the shoulders to the tip of the sharp-pointed bill, with no clear demarkation between head and neck, as in a duck.

Adult Loons acquire the nuptial plumage in early spring and one specimen taken on April 20, 1879, and another seen on May 18, 1924, had completed the molt. Young birds retain the grayish brown dress throughout the winter and the following summer and do not assume the breeding plumage until their second spring. One in the dull dress of immaturity was seen on the Harbor on May 27, 1929. Old birds in winter may be distinguished from young by their dark, almost black, backs. Walker Hand informed me that Loons frequently cross over the town in migration time singly or in bunches of two to five. Foggy and overcast days are their preference for migratory flights, in daytime, but judging from their occurrence here these flights are as frequently performed at night.

Fishermen are familiar with the cry of the Loon which is heard at Cape May, more especially in foggy weather, offshore, a tremulous, weird and mournful yell, once heard, never to be forgotten. Immediately before a storm they become particularly vociferous and uneasy and seafaring men always regard their calling as a forecast of bad weather. Loons passing over the town at night on migration sometimes give voice to their wailing cry, but the winter resident Loons that we see most frequently on sea and Harbor are generally silent. A Loon that I found on the beach on April 30, 1923, incapable of flight from an accumulation of oil on its plumage, uttered vigorous protests as I approached it. Its call as it caught sight of me was a long-drawn wailing cry *koo-lairrrrrrrrr'—ooo*, accented on the long drawn out second syllable and falling to a short low note at the end. It also gave a rapid *cac-cac-cac* cry, and a goose-like *go-la'rgle* as I poked it with a stick. It was very pugnacious and flew at my legs with wide spread wings and lunging bill. It rested partly on its breast and could not support itself on its legs, progress being made by shoving its heavy body forward with its widely divergent, flattened tarsi, while it made a succession of forward plunges. The curious trail through the sand showed how far the poor bird must have dragged itself, and although close to the water's edge it made no effort to take to it, probably aware of its disabled condition. It twisted its thick neck about so that it could dart its bill in any direction, the short velvety feathers standing on end like ruffled fur and the little eyes glittering as if with anger and hate.

It is hard to imagine a more perfect water bird than the Loon as we know him in his winter home on our New Jersey coast. Helpless if he be cast on the land and of labored flight through the air, he is a perfect master of the water, whether feeding or resting, swimming or diving. At Cape May we associate him naturally with the winter—with the ice cold waters of the ocean and Bay; with piercing winds coursing over bleak brown marshes and with fog banks out beyond the surf. Thence at intervals comes his wailing cry which seems to impress one with a feeling of utter loneliness, as if the bird were expressing a preference for solitude and the lack of any desire for communion with mankind or with other fellow beings.

RED-THROATED LOON

Gavia stellata (Pontoppidan)

The Red-throated Loon, which so far as general appearance and habits go, is a small edition of the other species, is found about Cape May throughout the winter months but scarcely any of the local fishermen or gunners recognize it as distinct.

So far as my experience goes we see it only in the plain gray and white plumage of winter and I have never seen a red-throated individual on the New Jersey coast while Charles Urner in his long experience in the Barnegat region has found but one, a belated individual on June 20, 1923, which had attained the full nuptial plumage. As we see it floating on the water the Red-throated Loon appears dull gray except for the snowy white breast, which is sometimes exposed as it rolls over on its side; there are no black markings whatever. Besides its generally smaller size the bill is noticeably smaller and apparently curves up more sharply on the under side than in the Common Loon, giving the head a rather different profile. This small Loon seems to be more a bird of the ocean, where it is frequently seen, and does not so often enter the Harbor as does the larger species. In midwinter we see a few individuals on almost every visit to the shore and we have records from mid-November to the end of April and single individuals as late as May 10, 1929, and May 7, 1933, but during the migratory seasons at the beginning and end of this period, and occasionally in winter, the Red-throated Loon seems to become truly gregarious and occurs in such numbers as to make one wonder if, after all, it is not the more abundant of the two. At such times the "ripps" at the entrance of Delaware Bay are a favorite resort and Walker Hand saw upwards of one hundred there on April 9, 1910, April 10, 1921, and April 24, 1914, while on another occasion he saw the Coast Guards' powerboat, returning from a wreck off the Point, put into a flock of Loons, evidently of this species, and force them to simultaneous flight. On March 17, 1927, I found thirty Red-throated Loons at the ripps busily engaged in feeding along the line of rough water. They were diving constantly, some emerging as others disappeared, while still others were taking wing only to fly a few yards and splash back again into the water. Both air and water seemed to be thronged with them. The wings of the flying bird appeared very narrow and the beats very rapid, while the head and neck were stretched out in advance and the feet behind, the neck being slightly arched. A month later (April 24) the flock had increased enormously and we estimated that it contained at least five-hundred birds. They were in three sections of approximately equal size and several counts of one section through the glass showed one hundred and forty to one hundred and fifty birds above water. The scene was one of wild activity as the birds dived and flew back and forth, keeping always in the smoother water but as close to the ripps as possible. On January 16, 1932, there were 175 birds present, individuals constantly rising from the water and flying for considerable distances so that there were always many in the air. They seemed to take wing much more easily and more quickly than does the Common Loon. On March 23, 1930, some of them came in so close that Edward Weyl was able to identify them with ease. None, however, had the red throat of the adult, nor had any of the fifty that were still there on April 5.

Like the other birds which feed at the ripps their numbers vary greatly from day to day; so on February 5 of the same year there were but five in evidence at the same spot, while on March 5 they had increased to fifty. Doubtless the condition of wind and tide and the consequent presence or absence of fish is the cause of these fluctuations. Of the birds observed at the ripps on November 28, 1931, some flew quite high above the water and splashed in as they came down diagonally while others flew close over the ocean. Again, I saw one flushed by a passing boat fly high up in order to pass over it and descending rapidly dived the moment it struck the water. In a good light these flying Loons showed a distinct whitish edging to the spread wing in contrast to the general gray color.

I have looked upon the large assemblages of Red-throated Loons in April as gatherings preliminary to migration and Julian Potter has seen them on April 22, 1923, and April 18, 1926, flying up the coast in bunches of five, ten and fifteen with many scattered individuals. In the autumn there is a similar migration and on November 15, 1931, he counted six hundred in about an hour, passing south off Long Beach farther up the coast accompanied by Gannets, of which 115 passed in the same space of time.

HOLBOELL'S GREBE

Colymbus grisegena holboelli (Reinhardt)

This bird, the greatest of our grebes, is essentially a large edition of the Horned Grebe. It is present at Cape May in small numbers throughout the winter months and we have records from November 28, in 1926, to March 23, in 1930. It doubtless occurs more abundantly than our experience would indicate but it attracts little attention and opportunities for continuous winter observations have been lacking.

Its yellowish bill and generally larger size are the best marks of identification as compared with the Horned Grebe. As we see it at Cape May it has always been in the winter plumage, dusky above and white below, but Julian Potter saw an adult on the Delaware River near Philadelphia on May 13 and 15, 1919, in full red-throated nuptial plumage. It was diving for small fishes and one was brought up at nearly every plunge which was dipped again into the water and shaken vigorously before swallowing.

Holboell's Grebes like the Red-throated Loons seem to assemble in flocks before migration as we have seen several together at such times. Farther up the coast, Julian Potter found them on the ocean off Brigantine on February 22, 1926, and Joseph Tatum saw them at the same place on February 22, 1930. After the severe cold weather of February, 1934, Tatum found ten dead on the coast of Long Beach and among the 120 frozen birds examined there by Charles Urner no less than twenty-four were Holboell's Grebes. There were some inland records at this time and it would seem that a flight of the birds from much farther north had occurred as a result of the cold. One was found as far inland as Lititz, Lancaster County, Pa., by Barton Sharp.

HORNED GREBE

Colymbus auritus (Linnaeus)

These little divers are seen throughout the winter on the Harbor, the sounds and thoroughfares, on salt pools along the coast, and, less frequently, out on the ocean. They act much as do the loons and swim about with their dark colored backs well arched above the water and their necks held vertically. Sometimes, however, especially when it has been disturbed, a Horned Grebe may sit so low in the water that only the very center of the back is exposed. As a rule they are leisurely in their actions but when danger approaches they quicken their pace and glide easily over the water. It is by diving, however, that they usually seek to escape and when excited the submergence is often so nearly instantaneous that we fail to see just how it is accomplished, although when the birds are undisturbed and busy feeding we can study the operation with ease. There is a spasmodic muscular effort of the whole body, a raising of the shoulders, a curving over of the neck, and, bill first, the bird goes under with scarcely a ripple to mark the spot. One observed on the Harbor, March 24, 1923, remained submerged for from thirty to thirty-five seconds, on a number of successive dives.

Another individual observed on March 11, 1923, by Julian Potter was feeding on the ocean just within the surf line and dived repeatedly into the on-coming waves, emerging each time in the trough of the sea just in time to plunge under the next breaker. On April 2, 1922, Walker Hand and I cornered a Horned Grebe in the shallow water of a blind branch of Cape Island Creek, back of the Harbor. It swam about in great concern with head very erect turning it now to the right now to the left, and once it reared up until its body was two-thirds out of the water, with the possible intention of taking wing, which was what we expected it must do. To our amazement, however, it suddenly dived into the mud and shallow water and kept submerged for some twenty yards when it came to the surface and immediately dived again for fifty yards more into the deeper water and so escaped.

I have never seen Horned Grebes flying overhead in daylight at migration time as loons so often do and it is possible that they migrate entirely by night. Indeed it is not often that we see them flying at any time and if they do take wing it is usually because of adverse wind or tide, which retard their swimming or diving, or due to fright. One bird that was swimming on the Lighthouse Pond (November 6, 1927) arose from the water and flew directly in the face of a strong wind that was blowing across the water. Its neck stretched out in front and its legs behind seemed to balance one another, while its rapidly moving wings produced a flicker of white. I thought that the mass of aquatic vegetation which filled the pond made it difficult for the bird to swim and it was compelled to take wing in order to reach another feeding place. Other individuals seen in flight for short distances had the neck slightly arched, the wings beating duck-like and showing large white areas, the feet stretched out at divergent angles and the big webbed toes conspicuous.

While we usually see Horned Grebes singly or in groups of two to four I found six swimming close together on the Harbor on February 22, 1926, and a flock of thirteen in the same place on March 31, 1929; while Edward Weyl saw forty-two at Beesley's Point on January 28, 1934, and on Barnegat Bay, farther up the coast, I saw a flock of twenty on December 27, 1928. Weyl's birds, he writes me, were definitely in a group but not in a dense group such as ducks form and were not diving when first seen. About an hour later they were widely scattered and were diving continually. On the larger bodies of water I think Horned Grebes are more likely to assemble in flocks and also to take wing, especially when disturbed by passing boats. The Barnegat birds did this repeatedly as our powerboat followed them, and as we looked down upon them the white on their wings was very conspicuous. Sometimes they failed to rise entirely clear of the water and simply plowed over the surface.

Horned Grebes exhibit two plumages as we see them at Cape May. In the winter and immature plumage, in which we most frequently find them, they are black and white, the upper parts mainly dull black and the breast appearing somewhat soiled white. The throat is also white and on each side of the head is a conspicuous white patch separating the black of the crown from that of the sides of the neck. This white spot is really a very conspicuous feature and is visible at a considerable distance. After constant diving the plumage appears very wet, sleek and glossy with the black and white markings strongly contrasted. Often, when the bird is swimming, the feathers of the breast overlap the closed wing so as to produce a white band just above the water line, on either side, and in one instance the effect on the sides of the body below the wings was as if they were finely barred or mottled with dusky, producing a steel gray appearance. In the fully adult breeding plumage of spring, as seen in the bird studied on April 2, 1922, the curious expanded head tufts were fully displayed; the neck was carried slightly curved back over the shoulders and the head somewhat lowered. The entire neck and back were black, breast and sides bright tawny-rusty—very bright in the sunlight; the eyes were bright scarlet and the bill black with a white tip. When this bird emerged from a dive, however, it presented a very different appearance; all of the feathers were plastered down on the body while the head tufts had disappeared so that the head and neck were much more slender. Another individual in full nuptial plumage was seen as early as March 23, 1925, which stood upright in the water to plume itself and fairly leaned over backward in the operation. I have never found a Horned Grebe in the molt. That it has two complete molts each year is certain in order to accomplish the change from the highly colored nuptial plumage to the modest black and white winter dress and back again, but just when the changes take place I have not been able to ascertain.

While we usually see only one to six Horned Grebes on a winter walk about the Cape many more may be seen from a boat on the sounds and in migration time they are everywhere more common. Earl Poole saw nearly a hundred on April 1, 1921. The bulk of our records fall between November 7 and April 1 but there are outlying dates as early as October 21, 1923, and October 22, 1922, and as late as April 18, 1926, May 3, 1914, when Julian Potter saw no less then twenty, and May 26, 1929, when Walker Hand and I saw a single individual on Skunk Sound.

The diving of the Horned Grebes attracts universal admiration, but we have little else in their winter home with which to associate them in memory. They rarely leave the water and their feeding activities are carried on beneath the surface while their keen sight when above water usually detects our presence long before we are aware of theirs and they are immediately on guard. There is, therefore, little opportunity for close acquaintance.

PIED-BILLED GREBE

Podilymbus podiceps podiceps (Linnaeus)

Pied-billed Grebes, while not so well known at the Cape as the other "Hell-Diver"—the Horned Grebe, are seen regularly in spring and autumn and less frequently in midwinter. They seem to be very rare in summer but may occur at this season more frequently than we suppose. The protective character of the bird's coloration and the admirable concealment afforded by the aquatic vegetation of our inland ponds may easily account for our failure to find them here in the breeding season. That they do nest a little farther north in the peninsula and across the Bay in Delaware is well known.

While preferring fresh water these Grebes are by no means confined to it during the migrations and we see them on the Harbor and on brackish ponds in the salt meadows. They do, however, seem to seek localities where aquatic vegetation abounds and are not so frequently seen on the broad expanses of open water where the Horned Grebe is at home; nor have I observed them on the ocean. When floating on the water the Pied-bill sits rather high with the neck somewhat curved backward and never held so erect as in the Horned Grebe, while the large size of the head is a character that will distinguish it at a considerable distance. The general color is dull brown without the contrasting white area on the sides of the head as in the other species, but the feathers of the posterior parts, which seem widely spread as the bird swims away, are rather conspicuously white. On April 2, 1922, a single bird swimming on the inner Harbor near the boat houses and in full nuptial plumage, gave me an excellent opportunity for study while I remained concealed. The head and neck seemed relatively very thick and heavy, while the jet black eye with its narrow white circle and the short heavy whitish bill with its median band of black, were clearly distinguishable with the glass. There was also a vertical band of black at the base of the bill where it joined the feathers. The high arched bill of this species, so different from the slender straight bill of the Horned Grebe, is responsible for the local name of "Hen-bill Diver" which, however, has come to be used indiscriminately for either bird. In winter and immature plumage the bill is duller in color while the underparts are dirty white.

There is also a conspicuous white triangular area on the throat which is distinguishable at a considerable distance. Birds in fully adult plumage were seen on March 5, 1896, and March 31, 1899.

On March 20, 1924, a Pied-bill was found floating on the Lighthouse Pond which was instantly identified by its profile, while still a long way off. It was diving at regular intervals when suddenly, apparently alarmed by the arrival of a Herring Gull, it took wing. Its head and neck were stretched out to their fullest extent, wings beating rapidly like those of a duck and the big lobed feet, clearly seen with the glass, stretched out behind. It traveled some thirty feet, low over the water, and then splashed in again, while going at full speed. Another individual, seen on the Lighthouse Pond, September 21, 1930, was plowing its way slowly through the mass of Utricularia and other plants which covered the surface. Only the middle of the back was exposed the head being thrust below the surface except now and then when it was raised for observation, this method of feeding being doubtless due to the density of the vegetation which rendered diving impracticable. Once the bird raised its body and flapped its short wings and again it attempted to fly but only succeeded in skittering along over the submerged plants being apparently unable to clear itself from them.

Only on rare occasions have I seen Pied-billed Grebes take wing and I have never seen them on protracted flights overhead. Their migrations I presume are performed at night.

Sometimes I have seen them swimming slowly and have noted the head moving forward and back in unison with the alternate strokes of the feet. They dive in the usual grebe fashion plunging over forward and sliding quickly beneath the water, and also by simply sinking below the surface with a slight backward movement. Two individuals seen on the Lighthouse Pond (September 26, 1926) when startled sank out of sight in this manner and upon emerging showed only the head and upper portion of the neck, the triangular white throat patch was, however, as usual, conspicuous. While we usually see Pied-bills singly, I have found three, four and five together on one or other of the ponds but have not found them associating closely in flocks or flying together as the Horned Grebes sometimes do.

On October 1, 1923, while searching the surface of Lake Lily for possible water birds my glass came upon a Pied-billed Grebe resting among the brown lily pads which its plumage so closely resembled. It was trying to swallow a medium sized frog which it had caught, picking it up and dropping it again repeatedly. It would succeed in getting its prey fairly into its mouth only to reject it again. Whether the frog was not entirely to its liking or whether the bird was merely trying to reduce it to a more pulpy condition before swallowing I could not determine. Presently I discovered three more Pied-bills close by, which had hitherto completely escaped me, so closely did they match the lily pads—two young of the year, like the first one, and one adult. The young birds were ruddy brown on the neck and breast, with a short unbanded bill, while the old one appeared to have its head very narrowly streaked with brown and white. They were sitting duck-like with their bodies well above the surface of the water, pluming themselves. They would raise one round wing and then the other, while they probed behind it or would poke the bill down among the feathers of the breast. Occasionally one of them would stretch a foot out above the water displaying the lobed toes widely spread, making it appear all the more out of proportion to the size of the bird. One individual scratched its head with its foot, bending the neck around on one side, and several times one would dive, remaining under the water fifteen seconds on the average. When I moved out from the bushes so that they saw me they instantly sank lower in the water and swam out toward the middle of the lake, diving repeatedly and emerging each time farther from the shore. They plunged over forward, like a loon, and when they came up only the head and a small portion of the neck—about two to six inches—were visible. Occasionally one would submerge by sinking. Once two of the birds reared up in the water and with both legs and wings going splashed over the surface for a few feet as if in pursuit of one another, but did not actually take wing. The bird that had the frog apparently had the rudimentary tail feathers and posterior plumage erect.

These Pied-bills later joined a Ruddy Duck which, like them, had remained undetected among the concealing lily pads and followed it about over the lake. Compared with it they held the neck much more erect, never bending it back sufficiently for the head to rest on the shoulders as ducks so frequently do, while the pointed bill gave them a totally different profile from that of the duck. The neck of the Grebe, too, seemed much more slender and the head heavy in proportion. The rusty brown coloration was beautifully protective against the water which was brown with submerged vegetation, and with the scattered lily pads, now brown and withering, and continually raised and lowered by the wind and the ripple of the water.

Pied-billed Grebes are seen about Cape May most frequently between March 20 and April 24, in spring, and September 21 and October 30, in autumn, and appear to be more abundant in the latter season. We have outlying dates for November 4, 1923; November 6, 1927; November 10, 1935; December 11, 1921; December 24 and 31, 1922; December 27, 1931; January 3, 1926; and January 8, 1928 (Ludlam's Beach—Potter); all single birds. They would indicate that some individuals occasionally spend the winter; while single birds seen on May 10, 1924, and August 8, 1926, may have been summer residents at some point not far distant. Indications of nesting or the actual discovery of nests are very rare in southern New Jersey. Gilbert Pearson has recorded these birds on a flooded bog a few miles from May's Landing on June 1, 1919, while inspecting a rookery of Great Blue Herons, and says: "two Pied-billed Grebes surprised me by calling among the lily pads." (Bird-Lore, 1919, p. 271.) On July 22, 1933, Victor Debes found an adult with eight young about three-quarters grown on the Tinicum marshes above Chester, Pa., after the region had been flooded by the waters of the Delaware River for some months and unusual opportunities for the nesting of various aquatic birds had been provided (Auk, 1934, p. 320). While this locality is quite outside of our area the occurrence is of interest in connection with our next record. This was at Greenwich, Cumberland County, not far from the Cape May line at the head of Delaware Bay. Here Herbert H. Mills had constructed a dam, largely for the accommodation of water birds, and upon this sheltered pond Edward Parry and William Pepper Jr., on June 20, 1934, came upon a floating nest of the Pied-billed Grebe, attached to partly submerged bushes a little back from the open water. They saw two downy young probably not more than a day out of the eggs both of which dived when they were still several feet away. One was later caught but when returned to the nest immediately struggled out again and dived out of sight. The storm of Labor Day washed out the pond but the dam was rebuilt a year later and we understand that the Grebes have returned.

While looking for birds on Price's Pond on the Bay side of Cape May above New England Creek, on July 6, 1936, in company with Ernest Choate, I discovered a single male Pied-bill floating on the water which is largely covered by aquatic vegetation of various sorts and flanked by extensive cattail marshes. I found the bird on nearly every visit to the pond during the summer. It was usually busy diving, bending over forward and sliding in leaving not a ripple to mark the spot. At other times it submerged backward and only once did it make a splash and that I think was when it had caught a fish in rather shallow water. Several times I saw a fair sized fish in its mandibles as it emerged from a dive and, after a slight adjustment, the prey was swallowed. Sometimes the bird would float well out of the water and at others would be partly submerged or perhaps only the neck and head would be in evidence. Hampton Carson and John Hess kindly waded out into the pond and worked their way well into the cattails but while they flushed eight American Bitterns and a Least Bittern as well as many Little Blue and one Great Blue Heron, which had been concealed there, they found no trace of a female Grebe or young. A similar search by Ernest Choate failed to discover any. On August 7, however, Conrad Roland and I found the female accompanying the male on the water and on several subsequent occasions the two birds were seen but no young. Finally on September 12, Choate found five Grebes swimming about together and as it seemed to be too early for the arrival of migrants from farther north he assumed that these birds were the summer pair and three of their offspring. James Tanner tells me that there were two still there on October 16.

Where there are wider expanses of water and marsh the Pied-billed Gerbe occurs in greater numbers especially at times of migration. Victor Debes saw fifteen at Delaware City in late March, in 1930, and there were ten at Brigantine on April 5, of the same year, while in autumn Julian Potter found twenty-five at Fish House on the Delaware River and Charles Urner twenty-eight at Barnegat Bay on October 8, 1927, all of which are of interest in comparison with the occurrence of the bird at Cape May.

SOOTY SHEARWATER

Puffinus griseus (Gmelin)

The slaty-black Sooty Shearwater is of frequent occurrence about the fishing banks off the coast in late May and June. So Wharton Huber was informed by fishermen going offshore at Corson's Inlet on June 3, 1923, who brought back several specimens to substantiate their statements, while Capt. John Taylor found them common on the banks off Five-Mile Beach on May 24, 1890, and obtained two specimens, now in the Public Museum at Reading, Pa.

Charles Urner writes me that "on the water the Sooty Shearwaters look a bit like Crows and on the wing have the flap and sail flight, traveling rather close to the water."

Most of our New Jersey records are of dead birds found on the beaches where they have been washed up by the tide. The following have come to our attention: Atlantic City, June 3, 1893, J. P. Remington. Sea Isle City, May 25, 1898, Theodore L. De Bow. Stone Harbor, June 17, 1923, Turner McMullen. Cape May Point, June 1, 1923, Witmer Stone. Stone Harbor, June 17, 1923, Richard Miller. Cape May, May 30, 1936, Frederick C. Schmidt. Farther up the coast they have been observed several times from the beach or from boats offshore in the Barnegat area: Barnegat, September 6, 1926, two, and July 10, 1927, one, and four at Seaside Park, August 8, 1926, all recorded by Charles Urner, and one seen off Barnegat Light by John Kuerzi. Fletcher Street while fishing for bluefish near Southeast Light Ship, June 24, 1934, saw two Sooty Shearwaters with a flock of fifty or more Greater and Cory's Shearwaters from which they were easily distinguished by their uniform dark coloration. He noted one of them dash into a school of arching bluefish and dive while among them as if after some morsel of food. While our records from the south Jersey coast are all in May or June the birds probably are to be found later than this as shown by the records of the Barnegat area and in other parts of the North Atlantic from which we judge that they start for their breeding grounds on sub-Antarctic islands in the south Atlantic in October.

AUDUBON'S SHEARWATER

Puffinus lherminieri lherminieri (Lesson)

I have but a single record of this southern shearwater on the Cape May coast. On August 2, 1926, I saw a bird just beyond the surf at the bathing beach making short flights and then resting on the water as if unable to maintain itself in the air for any length of time. Swimming out as near to it as possible I noted that its wings were shorter and broader than those of the Common Tern although it approached that species in size. When it flew it kept a uniform distance from the water following the curve of the waves with great nicety and disappearing now and then in the trough of the sea. When at rest it had the pose of a small gull with head resting close down on the shoulders. In color it was dusky black above and pure white below. It managed to keep out of reach and gradually worked its way out to sea. A few hours later it appeared again and was caught by the lifeguards who were out in their boat just beyond the surf. Through the intervention of Dr. T. S. Palmer, who was present, it was kept until I could return to the beach but died shortly after it was brought in. On dissection it proved to be much emaciated and its stomach was empty.

This seems to have been the first recorded occurrence of the species on the New Jersey coast and it was evidently blown north by the tropical hurricane which caused so much damage in the Bahamas and on the southern coast of Florida a few days before. The bird was a female, apparently a bird of the year and is now in the collection of the Academy of Natural Sciences of Philadelphia. (Auk, 1926, p. 536.)

Audubon's Shearwater, unlike the other species that occur off the New Jersey coast, breeds in the West Indies and the Bahamas and does not indulge in such long oceanic flights. Indeed under normal conditions it does not seem to migrate at all.

GREATER SHEARWATER

Puffinus gravis (O'Reilly)

This shearwater, similar in general appearance to Cory's Shearwater, has a more northerly range during its non-breeding season, occurring from Newfoundland and Cape Cod across the Atlantic to Scotland and northward. Its occurrence off Long Island is similar to that of Cory's and most of the observations or captures there have been from June 27 to October 31 when the birds were on their migrations to and from their nesting grounds on islands in the Antarctic.

Their occurrence off the New Jersey coast is doubtless similar but we have, so far as I am aware only two records from this part of the Atlantic. A single bird was seen off Barnegat Light on August 24, 1926, by John Kuerzi while Fletcher Street saw about thirty individuals near Southeast Light Ship in company with a few Sooty and Cory's Shearwaters on June 24, 1934. When they arose from the water he observed that they kept close to the surface and flapped and sailed alternately on arched wings.

CORY'S SHEARWATER

Puffinus diomedea borealis (Cory)

Like other shearwaters this species is pelagic, only occasionally coming within sight of the shore when blown in by storms. In its post breeding migration from the Azores, it reaches Cape Cod by early August and is to be found off our Atlantic coast as far south as Cape Hatteras from then until November when it retires again to its nesting area. Its range off our coast is almost entirely south of that of the Greater Shearwater and the two are supposed to occur together off the New Jersey seaboard only during October when the southward migration of the latter is under way, as has been pointed out by Wynne Edwards (Proc. Boston Soc. Nat. Hist., 40 p. 2333–46). Evidence presented below, however, shows that they occur together also in June but perhaps only occasionally. Cory's and the Greater Shearwater are closely similar in flight, shape and color, but the head and neck are said to look utterly different in the two species when seen in life even at a distance. In Cory's the dark cap extends right down the sides of the head, gradually fading into the white throat; while in the Greater there is an abrupt line of demarkation between the two colors just below the level of the eye. The upper tail-coverts of the Greater Shearwater are conspicuously tipped with white, forming a prominent patch visible at a distance, while in the Cory's these tips are narrow and smoky and useless as a field character. (Wynne Edwards.)

Ludlow Griscom states that Cory's Shearwater is an uncommon but regular visitor to the seas off the Long Island coast rarely approaching within sight of land. Most of the captures or observations seem to have been during the October migration. While it has in all probability occurred during storms, at several points on the New Jersey coast we have but one record and that not certainly of this species. On August 22, 1908, Dr. William C. Braislin observed shearwaters in considerable numbers all supposedly of this species, off the beach opposite Forked River. He writes: "They were migrating southward mostly in small flocks, maintaining a rather low, steady flight with even, deliberate wing strokes. Their dark upper and white under parts, manner of flight and a previous scraping acquaintance with the species led me to a rather positive identification. An easterly wind then blowing perhaps led them nearer than usual to the beach." (Cassinia, 1908, p. 42.)

On June 24, 1934, Fletcher Street cruised out on a fishing boat to within two miles of Southeast Light Ship which lies some twenty-three miles off Cape May. There he encountered a flock of about fifty shearwaters, two of which were easily recognized as Sooty Shearwaters and he had a good view of thirty of the others as they rested on the water about one hundred feet away. Two showed the blending of the colors of the upper and lower surfaces characteristic of Cory's Shearwater while the others were obviously Greater Shearwaters.

*ATLANTIC FULMAR

Fulmarus glacialis glacialis (Linnaeus)

There is but a single record of this northern petrel for New Jersey, an individual picked up in an exhausted condition near Ridgwood, Bergen County, in early December, 1891, following very stormy weather on the Atlantic (Henry Hales, Ornith. and Oöl., March, 1892, p. 39).

LEACH'S PETREL

Oceanodroma leucorhoa leucorhoa (Vieillot)

Small petrels in the characteristic black, white-rumped, plumage occur all summer on the ocean off the Cape May coast and represent the two species Wilson's and Leach's, the latter being apparently the more plentiful at least in late summer or autumn. While rarely coming within sight of the beach they may often be seen from the end of the "rock-pile," or jetty, which extends out from the mouth of the Harbor for nearly a mile seaward. Here on July 22, 1920, Fletcher Street saw twelve or more flying back and forth with feet dangling low over the waves, in the characteristic petrel attitude so often referred to as running on the water, William Yoder and Henry Gaede reported six present there on July 9, 1922.

We also have many observations of petrels on Delaware Bay, supposedly of this species, by fishermen going out from Reed's Beach, our records run from August 13 to September 2 although petrels undoubtedly were present both before and after those dates. Mrs. Alice K. Prince has seen them as far north as Cross Ledge Light where they were flying about the fishing boats on July 4, 1912, and June 11, 1913, and below Egg Island Light on August 14 and 27, 1914. Several were seen in Maurice River Cove on June 11, 1914, and she saw two in the lower part of Maurice River on August 23, 1915, driven in by storms. Dr. Samuel W. Woodhouse secured two Leach's Petrels on July 25 and August 1, 1860, off the mouth of the Bay in lat. 39° 30', long. 72° 15' and Norris DeHaven shot one on the sounds northwest of Atlantic City on August 24, 1893, which had been blown in by a storm. The memorable hurricane which prevailed at Cape May August 21 to 23, 1933, brought many petrels into the town. On this occasion all of the low grounds and many of the streets and gardens were covered with water and petrels were seen flying like swallows over the inundated Beach Drive and the flooded meadows. All that could be identified were Leach's Petrels. The next day several were still to be seen in the town and on August 25 I found two on the Bay shore north of Cape May Point which for several hours remained close to the beach beating back and forth |and|never ranging more than a hundred feet offshore. There was not sufficient wind to force them in to the shore line and I think that there must have been food of some sort floating on the water to hold them there, although I did not actually see them pick up anything. Viewed from close at hand these birds seemed larger than I had expected. Their general color appeared black in the sunlight with a grayish area over the bones of the wing and a pure white rump. They flew over the sandy beach as much as over the water but never alighted definitely on either nor did they at any time fold their wings. Sometimes they would sail down so close to the waves or sand that it was difficult to determine whether their feet touched or not. Occasionally one of them did rest its body on the water but only for a moment, with no attempt to draw in the wings which were always kept fully expanded. As the low waves came in the birds would float in the air above them with no apparent effort and when they found themselves close to the surface their feet would often dangle below their bodies and when moved would give the impression of "walking the waves."

This August hurricane saturated all the gardens and orchards for more than a mile inland with salt water and killed the foliage as if struck by a heavy frost. Besides bringing these petrels into Cape May it carried them inland as far as Reading and Lancaster in Pennsylvania and to many points in New Jersey. Julian Potter saw two at Camden on August 24, and P. B. Philipp two at the mouth of Shark River on the upper coast.

Leach's Petrels are now and then found dead on the beach among the trash washed up by the tide, some quite fresh, others little more than skeletons, every bit of flesh having been eaten away by the sand-fleas. Such occurrences have usually been in August. Unlike Wilson's Petrel Leach's breeds on our Atlantic coast digging its burrows from Maine and Penekese Island, Mass., to Greenland, and winters south to the Equator.

1936
CR

WILSON'S PETREL

Oceanites oceanicus (Kuhl)

While this species is known to occur off our coast with the Leach's Petrel it seems to stay farther from shore or to be most plentiful earlier in the season. At any rate, after the hurricane of August 21–23, 1933, when many petrels were blown far inland only one of the numerous records was of this species. This occurrence was at Reading, Pa., where Earl Poole found two Wilson's Petrels, although sixteen or more Leach's Petrels were examined and upwards of one hundred seen there. The two species are quite similar; both have white rumps but the square, not forked, tail of Wilson's, its much longer tarsi and the lack of grayish brown on the upper surface of the wings serve to identify it.

We have records of several dead Wilson's Petrels washed up on the beach at Cape May in August, while one was found at Atlantic City as late as September 12, 1898. Two collected by John Taylor off Five Mile Beach on May 24, 1890, are now in the Reading, Pa., Museum, and Ernest Choate saw one on the fishing banks off Cape May in July, 1936. Charles Urner tells me that they are common several miles offshore in the Barnegat Bay region, his extreme dates being June 20 and August 26. His only records for Leach's Petrel were after the hurricane of August 1933 when several were seen from August 25 to September 3.

It is interesting to find these two petrels mingling during the summer, one of which breeds on the New England coast while the other is a native of islands in the south Atlantic and comes north to "winter" during our summer as do the larger Sooty and Greater Shearwaters.

The Storm Petrel, *Hydrobates pelagicus* (Linn.) has been attributed to the New Jersey coast in error.

WHITE PELICAN

Pelecanus erythrorhynchos (Gmelin)

While mainly a bird of the Western States, the White Pelican has occurred occasionally in the East in former years. Titian Peale is quoted as having killed a pair of these birds on the Delaware "a few miles below Philadelphia" (Baird, Brewer and Ridgway, Water Birds of N. A., II, p. 137). Turnbull (1869) says: "Has been seen at rare intervals on the Delaware, and on the sea coast near Cape May," the former locality doubtless referring to Peale's record.

Dr. C. C. Abbott (Birds of New Jersey, 1868) states that he "saw three flying off Sandy Hook, in February, 1864, and has seen one mounted specimen said to have been killed near Tuckerton." We know of no recent records.

BROWN PELICAN

Pelecanus occidentalis occidentalis (Linnaeus)

The Brown Pelican is a southern species occasionally blown north by tropical storms as are the Man-o'-war-bird, Audubon's Shearwater, Purple Gallinule, Royal and Sooty Terns.

The earliest Cape May record of the Brown Pelican was one shot by a man named Souder on "Poverty Beach" just above where Madison Avenue now joins the Ocean Drive. Otway Brown saw the bird and tells me that the date must have been in the early eighties. H. F. Graves, of Cape May, tells me that in the early autumn of 1923 he saw two Brown Pelicans perched on pilings on Richardson's Sound. Their flight as they took wing northward was unmistakable and he had become well acquainted with the birds in Florida but a short time before. He again saw a Brown Pelican on the bank of the Harbor, about one hundred yards from the wharf at Schellenger's Landing, on September 20, 1928.

Farther up the coast one of these birds was shot at Ventnor, in May, 1902, and preserved by the late William H. Werner and another was wounded and captured by Walter Layton at Townsend's Inlet, May 5, 1909. Across the Bay to the south Joseph Tatum tells me of a Brown Pelican seen by R. W. Schofield, who is familiar with the bird in the Southern States, one mile off Indian River Light below Rehoboth, Delaware, on May 30, 1934.

GANNET

Moris bassana (Linnaeus)

Gannets are most frequently seen out over the ocean between South Cape May and the Point, doubtless attracted by the profitable fishing to be had off the entrance to the Bay, but they occur also at other points along the shore. Adult birds can be distinguished at once from the Herring Gulls that are likely to be present, by the whiteness of their plumage— whiter than that of the whitest gull—and by the jet black terminal portion of the wing which stands out in striking contrast, making the adult Gannet a very black and white bird. The tail appears relatively long and wedge-shaped, and the neck very thick, the two almost equal in length and balancing each other, as it were, at either end of the body. The wings seem to be attached at the bird's middle and are distinctly narrower than those of the gulls and sharp pointed. The head is usually deflected in flight at about 30° and the large heavy white bill is conspicuous. The wingbeats are similar to those of a gull but seem to cover a greater arc and the bird sails more frequently while the narrowness of the wings when set for sailing is particularly noticeable.

Ten observed on April 30, 1923, were traveling in single file beyond the surf, their every movement showing to advantage against the leaden gray sky and water. Flapping steadily for several minutes they arose thirty feet above the sea and then on set wings dropped back on a long slant until just above the waves when the flapping began again and up they went, pressing all the while northward along the coast in this rising and falling flight. Four observed off Seven Mile Beach, November 16, 1929, were constantly wheeling and turning, poising for an instant with one wing vertical and the other down, displaying for the moment their maximum of white.

Most of the Gannets seen off the Point are engaged in fishing. They fly at an elevation of twenty feet or more, flapping and sailing, when suddenly one of them will check its flight and with a motion as if to turn a somersault, plunge abruptly into the sea with a pronounced splash. It strikes the water with the wings only partly closed, usually descending at an angle but sometimes almost vertically. The diving bird completely submerges, but reappears in a moment or two and is almost immediately in the air again, throwing the wings far forward as it rises from the water, with labored beats. Up it goes to its former elevation and the short sails and flapping begin again followed by another dive. Sometimes instead of striking the water it glides off horizontally close to the surface and before mounting once more it may be temporarily lost to sight in the trough of the sea, flashing into view again in a startling manner.

Gannets are usually associated in small parties of four to six or may fish individually, scattered about over the sea; sometimes however as many as fifty are in sight at once and present an attractive picture, some sailing well above the water, others splashing into the waves and others, still, winging their way along in their characteristic rising and falling flight. On April 30, 1923, forty or more were distinguishable far out over the "ripps" feeding with the gulls, their snowy white plumage showing brilliantly in the sunlight which streamed over the waters from the west, while the splashing of the diving birds and the scattering spray could be clearly seen with the glass; one or two birds seemed to strike the water every half minute. Again on March 18, 1927, thirty were detected well offshore opposite South Cape May which were studied for some time. There were a number of Herring Gulls in attendance upon them and the contrast between the two species was marked; the pure white and black plumage of the Gannets and their quick wheeling and diving appearing so different from the duller grayer coloration of the gulls with their slower, heavier action. The Gannets were circling on set wings comparatively low over the water and were continually plunging almost vertically into it with a distinct splash of spray. Upon emerging they would rest for a moment on the surface and then rise again, flapping along on top of the water for several yards before getting under way. Once fairly in the air they set their wings preparatory to another dive. Meanwhile the gulls swarmed about them swooping down after every scrap of food that the Gannets might drop. Sometimes they seemed actually to plunge into the water but only the head went under as they snatched at some floating morsel.

When Gannets venture in as far as the surf line their plumage is seen to much better advantage and the deep orange buff wash on the head of the adults can be distinguished. All the birds that one sees, however, are not in the black and white dress of the adult. In both spring and fall, uniform drab gray birds are seen which represent the immature plumage but which are readily identified as Gannets by their shape and actions. Of the ten seen close inshore on April 30, 1923, five were adults, three uniform gray and two gray with a broad white patch across the upper part of the wings and shoulders. Presumably these last were changing to the adult plumage. William Yoder saw one still in immature plumage on March 22, 1924. The gray birds are thickly marked above and below with small creamy spots but these can only be detected when close at hand.

April would seem to be the time of the Gannet's greatest abundance in spring although we have records from March 11 to 28 and occurrences as late as May 16, 1910. In autumn we have a record of a single bird as early as October 13, 1928, while Charles Urner saw one at Barnegat on October 11, 1923. All of our other observations are in November, from the 7th to the 22d. The most extensive migration of Gannets seems to be in this month. On November 15, 1931, Julian Potter saw 115 passing south off Long Beach in the space of half an hour along with a flight of Red-throated Loons, while on November 12, 1927, Charles Urner saw two hundred pass along the beach at Point Pleasant. In 1928 the flight was observed by several persons; William Yoder counted 145 Gannets passing Island Beach on November 18 while Benjamin Hiatt saw forty-five off Long Beach on the 11th. At Stone Harbor John Emlen saw thirteen on November 4 and Julian Potter six on the 25th at Cape May. We have no records of flights at Cape May at all comparable to those seen on the more northern parts of the coast and it may be that the migrating Gannets fly farther offshore when passing the Cape. I am informed that they winter more or less regularly on the south side of Delaware Bay and stragglers occasionally cross over to Cape May, one having been recorded on December 23, 1923, and another on February 25, 1906, while the Christmas census of the Delaware Valley Club has shown a number present at the ripps, at the mouth of the Bay, on December 27, 1931; twenty-six on December 24, 1933; ten on December 23, 1934; two on December 22, 1935.

The only inland occurrence of a Gannet at Cape May of which we have record was one reported by Walker Hand. The bird had been driven into some bushes by a pair of Bald Eagles and eventually killed by a man with a club! Gannets occasionally come up the Delaware River and one was caught in a shad net near Salem by S. B. Irwin and J. H. Cullen on May 25, 1890, while another was found at Secane, Pennsylvania, some miles from the river on November 22, 1922, by Harold Held.

The Gannet shares with the terns the honor of representing the high divers on our coast, just as the loons and grebes and certain ducks are the exponents of surface diving, and the usually diagonal plunge with its conspicuous splash constitute one of its most striking field characteristics.

DOUBLE-CRESTED CORMORANT

Phalacrocorax auritus auritus (Lesson)

We see Cormorants at Cape May almost entirely as transient birds of the air, passing overhead in irregular V-shaped flocks, like geese. In fact this formation and their comparatively slow wing beats bear such a general resemblance to those of the Canada Geese that they fully warrant the popular name of "Nigger Geese" so frequently applied to Cormorants by local fishermen and gunners.

On their northward flight Walker Hand has recorded the first flocks from March 21 to April 22, during ten years observations, and they occur frequently in May. I have seen them still winging their way north over the Cape as late as May 22, 1921; May 23, 1918; May 25, 1922; May 26, 1923, May 26, 1929; May 27, 1931. Julian Potter saw a flock of six on Grassy Sound on June 20, 1923, while on July 9, 1916, Walker Hand and I saw two come flying in from the ocean at the mouth of the Harbor. We have three August records, five birds passing on August 5, 1922, and one each on August 9, 1924, and August 23, 1924. Whether these summer occurrences are to be regarded as early migrants or non-breeding birds that have been in the neighborhood all the time it is difficult to say. On their southward flight Double-crested Cormorants occur most abundantly at Cape May during September and October. Brooke Worth saw several flocks passing Seven Mile Beach on September 20, 1931, and during a ten day visit to the Cape in 1925 Julian Potter saw flocks passing on September

22, 23, 28, October 1 and 2, and three flocks on the 3d numbering forty-four, eight and fifty-two birds respectively, while on Ludlam's Beach on October 14, 1928, he saw flocks of fifty, sixty and seventy individuals. Fletcher Street saw some at the same place on November 5 and 6, 1926, and Ludlow Griscom two at Cape May as late as November 11, 1921. We have no record of their wintering. Some of these transient Cormorants travel up the Delaware River as far as Philadelphia or even farther and Julian Potter saw four perched on pilings at Gloucester on several dates from May 13 to 25, 1924, while in the autumn of 1890 Samuel Rhoads saw single birds while crossing on the ferry from Camden.

Looking at a passing flock of Cormorants one is impressed by the solid dark color of the birds and the length of the tail which is proportionately greater than that of a duck or goose and in a measure balances the long neck as does the tail of the Gannet. The flight is heavier and slower than that of a duck but not quite so slow as a goose and the wings are comparatively narrow. The birds fly at varying altitudes; one flock, the largest I ever saw, numbered one hundred and twenty-five individuals and crossed the town on April 30, 1923, scarcely more than twice as high as the houses, while another flock, seen on May 22, 1921, back of Two Mile Beach arose to a considerable height. The latter flew in a perfect V; the former in a long line, company front, a part of it now and then jutting forward making an irregular V. On May 26, 1923, a flock of twenty-two crossed low over the houses and the shape of their heads and their relatively long tails at once caught the eye as well as the sheen of their plumage as the sunlight struck them. Another flock of seventy-four passing north over the meadows at Townsend's Inlet on May 10, 1925, offered a splendid opportunity for study. We saw them first far away to the south when they appeared as a long, narrow, irregular line across the horizon. Later, as they came nearer, a distinct undulating motion, due to the rhythmic beating of the wings, could be detected. Then the individual birds could be discerned flying wing tip to wing tip, the line bulging forward now at one point, now at another. As they passed a little to our left we saw them in a distinct V formation one side of which lengthened until it included fifty-four birds to twenty in the other. Then the long line buckled and for a time there was a double V. Now one bird and now another missed a stroke or two and sailed on set wings resuming his wing beats without losing place. They passed on out over the ocean becoming less and less distinct until they were swallowed up in the haze. For only five minutes, all told, were they in sight. A similar flock of ninety-five birds was seen passing over the Harbor at Schellenger's Landing on May 26, 1929, which had almost gone by before it caught my eye. When we consider for how short a time a passing flock is visible, and how many flocks are seen on casual trips to the shore in spring and fall, we realize how many must pass unnoticed during the periods of migration.

When the Double-crested Cormorants fly low the brick red of the gular sac and lores can be clearly seen as they pass overhead, both in spring and autumn birds. In many flocks, too, a variation in plumage may be noticed, the adults being entirely black both above and below while the birds of the year have the under parts brown. Both plumages occur in spring as well as autumn and in a flock seen as late as May 16 there were brown-breasted birds. The molt of the young apparently does not take place until their second summer.

Opportunities to observe Cormorants, other than as passing migrants, are not frequent at Cape May since they do not often stop here to feed.

I have reports of single birds seen on the waters of the Bay presumably feeding or resting and one observed on August 23, 1924, may have been so engaged. It arose opposite Cape May Point and circled high in the air, flapping its wings a dozen times and then sailing for a short distance. It mounted higher and higher and finally flew off to the north. Another bird that I saw flying close to the water of the Harbor had apparently risen from the surface. It passed down the coast flapping continuously. In both of these birds the crest of the head was clearly seen and the former, as he turned in the air, spread his square tail. Wrecks stranded on the beach are always attractive to Cormorants which make use of the masts and spars as resting places but few such opportunities for the study of the birds at rest have occurred of late years on the Cape May coast. On one occasion a number of years ago a wrecked vessel at the mouth of the inlet was used by some of these birds and here Walker Hand tells me he saw his first Double-crested Cormorant which was shot from its perch. At Corson's Inlet a few years since a single Cormorant was reported to me resting on a telegraph pole along the beach.

The first one that I ever saw at rest at Cape May was on September 26, 1926, and I was surprised to find it perching with Herring Gulls on some pilings on the south side of the Harbor close to the shore. The body of this bird was held at an angle of about 45° with the tail at the same slope but the tarsi perpendicular. From the side the tail seemed very slender, like a narrow stick or reed, but from the rear it was seen to be partly spread but very flat, with the feathers all in the same plane. The breast and belly were brown in contrast to the black back and tail, while the back presented a shell-like pattern due to the glossy edgings of the feathers. The manipulations of the long, slender neck were interesting. The bird would bend it down close over the back to plume the feathers there and then poke it down into the plumage of the breast. Several times the head was bent far over and raising one foot the bird would scratch the side of its neck all the way to the bill, using the claw of the longest toe. Now and then the neck was craned up in a peculiar manner and twisted so that the throat was almost uppermost and when a gull approached it was stretched forward and kinked in the middle; the gull invariably retreating. In jumping from one post to the next the bird gave an awkward hop landing flat-footed in a clumsy manner. When a gull would hover over him in an effort to dislodge him and secure his perch, a scheme that never failed in the case of another gull, the Cormorant never budged. When I came out into full view the gulls immediately took wing but the Cormorant held his post, sitting a little more upright and craning his neck toward me. It was difficult to make him leave by throwing sticks or shouting but he finally sailed down to the water settling after several strokes of the wings, and swam slowly about. He struck the water tail first, submerging it entirely and rested with his back arched above the surface. The neck was held erect and the bill at an angle of 45° like a loon, but he kept continually twisting the neck right and left. Soon he flew over to join a bunch of gulls but they flew off the moment he alighted. He then took wing and mounting higher and higher made off toward the Bay in a straight, steady flight. As he left the water he seemed to kick with his feet in an effort to get under way. Local fishermen say that a Cormorant cannot fly until he gets his feet wet; in other words he always goes from a perch to the water before taking an extended flight, and this bird fulfilled the tradition.

On November 25, 1934, I came upon a Cormorant resting on the stern of a row boat on the shore of Lake Lily at Cape May Point. He was rather loath to leave but as I approached he dropped from his perch and was able to splash his feet in the water as he spread his wings for flight, thus fulfilling the requirements without actually alighting on the lake. The orange yellow on the base of the bill in this individual was very conspicuous. The barriers surrounding fish pounds are favorite roosting places for Cormorants in other parts of their range and doubtless they have the same habit here, as Julian Potter has seen them resting on pound stakes off Seaside Park farther up the coast.

While I have had reports of Cormorants resting on the waters of the Bay and fishing there I have only once seen them on the water. On May 9, 1931, I discovered a flock just beyond the breakers opposite the Coast Guard station at Cape May Point. There were twenty-seven in all and as they rose and fell with the waves, disappearing at intervals in the trough of the sea, I took them at first glance for a flock of Scoter Ducks. They were swimming slowly up the coast with necks drawn down so that they seemed abnormally short, which added still more to their duck-like appearance. They formed two flocks at first, sometimes clustered together as they rested, and then strung out in a line as they began to swim again. When they finally took wing they arose slowly and their great size as compared with ducks was at once apparent. Each flock circled twice over the beach gaining altitude and then joined forces and went off in a northeasterly direction gradually assuming a long shallow V formation. As in most spring flights there were a few brown-breasted birds but the majority were glossy black.

The spring line of flight seems to be along the sea coast but there is also a definite crossing of the peninsula in the vicinity of Cold Spring, by birds which have probably been feeding on the Bay. I saw a line of thirty-nine crossing at this point on April 25, 1931, and Otway Brown on the same date, in 1929, saw an immense flock stretching well across the sky while I have seen smaller flocks on several occasions.

There is something strangely fascinating about the sight of the long black lines of passing Cormorants coming from unknown waters far to the south of us and pushing on to breeding grounds far to the north, rarely pausing in our vicinity except occasionally far out on the Bay to rest on the water and dive for food. More than any other of the larger birds of passage they seem to me to exemplify the mystery and the spirit of migration.

* EUROPEAN CORMORANT

Phalacrocorax carbo carbo (Linnaeus)

The adoption in our earlier lists of the English name "Common Cormorant" for this bird has resulted in several erroneous records, for while it is the "common" Cormorant in Europe, the common Cormorant of America is the Double-crested species. The European Cormorant is now known to breed at only a few points in the Western Hemisphere, from Greenland to Nova Scotia. I know of but a single record of its occurrence in southern New Jersey and none for Cape May County although doubtless a few, probably this very individual, must have passed along the Cape May coast or across the peninsula.

Joseph Harrison shot a specimen in immature plumage near Salem, N. J., at the head of Delaware Bay, on October 21, 1929, which is now in the Academy of Natural Sciences of Philadelphia. It was in company with another Cormorant probably of the same species. He was not aware, at the time, that the bird was not a Double-crested Cormorant nor did he detect any difference between the two individuals. It is possible that the slightly larger size would be noticeable were both species present side by side and the more or less white mottling of the under surface would also be diagnostic in the immature plumage. Farther north along the coast Charles Urner states that several times he has seen winter birds about Barnegat Bay which, judged merely by size and manner of flight, were probably of this species. On February 23, 1931, however one was positively identified as a European Cormorant. It was a single individual flying from the Bay through the inlet. It was first noted by John H. Baker and as it came closer in good light he could see distinctly the white on the cheek, chin and flanks of the otherwise black bird. Urner truly adds that "since identification of single Cormorants in the field is so difficult unless the bird is in, or approaching, breeding plumage, or is seen very near at hand, this species is probably of more regular occurrence than the published records indicate." (Auk, 1932, p. 341.)

MAN-O'-WAR-BIRD

Fregata magnificens Mathews

This tropical species has occasionally been blown north to Cape May by hurricanes. Maynard in his "Birds of Eastern North America" states that one was shot on the meadows at Cape May Court House in the spring of 1877 but gives no details or authority for the record. This was the only reported occurrence on the New Jersey coast until the tropical storm which swept the West Indies and southern Florida on the first days of August, 1926. Immediately following this storm, on August 3, a Man-o'-war-bird was seen soaring over the board walk and bathing beach at Cape May by Mrs. Emlen H. Fisher who writes me as follows: "The bird I saw had a wing spread of at least three feet the wings tapering to a point and the tail long and like a king-crab's, only with a fork at the tip, possibly made by two long feathers crossing when the tail was closed. The bill was long and curved over at the tip. The color was slate gray all over, and the neck was either drawn in like a heron's or there was a sort of pouch or bulge below the base of the bill. The bird hung perfectly motionless facing the wind for fifteen or twenty minutes and did not move an inch in space nor move a feather except to turn his head and look down at the small group of people gathered below.

He finally flew off to the south and disappeared." Mrs. Fisher later examined a Man-o'-war-bird in a museum and at once recognized her Cape May bird. (Auk, 1928, p. 367.)

On September 15, 1935, Joseph Tatum saw a Man-o'-war-bird on Brigantine Beach, N. J., and although the locality is beyond the Cape May County boundary the bird must have passed Cape May. He reports that it "came in over the lower end of the beach, soaring in three hundred feet circles at one hundred and fifty to two hundred feet altitude, and gradually drifted off to the south. It was apparently very black with a sharply defined white breast. It had a wonderful long tail deeply forked which opened and closed continually. In the ten minutes that it was in sight it made only three quick, successive flaps of the wings, drawing them in close to the body and arching them deeply. As soon as it appeared all of the gulls on the beach mounted to about the same height but did not go near it." (Auk, 1936, p. 95.)

The fact that I had been on the beach several times on the day that Mrs. Fisher saw her bird and on every other day for a week or more shows how easily one may miss these rare stragglers to our coast and doubtless many more of this or other species go unrecorded.

GREAT BLUE HERON

Ardea herodias herodias (Linnaeus)

The Great Blue Heron is not only the largest of our Cape May herons but enjoys the distinction of being our largest wild bird as well—at least so far as height and wing area are concerned. Occasionally we see them passing overhead, their great broad pinions slowly flapping as the body sails majestically through the air, the head drawn back on the shoulders, the great neck contracted into a loop so that it is no longer conspicuous, and the long slender legs trailing behind. But more frequently we see them on the ground. Usually it will be on the great salt meadows which stretch away northward from the Harbor, until they meet the horizon and reaching westward to the fringe of low woodland which marks their inland limit.

Here we come to know the Great Blues as gray sentinels of the marsh, standing rigid like so many old pilings or weather worn stakes. More than once, indeed, in sweeping the marsh with my glass, I have debated long and seriously as to the nature of certain of these old posts and pilings until a chance movement revealed them as Great Blue Herons. On certain cloudy days when dark skies deaden all color distinctions and when thick fog rolls in from the sea to obscure and distort one's vision, the lines of

Great Blues on the meadows have been all but blotted out, yet through the mist the white area on the side of the head stands out with uncanny distinctness, an unfailing identification mark. On other days when the mirage is present on the meadows, when houses and trees rise to twice their height and take on fantastic shapes, these great herons appear like tall thick posts and their heads and necks become so broad and white that it is with great difficulty that we can convince ourselves that we are looking at birds. These herons of the marsh may be seen in action, too, when at low tide they feed on the exposed flats and do not hesitate to wade into water so deep that it all but touches the feathers of their bellies and they then appear as great long-necked, heavy-bodied birds, floating on the surface of the sound.

Another of their favorite haunts about Cape May is on the shallow ponds near the Lighthouse and in the more bushy meadows just above, through which flows Cape Island Creek, and in similar situations on the Pond Creek Meadows nearer to the Bay. The energetic ditching of the mosquito exterminators has, however, sadly depleted the numbers of the herons in these localities, an example of how a campaign directed against one form of life may seriously affect the welfare of another. In these latter localities, against a background of bushes and low woodland, with patches of marsh elder and great tufts of marsh grass and tussocks of sedge interspersed, the Great Blue Herons are much more difficult to see. Indeed so closely do their colors blend with those of their surroundings that despite their size they often merge into the background and vanish from sight. The sedges and rushes, moreover, have by late summer grown so tall that they conceal the legs and bodies of the big birds with the result that we see, at best, only their gaunt necks stretching above the grass and held rigid, like poles driven into the marsh. On one occasion while watching some Egrets several Great Blues stood directly in my line of vision and actually in the field of my glass yet they were not detected until they moved. It has seemed to me that birds that habitually adopt this rigid, stake-like pose have already sighted the intruder and take this position just as the smaller Green Heron freezes in its characteristic pose the instant it realizes that danger is near.

On one or other of their two principal feeding grounds I have come upon Great Blues in every month of the year but this does not imply that they are present continuously, nor that they are uniformly abundant at all seasons. As a matter of fact they are rather scarce during the months of May and June when activities at their distant nesting sites occupy their attention and it seems probable that their excursions for food at this time are limited to localities close to the rookeries, the nearest of which are in the vicinity of Pennsgrove, some fifty miles to the northwest, and in Delaware across the river. When the nesting is over and the young on the wing they resort at once to the great salt meadows where they spend the remainder of the summer and a distinct increase in their numbers is noticeable toward the latter part of July all along the coast; July 26, 22, 23, 17, 21, and 31 in seven of the years covered by my records; though it is usually early August before one can expect to see them regularly about the Lighthouse Pond and other resorts near to the town.

In these latter spots we see, as a rule, only one or two birds at a time, though as many as five may be present and on one exceptional occasion (September 3, 1920) there were no less than thirteen. On the broad meadows north of the Harbor, however, Great Blues assemble in much larger numbers and I have counted as many as twenty-seven on Jarvis Sound on September 4, 1920, and twenty-two on July 21, 1924, twenty-five on August 25, 1920, and twenty-seven back of Seven Mile Beach on July 31, 1927, while Walker Hand reported forty-seven at the former locality on August 27, 1903. The period from late July to mid-September seems to be the time of their greatest abundance but they are present in smaller numbers at other seasons, a few, even in spring, back of Seven Mile Beach and on Grassy Sound, and they will linger in winter so long as mild weather conditions prevail. I saw five Great Blues fishing on the ponds by the Lighthouse on December 12, 1920, and there were four present on the 26th and two on the 30th. Two were present at the same place on December 23, 1923, one on January 13 and 18, 1924, and January 23, 1927, and three on February 8, 1925. On the great meadows three were seen on January 10, 1922, nine on December 26, 1926, and February 15, 1933. The counts on the Delaware Valley Club's Christmas census are as follows:

1931.	December 27, seventeen.	1934.	December 23, thirty.
1932.	December 26, thirty-five.	1935.	December 22, thirty.
1933.	December 24, twenty-eight.	1936.	December 27, twenty-five.

All through the Spring Great Blues continue to occur as individuals or in parties of two or three and Walker Hand has reported a definite increase in some years about March 20, indicating the northward migration of birds that wintered farther south. While we generally suppose that Great Blue Herons move south when the meadows become ice-bound some of them evidently remain even in the most severe winters. Following the temperature of −11°, in February, 1934, Julian Potter found one of these birds at Brigantine so nearly starved that it could scarcely stand. The meadows and sounds were frozen over and the beach piled high with slushy ice.

After the nesting is over herons seem to be concerned only with the search for food and a suitable roosting place, and I am inclined to think that a Great Blue, finding a spot to its liking, will remain there for a considerable time, as we know is the habit of the white herons; and during their sojourn on a particular piece of marsh they probably roost in trees not far away. I have several times seen bunches of Great Blues and Night Herons fly from Jarvis Sound to roost in large oaks in the woods on the mainland back of Bennett. On September 3, when passing the spot, four Great Blues took wing and on October 11 there were twelve roosting there. When disturbed while feeding on some bush-bordered pond they may alight on the tops of one of the island thickets on the marsh, but this is only a temporary location and they soon drop back again to the ground. Migrant birds, too, have been found in autumn roosting in trees at Cape May Point close to the Bay shore, but they always passed on again in the morning.

That the Great Blues are active very early in the morning, if not sometimes at night, there is silent testimony in the great maze of footprints which covered the overflowed portion of the great sand flat formerly existing below South Cape May and which formed a channel to the sea, whenever the tide turned. Measurements of these indicated Great Blues, American Egrets and Little Blues, none of which, however, frequented the spot during the day. The occupant of a cabin nearby confirmed the evidence of the sand by stating that he had heard and seen a large flock of herons feeding there "sometime in the night."

The Great Blues are the wariest of our herons and are ever alert and ready to take wing at whatever seems to them the opportune moment. On the great meadows above the Harbor they seem to realize the safety that lies in distance and show no alarm so long as the intruder does not approach too close, or direct his attention too obviously upon them, but elsewhere they move on at the first appearance of possible danger even though it may be but a few yards at a time. It is therefore not an easy matter to approach them undetected and to study them at their activities unaware. I have made many attempts to stalk them but my successful efforts have been few.

On August 2, 1920, creeping up on a small pond where some pure white Little Blues were feeding, I saw the head of a Great Blue which was standing rigidly erect in the grass. He had evidently located me long before I had been able to see either him or the smaller birds by the water's edge. As I sank down behind a convenient bush he turned his head directly toward me and after gazing steadily for some minutes turned it slowly away again. Finally, apparently satisfied that I had gone, he stepped boldly out of the sedge into the water moving in a most stately manner, his head gliding back and forth in perfect unison with his steps and always maintained at a uniform level, with the big bill exactly horizontal. Seen across the intervening sedge the effect was as if some gigantic cobra-like serpent were gliding along the ground with head erect. The bird waded out until the water was up to the feathered part of his tibiae, moving always with the greatest deliberation. Now and then he would pause and with neck bent in an S-like curve make a short strike at something in the water, and once he flew up the pond for some ten yards to his former stand, allowing his legs to dangle as he flew. The appearance of the moving necks of these great birds when seen at a distance is still more snake-like than in the instance just described, the heads sliding forward some six or eight inches with each step. On August 5, 1920, I was watching a silvery-mantled adult Great Blue stalking with stately gait through a marsh pond when suddenly several others alighted beside him, the combined spread of great wings seeming to cover the whole surface of the water. Turning my glass to ascertain where they were coming from I discovered half a dozen more gaunt necks reared from the rushes which I had utterly failed to detect before. On December 12 of the same year, I was able to study five individuals fishing on the Lighthouse Ponds. They were stalking like Egrets with bodies bent low over the water and were much less wary than in summer. Occasionally one of them, apparently sighting prey, would stop short and remain rigid for several moments with neck stretched out to its fullest extent; then he would relax and, withdrawing the neck to its usual curve, proceed with his cautious advance. Another individual on Lake Lily (September 17, 1923) had waded into the water up to his body and would pause in the same way every few minutes with neck extended over the surface of the pond.

On September 3, 1920, I saw the largest assemblage of Great Blues of that notable heron season. There were thirteen gathered close together on the shores of a shallow pond back of South Cape May, with many white herons of both kinds. They were in all sorts of attitudes, a few apparently squatting on the ground, some standing with head well down on the shoulders and neck looped in front of the breast, others with necks fully extended, and others, still, stalking about with body crouched low and neck in an S-like curve, busily fishing. One bird partly spread its wings and then stretched its neck out full length, pointing it diagonally downward; after remaining in this position for a few moments it resumed its normal pose. The whole action reminded one of a dog stretching himself after a sleep.

The flight of the Great Blue is very deliberate and impressive, the wing-beats numbering but two to a second. The head is drawn back against the shoulders as soon as the flight is well under way, the neck projecting in front in a compressed loop, forming a prow, while the legs, dangling at first, gradually attain the horizontal plane of the body as the speed increases. They stick straight out behind with the toes compressed and apparently touching, though it is always possible with the glass to see between the tarsi of the flying bird. When the heron comes to rest the legs are thrown forward and dangle as he drops into the shallow water or onto a mud flat, while the neck is often stretched out to check the momentum, or to maintain the equilibrium as the flight ceases. Once, while concealed in the rushes on Pond Creek Meadows a Great Blue came head-on and setting his wings sailed to a point just in front of me. As he touched the ground and raised his wings slightly, to maintain his balance, the alula of each wing stood up clear of the other feathers and a white spot near or under it showed clearly. On Price's Pond on the Bay side (August 10, 1936) I saw two Great Blues fly down to the marsh from a high tree where they had been resting. They sailed and flapped for a hundred yards or more holding their necks straight out in front and slightly deflected, an unusual sight as the neck of a flying heron is almost always immediately drawn back on the shoulders.

I have several times seen Great Blue Herons flying high overhead which seemed to be on migration. At dusk on September 1, 1922, eight passed directly over the town, flying in single file with beautiful regularity of motion, like the compact little groups of Laughing Gulls that we see winging their way to their roost at night. Again on September 8, 1921, two pairs passed over the town, at short intervals, about 8 p.m., flying southeast and Julian Potter reported a bunch of ten going down the coast headed into a strong northeast wind which caused them to drift sidewise southward. On the same date in 1928 he saw seventeen in small bunches going south with a northeast wind. On October 11, 1925, I startled six Great Blues from the scrubby woods near the Point, where they had apparently passed the night. They began to mount in the air in irregular circles, the heads and necks appearing white as the bright sunlight struck them. One bird seemed to be the leader and the others followed his every move as best they could. After reaching a great height they went off to the south. Another bird passed, on October 18, high over head. The most notable migration that I have ever seen was on the morning of October 13, 1928, between 8 and 9 a.m., a dark gloomy day with an easterly wind and promise of rain. Happening to glance upward I saw a long line of great birds passing high over the town, heading southward. With long legs stretched out behind, heads hunched down on shoulders, and slow rhythmic beats of the broad wings there was no difficulty in identifying them as Great Blues. They were in V-formation flying with great regularity and I counted thirty-three in the flock. Soon after another V passed consisting of twenty-three birds, followed by others numbering twenty-six, and twenty or 102 in all. In one flock two birds were flying inside the V. No more were seen but how many had passed before I noticed them it is impossible to say. Curiously enough on the next day, below Corson's Inlet, about 4 p.m., Julian Potter saw a similar flight, nine flocks passing all in V-formation and numbering from five to thirty-six individuals in each,—sixteen, thirty-six, sixteen, five, twenty-five, fourteen, five, thirteen, five, 135 in all. On July 22, 1921, I saw three birds coming from the north at a considerable height which encountered a strong gale of wind over the Harbor and had great difficulty in maintaining their flight, turning, sailing and beating their broad wings in a most unusual manner. These, however, were probably local birds caught unexpectedly in the gale and were not on migration.

The Great Blues, while sufficient unto themselves, are often found associated with American Egrets and Little Blues but it always seemed that the initiative in seeking this association rested with the smaller species. Indeed on three occasions the big birds seemed to distinctly resent the intrusion of the Egrets. On August 2, 1929, one of them drew back his head in striking position and opened his bill as an Egret strode past and two days later I saw one advance upon an Egret with head drawn down closely on his shoulders and the Egret at once flew away. The same thing occurred on another occasion but this time the Great Blue crouched low, his head not as high as the Egret's shoulder, and with his neck bent in the form of a flattened S. The Egret as before flew to the far side of the pond. Compared with the Egret the Great Blue, while only a trifle taller, is of much heavier build while the neck is very much thicker and the body usually held in a more nearly horizontal position.

The color of the Great Blue Heron varies considerably. Young birds of the year, in late summer and autumn, are duller and browner, while the old adults are beautifully frosted with silvery gray on the back and wings, with the white and black markings on the head and neck much more sharply defined. When seen out on the sounds in bright sunlight they seem very highly colored, the entire head and neck appearing white at times. Young birds are seen on the meadows as early as July 12 and apparently retain their dull plumage through the following spring and summer. One individual which Otway Brown and I saw on the Fishing Creek marshes from the Bay Shore Road, May 2, 1936, was remarkable. It was slowly stalking prey not twenty-five yards away and was in good light. It appeared uniform chocolate or rusty brown throughout, but upon spreading its wings the inner webs of the feathers were seen to be of the usual blue gray coloration. We see a varying amount of rusty brown suffusion on all young birds of this species but this one had developed it to a surprising extent and it was hard to believe that the bird we saw was really a Great Blue.

On July 30, 1933, Charles Urner saw a curiously colored heron on the Tuckerton meadows associated with Great Blues, about half a mile away. He writes me that "it seemed very light colored in the good light falling directly upon it and its posture seemed somewhat different from the others, an indescribable difference which seemed to be based in part on position of the neck, held straight but thrown back out of perpendicular. The head and neck showed no dark coloring and the color of the body and wings was decidedly lighter than the Great Blues. It eventually flew directly over me and alighted in a pool teeming with bird life—Little Blues, Great Blues, Egrets and many gulls. I was again struck with its strange color pattern and its greater size than any of the Great Blues present and made careful note of the colors. Bill large and very bright yellow; entire head white; neck very light, just a trifle dirty looking on the sides but general impression white; back and shoulders very definitely lighter than any of the Great Blues, the difference quite striking; under parts whiter with some dark streaking on the sides of the breast; legs definitely lighter than any of the Great Blues, not black but a rather tawny color; size difference in comparison with the Great Blues very noticeably larger." A later examination of a specimen of Würdemann's Heron in the American Museum convinced him that it was that bird as it checked in every detail.

On September 6, 1933, while visiting the meadows back of Seven Mile Beach, in company with Ernest Choate and Conrad Roland, I saw a single heron about half a mile away which tallied very closely with Urner's bird but unfortunately it failed to take wing and we were forced to be content with a long distance view. Even so, however, the white head and neck were plainly seen through the glass. Whether this was the same individual that was seen by Urner and whether either or both were Würdemann's Heron, or merely albinistic Great Blues cannot be determined. Julian Potter and I studied another white-necked bird here on August 15, 1937, and could not be convinced that the coloration was due to light incidence as nearby birds were normal. But light plays strange tricks!

The Great Blues gather at their breeding places quite early in the spring and by March 30, 1925, I found over fifty at one of the large rookeries in a remote part of the pine barrens some sixty-five miles north of Cape May. The nests were located in a dense white cedar swamp bordering one of the streams which here traverse the barrens. As we approached an occasional heron could be seen majestically winging its way across the horizon or one would arise from the cedar swamp and flap about for a moment or two above the tree tops. Entering at a favorable spot, pushing our way through the low shrubbery and jumping from tussock to tussock of swamp sedge, we soon caught glimpses of the great birds taking wing beyond us in the dense forest of cedar trunks which stood close together like a barrier of bare poles surmounted by bushy tufts of foliage. Farther on in the tops of some dead cedars were a number of the nests, ragged masses of gray sticks, some of them sadly damaged by the winter's storms, others larger and more substantial and on these the birds perched, one to each nest. They stood with neck and body erect like sentinels, the big orange yellow bills and white crowns showing clearly against the gray sky. Other birds with necks crooked in S-like curves balanced themselves against the wind on the less substantial perches offered by the slender twigs of the cedar tops. With bodies bent over and legs crooked at various angles, they spread and folded their wings in a constant effort to maintain their equilibrium. Woe betide the bird that tried to alight on a nest already occupied. The guardian sentinel would spread his wings and, with crest erected in a vertical plume and neck feathers spread like bristles, he would lunge at the intruder with his powerful bill, his yellow eyes gleaming, and give voice with a gargling cry much like the call of an angry goose. Sometimes there was a great turmoil as several conflicts were in progress at once and then, each bird having apparently secured a nest, silence would again prevail. Nesting had not yet begun and all the birds that were present seemed to be males each of which, having secured a nest, was holding it until the female's arrival.

On May 10, 1928, when this rookery was again visited, it was evident that the birds had eggs as the females flushed reluctantly from the nests as we advanced through the swamp. The thickly crowded cedars made a dense and dark retreat and, at this early date, there was but little foliage in evidence on the undergrowth, only the spots of pale green formed by the opening leaves of the sweet pepper bush (*Clethra alnifolia*) and the vivid splotches of red where the ripening clusters of swamp maple seeds hung on the bare branches. Standing perfectly still among the tall straight trunks we awaited the return of the birds which had perched on tree tops near the edge of the swamp. We had not long to wait as the herons were eager to resume their incubation. A female came in and alighted on a nest. She would fluff up her breast and back plumes and shake her wings against her body, at the same time ruffling up all of her plumage. She would then reach down over the nest and poke with her bill apparently arranging the eggs and then settle gradually forward and slide into the cup, with head and bill resting on the rim, the latter projecting a little beyond it. When a bird comes in from a distance it tacks back and forth over the rookery on set wings, with legs dangling and partly spread. As it sails down preparatory to landing on the nest the neck is fully extended but if frightened by our presence it is at once drawn back into a curve, the wings are flapped with redoubled energy, and the bird beats a hasty retreat. As the males stand at rest on the tree tops the great orange yellow bill is most conspicuous. There is a striking patch of chestnut on each shoulder and a black spot above, on either side. The thighs are clear maroon. As the wind blows the long crest plumes stream out and the yellow eye with its black pupil gives a piercing and hostile expression to the bird's face. On June 29, 1930, when this same rookery was visited, it was found to be entirely deserted, the young having all taken wing a short time before and the entire colony departed for the coastal meadows, leaving only broken fragments of pale blue egg shells and great splashes of white excrement covering the ground and bushes as evidence of recent occupation.

Another Great Blue rookery was located near Pennsville not far from the Delaware River, about sixty miles northwest of Cape May. It was situated in a grove of giant pin oaks in the heart of a wet wooded swamp with dense undergrowth. Some of the trees were ninety feet in height and one of them held thirteen nests. Green and Night Herons also bred with the larger birds and when I visited the spot on May 2, 1896, all had eggs. On May 7 the eggs had hatched and the parents were busy feeding the young. Another rookery is located near Marshalltown which may possibly have originated from the last, and visits to it have shown that some birds were present there as early as February 25, in 1922, possibly wintering individuals which had used it as a roosting place, while there were eggs in the nests as early as April 28, 1918, and on March 29, 1936, it contained 126 nests. It was in this rookery, too, that the first American Egrets to return as breeding birds to New Jersey, were found in 1928. Several smaller heronries have existed in different years in the same general region.

Dr. T. Gilbert Pearson gives us an account of another Great Blue Heron colony at "Makepeace Reservoir," on the Elwood Road east of Weymouth in the heart of the pine barrens. A dam had been built in order to flood the adjacent cranberry bogs and a lake formed somewhat over a mile in length. This involved an area of white cedar swamp, many of the trees of which had been killed by the flooding. He visited the spot on June 1, 1919, and found a colony of sixty-six nests in the dead cedars. "They were collected in three main groups, each within sight of the others. The herons exhibited proper precaution and departed while the boats were yet several hundred yards away. On many of the nests young were observed, but upon close approach they showed the usual heron characteristic of crouching down out of sight. Five nests of Green Herons and three nests of Purple Grackles also were noted. These were supported by limbs growing out from the trunks only a few feet from the water, while the nests of the Great Blue Herons were situated at a height of from twelve to forty feet." White-bellied Swallows were nesting in holes of dead trees and stumps and one of these examined was found to contain large numbers of Great Blue Heron feathers." (Bird-Lore, 1919, pp. 272–273.) I visited this rookery with Arthur Emlen and others on May 17, 1936, and while the general conditions were still as described by Dr. Pearson, with three feet of water in the pond, the number of nests had decreased to seventeen. On May 4, 1935, Emlen had visited it and found twenty-one nests, many containing young.

For a great many years a breeding colony of Great Blue Herons has been located near Delaware City, Delaware, fifty-five miles northwest of the Cape, which has doubtless supplied many of the herons that frequent the Cape May meadows in late summer and autumn, as herons are well known to have a post-breeding movement, a search for food rather than a true migration. This colony, however, has had a precarious existence in later years. It has long been a mecca for "oölogists" who have gathered innumerable eggs for exchange or sale under the erroneous plea of advancing science, while the tall trees have proved desirable for firewood and their number yearly depleted. In 1930 it seemed to have almost regained its former dimensions as some seventy-five pairs of birds were reported present but in the spring of 1933 they had been reduced to twelve, while in 1935 the colony moved to another piece of woodland some two miles from the original site and was found by Julian Potter, on April 19, to number forty pairs. Turner McMullen furnishes the following data on the Marshalltown heronry in Salem County: April 15, 1917, twelve occupied nests sixty to eighty feet up in large trees, contained sets of five; April 10, 1921, three nests contained six, five and four eggs, respectively; April 8, 1922, twelve occupied nests contained sets of four or five. April 12, 1925, fifteen nests contained sets of from four to six. The Elwood heronry on April 9, 1927, contained twenty-seven nests each with from one to five eggs. In 1898 there was a breeding colony near Pitman Grove which Dr. William E. Hughes informs me was located in tall pines and the nests held eggs on May 1.

Whether the Great Blues ever nested nearer to Cape May than the colonies above mentioned I do not know, but I have been unable to find any evidence of it. It is as a post-breeding bird only that we know the Great Blue Heron at Cape May. Whether as a silent sentinel on some woodland pool or out on the broad salt meadows, spearing its hapless prey in the shallow waters of the sounds, it seems to tower above its associates and to possess a peculiar majesty of poise and action that mark it out as among the elite of the bird world.

AMERICAN EGRET

Casmerodius albus egretta (Gmelin)

One of the ornithological joys of the Cape May summer is the frequent presence of white herons. There are two kinds, the Little Blue which is white in its first summer and autumn, and the larger, less abundant, American Egret. To the local gunners and baymen the former is the "white poke" and the latter the "white crane." Of late years a third species, the Snowy Egret, has returned once or twice to its old haunts at the Cape, but it may still be regarded as accidental. The graceful snow white forms of these birds standing out against the green woods and marshes add a tropical note the charm of which can only be fully appreciated by those who follow the daily fluctuations of the passing pageant of summer bird life. We look forward eagerly every year to the day in late July when these beautiful exotic species will suddenly appear among the prosaic life of our familiar ponds and marshes.

The recent ditching and draining operations have been a serious menace to the presence of these beautiful birds in the immediate vicinity of the town and they now, for the most part, are compelled to seek the more distant sequestered ponds where they can still find conditions to their liking and where intruders are less able to follow them. Fortunately there seems now to be no disposition to actually molest them in the Cape May region, and those that occur are much tamer than they were some years ago. Out on the open meadows, north of the Harbor, where feeding conditions are entirely to their liking, and where they are free from molestation, they occur in yearly increasing numbers during the latter part of the summer and in smaller numbers at other times. Where twenty years ago there were only occasional migrants one may now see one hundred or more on a single day and often several dozen in one spot. The curious northward wandering of the heron tribe after the nesting season is over is better illustrated by the white herons than by any of the other species because they come into a territory where they do not occur as breeders and thus at once attract general attention.

In 1920, when a few American Egrets were present all through August on the shallow pond at the Lighthouse, I was able to study them almost daily. They associated constantly with the Little Blue and Great Blue Herons but were much more wary than the former and kept strictly to the large open pond or to a smaller one that was screened on the ocean side by a densely wooded marsh island and thus afforded them perfect seclusion. They never scattered over the meadows to feed as did the smaller species.

The difference in size between the two white herons when seen side by side is striking, the Little Blue scarcely reaches to the Egret's shoulder when both stand erect, while, when bent over in feeding, it could easily pass under the body of the larger bird. While not quite so tall as the Great Blue and by no means so heavily built, the Egret compares favorably with it in size. Like all white birds Egrets differ in appearance in different lights. Against a leaden sky they sometimes look quite gray while when wading and in a dull light they appear, in some instances, to have the head and neck gray, but the moment they turn about these parts suddenly become pure white again. The effect has been so deceptive at times that it was almost impossible to realize that the birds actually were pure white. On August 28, a day of continuous rain and drizzle, when the whole flock was seen on the wing, the individuals appeared alternately pearl gray and snow white according to the incidence of the light.

The actions of the Egrets were easily studied during that summer of 1920. On August 3, I watched two early arrivals with the glass at a distance of about a hundred yards. Their great size, as compared with the Little Blues which I had been studying only the day before, was noticeable and their great golden yellow bills at once caught the eye, so different were they from the short dull olive bills of the smaller species; also their jet black legs. They kept moving about continually in the shallow water with the same slow, stately motion that characterizes the Great Blues, although the pose was not quite so majestic. Compared with the Little Blues they were not nearly so quick and active in their movements. The neck was kept extended at full length most of the time and stretched out forward at an angle of 45°.

It strikes one as being unnaturally long and slender and we cannot but feel that it must surely bend from the weight of the head which appears top heavy, the impression being enhanced when the head is turned sideways. When viewed from the rear the neck does actually seem to bulge out of line either on one side or the other and when it is somewhat twisted it seems to be thicker in one part than in another, due of course to the marked lateral compression. When wading in a strong wind the neck is often blown considerably out of plumb and the bird has difficulty in making headway, being forced to advance diagonally like a boat making a tack. Now and then one of the birds I was watching would lower its body to an almost horizontal position, draw back its head so that the neck assumed an S-like curve, and make a sudden lunge at some prey in the water, resuming the former attitude immediately. One bird that was standing at rest on a sand bar had its neck twisted in a peculiar manner as if the U-like loop, which hangs down in front when the head is down on the shoulders in the normal position of rest, had been turned at right angles and flattened against the breast.

While the Egrets associated with the Great Blue Herons they were evidently more or less in awe of them and several times when threatened by the latter birds they immediately flew out of danger. The Egrets were never noticed to be the aggressors. On another occasion when two Egrets were joined by several Great Blues they immediately flapped their wings and dodged away from them and soon after, apparently realizing that they were not welcome, they flew off to another pond. Again when an Egret was forced to walk past a Great Blue he ran as soon as he got on the other side with wings flapping as if to get out of harm's way as soon as possible.

When in flight the Egret takes the same attitude as the Great Blue Heron, with neck looped out in front and head settled back on the shoulders; the long black legs stretched out behind with the toes touching but the tarsi clear of one another. As the bird begins to get under way the legs hang down for the first few strokes of the wings but are raised and stretched out horizontally as the momentum increases. The neck, at first loosely curved in an S, is quickly contracted, the head settling down by successive jerks onto the shoulders until the normal position is attained and the bird glides away with slow, rhythmic wingbeats. When short flights are taken, low over the water or marsh, the Egret uses the same deliberate wingbeats but the legs are allowed to dangle and are thrown forward, as the bird settles again, and lowered into the water, while the wings flap gently until the feet touch bottom. One bird flying in to join the main flock dropped right into deep water that reached up to his body, and immediately strode off in the characteristic manner of the species with neck stretched forward and upward, recalling the hissing attitude of a gander. Since the long legs were completely concealed, the body had the appearance of floating on the water. Later this same bird while still in deep water raised one foot and, bending down his head, scratched the back of his neck, the upper part of the tarsus remaining under the surface.

On August 26 in their usual haunt in the shallow pond by the Lighthouse, there were seven Egrets accompanied by one Little Blue and a Great Blue, and I succeeded in approaching to within two hundred feet of them. They presented a beautiful picture and covered, in their behavior, all of the late summer activities of the species. In the immediate foreground stretches a broad flat sand bar and at its farther edge a narrow strip of shallow black water in which the birds stand and which holds their reflection like a mirror. Behind them, making a striking background for their snow white plumage, is a dense growth of bright green spartina grass, with spikes of yellow bloom, and back of this again the dull green and brown of one of those island patches of low woodland that here and there dot the meadows. Some of the Egrets stand with necks outstretched in the usual attitude of wariness, their great bills shining like gold in the sunlight. Others have their necks contracted into the rather angular S so characteristic of the heron tribe. Some of them are walking about with body more nearly horizontal but neck still in the curved position, striking now and then to right and left as they advance. One of them suddenly becomes rigid, right in the middle of his stride, with neck well extended and body tilted slightly forward, and remains so for a couple of minutes finally making a stab at something in the water. Another bird bends his body to a horizontal position, low over the water, and with neck drawn into a close S rushes rapidly at one of his companions. Both jump a couple of feet into the air, spreading their wings half open. They repeat this several times but make no effort to strike one another. One seems to have the feathers of the neck all on end after the effort and proceeds to ruffle up the rest of his plumage, letting the wings hang loosely below the body before coming once more into repose. Another individual is busy preening its feathers, bending the long slender neck into a veritable pretzel in order to reach under the wings and over to the base of the scapulars. While watching these birds a second Great Blue joined the group which was now closely congregated in an area of fifteen feet square, further illustrating the social tendencies in herons, as in other community breeders.

From mid-July or early August until sometime in September we usually see a few Egrets in the vicinity of Cape May, in some years there may be a large flock while in others few are recorded. Walker Hand's records, prior to my studies, are as follows: 1903, three present in August until the 25th; 1905, August 2, two; 1908, seven on August 5 and two until September 8; 1911, several during August and September, five on September 5; 1917, August 12, three. In 1919 Julian Potter saw single Egrets on July 27 and August 10.

During my seventeen years of intensive study of the bird life of the Cape, 1920 was, perhaps, the most notable heron year because they occurred close to the town. Two Egrets appeared on the shallow pond at the Lighthouse on August 3; three more came on the 21st, and two more on the 26th, while eleven were present from August 28 to September 3. From two to eight could be found on the pond every day until September 20 when the flock had dwindled to seven and to five on October 3. When I next visited the spot, on October 17, none was to be seen. In 1921 and 1922, after the ditching there were no Egrets seen immediately about Cape May but we learned of the presence of some on the meadows back of Seven Mile Beach where Delos Culver saw several as late as September 23, 1922. On August 14, 1923, I saw three on a large drained dam at Dennisville where they had apparently been feeding all that month, and on July 22 and 25 single birds were seen flying over Cape May; there were five on the shallow pond at the Lighthouse on August 26. In 1924 five flew over on July 26 and single birds were seen on August 24, September 3 and September 20. In 1925 two were present on the waterworks pond near Rio Grande on the exceptionally early date of July 12 and single birds on the great meadows July 16, August 1, 8 and 12 with ten on August 23 and six on September 1. In 1926 two were seen back of Seven Mile Beach on August 29 and one flying overhead September 8. In 1927 three were found feeding on the meadows back of Jarvis Sound on August 13 and again in the same spot on the 31st. There was but a single record in 1928, a solitary bird which came in close to Schellenger's Landing on August 2. In 1929 Julian Potter saw one near Ocean View on July 7; two were present on the Lighthouse Pond on July 12 and three on the 15th. The summer of 1929 marked the beginning of a steady increase in the Egret population; in July a large roost was established near Cape May by the Little Blue Herons which is described in the account of that species and to this a number of Egrets resorted every evening. There were at least twelve there on July 30, six on August 13 and twenty-five on August 31, approximately that number having occurred there on the intervening evenings and up to September 6 when the roost was abandoned by the Little Blues. On that night two Egrets were the only birds to come in. During the occupancy of the roost some of the Egrets were always in the van of the evening flight which began regularly about 5:45 p.m. They came flapping along from the north in much more majestic pose than the Little Blues and pitched down into the trees where they remained conspicuously on the outer branches for some time finally retiring into the deep foliage. They always followed the narrow strip of woodland which borders the meadows on the west. The last Egret of that year was seen by Otway Brown on the meadows back of Ludlam's Beach on October 8.

In 1930, there were one or two Egrets to be seen on the great meadows from July 19 to September 27, the number increasing to seven on July 10 while at the shallow pool near the Lighthouse, which this year had been nearly grown up with cattails, three were seen on August 24. These were joined by others until on September 3 there were fifteen along with some Little Blues, while on August 30 there was a single Snowy Egret in the party. The summer of 1931 found American Egrets present on the meadows from July 27 to October 10, two on the first date and one on the last but increasing from the middle of August to late September—sixteen on August

15, thirteen on the 26th, eleven on the 31st, twenty-one on September 28 and ten on October 4, the only days upon which the region was visited. With the exception of the first day the birds were seen only on the meadows south of the Wildwood Road and curiously enough it was there that nearly all of the Egrets recorded for 1932 and many of those seen in subsequent years were found. In this year contrary to all precedent they occurred here in spring, seven being seen by Otway Brown on May 30, one in the same place on June 4, ten on the 11th and six on the 25th. They remained there through July and I saw sixteen in a close flock on July 13. By August 6 the flock had increased to fifty and one was seen as late as October 9. In 1933, the first record is of five on July 8 in the same section of the meadows, fourteen by July 22 and twenty-five by August 5. The last ones were noted on September 23.

In 1934 Egrets appeared in spring on the Bay side meadows and Otway Brown saw three there on May 30 and found them there regularly up to July 9, with as many as ten on June 24, while Julian Potter counted two hundred near Bridgeport on September 29. They were again common on the ocean side this year on the meadows south of the Wildwood Road where they appeared on July 28 with two or three each day thereafter. On August 13, on a very dark morning about 9 a.m. after a night of rain, twenty were seen back of Seven Mile Beach, the occurrence being in many ways unusual. The birds stood close together with a large number of other herons, and were clustered about a dry spot on the meadows. I was inclined to think that they had been there all night as they were not feeding. With the Egrets were eighteen adult Little Blue Herons, thirty-six in the white plumage and four Great Blues, while a swarm of Common Terns and some Laughing Gulls hovered overhead or came to rest alongside of the larger birds. The same number of Egrets with forty Little Blues were clustered at the same spot on August 17, under the same weather conditions and with a flock of terns in attendance as before. In 1935 they appeared back of Five Mile Beach on July 6 and remained until September 29 when I found eighty-six present, while the next day Nelson Pumyea counted 110. In 1936 they were in about the same numbers. Our latest American Egret record for Cape May is one seen on October 28, 1936, by James Tanner, on the meadows above the Harbor, but at Bridgeport, Gloucester County, Joseph Tatum found one on November 25, 1934, and Julian Potter saw what was probably the same bird on December 1, following. Potter also saw one at Marshalltown, Salem County, on November 22, 1936. Coincident with the increase of Egrets at Cape May they spread northward into various parts of New Jersey, Pennsylvania and other states during late summer and early autumn, especially where bodies of open water or marsh land attracted them. As described in the account of the Little Blue Heron they occurred in abundance on the Tinicum meadows below Philadelphia when that area was under water during a break in the Delaware River dykes. Here they were found not only after their breeding season in the south but to some

extent in the spring, and Brooke Worth saw twenty-eight Egrets there on
June 13, 1934, most of them with the delicate nuptial plumes still in evidence.
In the Barnegat region of northern New Jersey, Charles Urner recorded
twenty-one in 1928, fifty-eight in 1929, and 264 in 1930.

It was an open question, in the old days, whether the American Egret
ever had nested in New Jersey. Wilson stated, over one hundred years ago,
that it bred in the extensive cedar swamps in the lower parts of the state,
but gave no definite information nor any evidence that he had ever visited
a rookery. It may well be that his statement was merely an inference from
the same northern summer movement that we see today and which was not
understood in those days. I had always been in hope that the Little Blue
Heron, which was well known to have been a regular summer resident in
New Jersey in Wilson's day and which had suffered less from the millinery
interests than the Egrets, might under present protection return as a breeder,
but I had no hope for the Egrets. It was therefore a delightful surprise to
learn, in 1931, that these splendid birds were actually nesting in the state.

On May 6, 1923, and May 2, 1926, Turner McMullen had seen Egrets
in a marsh on the Bay side of Salem County which aroused his suspicions
as their previous occurrences had been limited to late summer and autumn.
On May 5, 1928, he had the satisfaction of finding one on a nest in a colony
of Great Blues near Marshalltown. The nests were in tall red maples and
sweet gum trees standing in several feet of water and surrounded by a dense
flooded woodland. The Egrets were seen on the nests on May 9, 1931;
May 2, 1933; May 19, 1934; June 23, 1935. The nesting was announced
by Julian Potter who had visited the spot on April 29, 1934 (Auk, 1934,
p. 368). He and I were there again on June 3 and could plainly see the big
white birds on their nests, while an occasional individual would be noticed
flying across the adjacent fields to the river marshes or returning, doubtless
with food for the young, and one was seen in a small nearby swamp busily
feeding during the entire time of our visit. About ten pairs of Egrets and
perhaps 100 pairs of Great Blues were nesting in the swamp that year and
both had young on June 3, the Egrets apparently hatching later than the
other species. Julian Potter estimated that twenty young were reared that
year, he had noticed the first incubating bird on April 29 and saw a young
one standing on a nest but able to fly on August 9. During the summer
breeding and non-breeding Egrets used the swamp for a roost and on July 27,
he saw no less than eighty come in between 5:30 and 6:30 p.m. Three days
later only fourteen were counted, the majority having moved to another
roosting place. In May, 1935, the colony remained at ten pairs, but in
May, 1936, the Egrets seemed to have deserted the locality.

In May 1932, a single nest of the Egret was found by Herbert Buckalew
in a colony of Little Blues and Night Herons near Milford, Delaware, which
at the time was thought to be the most northern breeding station. Then
on June 9, 1935, both Egrets and Little Blues were found nesting in a colony
of Night Herons in New Jersey farther north along the Delaware River.

I visited this heronry with Fletcher Street on June 6, 1936. It was in a low, but not wet or flooded, woodland composed of sweet gums and red maples about thirty to thirty-five feet in height. We estimated that there were perhaps 150 pairs of Night Herons present and their young were climbing about in the trees or jumping about on the ground, which was, as is usual in such places, well splashed with white excrement and scattered with feathers. Ten pairs of Little Blue Herons and four pairs of Egrets had nests, the former off in a group by themselves. The young of both species could be seen raising their heads up from the nests but they had evidently not hatched as early as the Night Herons.

In May 1936, Charles Urner found a pair of Egrets nesting in a Great Blue rookery near Tuckerton, the third breeding locality for the state. The increase in summer Egrets on the Bay shore about the time of their resumption of nesting was noticeable. The Little Blues had been of frequent occurrence there as early as 1912 but the Egrets had been mainly restricted to the ocean meadows. The occurrence of Egrets in spring or early summer began to be recorded at about the same time.

The only vocal effort that I have detected in an Egret at Cape May was on August 30, 1930, when one of those feeding near the Lighthouse Pond flew directly over my head and alarmed at my presence uttered a rather gutteral *gar—ouk, gar—ouk.*

Except when coming in to roost we usually see Egrets on the wing only when making short flights from one pond or feeding place to another, passing low down over the marsh, but occasionally we may happen to glance up and see one crossing the sky at a considerable elevation, either coming from the south in search of new feeding grounds or returning as the first crispness of autumn tingles the air, and sometimes we may be fortunate enough to see several flying together in close formation like ducks. There are few more beautiful sights than one of these migrant Egrets passing overhead. The great snow white wings beating in slow rhythm, head resting easily back on the shoulders and the edge of the looped neck cleaving the air, while the long legs stretch behind like a slender rudder. Now it passes directly over us and then as we follow it with the glass it gleams silvery white as the sun's rays strike it at a favorable angle. Now it comes into brilliant relief against a patch of blue sky and then for a moment is invisible as it passes a floating white cloud. Once again we catch the motion of its wings and hold it in the field of the glass while it grows smaller and smaller until it is swallowed up in the heavens.

EP

SNOWY EGRET

Egretta thula thula (Molina)

The first record of the Snowy Egret for New Jersey in recent years was a single bird seen with five Little Blue Herons on the meadows back of Seven Mile Beach on September 23, 1928, by Philip Livingston, Edward Weyl and William Yoder. Close inspection of the birds "showed that one individual was more agile and had a thinner, entirely black bill. When it took wing it showed clearly its black legs and yellow feet, in contrast to the dark olive legs and feet of its companions. At the close range afforded it was easily identified as a Snowy Egret." (Auk, April, 1928, p. 230.) Another was identified on the meadows back of Ludlam's Beach on August 9, 1930, by Charles Urner and David Baird.

On August 30 of the same year I found one on the shallow pond near the Lighthouse at Cape May Point. The pond had in this year largely grown up in cattails and in forcing my way through them to the edge of the water, to have a closer view of some Little Blue Herons which were feeding there, I found myself face to face with a Snowy Egret and not more than twenty feet away. It was standing on a mud lump and its toes spread out on the black mud were conspicuously yellow in sharp contrast to the black tarsi, but as the bird moved about I found that the latter, while solid black in front, were yellow on the back and part way around on the sides. The yellow color, while slightly tinged with olive, is distinctly yellower than the color of the tarsi of any Little Blue which are always olive with no tinge of yellow. As the bird flew past me a little later at very close range the contrast of the black and yellow parts of the tarsi was clearly evident. The bill of this bird was slightly longer and more slender than that of the Little Blues standing nearby and was entirely black, while in the latter only the terminal third appears black the basal portion being lighter colored.

When feeding the Snowy Egret was somewhat quicker in action than the Little Blues and its neck seemed a trifle more slender and was twisted about more than in that species. Its actions were more like those of the larger American Egret, two of which were in the party, and it was on the whole more stately than the Little Blues. The whole flock of twenty-one birds now flew to another part of the pond and upon again approaching them and studying each individual with the glass I was unable for some time to find the Snowy Egret when it suddenly appeared directly opposite to me on the far side of the pond. It was probably there all the time but with its back toward me and its feet in the water, in which position it would be very difficult to identify. The Cape May birds were all in their first summer plumage and of course lacked the beautiful plumes of the breeding season.

The Snowy Egrets nested on the coastal islands of Cape May County at least through the seventies of the last century. Alexander Wilson, writing in 1812, says "On the 19th of May I visited an extensive breeding place of the Snowy Herons among the red cedars of Somers' [= Peck's] Beach, on the coast of Cape May" and describes the heronry in some detail. William H. Werner told me that he found them nesting in numbers in 1872 on the beach, where Ocean City now stands, eight to ten nests to a tree and Mrs. Henry W. Hand, in 1926, when in her ninetieth year, told me of visiting one of the islands as a girl and climbing up on the prostrate and dwarfed cedars from the tops of which the whole colony of nesting white herons could be seen, spread about her on every side.

Snowy Egrets have wandered occasionally to northern New Jersey and Pennsylvania in recent years. Lester Walsh found one on the Troy Meadows near Ridgwood, N. J., on August 5, 1929, and has given an excellent account of its actions. He writes: "I had succeeded in crawling to within fifteen feet of the edge of a pool in which several species of herons were feeding. The actions of one small white heron in particular attracted my attention. Instead of searching for its prey in the manner of the Little Blue Heron, it seemed to deliberately roil the water with raking foot motions and then seize the food that had been disturbed from the bottom. As the light became stronger I noticed the bill was black except for a small yellow area at the base of the upper mandible. The bird was noticeably smaller than the Little Blue Herons in its company and when a blundering cow had put the

herons to flight I detected yellow toes against a background of dark legs."
On August 7 he and Charles Urner found "two Snowy Egrets feeding with
ninety immature Little Blues on the same pond and were soon able to pick
out the birds without the aid of glasses simply by observing their peculiar
feeding habits." (Auk, 1929, pp. 536–537.) Charles Urner gives the total
number of Snowy Egrets seen in coastal New Jersey in 1930, as nine,
while in 1933, he saw two at Tuckerton on August 13, one at Brigantine on
August 19, and one on September 10, while Julian Potter found one at the
former locality on August 1, 1932. In 1936 Urner tells me that he saw
one on the Newark meadows on July 25 and another on September 26,
while at Tuckerton Snowy Egrets were seen from August 8 to 30 and on
September 20.

J. F. S.

LOUISIANA HERON

Hydranassa tricolor ruficollis (Gosse)

On August 1, 1920, when the first flock of Little Blue Herons appeared from the south and was being studied on the marsh east of the Lighthouse, I was astonished to see among them a single Louisiana Heron. It afforded an excellent view as it flew over to an adjoining pond and creeping up on it, I had a still better view as it took wing a second time. As it turned in the air the pure white breast could be clearly seen as well as the maroon tints on the neck, while the latter appeared relatively longer and more slender than that of a Little Blue. Although the other herons remained all summer the Louisiana was never seen again. It had probably overshot the mark in its northward postnuptial flight and had returned again within the limits of its normal range.

On August 29, 1926, Julian Potter and William Yoder were fortunate to see another individual of this species. It was in one of the cedar islands on the meadows back of Seven Mile Beach where the Night Herons have a considerable rookery. They describe the bird as a "long, thin heron, with brownish neck and white under parts." On July 30, 1929, still another was seen flying with the Little Blue Herons into the roost opposite Bennett, which the latter species had established that year, while on August 26, possibly the same bird was found by John Emlen on the meadows back of Seven Mile Beach.

On August 20, 1932, I had the best opportunity of studying this heron that I have enjoyed at Cape May. The bird was on a shallow meadow pool on the Avalon Road back of Seven Mile Beach. No other herons were near and its only companions were some Killdeers and a Dowitcher. It stood by a patch of grass to which it usually returned after a quick run through the shallow water. As it ran it bent over and lunged its bill at the small fish that were present, drawing the neck into a curve and shooting it out again, while it flapped its wings once or twice and splashed the water with its feet. It was very active and slender. The neck was white in front, lavender purple behind and had a row of dark spots down the sides; the body was pale gray blue above, the belly white and the legs and most of the bill greenish yellow. I watched it from a car at a distance of one hundred feet and it showed no alarm, nor did it take wing when I approached closer for a better view. Frederick C. Lincoln and I saw another Louisiana Heron with Egrets back of Five Mile Beach on August 5, 1933. Charles Urner has one record for northern New Jersey, a single bird seen by him on the Newark Meadows, on August 13, 1930, and again by J. F. Kuerzi on August 24, following.

LITTLE BLUE HERON

Florida caerulea caerulea (Linnaeus)

The Little Blue Heron is peculiar among our Cape May herons in that the young birds in their first summer are pure white and do not assume the slaty blue plumage of the adult until the approach of their first breeding season. The white birds really have slight slaty tips to the outermost wing feathers but these are not distinguishable at a distance. Occasionally an individual will fail to molt completely and we may find a few white or mottled birds during the following summer that have bred in that plumage.

Our experience with the Little Blues at Cape May is largely limited to the white phase of plumage and mainly to the months of August and September. Like various other members of the heron tribe they have a northward migration—or better, a wandering in search of food, which begins after the nesting is over and, judging by the present species, in which old and young are so easily distinguished, this movement is limited, in the main, to the young. In recent years, however, there has been a steady increase in the number of adult birds and while, in 1920, it was a rare thing to see a bird in the slaty blue plumage there are today one or more in every flock and sometimes they are as numerous as the white birds.

While the first Little Blue Herons are not usually seen about Cape May until July 15 or even August 1, there have been earlier records, some of which evidently represent the normal northward spring flight of the birds from their winter home far to the south, individuals which have, so to speak, overshot the mark and landed north of their breeding area. John Carter reported one at the Point on May 30, 1922, and Walker Hand another on a pond near Rio Grande in May 1923. Others were seen on May 29, 1926; April 30, 1927; July 4, 1928; May 5, 1929; and two opposite Weeks' Landing on May 26, 1929, one adult and the other mainly white. On May 30, 1930, John Gillespie saw a flock of eight and since then there have been a few spring records every year, the earliest being on April 19, 1931.

In former years, probably up to 1880, the Little Blue Heron bred on some of the island beaches of Cape May County which have now been almost entirely converted into summer resorts. Indeed Charles S. Shick states that "a few pairs still hold out (1890) in a thick grove of cedar trees on the lower part of Seven Mile Beach." He was informed by Capt. William Sutton, an old resident in this locality, that "in former years there was a large heronry on this beach, which the residents of the mainland would visit every spring when they would secure hundreds of eggs. He stated that even after taking large basketfuls one could not notice a diminishing of nests. He was confident that several thousand pairs occupied the lower end of the beach." (Auk, 1890, p. 327.) It would seem probable, however, that Capt. Sutton's report had to do largely with Black-crowned Night Herons which have for many years been abundant breeders on this island, but curiously enough Mr. Shick says that they were not common in 1890. There have been plenty of suitable nesting sites on the mainland for the Little Blues and I had always entertained a hope that some of these beautiful birds would return to New Jersey to breed now that the slaughter for millinery purposes, which had been the main cause of their extermination, was a thing of the past. The increase in their numbers in recent years and especially their occurrence in spring lent strength to this hope.

On May 29, 1927, when members of the Delaware Valley Club conducted an ornithological excursion to Delaware we observed a steady flight of Little Blue Herons passing regularly from some spot to the west of Lewes, out to the coast marshes for food and back again. That these birds came from an occupied rookery there seemed no possible doubt, and its presence directly across the Bay from Cape May made it still more likely that this species might return to its old haunts in New Jersey. At any rate the presence of this breeding colony so near at hand in all probability accounted for the early presence of Little Blues at Cape May during the immediately preceding years and also for the recent increase in the number of adults.

On June 6, 1930, Charles Pennock visited a breeding colony of Little Blue Herons which had been discovered by Herbert Buckalew not far from Milford, Delaware, no doubt the one from which the Lewes birds were flying, although the exact location may have changed. He writes that it "was located in second growth pines and deciduous trees, the former predominating. They were perhaps forty feet in height and set very thickly on the ground with their tops closely interlaced. The nest trees covered an irregular area of about seventy-five by one hundred feet square and so dense was the foliage above that it was usually impossible to determine the number of nests in a tree or the number of birds in the colony, but we felt sure that there were fifty pairs of Black-crowned Night Herons and one hundred of the Little Blues while six nests were the maximum per tree. The two species were somewhat segregated, but the Night Herons seemed to have entered into the Little Blue territory though there was no sharp line of division. Many young birds were out in the tree tops, some still in the nests and several hanging dead where they had fallen and caught in the crotches of the limbs. There were apparently no eggs at this time although eggs had been seen in some of the nests a week previously." (Auk, 1930, p. 535.) I visited this heronry on May 30, 1934, with members of the Delaware Valley Club including Herbert Buckalew, who had originally discovered it, and we estimated that there were about one hundred pairs of Little Blues nesting there.

It was not until 1935, however, that my long cherished hope was realized and then not in Cape May County nor in any of the spots that I had regarded as possible New Jersey nesting sites! On June 9, 1935, Roger Peterson, in company with Julian Potter and others, found a few Egrets on the marshes adjoining the Delaware River some distance below Camden and thinking that they might possibly be nesting in the neighborhood a search of the woodlands farther inland was made with the result that a breeding colony of Black-crowned Night Herons was found in which not only a few Egrets but some Little Blue Herons were nesting. Fletcher Street visited the spot on June 22 and thought that there were about ten pairs of Egrets and four pairs of Little Blues. Under his guidance I saw the birds on June 6, 1936, when we found four pairs of the former and ten of the latter, with probably 150 pairs of the Night Herons. The wood was not one that we should have thought suitable for a heronry, the trees were not over thirty feet in height, mostly red maples, and while the ground was low there was no water whatever and the tract was surrounded by open farming country. We could see the young Little Blues standing in the nests covered with the long pure white down, and with their throats vibrating, as the day was warm. The adults were all in the blue plumage. The young Night Herons were farther advanced than either the Egrets or Little Blues and were climbing actively about the branches. It is idle to speculate but the finding of these birds nesting so far north in the state makes one wonder whether some of them, at least, did not breed in the Night Heron rookeries in years gone by and whether a few may not have relocated in such places soon after the "white herons" again became common summer visitors to New Jersey, and were not previously detected. At Cape May, however, the Little Blues still remain irregular migrants and late summer visitors, and it is as such that my studies of them have been made.

We have no records to show just how rare these herons became in New Jersey after they were driven from their coast island nesting places, but it seems probable that some came north every year in late summer and that there were years of exceptional abundance, notably 1902, when some flocks of over one hundred were seen (cf. William B. Evans, Cassinia, 1902, pp. 15–21). Since 1892, at least, there have been a varying number of Little Blues on the great salt meadows of Cape May, from mid-July or early August until the end of September of every year, as well as on the Bay shore and on certain of the inland ponds. These birds were all in the white plumage of immaturity, appearing to the casual observer as miniature American Egrets. The local fishermen, however, distinguish the two and have distinctive names for them, likening the smaller species to the Green Heron and terming it the "white poke."

We sometimes see Little Blues at Cape May passing overhead singly or in parties of five or six in close formation, or they may be standing like white stakes scattered about over the meadows conspicuous against the broad greensward, or perhaps they will be busy feeding on some tidewater creek or shallow meadow pond. They are then active and restless and frequently shifting from point to point. In sweeping the meadows with the glass, from some point on the mainland, for possible white herons, roosting Herring Gulls are sometimes mistaken for them, especially when there is a mirage on the marsh' which increases their stature, but I find that in brilliant sunshine the gulls appear pearly white while the herons are chalky white in comparison. It is not so easy to study the Little Blues on the meadows as the grass is usually short and there is no shelter behind which the observer may approach the birds; we are therefore forced to view them from a distance. On August 15, 1924, I watched a flock feeding back of Jarvis Sound. There were nineteen white birds and four blue adults. At first they were down in one of the drainage ditches and only their heads and necks could be seen as they occasionally stood erect and looked out over the marsh, ever alert for the approach of danger. When they stooped again to feed they disappeared entirely. Then they moved out to one of the shallow meadow ponds where they showed to advantage, clear of the obscuring vegetation. Some of them stalked gracefully about in the water, their bodies bent over and their necks held out at an ascending angle; others stood in erect poses on the bank and others still selected as roosting posts some short stakes that had been driven into the ground. This flock was present all summer and was augmented by additional arrivals until by August 31 it numbered at least one hundred birds which could be seen from the turnpike settling to roost about sunset in a grove of trees bordering the meadows. They were perching all over the branches and the air seemed full of them as they flew in from the north and south. Sometimes the entire assemblage would take wing for a few minutes and then settle down again. I was unable to study this flock after that date but fishermen informed me that the meadows as far as Avalon were covered with scattered white herons as late as September 20.

In certain years, before the Mosquito Commission destroyed their haunts, we have had a definite summer colony of Little Blues on some of the ponds close to the town where better opportunities were offered for a careful study of their activities. In 1917 there was a flock present on a shallow body of water in the upper part of Pond Creek Meadows which I first saw on August 12 and which presented a beautiful picture. The marsh with its varying tints of green stretched away on all sides, cut here and there by slightly elevated dykes marking the line of ditching operations, and bordered by low bayberry bushes and by rose mallows just now in gorgeous bloom. Away off in the distance there was a momentary flash of white above the sedge which we recognized as a white heron, then two slender white necks stood out clearly against the green background. Anxious for a nearer view of the birds we crept through the tall grass toward a projecting wood edge and followed its shelter to its very tip where we came out into a clump of bayberry bushes and arose slowly until we stood erect. A bare muddy flat, now nearly dry at low tide, lay some four hundred feet away and upon it stood the greatest array of white herons that I had, up to that time, seen at the Cape, to say nothing of the other birds that were mingled with them. There were twenty-three Little Blues in the white plumage and three Egrets, the latter standing twice the height of their lesser companions and maintaining a more dignified attitude. The Little Blues were dodging about with frequent spreading of wings, now and then making a feint at one another and jumping entirely clear of the ground, or flapping away for twenty feet or more only to alight again with a great swoop. There were Night Herons, too, a dozen or more, both old and young, standing stock still near the edge of a small pool and occasionally stalking over to another position with head extended and body arched over like a hunchback. Farther back in the taller grass stood three Great Blues, like sentinels on guard, only their tall necks showing above the sedge.

My companions, Walker Hand and Dr. Smallwood Eldredge, passed out into clear view but the birds were in no hurry to leave and a third of the distance that separated us was passed before any of them took wing. The Great Blues left first but went no farther than a wood edge that lay on the far side of the marsh, where they settled themselves in the top of a large oak. Then, one at a time, the Night Herons flew away but the white herons were still loath to leave, and after rising once or twice they settled again on the flat. As the intruders advanced, however, they at last took flight and settled temporarily on the tops of the oaks until the trees looked as if a number of small white sheets had been spread over their branches. This flock of herons along with Black Ducks and a variety of shore birds had apparently been located on this pond most of the summer, but with the advent of the shooting season, on August 15, they moved elsewhere.

The summer of 1920 was our greatest white heron season in the immediate vicinity of Cape May. On July 25 a single Little Blue was reported to me as present at Green Creek, on the Bay shore, and on the 29th I saw one fly southward from Pond Creek Meadows and drop with its characteristic twist, like a collapsing airplane, into the Lighthouse Pond. On August 1 a flock of five was flushed from the ditches east of the Lighthouse and from then on to the time of my departure, on September 6, these beautiful birds were to be seen daily in this vicinity, their numbers varying from three to six until August 26, when the flock increased to twenty-one, to be still further augmented on September 5 to thirty-one individuals. How rapidly they moved to the south I am unable to say but on a subsequent visit, on October 3, only a single bird was to be seen and by October 17, and doubtless sometime before, it too had departed. At first these herons kept pretty close to the open shallow ponds east of the Lighthouse, or to the nearby ditches, and when disturbed generally took refuge in a sheltered pond lying behind a thick growth of small trees and bushes which formed a sort of wooded island in the marsh. They were almost always in more or less intimate association with Egrets and Great Blues. Later in the season they would spread about over the meadows and the individuals would become more widely separated as they traversed the various ditches and salt creeks in search of food. As the marsh grasses became taller they were afforded much better protection and were usually completely concealed from view until they took wing. They were now much more easily approached as their necks were not long enough to reach above the level of the sedge, as in the case of the larger herons, and they could not detect the approach of an intruder. They were, however, the least wary of the three species and would hold their position in full view for some time after the others had taken wing. Toward the end of August the Little Blues came regularly as far east as South Cape May and later quite up to the edge of Cape May itself, where they found a roosting place on a wooded flood gate and dyke at the head of one of the salt creeks, within a stone's throw of the houses. They had another favorite perch on top of a pile of brush on the edge of the Lighthouse Pond, sheltered from the beach by a low thicket. On both of these roosting places they congregated, especially in the afternoons, in parties of from three to seven. Whether they roosted there at night I cannot say but it would seem that they fed very early in the morning on a sand flat back of the beach, judging by the maze of heron tracks of all sizes which was to be seen there, and the fact that they were never observed there in the daytime.

Compared with the Egrets, with which they were so frequently associated, the Little Blues seemed scarcely half as large, their heads, when held erect, reaching only to the shoulders of the Egrets. The upper mandible seemed often to be very dark and almost black while the lower appeared largely dull greenish white, the closed bill being relatively thicker at the base than that of the Egret, and proportionately less slender. On one occasion I was convinced that a certain individual had a wholly black bill but a turn of the head revealed the normal dark olive color and a possible Snowy Egret vanished as the light became clearer! Such are the vagaries of light and shadow which confront the observer. In another bird the bill actually was black but the color proved to be due to the black mud in which it had been probing, and later experience taught me that the slenderness of the Snowy's bill was as good, if not a better, character than its color. The tarsi of the Little Blue also appear to be black in certain lights but usually the olive color is easily detected.

Little Blue Herons are restless birds, much more active and quicker in their movements than the Egrets, and their wingbeats run thirty-five in ten seconds as against twenty for the larger bird. When on the wing they present a beautiful sight, flapping low over the dark green meadows or high overhead against the clouds and sky. Two occurrences in particular have impressed themselves on my memory. One at sundown on the evening of September 2, 1920, when eleven of the birds left their feeding grounds and winged their way to their roost on the Lighthouse Pond. They headed directly west and as their wings rose and fell they caught and reflected the rosy glow of the rays of the sinking sun. Again on a dull afternoon in August, with gray sky and a threat of rain, six Little Blues accompanied by five Egrets came in at a considerable elevation from the direction of Pond Creek. The peculiar light played strange tricks with the white birds. At one moment they disappeared completely melting away into the prevailing gray of the sky and then as suddenly they would flash forth again in silvery whiteness. When the flying Little Blue Heron arrives above the spot where it desires to alight, the wingbeats cease and it sails graduallv down on set wings, flapping again once or twice as if to hold its course. If low enough it makes a landing from the last sail, but more frequently, while yet at a considerable height, the bird gives a sudden twist, as if one wing had broken, or given way, and turning the tip of the other wing directly upward it hurls itself to earth, landing with a short swoop. During this operation the neck is reared from the shoulders into the characteristic S-like curve, and as the bird nears the ground the legs are swung forward, dangling below the body, while the wings flap lightly before they are finally folded and the bird comes to rest. The sudden pitching to earth of a flock of Little Blues is an impressive sight, but it is completely overshadowed by the greater spread of the Egrets' pinions should any of them be alighting at the same time. One Little Blue, which came sailing directly toward me, elevated the alula of each wing as it prepared to alight and they stood up like thumbs on a spread hand.

I had several opportunities to study the Little Blues at comparatively close quarters during that memorable summer of 1920. On August 2, I located several of them in a small pond near where the marsh joins the upland and by creeping slowly through the grass I was able to approach to within thirty-five yards of them. There were five lined along the far side of the pool, wading in the shallow water, with two Great Blues in the sedge farther back. The white plumage of the smaller birds shown like snow against the black mud and dead sedge of the bank. At first they were spread out at equal intervals but gradually they came close together. While they headed to windward their plumage was sleek and smooth, but as soon as they turned the scapulars and sometimes the feathers of the crown were ruffled and fluttered conspicuously in the breeze. Their movements were deliberate and they walked slowly, raising the feet daintily, bending the posterior leg and often holding it so for a moment before drawing it forward, the foot being just below or just above the surface of the water. So high did they step that the body was sometimes tilted over a little, while the neck, bent in the usual S-like curve, moved slightly forward and back with every step. When preparing for a strike it was suddenly drawn back into a tighter coil and loosened should the opportunity pass. When a strike was made it was executed with great quickness and the former attitude immediately resumed. I could often see the bird swallow its prey accompanied by a raising of the head and a gulping action of the throat. The water usually reached to the juncture of the tarsus and tibia of the wading bird and sometimes even to the feathered part of the latter. More rarely the birds stood out on the muddy banks and once or twice one of them would ruffle up its plumage and probe among its breast feathers with its bill. Occasionally one heron would make a short flight, for a distance of six or eight feet, with wings fully spread and legs slightly dragging. When they flew for greater distances the legs would be stretched straight out behind, to be brought forward again to the perpendicular as the bird settled, while the wings were held for a moment partly expanded until perfect equilibrium was attained.

When a bird became suspicious the neck was held up vertically and the curious bend or crook, caused by certain elongated vertebrae, was clearly seen. One individual held its neck out horizontally at full length forming a continuous curve with the normal slope of the back and "craned" it from side to side somewhat like the action of a goose. When one bird attempted to pass close to another it took several quick steps and at the same time flapped its wings as if to gain speed, much as chickens do when taking a quick run. The Little Blues seemed distinctly afraid of the Great Blues and beat a hasty retreat whenever one of the larger birds approached. A single Little Blue was, however, often associated with several Egrets or Great Blues and seemed to derive some benefit from the companionship, but he was always found far out at the end of the assemblage, as if fearing to approach too close.

As an illustration of how completely these white birds may be concealed by the vegetation, I was watching some Egrets, one day, and while I was attempting to withdraw without disturbing them I made some motion which caused them to take wing, along with a single Little Blue that was with them. Immediately, to my astonishment, twenty more Little Blues arose from immediately in front of me, where they had been feeding completely concealed by the tall grass. The twenty-seven snow white birds wheeled about for a few minutes and then settled on the tops of some low trees near at hand, some holding their necks stretched up to full length, others with them drawn down on their shoulders. Soon they dropped into a pond which lay behind the grove and upon creeping round the corner of the thicket I could see the long necks of the Egrets and of some Great Blues which had joined the party but the smaller birds were once more entirely concealed.

On September 3, 1920, there was another assemblage of herons at this same spot, twenty-one Little Blues and a larger number of Egrets than on the previous occasion. The smaller birds were scattered all about over the grassy meadows, feeding in the shallow ditches, with heads bobbing up and down, like a lot of white chickens. On September 5, there were no less than thirty-one Little Blues and six Egrets, the largest assemblage of white herons of the entire summer. In subsequent years we have had a few individuals present on the Lighthouse Ponds or about the ditches, but the activities of the Mosquito Commission in draining the meadows between Cape May and the Point have been responsible for driving these beautiful birds farther afield.

The occurrence of adults with the white juvenile birds has been much more frequent of recent years than in the past. I saw my first one on August 26, 1920, near the Lighthouse; but on August 15, there were four in a flock of nineteen on the meadows back of Jarvis Sound, and later there were six. In 1925 two were seen on July 13 and five on August 1. In 1926, one each on August 10 and September 1, and two on August 23. In 1927 seven with eighteen white birds were present all through August back of Seven Mile Beach. Six were seen at this locality on July 25, 1928, also nine on July 2 and three on July 4 on the Lighthouse Pond, and a single bird, July 28, 31 and August 31. In July, 1929, one fifth of the birds coming into the roost that existed in that year were adults—124 by actual count; on August 1, 1932, twenty adults were seen and on August 13, 1934, eighteen, and since then they are to be seen in every flock on the meadows often equalling the white young birds in number.

On July 13, 1925, I had an exceptional opportunity to study the adult bird at close quarters on the waterworks pond near Rio Grande. There were two of them accompanied by two in the immature white plumage and hidden in the bushes which border the pond I could see their every action. In good light the old bird was deep slaty blue on the back, wings and entire upper parts, with no trace of the silvery tints or lighter plumes that characterize the adult Great Blue and Green Herons. The neck and the narrow breast plumes were rich maroon while the forehead and bill were light blue gray, paler than the body plumage and the resultant "blue face" was very conspicuous. One bird flew up to a dead limb on a well foliaged tree where he stood on straight legs with neck held upright and not down on the shoulders as is customary in the Green Heron. After a time he returned to the pond and began to pursue one of the white birds wading rapidly through the water and forcing it to move from place to place, often with a flap or two of its wings to aid in its progress. Wherever it went he followed and now and then both took wing and flapped over the water for a few yards. After some twenty minutes the old bird desisted from the pursuit. On the meadows back of Seven Mile Beach, on August 1 of the same year, I studied a flock of Little Blues with the glass at a considerable distance. They were exploring the narrow creek bottoms most of the time and one could see only the head and neck of a bird as it straightened up from its feeding, but even so the peculiar light blue bill and forehead of the adults could be distinctly seen and seemed to be one of their most striking characteristics, although the habit of both old and young of holding the neck out at an angle of 45° when walking was distinctive. On the surface of the meadows, the adults sometimes appeared absolutely black and at other times pale blue or rich slate blue according to the angle of the light. In this flock was one bird with irregular splotches of white scattered over the slate blue plumage, evidently one of those exceptional birds that have not molted completely, somewhat analogous to the white-headed Laughing Gulls that we see in summer, but unlike these the piebald Little Blues some times breed in this plumage. Another of these birds was seen on May 26, 1929, in company with a plain slate blue adult, but it was almost entirely white with only a few blue feathers in the wings and on the back. Other mottled individuals were seen on the Lighthouse Pond on September 17, 1927; August 23, 1925; May 30, 1930 (several white primaries); and one on Price's Pond on July 15, 1936, the most mottled of all. There were also two flying back and forth at the Lewes rookery in Delaware, in May, 1927.

While the Little Blue in flight is not so much larger than the Green Heron in body bulk, there is little possibility of confusing them, as the long legs of the former trailing out behind will serve to identify it while its slower wing beats are also characteristic. The roosting of the white herons of summer time in the vicinity of Cape May has always occasioned much speculation. We had reason to think that those observed in 1920 about the Lighthouse roosted in nearby thickets and that those which fed on the great meadows back of Five Mile and Seven Mile Beaches resorted to some of the woodland bordering the mainland to the west, each individual flock by itself. On August 31, 1924, however, just before sunset, I saw from the shore road a great accumulation of these birds, apparently mainly Little Blues in the white plumage, flying about a grove of rather low trees bordering the meadows some miles below Cape May Court House. They were perching all over the branches and the air seemed full of them as they flew in from the north and south. Sometimes the entire assemblage would take wing for a few minutes and then settle down again. It, therefore, seemed likely that the birds all gathered together each evening at a common roosting place and this was abundantly proven in August, 1929. I am still undecided as to whether there is a common roost every year, and also whether such a roost is occupied all summer or only toward the end of the season.

On July 28, 1929, about sunset I saw a large flock of Little Blue Herons in the white phase settling in a low thicket to the east of Bennett, near to the border of the meadows. Two days later in company with Walker Hand I walked down the edge of the meadows from near Weeks' Landing to the spot where I had seen the birds, arriving there about 6 p.m. We saw only an occasional bird on the way and a few in some tall woodland, in a swampy spot, just to the north of the roost. Reaching there we found that many birds had already arrived having apparently come in from the western side of the tall woods and thus escaped our attention. Others followed quickly flying over the tall woods and dropping into the roost, a dense thicket of small trees and bushes, the tallest not over fifteen or twenty feet. One flock, that we had in good view and which was flying in a long straggling line, numbered fifty-seven individuals and when the entire assemblage suddenly took wing and settled again we counted seventy-five adult birds in the flock. All the herons seemed to be in the roost by 7 p.m. There were at least twelve Egrets which stood conspicuously on the highest branches of the thicket, the smaller birds being scattered all over it, most of them in plain view although they may have settled into the heart of the bushes later.

On August 13 I again visited the roost locating in a pasture field in full view of the spot and about one hundred yards away. At 4:40 three Egrets came in from the north and settled in the tall woods to the left, then the Little Blues and a few more Egrets came along, one or two at a time, nearly all of them stopping in the tall woods. At 5:45, twenty-eight of them arose and passed on to the south and six minutes later nineteen of them returned; at 5:56, thirty-three birds flew off to the south and after that none of them stopped at the old roost, but all kept on to the south from three to fifteen in a flock. I left at 6:20. One of the Little Blues was a mottled individual and I counted six Egrets in all. On August 22 Walker Hand and I spent an hour (4:55 to 6 p.m.) in the heart of the swamp where the tall trees grew but except for a Great Blue and two Black-crowned Night Herons, which seemed to be settling there for the night, only nine herons arrived. Going out to the pasture field we counted 148 Little Blues and three Egrets passing overhead from the north and disappearing over the hedge rows to the south, between 6 and 6:20. On August 26, standing in the next field I counted 146 herons between 6 and 6:20 passing on to the south as before. On the next evening taking my stand about half a mile farther south I located the roosting place that they had finally adopted in the last section of woods and thickets north of Cold Spring. Egrets and Little Blues to the number of 160 went in between 5:45 and 6:35, the bulk arriving from 6 to 6:20 and probably others came after I had left. On the 29th, in company with John

Emlen and others, I got to the roost by 5 p.m. It was a dark cloudy evening and the birds seemed to come earlier and in closer ranks. The first ones arrived at 5:30 and there was one flock of sixty. In twelve minutes time we had counted two hundred and the last one came in at 6:15. The total count was 522 herons of which at least five were Egrets and the rest Little Blues, while of the latter about one quarter were adults. One flock contained eleven adults and ten white birds and again we saw a mottled individual. With Walker Hand, Edward Weyl and Philip Livingston I was once more at the roost on August 31, prepared to make an accurate count of the birds; the result was 549,—twenty-five Egrets, four hundred white Little Blues and 124 adults. They came in between 5:40 and 6:20. On September 1, the count was 540 and on September 3, 440 with two flocks which we were unable to count. On September 6, after two days of stormy weather, I visited the roost and remained until dark but saw only two birds, both Egrets. Evidently it had been deserted.

I am of the opinion that the first roost was only a temporary one, and possibly the later one also, but it would be interesting to discover how the birds know where to go. They come on in steady flight without a pause until over the roost, when they pitch suddenly down and settle in the tree tops. On one evening I was secreted in the heart of the woods quite close to the birds. They seemed to alight at first on the upper branches and gradually work down into the thicket; the Egrets which perched at first on the higher branches eventually roosted low down also. I watched some of the Little Blues perched on some dead limbs directly over my head and could see them raising themselves slightly on one leg and then on the other, opening and closing their toes on the limb as if to stretch them. One bird scratched its head with its foot, fluffing up all the feathers as it did so; another uttered a cry *skauk, skauk, skéeee-augh*. There was much excitement amongst the roosting birds every time a new delegation arrived, guttural calling and changing of positions.

It seems incredible that so many of these herons could come from the nearby meadows when we consider the comparatively small numbers seen there during the day, and as almost all of them came from the north it would seem reasonable to infer that they represented birds from a wide area extending well beyond the Cape May district. This supposition is rendered more likely from the fact that we saw, on August 27, a flock of sixty-eight in close ranks and flying steadily south, pass over the Stone Harbor Road ten miles to the north. Great numbers of Little Blue Herons and American Egrets have spread over northern New Jersey of late years as well as over most of the eastern United States during July and August. Charles Urner counted one hundred Little Blues and two Egrets in the vicinity of Barnegat Bay on August 11, 1928, and on the temporarily flooded Tinicum meadows below Philadelphia, John Gillespie reported about 150 white herons on July 22, 1933, of which at least fifteen were Egrets. At the same spot on July 29 of the same year, Victor Debes saw about one hundred white herons fly over the river apparently heading for some roost in New Jersey, which supports my theory that the population of the Cape May roost was more than local.

I failed to locate another roost until 1936 when Frank Dickinson told me of one he had seen in a dense pine and oak woods south of Weeks' Landing. Upon visiting the spot, we found probably 150 Little Blues associated with as many Black-crowned Night Herons. The trees were rather low as in the case of the 1929 roost, but the deer flies were present in such enormous numbers that a detailed study of the colony was simply out of the question. The birds were still there on August 29. The Night Herons were roosting during the day and seemed to scatter onto the meadows at dusk as the white birds came in. Charles Urner has published an interesting account of the increase of Little Blues and Egrets in northern New Jersey during the years 1928 to 1930 (Cassinia, 1929–1930, p. 9), and describes roosts at Barnegat, Princeton, Troy Meadows and Newark Meadows, the first with 550 birds in 1930, was the largest. He found, as I did at Cape May, that the birds came in earlier on cloudy evenings, in fact some resorted to the roost when a heavy thunder cloud darkened the sky in the afternoon and left again when the sun came out.

Arrival dates at Cape May for the Little Blue Herons are:

1908.	August 1	1924.	July 28	1930.	July 11
1917.	August 19	1925.	July 12	1931.	July 19
1920.	July 29	1926.	July 21	1932.	July 9
1921.	August 18	1927.	July 14	1933.	July 8
1922.	August 10	1928.	July 2	1934.	July 1
1923.	August 4	1929.	July 12	1935.	July 13

Dates for the last one seen are:

1920.	October 3	1928.	October 9	1932.	September 7
1925.	September 3	1929.	October 5	1933.	September 6
1926.	September 27	1930.	September 27	1936.	October 5
1927.	October 10	1931.	August 31		

While we always think of the Little Blue Herons as birds of the late summer, our records show us that as a matter of fact they are now present during six months of the year "and we realize," as I wrote in 1934, "that the finding of a nesting colony somewhere in the peninsula, that goal to which the ornithologist of the Cape is ever looking forward, is by no means beyond the range of possibility." The discovery in the very next year of nesting birds a little farther north in the state justified my prediction. Let us hope that the birds may increase as breeders as they have as summer visitors, and that the nesting colonies of Night Herons which seem to attract them may be unmolested wherever they may be established and not subjected to wanton slaughter as they have in the recent past.

GREEN HERON

Butorides virescens virescens (Linnaeus)

The Green Heron is the most abundant of Cape May's herons and prior to the extensive draining operations of the Mosquito Commission there was not a day—scarcely an hour indeed, all summer long, that one or more could not be seen on the wing over the salt meadows which stretched between Cape May and the Point, or flushed from one of the shallow ponds or ditches which lay on either side of the turnpike connecting the two places. While considerably reduced in numbers in recent years, there are still Green Herons to be found in the Cape May area, especially about the ponds near the Lighthouse and along Cape Island Creek below the Harbor. Occasionally one will be seen flying directly over the house tops of the town and now and then, before the culvert was constructed, individual birds would follow up the course of the creek at low tide, feeding on the muddy bottom, until they were close to the outlying buildings.

On the great open salt meadows north of the Harbor they are still plentiful and may be seen on the wing, passing from one feeding ground to another or following the course of the larger thoroughfares. We may also catch them with the glass as they explore some shallow meadow pool or rest amongst the short grass with their long necks stretched up, on the lookout for danger. Every inland pond or wooded mill dam will have its Green Herons, too, which take wing precipitately with their familiar cry, as we come into view, or freeze into statue-like poses on the muddy banks where they may be feeding among the rushes and arrowheads.

When, in 1920, I made a special study of these herons in their old haunts between Cape May and the Point, they regularly increased in numbers as the summer advanced and by August 31, when I followed the creek banks back of South Cape May, after the spartina grass had attained its full growth and furnished good shelter for them, I could count no less than fifty individuals in a distance of a quarter of a mile. Green Herons are wary birds and as one crosses these marshes they are continually taking wing ahead, while one is still quite a distance away. Many of those seen on the wing, however, must take flight of their own initiative, as when watching them with the glass at a distance no cause of alarm could be detected when they flew. The Green Heron acts individually, and while as many as twenty may be flushed in crossing a section of marsh, they always take wing one after another and are scattered far across the flats, the first ones beginning to alight again as the last take wing. On but two occasions have I noticed anything like concerted action. On the mornings of August 7 and 8 parties of eleven and seven, respectively, were seen from a distance to arise together and travel like a bunch of Crows in rather close formation to some distant point. Possibly they migrate in a similar manner.

The call of the Green Heron is a harsh *kaup, kaup, kaup,* or *skeuk, skeuk, skeuk,* as it may sound in the case of different birds or to different ears. When come upon very unexpectedly and evidently much frightened, one individual gave a more prolonged cry: *skowk, skowk, skowk—owk—owk—owk.* On another occasion when following one of the smaller streams through the marsh grass I flushed a number of these herons at close quarters which seemed to have several low preliminary notes before the usual call, as if the bird were getting up steam; *puk—puk—puk—keuk, skeuk, skeuk,* they seemed to say, the first three notes close together, the latter separated by greater intervals. When first taking wing the Green Heron's neck is outstretched and the feathers of the crown erected, and when its flight is finished and it is about to alight on a tree or bush, this same general attitude is resumed, although the neck is then usually bent in an S-like curve while the legs dangle loosely, ready to grasp the perch at the first opportunity. During protracted flight, as one usually sees them, however, the head settles back on the shoulders and the neck is looped out in front while the legs are swung straight out in the rear touching one another at the toes and at the tibio-tarsal juncture. The wings move with a regular rhythmic swing about thirty beats in ten seconds. In flight the Green Heron has something of the action of a Crow and both birds appear equally black when silhouetted against the sky, but the heron holds more steadily to its course, with less rising and falling and one soon learns to distinguish it instantly. Once or twice I have seen a Green Heron carried along in a strong wind at a much faster pace than usual, flapping vigorously as if to keep from being blown over.

When come upon resting in an open pond the Green Heron sometimes flushes but usually becomes rigid or "freezes," as it has been termed, in an apparent effort to escape notice. It seems to me that the former action is taken when the bird has seen the intruder first and has plenty of time to get away safely, and the latter when it has been taken somewhat unawares and is not so sure about the safety of flight. If you continue to advance upon him he will take flight anyway as soon as he realizes that he has been recognized, but from the number of Green Herons that have taken wing from directly in front of me, and which up to that moment were undetected,

I am convinced that there were probably still others that remained frozen and escaped my notice entirely. When freezing, the heron draws its head and neck down and points the bill up at an oblique angle, the long feathers of the neck overlapping the fore part of the body so that the bird appears strikingly like an old stake or stump. The legs are usually somewhat spread and one bird, taken unawares while I was close at hand, assumed a most ludicrous bowlegged posture, not taking time to turn and adjust his feet although the head and neck were in the normal position. He remained rigid, his yellow eye fixed steadily upon me, and although the wind ruffled up his inner wing feathers, he never moved from his uncomfortable position so long as I stood still. At my first movement, however, he took wing.

Once I saw one of these herons freeze when I was completely concealed and some distance away, and although he retained his rigid attitude for some time I could see no reason whatever for his action. Birds feeding on open mud flats frequently assume this rigid pose, whether because they sight an intruder afar or to aid in securing their prey, I cannot say, but they look exactly like so many snags protruding from the mud, or like irregular masses of the mud itself, and it requires careful examination with the glass to ascertain their true character. On one occasion a bird froze when alighting on a tree although usually when settling on a perch the neck is kept well extended and the bird in a short time drops to the ground again.

Often when no danger threatens I have seen Green Herons stand on some post or limb of a tree with neck fully erect, as if making observations, and the same pose has been noticed in birds on the ground among grass, or when they were supporting themselves among reeds or coarse sedge, and grasped the stems with their feet with tarsi horizontal.   Indeed by carefully sweeping the ground with the binoculars a surprising number of Green Heron necks will sometimes be discovered, sticking up like stakes among the grass. When resting on a tree voluntarily, and not as the result of being frightened, the head is drawn down on the shoulders and the neck looped below on the front of the breast. While frequently alighting in trees in the neighborhood of their nesting grounds, these herons do not usually do so in other places, and the few times that I have found them in trees were during the migration when the individuals had probably just completed a long journey. Once (September 5, 1920) there were seven in the tall pines at the Point, and on the 13th of the same month, one was resting on bayberry bushes on the dunes below the town. Once or twice individuals have been seen resting and pluming themselves on stakes or boat landings.

Early in the summer the favorite feeding grounds of the Green Heron seem to be the open shallow ponds and wet marshes, but later in the season, when the marsh grasses have attained their full growth, the birds frequent the narrow tidewater creeks and ditches to a greater extent. When feeding in shallow water and entirely undisturbed the Green Heron crouches low and with head about half drawn back advances rather rapidly, but ever warily, raising its feet clear up and well out of the water and placing them down again with toes widespread, as it steps carefully along. Suddenly spying its prey it crouches still lower with neck outstretched and head raised slightly so that it is about on a level with the shoulders, while the neck sags a little below. It then advances, much as a cat would do, and with a quick lunge of the bill secures the luckless frog or small fish which it has been stalking. The original attitude is then resumed and once more the bird advances with rapid steps, sometimes all but running. The whole action is very cat-like and the body is bent so low over the mud and water that it is often partly concealed by tufts of low sedge and grass. One day while watching a stalking heron a large snapping turtle crossed its path, and I was struck by the similarity between them; the head and back of the reptile on the same level and the neck depressed, exactly as in the bird.

Sometimes the feeding individual becomes rigid at the sight of its prey. Then there is a stealthy step or two forward, out shoots the neck and the spear-like bill has secured the quarry. A bird that I watched on August 21, 1922, on the edge of the pond at the Lighthouse was unusually tame and allowed a close approach, flying only a few yards before alighting again. It ruffled up its crest feathers as it took wing and held its head erect for a few moments after alighting, then it drew in its neck and flattened its crest. In both attitudes there was a peculiar and constant flirting of the short tail feathers but this did not involve the upper tail-coverts which hung close over the base of the tail. The bird seemed to be a young one and this action was perhaps the result of a nervous indecision as to what should be its next move.

An entirely exceptional method of feeding was observed on August 25, 1924, when a Green Heron was noticed perching on a post which stood in a pool where minnows abounded. The perch was not over fifteen inches above the surface of the water. When a fish approached its stand the bird bent down slowly over the water holding tightly with its long toes to the top of the post until the entire neck and most of the body were below the level of the feet and the bill close to the surface. Then there was a sudden lunge, the fish was caught, and the heron recovered its position on top of the post and swallowed its prey. This performance was repeated again and again and it was evident that the water was too deep for the bird to reach the fish by wading.

Green Herons nest in the low woods which cover parts of some of the coast islands, especially Two Mile and Seven Mile Beaches as well as in thickets on the mainland or elevated spots on meadow islands where marsh elder or a few bayberry bushes furnish shelter. There may be only a single pair nesting in a certain clump of woods or there may be a sort of colony of half a dozen pairs or more. One of these in a cedar thicket on Seven Mile Beach contained twenty pairs in May, 1927, and the dense thickets of bayberry at Corson's Inlet and at the lower end of Two Mile Beach supported considerable assemblages of nesting birds as did some of the cedar thickets on the Bay side. There is much excitement when one intrudes upon one of these heron communities and the birds are continually flying out and back again or perching on the topmost branches of the thicket, with necks outstretched and crests erect as they balance themselves against the wind. Richard Miller, who observed the colony on Seven Mile Beach, from 1913 to 1923, tells me that the nests were usually from fifteen to thirty feet from the ground, although one placed in a tangle of vines was only three feet up. Full sets of eggs were found as early as May 13 and as late as June 17, but the average date seemed to be about May 30. The first young were found May 23 and by early June most of the eggs had hatched. I have record of one nest on an island back of Five Mile Beach which contained eggs as late as July 7.

Nests of the Green Heron examined by Richard Miller on Seven Mile Beach are as follows:

May 30, 1913, five eggs, five young, just hatched.
May 30, 1915, three sets of five, three of four and two of three, some pipped.
June 4, 1916, one of two, two of five.
May 30, 1919, three of five, two of four, one of three, pipped.
May 31, 1920, four of four, one of four young.
June 12, 1921, five of four, one of three, one of four young.
May 21, 1922, one of one, two of four.
June 17, 1922, four of four, one of three, one of two and one with three young.
June 8, 1923, one of three, pipped, one with three young.
May 18, 1924, only three nests, one, two and three eggs.
July 4, 1924, one of one, two of three, three of four.
June 7, 1925, one of three, one of four, one with three young and another with four.
July 3, 1927, one with three young.
May 26, 1928, one of three and two of four.
June 24, 1928, one of three.
June 2, 1929, one of one, one of two and one of four.
June 26, 1932, one with four, three with four young and one with three all about one third grown.

Turner McMullen on May 14, 1916, at the same colony found two sets of five, one of four and one of three.

Green Herons arrive at Cape May from the south about the twentieth of April according to the records of Walker Hand, whose dates run from April 10 to 29, while in 1922 he and I found a single individual on Lake Lily as early as April 2 and Julian Potter saw another on April 9, 1922. In the autumn most of these herons have left by September 18, although one was seen as late as September 27, 1926, September 25, 1927, and three on the 29th. Other individual birds were observed on September 23, 1929; September 28, 1930; September 26, 1931; September 25, 1932. We have several still later dates: October 1, 1924; October 2, 1925; October 5 and 15, 1935; October 11, 1936 (Tanner); and an entirely exceptional occurrence on November 11, 1906, reported by Walker Hand.

The plumage of the Green Heron varies greatly with age and the fine bluish slate back of the adult with its rich maroon neck plumes and orange yellow legs are in striking contrast to the dull bronze green of the bird of the year with its brown-streaked breast and belly. Indeed so completely is the green lacking in the full plumaged bird that it is really not a "green heron" at all, and those who are not personally acquainted with the Little Blue Heron and its characteristic action and appearance, have doubtless more than once identified an old Green Heron as the latter species. When the head of the bird is drawn down on the shoulders and the beak pointed diagonally upward in the freezing attitude, the lines and stripes on the head, neck and breast come into beautiful alignment, while the blue gray and maroon tints of the adult blend perfectly.

A generation ago Green Herons were considered legitimate game and along with Black-crowned Night Herons were extensively hunted, especially by boys, while their eggs were gathered and eaten and I am told by residents of the Cape that they were the best of any of the wild birds' eggs, surpassing those of the rails and gulls which were so universally used for food. Crows' eggs they tell me were also tried but would never cook properly and were inedible.

While held in more or less popular contempt and ridicule, and burdened with offensive local names wherever he occurs, the Green Heron seems nevertheless to fill an important niche in nature's scheme. His presence adds an indescribable something to the environment of salt marsh or woodland pool that we should not care to lose, while his coarse and explosive cry lends a touch of wildness to otherwise prosaic surroundings. When studied carefully, moreover, we find in him an interesting personality minding his own business, and carrying on his daily activities as fisherman and game stalker with a grace and effectiveness that enlist our damiration

BLACK-CROWNED NIGHT HERON

Nycticorax nycticorax hoactli (Gmelin)

While not so abundant as the Green Heron, the Black-crowned Night Heron may be found on any body of fresh or brackish water about Ca peMay as well as on the ponds and along the creeks of the great salt meadows. At dusk and in the early hours before dawn its hoarse call may be heard as it passes overhead both in the open country and above the town itself. As a few individuals may remain through the winter it may be looked for in every month of the year.

One of the haunts of the Night Herons, which is most associated with them in my mind, is the shallow pond to the east of the Lighthouse. Here on almost any day in summer or early autumn they may be found standing in the water or on the mud bars, looking like snags or short posts exposed at low tide. They usually remain stationary with the head well down on the shoulders so that the contour of the back forms a continuous curve and the bird's body seems oval and pointed at each end. At other times the neck will be deflected until the point of the bill barely touches the surface of the water or may be partly submerged. In deeper water they often stand so that

(148)

the legs are wholly submerged and the body is held horizontally immediately above the surface, with the neck stretched out in front. In these attitudes the birds patiently await the approach of their prey and are ready for an instant strike. During the daytime, at least, the Night Herons seem to prefer this way of procuring their food rather than actively walking after it. Along the ditches and creeks back of South Cape May I found several "stands" which the birds seemed to have used regularly, though probably more for resting places than for feeding. They were in the tall grass or under bunches of overhanging sedge and overlooked the water. Where the grass had grown to a height of two or three feet they were somewhat arched over above and were always plentifully splashed with excrement. One bird which I was able to stalk while occupying one of these stands, sat motionless as long as I watched it, with its head drawn closely down on its shoulders.

On mill ponds in the interior of the country I have seen Night Herons perched on stumps which projected a few inches above the water or on the edge of boat landings, but where the water is not too deep they prefer to alight close to it on the muddy shores or exposed sods.

While usually stationary in the daytime they sometimes have occasion to move from one stand to another and they then walk in a more or less humpbacked attitude, the neck and head held rigid and not moving back and forth in unison with the bird's steps as is usual with the larger herons, while the feet are raised rather high, often quite clear of the water if the bird is walking through it. On July 23, 1932, I saw a Night Heron fishing on a shallow pool on Seven Mile Beach; an adult bird with plumes. It stood crouching with its yellow legs bent at the heel in an apparently uncomfortable position, and its red eyes staring into the water. It would move one foot forward very stealthily spreading the toes wide apart and when in the required position made a thrust but sometimes waited a long time before taking this action and it missed its prey about half the time. When changing location it ran with a clumsy and ridiculous gait sometimes jumping from one mud lump to another, sometimes making several flaps of the wings, but always landing on a lump of mud and avoiding stepping into the water. Its object may have been to avoid alarming the fish. It was seen in the same spot the following day.

Another habitat of the Night Heron, none the less characteristic, is on the open salt meadows, north of the Harbor. Here I have seen them in mid-May winging their way out from the mainland at dusk, usually one at a time, apparently to feed. They would come over at a considerable height with steady wingbeats and then sail down on set, decurved wings and dangling legs, flapping again slowly as they settled into the grass. Again at sunrise I have seen them returning, having presumably spent the night feeding on the salt ponds or on the tidewater creeks although some of them feed there regularly during the day. Birds that arrive on the meadows just before sunset stand out very conspicuously against the green of the marsh as the horizontal rays of light strike them, and their pale gray plumage brightens up until it would appear that their breasts were snowy white. Once or twice I have seen these marsh birds begin to feed, sneaking stealthily along the border of a pond with neck drawn in below the shoulders and bill pointed diagonally downward ready to strike. Probably there is more action in their feeding habits at night than in the daytime.

Night Herons are more given to roosting in trees during the day than are the other species of herons and every year about the 20th of August I begin to find gatherings of them in the pine woods at Cape May Point. Probably these are migrants from farther north as about the same time we hear more Night Herons calling as they pass overhead in the dark. As evidence of the night flight of this species, besides the calls of passing birds, we have the observation of Julian Potter of seventeen birds flying south at the time of a fire along the beach front, at 3:30 a. m., on September 30, 1925.

While usually acting as individuals, Night Herons will sometimes, perhaps from necessity, move in small flocks, especially in late summer or early autumn, when they are preparing to migrate. On September 5, 1921, I saw fourteen arise from the Lighthouse Pond and cross over to a nearby thicket where they had been in the habit of roosting, and which was strewn with their gray feathers. On September 17, 1923, a group of twenty flew from the same spot, while on May 28, 1923, there were fifteen feeding there and later in the season we can frequently find from twelve to twenty-five present. Of scattered birds that may be flushed here and there along the ditches, or on ponds farther inland, we rarely record more than six to ten in a day's walk, though on September 4, 1920, I flushed thirty on the marshes between Cape May and the Point, all of them taking wing independently as I advanced. William Rusling, Audubon Society warden at the Point during the southward flight of 1935, has given me a résumé of his observations of the Night Herons on migration. He tells me that they cross the Bay on a northeast wind often in the rain and mainly at night but sometimes while it is still quite light. They congregate in the pines east of Lake Lily where some flocks often roost during the day. They begin calling soon after sundown until there is quite a chorus of squawks. They arise one or two at a time until there is a loose flock of twenty or thirty and begin to circle higher and higher until they reach an elevation of several hundred feet and then in a loose V they start across the Bay. If they leave while it is still light enough to see, they call but little, but after dark they keep calling until out of hearing. Some of his daily counts are as follows:

September 2—250	October 15— 175	November 3—136
September 7—500	October 23—1,050	November 15—125
September 24—130	November 1— 588	November 16— 3
	November 2— 450	

James Tanner, Audubon warden for 1936, recorded migrating birds from September 1 to November 12, with the largest flight on October 28.

On the wing the Night Heron resembles most closely the Green Heron, although its larger size and lighter coloration will always distinguish it. Its legs, moreover, are shorter in proportion than those of any of our other herons, and in flight the toes project only a short distance beyond the tips of the tail feathers. It thus has a sort of bobtailed appearance for a heron, while the shorter neck forms less of a prow than in the other species. At a distance, especially when seen in silhouette, a group of flying Night Herons bears considerable resemblance to a flock of large-sized crows. When they come to roost in a tree the wings are strongly decurved and the neck is stretched out in front as the bird reaches its perch.

The call of the Night Heron is a *quonk* or *quawk* as it sounds to different ears, and is uttered as the bird is flushed from a roosting place or as it passes overhead in the night, either in migration or when returning from its feeding grounds. One that I flushed from a ditch, almost at my feet gave a goose-like cry, *quongle-gogle*, which was doubtless exceptional and due to excitement.

At several places in lower Cape May County the Night Herons have nesting colonies and some associated with the Great Blues in the rookery at Pennsville. At Cape May Point small colonies of two to six pairs are said to have nested but usually they prefer to form much larger rookeries. One of the best known of these was on Seven Mile Beach where both Night and Green Herons occupied a grove of cedars and hollies back of the sand dunes. Here the Night Herons built their nests from twenty-five to thirty feet from the ground and had sets of four eggs by May 30 and some pairs at least a week earlier. The first young were hatched about June 10 although eggs were still to be found as late as June 16.

Recent developments incident to the conversion of this beach island into summer resorts greatly disturbed the herons, and when the tops of most of the holly trees were cut out by vendors of Christmas greens the birds departed. Doubtless it was the successor of this same colony that was later located in a long straggling growth of red cedars, a quarter of a mile back on the salt meadows, which occupies a slightly elevated sandy ridge or island some two hundred yards in length. At the height of the breeding season when the young are in the nests this grove is anything but an attractive spot. The trunks and branches of the cedars, as well as the dense thickets of poison ivy and the masses of dead and tangled branches on the ground are all thickly splashed with white excrement, while pieces of dead fish and other food dropped from the nests lie scattered all about. These attract swarms of carrion flies, while mosquitos and green-headed flies everywhere abound. The all pervading stench and the heat of the meadows do not add to the pleasure of a visitor. Several pairs of Fish Hawks which have nests in the cedars circle constantly overhead whistling shrilly, and many Fish Crows, also with nests, keep up a continual harsh chorus of protest so long as we remain, while the young herons utter their peculiar and characteristic clicking sounds.

Julian Potter visited this colony on June 13, 1926, and tells me that at that time some of the nests contained eggs, while there were young in various stages of growth. The nests were scattered about on the limbs of the cedars at varying distances from the ground, the lowest one being in a holly bush only five feet up. The nesting trees were none of them over twenty-five feet in height but a number were old and stunted their trunks measuring from twelve to fifteen inches in diameter. Many of the unfeathered young were bleeding from the constant attacks of mosquitos and flies. The older fledglings were standing about on the nests or on neighboring limbs and when the intruders attempted to climb up to them they scattered to all parts of the tree, for although unable to fly they were expert climbers. They went out to the very tips of the branches where they hung on with feet, wings and bill. The bill was used in two ways; the lower mandible was hooked over a limb or the small twigs were grasped between the mandibles. In all their scrambling not a single youngster was seen to fall; and a fall at this stage of their existence probably meant death, as no young were found on the ground except a few dead ones. When captured the young herons protested with faint squawks and then after a series of neck contortions disgorged the contents of their stomachs consisting mainly of entire shrimps and macerated and partly digested killie-fishes, an altogether nauseating mess. On June 27 Potter again visited this rookery and states that many of the young were now fully grown and able to fly awkwardly from tree to tree, alighting however with much difficulty. Some fell to the ground with a loud thump and made off through the bushes at a rapid rate. One bird had something projecting about two inches from its bill which, upon being disgorged, proved to be a fourteen inch eel. While the head was partly digested the tail was still intact, showing that the bird, unable to swallow it all at once, was digesting it by degrees. Another young bird had hastened out on the marsh and entering a rather deep pool swam around the edge of it like a duck. The clicking note so characteristic of the rookeries of the Night and Great Blue Herons was now heard only occasionally, from the few nests that contained quite young birds.

When I visited this rookery on May 15, 1928, the birds were just beginning to nest, while on July 3, 1927, some nests still contained young although most of them were deserted and the young either on the wing or climbing about the branches. On July 31, 1926, all of the young took wing with the adults. On all of these occasions, as well as when others visited the colony, the birds that could fly, both old and young, at once left the trees and settled on the marsh a hundred yards or more from the thicket, ranging themselves in more or less regular ranks, eventually crouching down in grass, and there they would remain until they thought danger had passed. Some of the adults which still had young in the nests would fly over the rookery at intervals and sometimes alight on the tree tops, in evident concern for their broods.

John Gillespie, with the true spirit of the bird bander, banded a number of the young in this colony regardless of the surroundings and one banded June 13, 1925, was shot at Millville, N. J., July 28, indicating that they do not travel far during the summer.

In 1927 part of this colony moved farther north to a more extensive thicket of larger trees and more dense underbrush and the next year the majority of the birds had established themselves there. Data furnished by Richard Miller on nests in the old Seven Mile Beach heronry are as follows:

May 30, 1915, nests with two, three, four and four eggs.
June 4, 1916, two nests each with one three and four eggs and one with three young.
June 16, 1918, four nests with three eggs, three with four.
May 31, 1920, three with two eggs.
June 12, 1921, two nests with three eggs, two with four well grown young.
May 21, 1922, one with two eggs.
July 4, 1924, one nest with three eggs; many with young half to three quarters grown

Turner McMullen tells me that the earliest eggs ever found here were a set of four on May 10, 1931, and that on May 12, 1934, there were thirty-six occupied nests containing from one to five eggs each. Back in the interior the Night Herons seem to nest earlier and in a heronry not far from Camden he examined fifteen nests on April 5, 1930, all of which contained five eggs. An earlier visit to this heronry by Julian Potter on May 20, 1923, found that at this date most of the eighty nests that it then contained were burdened with well-grown young. This rookery was in a grove of tall oaks and the nests placed in the treetops were well out of reach while the undergrowth of poison ivy was doubtless a protection. There were still fifty pairs nesting here in 1936. Another colony a few miles south contained about one hundred and fifty pairs of birds in June of the same year. It is surprising how many of these birds may congregate in the vicinity of towns and villages without attracting attention.

Black-crowned Night Herons present two very different plumages. That of the adult is plain pearl gray, lighter on the underparts, with a greenish black saddle and crown patch and with several slender white hair-like plumes drooping from the occiput onto the shoulders. The young bird of the year is brownish gray throughout streaked with buffy, which is most conspicuous on the neck and sides of the body. Birds in this plumage are not usually seen before July 15, as the young do not stray far from their rookeries until this time and often not until later. One was seen, however, on July 8, 1922, and others on July 6 and 13, 1925, both of the latter being accompanied by adults.

An adult bird with plumes was seen on July 23, 1932, which had the usual green back and cap replaced by darker gray. On September 9, 1924, I found a Night Heron in the brown plumage of the first summer, roosting in a low tree in a thicket at the Point, which was fast asleep and I was able to approach to within arm's reach of it before it took wing. It was perched about four feet from the ground, resting on one leg with its head and neck bent over in front, the head being poked under the left wing and completely concealed by the plumage. These young birds seem to be possessed of little sense of fear and I saw one walk slowly and deliberately from the shore of the lake at the Point through the pine woods and directly into a temporary camp that had been pitched there, to the astonishment of the campers.

Another of these brown young birds was found feeding on the shore of Price's Pond on August 15, 1935. I walked up to within eight feet of him but he continued his fishing without paying the slightest attention to me. He held his mandibles partly open and his throat vibrated rapidly as it was a very warm day.

The spring flight of the Night Herons reaches Cape May by early April, probably earlier, in some years at least, but birds were heard going overhead on April 5, 1926. They remain through October to mid-November while occasional birds pass the winter if the season be open. We have records for November 30, 1909; December 14, 1917; December 2, 1923; December 13, 1929; while the counts on the Christmas census of the Delaware Valley Club are as follows:

December 28, 1930—one	December 23, 1934—one
December 27, 1931—three	December 22, 1935—two
December 26, 1932—seven	December 27, 1936—eighteen
December 24, 1933—seven	

Single birds were also seen on January 30, 1922, and December 23, 1919, the last mentioned was found by Walker Hand seeking shelter in the shrubbery of the Physick garden in the heart of the town.

In Louisiana the Night Heron has long been shot for food and is recognized as a game bird under the name of "Grosbek," and in days gone by it was hunted in Cape May, usually in its roosts or breeding places where numbers of the birds could be killed with little trouble. That this practice still prevails, to some extent at least, is evidenced by the finding in June, 1932, of a pile of bodies of Night Herons in the rookery on Seven Mile Beach where they had been left by the gunners who for some reason seemed to have lost their appetite for the birds. Perhaps the surroundings of the breeding colony already described may have had something to do with it!

YELLOW-CROWNED NIGHT HERON

Nyctanassa violacea violacea (Linnaeus)

   This southern species apparently occurs sparingly in several of the colonies of Black-crowned Night Herons which nest in southern New Jersey and possibly has done so for many years although the fact has only recently been determined.

   On June 13, 1926, while visiting the rookery of Black-crowns in the cedar grove back of Seven Mile Beach, Julian Potter and John Gillespie discovered a single Yellow-crown among the ranks of the commoner species and it was seen again on June 27. (Auk, 1926, p. 538.) On July 24, 1926, Richard Erskine saw two Yellow-crowns feeding in a small pond close to the roadway opposite the rookery. All of these birds were adults. On July 31, I visited the colony and saw one of the Yellow-crowns fly from the cedar trees out into the marsh. The clear dark bluish slate color of the plumage could readily be noted as well as the white patch on the top of the head and the white stripes on either side. The general tone of the plumage seemed almost exactly that of the adult Little Blue Heron. A gray bird, presumably its young, followed it out of the grove. Later on a Yellow-crown came into the thicket and perched on top of one of the cedars. It looked more slender than the Black-crowned Night Heron and acted differently. As it settled on its perch it stretched out its neck, but kept it more or less crooked, and flapped its wings several times rapidly exactly as the Little Blue Heron does. When the whole flock of Night Herons left the rookery and settled out on the meadows, the Black-crowns all crouched down in the grass so that they were invisible but the four Yellow-crowns that we now saw among them held their necks clear above it, so that the black and white pattern of the head could be distinctly seen, while the plumage of the occiput looked truncate, as if the feathers had been clipped.

On June 6, 1927, Benjamin Hiatt and Charles Doak visited the rookery and by carefully marking the trees from which a Yellow-crown flew and waiting patiently until the birds returned to the grove, located two nests of this species; one containing five eggs and the other eight—the latter unusual for any heron. On the same day practically all of the nests of the Black-crowns contained well-grown young. On June 12, three Yellow-crowns were seen and the two females were observed brooding on their nests. Five of the eight eggs had hatched and the young could easily be distinguished from those of the other species. They were grayer and had the hair-like feathers on the back coarser. Julian Potter tells me that on June 26 one of the young, now sixteen days old, had pin feathers just bursting out. The skin and soft parts were yellow as compared with the greenish color of the young Black-crowns; eyes orange instead of yellow, and the down on the tips of the feathers white instead of brownish. On July 3, 1927, when I visited these nests the two adults were very solicitous for their safety, quite unlike the usual attitude of the Black-crowns, and returned again and again to perch on the nearby trees while I was examining the young. Their posture was more like that of the true herons than of the Black-crowned Night Heron. The neck appeared proportionately much longer and more slender than in the latter species, and was kept loosely curved as the bird made short flights from one tree to another or perched on the topmost branches. The uniformity of the bluish slate plumage was especially notice-able as was the contrast between the slaty blue neck and the black and white head. From the side there appeared two broad black bands from the forehead and bill, respectively, to the nape, and between them a white stripe. As the bird faces you the broad buffy white crown is splendidly displayed and the white bands running back from the base of the bill on either side. The latter is black tinged with greenish at the base and appears very thick and short. There are fine plume-like hairs projecting from the occiput while some slender plumes from the lower back project slightly beyond the tip of the tail.

On July 24 there were still some young in the nests but they were evidently about ready to leave, as they crawled actively out on the limbs. They were now distinctly darker colored than the young of the other species, especially on the back, and were more slender, with longer neck and legs and the markings on the back smaller and more scattered.

There was about a quart of fiddler crab shells under the nests showing that these crustaceans were a favorite food of the birds. On May 15, 1928, I saw a single Yellow-crown at the rookery but so far as I know no nests were located. On June 3, 1930, Julian Potter saw two individuals in the colony and three on July 28, 1935.

Otway Brown tells me that one of these herons was killed by a boy at Cold Spring about 1921, but we have no further evidence of its earlier occurrence in the county, although I have examined a specimen killed at Browns Mills farther north in the state on October 5, 1911, by Samuel Brown. On August 16, and September 2, 1922, Charles Urner flushed a young Yellow-crown near Elizabeth, N. J., and by comparing it with young Black-crowns that were present was able to determine the distinguishing characters so as to make identification in the field certain—darker and bluer upper surface, darker crown, more prominent whitish spots on the wings, longer and darker legs, and different, more erect, pose. Subsequently a number of immature individuals were identified in the Barnegat and Brigantine areas and one or two adults. The source of these birds was a good deal of a mystery until the discovery by Charles Urner and C. Brown of breeding birds in a colony of Black-crowned Night Herons near Absecon in the summer of 1936. The numbers of immature birds at Brigantine had been increasing regularly and in this year as many as twelve were seen at one time in the bayberry roost on the outer beach. The records are as follows:

Barnegat Bay.	August 9, 1925, two immature birds. Charles Urner.
Absecon.	September 4, 1927, one. Charles Urner.
Brigantine.	July 22, 1928, one immature. Charles Urner.
Barnegat Bay.	August 11, 1928, one immature. Charles Urner.
Troy Meadows.	June 16, 1929, one adult. Lester Walsh.
Troy Meadows.	August 5, 1929, one immature. Lester Walsh.
Brigantine.	August 3, 1930, two immatures. Julian Potter.
Brigantine.	September 14, 1930. John Gillespie.
Brigantine.	July 29, 1931, one immature. Clifford Marburger.
Brigantine.	July 27, 1932, one adult. Julian Potter.
Brigantine.	August 1, 1932, one immature. Julian Potter.
Brigantine.	August 12, 1932, one. Julian Potter.
Brigantine.	August 25, 1934, seven. Julian Potter.

19 36

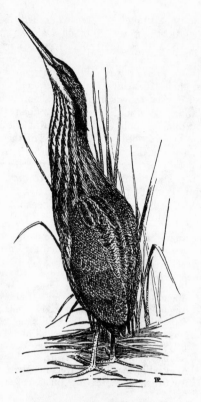

AMERICAN BITTERN

Botaurus lentiginosus (Montagu)

The Bittern, like the rails, is not a bird that offers us many opportunities for study. Crossing one of our fresh or slightly brackish marshes covered with a growth of grass and sedge two or theee feet in height, a bird of the general build of an immature Night Heron springs suddenly from almost under our feet, flies for some distance, and then drops back again into the cover. We note the rich rusty brown of the plumage, the narrow streaks of buff, and the black edgings to the wings, all quite different from the ashy gray of the young Night Heron. We see the short olive yellow legs stretched straight out behind and the neck bent down from the shoulders and curved up rather abruptly causing the bill to point diagonally upward. That is about all we usually see of a Bittern.

There are, however, occasions when we are able to creep up on one of them at rest or by chance we come suddenly upon one before he catches sight of us. One that I saw on the shores of Lake Lily, April 3, 1923, had frozen the moment he saw me and stood rigid in the scant growth of grass and sedge, with neck contracted and head nearly on his shoulders, his yellow eye very conspicuous and his bill pointed up at an angle of 45°. When he realized that he was discovered he at once took wing. Another I came upon was on the edge of Cape Island Creek, within the town limits. I was crossing on an old tressel and he did not see me until I was directly over him. He adopted a rigid attitude glaring up at me with his yellow eyes and we stood for some minutes looking steadily at one another; then a slight movement on my part caused him to take wing. One that I flushed from a field of weeds and grass back of the Lighthouse Pond, over a hundred yards from water, on September 17, 1923, had a regular form like that in which a rabbit rests, and apparently had been sitting there for some time.

On October 15, 1928, while following a path on the Fill which crossed a damp depression thickly covered with vegetation, I came upon a Bittern walking slowly ahead of me. He was hunched up in a rail-like attitude stepping carefully and deliberately and darting out his neck as he apparently picked insects of some sort from the stems of the grasses. On June 25, 1932, I flushed an immature bird on a pond on the Bay Shore Road which froze in the usual way and when holding its head erect displayed long tufts of yellow downy filaments standing up from the crown, remnants of the natal plumage. Upon further approach this bird ran like a rail crouching low and disappeared in the grass.

The Bittern is a summer resident of the Cape May district and formerly nested rather commonly on the Pond Creek Meadows where some may still breed if they have not been entirely driven away by the draining. They also occurred in the breeding season among the cattails and sedges which bordered the old Race Track Pond and other ponds to the north, all of which have now been drained. They still remain, however, on Price's Pond, with its extensive undrained cattail marshes. A few also probably bred on the edges of the island beaches along the coast for, on July 4, 1927, I found one on Elder Island north of the Harbor, which had apparently been carried there from Two Mile Beach by a heavy northerly gale that was blowing. When the bird tried to make headway against it he was unable to advance and was blown down again into the bushes. Another individual was flushed from a marshy spot on Two-Mile Beach on July 9, 1916, and while occupying a gunning shack on the marshes nearby, in May, 1922, we heard one "pumping" somewhere over on the island.

The Bittern's nest is placed on the muddy bottom of the marsh in thick patches of grass and sedge from eighteen inches to three feet in height, some of the stalks often arching over it. Richard Miller found one on the Pond Creek Meadow on May 22, 1921, which contained five eggs and another on May 30, 1922, with four eggs heavily incubated, while a third found May 30, 1921, contained two nestlings about a week old. Turner McMullen examined one at Fortesque, Cumberland County, on May 19, 1934, with four eggs.

The peculiar pumping "song" of the Bittern is frequently heard about Cape May during the spring when the birds are mating but I have never had an opportunity to study the bird while going through the performance.

Most of the Bitterns that we see are in the marshes or about the ponds where they nest but during the migrations they may be found anywhere in grassy fields, on the salt meadows, or even immediately back of the sand dunes, dropping down at the end of their night flight to seek rest and shelter during the day. These migrants are entirely unacquainted with their surroundings and one of those flushed from the sand dunes flew over the housetops in a bewildered manner and came to rest close to the old railroad station.

Migrant Bitterns arrive at Cape May according to Walker Hand in early April, his dates for a number of years (1905–1916) running from April 6 to 17 while I have first records of April 15, 1920; April 10, 1921; April 14, 1924; April 24, 1927; April 7, 1929; April 30, 1932.

They seem to increase in numbers in September but whether due to the arrival of migrants or to the greater activity of birds reared in the neighborhood I cannot say. We have a number of October observations usually to October 25, also two on November 7, 1920; one on November 14, 1926; one on December 20, 1929; and two on December 5, 1916, which were caught in muskrat traps. That some Bitterns remain through the winter is indicated by the Christmas census of the Delaware Valley Club but as we have no records for January, February and March, it may be that they move south as the weather grows colder or perish. The Club Christmas census records are as follows; single birds unless otherwise stated:

December 27, 1919	December 26, 1932, four
December 28, 1921	December 23, 1933, three
December 26, 1926	December 23, 1934
December 25, 1929	December 22, 1935, four
December 27, 1931, two	

We often find Bitterns in pairs but never in flocks although several may be flushed individually from the same marsh, probably members of one family. Walker Hand flushed six, one after another, while crossing the meadows east of the Lighthouse on August 8, 1920, and from the cattail marsh surrounding Price's Pond near New England Creek, on the Bay side, Hampton Carson and John Hess flushed eight Bitterns on July 15, 1936, all of which immediately settled again a little farther on. Their presence was not suspected until the cattail belt was entered.

LEAST BITTERN

Ixobrychus exilis exilis (Gmelin)

This miniature of the American Bittern has much the same secretive habits but appears to be a much rarer bird about Cape May, although rather more common on the marshes farther north along Delaware Bay. It seems to be mainly restricted to fresh water marshes well covered with cattails or tall sedges. In flight the appearance of the bird is similar to that of the American Bittern, with neck stretched out in front, bent down at the shoulders and then sharply up again, and with the bill pointing diagonally upward. The prevailing yellow buff color of the wings is very conspicuous as they are spread in flight and in strong contrast to the dark glossy green back. The wing beats are rather rapid but not so quick as those of a rail.

On July 12, 1925, I watched a pair on the marshes on the Bay side at South Dennis, which flew repeatedly back and forth over the sedge from the edge of the creek to a thicket of cattails which evidently sheltered a brood of young. When not on the nesting ground, however, the Least Bittern does not often take wing except when disturbed by an intruder.

The only spot where I have found Least Bitterns immediately about Cape May is along Cape Island Creek back of South Cape May. Here in 1891 they were of frequent occurrence during August, flying up and down the stream or resting in the marsh elder bushes on the banks. They would take up a position with neck and head held vertically and remain rigid so that the whole bird with its slim body, seemed like the trunk of one of the bushes and when the wind blew it seemed to sway back and forth in unison with the surrounding reeds. Doubtless these birds nested in the great cattail swamps which lay just to the north of the creek.

The draining of this entire area has sadly depleted the number of the Least Bitterns as well as other forms of bird life. Several were found here as late as 1903, and on August 31, 1920, I flushed one from a small tract of coarse sedge along the creek, immediately back of South Cape May, which flew a short distance and dropped back into the vegetation. After once settling I was unable to make it rise again and I presume that, rail-like, it skulked away among the stems of the sedge until it reached a spot sufficiently remote for safety. On May 30, 1921, H. M. Harrison found a nest of the Least Bittern in a thick growth of narrow-leaved cattails bordering one of the ponds in West Cape May. It was the characteristic flat structure supported among the cattails about a foot above the level of the water and contained five eggs.

During the last two or three years the growth of cattails about the ponds at the Lighthouse has increased to such an extent that it furnishes excellent shelter for rails and bitterns and as it is more or less inaccessible, it is probable that many may nest in its recesses. Additional draining, however, will ruin it as a resort of any water bird. On August 13, 1928, I saw a Least Bittern flying over the top of this growth which plunged down into it. The characteristic buffy yellow wings and the touches of mahogany and bronze green on the back shown out clearly against the background of the cattails. By 1930 the cattails had virtually covered the shallow pond and the swampy area behind it and here on July 31 and August 7 I saw single Least Bitterns which I infer had nested there.

Price's Pond above New England Creek on the Bay side formerly rather inaccessible has recently been made available to visitors by the opening of several roads for which there was no possible use except as they furnished work for the unemployed. Here during August, 1936, I saw several Least Bitterns and have no doubt but that they nested in the broad cattail swamp adjoining the pond on the east.

The birds breed in several swamps in Salem County to the northwest of Cape May and Turner McMullen examined two nests there on May 29, 1921, each containing five eggs and another with four eggs on May 28, 1932.

As further illustrating the nesting of the Least Bittern, Julian Potter has recorded a nest at Camden, N. J., finished on May 11, 1919, in which the young hatched on June 1 and climbed out on June 8. Another nest found on May 27, 1922, contained young just hatched. Turner McMullen has given me the following data on nests that he has examined:

Camden County—
    June 6, 1935, five eggs; four eggs.
    May 27, 1922, three eggs; three young.
    May 25, 1935, four eggs; five eggs;
        five eggs.
Marshalltown, Salem County—
    May 29, 1921, five eggs; five eggs.
    May 28, 1932, four eggs.

Woodbury, Gloucester County—
    June 5, 1915, five eggs; four eggs.
    May 27, 1916, five eggs; three eggs.
Palmyra, Burlington County—
    June 14, 1930, one egg; four eggs.
    June 19, 1930, six eggs; four young.

WOOD IBIS

Mycteria americana (Linnaeus)

The occurrence of Wood Ibises at Cape May was one of the outstanding features of my bird studies on the peninsula.

August 10, 1922, when I saw the first one, was a day of strong wind from the north-northeast, with fleecy clouds on the horizon and blue sky overhead. Over the pine woods at Cape May Point I had noticed six Broadwinged Hawks, soaring high in the air; now sailing into the wind, now floating off with it and then turning to face it once more. While I was watching their evolutions through my glass some thirty Turkey Vultures came into view, strung out in a long line reaching toward the mouth of the Bay, and all facing the wind. As I turned my attention to these master soarers I suddenly realized that there was quite a different bird soaring with them away up under the canopy of the sky. It appeared white below and, unlike any heron or other long necked bird of my acquaintance, held its neck stretched straight out in front, while its broad wings were motionless except for an occasional slight undulatory movement. It pressed right into the wind going higher and higher and occasionally soaring in a circle. I kept the glass focussed on it until it had mounted so high that it looked no larger than a swallow when suddenly I lost it in the sky.

I had given up all hope of ever seeing it again when to my astonishment, about half an hour later, it came sailing swiftly over the tops of the trees headed for the Bay but before I could get clear of the woods it had once more disappeared. Shortly, however, it returned much more slowly, against the wind, and lower down than on either of the previous occasions, headed this time for the Pond Creek Meadows. I could now see clearly the white under wing-coverts and the tips of the black primaries distinctly separated from one another. Its head appeared bare and somewhat vulturine in character and it was turned slightly from side to side as the bird passed overhead. The tail seemed to be square at the end and not very long so that the black legs projected clear beyond it. When the bird circled, as it did once or twice, it tilted up sideways so that the upper side of the body came into view showing that the back and fore part of the wings, everything in fact except the flight feathers, were white.

Although I was now sure that I had seen a Wood Ibis, I was anxious to make a closer acquaintance with it, but search as I could that day and others I failed to find it.

I had no expectation that it would return the next year but on July 7, 1923, David Baird told me that about 8:30 that morning he had seen two large black and white birds, sailing toward Cape May Point, over the meadows and that from my description of the Wood Ibis of 1922 he thought they must be Wood Ibises. About 3:30 that afternoon with Julian Potter I accompanied Baird to the spot but we saw nothing of the birds until we reached the shallow pond just east of the Lighthouse when to our amazement four of these splendid storks—for they are really not ibises at all—came sailing from behind us right over our heads as we stood screened from view by a thicket. One flew in front and the three others abreast a little way behind him, all with wings set as if intending to alight on the pond, but, seeing us, they kept on past the Lighthouse and apparently settled at the head of the deep pond to the west.

We circled this pond and caught a glimpse of the birds far back beyond the growth of cattails with which it is surrounded so we continued on and in the shelter of the cattails crept up to within 200 feet of them. They were feeding on the soft mud bottom of a shallow pool. As they walked about they held the body horizontal, the head bent over and the heavy bill down in the mud and water, nearly to the gape, but the mandibles were parted so that one could see between them even at the base. The birds' steps seemed enormous and their legs were well bent at the "heel" so that the tarsus seemed often nearly horizontal as they waded in the mud and water, and as they stepped the heels kicked out and up behind the tail. At each step one foot was brought forward beside the submerged bill and wiggled slowly back and forth, apparently to stir up the fish or frogs that formed the bird's prey. This was done regularly, first the right foot, then the left, as the bird advanced, and as one foot was sunk down in the mud after the stirring process the other was brought forward and shaken through the water in the same manner.

Now and then one of the wings was extended to the full to balance the bird and occasionally both were raised over the back when the feet sank too deep in the mud. When in its progress the bird stepped from the mud onto a sand bar or piece of firmer ground it seemed to suddenly loom taller. The four followed one another in rather regular fashion and finally all waded out on firmer ground and came to rest. The bill now pointed straight down over the breast, a pose that recalled that of the Adjutant Stork in the Zoo.

Viewing the birds as we did at comparatively close range, the body plumage appeared dirty white, or white with a dull pinkish or buffy tinge, not snow white like the Egret, and with the wings closed the black of the flight feathers showed only as a narrow line on the edge of the outer primary. The black tail was likewise concealed by the white coverts which overhung it. The skin of the head was blackish and was covered by scattered fuzzy, down-like feathers. The big bill was light horn color and the breast soiled white. One bird had a white spot on the forehead like polished bone. As they plumed themselves we could distinctly see the curved tips of the mandibles open to pick up the end of a feather and every now and then one of the birds would snap its mandibles together rapidly in the air.

As they seemed a little suspicious and restless we withdrew without disturbing them but on our way back we saw them on the wing again over the meadows going as far as South Cape May where they wheeled about and once more sought the pond by the Lighthouse. They flapped their wings several times in unison, usually from two to twelve beats and then sailed a short distance on set pinions, and then again the flapping, the two alternating regularly. As the sun struck them in flight their colors seemed much brighter and purer—jet black and white in strong contrast.

Later in the afternoon as we stood behind South Cape May two of the Ibises came up the meadows again, flapping and sailing in the regular way. The head and long bill in front and the feet and legs behind appeared to balance one another and the wings seemed to be about in the middle. The great spread of wing was impressive. As a bird came nearer to us its head could be seen to be slightly deflected and was turned a little from side to side as it flew, producing a decided vulturine effect that was enhanced by the bareness of the head, which was much more striking now than when the bird was on the ground.

Subsequently these four birds were seen nearly every day when they were looked for until August 18; either on the shallow pond at the Lighthouse, on the wing over the meadows, or soaring high up in the sky. Once or twice they were seen soaring east of Broadway over the town and on one occasion they followed the meadows up to the golf links which they crossed and flew almost to Schellenger's Landing where they turned and retraced their way to the Point. Again they were seen soaring at a great height with Turkey Vultures above Erma, some five miles from their favorite pond. Once again I saw them ascend from the pond at 10:30 on the morning of July 26, and circling slowly without the slightest movement of the wings they mounted up and up until they were barely discernable, finally disappearing in the east. Next morning, however, they were back on the pond.

Every evening about sunset they seemed to seek the small cattail-bordered pond where we first saw them and probably spent the night there or else in the grove described below.

On July 30, we watched two of them feeding on the shallow pond and one varied the usual method by walking quite fast, poking his bill into the mud with every step but never entirely removing it from the water. He reminded one of a man walking with a cane the tip of which he dragged along the ground, poking it down with every step. A study of the birds as they were about to take wing showed that they gave several successive jumps with both feet together as they began to flap their wings, the number of jumps depending on the time required for the wings to lift the bird clear of the ground.

On August 5, I saw one of the Ibises disappear behind a strip of woody thicket on the marsh and I inferred that he must have settled on a small pond located there and sheltered from view on the ocean side, a spot where herons of various kinds like to gather. Creeping through the thicket I found all four of the birds there but instead of being in the pond they were roosting on the branches of some dead trees close by. With the bushes as shelter I was able to approach quite close to them. They sat upright with their bills held straight down against their breasts and their feet close together. The plumage appeared dirty white, becoming purer and brighter whenever the sun shown upon it. The edge of the wing showed narrowly black where a portion of the outermost primary was visible from the side, but from the front, where the edges of all the folded primaries could be seen, as the wings hung loosely away from the body, the black band appeared much wider. The head was grayish, the bill orange horn-color, and the tarsi and toes black.

When the birds changed their position on the limbs they depressed the head and neck so that they were below the shoulders which gave them a humpbacked appearance. Sometimes on taking a perch on a new limb they would test its strength with one foot before resting the weight of the body upon it, at the same time raising the wings to help maintain their balance. Sometimes they would hold the neck out at an angle of 45° sloping down toward the ground and at others at an ascending angle, the latter apparently when nervous and on the lookout for danger. Once when the sun shown strongly one of the birds spread its wings to their full extent and held them so, in the manner of a Turkey Vulture airing itself. In this position the anterior border of the wings for several inches was white but the greater part, composed of the primaries and secondaries, was solid black. Another individual held its wings about half spread and drooping. Finally, disturbed by our closer approach, all four flew off, their legs dangling for the first few strokes of the wings after which they straightened out behind in the normal position of flight. Fletcher Street, who had approached from another angle, succeeded in photographing them.

On August 13 three of the birds were feeding in the shallow pond at the

Lighthouse, which had become quite full of water after recent rains; they waded about, their legs entirely submerged and with the bill in the water up to the eyes. The foot shaking went on as usual where the water was shallower and we could see the bill open and shut. One Ibis seemed to have something stuck in his throat as he shook his head violently from side to side and made several retching motions as if to vomit. This was repeated several times as he stood still in one spot.

On August 18, all four Ibises were high in the air soaring with a party of Turkey Vultures and on the 22d they were seen by the Coast Guards on the pond. The next day a severe northeast storm prevailed and when it had cleared up the Ibises had gone nor were they seen again although a careful watch was kept for them until early in September. They had evidently drifted back to some spot far in the south from which they had come, but why they should stray so far afield and spend the summer in such a restricted area is a difficult problem to solve.

None of the earlier writers on the bird life of New Jersey mentions the Wood Ibis although its post-breeding flights to the north have been known for some time and individuals have been shot in interior Pennsylvania, even in the high mountains of Sullivan County.

J. F. S

GLOSSY IBIS

Plegadis falcinellus falcinellus (Linnaeus)

One of the rarest of the casual southern visitors to the Cape May coast is the Glossy Ibis and the only definite record of its occurrence was of one secured on May 7, 1817, by Mr. Oram of Great Egg Harbor and sent to Thomas Say. George Ord in describing it (Journal Acad. Nat. Sciences, Philadelphia, Vol. I, pp. 53–67) says "it was considered a curiosity at Egg Harbor and was unknown there even by name. I have often been on the coast of New Jersey, in the spring and autumn, and was equally unacquainted with this bird." Charles Lucien Bonaparte, in his "Ornithology," tells us that this specimen was "carefully preserved in the Philadelphia [i. e. Peale's] Museum" and failing to satisfactorily identify it with any described ibis he later proposed for it the name of *Ibis ordi*. Turnbull states that John Krider shot one in 1866 on the meadows below Philadelphia and that it has been seen at long intervals on the Delaware River and at Egg Harbor but gives no details. Those supplied by Krider in his "Forty Years Notes," cannot be taken seriously as his memory had failed by that time as shown by many curious statements. Ord mentions another specimen secured near Baltimore and two shot in the District of Columbia, all apparently in May, 1817.

No Glossy Ibis has since been identified in Cape May County but some have been seen not far away. One was observed in a shallow boggy pond on the southern outskirts of Wilmington, Delaware, on May 27, 30 and 31, 1927, by three parties, all members of the Delaware Valley Ornithological Club, on Memorial Day field trips. John Emlen and Ben Hiatt saw a bird actively feeding in the pond on May 27, about the size of a Green Heron but with a long decurved bill. By wading into the mud and water up to their waists they were able to approach to within twenty-five feet of it and study it with glasses. The chestnut brown of the head and neck and the greenish luster of the wings and back were clearly seen and a line of white skin at the extreme base of the bill. As the neck was stretched out in flight the curved bill was very conspicuous.

Richard Erskine and I, unaware of the previous observation, discovered the bird there on the afternoon of May 30, when we stopped to watch the Florida Gallinules that frequented the pond. It was satisfactorily identified and studied for half an hour but in the poor light occasioned by an overcast sky it sometimes appeared entirely black resembling a black curlew, a name which has been bestowed upon it by some writers. William Baily and other members of the Club came upon it a little later and reported it as "feeding in the shallow water gracefully probing in the mire with its long down-curved bill and occasionally taking wing for a few feet as if some tidbit a little farther away had caught its eye." Its neck was somewhat curved in flight but fully extended and the feet stretched out behind. Norman McDonald, who was with this party visited the spot next day and found the Ibis still there (Auk, 1927, p. 417). In 1932, we have a record of one of these birds at Laurelton on the Metedeconk River, north of Barnegat Bay, on May 1, seen by Charles Urner and J. L. Edwards. It was recognized promptly "by its size (a bit smaller than a Little Blue Heron), by its down-curved bill, by the dark rusty of the head, neck and shoulders, and by the dark cast of the remainder of the upper parts. The bird was tame and permitted close approach, near enough to see the body color more distinctly and to note the dark color of the bill, legs and feet. It flew with neck extended, uttering as it went a rather annoyed *squawk*. While feeding on the mud flat the bird could be recognized as other than a heron, even at a distance too far to see the bill shape, by its manner of searching for food in the mud. It 'mouthed over' the mud, sometimes with mandibles partly open." (Auk, 1932, p. 459.)

On June 6, 1937, Julian Potter and others found a Glossy Ibis in a flooded marsh at Paulsboro, N. J. At twenty yards his mandibles were seen to open and shut as he fed in the shallow water. On June 13 Robert Haines found three of the birds present and Fletcher Street saw four on July 5. They remained until the 7th and roosted in a nearby woods.

WHITE IBIS

Guara alba (Linnaeus)

Audubon, writing in 1835, says "a few individuals of this species have been procured in Pennsylvania and New Jersey" but gives no particulars. Turnbull states: "I shot one at Great Egg Harbor in the summer of 1858" and an old mounted specimen in the collection of the Academy of Natural Sciences of Philadelphia labelled "New Jersey" may be this same bird.

We have no further evidence of the occurrence of this southern ibis in the state but on Long Island Giraud mentions specimens taken at Raynor South in the summer of 1836 and at Moriches in March 1843.

CAPE MAY DUCKS

The waters of Cape May have never been notable for their ducks and while they furnish excellent sport for local gunners the wildfowl do not gather there in sufficient number or variety to attract many sportsmen from elsewhere. The Black Duck is everywhere the most abundant species and such diving ducks as the Whistler (Golden-eye), the Shelldrake (Red-breasted Merganser), the Broad-bills (Scaup) and the South-southerly (Old squaw), with some Canada Geese, are regularly present on the sounds throughout the winter, sometimes in considerable flocks. Other diving ducks are irregular or rare and are more likely to occur during the migrations. These same species are to be found also on the open waters of the Bay while the Scoter Ducks or "Coots" abound both on the ocean and on the Bay during migrations and through the winter.

The Bay side ponds and meadows, notably the Pond Creek Meadows, are —or rather were—the rendezvous during the autumn and spring flights, and to a less extent during the winter, of such surface-feeding ducks as the Pintail and Teal, and, of course, the Black Duck which is as plentiful there as on the sounds, while other surface-feeding species have occurred there occasionally or rarely. Of late years, however, the activities of the Mosquito Commission and Federal Agencies providing work for the unemployed have carried ill-advised ditching and draining operations to such a pass that these resting places for migrant ducks have been almost ruined and Cape May's wildfowl population has been seriously depleted, the birds passing on to areas where suitable feeding and resting grounds have been maintained. The details of the relative abundance and time of occurrence of the several species of ducks at the Cape are presented beyond.

Within comparatively short distances fom Cape May—short that is to the flying duck or goose—are several notable places of assembly for transient or wintering waterfowl and these should be considered in connection with a study of Cape May's duck population. Barnegat Bay seventy miles north along the coast is the most noted of these duck centers and has, from the earliest times, been famous among duck hunters and more recently among those who prefer the study of ducks in life to mastering the technique of killing them. Fortunately Barnegat still remains a Mecca for all who take an interest in wildfowl—be they sportsmen, naturalists or artists. Here on the largest stretch of coastal waters in the state is one of the great winter resorts of the Brant while the Black Duck, Scaup, Canada Goose, Golden-eye and Red-breasted Merganser occur in abundance together with the Scoters which inhabit the bay as well as the ocean. The Canvas-back, Red-head, Buffle-head, Baldpate, Old-squaw and Pintail are less abundant with nine others still less common and eleven of only sporadic occurrence—thirty-six species in all, according to records kept by Charles Urner for the past twelve years. Through most of this time Mr. Urner has made weekly trips to the Bay and has counted the numbers of each species of duck that was present and by adding these together he obtains a pretty accurate estimate of their *relative* abundance but as the same birds are often counted over and over again on different days the figures are far in excess of the actual number of birds present. These figures are given below while Mr. Urner's dates for the period of occurrence of each species on Barnegat Bay are included in the detailed accounts of the several species which follow. His records are the most important duck records that have been obtained for New Jersey.

Total counts for all trips to Barnegat Bay for the twelve years, June 1923 to June 1935:

Brant,	454,264	American Merganser,	536
Scaup (both species),	203,347	Ruddy Duck,	337
Black Duck,	192,313	Mallard,	421
Canada Goose,	125,688	Blue-winged Teal,	417
Golden-eye,	19,739	Wood Duck,	295
Red-breasted Merganser,	14,865	Ring-necked Duck,	204
White-winged Scoter,	10,780	Hooded Merganser,	95
Surf Scoter,	10,583	Shoveller,	15
Canvas-back,	8,665	King Eider,	12
American Scoter,	8,660	American Eider,	7
Buffle-head,	5,556	Snow Goose (Greater?),	5
Old-squaw,	5,244	Gadwall,	4
Baldpate,	4,648	European Widgeon,	4
Pintail,	3,567	Harlequin,	4
Redhead,	3,360	White-fronted Goose,	2
Mute Swan,	1,601	Blue Goose,	2
Whistling Swan,	708	European Teal,	1
Green-winged Teal,	612		

Mr. Urner further informs me that it was possible in the past, at the height of the duck season, to obtain daily lists of twenty or more species of wildfowl by visiting the ponds to the north of Barnegat Bay, covering the Bay itself in a boat, and searching the ocean with the binoculars. His best week end list was for December 26–27, 1931, when he identified twenty-eight species of swans, ducks and geese. This list follows:

On December 27, 1931: 11,166 birds, 24 species

Whistling Swan,	3	Old-squaw,	26
Canada Goose,	350	American Eider,	1
Brant,	2,700	White-winged Scoter,	40
Mallard,	4	Surf Scoter,	14
Black Duck,	3,500	American Scoter,	9
Baldpate,	6	Ruddy Duck,	1
Pintail,	5	Hooded Merganser,	1
Green-winged Teal,	2	American Merganser,	6
Wood Duck,	1	Red-breasted Merganser,	15
Redhead,	3		
Canvas-back,	1	Additional species, December 26	
Greater Scaup and Lesser		Mute Swan,	25
Scaup,	3,700	Shoveller,	1
Golden-eye,	600	Blue-winged Teal,	1
Buffle-head,	150	Ring-necked Duck,	1

Other interesting duck counts furnished by Charles Urner and his associates follow, in all of which it is assumed, in stating the totals, that both species of Scaup were present.:

February 27, 1926: 4753 birds, 12 species

Canada Goose,	1,500	Scaups,	1,000
Brant,	1,000	Golden-eye,	a few
Mallard,	3	Old-squaw,	100
Black Duck,	1,000	American Merganser,	a few
Redhead,	150	Red-breasted Merganser,	a few
Canvas-back,	a few		

October 31, 1926: 643 birds, 12 species

Whistling Swan,	6	Old-squaw,	1
Mute Swan,	8	White-winged Scoter,	30
Canada Goose,	11	Surf Scoter,	30
Brant,	300	American Scoter,	100
Redhead,	6	Ruddy Duck,	1
Scaups,	150		

December 10–11, 1927: 15,741 birds, 17 species

Canada Goose,	470	Black Duck,	2,620
Brant,	3,250	Baldpate,	9
Mallard,	26	Pintail,	2

Green-winged Teal,	5	Buffle-head,	200
Redhead,	56	Old-squaw,	40
Canvas-back,	35	White-winged Scoter,	1
Scaups,	8,580	Surf·Scoter,	65
Golden-eye,	335	Ruddy Duck,	47

November 10–11, 1928: 26,589 birds, 23 species

Mute Swan,	19	Scaups,	13,136
Canada Goose,	127	Ring-necked Duck,	3
Brant,	9,190	Golden-eye,	63
White-fronted Goose,	2	Buffle-head,	77
Mallard,	8	Old-squaw,	39
Black Duck,	2,920	American Scoter,	1
Baldpate,	108	White-winged Scoter,	20
Pintail,	5	Surf Scoter,	55
Green-winged Teal,	1	Ruddy Duck,	31
Redhead,	620	American Merganser,	8
Canvas-back,	72	Red-breasted Merganser,	85

January 13, 1929: 18,080 birds: 18 species

Canada Goose,	1,300	Scaups,	5,000
Brant,	5,600	Golden-eye,	575
Mallard,	1	Buffle-head,	80
Black Duck,	4,750	Old-squaw,	60
Baldpate,	20	White-winged Scoter,	300
Pintail,	2	Surf Scoter,	15
Green-winged Teal,	1	American Scoter,	4
Redhead,	52	Red-breasted Merganser,	20
Canvas-back,	300		

January 11, 1931: 25,773 birds: 22 species

Mute Swan,	1	Buffle-head,	22
Canada Goose,	100	Old-squaw,	1,000
Brant,	13,100	American Eider,	1
Mallard,	4	White-winged Scoter,	800
Black Duck,	1,900	Surf Scoter,	200
Baldpate,	1	American Scoter,	1,000
Pintail,	17	Ruddy Duck,	1
Redhead,	4	Hooded Merganser,	1
Canvas-back,	500	American Merganser,	6
Scaups,	6,300	Red-breasted Merganser,	65
Golden-eye,	750		

Another count, or estimate, of interest was made by Dr. J. K. English on January 24–26, 1927, while shooting far out on the Bay where the birds could not be identified from the shore. His figures are as follows:

Canvas-back,	2,000	Canada Goose,	20,000
Scaups,	6,000	Baldpate,	100

Mr. Urner also furnishes me with the following seasonal list of Barnegat ducks exclusive of exceptional and erratic occurrences:

June.   Red-breasted Merganser and the three breeding species—the Black Duck, Wood Duck and Mute Swan.

July.   The same with the addition of the Blue-winged Teal.

August.   The same with the addition of the Mallard.

September.   In addition to the above the Baldpate and Pintail occur along with scattering Scoters of all three species.

October.   The following additional species arrive: American Merganser, Green-winged Teal, Redhead (formerly), Scaups, Ring-neck, Buffle-head, Ruddy Duck, Canada Goose, Brant and Scoters in force.

November.   Additional species arriving: Hooded Merganser, Canvasback, Golden-eye, Old-squaw and Whistling Swan.   The Wood Duck and Blue-winged Teal, however, usually pass on to the south.

December.   The same species present with an occasional Eider.

January.   The Hooded Merganser, Green-winged Teal, Pintail, Ringneck, Ruddy Duck and Mute Swan usually move south.

February.   There is usually an increase in American Merganser, Baldpate, Green-winged Teal, Pintail and Redhead, as the northward flight begins.

March.   There is a decrease in the numbers of most species but an increase in the Red-breasted and Hooded Mergansers, Green-winged Teal, Pintail, Wood Duck, Redhead, Ring-neck, White-winged and Surf Scoters, Ruddy Duck and Mute Swan; both changes being due to the progress of the northward migration.

April.   The numbers of all species fall off noticeably except the Red-breasted Merganser, Blue-winged Teal and American Scoter.

May.   The same species are present as in June with a few scattering records of late migrants of some of the others.

An interesting account of duck hunting on Barnegat Bay and adjacent waters by the late Norris De Haven will be found in "Cassinia" for 1909, pp. 11–18.

Another great rendezvous for transient waterfowl, which is occupied as late in the autumn and as early in spring as the presence or absence of ice may determine, is on the lower reaches of the Delaware River in Salem County.   While Barnegat Bay is populated mainly by Brant, geese and various diving ducks, and the ubiquitous Black Duck, this area is frequented almost entirely by surface-feeding species with the Black Duck again prominent.   A portion of this region seems to have been saved from the devastating activities of the Mosquito destroyers and similar agencies by the establishment, in February 1934, of a Federal sanctuary, the Killcohook Migratory Bird Refuge, embracing 1440 acres surrounding Fort Mott and including much of the New Jersey and a portion of the Delaware shore.   A large part of the area is enclosed by a dyke and overgrown with wild rice, while another section consists of mud flats at low tide with extensive marshes covered with cattails, reeds and coarse grasses. Julian Potter, who is thoroughly familiar with the region, tells me that the Black Duck is present at all seasons although few are to be seen in midwinter when both river and marsh are icebound, while the other species occur only in spring and fall. To give some idea of the duck population I quote from the records of Julian Potter and Fletcher Street giving their estimates of the numbers of ducks seen by them on different visits to the spot:

October 6, 1929. Black Duck, 10,000; Pintail, 30,000; Blue-winged Teal, twenty; Green-winged Teal, one; Ruddy Duck, one. The flock on this occasion was a mile in length and from twenty-five to fifty yards wide.

October 21, 1934. Black Duck, 2500.

September 28, 1935. Black Duck, 10,000; Pintail, 5000.

October 12, 1935. Black Duck, 1000; Pintail, 5000; Green-winged Teal, fifty.

October 20, 1935. Black Duck, 3000; Pintail, 2500.

November 3, 1935. (Street). Black Duck, 20,000; Pintail, 12,500; Mallard, fifty; Baldpate, twenty-five; Green-winged Teal, 2500; Shoveller, twenty-five; Ruddy Duck, one; Gadwall, one.

Fletcher Street writes me "when we arrived on the evening previous we expected that the flight of the ducks from the marsh, which does not take place until after sundown, would be up stream but when they began to pass, generally in that direction, they moved so fast that if was difficult to count them. The counts during five minute periods were as follows:

4.50–4.55,	13	5.25–5.30,	9,910
5.00–5.05,	3	5.30–5.35,	5,885
5.20–5.25,	465	5.35–5.40,	203

There was little movement south or east and many ducks, especially the teal, remained on the marsh.

November 17, 1935. Green-winged Teal, eight hundred; Shoveller, twenty-five.

December 1, 1934. Black Duck, 10,000.

December 1, 1935. Black Duck, 70,000; Pintail, five hundred; Shoveller, ten; Green-winged Teal, 2,500; Mallard, four. On this occasion the Black Ducks were not only in the Refuge but extended in one continuous flock from Fort Mott down into Salem Cove.

January 5, 1936. Black Duck, one hundred. River and marsh ice-covered.

March 15, 1936. Black Duck, 3000; Pintail, 3000; Green-winged Teal, seventy-five.

March 29, 1936. Black Duck, 2000; Pintail, one hundred; Green-winged Teal, forty; Scaup, four; Baldpate, two; Shoveller, four; Canada Goose, ten.

May 24, 1936. Black Duck, three hundred.

July 19, 1936. Black Duck, thirty.

What might be called another section of this Delaware River area is Riggin's Ditch and the adjoining marshes in Cumberland County not far from the Cape May County line, the only spot in south Jersey so far as I know, where the Blue-winged Teal nests. Julian Potter furnishes the following lists of characteristic species of this area:

September 2, 1935.   Black Duck, twenty; Green-winged Teal, one; Blue-winged Teal, eighty; Pintail, two.

September 25, 1932.   Black Duck, six; Blue-winged Teal, six.

October 12, 1932.   Blue-winged Teal, four; Pintail, six.

March 9, 1933.   Black Duck, ten; Blue-winged Teal, ten.

April 3, 1932.   Black Duck, ten; Green-winged Teal, one; Pintail, two.

May 1, 1932.   Black Duck, two; Blue-winged Teal, four.

June 7, 1936.   Black Duck, six; Blue-winged Teal, twelve; two females with broods of downy young.

July 28, 1932.   Black Duck, twelve; Blue-winged Teal, twelve.

A third resting ground for ducks, mainly fresh water or surface-feeding species, is the Delaware City region across the river from Fort Mott. This consists of extensive flooded and dyked meadows or swamps which we hoped might be preserved in the interests of the muskrat industry which furnishes pelts to the fur trade and a supply of meat, known as "marsh rabbit," which is held in high esteem by a certain portion of the population. It seems, however, from later advices, that the agencies of destruction have reached these marshes as well as those of New Jersey and if the draining operations are carried on as planned they too will be sacrificed.

These areas furnish ideal feeding grounds for waterfowl and many bird students go there to study these fascinating birds. The records of the Delaware Valley Club, especially those furnished by Victor Debes, show that the following species have been found there and indicate the times of their occurrence.

Whistling Swan.   March 28, 1928, six; April 14, 1933, three; April 28, 1934, three; April 4, 1936, two.

Canada Goose.   March 15, 1930, eighteen; March 11, 1934, thirty.

Mallard.   October 13 to November 20 and April 13 to May 14; one to ten per trip.

Black Duck.   October 13 to October 26 and March 2 to May 30; two to one hundred per trip.

Baldpate.   October 13 to November 20 and February 22 to April 19; two to two hundred per trip.

Pintail.   February 22 to April 19; ten to 5000 per trip. Also records for November 20, 1932, two; and January 24, 1932, thirty-seven.

Green-winged Teal.   October 13 to November 20 and February 22 to April 19; two to three hundred per trip.

Blue-winged Teal.   February 22 to April 17; two to two hundred per trip. Also May 30 nesting.

Shoveller. February 22 to April 14; two to twenty on nine different days.

Wood Duck. March 29 to May 8; one to six per trip.

Canvas-back. April 6, 1928, one; April 8, 1934, one; April 10, 1935, two; April 4, 1936, five.

Scaup. January 24, 1932, 11; March 27, 1933, thirty-six; May 14, 1933, one; April 8, 1934, ten; March 27, 1933, thirty-six; March 7, 1936, eighty.

Ring-neck. March 28, 1928, twenty-nine; April 14, 1931, four; April 13, 1932, fifty-four; March 5, 1932, thirty-six; April 17, 1932, fifty; April 8, 1934, ten; March 24, 1935, fifty; April 13, 1935, eighty; March 7, 1936, ten.

Golden-eye. March 15, 1930, seven; March 27, 1932, nine; March 7, 1936, fifty.

Buffle-head. February 25, 1933, two.

Ruddy Duck. March 23, 1930; March 27, 1932, one.

Hooded Merganser, April 3, 1931, three; November 20, 1932, two; February 25, 1933, one; February 25, 1934, three.

American Merganser. March 2 to April 14; five to fifty per trip. Also a record for November 20, 1932, one.

Red-breasted Merganser. March 29, 1929, twenty; April 3, 1931 ten; March 27, 1933, four; April 4, 1933, three.

The above summary is the result of only twenty-five trips and is therefore by no means complete and contains, for the most part, only spring data. It serves, however, to show the character of the duck population of the Delaware City region.

To show what ducks may be seen here on single day trips the following are presented:

March 30, 1923. American Merganser, two; Black Duck, one hundred; Pintail, five hundred; Shoveller, four; Wood Duck, three; Whistling Swan, one.

April 10, 1925. Mallard, two; Pintail, one hundred; Shoveller, two; Wood Duck, one; Blue-winged Teal, fifteen; Canvas-back, two.

March 17, 1928. European Widgeon, one.

March 18, 1928. Pintail, 5000; Shoveller, two.

March 28, 1928. Whistling Swan, six; Shoveller, twelve; Ring-neck, twenty-nine.

April 1, 1928. Green-winged Teal, four; Blue-winged Teal, twenty; Shoveller, twenty.

April 6, 1928. Green-winged Teal, three hundred; Blue-winged Teal, two hundred; Shoveller, fifty; Baldpate, fifty.

April 3, 1932. Mallard, eight; Black Duck, thirty; Baldpate, seventy-five; Pintail, thirty; Green-winged Teal, forty; Shoveller, twenty; Wood Duck, two; Ring-neck, fifty-four; American Merganser, ten.

The ducks occurring at the last three localities on the Delaware River, or at the head of the Bay, must pass Cape May on their migrations unless they follow the western shore line, while many of those leaving Barnegat for localities farther south must fly along our ocean front or well offshore. In any case the number that stop off with us depends entirely upon the conditions of water and food provided by nature and upon the extent to which man has altered or destroyed nature's plan. Cape May waters never were able to compete with Barnegat Bay or the great sounds of the Carolinas, to the south, in their lure for ducks and geese, and unfortunately the recent draining and poisoning operations, connected with "developments" and mosquito control work, have sadly affected their attractiveness for the waterfowl. The variation in the abundance of ducks and the alarming scarcity of certain species in the last few years are, however, not due to local causes. One must go far away to the northern breeding grounds of the birds to find an explanation, and efforts toward duck conservation must be based upon extended investigations in all breeding grounds and winter quarters. The evidence from one limited area is misleading and has little bearing on the general problem.

Our waterfowl may be roughly divided into two groups as follows: (John C. Phillips, Auk, 1932, p. 441)

Eastern Breeders	Western Breeders
Black Duck	Mallard
Pintail	Gadwall
Green-winged Teal	Baldpate
Blue-winged Teal	Shoveller
Wood Duck	Redhead
Golden-eye	Ring-necked Duck
Mergansers	Canvas-back
Canada Goose	Scaups
Brant	Buffle-head
	Old-squaw (far north)
	Scoters (mainly far north)
	Ruddy Duck

The eastern group breeds in eastern Canada, Greenland and the eastern United States, and it will be noticed that all the common ducks of Cape May with the exception of the Scaups, Scoters and Old-squaw are included in this list. The other group breeds in western and central Canada to Alaska and formerly in the north-central United States, where some still persist. While the decrease in the numbers of Brant in the winter of 1931–32, and in later years, was due to the disappearance of the eelgrass which constituted their principal food, it has been the second group, except Scoters and Old-squaw, that have snown the alarming decrease noted in recent years, which forced the Government to place the Ruddy Duck and the Buffle-head on the protected list in 1932 and the Redhead and Canvas-back in 1936. Spring shooting of waterfowl was abolished in 1913 and at the same time the Whistling Swan was given absolute protection, the Wood Duck in 1918 and the Snow Geese of the Atlantic coast in 1931. These three species had been so reduced, through no apparent cause but overshooting, that their existence was threatened. The decrease in the others was due to the widespread drought which affected the breeding grounds of the birds and prevented the raising of their broods. This was, however, but the climax to what had been going on for years. To quote from Frederick C. Lincoln (Bull. New York Zool. Soc., Sept.–Oct., 1936) "Even as recently as twenty-five years ago, the numbers of ducks and geese that reared their broods in the innumerable marshes and sloughs of the Western States and the vast well-watered regions of Canada and Alaska seemed incalculable. No one knows what the continental waterfowl population was, but figures of one hundred million and even one hundred and fifty million have been suggested and it is probable that both are conservative.

"As soon, however, as the first furrow was plowed in the prairies, the doom of important breeding grounds was sealed. Quickly following the plow came the dredge and in 1922 it was estimated that organized drainage districts in the United States included approximately seventy-five million acres." Coincidently "Canada was developing her granary in Manitoba, Saskatchewan and Alberta resulting in the permanent and complete destruction of additional prairie marshes, sloughs and pot-holes.

"While a progressive diminution of the numbers of ducks and geese has been noticeable during the last generation, it required a major calamity to bring their critical situation so forcefully to attention that remedial measures would be taken. The long period of deficient precipitation began in 1915 over hundreds of thousands of square miles of the finest breeding territory in the North Central States and the prairie Provinces of Canada. With the exception of a slight improvement in 1920 the shortage of rainfall continued to 1934, when all existing records were broken for duration, severity and extent of territory affected." After a slight improvement in 1935 the drought conditions by midsummer of 1936 were about as bad as ever. Meanwhile, however, large areas of land in the North Central States have been acquired and turned back into marshes or small lakes with favorable responses on the part of the ducks. There has been great difference of opinion as to whether duck shooting should be abolished altogether, limited to eastern species or continued with additional restrictions. The last course was adopted by the Government; whether it will prove successful or whether many of our ducks are doomed to go the way of the Buffalo and the Wild Pigeon remains to be seen.

*MUTE SWAN

Sthenelides olor (Gmelin)

Ludlow Griscom, writing in 1923, says that the common domesticated swan of England, the Mute Swan, had been introduced on the Hudson River near Rhinebeck and at the South Side Club near Oakdale, Long Island, and possibly it has been liberated elsewhere since his "Birds of the New York City Region" was published.

On October 24, 1916, a young bird was picked up exhausted at Elizabeth, N. J., and reported by Charles Urner. Since then he states that the species "has become completely naturalized and a number of pairs breed in a wild state in suitable ponds along the coast from the vicinity of Asbury Park to Bayhead." At Point Pleasant for a number of years there has been a breeding pair on a pond just south of Arnold Avenue. "The birds, young and old" writes Mr. Urner "gather in the fall into flocks, sometimes containing as many as thirty-five individuals, and fly about seeking feeding grounds.

"As the smaller ponds freeze some are trapped and wing-clipped, but a number fly south, and it is then not uncommon to see companies of these birds feeding about the northerly third of Barnegat Bay (north of Seaside Park) where they remain as long as there is open water." One was picked up dead on the marshes back of Long Beach by Charles Beck on January 8, 1920 (Auk, 1921, p. 273) and as many as eight were seen by Urner on October 31, 1926.

The question naturally arises: How many of the occasional swans reported flying along the shore are Mute Swans? and we are unable to say. However in Charles Urner's experience all swans going down the coast near enough to shore for indentification were Whistling Swans and it seems probable that the Mute does not migrate to any great distance. Most observers do not realize that there are two species to be distinguished and at a distance they are by no means easily recognized. The most striking difference is that the Mute Swan has a prominent knob at the base of the bill above, less conspicuous in the female and young, while the greater part of the bill is red or reddish orange; dusky in the young. The Whistler, on the other hand, has no knob and the entire bill is black except for a small yellow spot near the base. The Mute Swan when resting on the water curves the neck and points the bill at a downward angle while the Whistling Swan, except when feeding, usually holds the neck stiffly erect and the bill horizontal.

While we have no record of this bird from Cape May it may very well occur, especially when severe weather may force the north Jersey individuals southward.

J. F. S.

WHISTLING SWAN

Cygnus columbianus (Ord)

The headquarters of the Whistling Swans, during their stay in this latitude are on Chesapeake Bay. Some apparently occur there throughout the winter and in February they move to the upper bay and Elk River where they are joined by the great flock from Currituck Sound and all take off together for their northward flight. Such swans as occur at Cape May or on Delaware Bay are stragglers from the main flock. They are, however, by no means common on the New Jersey coast.

Walker Hand reported one shot at Cape May on February 1, 1917, and I found a dead one back of Seven Mile Beach on May 1, 1932, which had probably been killed a week or two before. Small flocks also have been seen passing, some of them high overhead, others off the jetty at the mouth of the Harbor. Two bunches of swans were seen on November 6 and 8, 1919; four flew past the jetty on April 25, 1923; a flock of seven on December 16, 1926; another of twelve on December 23, of the same year; a single bird on November 11 and 13, 1928; another December 1, 1935. Six were seen flying over the ocean on October 30, 1936, by James Tanner. The possibility of some of the passing swans of Cape May being the introduced Mute Swan must be considered.

Whistling Swans occur occasionally on Barnegat Bay on the upper New Jersey coast, where, however, the resident Mute Swan is more frequently seen. Charles Urner has recorded six on October 31, 1926; seventy-seven (flocks of fifty, nine and thirteen) on March 28, 1928, too far out to be absolutely sure of the species; four on February 22, 1931; and William Yoder saw nineteen on November 18, 1928. Across the river at Delaware City swans have several times been observed and among the records are several on November 13, 1927; six on March 28, 1928; three on April 14, 1933.

On March 25, 1928, I saw the only Whistling Swan of my many years' experience at Cape May. As I came over the sand hills above the Lighthouse where I could obtain a view of the shallow pond then located back on the meadows, and where we always looked for water birds, I was amazed to see a great white swan floating serenely on the still water accompanied by a pair of Mallards and a Green-winged Teal. It was a young bird with ashy neck and head, darkest on the head and becoming lighter downward. It also had a trace of ashy on the back and wings. The bill was black and very prominent and there was a whitish area on the lores. It swam farther out in the pond as I approached and called once, dropping the lower mandible as it did so. As it sat on the water the back curved over in front to its juncture with the neck, while the latter was held erect or slightly tilted back, but not curved.

After the ducks had taken wing, and when I was about twenty yards away, the swan flew, stretching its neck straight out, flapping its wings, and paddling with its feet, the wingbeats being about as fast as those of a goose. In the flying bird the juncture of the neck and body was about midway in its total length. It called while on the wing and circled over the beach and then back to the deep pond where it remained. I saw it feeding there later; it curved its neck and poked its head under the water, sometimes submerging the entire neck, so that is must have been taking food from the very bottom of the pond. When it swam the black feet and tarsi appeared with each stroke, just above the water close to the base of the tail. This bird was present on the same pond on April 1 but was not there on April 2 or 6 and had presumably passed on to the north. Two full-grown young swans were seen by Raymond Otter on the same pond in November, 1928, feeding at daybreak along the edge of the growth of cattails which covered part of the water.

On the wing the appearance of the swan is extraordinary, the great white neck stretching ahead fully twice the length of the body so that it appears to be drawing the flapping wings after it.

By way of comparison with the occurrence of the Whistling Swan at Cape May a few words as to its occurrence on Chesapeake Bay may not be out of place. The majority of these birds seem to go south to Currituck Sound for the winter but they stop on Chesapeake Bay on their northward and southward flights and apparently many remain there through some winters at least. They are seen to best advantage from Stump Point, Perryville, Maryland, where the Susquehanna River broadens out to form Havre de Grace flats long famous as a feeding ground for waterfowl. Julian Potter states that the best time to see them is in March, just before they take wing for the vicinity of Niagara Falls where they again come to rest.

He writes me "If large numbers are present, their 'clanking' can be heard plainly as you enter the grounds of the Government Hospital located on the Point. As you near the tip of the Point the dense white elongated flock of swans appears out on the bay. Some of the birds are sitting on the low islands, others are floating lazily about on the water. The 'clanking' is continous and now and then a portion of the great flock will rise in the air, and forming long glistening snow white undulating lines, fly up or down the bay. The numbers vary from year to year and if the bay does not freeze over some remain through the winter; a few linger on quite late in the spring. The following estimates will show their approximate abundance:

March 3, 1929,	5,000		March 13, 1932,	2,500
March 2, 1930,	3,300		March 29, 1932,	1,000
March 8, 1931,	10,000		April 1, 1932,	300
March 15, 1931,	15,000		December 18, 1932,	1,500
January 24, 1932,	5,000		March 12, 1933,	7,000
February 22, 1932,	2,000		April 9, 1933,	8

Of the last lot four were feeding close to the shore, dipping like surface-feeding ducks. They held themselves almost perpendicularly downward paddling with their broad dark feet. To all appearances they were adults."

Twelve Whistlers that I watched on Elk River, Maryland, with Dr. W. L. Abbott, on February 19, 1927, during a heavy gale rode the waves like corks. They rested in a long line, one directly behind the other and were majestic in their pose, holding the neck somewhat drawn down and bent over in front, so that the bill was near the water's surface, though now and then a bird would stretch out its neck at an angle of 45° bringing the black bill, which looked disproportionately large, into still greater prominence. Several of this flock were gray-necked birds of the year, the others adults and snowy white except for their black flight feathers.

The Trumpeter Swan now reduced to a few individuals and an inhabitant of the western United States and British America is said to have been shot on both Chesapeake and Delaware Bays according to Turnbull (1869). He made his statement "on the authority of reliable sportsman" but states further that "it must be a rare straggler and I have not noticed it in the Philadelphia market, where the Wild or 'Whistling Swan' is so frequently seen every winter." When we consider the difficulty of distinguishing the two species and the absence of any definite records or specimens from the Atlantic Coast its occurrence either on Delaware Bay or on the New Jersey shore seems very doubtful.

CANADA GOOSE

Branta canadensis canadensis (Linneaus)

As several Cape May gunners have told me Cape May is not, and never has been, "goose country" and yet Canada Geese may be seen there at any time from the end of October to the first of April. It is not to be inferred, however, that they are present continuously and most of those seen after January are migrants passing overhead. A flock is often established on the great sounds a little to the north or on Delaware Bay but even there there are periods of a few weeks in midwinter when they are absent, or exceptional seasons with much ice when they will move farther south. Julian Potter noticed this at Barnegat Bay farther up the coast and recorded a flock of sixty geese flying south on December 21, 1919, when the bay was frozen over. Their occurrence at any time is dependent on weather conditions; raw overcast days seem to suit them best and it is at such times, when rain is threatening, that flocks will most likely be moving. Raymond Otter tells me that the Canada Goose is the only waterfowl that migrates when the wind is blowing in the direction in which it desires to go.

Canada Geese are most frequent about Cape May during the migrations in February and March, in spring, and October and November, in autumn, and it is then that we see them passing directly over the town in flocks of varying size for it is rarely that we see a lone goose. I have heard them in March going past overhead when everything was wrapped in fog. Not a goose could be seen but their honking came down to us like the baying of a pack of hounds, faintly at first then louder and louder as they came nearer until directly overhead, and gradually fading out as they passed on their way.

The earliest dates that I have secured are October 11, 1925, a flock of five; October 14, 1928, a flock of twelve; October 26-28 of the same year many flocks of from six to fifty birds; October 17, 1931 a flock of thirty-two. Dates of last occurrence are April 1, 1921, a single bird; April 20, 1932, a flock of forty; May 8, 1932, eight. Usually, however, they depart by the end of March and in some years we have as latest dates, March 25, 1923; March 27, 1932; March 24, 1929. We have records of wintering birds in December 1911, 1926, 1927 and 1931 and in January 1924, 1925 and 1926; and of birds evidently migrating north in February of nearly every year. One exceptional occurrence was a single bird, probably sick or wounded, which was seen about the neighborhood during the spring of 1927 and which I last saw on July 16 at which time it could fly perfectly well.

A flock observed by Julian Potter on November 21, 1920, was typical of the occurrence of Canada Geese at the Cape. They came from over the land and were first observed while still far away as a thin black streak stretching across the sky, the birds being strung out in a long line. As they drew nearer the rear ranks broke and moved in a compact mass up toward the front and spreading out to one side formed an irregular V. They proceeded thus for a short distance and then again the flock became more or less broken and irregular and once again the long line was formed. As they passed directly overhead about half of those at the front advanced far ahead of the rest and they proceeded in this manner out over the Bay where they were soon flying low over the water, in single file, and finally about a mile from shore the entire flock came to rest. There were 150 birds in this flock while two others which came down the coast the same morning numbered 125 each. They also settled out on the Bay.

On January 3, 1926, Walker Hand and I saw eleven Canada Geese, coming across the Harbor from the north flying tandem, which passed directly over our heads as we sat on the shore. They showed to perfection the deliberate wing motion and the elevation of the neck at a considerable angle, while the last in line, probably an old male, had the curious crook in his neck which is another characteristic of a goose. One of these birds kept calling *au, au, au, au,* as he flew. A flock noted by Hand on January 2, 1924, was traveling north, and during the winter they may be seen flying in any direction irrespective of migration. When the migration is on, however, and the day propitious for goose movement, many flocks will often be seen passing in close succession. On November 7, 1920, Walker Hand saw four hundred geese flying over and on the 21st over one thousand passed during the day, eight flocks being in sight at one time. Julian Potter saw six flocks in an hour on February 22, 1928 and counted 237 individuals, and Hand reported five in fifteen minutes on March 24, 1929.

On March 26, 1932, I witnessed the largest flock of Canada Geese that I

have ever seen at the Cape. About 8 o'clock in the morning a flock of fourteen birds passed overhead on their way to the Bay, flying northwest. They
were followed immediately by another flock of forty, then upon turning my
head toward the coast I saw an immense assemblage of approximately four
hundred geese coming over, flying in a number of intersecting V's and straggling lines; for the moment they seemed to cover the sky but in a couple of
minutes they were passing out of sight.

It is not often that Canada Geese will come to rest on the small ponds
about Cape May as they prefer the larger bodies of open water—the Bay,
the sounds, and the ocean, but occasionally some will drop into Lake Lily or
the deep pond at the Lighthouse. Two were seen at the former spot on December 11, 1920, which honked loudly as they took wing and on October 11,
1925, I found five on the Lighthouse Pond. A terrific northwest gale had
prevailed on the previous day sweeping the entire Atlantic coast and doubtless affecting many migrant birds and I presumed that these geese, the first
to arrive from the north, had been driven to seek harbor on this pond.
They were sitting quietly on the water at the far end of the pool as I came
suddenly around a clump of bayberry bushes and for a few moments I had
an excellent opportunity to study them. Their characteristic colors were
clearly defined against the dark background of low shrubbery—the mottled
gray backs, jet black necks and white cheeks and flanks. Three of them were
feeding, leisurely dipping their heads into the water, while the other two,
evidently the parents of the family, sat on guard with necks erect. Presently
they saw me and took wing and with necks outstretched at an upward angle
they circled once and, constantly gaining altitude, flew off over the ocean
where eventually they came to rest. I could see them gracefully riding the
waves in single file, now coming clearly into view, now disappearing in the
trough of the sea; their black necks always visible but the white cheeks not
distinct in the poor light of an overcast day. Their wingbeats were char-
acteristically slow—slower than any duck and about as fast as those of a
Green Heron. These birds remained in the vicinity all day and about two
hours later I saw them crossing high in the air apparently with the idea of
settling again on the quiet waters of the pond. They still maintained the
typical goose attitude with head and neck directed diagonally upward as
they breasted the wind and occasionally set their wings for a few moments
and sailed into it. The white of the belly and flanks shown brilliantly in the
sunlight as they flew dead away from me at a great distance and then, by one
of those curious optical illusions, they suddenly seemed reversed appearing
exactly like dark-bodied birds coming toward me, with short snowy white
necks (the flanks) and round heads (the tail). It was really difficult to convince myself that the birds were really going away and that they had long
necks which were now concealed by their bodies—such are the tricks that light
and shadow play. I passed the pond later in the afternoon just in time to
see this same family of geese preparing to alight. I squatted immediately
in the dead grass and they came directly toward me on down-curved wings

and feet widespread. When about forty-five feet away they saw me and with several cries *gargle, gargle*, and a sudden increase in the rapidity of their wingbeats, they veered off again. I lay perfectly still and after circling widely they came back and settled on the far end of the pond where I had first seen them, splashing into the water, one old goose in front and the other in the rear, with the three young ones between them. The latter at once began to duck their heads under and to pull up grass while the old bird in front did the same but the one in the rear, doubtless the gander, remained ever vigilant, with neck rigidly erect swinging slightly to and fro as he swam along. The bodies of the birds seemed very broad in front as they rested on the water and as they flew over me, the demarkation between the color of the breast and the belly was marked. One of the young birds flapped its wings once as it swam and disclosed the gray under surface.

On February 6, 1932, I saw a flock of thirty-seven Canada Geese alight in the meadows back of Hereford Inlet and on March 6 of the same year there were eight feeding on the meadows back of Seven Mile Beach. In the latter case there had been a high tide and the meadows were very wet. The light-colored under surface of the birds' bodies shown very clearly in rather poor light but the black necks did not stand out against the dark marsh. Another was flushed on the next day during a heavy gale just back of the town. We have also seen them on Gull Bar and other sandspits out in the inlets.

On May 8, 1932, there were eight on a small pool at the head of Pond Creek Meadow which were seen early in the morning following a rainstorm and doubtless they had spent the night there. They were very wary and as we approached stretched out their necks and slowly took wing, all together, with very deliberate wingbeats. They circled several times and changed their relative positions finally taking off to the north.

I have only once seen Canada Geese alighting in fields about Cape May. On February 7, 1932, a flock of eleven was observed resting on the waters of Sluice Creek not far from the Bay and a few minutes later they arose, and passing over a hedgerow, alighted in an adjoining field where they walked about and fed among some green grass or grain that the mild winter had kept very fresh. Their necks were crooked as is usual with feeding geese and occasionally one of them would flap its wings. It is worthy of note that all of these occurrences of geese on the ground were in the same year and apparently many of the birds had located in the Cape May region; Walker Hand said that they had been "trading up and down" all winter.

Raymond Otter tells me that he has seen small flocks alighting in corn fields on several occasions and almost invariably when a flock alights in an upland field or on inland ponds, during the gunning season, they will stay around until all have been killed. He regards these as young and inexperienced birds. Those that frequent the sounds seem to be much more wary.

Over on the Delaware-Maryland Peninsula across the Bay, this feeding in the fields is a common occurrence. On February 26, 1927, Richard Erskine and I came upon thirty-nine of these birds walking slowly across a wet wheatfield by the roadside, near Cecilton, Md. So closely did their general coloration match that of the plowed ground that we had not noticed them at first. They were strung out in a long line, waddling somewhat from side to side as they advanced, with necks bent over forward and often humped or kinked near the middle. They were feeding as they went pick-ing at the sprouted wheat or at the fragments of corn fodder that were scat-tered about. When alarmed they held their necks straight up and the "kink" would disappear at once and now and then a feeding bird would rush at another one which had apparently trespassed upon his domain, with neck extended and head down. The white belly and gray sides and breast were in strong contrast when one got a front or side view of the birds. The black neck was still more conspicuous, but not until the birds raised up their heads in alarm, when the whole row of necks looked like black stakes rising from the ground, the cheek patches forming a dash of pure white at the top of each.

When approached these geese showed no inclination to fly but simply stopped feeding and walked away, heads up, at a rapid pace, now and then switching the tail nervously from side to side. Presently we discovered an-other group of seventeen farther on in the same field and thirty more in an old pasture not far from a house and, as we watched, another flock of thirty-two decoyed to them. They came in company front in a long line or shallow V, necks all straight out in front and heads slightly raised. As they neared the field they all stopped flapping and sailed down to join those already on the ground, their wings set at all degrees of curvature according to the angle at which the wind struck them; some held nearly flat and others strongly bowed with the tips pointing almost vertically downward. Now we saw additional flocks coming over from the Bay far out on the horizon and others passing on to the southward.

The smaller bunches, in the distance, resembled airplanes, while in the larger flocks the apparent congestion in front caused by our viewing the V from an angle, or from one extremity, appeared as a plane or a gigantic bird leading a flock of little ones, just as we see it in the case of the Scoters flying in dense masses far out over the ocean.

In season the farmers of this region go goose shooting on the wheatfields and have decoy Canada Geese to attract the wild birds. Of late years the ground has been baited and the geese return year after year to the places where they have been fed, which accounts for their abundance and tameness. While visiting Dr. William L. Abbott at his home on Elk Neck, Maryland, in November, 1935, flock after flock of Canada Geese were seen passing over every evening and back at dawn from their feeding grounds to spend the night on Chesapeake Bay. This region is the real wintering ground of the geese in this latitude and while we see many migrant flocks at Cape May it is not their true home, in any sense, so that the sight of geese anywhere but in the air is noteworthy. Even the flying birds of late fall and winter are but irregular visitors not following any definite schedule, like those of Elk Neck, and less frequently giving voice as they pass. It is the vocal chorus of the geese that adds so much to the wild charm of the Chesapeake waterways and Dr. Abbott was wont to depend upon their morning and evening honking as upon a timepiece; greeting the sunrise and foretelling the coming of the dusk.

In the Barnegat Bay region Charles Urner furnishes the following summary of extreme dates of occurrence of Canada Geese for several seasons:

1929–30.	October 12–April 13.	1932–33.	August 21–April 29.
1930–31.	November 30–April 12.	1933–34.	November 5–April 8.
1931–32.	September 27–July 4.		

As in the case of the Brant and several other species of waterfowl he has observed scattered individuals in every month of the year but the period from December to March covers the time of their regular occurrence in any numbers.

The disappearance of the eelgrass described under the Brant had a disastrous effect upon the Canada Geese which also fed on it to a considerable extent and Charles Urner's counts for a number of winters show very clearly the decrease in these splendid birds as a result of the scarcity of this food plant:

1927–28, 7,640	1930–31, 2,752
1928–29, 5,821	1931–32, 3,952
1929–30, 5,612	1932–33, 1,884

His total counts for all trips show this decrease even more strikingly. From June 1923 to June 1933 the total reached 120,510 while from June 1933 to June 1935 it decreased to 5,178. These figures, as already explained, do not mean the number actually present but the sum of all the daily counts and while they exaggerate the number present they are perfectly satisfactory for comparisons.

Alexander Wilson who was the first to give us an account of the Canada Goose on the New Jersey coast presents a good picture of the bird as he knew it one hundred and twenty-five years ago. He writes:

"Their first arrival on the coast of New Jersey is early in October and their first numerous appearance is the sure prognostic of severe weather. Those which continue all winter frequent the shallow bays and marsh islands; their principal food being the broad tender green leaves of a marine plant which grows on stones and shells, and is usually called sea-cabbage; and also the roots of sedge, which they are frequently observed in the act of tearing up. Every few days they make an excursion to the inlets on the beach for gravel. They cross indiscriminately, over land or water, generally taking the nearest course to their object; differing in this respect from the Brant, which will often go a great way round by water, rather than cross over the land. Except in very calm weather they rarely sleep on the water, but roost all night in the marshes. When the shallow bays are frozen, they seek the mouths of the inlets near the sea, occasionally visiting the air holes in the ice; but these bays are seldom so completely frozen as to prevent them from feeding on the bars. The flight of the Wild Geese is heavy and laborious, generally in a straight line, or in two lines approximating to a point; in both cases the van is led by an old gander, who every now and then pipes his well known *honk*, as if to ask how they come on, and the honk of "all's well" is generally returned by some of the party. When bewildered in foggy weather they appear sometimes to be in great distress, flying about in an irregular manner, and for a considerable time over the same quarter, making a great clamor. On these occasions should they approach the earth, and alight, which they sometimes do, to rest and recollect themselves, the only hospitality they meet with is death and destruction from a whole neighborhood already in arms for their ruin." (American Ornithology, VIII, p. 52, 1814.)

*HUTCHINS' GOOSE

Branta canadensis hutchinsi (Richardson)

This small edition of the Canada Goose occurs regularly in migrations in the Mississippi Valley and occasionally an individual strays to the Atlantic coast. "Homo" writing to "Forest and Stream" (March 2, 1882) states that he had secured a specimen at Tuckerton and that the bird was well known to gunners on Barnegat Bay as the "Sedge Goose," but gives no further details. Turnbull (1869) says that it is occasionally seen on Chesapeake Bay, and Dr. Henry Tucker tells me that a few years ago he secured one on his place on the Elk River, Maryland.

It seems to be very rare on the New Jersey coast.

AMERICAN BRANT
Branta bernicla hrota (Müller)

The Brant has always been a scarce bird about Cape May, its main rendezvous in New Jersey being Barnegat Bay, some seventy miles to the northward, where it has been a notable game bird as far back as our shooting records extend. Farther south in the state the bodies of water are apparently not large enough or not of the right character for its needs and so the Brant is absent. Even during the migrations it is exceedingly rare at the Cape and it would seem that the birds that winter farther south must follow a course well out to sea as they travel to and from the sounds of North Carolina. Such at any rate, is the belief of the old time gunners who say that once the Brant start from Barnegat, where the south-bound flocks are supposed to pause, they do not stop until their destination is reached. They also call attention to the fact that Brant will not travel over land and Raymond Otter tells me that he has seen them follow a water route several miles long to arrive at a point that they could have reached in a quarter the distance by flying over the meadows.

In his experience a few Brant can usually be found every winter on Great Sound and John Mecray states that there was a flock of some three hundred established there during the winter of 1935–36 but for the most part we see only stragglers at the Cape; usually in migration. Walker Hand's first record for Cape May was a single bird killed on January 28, 1924, while on December 3, 1926, two were seen, and nine were shot on December 7 and 28, 1930. Julian Potter saw two flocks of thirty birds each going up the coast on April 18, 1926, and a few were seen passing on December 23, 1920, and December 22, 1929, which completes our records, although doubtless others have been seen or shot and not recorded. Some of the flocks which have occurred at Cape May in midwinter are supposed to have been driven out by the freezing of Barnegat Bay and doubtless returned there later. The Cape May sounds are said to remain open when Barnegat is frozen over.

Barrel Island out in Little Egg Harbor Bay at the lower end of Barnegat is an excellent spot from which to watch the Brant and here on December 27, 1928, I spent some time with Edward Woolman studying their actions. There is a thicket on the higher part of the island which seems to be clear of the high tides and here the bayberry bushes and marsh elders reach the proportions of trees, rising ten to twelve feet in height with trunks six inches in diameter at the base, those of the former being smooth and gray like beach or hornbeam, while the latter are deeply corrugated and furrowed with undulating ridges of rough bark. Extending close up to this wooded ridge on both sides is a great bed of dry eelgrass one to two feet deep; the accumulation of many high tides, packed down into matted cushions and bleached by the sun.

Lying there in the shelter of the thicket one gazes out over the broad waters of the bay with the inlet far away on the horizon. The surface is simply paved with Brant floating safely beyond the range of the blinds which are located on smaller islands and sand bars. They are always alert to take wing at the approach of any suspicious object and are constantly coming in from farther up the Bay, possibly seeking safety on the wider reaches of water. Flying head-on with their pointed wings in rapid action they appear uniform black, then suddenly as they wheel the white of the bellies and under tail-coverts flashes out, forming a surprising burst of color when the whole flock moves in unison.

A wind has now arisen and the Brant on the water rise and fall in rhythmic motion with the undulation of the waves, presenting the effect of a series of little white caps as the white flanks and black backs of the birds are alternately exposed.

They all face one way and tilting up as they feed make a constant movement all along the line. Now and then something disturbs one portion or another of the great raft and they arise and stream away, alighting again farther on, and once a passing airplane caused them all to rise high in the air until they appeared more like swallows than Brant.

Again and again I counted sections of the flock as they rested on the water and by comparison arrived at a conservative estimate of 10,000 birds in this single gathering.

Now the sun begins to sink behind a bank of lead blue clouds, whose upper borders are soon edged with crimson and this brilliant color is reflected in the water and on the wet mud flats that are left by the ebbing tide. As soon as the tide begins to shoal the Brant drift away to deeper waters and the Herring Gulls come out to feed on the flats, gaunt silhouettes in the fading light, while great flights of black Coot Ducks (Scoters) cross the inlet like flocks of Starlings going to roost.

The Brant remain on Barnegat Bay all winter long and stragglers are noted as late as May (May 15, 1927)

On March 13 of that year Richard Erskine and I found several large flocks still present. They rested out in the middle of the bay and one flock was busily engaged in feeding. The birds ranged in a long line, tandem, swimming or drifting over the shallows at a fair speed. They carried their necks stretched out in front at an angle of forty-five degrees, frequently bending over until the bill touched the surface of the water. Every now and then one bird or another would raise itself for an instant and flap its wings vigorously while all along the line birds were constantly tilting up to feed and righting themselves again. Brant do not dive but, like the geese, stretch their necks down to the bottom where they can reach the eelgrass or sea cabbage, and in so doing elevate the posterior part of the body. While the head and neck are submerged the body, tail up, bobs like a cork on the water; the tips of the wings and tail stand out clearly while the white flanks flash in the sunlight.

Several other flocks all of over a hundred birds were seen moving about the bay, sometimes flying short distances over the water and settling again, coming slowly down to the surface and alighting without a splash.

We saw another flock of 150 birds flying down the beach, high in air. They were stretched out in company front in an irregular line, a few scattering individuals ahead and a few others bringing up the rear, while in the middle the flock appeared congested. The neck and breast of the Brant are jet black but the neck appears distinctly shorter in relation to the size of the body than in either the Canada Goose or the Black Duck. The relative shortness of the Brant's neck is even more emphasized in flight than when on the water and makes the bird look better proportioned and not "neck-heavy" as are the geese, swans and many ducks. The neck is carried straight out in front, the wings are more pointed than in a goose and the beats are intermediate in rapidity between those of a goose and a duck.

The black breasts and light underparts are well seen as the flock goes overhead. It is curious that with a color pattern so like that of the Canada Goose the Brant is so different in proportions, in actions and in pose.

Charles Urner, who has studied the Brant on Barnegat Bay for many years, gives the following schedule of extreme dates of its arrival and departure for several seasons:

1929–30.	October 12–July 13.	1932–33.	November 6–May 7.
1930–31.	November 9–May 26.	1933–34.	November 5–April 8.
1931–32.	November 8–April 17.		

He has, however, records of its occurrence for every month of the year although it is very rare in May and is usually present in numbers only from November to March.

In earlier times Charles S. Westcott ("Homo"), writing in "Forest and Stream," records Brant in 1884 as arriving about November 16 and in 1885 on October 17

The principal food of the Brant has been the eelgrass (*Zostera marina*) which grew in abundance all along our seacoast bays. According to Charles Urner, whose account we quote (Abstract Proc. Linnaean Soc. New York, No. 43–44, pp. 37–39), it controlled the shifting of sand in the bay, sheltered a vast amount of life—food both for fish and fowl, and constituted almost the sole diet of the wintering Brant as well as an important part of that of the Canada Goose and certain ducks. As early as 1929 he noticed that the dense beds of floating eelgrass in the coves which furnished attractive feeding grounds for many shore birds were becoming less extensive each summer. It developed that the plant was affected by a species of blight which eventually destroyed it and by the winter of 1931–1932 the condition became acute. Flocks of Brant on their southward migration dipped low as usual upon reaching the bay but seeing nothing but white sand on the flats and shallows which had been formerly green with a thick growth of the plant, they passed on in search of better fare. On February 22, 1925, in company with Ludlow Griscom and J. A. Weber, Charles Urner estimated eighty thousand Brant present in great rafts, miles in extent, between Egg Harbor and Manahawkin Bay. In the winter of 1932–1933 the high count for Brant was only eighteen hundred. Some of the birds formerly wintering in the Barnegat area had taken up winter quarters on Absecon Bay and some ten thousand wintered there in 1931–1932 and in the following winter, though in 1933–1934 there were not more than two thousand. Their chief food seemed to be the sea cabbage (*Ulva luctura*) but many of them were forced to go out on the meadows to pull up the roots of various marsh grasses for food, something that they never did in the past. To further illustrate the Brant calamity we quote Mr. Urner's counts:

1927–28, 28,800	1930–31, 11,550
1928–29, 25,090	1931–32, 5,400
1929–30, 17,603	1932–33, 2,320

His total count of Brant seen on all trips from June 1923 to June 1933 was 403,724 while for June 1933 to June 1935 it was only 50,540, yet a great many more trips per year were taken in the latter period. These totals, as explained elsewhere, do not represent the actual number of birds present as the same flocks are probably counted again and again on successive trips but are perfectly satisfactory for purposes of comparison.

That Brant can survive very cold weather so long as food is available is shown by the presence of a flock seen by Julian Potter in the open water at the edge of the Barnegat Inlet on February 11, 1934, after a temperature of −11° a few days before and another bunch feeding on sea cabbage close to the highway bridge, while on January 24, 1925, after a temperature of −6° and sixteen inches of snow he counted no less than 150 on the bay.

On April 8, 1934, Joseph Tatum saw three flocks of Brant arise from the bay and fly off to the north, doubtless the beginning of their migration.

After all these years it is of interest to read what Alexander Wilson had to say of the Brant on Great Egg Harbor about the winter of 1812–13. He was not familiar with Barnegat Bay nor did he make any attempt to count or estimate the numbers of the birds that wintered on the coast which would have been very interesting for comparison with present day conditions: "The Brant is expected at Egg Harbor on the coast of New Jersey about the first of October, and has sometimes been seen as early as the twentieth of September. The first flocks generally remain in the bay a few days, and then pass on to the south. On recommencing their journey, they collect in one large body, and making an extensive spiral course, some miles in diameter, rise to a great height in the air, and then steer for the sea, over which they uniformly travel; often making wide circuits to avoid passing over a projecting point of land. Their line of march very much resembles that of the Canada Goose, with this exception, that frequently three or four are crowded together in the front, as if striving for precedency. Flocks continue to arrive from the north, and many remain in the bay until December, or until the weather becomes very severe, when these also move off southwardly. During their stay they feed on the bars at low water, seldom or never in the marshes; their principal food being a remarkably long and broad-leaved marine plant of a bright green color, which adheres to stones, and is called by the country people sea cabbage; the leaves of this are sometimes eight or ten inches in length; they also eat small shellfish. During the time of high water they float in the bay in long lines, particularly in calm weather. About the fifteenth or twentieth of May they reappear on their way north; but seldom stop long, unless driven in by tempestuous weather." (American Ornithology, VIII, p. 131, 1814.)

BLACK BRANT

Branta nigricans (Lawrence)

Curiously enough this Pacific coast bird was first described from a specimen procured at Egg Harbor, N. J., in January 1846, by George N. Lawrence. He states that "when on a shooting excursion some years since, at Egg Harbor, I noticed a bird flying at some distance from us, which our gunner said was a Black Brant. This was the first intimation I had of such a bird. Upon further inquiry of him, he informed me he had seen them occasionally, but they were not common. I have learned from Mr. Philip Brasher, who has passed much time at that place, that speaking to the gunners about them, they said they were well known there by the name of Black Brant, and one of them mentioned that he once saw a flock of five or six together." It was several years, however, before Mr. Lawrence was able to obtain a specimen but he states that later in the spring of 1846 two others were shot at Egg Harbor of which he secured one. (Ann. New York Lyceum of Nat. Hist., IV, p. 171, 1846.)

John Cassin (Birds of California and Texas, p. 52 (1856)) states that "To gunners of Philadelphia this bird is known by the same name, and we have seen several specimens which have been shot in Delaware Bay, and at various points on the seacoast. Our friend Mr. John Krider, Gunsmith, whose establishment is a favorite place of resort of the Ornithologists and gunners of this city, has had several specimens of this Brant brought to him within the last two or three years." In addition to these records we have only that of W. E. D. Scott (Bull. Nuttall. Ornith. Club, 1879, p. 226) that he "saw two specimens which were taken on Barnegat Bay by gunners on April 5, 1877," and a statement by "Homo" (Forest and Stream, for March 2, 1882) who states that he had secured Black Brant on Barnegat Bay, but gives no dates. In view of the statements of Lawrence and Cassin it seems remarkable that specimens of this bird have not been obtained in the East in recent years and we wonder whether the gunners from whom they received their information may not have been mistaken in some, at least, of their identifications.

J. F. S.

GREEN-WINGED TEAL

GREATER SNOW GOOSE

Chen hyperborea atlantica Kennard

Snow Geese are rarely seen east of the Maurice River, at the head of Delaware Bay, and are consequently very rare at Cape May and usually seen only as stragglers flying overhead during periods of migration. From the earliest times these "Bald Brant" or "Wavies," as they are locally named, have been known to frequent the marshes in the neighborhood of Fortesque and Bivalve and back of Egg Island on the upper Bay, where they come in at night to feed on the roots of the sedges which they dig out with their powerful bills. Many acres of marsh have been so thoroughly worked by the geese that all the sedge has been exterminated and they have become areas of soft mud. The Swedish settlers on the Delaware came down to the mouth of the Maurice River to kill geese "for their feathers leaving the carcasses behind them" and it would seem most likely that their birds were Greater Snow Geese. (Gabriel Thomas, History of West Jersey, 1698).

Alexander Wilson, writing in 1814, says: "This species, called on the sea coast the 'Red Goose,' arrives in the river Delaware from the north early in November sometimes in considerable flocks, and is extremely noisy, their notes being shriller and more squeaking than those of the Canada, or common Wild Goose. On their first arrival they make but a short stay, proceeding as the depth of winter approaches, farther to the south; but from the middle of February until the breaking up of the ice in March, they are frequently numerous along both shores of the Delaware, about and below Reedy Island, particularly near Old Duck Creek, in the state of Delaware. They feed on the roots of the reeds, tearing them up from the marshes like hogs."

His specimens were shot on the river, below Philadelphia, on February 15. He adds: "Among thirty or forty there are seldom more than six or eight pure white or old birds. The rest vary so much that no two are exactly alike." He considered the "Blue-winged Goose" the same as the Snow and the individual which he figures as the young female of the latter is in reality an adult Blue Goose!

Definite reference to Snow Geese on Delaware Bay in later years are few although certain works on wildfowl shooting seem to assume that it was generally known that the birds were to be found there. Five specimens in the collection of the Academy of Natural Sciences of Philadelphia were obtained by Dr. Wm. Louis Abbott on Delaware Bay, four of them on March 5, 1879, and one in January 1882, but these are the only specimens that we know of from this region.

In an issue of the "Smyrna (Delaware) Times" published early in January, 1884, it is stated "that a large number of white wild geese have been spending the winter in the coves at Bombay Hook, especially near Collin's Beach. It is no uncommon thing for them to spend the winter or part of it in the Bay but heretofore they have confined themselves to the Jersey side. In the past eighteen or twenty years Mr. Benson in charge of the lighthouse here tells me that he has not seen them so numerous." The writer goes on to explain the difficulty of approaching the flock and how gunners when the ice is breaking up cover their skiffs with ice and themselves with sheets and lying down in the boats are able to drift right in among the birds. "Captain A. A. Clay of Philadelphia, a season or so ago, managed to secure some by decoying them on the marshes where they go to feed. (Forest and Stream, January 12, 1884.) Charles S. Westcott writing in "Forest and Stream" under the pseudonym of "Homo," presents much information regarding the Snow Geese of Delaware Bay. He tells us that they were present in March 1882 and again in early March 1883, when they occupied the extensive meadows below Bombay Hook, tearing up the grass in order to get at the tender roots and making the ground look as if a herd of swine had been rooting there. "It is difficult to get at them as they sit out on the Bay the greater part of the time when not feeding and when they go to feed they start in two or three bodies so that if the gunner has made a blind where he expects them to visit and they do not come to him the first shot or two will drive them off to another portion of their extensive feeding grounds where they remain until ready to go out on the Bay again, and cannot be approached."

He reports them present again on March 13, 1884, when at "almost every poulterer's stand in Philadelphia more or less Snow Geese can be seen hanging up for the first time this season. They are termed 'Brant,' 'Bastard Geese,' 'Gull Geese' etc. A pair purchased were very cheap and were excellent eating. Very few were solid white the majority being in the sooty white plumage of the young bird." They were probably shot near Smyrna, Delaware. On November 30, 1884, he states that the usual flocks had made their appearance below Bombay Hook and were using the same meadows for feeding that they resorted to the year previous. This information was obtained from the crews of the oyster boats coming up the river from the Bay. On December 18, he states, that "the oystermen had taken two shots into the flock with a big shoulder-gun and had killed fifteen or twenty. They were sold in the city and brought good prices. There are more of the immature birds than usual which may account for the oystermen getting within range of them." On March 19, 1885, he again reports the geese present: "W. H. Childs went down Delaware Bay and saw immense flocks of Snow Geese which he was informed had been there about a week." He adds that the birds occur there regularly every spring and fall, "in large numbers and not as occasional stragglers."

On Thanksgiving Day in 1936 Lester Walsh flew down the Bay and while he saw no Snow Geese at Egg Island Point he counted 4500 from the air between Cedar Island, Parmour Island and Cape Charles.

In early February, 1928, Wharton Huber and I made a short cruise over the waters of the upper Bay which resulted in the discovery of a few single birds passing high overhead, but the main flock was not to be found and was supposed by our boatman, who was familiar with the movements of the geese, to be over near the Delaware shore. We were shown a marshy spot with a pool in the middle on Egg Island, where he said the Snow Geese came to feed and where they were often shot, sheets of paper being stuck up on the low bushes to serve as decoys. On March 26, 1933, Richard Pough and other members of the Delaware Valley Club visited Fortesque and saw between seven and ten thousand Snow Geese. Two Bald Eagles kept the geese in almost constant motion but although the birds were watched for an hour no damage was done and the geese, as well as many ducks that were present, seemed able to outdistance the Eagles with ease. Richard Pough, who has made other visits to the locality, tells me that the geese occur late in October and rest out on the Bay to the west of Egg Island Point.

Another trip was made to study the Snow Geese on March 28, 1936, by Fletcher Street, Charles Urner and others. Street tells me that upon arriving at Fortesque they discovered a large raft of the birds well out in the Bay and securing a boat found them to be about a mile and a half offshore and succeeded in approaching to within a hundred yards of them before there was any commotion. Then the rear birds began to fly up to the head of the line which action continued until the boat startled the entire flock which took to the air at once. The birds flew low over the water for about three hundred yards before alighting. The formation was triangular; a narrow point close to the water while the height of the mass increased toward the rear. While the birds were on the wing they counted seven Blue Geese in the flock. It was an inspiring sight to see as many as ten thousand of these snow white birds in the air at once drifting along the surface of the water like a great cumulus cloud and to realize that they were viewing, perhaps, at one time, a majority of the species, undoubtedly concentrated there on their northward migration.

Charles Urner describing the same trip says "We found them just a thin audible line a mile or more offshore. We hired a boat and in no great time were upon them. Never shall I forget the picture as this compact mass of probably five thousand black-tipped white birds, necks stained from months of dabbling in the Jersey mud, took to the air—a puff of great noisy snow flakes that quickly settled into layer upon layer of drifting white, all but hiding within their depth a scattered few, dark-bodied, white-necked Blue Geese" (Bird-Lore, 1936, p. 333). The two accounts illustrate the variation in estimates of numbers in waterfowl even among trained observers!

Charles Nichols states that on April 2, only one small flock was to be seen and on April 7, all had gone. Various views have been expressed as to the nature of the occurrence of Snow Geese on Delaware Bay. Some claim that a certain number of the birds winter here; others that all of them eventually go south to Currituck Sound where the main winter flock is located and that they return in February to congregate at the head of Delaware Bay preparatory to their non-stop flight to the St. Lawrence River at Cape Teurmente east of Quebec, where between 8000 and 10,000 rest after coming from their breeding grounds in the Arctic and before resorting there in the spring. From the evidence presented it would seem that the bulk of the birds do go south but that not only stragglers but flocks of considerable size remain at least as late as mid-December. But the fact that Richard Pough made an airplane trip over the upper part of the Bay on January 29, 1933, and failed to sight any Snow Geese shows that it is not a regular thing for them to tarry so far north. Then again their habits may have changed as conditions are not so favorable for their wintering as in times past. A flock of about one hundred passing over Troy Meadows in Morris County, in northern New Jersey, on April 2, 1933, and observed by Charles Nichols, and another of thirty-seven seen by Donald Carter passing Boonton, not far away, on April 6, 1924 (Auk, 1924, p. 472; 1933, p. 352), would seem to represent these Delaware Bay birds on migration.

Our records for the immediate vicinity of Cape May, where the bird is of only casual occurrence, are as follows. Walker Hand reported five seen flying over the town on February 25, 1917, and eighteen on January 27, 1925. The latter were travelling in a southwesterly direction and could be heard calling as they went. There was a slight flurry of snow in the air at the time. Hand also reported a single bird shot by a gunner on January 28, 1924, and a flock seen off the jetty on December 3, 1926. On October 11, 1925, after a notable gale from the north, I found a single Snow Goose on the Lighthouse Pond in company with five Canada Geese. He was busily engaged in pulling up grass and sedge by the roots from a black mud bank on the side of the pond and when disturbed swam out to join the other birds and eventually flew off with them. In flight his wings were rather more pointed than those of the Canadas and the strokes more rapid, while his shorter neck was characteristic. In all these respects he resembled the Brant and this doubtless is responsible for the gunners' name of "Bald Brant" so generally bestowed upon the Snow Goose. On December 27, 1931, Joseph Cadbury reported two seen flying down the coast, while on November 20, 1933, Otway Brown saw several out on the Bay where Richard Pough found three on December 24, following. In the autumn of 1934, Raymond Otter tells me that he saw a great flock of Snow Geese over Jenkins Sound, near Cape May Court House, so large that when the birds raised against the sky they looked like a great white snowbank. This was very unusual and is in fact our only record of a large flock on the ocean side of the peninsula. In November 1934 Socrates McPherson obtained a wounded Snow Goose which he kept in captivity until its wound had healed and then turned over to his nephew Michael, of Cold Spring, who had a Blue Goose which had decoyed to his ducks. The two geese associated at once and have since been going about the farm wing-clipped. On May 2, 1936, I saw a Snow Goose sitting on the beach near Broadway which I learned was later captured by some boys and added to the goose group above referred to which now contains a Canada Goose which had decoyed to the others and was caught and pinioned.

In the Barnegat Bay region Charles Urner tells me he has occasional records of Snow Geese in October and November and more rarely in April. His dates are: November 24, 1925, one shot; November 30, 1930, one seen by Lester Walsh; April 1931, one caught alive; November 13, 1933, flocks of thirty and forty; November 14, 1933, several flocks, one of two hundred birds; October 14, 1934, two; November 11, 1934, one.

The Snow Goose is a very conspicuous bird both on the water and in the air. Its plumage is entirely snowy white except for the jet black wing feathers which show to advantage only when in flight. Those that we saw flying overhead at Egg Island made a beautiful picture as the late afternoon sunlight lit up their brilliantly contrasting plumage. The rusty stains which sometimes envelop most of the head seem to be due to iron deposits in the mud in which they dig (Kennard, Auk, 1918, 123).

*LESSER SNOW GOOSE

Chen hyperborea hyperborea (Pallas)

This goose, a native of north-central North America, migrating down the Mississippi and to California, is a slightly smaller and more slender edition of the Snow Goose of the Atlantic Coast (cf. Kennard, Proc. N. E. Zool. Club, IX, pp. 85–93). It occasionally straggles eastward but cannot be distinguished in life from its larger relative unless the two are side by side and close at hand. Therefore more of them may have occurred on eastern waters than we suppose. Charles Urner records one specimen which could not be preserved and which was taken on the salt meadows near Elizabeth, N. J., on October 29, 1916, with two others and which had a wing length of 14.75 inches. Turnbull (1869) tells us that "John Cassin procured in the Philadelphia market, two pairs in the course of twenty years, of this inhabitant of Northwest America." At the time that these specimens were secured they could not have been shipped from the west and must have come from Delaware Bay or the New Jersey coast. As Cassin was the first to distinguish the two forms of Snow Geese he knew what he was talking about and furthermore it was one of the above birds upon which he based his name *albatus* and which, an unquestioned Lesser Snow, is still in the collection of the Philadelphia Academy. The Blue Goose, a close associate of the Lesser Snow, both in the breeding area and on migration, has occurred as a straggler on our coast but its conspicuous coloration makes it easy to identify while a straggling Lesser Snow would be easily overlooked. Two Snow Geese seen by Graham Rebell at Tuckerton, on November 11, 1934, in company with two Blue Geese might very well have been a Lesser (Auk, 1935, p. 182).

BLUE GOOSE

Chen caerulescens (Linnaeus)

The only record of this goose for Cape May is one which decoyed to some Pekin Ducks on a pond on the farm of Michael McPherson at Cold Spring in October, 1934, and was caught in a crab net. It was later pinioned and turned loose but made no effort to leave the premises. It was in the immature dark-headed plumage when caught but the following spring assumed the typical white head. It has later been joined by two Snow Geese which were captured along the shore, having been wounded by gunners as already explained.

The only other New Jersey records of the Blue Goose are one associated with two Snow Geese seen by Graham Rebell at Tuckerton November 11, 1934 (Auk, 1935, p. 182) and five seen by J. L. Edwards and Fletcher Street at Fortesque, on March 29, 1936, in a great flock of Snow Geese. A specimen in the collection of the Academy of Natural Sciences of Philadelphia was obtained by Dr. W. L. Abbott in the Philadelphia markets and doubtless came from the head of Delaware Bay. The specimen figured by Alexander Wilson in his "Ornithology" (1814) as the young of the Snow Goose came from the same spot where Wilson tells us the Snow Goose was common.

* WHITE-FRONTED GOOSE

Anser albifrons albifrons (Scopoli)

We have no record of the occurrence of this western goose at Cape May. Turnbull (1869) states that it is rare in New Jersey and C. C. Abbott (1868) reports one shot on Barnegat Bay years ago. Of late years we have two records for the same locality; two seen by Charles Urner and J. L. Edwards on November 10 and 11, 1928, and another seen by Ludlow Griscom on November 28, 1926.

*WHITE-FACED TREE-DUCK

Dendrocygna viduata (Linnaeus)

A specimen of this South American duck was shot by Hon. John W. Griggs on the Hackensack Meadows early in October, 1912. While it showed no signs of being a captive it seems hardly possible that it should have found its way so far north by normal flight, and it probably escaped from some zoological garden (G. B. Grinnell, Auk, 1913, p. 110).

*RUDDY SHELDRAKE

Casarca ferruginea (Pallas)

One of these Old World ducks was shot on Barnegat Bay on November 14, 1916, by W. H. Eddy who had it mounted. I examined it on two occasions and could find no evidence of its having been in captivity. It has occurred also in Greenland and North Carolina (cf. G. B. Grinnell, Auk, 1919, p. 561).

MALLARD

Anas platyrhynchos platyrhynchos Linnaeus

Mallards occur on the Bay shore meadows and adjacent freshwater ponds but are not nearly so common as in years gone by; on the sounds and meadows on the ocean side of the peninsula they have always been rare. Raymond Otter tells me that the reduction in their numbers has been very marked since the Mosquito Commission has been draining the meadows along the Bay and that a gunner today who gets a pair in a season's shooting is doing remarkably well. The last large flock of Mallards that he saw was in the autumn of 1916 when some five hundred remained on the Pond Creek Meadows for two weeks in October. John Mecray says that the few that come into the sounds associate with the Black Ducks. Migrating Mallards occur at Cape May from mid-March to mid-April while in the autumn they arrive in late September or early October and remain until the first freeze drives them south.

Walker Hand received information from a man well acquainted with our wildfowl that a pair of Mallards bred on the New England Meadows in the spring of 1927 and that he saw them several times with their brood of young, while Richard Harlow found a nest on the coastal marshes of Burlington County on June 16, 1918, which contained ten hatching eggs. However, at Delaware City, semi-wild Mallards occur commonly and are said to nest but they would seem to owe their origin, in part at least, to domestic birds and it is difficult to decide which are really wild individuals. The nesting birds of New Jersey may have a similar history.

Pond Creek and New England Meadows have been the best places to find Mallards immediately about Cape May. I came upon a fine drake in the former locality on April 4, 1924, when the entire meadow was well flooded and offered ideal duck conditions. He was swimming about by himself and made no effort to join the numerous pairs of Black Ducks that were present and apparently mated. One of the latter swimming with his mate near to the Mallard made a vicious attack upon him and drove him away. Even across the meadows the colors of this drake could be clearly seen with the glass—the bright glossy green head, light gray back and sides separated by a longitudinal black line, and the black and white spot on either side at the base of the tail. The curled feathers of the lower rump and the pale green bill were also noticeable. A pair was present on the shallow pond near the Lighthouse on March 25, 1928, in company with a pair of Green-winged Teal, and a single drake on April 11, 1926. Philip Laurent obtained two on Five Mile Beach, where they were always regarded as rare, in winter in 1891 or 1892, and Fletcher Street saw one on November 6, 1926, on Ludlam's Beach. At Delaware City we have records of two seen on April 10, 1925, and eight on April 3, 1932. At Fort Mott on the Delaware fifty were seen on November 3, 1935. In the Barnegat Bay region Charles Urner has seen Mallards in every month of the year, but whether the summer birds are truly wild may be open to question. The time of greatest abundance there is from November to April. On December 13, 1885, Charles S. Westcott ("Homo") writing in "Forest and Stream" says that many Mallard still remain with the Black Duck in Salem Cove, on the lower Delaware, in spite of the very cold weather. The year previous they had left long before this date although the weather was then much milder.

Hybrid Mallard × Muscovy Ducks occur occasionally on Delaware Bay and elsewhere probably bred on some private duck pond. Some of them closely resemble the Mallard in plumage but are twice as large. C. S. Westcott records two shot on the lower Delaware, on December 21, 1885 (Forest and Stream, Dec. 24, 1885).

COMMON BLACK DUCK

Anas rubripes tristis Brewster

The Black Duck is the common wild duck of Cape May. Present throughout the year, it is the only duck to be found regularly in the summertime, though its activities, as we see them at this season, are merely the shifting of pairs or family flocks from one feeding ground or shelter to another. In winter the Black Ducks increase in numbers augmented as they are by migrants from farther north and great flocks gather on the sounds and daily pass out to sea and back again. After the gunning season, however, when they are free from molestation, they are glad to remain on the shallow waters of their feeding grounds. They occur also in large flocks in autumn and winter on Delaware Bay and nest on the Bay side meadows and occasionally on the ocean side too. The marshes around Fort Mott, where the Killcohook Bird Refgue has been established, are a favorite resort and it was estimated that ten thousand birds were present there on September 28, 1935, and October 6, 1929, and two thousand on November 3, 1935. Black Ducks are also abundant about Delaware City across the river to the west and breed there, while in the Barnegat Bay region to the north, Charles Urner has observed them in every month of the year. The principal migratory movements, however, take place during August and March with the peak of abundance from November to January.

Ducks are difficult subjects for intimate study as their extreme wariness enables them to detect the presence of the observer before he discovers them, so that they are ever alert and ready for instant flight. For countless generations their training has been directed toward matching their wits against those of the gunners who have ever pursued them, and thus has developed that suspicion of mankind which self-preservation has rendered necessary and which is today one of their chief characteristics. Then, too, their habitat is usually the open water of the sounds or the Bay, where they cannot be approached very closely. Our acquaintance with the Black Duck is thus, for the most part, limited to birds in the air or to those which are resting on the water at a safe distance.

Usually the flying birds of summer are passing from haunts on the Bay shore to Jarvis Sound or other bodies of open water on the ocean side of the peninsula, but I have also seen them crossing over the town and putting out to sea bound, apparently, for the Delaware coast. Early one morning (August 8, 1920) a bunch of seven, probably a family group, came down the coast at a considerable height and swerved off to the west in the direction of Pond Creek Meadows. As no migration was yet in progress these flights would seem to indicate considerable latitude in the selection of feeding grounds, and with the advent of the Yellow-legs season, when the shooting of these birds was permitted, the fusilade of guns in the favorite resorts of the Black Ducks scattered them still more widely. On these same meadows of Pond Creek, on August 16, 1920, the first day of the shore bird shooting season, they were greatly disturbed, taking wing again and again, quite at a loss where to find shelter as the bombardment continued. Fortunately with the abolishing of all summer shooting this menace to their peace has disappeared.

On the wing Black Ducks, like other Anatidae, fly in more or less of a V although the big flocks of midsummer, composed, perhaps, more largely of young birds, have a more massed and irregular formation. The neck of the flying Black Duck is stretched straight out in front and slightly elevated as if the head of the bird were straining to the utmost, and it always gives me the impression that the heavy body is being dragged along by the neck, the rapidly beating wings being set so far to the rear that it does not seem possible that they can be the motive power. The tail and legs of the flying duck are so short that they cut but little figure and the bird seems to lack balance. When studying a Black Duck on the wing with the glass, it is possible to note the lighter color of the neck, the narrow brown and black striations contrasting with the solid black brown of the body, also the light under coverts of the wings which appear almost silvery white when the bird turns into the sunlight.

The usual summer flocks a few years ago consisted of from three to twelve birds and sometimes single individuals were seen. In recent years, however, they collect in larger bodies and on July 1, 14, and 17, 1924, a flock of forty was observed. On August 10, 1927, too, while following the windings of a salt creek through the meadows east of Cape May Court House, I flushed thirty Black Ducks in bunches of four and five, some taking wing at almost every turn in the stream where they had been concealed by the marsh elder bushes and the overhanging banks, which had likewise shielded my approach from them. They were obviously old and young of several families. During August, 1936, fifty or more came every evening at dusk to the marsh at the mouth of New England Creek.

I have frequently found Black Ducks resting at the upper end of the Lighthouse Pond where, flanked by the dense growth of cattails, they are safe from approach and stand on the muddy shores deliberately pluming themselves oblivious to observers across the water. They were also sometimes seen on the old Race-track Pond, now drained, and on smaller ponds on the marsh back of South Cape May, and are frequent on the shallow pond east of the Lighthouse and on pools left by the sand shovels among the Bay shore dunes, as well as at Price's Pond a little to the north. On the latter ponds, where they can be more closely approached, they are very wary and ready for instant flight.

Sometimes I have been fortunate enough to stalk Black Ducks on the smaller ponds, where bushes along their edges offered a chance of concealment. On July 14, 1920, I approached several which had settled on one of the lily ponds in West Cape May, and coming suddenly over the dyke-like bank found myself almost on top of them. Fourteen birds immediately took wing in the greatest confusion, with much splashing of water and showers of spray. One broke into full cry and kept it up until the flock was well under way. The cry, when heard thus close at hand sounded more like, *skeag, skeag, skeag,* than the conventional *quack, quack, quack.* On August 3 of the same year, under cover of bushes and tall grass, I crept up on seven Black Ducks on the old Race-track Pond and was able to study them for some time unseen. They were clustered together in the middle of the pond which was shallow throughout, with scattered patches of low sedge and some mud flats. Most of the birds had their heads under the water poking vigorously in the mud and aquatic vegetation. Two or three stood on the exposed hummocks of mud and with heads erect and bodies rigid looked at a distance exactly like black stumps. Indeed, owing to this resemblance, it was some time after I caught sight of the group before I realized that they were ducks. Through the glass the light yellowish basal portion of the bill could be clearly seen as well as the brilliant green speculum of one bird, probably the drake. They flushed suddenly, while I was recording my notes and were out of sight when I again raised my head. Again on the shallow pond at the Lighthouse (July 15, 1929) there was a flock of nineteen, two large ones, evidently the parents, standing erect on the shore and the others, apparently only two-thirds their size, feeding on the shallow water with heads submerged. So persistently did they feed in one spot that for some moments I took them for lumps of black mud until some of them raised their heads. With that mysterious flock action they all took wing simultaneously and despite their small size the young birds could fly as well as the adults. At the Race-track Pond, on another occasion, four Black Ducks came in and crouching down in the grass I could see them circle a couple of times as if to assure themselves that all was well. Then they came dropping down on rapidly beating wings, heads up and feet extended, until they splashed into the water and began swimming leisurely about. Another bunch, a few days later, came in in the same way but though they circled again and again, and made as if to settle, they were evidently suspicious and eventually gave it up.

On July 9, 1923, I worked my way through the dense growth of cattails at the head of the Lighthouse Pond almost to the edge of the water, where a narrow channel flows through, and as I stood perfectly still a family of Black Ducks swam past along this thoroughfare within six feet of me. Two adults and six young birds scarcely full grown—all in single file. A little farther on they waddled out of the water and disappeared among the cattails leaving one old bird on the edge of the channel as a sentinel. They were entirely unaware of my presence. While on Two Mile Beach on July 4, 1927, a similar string of young, nearly full grown but apparently unable to fly, came past me following a little ditch from the marsh up into the thickets of bayberry. The female suddenly spied me and deserting her flock flew back almost to my feet and flopped about on the short marsh grass with wings extended. Then she arose and after several wing strokes dropped back again into the grass. As I did not follow her she began to walk about, her neck held erect so that her head was just above the level of the grass, and watched my every movement. The young meanwhile had disappeared into the dense shrubbery and I was unable to locate them. The duck circled around several times but always returned to where she could watch me, and upon approaching her she once more endeavored to lead me away, even to the extent of taking short successive flights out onto a nearby salt pond, settling each time on the water a little farther out, as if she expected me to follow her there. I was unable to determine whether the other parent was with the young or not. Every time this bird dropped to the water she threw her feet forward, just as if alighting on the ground, so that they struck the surface before her body. Even in cases where the young ducks were fully grown I have seen the female drop back on the meadows when they flew away and walk about keeping strict watch on me for several minutes before she followed her brood.

Family broods like these join forces as the summer advances and I counted over sixty birds gathered on a sequestered pool on the Pond Creek Meadows on August 12, 1917, where many white herons, Yellow-legs and countless smaller sandpipers had sought shelter, and on July 14, 1924, I found a flock of forty feeding on Jarvis Sound, where they swam through the shallow water which was almost clogged up with seaweed. They formed a long line, for the most part in single file but sometimes with two or three abreast, some birds bending their heads down below the surface fed as they swam slowly along, while others were tilting completely over with tails erect, bobbing up and down like giant corks and gave a varied and ever changing aspect to the procession. A few days later I saw presumably the same flock, numbering this time thirty-eight, flying over Cold Spring headed for Pond Creek. On August 18, 1925, a similar flock was seen crossing the peninsula in the same way, while on July 21, 1926, another string of birds crossed against the glow of the sunset, going in the opposite direction, at 7.30 in the evening. A flock of young and adults of two or more families frequented the shallow pond east of the Lighthouse in the summer of 1928 and was found on July 4, 7, and 31 to consist of twenty, twenty-five and twenty-three birds respectively.

Some of them were bathing and I was able to approach them quite closely in the shelter of some bushes. They ducked the head completely under and churned up the water with the wings shaking the tail vigorously from side to side and now and then rising straight up in the water to flap the wings, displaying conspicuously the white under surfaces.

Augmented by migrants from the north these family gatherings form the winter flocks of the sounds which number 1000 to 1500 individuals and which feed there, usually moving out to rest on the ocean during a good part of the day. While on the beach below Corson's Inlet, on January 1, 1926, I saw two immense flocks of Black Ducks arise from the sounds and fly out to sea and again at 5.30 in the evening I could see them in a great raft resting just beyond the surf line, rising and falling with the waves and apparently awaiting nightfall before venturing back to their feeding grounds. On January 23, 1927, I saw a similar raft lying off Stone Harbor which I estimated to contain 1200 individuals while there were 1000 in the same place on February 22, 1928, and again on March 5, 1932, when they flew seaward in the morning and disappeared beyond the fishpounds a mile offshore.

Julian Potter saw six great rafts of Black Ducks off Ludlam's Beach on December 4, 1921, each containing from five hundred to one thousand birds and a large raft lay a mile off of Townsend's Inlet on January 3, 1931.

In exceptionally cold winters, when the sounds freeze over, most of the Black Ducks seem to move farther south where they can find open water until the weather moderates, but they are nevertheless able to survive our coldest weather. On February 11, 1934, after a fall to −11° on the 9th when all of the surrounding waterways were ice-covered Julian Potter found a number of them at Brigantine sitting in groups on the frozen meadows and the following winter on February 3, after a −6° temperature on January 28 and sixteen inches of snow on the 24th, he found two hundred Black Ducks with one hundred and fifty Brant and one hundred Scaup sitting on the ice at the same spot. Again on December 21, 1919, when very cold weather prevailed there were twelve Black Duck, an equal number of Scaup and two hundred Herring Gulls resting on the Brigantine icefields.

Considerable flocks remain as late as early April, which must be migrants bound north as birds have been found apparently mated by this time back on the meadows. On April 2, 1922, Walker Hand and I found a flock of sixty-seven Black Ducks accompanied by two Blue-winged Teal on a shallow pool on the Fill below the airdrome which took wing as we approached and flew north in the direction of Jarvis Sound. Whether the local breeding birds remain through the winter or arrive in the spring from farther south I cannot say but by March 20 I have found Black Ducks evidently paired. On this date in 1924, I saw no less than fourteen pairs on the Pond Creek Meadows, which that year were well under water and formed admirable feeding grounds for all sorts of surface-feeding ducks. They were scattered all about busily feeding, moving slowly among the submerged tussocks of sedge, now and then disappearing behind some tall bunch of grass or swimming out into the open channels. The birds of each pair kept close together; the drake, distinguishable by his larger size, always in the rear. When they occasionally raised their heads the greenish yellow bills of the males were conspicuous. Now and then a pair would take wing and fly for a short distance and either return immediately to the water with a splash, or mounting higher in the air, circle widely, the male close behind the female. These birds traveled rapidly on their customary short-arc wingbeats and then on set down-curved pinions sailed for a short distance, flapped and sailed again, coming lower and lower until finally they dropped straight down with feet dangling and wings beating and plumped into the water. On April 4, 1924, about twenty pairs were present in the same spot, one male was resting with his mate on some sods left by the ditchers when another pair alighted nearby; he immediately made for them and all three took wing. He pursued the pair to the far end of the meadows and then returned to his mate. In another patch of water far across the marsh several splashes seemed to be the results of conflicts between the males that were feeding there and two were clearly seen through the glass pursuing one another through the narrow channels, twisting and turning in every direction.

Until the abolition of spring shooting the Black Duck had been all but exterminated as a breeding bird in southern New Jersey. One nest that William Baily found at Ocean View on March 30, 1901, was ready for eggs when a gunner flushing the female bird shot her, and no doubt most of the attempts at local nesting met with the same fate. Robert T. Moore, in an excellent picture of the marshes of Great Egg Harbor River (Cassinia, 1908, p. 37) describes the discovery of a nest with nine eggs on March 22, 1908, and doubtless the birds never ceased to nest in small numbers in this and other remote spots in south Jersey. Indeed there is reference to the apparent nesting of perhaps four pair in the Barnegat Bay region in this same year (Forest and Stream October 24, 1908), while in a much earlier issue of the same journal (August 2, 1884) Charles S. Westcott states that two broods were raised in the same vicinity and regrets that owing to lack of any protection they will all probably be shot before the real duck season begins.

No sooner had spring shooting been stopped and the Migratory Bird Treaty gone into effect, than we heard report after report of the nesting of Black Ducks until now the species is re-established as a regular breeding bird of the Cape May district. Many nests have been reported along the edges of the Pond Creek Meadows and in the spring of 1924 two pairs nested on Two Mile Beach. One of these nests I found in a hollow just in front of the high sand dunes under some bayberry bushes at the lower end of the island. It was in a dry sandy spot with no water near and was made of down, dead bayberry leaves and some dry grass. The rim stood up several inches from the ground and was two inches thick, the inside measurement of the nest being seven by nine inches and its depth three inches. It contained six eggs. The female bird flushed every time I came by yet only once was I able to get near enough to see her leave the nest. She ran with a waddling gait from under the bushes and immediately took wing flying out toward the ocean and circling back again in a long curve. She passed over the dunes and alighted somewhere back on the salt meadows. The male bird I never saw.

Another nest was shown to me by W. A. Squires on May 9, 1931, who had just found it back of the sand dunes a short distance above the Coast Guard Station at the Point. It was placed under bayberry bushes in tall dead grass and was lined with feathers and down and contained seven eggs. Two nests reported to Walker Hand in April, 1928 and 1929, were in low woods half a mile from water. Additional nests found in Cape May County and reported to me by Turner McMullen are as follows:

May 19, 1924, Seven Mile Beach, one with seven eggs and another with four.

May 5, 1929, Cape May Court House, eleven eggs.

May 18, 1930, Cape May Court House one with seven eggs and another with eleven.

In Salem County, where the Black Duck seems to be much more plentiful, especially as a nesting bird, he has examined a number of nests of which the date and number of eggs of each are appended:

April 29, 1922, nine; seven.
April 23, 1923, six; nine; ten.
May 6, 1923, three; nine; ten; eleven.
April 20, 1924, ten.
May 24, 1924, ten.
April 12, 1925, thirteen.
April 13, 1925, nine.
May 2, 1926, ten.
April 22, 1928, eight.

April 22, 1928, ten; ten; eleven; ten downy young.
May 4, 1930, nine.
April 18, 1931, eleven; ten.
April 25, 1931, ten; ten.
May 7, 1932, ten; ten; one.
April 3, 1933, twelve.
May 2, 1933, eight.
May 30, 1933, nine.
April 14, 1933, nine.
April 21, 1935, four; ten; twelve.

The young Black Ducks hatch late in May or early in June as some were found at Corson's Inlet only a few days old on June 8, while on May 26, 1929, and again on May 25, 1934, Otway Brown saw an old duck cross the the Bay Shore Road near Cold Spring followed by a string of nine half-grown young in single file.

Walker Hand tells me that in the stomachs of Black Ducks that he has shot he has found a wide range of food, small minnows, weed seed, little mussels, which seem to be digested shells and all, and in midwinter quantities of periwinkles, which seem to constitute an important item at this season when other food is apparently scarce. Where they have been feeding he has frequently noticed much eelgrass floating on the water, which had been pulled up by the roots, and as he had never found this plant in the stomachs of the ducks he presumed that they had been seeking some sort of food that was to be found amongst it.

The return of this splendid bird as a common summer resident of the Cape is one of the great benefits that we have reaped from the migratory bird legislation but the senseless draining of the meadows and marshes in recent years threatens to destroy both its feeding and nesting grounds which can only result in the disappearance of the ducks.

It is always inspiring to see a pair of Black Ducks winging their way over the meadows, for there is nothing that emphasizes the natural beauty of a spot more than the presence of wildfowl and one of the greatest delights of summer ornithology at Cape May is to follow up the Black Ducks and to match our wits against theirs in our efforts to learn more of their home life and actions.

RED-LEGGED BLACK DUCK

Anas rubripes rubripes Brewster

Raymond Otter and John Mecray, two leading sportsmen of Cape May, tell me that the gunners of the region recognize the two kinds of Black Duck, the resident bird with legs ranging from green to brown and which weighs from two and a half to three and three-quarters pounds, and the larger Red-legged Black Duck which, according to Otter, arrives on the sounds and meadows about mid-November and leaves on the first sign of spring. The latter has red legs and weighs from three to three and a half pounds while one shot by Logan Taylor reached four pounds. He adds that these birds have the straight "up spring" of the Mallard as they leave the water in flight and he therefore terms them "Black Mallard." I saw one of these big Black Ducks on Pond Creek Meadows as late as April 4, 1924, and his coral red legs were conspicuous as he stood on a mud lump while I watched him from the sand dunes to the south.

GADWALL

Chaulelasmus streperus (Linnaeus)

Raymond Otter tells me that he shot two of these ducks in the autumn of 1924, which were indentified by Walker Hand. He secured another in October, 1925, one in 1928 and two in 1929. This is all the information that I have been able to secure regarding the occurrence of the bird at the Cape where it is evidently very rare. There is a record of one killed on Delaware Bay in March, 1900, and one was seen at Fort Mott by J. L. Edwards on November 2, 1935, easily recognized by its conspicuous white speculum.

Charles Urner has four records of single birds for Barnegat Bay: April 8, 1928; December 26, 1932; September 29, 1935; January 25, 1936.

C. S. Westcott records "Gray Ducks" on Delaware Bay on October 11, 1885, and October 16, 1884, associated with teal.

*EUROPEAN WIDGEON

Mareca penelope (Linnaeus)

While we have no record of the occurrence of the European Widgeon at Cape May it undoubtedly must occur there at least in its flights from localities to the north and south, where a few of these birds are to be found nearly every year.

Charles Urner and others who study the birds of the Barnegat Bay region have reported one seen on February 16, 1924; another in a flock of 150 Baldpates on March 14–15, 1925; a third on December 9, 1928; and still another on November 9, 1930. In earlier times Charles S. Westcott ("Homo") who contributed many notes on New Jersey waterfowl to "Forest and Stream" states that Samuel Smith of Tuckerton told him that several European Widgeon had been killed on Tuckerton and Barnegat Bays near Little Egg Harbor Inlet. (F. and S., March 2, 1882.) On March 17, 1928, Benjamin Hiatt saw one at Delaware City across the Delaware River from Fort Mott while Julian Potter identified one at Perryville, Maryland, on January 24, 1932. The last locality is entirely out of our region but the date is interesting for comparison.

J. F. S.

RED-BREASTED MERGANSER

BALDPATE

Mareca americana (Gmelin)

The Baldpate, the "Widgeon" of the Cape May gunners, has never, from the information that I have obtained, been a common bird about Cape May. It occurs most frequently on the Bay shore meadows and has suffered severely from the draining of its favorite haunts. On the sounds it is rare and quite unknown to many of the gunners who shoot there. Curiously enough, as Raymond Otter tells me, while the Baldpate is more common on the Bay side the largest flocks have been seen on the sounds or adjacent creeks. He thinks that the scattered birds from the Bay bunch together to fly into the sounds and spread out again when they return. Most of the birds that are shot, he tells me, are immature and only occasionally is a full plumaged male reported. He gives October and December as their time of most frequent occurrence. On the sounds, however, I have mainly mid-winter records. Two were shot from a flock of about thirty on January 31, 1925, and another on December 5, 1929, while an adult drake was killed at Corson's Inlet on January 1, 1928, and Julian Potter saw a flock of six back of Ludlam's Beach on January 8, 1928.

It is on fresh water ponds about Cape May Point that I have usually found the stray Baldpates that have come under my personal observation. On March 20, 1924, I found a pair on the flooded meadows of Pond Creek. They were swimming about among the submerged grass and sedges and kept close together, the female always in the lead. Once they took wing and after a wide circuit of the marsh returned and alighted at almost the exact spot from which they started. Through the glass the male appeared very brilliant with his white crown and conspicuous white spot on the bill, while the pinkish tint of the breast was also noticeable. The female was much duller and browner. They were feeding close to a bunch of Black Ducks and finally took wing with them flying close together at the head of the flock, easily distinguishable by their smaller size and generally lighter coloration.

Two Baldpates were flushed from the shallow pond at the Lighthouse on February 22, 1926, and a female on the deep pond November 6, 1927, while a solitary individual was seen there on September 27, October 7 and 21, and four on October 5, 1929.

At Delaware City, we have records of fifty on April 6, 1928, and seventy-five on April 3, 1932, and at Killcohook Reservation, at Fort Mott, twenty-five on November 3, 1925, which give a further idea of the time of their migrations.

In the Barnegat Bay area Charles Urner's inclusive dates of occurrence of the Baldpate for several winters are:

1929–30.	October 12–April 13.	1932–33.	October 9–April 9.
1930–31.	September 14–March 23.	1933–34.	October 8–March 16.
1931–32.	October 11–April 17.		

He has extreme dates of September 7, 1925, and April 30, 1927, and finds it most abundant in October and April.

It is interesting to read in the classic "American Ornithology" of Alexander Wilson, published over one hundred years ago, that in the early part of the last century the "Widgeon" was "common in the winter months along the bays of Egg Harbor and Cape May, and also those of the Delaware" and that "they leave these places in April." George Ord in his later edition of the work tells us that on the thirtieth of April, probably 1815, he observed a large flock of them accompanied by a few Mallards and Pintails, feeding upon the mud flats at the lower end of League Island below Philadelphia. Still more interesting is his statement that "a few of the birds breed annually in the marshes in the neighborhood of Duck Creek, in the state of Delaware. An acquaintance brought me thence, in the month of June, an egg, which had been taken from a nest situated in a cluster of alders. The nest contained eleven eggs." We fear, however, that the nest must have belonged to a Black Duck as there is no other indication of the Baldpate ever nesting so far south.

J. F. S.

PINTAIL

Dafila acuta tzitzihoa (Vieillot)

While occurring rather frequently on the Bay side meadows of Cape May County the Pintail is very rare on the sounds as is the case with most fresh water ducks and John Mecray tells me that he thinks their occurrence there is due to strong westerly winds which drive them across the peninsula but that they do not remain. On the lower Delaware River and at the head of the Bay they occur on the wet meadows in great rafts and Raymond Otter tells me that Salem Cove and the surrounding waters were literally covered with them in the spring of 1935. He further states that they began to decrease in the Cape May area after 1930 and since then the numbers arriving seem to grow less each fall due, he thinks, to the drainage of the Bay shore meadows which has ruined their natural feeding grounds. In his experience they arrive about October 1 and have usually left by December 1. I have, however, seen them as early as September 27, 1928, and we have a record of a single bird on December 23, 1934, while in spring I have seen them during the latter half of March and early April.

In the spring of 1924, when the Pond Creek Meadows were flooded, an exceptionally good opportunity was afforded for studying the fresh water ducks and on March 20, while watching the Black Ducks that had gathered there I found a pair of Pintails. The male with the light line down the side of his neck was very conspicuous and his elongated tail feathers, standing clear of the water as he swam about, formed another identification mark. Returning, on April 3, I found three males and two females feeding in the same place. Through the glass the brown head and neck of the males could be seen clearly and the fine white line down the side of the latter. A conspicuous white patch was visible on the flanks and a black spot near the middle of the wing. They fed continuously with heads submerged so that they appeared like so many lumps of mud or sod until a movement disclosed their identity. At long intervals they raised their heads and looked about them and then resumed their feeding. They huddled close together and their heads and shoulders appeared very wet and shiny when raised from the water. On one occasion one of the males rushed at another and for a few minutes all swam with heads erect and bills pointed diagonally upward. The next day there was but one pair present.

Three Pintails were seen on March 18, 1927, one on October 11, 1925, and one on November 16, 1929, on Lake Lily, and David Baird saw four on the Lighthouse Pond on October 28, 1928, while on March 14, 1931, I found five males and seven females clustered close together on a small pond near Goshen.

At Delaware City, Delaware, members of the Delaware Valley Club have recorded Pintails as follows: Thirty on April 3, 1932; five thousand on March 18, 1928; one hundred on April 10, 1925; ten thousand, including one albino, on March 2, 1930. At Fort Mott, across the river on the New Jersey side, there were fifteen thousand Black Duck and Pintails on September 28, 1935, and on October 6, 1929, Julian Potter estimated that Pintails to the number of thirty thousand were present, the rafts extending for a mile down the river with an average depth of from twenty-five to fifty yards. On November 3, 1935, there were twelve thousand five hundred present. Years before, on March 15, 1883, Charles S. Westcott ("Homo") says in "Forest and Stream" that the Delaware shore on the lower stretches of the river was black with "Sprigtails" which were very hard to approach on the bare marshes. The only successful method was to sink a box in the mud and wait for them. "The job," he adds, "is a dirty and a hard one and scarcely repays the trouble as the birds soon learn to avoid the place where the hide is made."

In the Barnegat Bay region Charles Urner has dates of extreme occurrence of the Pintail as follows:

1929–30.	October 12–April 13.	1932–33.	October 11–March 12.
1930–31.	September 14–March 22.	1933–34.	September 17–March 18.
1931–32.	September 27–April 17.		

Unusual dates are July 26, 1925 and April 19, 1925. It is found chiefly from September to November and in February and March with only a few during the winter.

GREEN-WINGED TEAL

Nettion carolinense (Gmelin)

The Green-winged Teal has the same distribution as the Blue-wing about Cape May. It was formerly much more plentiful than now on the Bay side meadows especially in the autumns of 1914 and 1918 when large numbers visited Pond Creek Meadows according to Raymond Otter and he states that the draining of the meadows are in all probability the cause of the disappearance of the Teal. They usually arrive as early as September and sometimes remain until January even in freezing weather. They occur occasionally on the sounds but are not at home there as they are distinctly fresh water ducks. Walker Hand reported a pair shot while sitting on the sods on the north side of the Harbor on December 12, 1910. My only spring record is of a pair seen on the shallow pond at the Lighthouse on March 25, 1928, accompanied by a pair of Mallards. When they flew each pair kept by themselves the male close behind his mate.

On October 11, 1925, after a terrific northwest gale which swept the whole Atlantic coast, I found three ducks feeding on Lake Lily at Cape May Point—a Black Duck, a female Pintail and a male Green-winged Teal. They remained in the vicinity all afternoon and I saw them on the wing several times. The Teal was always flying in the lead and if he happened to be the last to flush he had no difficulty in immediately passing his companions, thus upholding the reputation of the species as a speedy flier. Later on the same day I found another Teal associated with five Canada Geese on the Lighthouse Pond and every time they took wing he completely outflew them. Another was on this pond in company with a Blue-wing on September 25, 1927, while Ludlow Griscom found a single female there on November 11, 1921, and Julian Potter saw four on the same pond on December 23, 1928. Another was shot on Delaware Bay on December 16, 1921, and James Tanner saw five on Pond Creek Meadow on November 7, 1936.

A very exceptional occurrence was three individuals on the Lighthouse Pond on August 12, 1930, associated with Black Ducks. They fed with the latter, submerging their heads as they swam about, but when the flock took wing they at once separated from their companions and flew together neither mingling with nor following the larger birds.

The grayish color and conspicuous green speculum are the best identification characters of the Green-wing in the dull plumage in which we usually see it at Cape May.

At the Killcohook Reservation at Fort Mott, on the Delaware, a Green-wing was seen by members of the Delaware Valley Club on September 2, 1935; fifty on March 15, 1935; twenty-five hundred on November 3 and December 1, 1935; seventy-five on March 15, 1936, and forty on the 29th. At Delaware City across the river our records are: four on April 1, 1928; three hundred on April 6, 1928; forty on April 3, 1932.

In the Barnegat Bay region Charles Urner has dates of occurrence for several seasons as follows:

1929–30.	November 10–April 13.	1932–33.	October 9–April 9.
1930–31.	October 12–March 23.	1933–34.	October 8–March 18.
1931–32.	October 11–May 1.		

Extreme dates are August 26, 1934 and May 1, 1932 and the months of greatest abundance, October and April.

*EUROPEAN TEAL

Nettion crecca (Linnaeus)

Occasional European Teals have been seen in the eastern United States and one was positively identified by Charles Urner in the Barnegat region on November 12, 1933. So far as I know there is no record for Cape May although a stray individual in a flock of Green-wings would easily escape notice, indeed identification is difficult unless the bird is within reasonable distance, so that the absence of the vertical white bar in front of the wing may be noticed. The females cannot be distinguished in life.

There are other records for northern New Jersey. One seen at Boonton, N. J., on the Jersey City Reservoir, on February 27 and April 3, 1932, and one on Troy Meadows, Morris County, April 3, 16 and 30, 1932 (J. L. Edwards, Auk, 1932, p. 460).

BLUE-WINGED TEAL

Querquedula discors (Linnaeus)

This fresh water duck occurs in small numbers on the Bay side ponds during the migrations, usually in September and October and in March and April. Raymond Otter states that there were many of both kinds of Teal on the Pond Creek Meadows in the autumns of 1914 and 1918 but that today he rarely sees one. John Mecray tells me that they occur occasionally on the sounds but are very rare on the ocean side of the peninsula, while I have seen them there once or twice associating with migrating Black Ducks. On April 2, 1922, Walker Hand and I saw two in a flock of sixty Black Ducks on a moist spot on the Fill where, in winter and early spring, a considerable shallow pond used to form. The Teal were not detected until the flock took wing when they were at once recognized by their smaller size and general light gray appearance, in sharp contrast to their larger, darker associates, which seemed to be at least twice their bulk. They showed much white on the wings in flight. On March 22, 1930, another was seen on Fishing Creek near the Bay in company with six Black Ducks and a pair on April 7, 1930. Most of my records of Blue-winged Teal, however, are of birds seen on the deep pond at the Lighthouse. There was one there on September 26, 1926, swimming leisurely about close under the dense cattail growth which covers the sides of the pool. Three were there on September 27, 1928, two on the 30th and four on October 7 of the same year, while another was seen on September 25, 1927, accompanied by a single Green-winged Teal. These two birds remained, apparently all day, in a small patch of clear water in the middle of the pond which was otherwise completely covered with a dense mass of submerged vegetation. The three seen on September 27 were associated with two female Baldpates and a male Wood Duck.

Along the Delaware River, Charles S. Westcott states that they were plentiful in the eighties arriving in force October 4 to 11, 1885, and October 16, 1884.

Raymond Otter has seen Blue-winged Teal at Cape May as early as August and Henry Collins found two at Brigantine, on August 28, 1932, but these may not have been migrants as the Teal have of late years resumed nesting in portions of Maryland, Delaware and New Jersey. Directly across the bay from Cape May, at Cedar Beach, Delaware, Herbert Buckalew found a nest with eight eggs on May 11, 1933, and on May 30, 1934, Turner McMullen found another in the same locality containing ten eggs. In June, 1934, I found a pair of Blue-wings on Cedar Swamp Creek, near Petersburg, which no doubt were nesting nearby. On Riggin's Ditch, which flows into Delaware Bay near Heislerville, in Cumberland County, Julian Potter found four of the birds on May 1, 1932, twelve on July 28, fifteen on September 25, and four on October 12, of the same year. On September 2, 1935, there were eighty present and on April 19, 1936, fifty while on June 7, 1936, he found twelve as well as two females each accompanied by a brood of downy young which confirmed his suspicion that they were breeding in the immediate vicinity. These nearby nestings are no doubt responsible for the increase in Blue-winged Teal at Cape May in the past few years. In August, 1936, they appeared on Price's Pond on the 29th and on almost every day thereafter through the early autumn a flock could be flushed from the extensive cattail marsh to the east of the open water, sometimes as many as thirty would arise together and fly over to the marsh along New England Creek, usually returning when they thought that the coast was clear.

Across the river at Delaware City, counts of Blue-winged Teal by members of the Delaware Valley Ornithological Club are as follows: April 12, 1925, sixteen; April 1, 1928, twenty; April 6, 1928, two hundred; while at Pennsville, on the New Jersey side, two were seen as late as May 13, 1934, and at Fort Mott, twenty on October 6, 1929, and eighty on September 2, 1935.

In the Barnegat Bay region Charles Urner's inclusive dates for several seasons are as follows:

1929–30.   July 23–October 13 and April 13.
1930–31.   September 7–14 and April 12.
1931–32.   August 22–December 26 and April 6–May 1.
1932–33.   August 28–November 11 and February 26–April 22.
1933–34.   September 17–October 8.

His extreme dates are July 22, 1927, and May 3, 1928. He has never seen one between December 26 and February 26 and it is very rare in winter.

SHOVELLER

Spatula clypeata (Linnaeus)

The Shoveller is a very rare bird at Cape May and Raymond Otter has but a single local record, a pair shot by Edward Roth on Bennett's Creek Pond, near Erma, on November 25, 1929. The birds were identified by Walker Hand. While there have been one or two sight records they do not seem convincing. Back of Five Mile Beach a single Shoveller was shot by John Taylor in the winter of 1888–89 and Norris DeHaven shot one at Atlantic City in the same winter. We have several records of Shovellers shot on the lower Delaware near Salem where the fresh meadows would be more to their liking. One was shot by Dr. Henry R. Wharton on September 23, 1904, at Salem, twenty-five were seen at Fort Mott, on November 17, 1935, ten on December 1, 1935, and four on March 29, 1936, by Fletcher Street and Julian Potter. At Delaware City they have been observed on:

April 10, 1925, two.	March 24, 1930, nine.
April 10, 1927, two.	March 7, 1931, eight.
March 18, 1928, two.	February 22, 1932, eight.
March 28, 1928, twelve.	March 5, 1932, twelve.
April 1, 1928, twenty.	April 3, 1932, twenty.
April 6, 1928, fifty.	February 25, 1933, four.
April 27, 1929, two.	

Charles Urner's records at Barnegat Bay are mainly in the autumn. They are scattered from September 14 (in 1924) to December 26 (in 1931); also January 3, 1932, March 12, and April 9, 1933.

WOOD DUCK

Aix sponsa (Linnaeus)

The Wood Duck, as its common name Summer Duck would imply, is strictly a summer resident in the Cape May Peninsula arriving at the end of March or early April and remaining until October, our actual dates for recent years running from April 10 to October 12, while Walker Hand has told me of one shot November 7, 1930. It is very rare immediately about Cape May, occurring only as a straggler in migration about ponds at the Point. Its true home is on the wooded streams and ponds farther up in the peninsula, about Ross's Mill Pond, Price's Pond and on the headwaters of Dias, Fishing and Green Creeks.

In the latter locality Walker Hand tells me the birds were accustomed to leave the woodland between sundown and dusk and seek the open meadows, apparently to feed. They came singly, in pairs or in bunches of from three to six and had a remarkably regular line of flight, so regular in fact that gunners were able to conceal themselves in the cattails with the certainty of having the birds pass close to them. They were hunted in September and so persistently that the entire local flight was frequently killed before the birds were ready to leave for the south. The same conditions prevailed at Dennisville and South Dennis at the head of the peninsula where the birds bred in the wild recesses of the Timber and Beaver Swamp and came forth at dusk to the extensive fresh meadows leading out to the Bay. Of course the shooting has ceased since the Wood Duck has been afforded complete protection by law and it seems now to be increasing in numbers. John Mecray tells me that it occurs as an accidental straggler on the sounds but is very rare there.

While we have no record of a nest actually found in Cape May County the late William B. Crispin found one in Salem County on April 25, 1908, in the woodland where the Great Blue Herons have their rookery. It was in the hollow trunk of a black gum tree which had been broken off about forty feet from the ground and was situated about five feet from the top of the stub, the birds passing in and out through the chimney-like opening. There were sixteen eggs which were somewhat piled up on one another as there was not room for them all on the floor of the nest. Walker Hand found a pair of Wood Duck at the head of Dias Creek on April 10 some years ago which judging from the date must have had a nest nearby.

Turner McMullen has furnished me with data for a number of nests of the Wood Duck that he has examined in Salem County and a few found by Richard Miller are listed under his name.

May 16, 1920, eighteen eggs.
April 17, 1921, sixteen eggs.
May 1, 1921, eight eggs (Miller).
April 27, 1924, two eggs.
April 12, 1925, one egg.
April 19, 1925, one egg; twelve eggs; fourteen eggs.
April 10, 1927, sixteen eggs (Miller).
April 7, 1928, eleven eggs; twelve eggs.
April 14, 1928, six eggs; sixteen eggs; twenty-four eggs; twelve eggs.
April 29, 1928, thirteen eggs; thirteen eggs.

April 13, 1929, seven eggs; seventeen eggs; twenty eggs; thirteen eggs; fifteen eggs.
April 18, 1929, eighteen eggs (Miller).
April 27, 1929, ten eggs.
April 6, 1930, twenty-two eggs.
May 4, 1930, four eggs.
April 15, 1931, four eggs; nine eggs; fourteen eggs.
April 15, 1931, twenty-four eggs.
May 31, 1931, eleven eggs.
April 12, 1933, six eggs.
April 19, 1933, one egg.
April 21, 1934, eighteen eggs.

I saw a female Wood Duck flying about the shallow pond at the Lighthouse on August 27, 1925, which may indicate an early migration southward. Its uniform dark brown appearance both above and below, with its brilliant green speculum served to distinguish it. An adult male in full plumage was seen by Julian Potter on the deep pond nearby, on October 1, 1925, while I saw another beautiful drake on the same pond on September 27, 1928. This bird was in company with three Blue-winged Teal and two Baldpates, the only time that I have seen the Wood Duck associating with other species. A female was seen on this same pond on May 14, 1930, and another on October 12, of the same year.

Apparently a pair of Wood Ducks must have nested nearer to Cape May than usual in 1930 as I found a family of ten on a small pond in the meadows east of the shallow pond at the Lighthouse, where several interesting water birds have been seen, but which like all our other ponds seems doomed by the draining for mosquito control. These birds were associated with six Black Ducks. When they flew their small size and very dark brown coloration were noticeable as well as the very short, apparently slightly upcurved, bills. The whitish bar on the wings was also discernible through the glass. On September 27 ten Wood Ducks were feeding on the Lighthouse Pond, probably the same flock. As they flew directly past me their white bellies could be seen as well as the white wing bars and white on the cheeks. The general dark brown coloration, however, was the most striking character. The next day there were eleven present and, as I lay concealed in the cattails at the head of the pond, I had an even better view of the birds. All the characters mentioned were clearly seen as well as the very dark color of the breast, while the brilliant glossy plumage and long drooping crest feathers of the males were conspicuous.

On August 7, 1936, in company with Conrad Roland, I saw four young Wood Ducks accompanied by their parents on the far side of Price's Pond on the Bay side feeding close to the base of the cattails which border the water and extend for several hundred yards to the east. I had seen several Wood Ducks flying about the vicinity at dusk all through July and thereafter at about sundown we flushed from eight to twelve of the birds from the pond, probably an earlier brood. Even in the gathering dusk when colors could not be readily distinguished we could note the much smaller size of the birds as compared with the Black Ducks that were present as well as the light terminal border to the secondary feathers and, when they came very close, the white bellies. The wet woodland immediately behind the marsh doubtless furnished satisfactory hollow trees for their nesting. At Delaware City, Wood Ducks have been seen frequently by members of the Delaware Valley Club from April 3 to 30 of nearly every year, from two to six at a time.

In the Barnegat Bay region Charles Urner tells me that the Wood Duck nests and doubtless occurs in winter although he has no January record. His observations run from February 26 to December 27.

Alexander Wilson gives us an interesting account of a visit to a Wood Duck's nest over one hundred years ago: "On the eighteenth of May I visited a tree containing the nest of a Summer Duck, on the banks of the Tuckahoe river, New Jersey. It was an old grotesque white oak, whose top had been torn off by a storm. It stood on the declivity of the bank, about twenty yards from the water. In this hollow and broken top, and about six feet down, on the soft decayed wood, lay thirteen eggs, snugly covered with down doubtless taken from the breast of the bird. On breaking one of them the young bird was found to be nearly hatched, but dead, as neither of the parents had been observed about the tree during the three or four days preceding; they were conjectured to have been shot. The tree had been occupied probably by the same pair, for four successive years in breeding time." (American Ornithology, VIII, p. 97, 1814.)

REDHEAD

Nyroca americana (Eyton)

The Redhead is of irregular and rather infrequent occurrence about Cape May. There are occasional flocks on the sounds or some may associate with the Broadbills that are wintering there, while stragglers on migration may be seen occasionally on the Lighthouse Pond or similar small bodies of water. Walker Hand told me of a flock on the sounds, December 20–23, 1919, from which a number were shot and Raymond Otter reports several flocks present during December and January in the winter of 1922–23. Since then only single birds or small bunches have been seen, and in the past six years only two have been shot. In November, 1928, he and Cecil McCullough shot three on the Lighthouse Pond, and I saw single birds there on October 27, 1928, and October 10, 1929.

Julian Potter saw a flock of twenty at Corson's Inlet on November 28, 1926. The scarcity of Redheads as well as several other ducks at Cape May is attributed to the lack of suitable food.

In the Barnegat Bay area Charles Urner tells me that there has been a marked decrease in the number of this duck since 1928. His records of occurrence in subsequent winters follow:

1929–30. November 10–February 9.	1932–33. October 16–March 12.
1930–31. November 9–February 22.	1933–34. November 12–April 8.
1931–32. November 1–February 14.	

Extreme dates are October 3, 1928, and May 22, 1926.

RING-NECKED DUCK

Nyroca collaris (Donovan)

Raymond Otter tells me of a pair of these ducks shot near Jenkins Sound in the autumn of 1935, by Charles Bellangey. So far as I can ascertain this is the first record for the Cape May district. The heads were examined by Otter and positively identified. On October 25, 1936, James Tanner saw a single female on Lake Lily at the Point. At Delaware City, the Ringnecks seem to be much more plentiful and the Club has records of twenty-nine on March 28, 1928; fifty-four on April 3, 1932; fifty on March 24; eighty on April 13, 1935.

In the Barnegat Bay region Charles Urner states that they are increasing in numbers. He has records of occurrence as follows:

1929–30. November 25.	1932–33. October 30–April 1.
1930–31. December 19.	1933–34. November 5–March 18.
1931–32. November 1–December 26 (thirty-one seen).	

Extreme dates are October 12, 1928 and April 1, 1933.

CANVAS-BACK

Nyroca valisineria (Wilson)

The Canvas-back is even rarer than the Redhead at Cape May due to the lack of fresh water and of its essential food the wild celery (*Valisneria*). Walker Hand told me that an occasional one was shot among the Broadbills on the sounds and had a definite record for December 5, 1929; old time gunners speak of birds having been killed "once in a while." Raymond Otter has heard of but three in his time, two killed by Morris Brown on the Harbor in the winter of 1932–33 and the other shot by himself on Cape Island Creek during a freeze in the following winter. Elsewhere along the coast Fletcher Street has reported twelve back of Ludlam's Beach on November 5, 1926, and there is a record of two at Delaware City on April 10, 1925. In the Barnegat Bay area, Charles Urner has recorded them in:

1929–30.	November 10–February 9.	1932–33.	November 6–January 23.
1930–31.	December 7–February 23.	1933–34.	November 5–April 8.
1931–32.	November 1–February 14.		

Extreme dates are October 5, 1928, and April 25, 1926, most occurrences have been in December and January often only a few and seldom in very large flocks although Dr. J. K. English reported two thousand present far out on the bay on January 24, 1927, in company with Canada Geese and large numbers of Scaup. Alexander Wilson, who first described and named the Canvas-back tells us that in the autumn, about 1812, a vessel loaded with wheat was wrecked at the entrance to Great Egg Harbor and went to pieces, the grain flowing out over the water. This attracted great quantities of Canvas-back Ducks, a species which was quite unknown to the native gunners who termed them Sea Ducks and gathered in numbers to shoot them. The birds remained for about three weeks and the slaughter was enormous. "During the greater part of the time a continual cannonading was heard from every quarter," and 240 were killed in a single day. The Canvas-back has now become so scarce that its shooting was stopped in 1936.

GREATER SCAUP DUCK

Nyroca marila (Linnaeus)

LESSER SCAUP DUCK

Nyroca affinis (Eyton)

While the local gunners distinguish between the two species of Scaup Ducks when shot, they refer to them collectively as "Broadbills" and it is hard to obtain separate information on their relative abundance or times of occurrence. Indeed, unless very close at hand, it is difficult to distinguish them in life. Walker Hand told me that the sounds were their center of abundance during the winter months although there were also large flocks on the Bay. Each sound used to have its own flock which stayed there throughout the season unless the weather became excessively severe and only rarely would a sound's flock go out to sea. Raymond Otter and John Mecray both tell me that the Broadbills are much scarcer than they were ten or fifteen years ago and they attribute this to the custom of baiting the waters of Barnegat Bay to the north with grain, in order to hold the flocks there and draw them back again when freezing temperatures drive them south to the Cape May sounds to find open water. But the main reason seems to have been the decrease in breeding due to the drought. In the migrations small bunches of Scaup will be found on the Harbor and scattered individuals on the ponds by the Lighthouse or on small bodies of water elsewhere. While most abundant on the sounds during December, January and early February, I have records elsewhere as early as October 25 and until April 8 in the spring. Alexander Wilson in his "Ornithology" (1814) says that these ducks are better known by the name of the "Blue-bill" and are "common both to our fresh water rivers and seashores in winter. It is sometimes abundant in the Delaware, particularly in those places where small snails, its favorite shellfish, abound; feeding also, like most of its tribe, by moonlight. They generally leave us in April, though I have met with individuals so late as the middle of May, among the salt marshes of New Jersey."

Walker Hand has told me that the causeways and bridges across the meadows that have been built in recent years to connect the resorts on the ocean beaches with the mainland, have disturbed these birds as well as the Old-squaws not a little as they do not like to pass them and consequently do not use the waterways of the meadows as freely as in times past.

The Broadbills can easily be recognized, as they fly overhead, by their small size, as compared with the Black Duck, the shorter neck, dark breast and white belly, and by the constant flicker of white from their rapidly beating wings. A flock that passed high in air near Egg Island at the head of Delaware Bay, in February 1928, was particularly memorable as it came by just before sunset and the horizontal rays of light brought out with remarkable distinctness the colors of the birds and after they had reached a distance too great to distinguish the individuals, even with the glass, the twinkle of the pulsing wings was still visible.

The Broadbills are very hardy birds and endure our coldest winters. On December 21, 1919, Julian Potter found twelve resting on the ice on Barnegat Bay along with some Black Ducks and two hundred Herring Gulls and on February 8, 1935, after a temperature of − 6°, there were one hundred on the ice-bound waters of the Bay.

Christmas census records of the Delaware Valley Ornithological Club are as follows:

December 26, 1932, thirty-one.	December 23, 1934, fifty.
December 24, 1933, two thousand.	December 22, 1935, twelve.

In the Barnegat area Charles Urner has records of occurrence as follows:

1929–30.	September 7–April 13.	1932–33.	October 16–April 23.
1930–31.	October 12–April 12.	1933–34.	October 8–April 8.
1931–32.	October 25–April 17.		

It is most abundant there from November to February but he has seen scattering birds in every month of the year.

Mr. Urner makes no attempt to separate sight records of the two species of Scaup and I quite agree with him that only in exceptional instances is it possible to do so with any pretense to scientific accuracy. The Greater Scaup differs from the Lesser in its somewhat larger size; in the greenish instead of purplish reflections on the top and sides of the head; and in the faint, instead of strongly marked, dark vermiculations on the sides of the body. The females differ only in size.

In large collections that I have examined from along the Atlantic coast, from New Jersey to North Carolina, the Lesser Scaup has always been the more numerous.

GOLDEN-EYE

Glaucionetta clangula americana (Bonaparte)

The Golden-eye, or "Whistler" of the local gunners, is distinctly a winter duck and restricted to the sounds and adjacent creeks on the ocean side of the peninsula, where it occurs every winter in considerable flocks, and to the open waters of the Bay. It is very wary and exceedingly hardy. Walker Hand has told me that they can live perfectly well about an airhole in the ice when the freezing has driven other species to seek elsewhere for open water. Raymond Otter considers that the Whistler along with the Shelldrake (Red-breasted Merganser) are the most common of the diving ducks today. They arrive as soon as winter sets in, about December 1, and remain until March. The Christmas census of the Delaware Valley Ornithological Club shows the following counts although the sounds were not examined:

December 26, 1932, twenty-two.
December 24, 1933, sixty-nine.

December 23, 1934, forty-five.
December 22, 1935, eleven.

and a bunch of twelve was seen on Richardson's Sound as late as March 23, 1930.

On Great Sound, where there is a stretch of water two miles across, the Whistlers find congenial quarters and I have seen a flock of over one hundred passing from there out to the ocean on February 22, 1928, crossing Seven Mile Beach at a considerable elevation. They came in detached bunches and were passing for several minutes, their brilliantly contrasting plumage— the glossy black heads, white bodies and black and white wings of the drakes, and the brown and white of the ducks, shown to perfection as they flew over our heads in the sunlight. In the Barnegat region Charles Urner's records for several winters are as follows:

1929–30. November 10–April 13.	1932–33. November 6–April 23.
1930–31. November 9–May 12.	1933–34. November 5–May 16.
1931–32. November 8–April 17.	

Extreme dates November 3, 1925, and May 30, 1929, with one on April 30, 1927; greatest abundance from December to February.

J. F. S.

BUFFLE-HEAD

Charitonetta albeola (Linnaeus)

The "Butter-ball," as it is locally called, is a rare duck at Cape May to-day but according to Raymond Otter was quite common twenty years ago when it came in November and remained regularly until March, frequenting the sounds and to a less extent the interior ponds. He tells me that it decoys very easily and being a fine duck for the table its numbers were depleted until it is today threatened with extermination unless given the absolute protection that it requires.

Walker Hand has told me that it frequented the salt creeks and ponds along the edges of the larger sounds and I have found it more commonly back of Seven Mile Beach than elsewhere. My records run from November 28 to the latter part of March but in the past four years I have not seen any.

Julian Potter found a flock of twenty-three back of Ludlam's Beach on January 8, 1928, and in 1932 I saw small flocks of as many as eight from March 5 to 26, on every visit to the meadows back of Seven Mile Beach.

A drake Butter-ball, evidently a disabled individual, was seen by Richard Miller on the Lighthouse Pond on June 8, 1919.

At Barnegat Bay and vicinity Charles Urner's dates of occurrence are:

1929–30.	November 10–April 13.	1932–33.	October 30–April 9.
1930–31.	November 9–March 23.	1933–34.	November 5–April 8.
1931–32.	November 8–April 17.		

Extreme dates are October 30, 1932 and April 30, 1927; most abundant from November to March, with peak in December. Once seen in August.

The Butter-ball is another species which has decreased in numbers to such an extent, due to the droughts in its breeding area, that it was placed on the protected list in 1932.

OLD-SQUAW

Clangula hyemalis (Linnaeus)

The "South-southerly," as the Old-squaw is called at the Cape and elsewhere along the New Jersey coast, is strictly a bird of the sounds and the sea although many of them round the Point and may be found regularly in Delaware Bay. Walker Hand has told me that they have decreased very greatly in numbers since his earliest recollections and John Mecray says the same thing. Hand stated that these ducks are accustomed to winter on the thoroughfares and that they persisted in flying against and wind and would go deliberately past a gunner rather than leave a waterway and cross the meadows. The result was that they were easily killed by anyone who persistently pursued them and large numbers of them were shot even though they are not generally regarded as edible. This practice has apparently been abandoned for the most part today as Raymond Otter tells me that they are still here in numbers.

In recent years members of the Delaware Valley Ornithological Club have seen them not infrequently during the winter, resting on the waters of the Harbor, or off the end of the jetty, flying along the coast just beyond the surf line. Usually there are only two or three birds together but on several occasions bunches of five or six have been seen and on December 26, 1926, Julian Potter saw a flock of fifty-four while seventy-five were seen to pass along the coast of Seven Mile Beach on the morning of February 22, 1928, in several detachments.

On January 3, 1926, I found a flock of five on the Harbor close to the southern shore, where they could be studied from the shelter of the marsh elder bushes. They were swimming about and diving repeatedly. There were two males and three females, the more brilliant coloration of the former making them conspicuous; their black ear patches, jet black wings, and white flanks, combined with the gray and white back and mottled neck, were prominent features. The long tail feathers held clear of the water at an ascending angle were also characteristic. They kept close together and whenever one emerged from a dive it at once swam over to the others if they happened to have gone on a little in advance. On the wing the large amount of white in the plumage of the Old-squaw will distinguish it from any other duck. The peculiar call, uttered both on the wing and while at rest, can be heard for at least a mile when a number of birds are calling together—*ow-owdle-ow, ow-ow-owdle-ow.*

Our records of Old-squaws in recent years run from October 28, to April 8, with three exceptional occurrences probably of birds that were non-breeders or had been injured and prevented from migrating. One of these I saw near Brigantine, farther up the coast, on July 17, 1921; another was seen by Julian Potter on June 18, 1923, flying off the jetty; and one by Brooke Worth on Seven Mile Beach on June 10, 1928. The first was resting under an overhanging mud bank and was a male in full nuptial plumage, very different from the white and black bird of winter. He squatted at the edge of the water with his long tail elevated against the bank. Every moment or two he would switch his tail to the right or left as our boat was propelled toward him and he finally waddled off his perch and took to the water. He swam easily and held his tail up at an angle above the surface, continuing to switch it from side to side at intervals. Soon he dived and emerged well out in the channel, but showed no inclination to fly, which led us to suppose that he had suffered some wing injury. He was well camouflaged as he sat against the black mud bank. As he faced us the large white areas on the sides of his head were conspicuous, appearing like immense spectacles, separated by a narrow black line down the crown. The bill was dull pinkish red and black and the long spine-like tail feathers jet black. The black breast and black and brown of the upper surface matched exactly the dark mud and the white areas looked like the white shells sticking there. Doubtless he had been passed by, many times, unseen. Julian Potter saw another summer Old-squaw off the jetty on June 18, 1923.

Our earliest record in the autumn is of four that I saw on the Harbor, October 28, 1928, and the latest, six at Corson's Inlet, seen by Julian Potter on April 8 of the same year. As the former, still in winter plumage, took wing all together it was possible to see the coloration from above and the black running down over the rump and out on the wings gave the impression of a nearly black bird with a white head, quite different from the impression that we get when viewing them from below.

Alexander Wilson writing of this duck a century and a quarter ago says: "On the coast of New Jersey they are usually called Old Wives. They are chiefly salt water ducks, and seldom ramble far from the sea. They inhabit our bays and coasts during the winter only; are rarely found in the marshes, but keep in the channel, diving for small shellfish, which are their principal food. In passing to and from the bays, sometimes in vast flocks, particularly towards evening, their loud and confused noise may be heard in calm weather at the distance of several miles. They fly very swiftly, take short excursions, and are lively restless birds."

The Old-squaws are most abundant from the end of November until February. They are probably as well fitted to withstand our coldest winters as any of our winter visitants and Julian Potter saw two sunning themselves on a small iceberg floating in Absecon Inlet during the phenominal cold of February 1934, when the temperature went to $-11°$.

The counts of the Delaware Valley Ornithological Club on the Christmas census for the past several years are:

December 27, 1931, sixteen.	December 23, 1934, eighty-four.
December 26, 1932, seventy-five.	December 22, 1935, sixty-four.
December 24, 1933, 104.	

This however covers only some of the smaller sounds and the ocean front.

In the Barnegat Bay area the records of Charles Urner give the following range:

1929–30.	November 10–April 13.	1932–33.	November 6–April 9.
1930–31.	November 9–May 25.	1933–34.	November 5–April 8.
1931–32.	November 8–March 26.		

Extreme dates October 14, 1928, and May 26, 1928; most plentiful from December to February; peak of abundance January.

AMERICAN EIDER

Somateria mollissima dresseri (Sharpe)

Cape May gunners are not familiar with Eider Ducks of either species and neither Walker Hand, Raymond Otter nor John Mecray ever saw one here. Turnbull in his beautiful little brochure on the "Birds of East Pennsylvania and New Jersey," published in 1869, states that both the American Eider and the King Eider have been seen occasionally in severe winters at Great Egg Harbor.

The only records that we have were made by the Audubon Association wardens stationed at the Point. A female or immature male was seen by William Rusling on the ocean off Cape May Point near the jetties in front of the Villa Maria. It remained there from October 10 to 14, 1935, and he watched it several times a day for four days having it in good view, once within twenty-five feet. When caught by the tide and carried out from shore it immediately swam back to the jetties. He identified it as an American Eider.

Three were seen off the end of the jetty at the entrance to the Harbor, on November 3, 1936, by James Tanner.

In the Barnegat Bay region several of these ducks have been seen. Charles Urner sends me the following list: One on November 3, 1930 (Harry Ridgway); a drake, January 11, 1931; one on December 27, 1931 (Lester Walsh and Charles Nichols); two on December 17, 1932 (Oscar

Eyre); one drake, on December 26, 1932 (Urner and others); one young drake, February 18, 1934 (Urner).

Our earliest Cape May record is Alexander Wilson's (1814) that "they are occasionally seen in winter as far south as the capes of Delaware." and John Krider in his "Forty Years Notes" (1879) says that he met with it only once when he obtained four full plumaged birds at Barnegat in February.

KING EIDER

Somateria spectabilis (Linnaeus)

Turnbull in recording this Eider from Great Egg Harbor (Birds of East Pennsylvania and New Jersey, 1869) states that such specimens as have been secured were generally young birds. We have no recent records for Cape May County but a specimen in the collection of the Academy of Natural Sciences of Philadelphia was shot on Great Bay on November 29, 1928, by W. A. Barrows, III. It was a young male, just beginning to molt into adult plumage. Another specimen, an immature female, was secured by L. I. Smith, Jr., on the lower Delaware River on December 4, 1900, and is now in the Club collection at the Academy.

Charles Urner sends me the following list of individuals observed in the Barnegat Bay region by him or his associates: One December 15, 1924, (Watson); one March 3, 1929 (Jacques); two November 9, 1930; one January 14, 1934; two February 4, 1934; four February 18, 1934. The 1934 birds were seen immediately following the extremely cold weather of that winter which was undoubtedly the cause of their southward wandering.

HARLEQUIN

Histrionicus histrionicus histrionicus (Linnaeus)

This is another rare visitant to the New Jersey seashore in severe winters and is so recorded by Turnbull (1869) but without any further details. No doubt his record was based on a Great Egg Harbor occurrence. Of late years we have no record for Cape May and only a few for the New Jersey shore. A single individual was seen by Charles Urner at Barnegat Bay on February 18, 1934, and two on February 25 following. One was also seen on the latter day by J. L. Edwards and others and one on March 4, 1934, by J. L. Kurzi. Just how many birds these observations covered it is impossible to say but evidently a few Harlequins were driven south by the extreme weather that prevailed in early February of 1934.

LABRADOR DUCK

Camptorhynchus labradorius (Gmelin)

This extinct duck seems to have been a rare bird even in the time of Wilson and Audubon and we have practically no information as to its habits nor any definite records of its occurrence on the New Jersey shore. Wilson says that "it is called by some gunners the sand shoal duck from its habit of frequenting sand bars and that early in the month of March a few are observed in our [Philadelphia] market"; he apparently never met with it in life. Turnbull (1869) says "rare, a few are seen every season." He too probably had no personal experience with it and Audubon apparently knew nothing personally of its occurrence on the New Jersey coast, although he says that it occurs there "in greater or less numbers every year." He adds "A bird-stuffer whom I knew at Camden had many fine specimens, all of which he had procured by baiting fish-hooks with a common mussel, on a 'trot-line' sunk a few feet beneath the surface, but on which he never found one alive, on account of the manner in which these Ducks dive and flounder when securely hooked. All the specimens which I saw with this person, male and female, were in perfect plumage."

The last specimen of the Labrador Duck was shot in the autumn of 1875 by J. G. Bell on Long Island and is now in the U. S. National Museum collection. A female and immature male in the collection of the Academy of Natural Sciences were probably obtained on the New Jersey coast, perhaps in the fifties, as they resemble in preparation other specimens from the collection of John Krider, but this is merely supposition. An adult male in the same collection I found in the private collection of the late George W. Carpenter, formerly of Mount Airy, Philadelphia, where he maintained a small museum on his estate. This bird also may well have come from the New Jersey coast but no data whatever were attached to it (cf. Auk, 1893, p. 363).

These early records must almost certainly have referred to the Great Egg Harbor or Cape May regions as those are the localities where Philadelphia gunners of that day did their shooting.

WHITE-WINGED SCOTER

Melanitta deglandi (Bonaparte)

This is the largest of the Scoters and the one most easily recognized at a distance on account of the flicker of white on the wings caused by the pure white speculum. The knob at the base of the bill of the adult male White-winged Scoter is black and the fore part of the upper mandible orange or purplish red. While seen regularly off the beaches it is probably not so abundant as the Surf Scoter or it may occur in greater numbers farther out at sea. The presence of favorite food doubtless has much to do with the distribution of these ducks and the destruction of mussel beds causes them to disappear from areas where they were formerly of regular occurrence.

While we have records of long flights of White-winged Scoters following the coast line during the migrations we have not seen many large rafts of them near shore. One of the most notable occurrences of this kind was observed by Julian Potter on April 6, 1930, when he found an area estimated to cover several square miles of Delaware Bay literally covered with them. The passage of an airplane caused them to arise all together and the water seemed fairly to explode with ducks. There was another enormous gathering of White-wings off of Gull Bar on October 8, 1932.

As in the case of the Surf Scoter we have records of individual birds in midsummer: two off the bar at Anglesea on June 20, 1923, and one on July 24, 1924, flying up the coast; also one perched among the grass on the edge of the meadow on the Harbor on July 8, 1933, which took to the water as we approached in a boat and immediately dived some forty feet and then swam away. Another was seen at the same spot on July 19, 1930.

Charles Urner's records of the occurrence of the White-winged Scoter in the Barnegat Bay region are as follows:

1929–30.	October 12–April 13.	1932–33.	October 16–March 26.
1930–31.	October 12–March 23.	1933–34.	November 5–April 8.
1931–32.	November 15–February 14.		

He has records for every month except July but the time of greatest abundance is from October to April.

SURF SCOTER

Melanitta perspicillata (Linnaeus)

Scoters or "Coots," as they are called by the local gunners and fishermen, are preëminently the ducks of the sea, and rarely do we see them close to Cape May either on the Harbor or on the thoroughfares. They do occur plentifully on Delaware Bay but with its stretch of twenty miles in its lower reaches it is really but an arm of the ocean. Raymond Otter tells me that on certain sounds directly connected with an inlet he has found flocks of this species, the "common Gray Coot," and more rarely the other species. All three Scoters—American, Surf and White-winged, occur off the Cape May coast during the winter and in the seasons of spring and autumn migration, but so closely do they agree in habits, and so difficult is it to distinguish them when flying far offshore, that it is almost impossible to differentiate the several species so far as peculiar habits, relative abundance, or actual time of occurrence are concerned, and during their presence on our coast they often mingle more or less. Such general remarks as we have to make regarding them refer as much to one species as to another. At the times of greatest abundance, from mid-March through April and from mid-September through October, we see Scoters from the beach passing constantly far out over the ocean, in long jet black lines, conspicuous against either sky or water. They literally "stream" along over the surface like slender wisps of cloud or mist, drifting with the wind. Now one of these whisps seems to swell out in the middle as the birds gather more closely together, and then it thins out and lengthens, then once again the congestion develops at the head or rear of the column. Now there are little knots formed at several points along the stream or perhaps it breaks up into small "clouds" which later drift together and form again the long slender line. The formation is ever changing but the streams of birds are always pushing steadily ahead as if driven by some unseen power behind them. Now and then a flock will rise twenty feet or more above the surface of the water and then drift down again low over the waves.

With the aid of glasses we are able to resolve the moving streams into individual birds and we see them in long lines, like strings of black pearls, each bird following close after the one ahead, then the rear guard presses forward and there are several parallel lines at the head of the column. As we see these from the side they appear like a single congested mass of birds while to the naked eye the impression is often of a great grotesque bird leading a string of little ones. These strings of flying Scoters may be in view all day long passing at frequent intervals out on the horizon with smaller detached strings nearer to the shore. The flocks may vary from half a dozen to twenty-five or from fifty to several hundred but the birds always fly in strings bill to tail in a regular follow-the-leader formation. It is very difficult to estimate the number of Scoters that are passing offshore and an actual count from the beach is impossible. Julian Potter states that he was sure that he saw 5000 passing south during part of the day on October 11, 1925, while on November 19 there were thousands. I have seen them resting on the sea when they covered very large areas reaching away to both right and left but I could not guess at their numbers. In these assemblages probably all three species were represented.

Scoters have relatively short necks and hold them straight out in front when in flight. Their wings are pointed and appear rather narrow while the wingstrokes are exceedingly rapid—so rapid in fact as often to produce the curious optical illusion of two pairs of wings to a bird instead of one. The head is rather large and the bill more or less swollen at the base; in the adult male Surf Scoter, the upper mandible is red above and white on the sides at the base, but this can only be seen when the birds are reasonably close at hand; and the pure white patches on the forehead and nape are the best identification marks.

It is local tradition at Cape May that the "Coots" always fly south no matter what the season, and up the Bay but never down. While this is not absolutely true, it nevertheless is the case in a vast majority of instances and Walker Hand, who gave the matter much study, was of the opinion that ninety percent of the Scoters that we see are flying in this direction. While he cruised pretty well all over the Bay and saw thousands of Scoters going up as far as Egg Island he saw scarcely any travelling back again. Whether they go by night or overland, which is very unlikely, and whether their regular northward line of migration lies farther out to sea we have not been able to determine. On April 3, 1922, to cite one instance out of many, there were thousands of scoters streaming down the coast and up the Bay, as far as I could see from Cape May Point, but only a few scattered individuals going in the opposite direction. While we see them more often on the wing, these ducks rest frequently on the ocean especially during migrations and great rafts of them fairly cover the water for considerable areas. Sometimes they will be asleep with their heads tucked over under the feathers of the back, while at others they are busy pluming themselves, rising up in the water with rapidly flapping wings and settling again or rearing up, as many ducks do, without using the wings, as if treading water. These actions are individual and occur here and there throughout the flock producing the appearance of constant motion.

We have records of Surf Scoters as early as September 4, 1926, a string of seven and a pair flying north; September 5, 1926, a flock, and immense numbers all through the autumn thereafter. There are several May records, one as late as the 30th, as well as seventy-five birds in a close flock that I saw back of Little Beach Island farther up the coast. Our midsummer records, apparently non-breeding birds that did not go north, are the following: June 18, 1923, ten flying off the mouth of the Harbor, and an adult male on the channel back of Two Mile Beach; on July 6, 1922, one beyond the surf at South Cape May which dived on our approach but did not fly; July 11, 1931, an adult on Jarvis Sound; July 25, 1930, a flock of eleven seen by Julian Potter; July 27, 1922, one on the Bay at Town Bank; July 31, 1922, one on the Bay above Cape May Point, which swam near shore with tail erect and reared up several times to plume itself; August 1, 1931, two off the beach; August 6, 1927, three flying north.

While all three species of Scoter have been recorded on each of the several Christmas censuses of the Delaware Valley Ornithological Club for the Cape May district, they have usually been in comparatively small flocks or individuals, and large flocks, if seen at all, have been well out on Delaware Bay. This emphasizes the fact that these birds are far more abundant during the migrations than as winter residents.

Charles Urner tells me that the Surf Scoter is the most plentiful species on Barnegat Bay and on Great Bay as well, where it feeds and lives throughout the winter. He thinks that the other species may be more abundant farther out to sea. This corresponds with my experience in the Cape May area but I have not felt that my estimate of numbers was sufficiently accurate to be depended upon.

In the Barnegat Bay region his records show this Scoter to be present during recent years between the following dates:

1929–30.	October 12–April 13.	1932–33.	October 16–April 9.
1930–31.	September 7–April 12.	1933–34.	October 8–April 8.
1931–32.	November 16–March 16.		

It occurs commonly from October to January but has been observed in every month in the year except July.

The Scoters are perfectly adapted to a life on the ocean and probably seldom if ever voluntarily come on shore during their long sojourn with us, feeding, sleeping and pluming themselves as they float on the waves, as truly sea birds as the winter loons.

AMERICAN SCOTER

Oidemia americana (Swainson)

While the great migratory streams of Scoters seen well offshore are doubtless largely made up of this species it is only when flocks of them come in close to the beach that we can with certainty distinguish them from the Surf Scoter and even under such circumstances the brownish females and young of the two species cannot be satisfactorily separated. The plumage of the old male is entirely black lacking the white head markings of the Surf Scoter while the swollen base of the bill is yellow or orange.

On August 1, 1925, an unusual time of year for such an occurrence, I saw a raft of over one hundred American Scoters sleeping on the ocean off the lower part of Seven Mile Beach, rising and falling with the waves like so many corks, now raised into prominent view, now sinking into the trough of the sea. Two others were seen at the same time on the thoroughfare nearby accompanying two Red-breasted Mergansers, and Julian Potter saw a summering individual at Cape May on June 18, 1923.

On April 1, 1922, I watched a flock of a dozen immediately off the pier which were swimming leisurely about, taking wing for short distances and settling again at once. Then on April 29, 1923, two jet black males and two females, with their dull brownish plumage and whitish ear patches, were found on the sea just beyond the surf line. They rode the waves beautifully, disappearing in the trough of the sea and floating up over the next wave crest. Now and then with a heave of the shoulders and a ducking of the head they would dive. One male rolled over on his side and plumed himself for some time, holding one black foot aloft like a balancer. Once he rose up and flapped his wings while several times he stood up vertically on his tail, as it were, with no action of the wings whatever, apparently looking for his mate. Then swimming close to her he would spread his tail feathers until they stood erect like a semicircular crest of bristles. After diving, too, he would shake the tail vigorously from side to side.

(245)

On April 2, 1924, there were five swimming and diving just off Poverty Beach. They usually went under the water all together but when one of them became separated and realized that the others were under he dived at once. They bobbed lightly about in the rough sea, now in sight now concealed, and then in a moment down they would go apparently for some choice food on the bottom. When a flock rises from the water and starts on a flight the birds move from their more or less scattered positions into a regular line, those farthest away rapidly catching up with the others and taking their places, until the regular formation is attained with each bird the same distance from his neighbor.

On April 4, 1930, Julian Potter noticed a flock of ten American Scoters about twenty-five yards off the Bay shore and heard them calling, a rather prolonged whistle much like the note of the Piping Plover; both male and female were heard making the call. On April 12, following, I found a pair on the ocean beach near the Lighthouse, the male standing up while the female squatted on the sand. When I approached, instead of flying, as I had expected they would, they ran for some distance to the water and swam out of reach. These birds were apparently sick from the oil which they had encountered on the water, for, a little later, I found six dead birds on the Bay shore and eight still alive but evidently far from well. They made their way with difficulty to the water, some of them shaking their tails from side to side and all pluming their breast feathers, which were apparently caked more or less with oil. We have only one record of a bird of this species on the Harbor, a single one seen May 16, 1924, and only the two instances already mentioned of the occurrence in summer of non-breeding individuals.

Charles Urner's records of the occurrence of the American Scoter on the coast in the vicinity of Barnegat Bay for several seasons are as follows:

1929–30.	October 12–January 12.	1932–33.	September 17–April 9.
1930–31.	October 12–March 23.	1933–34.	September 8–March 18.
1931–32.	September 27–April 17.		

He has records for every month in the year but it is most abundant in October and November and in January and April.

RUDDY DUCK

Erismatura jamaicensis rubida (Wilson)

The Ruddy Duck seems to be a decidedly rare bird about Cape May and for some years past we have no records at all. All of our observations have been made on the fresh water ponds about Cape May Point and, so far as I know, none has ever been seen on the sounds.

On October 1, 1923, a brown individual with dusky crown, evidently a female or immature male, was swimming about on Lake Lily with four Pied-billed Grebes. It dived occasionally and plumed itself reaching far back over the shoulder to the under tail-coverts, which, exposed by the operation, flashed out pure white. While swimming the duck's short spiny tail was spread and held up vertically. On October 7, following, Julian Potter saw one in the same place, probably the same bird. On March 20, 1924, there was one on the flooded Pond Creek Meadows, apparently a female, swimming about among the submerged grass, and on November 16, of that year, Julian Potter saw two. Another individual seen on Lake Lily on April 2, 1921, was in about the same plumage as the one above described and went through the same actions. It also bent its head over its shoulder and apparently rubbed its cheeks against the scapular feathers. It scratched its head with its foot and once raised itself in the water and flapped its wings vigorously. Potter saw another here on October 21, 1923, and two back of Ludlam's Beach on December 4, 1921, our latest date.

Raymond Otter tells me that he has never secured more than five Ruddies in all his experience in duck hunting at the Cape. Two on a pond back of South Cape May in November, 1927, one on the Pond Creek Meadows in the same month and two on the Lighthouse Pond in 1928.

Julian Potter and Edward Weyl saw a single bird at Fort Mott on October 6, 1929, and Fletcher Street another on November 2, 1935.   In the Barnegat region Charles Urner's records are as follows:

1929–30.	November 10.	1932–33.	October 9–January 22.
1930–31.	November 9–January 11.	1933–34.	November 5–December
1931–32.	November 8–January 17.	10.	

His extreme dates are October 3, 1928, and January 22, 1933, with a spring flight March 10 to May 30, 1929, and twenty-five on March 12, 1933.

The Ruddy Ducks that we see in New Jersey seem to be stragglers from the main flight which, as in the case of many other species which breed in the Northwest, migrates in a southeasterly direction reaching the coast, for the most part, south of Delaware Bay.   Alexander Wilson who first described the Ruddy Duck regarded it as extremely rare.   George Ord however reports a number on the Delaware below Philadelphia in October 1814 and 1818 and one in April 1819.   C. S. Westcott ("Homo") writing in "Forest and Stream" says that there was a great flight about October 17, 1884, after a severe storm and Howell's Cove, below Gloucester, was full of them but they were in such poor condition that he thought it a shame to shoot them.   They continued plentiful all month "beyond anything seen for years" the "stupid little Stiff-tails" being everywhere abundant and many were plucked and palmed off as teal in the Philadelphia markets. While never regularly common on our coast the Ruddy Duck has become very rare everywhere since the disastrous droughts on its breeding ground, so rare in fact that its shooting was prohibited in 1932.   While practically all of our observations have been of immature birds, Julian Potter saw an adult male at Camden on October 9, 1932.

J. F. S.

BUFFLE-HEAD

HOODED MERGANSER

Lophodytes cucullatus (Linnaeus)

The Hooded Merganser, or "Hairy-head" as it is known to the gunners of the Cape, is a frequenter of the salt creeks and ponds of the meadows, especially on Old Man's Creek back of Seven Mile Beach, where gunners in the past could almost always find this duck in season. It occurred also on the Bay side meadows and occasionally on small ponds elsewhere. They have decreased very much in recent years but Raymond Otter still finds some on the sounds from October until early spring. As in the Shelldrake, the females and immature birds outnumber the old males.

On November 7, 1931, I found a pair swimming on Lake Lily at Cape May Point, their crest feathers appearing very much plastered down when they emerged from their dives. On December 23, 1934, I found five in the thoroughfare back of Seven Mile Beach opposite Stone Harbor. One was shot at Corson's Inlet by W. G. Caruthers on January 1, 1920.

Charles Urner has found this duck in the Barnegat Bay region as follows with one on August 1, 1936:

1929–30. —— to March 9, 1930.	1932–33. November 20–March 26.
1930–31. December 21–May 12.	1933–34. November 12–December
1931–32. October 11–March 26.	24 (thirty-three).

AMERICAN MERGANSER
Mergus merganser americanus (Cassin)

This, the largest of the "Shelldrakes," is distinctly a fresh water bird and is therefore not nearly so common as the Red-breasted species about Cape May. It occurs occasionally on the sounds and on the upper portion of Delaware Bay. Arriving from the north in November it may be found in December and January, while stragglers occur both earlier and later. Wilfred Swain shot one on November 18, 1933, and Raymond Otter got two in December, 1935. Others have been seen on April 13, 1919, by Walker Hand and on May 2, 1926, by William Yoder, while Norris DeHaven got one as late as March 7, 1896, back of Atlantic City. Dr. Henry R. Wharton shot one near Salem on the Bay side on January 4, 1915, and ten were seen on April 3, 1932, at Delaware City, a region where they are naturally more plentiful. On Barnegat Bay, Charles Urner has recorded the presence of the American Merganser for a number of winters as follows:

1929–30.	January 12 (thirty)–April 13.	1931–32.	November 15–March 26.
		1932–33.	December 26–May 28.
1930–31.	December 21–May 14.	1933–34.	November 18–May 16.

Extreme dates are August 20, 1927, and May 16, 1934.

RED-BREASTED MERGANSER

Mergus serrator (Linnaeus)

PLATE 20, p. 177

This so-called "Fish Duck," the "Shelldrake" of the Cape May gunners, is the most abundant duck on the sounds and thoroughfares of Cape May during April and May while in October and November it occurs again but apparently not so plentifully and some remain throughout the winter. Hand's dates of arrival of migrants in spring range from March 23 to April 13, and many remain well into May. It is not unusual, too, for scattered individuals to occur on the sounds during the summer, probably barren birds, and we have such records for June 20, 1923 (six); June 5, 1925 (two); July 8, 1923 (three); August 1, 1925 (two); August 7, 1930 (one) August 23, 1930 (one). Of late years, however, they have been very scarce at this season. During spring and autumn we usually see these Shelldrakes in small flocks of from six to twelve, or in pairs, but they often gather on the larger sounds or inlets in considerable numbers and I have counted flocks of fifty to one hundred, especially in May.

As they fly along the coast, the white wing speculum is their most conspicuous characteristic but we can also detect the crested head and narrow bill at considerable distances. I have, however, become better acquainted with them from a skiff anchored under the bank of a thoroughfare or from the landing of one of the little baymen's cabins which here and there dot the salt meadows, raised on stilt-like supports above the reach of high tides.

On May 22, 1922, a pair came swimming up the thoroughfare directly past me as I lay in my skiff. They kept close in under the muddy bank passing in and out around every bit of fallen sod, ploughing through floating masses of bladder-weed and sea-cabbage and amongst the sedge that grows on the low mud bars. They stretched their necks straight out in front with the head submerged most of the time so that only the back of the neck and upper part of the back were visible. In deeper water they pursued one another holding the neck in the same position. Now and then they would rear up in the water and flap their wings vigorously. At such times the plumes on the head were thoroughly wet and no crest was apparent. Next day there were five of them just outside the cabin floating on the still water of the thoroughfare and occasionally diving for food.

On May 16, 1927, I watched a bunch of five Shelldrakes in flight, one male and four females. Their necks seemed shorter than those of Black Ducks as they passed me and against the light the head and shoulders of the male appeared black. Presently they set their wings in a bow and sailed down to the water, some of them dropping in suddenly, others skimming the surface and gliding in gradually with scarcely a splash. Again on May 11, 1929, a dozen of them gave a fine exhibition of diving, sometimes all of them being under at the same moment.

The adult male is always much more conspicuous than the female with his brilliant metallic green and white back and ruddy black-streaked breast, but in the case of three midsummer birds seen on a bar on July 8, 1923, the male was so much more beautiful than his dull gray and white companions, that we wondered if the latter had not failed to molt and were still in the old bedraggled plumage of the previous year.

In one of the larger late spring gatherings in Turtle Gut Inlet on May 19, 1924, I counted thirty-six all dull colored except one old male. They kept working farther and farther from shore as I approached and finally all took wing with a great show of white wing flashes. Whether all of the dull colored individuals were females or whether the males take two years to acquire the full plumage could not be determined, but the preponderance of dull colored birds was suggestive. In a similar flock numbering eighty-seven, resting on the channel back of Seven Mile Beach, on May 9, 1925, I could count only six old males. This flock kept stretched out in a long line bobbing up and down with the ripples. Usually they kept the head close down on the shoulders but every now and then one of the birds would stretch his neck out at an angle of 45°, often swimming out in front of another bird and facing it as he made the motion, possibly a courting action. Raymond Otter comments on the scarcity of full-plumaged male Shelldrakes and tells me that in all his gunning he has never been able to shoot one, all his bags consisting of dull-colored birds.

The counts of the Christmas census of the Delaware Valley Ornithological Club are as follows, although the men did not cover the larger sounds:

December 27, 1931, twenty-one.   December 23, 1934, twelve.
December 26, 1932, twenty-one.   December 22, 1935, sixteen.
December 24, 1933, fourteen.

At Barnegat Bay, Charles Urner has seen this Shelldrake in every month of the year but finds it most abundant from November to April.

TURKEY VULTURE

Cathartes aura septentrionalis Wied

All through the year Turkey Vultures are present about Cape May. There may be some migratory movement, as they seem most abundant in autumn and least so during the breeding season, but the daily—even hourly fluctuation in their numbers at any given spot is so pronounced that it doubtless obscures any seasonal variation. Sometimes there may be anywhere from a dozen to a hundred in sight any time that we look aloft and again, for a period of a week or more, especially in summer, none at all can be seen. In autumn, too, there may be a large flock directly overhead and in a few minutes not a single bird is in sight. These variations in abundance are most pronounced in the immediate vicinity of the town and the lowermost portion of the peninsula; farther back in the country the birds are of more regular and continuous occurrence.

We most frequently see Turkey Vultures, or "Buzzards" as they are universally known at the Cape, sailing majestically high in the air, driving directly into the wind, or soaring slowly in interlocking circles as they drift across the sky. All the way from the Harbor mouth westward along the coast to the Bay shore their occurrence is usually of this nature and it is only occasionally that we see them on the ground. About the Harbor and on the meadows to the north, however, while they also may be seen soaring there will usually be some individuals standing on the edge of the flats or channel banks feeding on the refuse washed up by the tide—black blotches against the uniform green or brown of the marsh; while in winter they may often be seen farther back, near the upland, resting on the ground in the lee of some thicket of marsh elder where, well protected from the wind, they will remain all morning basking in the sunshine.

Turkey Buzzards are seen to best advantage, perhaps, in the autumn, when a strong wind is blowing from the northwest or southwest across the tip of the peninsula. Across the sky come the great birds in small detachments sailing out over the marshes and circling about to form several groups which are constantly augmented by individuals or other groups coming from far back beyond the horizon. Then the several groups drift together in one great maze of soaring birds, layer upon layer, one above another; then suddenly the circling ceases and all are facing the same way and off they go sailing majestically into the breeze on set wings like a fleet of airplanes in battle formation. Glancing upward a few moments later we find them once more in close order soaring about in an apparently hopeless entanglement and yet there is no collision, no interference, and during the whole perform-ance there is scarcely a wingbeat—the perfection of soaring flight. It is astonishing how quickly the formation of the flying Turkey Buzzards changes and how rapidly the whole concourse will drift away and disappear only to assemble again when next we look aloft as if out of the sky itself. There is no apparent reason for these gatherings or these intricate evolu-tions and they seem to be instigated by the sheer joy of flight. Back in the uplands and across the Delaware in Pennsylvania we see these large as-semblages of Vultures only when attracted by some large dead animal but these flying autumn hosts on the seashore have no such cause although they often may be migratory bands pausing on their way south.

In winter when the wind is stronger and approaches gale proportions Turkey Buzzards do not have such an easy time of it. On January 7, 1922, a flight came on from the east, down over the marshes toward the Bay, and kept in the air all morning over Cape May Point. At first there were forty-seven birds but later on they had increased to ninety. They assumed all sorts of formations, now facing the strong northwest wind in regular battalions and now trying to circle, but the circling birds were always carried back through the ranks of their fellows as soon as they turned away from the wind and only regained their poise when they faced it again. Occasionally a bird would be all but capsized and would flap his great wings desperately until his equilibrium had been regained. Subsequently, when the force of the wind moderated, they circled at will soaring round and round, up and up, until almost out of sight and then, veering off, they would, in a few moments, be lost to view beyond the horizon. There was a precisely similar exhibition on February 14, 1925, and on many other occasions when the birds gathered for their aërial evolutions. Such gatherings seem to be most frequent in September and October but have occurred in all months except May and June when nesting is in progress. The largest numbers that I have seen at one time were one hundred on February 8, 1925; one hundred and fifty on January 16, 1929; seventy-one on November 8, 1930; seventy-two on February 22, 1931; seventy on July 19, 1924. The number of Turkey Buzzards present in midwinter is shown by the Christmas census counts of the Delaware Valley Club which are as follows, although they undoubtedly contain some duplications:

December 26, 1927, 100	December 26, 1932, 80
December 22, 1929, 250	December 24, 1933, 150
December 23, 1928, 344	December 23, 1934, 250
December 28, 1930, 19	December 22, 1935, 155
December 27, 1931, 200	December 27, 1936, 118

In summer the Fish Hawks often join the soaring Buzzards and in the autumn migrant hawks of various kinds associate with them; flocks of Broad-wings or Sharp-shins, an occasional Red-tail, or more rarely a mighty Bald Eagle with his wings stretched straight away from the body in the same plane, in contrast to the V-shaped position of the Buzzard's wings and the decurved wings of the Fish Hawk. The only birds to out-soar the Buzzards, however, were the Wood Ibises that sojourned with us in the summer of 1923. On September 25, 1927, twelve Buzzards were accidently detected with the glass far above the limit of natural vision. Their wings appeared pale brownish and semitransparent due doubtless to thin cloud strata below them and the strong sunlight playing upon them from above.

Sometimes just before a storm, soaring Buzzards are driven before the wind low over the housetops of the town, pitching and tossing, this way and that, their great widespread wings quivering as they pass. I have seen single birds, too, soaring above the Admiral Hotel, maintaining exactly the same position for many minutes at a time, evidently supported by the upward currents of air caused by the wind striking the sides of the building. More recently the same thing has frequently been observed, to even better advantage, at the great airdrome where many Buzzards could be seen soaring in a very limited area to the leeward of the structure for considerable periods. Later, apparently tiring of the sport, they would alight on the roof and arrange themselves in long rows on the ridge pole.

Turkey Buzzards about Cape May find an abundance of food in the dead fish, horseshoe crabs and other refuse washed up on the beaches and meadows and along the banks of the Harbor, but with the decrease in farming there are fewer large dead animals than formerly and consequently fewer large feeding assemblages of Turkey Buzzards. Dead hogs carted out to some old field or wood edge are their principal "game" and even these are not now very frequent. On March 21, 1925, I found twelve Buzzards perched on dead trees on the Bay side dunes near the skeleton of a hog which had evidently furnished food for a much larger number as the sand for thirty feet around the carcass was marked with thousands of footprints of the birds and flecked with white downy feathers lost in their contests for choice bits of carrion. Again on September 1, 1929, I counted over one hundred Buzzards circling low over a bit of woodland back of Rio Grande where carrion of some sort was evidently present. The birds when gathering for a banquet do not act like the flying squadrons already described but circle round lower and lower in a business-like way, alighting either at the scene of the feast or on nearby trees or fence posts awaiting an opportunity to begin their repast. On July 10, 1931, Otway Brown reported a gathering of about one hundred assembled near Cold Spring and the morning after the feast nearly every fence post for some distance along the road had its satiated Buzzard, resting from the activities of the day before.

When in flight the tips of the primary feathers of the Turkey Buzzard's wing are widely separated and apparently turn slightly upward according to the strength of the air pressure. When the bird is soaring transversely to the wind the tips on the windward side are turned up very noticeably. The extent to which the wing is bent at the wrist also varies according to the strength and direction of the wind; in a light breeze there is little or no bend but under other circumstances the tip of the wing may appear "furled," the outer primaries being bent back at a distinct angle with the front of the wing. The size and shape of the tail seem to vary individually when the birds of a soaring flock are compared, but it is soon apparent that this is dependent upon their position with reference to the air currents. When flying with the wind the tail seems long and narrow, the feathers being closely compressed; but when in other positions they are widely spread and the tail in consequence is fan shaped and appears shorter. In a bird soaring in circles these changes in the shape of the tail can be seen as it successively comes into the wind and drifts away again. The black under wing-coverts of the flying Buzzard stand out conspicuously against the dull brown of the flight feathers a contrast that is emphasized when the bird is in bright sunlight and soaring at a considerable altitude. This may account for some of the alleged observations of Black Vultures by those who are unacquainted with this southern species in life, and who have read that the wings show white below. The legs are carried straight out behind, close under the tail.

During periods of rain Turkey Buzzards do not usually take to the air but remain all day at their roosting places and when overtaken by unexpected rain or snow they resort to fence posts temporarily and later beat their way heavily to their roosts.

Certain groves of trees, especially along the edge of the meadows, or where the ground is wet or boggy, seem well adapted to the needs of roosting Buzzards and for some years a row of large oaks on the east side of Pond Creek Meadow was regularly occupied by a dozen or more of the birds but most of them retire farther back in the interior. A favorite roost for many years past is a line of trees following a small stream south of Cold Spring where many Turkey Buzzards resort every night, especially in winter, and the splotches of white excrement on the bushes below and the scattered white down feathers and occasional flight feathers are evidence of their occupancy. Birds that have come some distance to feed on carrion often roost temporarily in trees near the scene of the feast and on one occasion I found seven occupying one small tree near Bennett, at sunrise, having evidently spent the night there. Particularly tall trees in the heart of dense woods are also selected as roosting places as shown by the telltale down feathers and excrement. When resting from a feast Buzzards will roost on fence posts and I have seen them perched on posts on the meadows at Jarvis Sound and on pilings along the Harbor, while at Sea Isle City on May 10, 1925, and again at Cold Spring on August 11, 1934, some individuals repeatedly rested on telephone poles.

On the ground Turkey Buzzards are as clumsy and repulsive as they are graceful and inspiring in the air. On October 3, 1921, on the beach below South Cape May, I came upon a gathering of twenty-two of them; some were feeding on a dead tern and numerous small dead fish stranded by the tide, while others perched on the rail of a rough footbridge over a nearby ditch. Some of the latter had their wings spread facing a southwest breeze and from the rear a white spot was visible at the bend of the wing, apparently white down exposed by the elevation of the alula. The birds on the ground walked about like chickens with wings usually held well up against the body, the tips being visible above the base of the tail, and the·feathered thighs in full view. One or two drooped their wings a little and seemed to raise the tail a trifle. Once in a while one individual would run rapidly at another with head held low and both would leap up in the air with wings partly spread and held aloft over the back. When feeding they grasped their food with the beak and often tore at it with their claws. These birds varied much in color, some being very black while others, probably older individuals, were dull brownish. The dark birds had leaden gray, instead of pink, heads and the wing coverts were tipped with buff.

On March 22, 1925, I found twenty-five Buzzards roosting about a pigsty on the edge of the meadow on Pond Creek and I was able to approach very close to them. The naked heads of the birds when viewed near at hand seemed very narrow and the ruff of feathers was seen to stand high above the forehead and to project below the chin, as the head was drawn well back on the shoulders. The effect was much like a rolled collar drawn over the head from behind like a sort of mantilla. All of these birds had the head bright pink and the nail or hook at the end of the bill white, except one individual in which it was black. The tertial feathers of one bird were quite gray. In some others, which I found perched on stakes along the edge of the Harbor (June 28, 1930), the heads seemed ridiculously small as compared with the size of the ruff, and the effect was emphasized when they reached back over the shoulders to plume themselves. On April 4, 1924, twelve Turkey Buzzards were feeding with Crows on a refuse heap near a hogpen on the edge of the meadow back of South Cape May and as they walked about on the ground they looked very brown and dull colored although the neck was blacker than the body and wings. The heads of all were red. It seems likely that the Turkey Buzzards frequent pigpens, which are usually located out on the marsh or on the edge of woods, more for the sake of the refuse that is brought there than for possible dead hogs.

A single Buzzard which I was able to approach closely on August 31, 1920, was feeding on a dried up carcass of a dead cat on a railroad siding near the golf links. By creeping up behind a tool house I could see him apparently trying to pluck out the eyes of the animal and to tear open the abdomen. He would stand on the body and tear at it with his beak with the result that he pulled it end over end several times, but he could make no impression upon it. When he saw me he took wing and in a few minutes had disappeared utterly in the sky. The habit of plucking out the eyes of dead animals is well known and Otway Brown tells me that he has seen Turkey Buzzards extract the eyes of a living leather-back turtle stranded on the beach.

On July 5, 1922, during a rain storm, a Turkey Buzzard was found feeding on a dead dog on the roadside near the town. There was no apparent odor to the animal but it was quite conspicuous, as in the case of the cat, which had no odor whatever. There is a popular saying at Cape May that Buzzards will not eat dead dogs but I have found them doing so on a number of occasions both here and elsewhere. I have, however, one instance where a good sized dog in a rather exposed position was not touched by the birds and eventually dried up. This was in the late autumn and decomposition did not take place to the usual extent. Again on March 17, 1927, I found a dead black snake under some bushes in the woods at Cape May Point. The odor was quite perceptible at some distance yet no Turkey Buzzard had found the carcass. I threw it out into the middle of the woods path and next day found a Buzzard busily engaged in devouring it. He stood on the snake holding it down with both feet while he tore off shreds of skin and meat with his bill leaving the skeleton picked clean.

At close quarters the bird appeared "hollow-eyed," the skin projecting out from the brow, and it was possible to see clear through the nasal apertures from one side to the other. The face was pale pink and the rest of the head fuzzy with scattered hairs and blackish, with no bright color. The body plumage was brownish with lighter edgings. I had approached to within twenty feet of the bird before he saw me and then in his efforts to escape he got entangled among the underbrush and could not spread his wings, making several efforts before he was able to clear himself and get under way.

Out on the meadows where the Buzzards find dead fish and other refuse they tear at them with their bills and once, in the summer of 1929, I found ten busily engaged in devouring crabs and frogs, which had been left stranded and had died when ponds at the Point went dry during a prolonged drought. In severe winters Buzzards must have a very hard time of it as their normal food supply is buried in the snow or frozen in the ice and sleet, and the occasional dead Buzzards that we find have doubtless starved to death. The wonder is that there are not more. In February, 1934, when the temperature fell to ten degrees below zero, they came with Crows into the yards of the farmhouses, but it would seem that many of them must move off to the south, at such times.

While Buzzards have much contention among themselves over their prey they associate freely in an apparently amicable manner with both Common and Fish Crows when feeding on the meadows. On July 19, 1926, twenty Buzzards were settled on the meadows near Schellenger's Landing and while not feeding, so far as I could judge, they would flap their wings and rush at one another every now and then as they do when striving for choice bits of carrion. After a time some of them spread their wings out over the grass as if to dry or air them, as I have often seen them do when roosting on trees. While feeding harmoniously with Crows I have seen Turkey Buzzards sitting near a Fish Hawk that was devouring a fish as he perched on a stake on the meadows, but they made no effort to approach closer and were apparently willing to take what he might leave.

Several times I have seen Common Crows pursue a Turkey Buzzard viciously probably because he approached too near to their nest and Fish Crows, Red-winged Blackbirds and Kingbirds do the same, while woe betide the Turkey Buzzard that, unsuspectingly and with no thought of evil, crosses over a breeding colony of Common Terns or Laughing Gulls. He is relentlessly pursued by a swarm of birds and driven far from the nesting island and during his retreat will often disgorge his last meal possibly as a defensive measure but quite likely only a nervous reaction inspired by fear.

Turkey Buzzards nest in bushy thickets in swampy localities in southern New Jersey, especially under fallen trees. One that I found near Pennsville (May 2, 1896) was under the dead foliage of a windfall white oak and Walker Hand tells me of another in the upper part of Cape May County which was under the bole of a fallen tree near the base. I have but one record of a nest near the town, which was under a dense mass of vegetation on the sand dunes north of Cape May Point, which formed a convenient sheltering canopy for the setting bird. This was found about 1910. There is no actual nest the eggs being simply laid on the dead leaves which cover the floor of the woods and the eggs in my experience are always two. Turner McMullen has furnished me with data on seventy-one nests of Turkey Buzzards which he or his friends have examined in Salem County where the birds nest abundantly in low swampy woods. Ten nests contained a single egg, probably incomplete sets, with dates running from April 10 to May 2; fifty-seven contained two eggs, dates from April 4 to May 8 (average April 12); two nests April 13, 1930, and April 26, 1925, contained three eggs; and two April 30, 1932, and May 4, 1924, contained four. The last two may, of course, have been due to two pairs nesting together. The great majority of the nests were under fallen tree trunks or tree tops or close to fallen trunks, many in tangles of greenbrier or Japanese honeysuckle vines, and some under piles of cut logs or among the roots of trees. One was under the roof of a fallen down shack and another under the floor of a deserted house.

The presence of so many Turkey Buzzards about the Cape in May and June compared with the few recorded nestings would indicate that some individuals each year do not breed, or else that these coast birds come from some miles back in the interior for food, and considering their splendid powers of flight, this would not be surprising. Certain it is, however, that the Buzzard population of autumn and winter is greater than that of the nesting season and that there must be a considerable migratory movement.

No other Cape May bird possesses such a striking dual personality as the Turkey Buzzard. Well as we know him it is hard to realize that the graceful soarer, floating on motionless wings, high above the stretches of salt meadows and upland farms, on some listless day of midsummer, is the same as the repulsive scavenger, with his little naked red head, his sunken eye and heavy, dingy dress, like an overcoat many sizes too large for him. And yet in either guise he attracts our attention and presents problems that are not yet entirely solved—just how does he soar, and how does he locate his food—by scent or sight? Furthermore he is perhaps our most characteristic southern bird and the presence of many Turkey Buzzards marks our entrance into the Austral Life Zone with the possibility of many other forms of both plants and animals unknown in the colder boreal regions to the north.

*BLACK VULTURE

Coragyps atratus atratus (Meyer)

While none of the older writers on the birds of New Jersey mentions the Black Vulture of our Southern States it has always seemed strange that occasional individuals do not stray this far to the northward and all students of Cape May birds have been on the lookout for them but with little success.

The only New Jersey record of a specimen secured in the state so far as I know is one "shot at Sandy Hook during the spring of 1877. It was feeding upon the carcass of a pig, and was easily approached." It was in the collection of Robert B. Lawrence, who reported the capture (Bull. Nutt. Orn. Club, 1880, p. 117). A specimen found dead at Coney Island about 1881 and another shot at Plum Island on May 19 or 20, 1896, while feeding on a dead sheep, constitute the only records for Long Island, although there are several "sight records"; two of which Ludlow Griscom regards as worthy of consideration.

In New Jersey we have had several sight records but only two of these were made by an observer who was thoroughly acquainted with the bird and the characters necessary to its identification. On May 22, 1930, while driving through the Pine Barrens to Vineland, Charles Urner saw large numbers of Turkey Vultures circling over the burnt areas devastated by widespread spring fires. The birds evidently found the carcasses of mammals killed by the blaze an attraction. "Near New Egypt," he writes, "I noticed one Vulture with a smaller span than the others near at hand. Its wings were shorter and broad for their length and on the outer half of the wing was an area or patch of whitish, showing both above and below. The tail appeared shorter, not extending so far beyond the rear "wing line," this being very noticeable as the bird soared overhead. It soared frequently but when flying its wingbeats were more rapid than those of the Turkey Vulture. The wings when the bird soared, while curved slightly upward at the tip, were not lifted as high as in the soaring Turkey Vulture or in the Marsh Hawk.

"I watched this bird for some time at rather close range and in varying light. At times the underparts seemed lighter than the upper surface and at other angles the different pattern of the under parts was evident, lacking the diagonal line of demarkation which divides the darker and lighter areas in the Turkey Vulture." (Auk, 1931, p. 116.)

Another individual was observed by Urner at Colt's Neck, on April 8, 1934 (Proc. Linn. Soc., N. Y., No. 47, p. 116), and the same or another at Shark River on April 15 (Bird-Lore, The Season, 1934, p. 181).

For anyone familiar with the Black Vulture in the South there should be no trouble in recognizing it at any time or place but to identify one in the North from book descriptions alone is a very different matter. One point that the average observer does not know is that the head in first year Turkey Vultures is leaden or nearly black. Light and shadow moreover have a great deal to do with the appearance of white on the wings. The rather frequent, and always quick, beating of the wings, narrower circles in soaring and relative shortness of the tail are the best field characters. That the bird is extremely rare in New Jersey is, I think, proven by the careful study of hundreds of Turkey Vultures that I and others have made without detecting a single Black.

THE AUTUMN HAWK FLIGHTS

One of the notable sights at Cape May Point is the autumnal hawk migration when literally thousands of hawks of some twelve species arrive from the north and eventually cross Delaware Bay on their way to winter quarters in the South. The first individuals are seen about August 20 and they continue to pass until about November 15 although the bulk of the movement is during the months of September and October.

The great flights are always coincident with a strong wind from the northwest and a falling temperature, on tingling mornings when we feel the first touch of autumn in the air. Many of the birds will be circling about overhead but others, lower down, will be found to be flying, for the most part, directly into the wind in exactly the opposite direction from that in which the migration as a whole is supposed to be progressing. Precisely similar conditions of wind and temperature bring the notable autumn flights of Flickers, Woodcock and miscellaneous passerine birds, but in their case the movement into the teeth of the wind is more marked and when an autumn passes without these days of northwest winds there are no flights. In the case of the hawks, however, there seem to be flights every autumn although they are more frequent and the birds more numerous in some years than in others.

With winds unfavorable for the usual flights I have often seen, with the aid of binocular glasses, large numbers of hawks circling high overhead, so high indeed, that they appeared like small swallows or even insects, and gradually drifting off to the south. William Rusling, Audubon Association warden for that year, saw the same thing during the autumn of 1935 and states that at 8:00 a. m. on one September day, with the wind east and southeast, he saw the birds circling at an estimated elevation of 3000 ft. and as they reached the beach they arose 500 ft. igher hand then slowly crossed the Bay moving in great circles at about the limit of vision.

Again on September 17, 1935, with the wind northeast he saw a great flight of Broad-wings moving in circles and gradually drifting across the Bay toward the Delaware shore. On days when there was little or no surface wind both he and I have detected, with the glass, hawks passing over southward beyond the limit of vision.

As a result of many years observation I have come to the opinion that the normal southward flight is at a great height, just as I saw it fifty years ago in the vicinity of Philadelphia (Auk, 1887, p. 161), and that it is only when the northwest gales threaten to drive the birds out to sea that they descend and head into the wind to save themselves and, incidentally, cause the well known visible flights; at other times they are passing regularly, but beyond the limit of man's vision. If this is so it follows that the hawks counted during a flight are only a relatively small proportion of those actually on migration and that perhaps the number killed by gunners is relatively smaller, in proportion to those passing through, than we had supposed, which would account for the fact that the several species of hawks have not yet been exterminated although greatly reduced in numbers!

Whether the migrating hawks scatter as the day advances, or are less active, I do not know, but it has always been claimed by native gunners that they are more abundant and lower down in the early morning than later in the day. They are, of course, busily engaged in seeking food from the host of smaller birds which are often in congested migration at the same time, or from the constant supply of meadow mice, grasshoppers and dragonflies.

In the spring there is no visible return migration of hawks; only the gradual withdrawal northward of the individuals which have wintered in the vicinity. There is the same absence of Flicker and Woodcock flights in spring and, it should also be remarked, of the strong northwest winds which are so prevalent in autumn. Whether there is a northward hawk flight in spring at a great altitude I cannot say, but if not how and by what route does the great autumn host return? The correspondence of visible flights with northwest winds is admirably shown by William Rusling's table for the autumn of 1935, kindly furnished to me with records of his other observations by the National Association of Audubon Societies. (See also Allen and Peterson, Auk, October, 1936, pp. 393–404.)

		Wind	Hawks Counted	Hawks Killed	Gunners
August	23	N W	74	0	0
	29	N W	0	0	0
September	6	N E, rain	33	0	0
	8	Variable	35	0	0
	10	Overcast	154	3	7
	11	N E—S E	173	2	3
	12	E	30	0	2
	13	Variable	258	7	3
	14	N E	232	37	7
	15	Variable	73	0	10
	16	N W	1392	258	17
	17	N E	471	31	7
	18	S	10	0	0
	20	N W	529	17	4
	21	Variable	522	27	8
	22	N W	424	0	0
	23	N W	655	135	12
	24	S	74	0	0
	29	W N W	64	0	0
	30	N W	280	56	24
October	1	S W	692	142	14
	2	N W	317	108	21
	3	S	366	48	10
	4	N W	782	107	8
	5	N W	120	18	17
	7	N W	328	69	10
	8	N W	79	1	5
	9	N E	74	5	4
	10	E	107	0	0
	12	N E	137	3	4
	13	S E	119	0	0
	15	N W	1346	65	9
	16	N W	245	9	5
	29	N W	90	1	1
	24	N W	202	5	4
	25	N W	240	12	5
	27	N & W	67	0	0
November	1	N E	35	0	0
	2	N E	55	0	0
	3	N E	282	0	0
	4	E	10	0	0
	6	N W	204	9	2
	9	N E	30	1	2

(For further discussion of migration at the Point see Introduction, p. 38.)

The hawk flights at the Point have been in progress as far back as the recollections of the oldest residents extend and doubtless were always a feature of the autumn although there is a lamentable lack of published records. For some time a bounty was offered by the State for the destruction of the birds but in 1885 this was reduced from fifty cents to twenty-five according to Charles S. Westcott (Forest and Stream, December 3, 1885), who, in spite of his usually commendable attitude on conservation, regarded this as a great mistake. This bounty, of which we find no other record, may have started the slaughter at this spot, although it had been abolished by 1903 and the beneficial hawks given protection. At any rate the shooting of hawks during the autumn had become a regular pastime among certain so-called "sportsmen" of Cape May and vicinity. Ten years ago, or more, one used to see a small army of men scattered over the sand dunes at the Point, or lined up along the roads in the neighborhood, armed with all sorts of weapons of various ages and vintage, everyone bent on the slaughter. Then with the advent of the automobile their numbers increased and recruits came from farther north in the state as well as from Pennsylvania, and by daybreak on the morning of a flight one might see lines of cars parked along the Cape May Point turnpike, which divides the wooded area, near the Bay, over which the hawks fly at an elevation but slightly above the tree tops. While the number of shooters has decreased in recent years the situation remains about the same. The bombardment begins as soon as it is light enough to see and we find in the assemblage Italians, negroes, boys and "sportsmen," the last group ranging from baymen and farmers to business men and men of leisure, participating in the sport. The fusilade continues through the greater part of the morning and so frequent are the discharges that it actually sounds as if some sort of engagement or mock battle were in progress. The most abundant hawks and those which constitute the bulk of the "bags" are the unprotected Sharp-shinned and Cooper's Hawks which are legitimate "game" according to the present State laws. The "army" proceeds to kill and maim the poor birds as they pass over, and, up to a few years ago, either through ignorance or carelessness, a large number of protected birds were killed. I have picked up dead Sparrow Hawks, Pigeon Hawks, Broad-wings, Marsh Hawks, Owls of several species, Ospreys, Whip-poor-wills, Nighthawks, Kingfishers etc., all victims of the gunners. Most of these birds were left where they fell and only the Sharp-shins and Cooper's Hawks, which the shooters knew were within the law, were carried away. Most of these hawks are used for food, much to the amusement of those not familiar with the gastronomic ability of the hardy residents of the shore, although I am assured that, properly prepared, hawks are very good eating. (cf. plate 26, p. 237.)

The slaughter was far greater and more indiscriminate in the past than at present and apparently there were more birds passing, the decrease being due no doubt to the killing of hawks which has been going on all over eastern North America, as well as at Cape May, and which is inspired by the manufacturers of firearms and ammunition, and by misguided sportsmen's organizations. The latter either have been ignorant of the findings of the

scientific investigations on the food of the hawks, carried on by National and
State departments of agriculture and by similar bodies abroad; or prefer to
place their individual opinions in opposition to those of trained investigators.

In past years I have secured a few counts of hawks killed at the Point
which are of interest. On one day in September 1920, 1400 were shot; on
September 8, 1921, one man shot 140 hawks; and on September 14, 1923,
800 dead hawks were counted; while from 1920 to 1925 individual "bags"
of fifty or sixty were frequent and I have seen single gunners carrying home
one or two peach baskets filled with dead hawks.

Beginning with 1931 the National Association of Audubon Societies, in
an effort to eliminate the killing of protected birds, has had a special warden
or a regular State game warden at the Point during the shooting season to
help the gunners to recognize the hawks that may legally be killed.   The
gunners, for the most part, have cordially coöperated with the wardens and
representatives of the Audubon Association and the results have been very
beneficial so that scarcely any protected birds are killed in recent years.
To warden Groves and to George B. Saunders, Robert P. Allen, William J.
Rusling and James T. Tanner, who represented the Association in 1931,
1932, 1935, and 1936, respectively, much credit is due for present conditions
and for the collecting of much information on the hawk flights which has generously been placed at my disposal.   The work of the Audubon Association
has culminated in the establishment of the "Witmer Stone Wild-Life Sanctuary" at the Point through the generous coöperation of Mr. Ralph Stevens.
This provides a refuge for birds of all sorts and to some extent limits the
hawk shooting.

Recent figures from the records of the Audubon Association give some
idea of the numbers of hawks that pass and are killed at present at the
Point.   In 1931 between September 29 and October 15, George Saunders
estimated that 10,000 hawks of all kinds went by on migration of which
some 925 were killed; in 1932, a poor year, Robert Allen counted 5765
hawks between September 15 and October 29, of which 366 were shot, while
in 1935 William Rusling, who was present continuously from August 23 to
October 29, counted 13,452 hawks passing south of which 1080 fell victims
to the gunners.   The varying ability of the latter is shown by the fact that
on one of the flight days in 1932 one man used 62 shells to kill 13 hawks
while in 1933 another sportsman killed 180 on a single day using eight
boxes of shells!

In order to illustrate the make-up of the hawk flights we quote from
the careful record kept by William Rusling in 1935:

Sharp-shinned Hawk,	8,206 passed	1,008 shot	
Cooper's Hawk,	840 "	62 "	
Sparrow Hawk,	777 "	1 "	
Osprey,	706 "	1 "	

Pigeon Hawk,	402 passed	3 shot
Broad-winged Hawk,	367 "	3 "
Marsh Hawk,	274 "	1 "
Bald Eagle,	60 "	0 "
Duck Hawk,	56 "	1 "
Red-tailed Hawk,	50 "	0 "
Red-shouldered Hawk,	12 "	0 "
Rough-legged Hawk,	2 "	0 "

The great majority of the migrating hawks are birds of the year and it is only occasionally that we see an adult. One considerable bag contained not one old bird and in several peach baskets full of Sharp-shins I failed to find a single adult, while William Rusling's count of passing Sharp-shins in the autumn of 1935 showed 26 adults as against 8180 birds of the year.

In 1932 Robert Allen estimated that of the gunners who took part in the hawk shooting 40% were natives who used the birds for food; 55% were Italians who made a similar use of them; while 5% were sportsmen who killed the birds for sport and as good practice for wing shooting. Members of the last group were the most flagrant violators of the law in killing protected species; nearly all of the others coöperating with the wardens when the law and its objects were explained to them.

The investigations of the contents of hawk stomachs by the U. S. Department of Agriculture and similar agencies has proved to the satisfaction of all reasonable and unprejudiced persons that the Broad-winged, Red-shouldered, Red-tailed, Sparrow and Pigeon Hawks are of the greatest service to mankind by destroying enormous numbers of field mice and similar harmful rodents as well as grasshoppers and other destructive insects. These species are therefore protected by law in most of our states. As to the Sharp-shinned and Cooper's Hawks it is well known that they feed largely upon small song and insectivorous birds, and it is on this ground that they are denied protection and that the hawk gunners justify the slaughter at Cape May Point. While a few young game birds are killed by hawks in their northern breeding grounds this is not the case on the migrations. Game birds are too large in autumn to be caught by Sharp-shinned Hawks and not a trace of game birds has been found in the stomachs of hawks killed at the Point. It should be remembered that the killing of small birds by these hawks has been going on since bird life began without causing any diminution in the numbers of the former, The hawks are created for this life and know no other; they are Nature's provision for the elimination of the weak and diseased song birds which naturally constitute the majority of those caught and in this way the vigor of the song bird strain is maintained, as the strongest and most active individuals of each species are left to propagate the race, on the same principle that stock and poultry raisers practice selective breeding. Nature is not so foolish as some persons would have us believe!

Our laws permit the killing of such hawks as the Sharp-shin not because they are "cruel bloodthirsty birds" but in order to allow the farmer and poultry raiser to shoot hawks that are actually destroying his stock. They never contemplated the *extermination* of hawks, yet this will eventually be accomplished if there be no check upon senseless hawk shooting, and with disastrous results. Not only will the strain of our song and insectivorous birds be sadly weakened, so that they will readily fall victims to disease, but the reduction of protected hawks, which some short-sighted persons still demand, will flood the country with rats and mice against which these hawks are Nature's check. It is a dangerous thing to upset Nature's balance by exterminating any species especially when no benefit ensues or necessity exists, and we should be loath to think that any citizens of Cape May are yet dependent upon hawk meat for their subsistence!

Wild life belongs to all the people, not only to sportsmen and gunners, and already numbers of nature-lovers are coming to Cape May to view the remarkable migrations of birds which its peculiar situation makes possible and the destruction of the great spectacle that these migrating hawks presents on their flight from the Northern States and Canada to the Southland during the crisp days of early autumn would be a sad mistake, not only from the viewpoint of the nature-lover but from that of the hotel and business interests of the town, which could be vastly enhanced if these hawk flights and other natural attractions of the Cape were advertised as they deserve—to nature-lovers, not to gunners. (cf. McDonald, Bird Lore, 1931, p. 97; Stone, Auk, 1922, p. 56.)

MISSISSIPPI KITE

Ictinia misisippiensis (Wilson)

On May 30, 1924, while on the Memorial Day outing of the Delaware Valley Ornithological Club, twelve of us, including Julian Potter, William Baily, and myself saw a Mississippi Kite flying overhead on the road from Higbee's Beach to the turnpike, about three quarters of a mile from the Bay. The bird remained for nearly half an hour over an open space of several acres, giving us ample time for careful observation, and when it finally passed on it disappeared over the woodland to the north.

There was a strong northwest wind blowing and the bird rose and fell through distances of fifty and one hundred feet. It held itself steadily against the wind, now and then tilting the tail up until we could see the entire upper surface and again drawing in the wings at the shoulders, sometimes simultaneously, sometimes one at a time, to "trim sail" as it were and keep its head to the wind, it also bent the tail to the right and left using it as a rudder. When it first appeared far off to the south the light color of the upper parts made us think of a tern but we soon saw that it was a bird of prey and its method of flight indicated a kite. The details of its color pattern when they came into view a little later left no doubt as to its identity. The tail was square cut and black, in strong contrast to the lighter color of the rest of the plumage. The head was pale gray looking almost white in the strong sunlight. The wings were bicolored, the fore part dark slate and the posterior portion (secondaries) and the entire under parts pale gray.

This seems to be the only record of this southern species for New Jersey. (Auk, 1924, p. 477).

SWALLOW-TAILED KITE

Elanoides forficatus forficatus (Linnaeus)

This straggler from the south has occurred several times in northern New Jersey but the only South Jersey occurrence, so far as I know, was an individual seen by J. Harris Reed flying high overhead in Cumberland County, close to the Cape May line, on June 4, 1893.

The other records are:

John Krider, New Jersey, one shot near Philadelphia, 1857 (Field Notes, p. 10)

Harold Herrick, Chatham (Forest and Stream, XII, 1879, p. 165)

C. C. Abbott, Bordentown, one seen November 1883. (Science II, No. 29, 1883, p. 222, and Birds of Mercer County)

E. Carleton Thurber, two seen by L. P. Sherrer and George Hild, near Horse Hill, Morris County, September 18, 1887 (True Democratic Banner, Morristown, N. J., November 10, 1887)

C. F. Silvester, one seen at Princeton, "some years ago" (Babson, Birds of Princeton, p. 46, 1901)

GOSHAWK

Astur atricapillus atricapillus (Wilson)

While an occasional Goshawk may come south in winter as far as New Jersey in any year, there are periodic invasions of them accompanying flights of the Snowy Owl in seasons when the supply of snowshoe rabbits and other arctic mammals is unusually meager. The underlying cause of these diminutions in food supply and consequent migrations is not understood but they occur with reasonable regularity every seven to ten years.

Goshawks have doubtless been seen or shot in Cape May County on every such flight but since the average sportsman or farmer is unable to distinguish one hawk from another we have but few records of their occurrence. A specimen in the collection of the Academy of Natural Sciences of Philadelphia was shot by Dr. W. Louis Abbott, probably near Dennisville, on January 22, 1879; one was seen by William Yoder at the Point on December 26, 1926; a third was seen by James Tanner, Audubon Association Warden in the act of attacking some chickens near Alexander Avenue, Cape May, on November 6, 1936; and three were seen on the Christmas census of the Delaware Valley Ornithological Club on December 27 of the same year.

SHARP-SHINNED HAWK

Accipiter velox velox (Wilson)

The Sharp-shinned Hawk makes up the bulk of the notable hawk flights which occur at Cape May Point during September and October and probably four-fifths of all the hawks present on these occasions belong to this species and its ally the Cooper's Hawk. In the autumn of 1935 William Rusling, who has made the most careful continuous observations on these birds, counted 8206 Sharp-shins as against 840 Cooper's and 2706 of all other kinds.

Early on the morning of a hawk flight the Sharp-shins are low down seeking out warblers and other small birds which are migrating from the north at the same time and which at this season form their natural food. No matter how early we may repair to the Point the hawks are there, having come in the night, or descended from a great height when the wind shifted to the northwest. We see them dashing through the bayberry bushes and dune thickets and darting along the shaded wood roads moving silently and at lightning speed and weaving their way through the tangled branches. Sometimes the head will be raised above the level of the back and the bird glances to right and left in the hope of sighting some quarry. The flight in the open is very different from the flicker of the Sparrow Hawk, the wingbeats are fewer and the sail longer though when making off in fright the Sharp-shin may flap more or less continuously. He will follow the undulations of the bushes after the manner of the Marsh Hawk, or sail low down across the meadows, but upon sighting a small bird in a bayberry thicket or in the woods he will swoop directly after it dodging in and out in a marvelous manner.

As one walks through the low woods and thickets the Sharp-shins are flushed on every side, some from low limbs, others from bushes, and still others from the ground, and nearly every one has secured his prey and has come to rest to devour it. One seen on September 13, 1921, dropped what remained of his breakfast and upon picking it up I found only the wing bones and shoulder girdle of a warbler, every bit of flesh and feathers having been neatly picked off. Another bird sailing overhead on August 21, 1920, had the remains of a small bird grasped firmly in the claws of one foot and every now and then he would bring the foot forward and bend the head over, so that he could pick at his quarry while on the wing. Other individuals carried small birds in their talons holding them close up under the tail, awaiting a resting place before beginning their meal. As a matter of fact, however, the number of song birds killed on any one day is an insignificant fraction of the numbers present and there are many days, too, when birds are scarce and the hawks must go hungry, furthermore those that are caught are in most cases the weaklings of the species as explained elsewhere (p. 267).

Later in the day the Sharp-shins mount high in the air and circle about in an intricate maze. Their short rounded wings and long straight tail—not fan-shaped as in the buzzard hawks—are characteristic, as are also their easy aërial maneuvers. There are several quick light flaps of the wings, which seem to be merely tapping the air, almost like the wings of a butterfly, and then a circular sail and again the quick wingbeats. There seem to be perhaps one hundred circling about as we gaze upward, then, as our eyes become focussed upon them, we see that there are others above them seeming no larger than martins or swallows, and now as we study them through the binoculars we are amazed to find others still higher up and more and more come into view as we shift the range, their wings flashing for a moment as they catch sunlight. We realize that there are incredible numbers layer upon layer far above the limits of our vision and the whole assemblage drifting slowly along in the wind.

Other hawks associate with the Sharp-shins although the latter are in the great majority. Early in the autumn there will be Broad-wings, Ospreys, and Pigeon Hawks and on occasion Marsh Hawks, attracted aloft by the unusual gathering, and later Cooper's Hawks and some Red-shoulders and Red-tails. Even though there are but a few birds in the air they always tend to associate, regardless of species, and at one time I saw twenty Sharp-shins intermingling with twenty-four Turkey Buzzards while a little later they deserted the latter and joined a solitary Bald Eagle which had come into view from up the Bay. Later still they all joined forces with new recruits until we counted fifty Sharp-shins, forty-three Vultures, the Eagle and a Fish Hawk, the formation changing with kaleidoscopic bewilderment. Where they came from and whither they go is still very much of a problem.

While the main flight of Sharp-shins is always over Cape May Point and along the Bay shore there are often scattered hawks in Cape May as well, where, losing all sense of fear, they will dash through yards and across streets in the heart of the town, in pursuit of small birds. On September 1, 1922, I saw one pursuing a flock of Starlings on the outskirts but he could not gain a foot upon them as they wheeled and turned in perfect unison and after a short chase he gave it up.

When perching undisturbed the Sharp-shin sits absolutely erect with the head horizontal and the tail hanging straight down below, making it appear very long and slender. One that I watched for some time would occasionally bend over and scratch its head with its foot and jerk its tail vigorously from side to side. The Sharp-shin seems always to roost on trees or bushes, usually right in the woods, and never out in the open on telegraph wires or poles as does the Sparrow Hawk. Two that I saw on August 26, 1920, evidently just arrived from the north, were chasing one another, diving and quartering with wonderful ability. Suddenly they darted into a thicket on the edge of the marsh and perched on low branches not more than three feet from the ground but as soon as one of them took wing the other was after him again. On September 17, 1922, Julian Potter saw a Duck Hawk pursuing a Sharp-shin. Both pitched down from a considerable height and the Sharp-shin was apparently struck just as he was diving into a bunch of bayberry bushes.

The first south-bound Sharp-shins arrive about the 20th of August:

1920.	August 21.	1925.	August 27.	1930.	August 30.
1921.	August 15.	1926.	August 28.	1931.	September 1.
1922.	August 21.	1927.	August 22.	1932.	August 14.
1923.	August 21.	1928.	September 1.	1933.	August 27.
1924.	August 13.	1929.	August 25.	1935.	August 23.

The first great flights occur in September, our dates from 1921 to 1935 running from the 6th to the 14th, with later flights at frequent intervals through this month and October; their number and size depending upon the presence of strong winds from the northwest.

Most of the Sharp-shins have passed on by November 1 and there are left only those that propose to winter in the neighborhood. Winter lists usually contain records of single birds but on the Christmas census of the Delaware Valley Ornithological Club, when several men participate and cover the country pretty thoroughly, more are seen.

December 23, 1923, two.	December 27, 1931, six.
December 26, 1926, three.	December 26, 1932, six.
December 26, 1927, three.	December 24, 1933, four.
December 23, 1928, five.	December 23, 1934, three.
December 22, 1929, one.	December 22, 1935, two.
December 28, 1930, three.	December 27, 1936, four.

January and February records are fewer, many of the wintering birds being shot when they come close to farm houses in severe weather in pursuit of English Sparrows. On such occasions they become very bold and must suffer for want of food as we have seen them darting across gardens in the town and in and out among the sheds and bathing houses on the beach front, sometimes alighting directly on the ground possibly in pursuit of a mouse or rat. In the country a single bird will often take up a winter residence in an old orchard and may be seen every day darting about the barn or outbuildings or coming directly up to the door of the house in hope of quarry of some kind. Such birds have been seen to plunge directly into the privet hedges which furnish winter shelter for the sparrows, in a desperate effort to secure one of them. During spring we have records of single birds for April 24, 1921; April 12, 1930; April 2, 1933; May 6, 1928; May 30, 1928; May 1, 1926. Some of these would seem to indicate breeding birds and we have one record of a nest found by David Harrower near Dennisville. Some of them, too, may be migrants bound north but as already stated we have no evidence of a spring hawk migration. Possibly it takes place at a considerable altitude and the usual absence of north winds in spring permits the birds to pass without descending to lower levels as in the autumn.

COOPER'S HAWK

Accipiter cooperi (Bonaparte)

Essentially a large edition of the Sharp-shinned Hawk, and with similar habits, Cooper's Hawk occurs regularly in the southward migration during September and October and is one of the predaceous hawks which may be shot legally. According to the Audubon Association's records George Saunders estimated that five hundred passed the Point from September 15 to October 15, 1931, of which thirty-five were killed; Robert Allen counted 1222 from September 15 to October 29, 1932, of which forty-two were shot; and in 1935, between September 6 and November 6, William Rusling counted 840 of which sixty-two fell victims to the gunners.

While September and October are the months of greatest abundance some individuals arrive in the latter part of August and I have the following records:

1921. August 19.	1926. August 14.	1930. August 20.
1923. August 31.	1928. August 31.	1932. August 30.
1924. August 29.	1929. August 17.	

A few Cooper's Hawks remain through the winter and from one to three have been observed on the annual Christmas census ever since it was undertaken, while eight were seen on December 26, 1932; five on December 24, 1933; ten on December 22, 1935. I have five records for January and a single bird was observed perched on a low tree not more than four feet from the ground from February 7 to 22, 1932. Most wintering birds are probably shot, eventually, which makes them appear rarer as the season advances. In spring I have records for April 2 and 3, 1922; April 30, 1927; April 19, 1931; May 16, 1928; May 21, 1922. The last bird was observed while I was living in a shack on the meadows back of Two Mile Beach and came flying across from the sea toward the mainland early in the morning, flapping vigorously with only short sails, as if much fatigued. The bird of April 3, 1922, was in a yard on Washington Street gliding about among the trees to the consternation of the Grackles which were feeding there.

We have no record of the Cooper's Hawk breeding in the county but in Salem County, to the northwest, Turner McMullen has found two nests with four eggs on April 11, 1924, and May 5, 1928, both in pin oak trees thirty and thirty-eight feet from the ground, while William Crispin found two with four eggs on May 2, 1905, and May 4, 1912, and three with five eggs on April 29, 1906, May 12, 1911, and May 8, 1912.

J. F. S.

OSPREY'S NEST IN LOW HOLLY TREE

RED-TAILED HAWK

Buteo borealis borealis (Gmelin)

The large Buteos or buzzard hawks have never been as conspicuous in Cape May hawk life as the Accipiters or bird hawks, the Harrier or the Falcons. The Red-tails are the largest of the group, big round-winged birds sluggish in action and loath to take wing when they have found a convenient perch from which to watch for mice and other prey. We know them best as migrants of autumn slowly soaring overhead, often associated with Turkey Vultures, or as birds of winter, resting for hours at a time in the early morning on the top of some lone tree in the open or on the edge of the woodland, with plumage ruffled up gazing intently over the brown fields and meadows, or apparently dozing on their perches.

These wintering individuals, once such a picturesque feature of the landscape, have, however, been sadly depleted in recent years and we now know these Buteos chiefly as minor components of the great hawk flights of autumn. A few are always seen when we visit the shore in October, or even earlier, our earliest record being September 2, 1920, when an adult and a bird of the year were seen soaring high over Cape May Point, the bright rusty red tail of the old bird and the dusky barred tail of the immature being well contrasted against the sky.

In 1935 William Rusling, who was present through the autumn in the interests of the Audubon Association, saw Red-tails at Cape May Point on sixteen days from September 12 to November 9, fifty in all. Usually there were from one to five passing in a day but on November 9 no less than eleven were seen. The Delaware Valley Ornithological Club's Christmas lists from 1931 to 1935 show Red-tails present on:

December 27, 1931, five.	December 23, 1934, four.
December 26, 1932, three.	December 22, 1935, eight.
December 24, 1933, four.	December 27, 1936, five.

Like other of our winter hawks the Red-tails seem to decrease from January to March and we have no marked spring movement although there is one record for April 7, 1922, and another for April 19, 1936, which may very likely have been north-bound migrants. I have never heard of one being seen at the Cape in the summer. In Salem County, however, Turner McMullen tells me of a single nest fifty feet up in a red maple tree which contained two eggs on April 6, 1930. A pair of the birds was seen in the same vicinity on April 19, 1933, and May 30, 1934, and probably nested there. This is the only nesting record anywhere in southern New Jersey. Joseph Tatum and others saw a pair of Red-tails at Egg Harbor, on May 22, 1932, which from the late date probably were nesting in the vicinity.

While banding of hawks has not been carried on extensively, owing to the difficulty of trapping the birds, there is a record (Forest and Stream, March 26, 1885) of a "large hawk" with a spread of three feet ten inches, possibly a Red-tail, which had been "tagged" if not banded. Charles D. Lippincott, the well known botanist of Swedesboro, N. J., announced the capture of the bird in a steel trap, about March 15, and states that it had a bell attached to its neck, measuring two and a half inches across, engraved "L. Perry, Athol, Mass." I can find no further information about it. Mr. Lippincott died some years before the record came to my attention.

RED-SHOULDERED HAWK

Buteo lineatus lineatus (Gmelin)

The occurrence of this Buteo at Cape May is essentially the same as that of the Red-tail. I had supposed from my scattered observations on visits of a day or two that it was more plentiful in the autumn migration than the latter but William Rusling, who made a daily study of the flight during 1935, tells me that he identified only twelve Red-shoulders from August 12 to October 27 as against fifty Red-tails, and saw them on seven days only. It is no easy matter to distinguish between the immature birds of the two species nor to identify every individual, so that the question of their relative abundance is, I feel, still unsettled. Julian Potter records an interesting flight of Red-shouldered Hawks at Fort Mott on the lower Delaware on November 10, 1929, the birds soaring higher and higher until they disappeared, doubtless a migratory movement.

The Red-shoulder is well known to nest in Salem County and may do so also in the upper part of the Cape May Peninsula. Certain it is that we have earlier records of its occurrence than we have for the Red-tail—August 15, 1921; August 26, 1923; August 6, 1927; August 12, 1930; July 16 and 20, 1925; July 5, 1928; while spring records are also more numerous, March 20, 1928; April 30, 1927; May 26, 1929; May 27, 1931; and this is just what might be expected from the proximity of Cape May to the known breeding range of the bird.

I have had few opportunities to study this hawk at close hand. On February 14, 1925, an adult flew across the railroad embankment north of the town and swooping over the swamp below me entered a thicket not more than three feet from the ground. It showed to perfection the narrow white bars of the tail and the buffy white blotches on the back and on the outspread wings, while the underparts, as the bird faced me, were rich rusty. Another individual seen the same day near the Lighthouse was being pestered by a flock of Crows which at length forced him to take refuge in a clump of bayberry bushes.

The numbers seen by the Delaware Valley Ornithological Club on its Christmas census from 1930 to 1936 are as follows:

December 27, 1931, seven.   December 23, 1934, five.
December 26, 1932, three.   December 22, 1935, seven.
December 24, 1933, six.   December 27, 1936, seven.

In Salem County Turner McMullen has found one nest with two eggs on April 26, 1925, located forty-eight feet up on a tulip poplar tree while W. B. Crispin has recorded a nest with one egg, April 22, 1911; three with three eggs March 30, 1899, April 10, 1910, and May 7, 1912; and an exceptional nest with five on April 14, 1907. George Stuart found a nest with two eggs and a runt in the same vicinity on April 28, 1917. The nest of April 10, 1910, was robbed and the birds had another clutch of three on May 7. It seems probable, therefore, that the other nest of May 7 was also a second breeding.

BROAD-WINGED HAWK

Buteo platypterus platypterus (Vieillot)

The Broad-wing is a summer resident in the woodland some miles north of the Cape and stragglers occur in the pine woods of Cape May Point at various times during the summer but, as in the case of the other hawks, it is during the great flights of autumn that they become really common. I have seen them in spring on May 20, 1924; May 8, 1927; May 10, 1928; May 10, 1931; May 7, 1932; probably birds that had just arrived from the south as they disappeared later in the season. Through June and July, when I was usually in the field daily, only occasional individuals were seen or sometimes a pair, as on June 28, 1928, at Cape May Point. Toward the latter part of August the southward flight begins and continues on favorable days through September and early October, our latest dates being October 21, 1923; October 12, 1930; October 15, 1932, while William Rusling saw one on October 29 and two on November 9, 1935. The Broadwings winter mainly in South America and the published winter records for New Jersey are all, I think, referable to Red-shouldered Hawks.

The Broad-wings are among the first of the autumn hawks to appear in numbers and when the first flight arrives there may be actually more of them in sight at one time than Sharp-shins, the latter being more active and moving about continually. Sometimes we may see fifty or more Broad-wings soaring round and round high overhead and easily recognized by their fan-shaped tail and broad rounded wings. On September 30, 1928, Julian Potter saw seventy-five of these fine hawks at the Point, sixty on October 3, 1920, and no less than 110 on September 28, 1930, while William Rusling counted 138 on September 7, 1935, twenty-nine on September 16, sixty-two on September 21, and eighty-eight on October 1 of the same year. George Saunders estimated that two thousand Broad-wings passed the Point during the autumn flight of 1931 but William Rusling counted only 367 in 1935 and Robert Allen 1177 Buteos of all three species in 1932. As all three men were present every day during the flight the figures give some idea of the variation in numbers due to weather and wind conditions. Conrad Roland saw a flock of Broad-wings on September 19, 1934, soaring so high that he was able to count fifty in the field of his glass at one time. The Broad-wing while probably the most beneficial of our hawks, feeding as it does on meadow mice etc., has, on account of its small size, been frequently mistaken by the gunners for the Sharp-shin and has suffered accordingly. On more than one occasion I have come upon hawk shooters plucking the plumage from Broad-wings so that they would not be recognized by the wardens.

When not disturbed the Broad-wing is a rather sluggish bird and I have on several occasions been able to approach quite close to roosting individuals. One bird standing on a telephone pole was observed scratching its head with its foot while another flew into a patch of woodland and alighted upon a tree directly over my head. He looked down and inspected me very carefully but made no move to depart, so long as I remained rigid, but with my first movement he was off. Another found in the woods east of Bennett was much disturbed at my presence and flew about calling continually so long as I remained, acting as if it had a nest of young in the vicinity although the date, July 14, was too late for that. On two occasions birds flushed from trees carried garter snakes in their claws.

Turner McMullen has given me data on two nests found in Cape May or Salem County by him or his friends, as follows: May 30, 1923, in a black oak tree forty-five feet from the ground and May 11, 1925, in a red maple at about the same elevation, while W. B. Crispin has recorded in the latter county three nests May, 1898; May 20, 1900; May 23, 1909; all of these contained two eggs.

ROUGH-LEGGED HAWK

Buteo lagopus s.-johannis (Gmelin)

I have seen but three Rough-legs at Cape May, all of them close to the beaches. On October 31, 1920, a fine specimen came soaring past the Lighthouse at Cape May Point. Its large size, the square black patches on the under side of the wings near the bend, and the black belly all stood out clearly. Again, on January 8 to 10, 1922, and doubtless for a much longer period a Rough-leg was to be seen over the great Fill where he apparently found an abundance of field mice. Sometimes I would see him soaring steadily round and round and at others he would pause and flap his wings rapidly over one spot, even throwing his feet forward clear of the body as if getting ready to drop on some prey but did not do so. His favorite perch was on one or other of the line of telegraph poles which runs parallel to the beach front. He would come flying across the Fill and then on set wings would sail for several hundred yards dropping lower and lower all the time until he came to rest on one of the poles. When disturbed he alighted again a little farther on. As he perched his head showed white streaks and much white was visible on the upper side of the body while the black belly was conspicuous. When he flew the jet black patches on the under side of the wings came into view and the dull white of the tail with its dusky border. My third Rough-leg was seen soaring and flapping over Pond Creek Meadow, on December 21, 1935, pausing now and again like a Marsh Hawk to hover over some prospective prey.

William Rusling during his study of the hawk flights in the autumn of 1935 saw two of these birds passing southward on October 27, and Julian Potter saw two at Port Norris as late as March 19, 1933. About Barnegat Bay, to the north, Rough-legs seem to be much more plentiful winter residents and Charles Urner has many records. Edward Weyl saw two there on January 19, 1930, and four on February 2, 1930, while Ernest Choate found one on February 12, 1935.

The Rough-leg is a winter visitant at the Cape and is here close to the southern limit of its range.

GOLDEN EAGLE

Aquila chrysaëtos canadensis (Linnaeus)

The Golden Eagle is a rare bird in southern New Jersey and on account of confusion with the immature Bald Eagle some at least of the recorded instances of its occurrence are open to doubt.

Charles Voelker has recorded one shot at Cape May on October 20, 1892, by J. Milford, which he mounted, while there are other records of specimens secured in the state as follows:

Vineland, February 19, 1868. (Collection of John H. Sage. Auk 1895, p. 179.)
New Egypt, 1893. (Collection of Charles A. Voelker.)
Moorestown, November 8, 1901. (Collection Moorestown Natural History Society.)
Crosswick's Creek, Autumn, 1888. (Dr. W. C. Braislin, Auk, 1896, p. 81.)
Rocky Hill, near Princeton, March, 1881. (Collection William C. Osborn.)
Browns Mills, mid-December, 1921. (N. DeW. Pumyea, see Bird-Lore, 1922, p. 99.)
Long Branch, August, 1897. (F. M. Chapman, Auk, 1898, p. 54.)

The Browns Mills bird was shot by a deer hunter who claimed that the bird was about to attack him and he was compelled to shoot it in self defence but, as Julian Potter says in reporting the occurrence, "he probably made up the story in self defence." The Vineland bird had gorged itself upon the carcass of a deer and was killed with a club.

BALD EAGLE

Haliaeetus leucocephalus leucocephalus (Linnaeus)

Eagles as we see them about Cape May are usually high overhead, flying from some inland point out to sea, and are most frequent at Cape May Point when they come down the Bay from nesting sites in upper Cape May and Salem Counties, or on the Delaware shore near Bombay Hook, to soar by the hour over the tip of the peninsula.

I have seventy-seven Bald Eagle records in the seventeen years covered by my intensive study of the local bird life but considering the small number of days afield during the winter, spring and autumn, this probably represents but a small proportion of the actual occurrences of the bird in the region. On the other hand we have no way of telling how many duplications there may have been and how many individual Eagles were present, I have no doubt, however, but that a Bald Eagle could be seen every day from mid-August to the end of October if one were on the watch all of the time. The occurrences during September and October are probably in large part migrating birds on their way south. The records kept by the Audubon Association's wardens at the Point, during the hawk shooting season, are interesting as the men were present most of the time and the birds they saw were apparently going regularly down the coast, but even their counts doubtless included duplications. From September 1 and October 27, 1935, William Rusling saw sixty Bald Eagles; in the autumn of 1931 George Saunders counted forty; while Robert Allen, from September 15 to October 29, 1932, saw only ten; but he was engaged entirely upon watching the hawk killing and did not pay particular attention to the Eagles. Rusling counted seven on September 16, six on September 17, four on September 14 and October 19, and three on September 3. I saw three at once on September 16, 1921, and September 23, 1933, Walker Hand, three on April 26, 1918, and Julian Potter the same number on September 29, 1925, and September 29, 1929, while he saw four on September 17, 1922. Eagles are seen regularly through the winter and the number of individuals resident at that season immediately about the Cape is indicated by the Christmas census of the Delaware Valley Ornithological Club:

December 28, 1930, three.	December 24, 1933, five.
December 27, 1931, four.	December 23, 1934, four.
December 26, 1932, two.	December 22, 1935, seven.

On October 13, 1924, Julian Potter saw two adult Eagles, doubtless a pair, performing interesting evolutions in the air. They flew with flapping wings, never soaring, one slightly back of and above the other. The upper bird would suddenly dip toward the lower with claws extended but just as they almost touched one another, the lower bird would turn completely over and after hesitating for a moment, with back down and feet extended to meet the other, would right himself and the flapping flight would be resumed. These unusual actions were repeated a number of times until the birds passed from view. The whole performance was executed without apparent effort and seemed to be in the spirit of play rather than actual combat.

We see both immature and adult Eagles about Cape May, the former dark colored except for the whitish tinge at the base of the tail, the latter with head and tail snowy white although there is sometimes a mottling of dusky on the tail. Thirty-six of the sixty seen by William Rusling in the autumn of 1935 were immature. In my experience the brown birds have far outnumbered the white-headed adults.

The most characteristic feature of the flying Eagle, as we look up at him from below, is the absolute right angled position of the wings—straight out from the body without a slant or curve from shoulder to tip, and of nearly uniform width throughout, so different from the gently up-curved V-like wing-set of the Turkey Vulture or the downward arched pinions of the Fish Hawk. Another Eagle character is the prominence of the head and neck which seem to project farther out than in any of our other birds of prey, contrasting especially with the small naked head of the Turkey Vulture. The big white bill is also conspicuous and the widely divergent tips of the primary feathers. In young birds the under side of the wing is often more or less mottled and sometimes the head is streaked with white. When flapping the wingbeats of the Bald Eagle appear more labored and heavy than those of either the Turkey Vulture or the Fish Hawk, somewhat more like those of the Great Blue Heron. Bald Eagles soar a great deal, sometimes in circles sometimes straight away with the wind and flap only occasionally although now and then I have seen one flap continuously for considerable distances.

On September 2, 1920, an immature bird came sailing across on the northwest wind and continued on out to sea. It appeared very black below with the exception of an oblong area on the front margin of the wings. The tail, which was spread wide all the time, was largely black and was slanted now on one side now on the other, to aid in holding the bird's position in the gale. On October 3, 1920, another Eagle in about the same plumage flew from the ocean constantly flapping nor did he indulge in any sailing until he was well in over the land. He, too, looked very black against the sky. Approaching the Point on September 16, 1921, and glancing up I saw three Bald Eagles soaring around so close together that I was able to get them all in the field of my glass at once. The uppermost bird appeared very ragged, apparently molting, and was much mottled with white below and on the shoulders. The others seemed blacker but all had dark heads and tails.

The two latter were apparently chasing the first one and kept close to one another, flapping a good deal in a heavy, labored way. On April 2, 1922, another Eagle was flying in a strong northwest gale, flapping to hold its position. It would beat its way in the face of the wind for a time and then turn and sail with it and later return again to windward. Its tail, hardly ever fully spread, was often nearly closed and held straight out behind like a rudder. On October 1, 1924, a very ragged young bird came over the Point ahead of a light wind but at a slight angle to it and flapped continuously.

Hawks often associate with soaring Eagles during the autumn flights and later. On December 12, 1920, there were two Eagles accompanied by a single hawk circling over the Point, when there came a flight of twenty-three Turkey Buzzards from out of the north with six hawks in attendance. After a short time all of the hawks deserted the Buzzards and attached themselves to the Eagles. There was no evidence of any desire on the part of the latter for their company nor was any resentment shown.

On October 11, 1927, I saw an immature Eagle near Cold Spring, swoop repeatedly at a Turkey Buzzard, both birds twisting and flapping with a string of six Crows in attendance which attacked the Eagle when he ceased his offensive and chased him out to sea. I presume that the Vulture had innocently intruded upon the Eagle-Crow controversy as I had never before seen an Eagle attack one of these birds. Crows have the same antipathy to an Eagle as that which they evidence toward a hawk or owl but in less degree. John Carter writes me: "In one instance my attention was attracted to a

Crow which was giving the rally call and flapping vigorously upward in a spiral course. Looking far up in the sky I saw the cause of his behavior in an Eagle, calmly circling on his way. The Crow's enthusiasm soon lapsed as his comrades refused to rally and it was rather uphill work when carried on alone." Other cases of Crow persecution have come to my attention and on one occasion enabled me to identify an Eagle soaring with a party of Turkey Buzzards, which I had not previously noticed. William Baily tells me of two Eagles on Seven-Mile Beach which were being pursued by a flock of thirty Crows on November 10, 1895, and Otway Brown came upon a single Eagle sitting on a fence at Cold Spring, July 7, 1933, which was being pestered by some Crows, while the week previous probably the same Eagle was being followed from perch to perch by a whole flock of these birds which seemed to show no fear of him whatever. While the Crows make a great disturbance when engaged in tormenting any victim they do not seem to bestow their choicest invective upon an Eagle but reserve it for some misguided Great Horned Owl.

Various birds attack Eagles if they approach too close to their nesting grounds or young. I have several times seen Laughing Gulls pursue Eagles that have sailed innocently out over the meadows and on May 22, 1922, a single gull forced an Eagle to turn back and beat a hasty retreat to the mainland, flapping vigorously all the way. Even as late as July 23, 1922, and July 14, 1924, I have seen Eagles which flew over Jarvis Sound attacked by gulls that were feeding there, although none of the gulls nested near the spot. Again on May 1, 1932, I saw an immature Eagle on the meadows, back of Five-Mile Beach, pursued by a flock of Herring Gulls which forced him to drop something that he was carrying, apparently a dead fish upon which they had been feeding. He recovered it however and paying no further attention to his pursuers went flapping vigorously away to the west. On August 1, 1936, a couple of Black Skimmers viciously attacked an Eagle that flew over their nesting grounds although all the nests had been washed out earlier in the season and there were no young anywhere about.

The Fish Hawk is the only bird that the Eagle seems regularly to molest and the account of the Bald Eagle compelling the Fish Hawk to relinquish his hard-earned catch is an ornithological classic.

Mark Catesby in his famous "Natural History of Carolina," published in 1731, gives us one of the earliest versions: "The manner of fishing (of the Fish Hawk) is (after hovering awhile over the water) to precipitate into it with prodigious swiftness, where it remains for some minutes, and seldom rises without a fish; which the Bald Eagle (which is generally on the watch) at once spies but at him fiercely he flies; the Hawk mounts screaming up but the Eagle always soars above him, and compels the Hawk to let it fall; which the Eagle seldom fails of catching, before it reaches the water. It is remarkable that whenever the Hawk catches a fish; he calls as it were for the Eagle who always obeys the call, if within hearing."

Alexander Wilson (1811) does it better, describing the incident as he saw it enacted many times on the shores of Great Egg Harbor: "Elevated on a high dead limb of some gigantic tree, that commands a wide view of the neighboring shore and ocean, he seems calmly to contemplate the motions of the various feathered tribes that pursue their busy avocations below; the snow white Gulls, slowly winnowing the air; the busy Tringae, coursing along the sands; trains of Ducks, streaming over the surface; silent and watchful Cranes, intent and wading; clamorous Crows, and all the winged multitudes that subsist by the bounty of this vast liquid magazine of nature. High over all these hovers one, whose action instantly arrests all his attention. By his wide curvature of wing, and sudden suspension in air, he knows him to be the Fish Hawk settling over some devoted victim of the deep. His eye kindles at the sight, and balancing himself, with half-opened wings, on the branch he watches the result. Down, rapid as an arrow from heaven, descends the distant object of his attention, the roar of its wings reaching the ear as it disappears in the deep, making the surges foam around! At this moment the eager looks of the eagle are all ardor; and levelling his neck for flight, he sees the Fish Hawk once more emerge, struggling with his prey, and mounting in the air with screams of exultation. These are the signal for our hero, who, launching into the air, instantly gives chase, soon gains on the Fish Hawk, each exerts his utmost to mount above the other, displaying in these rencounters the most elegant and sublime aërial evolutions. The unincumbered eagle rapidly advances, and is just on the point of reaching his opponent, when, with a sudden scream, probably of despair and honest execration, the latter drops his fish; the Eagle poising himself for a moment, as if to take a more certain aim, descends like a whirlwind, snatches it in his grasp ere it reaches the water, and bears his ill-gotten booty silently away to the woods." (American Ornithology, Vol. IX, p. 129.)

Walker Hand, in his long experience, has many times seen the Bald Eagle drop from a great height to force the Fish Hawk to drop his prey, crying his protest all the while, and he tells me that the rush of the air through the wings of the descending Eagle is like the roar of a falling forest tree and can be heard for a considerable distance.

With the diminishing number of resident Eagles this act of piracy is not often seen today immediately about Cape May but, on April 24, 1926, David Baird, hearing the continual calling of a Fish Hawk directly over Washington Street in the heart of the town, looked up and saw an Eagle pursuing it in the regulation manner, and on another occasion, shortly after, I saw a Fish Hawk mounting high in the air and screaming constantly and at a little distance I saw an Eagle soaring, but it apparently paid no attention the hawk nor had the latter any fish.

Whether the Eagle's success is not what it was in times past or whether we have inferred from the graphic accounts of the older writers that he was more universally successful than was the case, I do not know, but Dr. J. Percy Moore tells me that while he has frequently seen Eagles pursue Fish Hawks near the fish pounds of the Chesapeake he has never seen them succeed in plundering them of their catch. At Ocean City, Maryland, Henry Fowler has witnessed the whole proceeding practically as Wilson described it, but the Eagles that he saw were by no means always successful. At this spot, he tells me, there are a number of deep sea fish pounds located some four miles out in the ocean. When he visited these on August 19, 1913, he saw numbers of Scoter Ducks and petrels the latter skimming the water with feet pendant, as if attempting to run over the surface of the waves. As he approached the pounds numerous gulls were to be seen flying about over them while Fish Hawks perched on the pound stakes. Every now and then one of the latter would dive and secure a fish with which it made off for the shore. Then from somewhere overhead, where he had been sailing, came the Eagle close on the trail and now and then he succeeded in so harassing the Fish Hawk that the latter dropped its prey and the Eagle promptly snatched it up before it could reach the water.

The natives of the Bay shore have an explanation of this curious habit which constitutes a pretty piece of folklore. As told to Fletcher Street by Chucky Rose, a picturesque trapper and "proger" of middle Delaware, it is as follows: "The Eagle years ago agreed to show the Fishing Hawk how to build a nest if the hawk would keep him supplied with fish. After receiving the desired information, however, the hawk failed to keep his end of the bargain, so henceforth the Eagle made a practice of worrying and robbing the Fish Hawk every time he found him carrying fish." The Eagle however is not always the aggressor in their conflicts. On August 8, 1929, an immature Eagle settled on the meadows opposite Bennett, a few miles north of Cape May, coming to earth with a thud not far from where I was standing. He rested for a few minutes and finally flew off to the northwest. As he approached the wood edge two Fish Hawks which had a nest there promptly came to its defence and swooped at him again and again. This however was their regular nest defence and had nothing to do with the fishing problem although it does show that the Fish Hawks are not really afraid of an Eagle.

Bald Eagles feed extensively on carrion and those individuals that we see out on the salt meadows, often in company with Turkey Buzzards, are usually feasting upon dead fish or other refuse washed up by the tides, while dead animals back in the interior of the peninsula also attract them. Otway Brown tells me of two that he saw feeding with Vultures at Cold Spring on April 23, 1932, and on July 30, 1927, there was one on the meadows back of Two-Mile Beach, which having fed on some dead fish rested directly on the ground for some time. He was a particularly fine adult with snowy white head and very black body, the white extending far down on his neck and nape. He stood out in striking contrast to the green meadows. A similar individual came down to feed on something on the meadows near Barrel Island in Barnegat Bay on December 28, 1928, and as I watched him he held his wings aloft for a moment or two after alighting.

On July 6, 1931, while crossing the meadows back of Five-Mile Beach in an automobile we flushed an Eagle from the ditch at the side of the road which had been so busily occupied with a dead chicken that he did not see us approach and in his effort to escape from the narrow space between us and the railing at the roadside he all but got one wing into the car! Eagles also feed upon ducks, though if our information were more accurate we should probably find that the majority of their catch were crippled or weakling birds. Henry Fowler tells me that thirty years ago when he used to visit the shores of the Chesapeake and Elk River, Maryland, before the fish pounds were established, fishermen drew their seines up on the beaches and sorted their catch leaving great masses of smaller fish to rot in the sun. These bountiful repasts attracted birds of all sorts as well as other scavengers —hosts of Turkey Buzzards, crows, gulls etc., and among them frequently one or two Eagles.

Norris DeHaven during his long experience on the New Jersey coast and on Delaware and Chesapeake Bays, has told me that he never saw an Eagle catch a live fish, as has been claimed by some writers but has often seen them pick up wounded ducks. Once he saw an Eagle pursue an uninjured Buffle-head Duck but without success as it was too skillful a diver. With surface-feeding ducks they have better success. Once or twice, too, he had seen Eagles, in misty weather, swoop down upon his wooden decoys and in one instance actually seize one of them.

On March 26, 1933, Richard Pough found two Bald Eagles on Delaware Bay, at Fortesque, N. J., pursuing the ducks and Snow Geese which had assembled there in large numbers and which were kept in almost continual motion by the activities of the big birds. Although they were kept under observation for an hour not a single duck or goose was caught; they seemed able to outdistance the Eagles with ease. In Joseph Tatum's experience, near Brigantine, on March 5 of the same year, the Eagle was more successful. He had forced a large flock of Brant to take wing and selecting a laggard he gave chase. The Brant was forced in over the land and after several misses the Eagle turned over below his prey and struck upward into its breast. It fell to the ground and flopped about and the Eagle executing a quick turn dropped on his prey and with some difficulty bore it off to a sod bank in the marsh where he proceeded to enjoy his feast. It was quite evident that the Brant was in a weakened condition which accounted for the Eagle's success.

As to preying upon domestic stock or poultry we have little authentic evidence. When practiced, however, it is in cases of dire necessity but the widespread prejudice of farmers and sportsmen against any bird of pray, a prejudice which increases in proportion to the size of the bird, prompts the farmer to believe any story reflecting on the Eagle and to blame him for crimes of which he is entirely innocent. In Delaware the Eagle is blamed for killing young domestic ducks and doubtless this is true since ducks are wont to stray out on the marshes far from the protection of farmhouses and would be taken by the Eagle for wild birds while they would doubtless prove much easier prey and are available all the year round. While we have at least one instance of a chicken being found in an Eagle's nest there is no proof that it was killed by the bird and it might just as likely have been picked up dead.

Alexander Wilson published an account of "a woman who happened to be weeding in the garden, near Great Egg Harbor, and had set her child down nearby to amuse itself while she was at work; when a sudden and extraordinary rushing sound, and a scream from her child alarmed her, and starting up, she beheld the infant thrown down and dragged some few feet, and a large Bald Eagle beating off with a fragment of its frock." This is the only case of the sort reported in New Jersey by a reliable person and even here the story is apparently second-hand and, granting it to be correct, it is quite likely that the bird was gathering nesting material and took the child's clothing as something suitable for its purpose. In similar reports from elsewhere there is always a lack of careful observation and a great deal of unwarranted deduction.

We see Eagles at other places about Cape May besides the Point or the sounds. I have seen them passing out to sea at Schellenger's Landing and several times, both in August and in midwinter, high in the air over Erma and Cold Spring, respectively, while Walker Hand and I found a fine whiteheaded adult perched on the top of a dead tree on the edge of a remote landlocked pond back of Green Creek. In the vicinity of their nesting place they often have a regular perch from which they look out over the surrounding country. There was one Eagle that used to fly from Timber and Beaver swamp near South Dennis to a tree on the edge of the marsh where, sheltered by the foliage, he had a clear view far out to the Bay. Another had a perch in a well foliaged tree at the head of Bidwell's Ditch from which he had a similar view and where I have seen him from the Bay Shore Road many a time although unless one knew where to look he would easily have escaped notice.

Eagles have long been known to nest in various localities in Cape May County as well as in Salem and Cumberland Counties to the northwestward. Alexander Wilson has presented the first account of the nesting of the Bald Eagle in the state. He says, "In the month of May, while on a shooting excursion along the sea coast, not far from Great Egg Harbor, accompanied by my friend Mr. Ord, we were conducted about a mile into the woods to see an Eagle's nest. On approaching within a short distance of the place, the bird was perceived slowly retreating from the nest which we found occupied the center of the top of a very large yellow pine. The woods were cut down, and cleared off for several rods around the spot, which, from this circumstance, and the stately erect trunk, and large crooked wriggling branches of the tree, surmounted by a black mass of sticks and brush, had a very singular and picturesque effect. Our conductor had brought an axe with him to cut down the tree but my companion, anxious to save the eggs, or young, insisted on ascending to the nest, which he fearlessly performed, while we stationed ourselves below, ready to defend him in case of an attack from the old Eagles. No opposition, however, was offered; and on reaching the nest, it was found, to our disappointment, empty. It was built of large sticks, some of them several feet in length; within which lay sods of earth, sedge, grass, dry reeds, &c., &c., piled to the height of five or six feet, by more than four in breadth; it was well lined with fresh pine tops, and had little or no concavity. Under this lining lay the recent exuviae of the young of the present year as scales of the quill feathers, down &c. Our guide had passed this place late in February, at which time both male and female were making a great noise about the nest; and from what we afterwards learnt, it is highly probable it contained young, even at that early time of the season.

"A few miles from this is another Eagle's nest, built also on a pine tree, which, from the information received from the proprietor of the woods, had long been the residence of this family of Eagles. The tree on which the nest was originally built had been for time immemorial, or at least ever since he remembered, inhabited by these Eagles. Some of his sons cut down this tree to procure the young, which were two in number; and the Eagles soon after commenced building another nest on the very next adjoining tree, thus exhibiting a very particular attachment to the spot. The Eagles he says, make it a kind of home and lodging place in all seasons." George Ord adds a note in his edition of "Wilson" to the effect that on March 1, 1913, a set of three eggs was taken from the first nest described above in which the young were perfectly formed. The old bird remained on the nest until several blows of an axe had been given to the tree.

Most of the Cape May County Eagle nests have been located in more or less inaccessible spots in the heart of deep swamps or dense stretches of woodland, due perhaps to the antipathy of the farmers who regard the Eagles as enemies to the poultry yards and cut down the Eagle trees when they come upon them. The persistence of egg collectors, too, has had much to do with the difficulties that the birds have experienced in raising their broods. A set of Eagle eggs has always been a great desideratum to a collector although, with our present knowledge, no scientific value attaches to the collecting of any additional specimens. In March, 1935, three of the occupied nests in Cape May County were robbed by "oölogists."

In the winter of 1935-36 Fletcher Street and Turner McMullen, of the Delaware Valley Ornithological Club, endeavored to locate all of the South Jersey Eagle nests with the object of giving them adequate protection and no less than six were found in Cape May County, of which four were known to have been occupied during the previous nesting season; while four more were located in Salem and Cumberland Counties.

The nest trees were posted, with the coöperation of the Audubon Association, and as a result all of the four occupied Cape May nests produced young —one, two, three and three respectively. Details of the earlier history of these nests have been furnished to me by Turner McMullen.

Cape May County nests:

Nest No. 1. Located fifty feet up in a large sweet gum tree. On June 6,

1920, it contained two young about ready to fly with young in the same condition on June 15, 1924.

Nest No. 2. Seventy feet up in a pine tree, six miles distant from the preceding nest. Contained three young on May 3, 1936.

Nest No. 3. Located sixty feet from the ground in the extreme top of a sweet gum tree June 15, 1924, contained two young ready to fly; May 27, 1928, two young visible from the ground; March 3, 1935, four eggs; March 8, 1936, three eggs.

Nest No. 4. Located seventy-two feet up in a sweet gum eight feet in circumference; the nest which was measured was eight feet in height and eight feet across the top. On March 9, 1930, it contained a single egg and a large bone and on March 22 two eggs with the bone still there. The following autumn the top of the tree was cut off and the nest lowered into a truck and sold with the bone still in place! Probably the same pair of birds built a nest about a quarter of a mile from the site of the last in a pine tree about fifty feet from the ground. On March 3, 1935, it contained one egg and on March 8, 1936, three eggs. On April 19, 1936, it held one young about two days old and two eggs. On May 3, only the fledging remained accompanied by the portions of three eels!

Nest No. 5. Located forty-five feet up in a dead pine on March 22, 1930. On January 25, 1936, a new nest was located one hundred feet south of the last, in a living pine forty feet from the ground, and birds seen in the neighborhood. On April 19, 1936, it contained two young.

In Cumberland County:

Nest No. 6. In the dead top of a tulip poplar, ninety feet up examined March 13, 1932. On March 8, 1936, another nest was built apparently by the same birds, one hundred yards farther south, also in a poplar tree seventy-five feet from the ground, with birds in the vicinity. It contained two young on April 19.

In Salem County the located nests are as follows:

Nest No. 7. Eighty feet up in a large white oak. Was visited March 6, 1921, April 1, 1923, and March 14, 1926. This nest was destroyed and by 1932 another was built in a willow oak close by which was visited March 19, 1933, April 14, 1934, March 9, 1935, and February 1, 1936. While the birds were present "oölogists" had robbed the nest nearly every year and no young were seen.

Nest No. 8. In a Red Maple in a swamp two miles north of the last. Old bird was on the nest on April 21, 1929, and February 21, 1932. This nest fell down and another was built by March 9, 1935, sixty feet up in another white oak two hundred yards south.

Nest No. 9. Forty feet up in an oak, two miles and a half north of the last. On March 15, 1930, this nest had fallen and another was built in a red maple half a mile away and forty feet from the ground. It contained a single egg on this date and birds were seen there on March 19, 1932.

Nest No. 10. In a white oak sixty feet up. Was in use in 1935, and in 1936 had been robbed by May 2. We have less definite information regarding three other nests in these three counties.

One nest that I examined some years ago (No. 5 above) was in a dead pine in the center of a clearing and was about forty-five feet from the ground. It was built in March, 1927, but in the spring of 1928 Walker Hand, upon visiting the spot, found a Great Horned Owl in possession which crouched down in the nest with its head and "ears" conspicuous over the side. It was apparently incubating eggs. All of the nests that I have seen were about six feet in height and from four to five feet across and were flat on top with only a small concavity large enough to hold the eggs.

My personal experience with nesting Eagles has been mainly at Bombay Hook in the state of Delaware. The Delaware River is here bordered by broad marshes and not infrequently outlying bits of higher ground are separated from the mainland proper by other extended areas of marsh. On all the higher ground are scattered trees, while farther back are fringes of woodland sometimes of considerable extent. Where the woods have extended to the shore they have been drowned out by numerous high tides and many of the trees killed while later the winds have twisted and broken them until they now present a forest of tall stubs and bare trunks. The morning of my visit to this region, on March 8, 1919, was one of mist and fog so dense that we could scarcely distinguish objects twenty yards away. Sounds carry well in a fog and as we approached the bordering woodland we heard many bird voices although the birds were quite invisible. There came the call of a Killdeer, the first of the season, as the bird passed overhead, and then from out over the marsh came the honking of Canada Geese, a wild desolate sound in keeping with the morning. On they came like a pack of baying hounds in full chase and now they are directly over us, but invisible as our eyes could not penetrate that curtain of fog.

We soon came to a felled tree lying on the ground with an Eagle's nest still in its forks. This was the one described to me by Fletcher Street and to which he had climbed the year before. The farmer had cut down the tree to avenge some imagined misdeeds of the Eagles. The nest had been built in the three-forked crotch of a sour gum tree the rim being forty feet from the ground; its total height was six feet and its diameter four and a half. It was built of dead branches one or two inches thick and a number of them three feet in length; they were interwoven with sods and tufts of grass and the upper surface was practically level with a cup in the middle six inches deep and twelve across, lined with fine grasses and dry corn fodder, while there was an eight inch border of the latter around the cup in which the two eggs rested. A few downy feathers were scattered about. Another nest also described by Street was forty-three feet up and measured five and a half feet across. In it sedge and sod were substituted for the corn fodder. It likewise was flat on top as were all the nests that he had seen. Upon crossing a little clearing in the flooded woods we came upon an Eagle's nest in the fork of a sour gum the successor to the one cut down the year before. It tallied well with the one on the ground—a great mysterious mass ill-defined in the gray haze, while the two birds dim and ghostly, perched on two broken tree trunks, like grotesque cappings to wooden totem poles. This nest proved to be empty.

As the view cleared we could see far over the marsh where on patches of higher ground nearer to the Bay were other stretches of woodland and other great nests—some Fish Hawks' and some Bald Eagles'. One that particularly attracted our attention had originally been built by a pair of Fish Hawks and last year had harbored a pair of Great Horned Owls. They had been content to take it as the Fish Hawks had left it but the Eagles had built it up according to their standard and had added a mass of material to its bulk. The female Eagle was on the nest as we approached and the male perched on a treetop nearby but later she flew off and joined him. They flew back and forth past the nest tree with rapid wingbeats, now and then sailing a short distance and all the while uttering a series of short squeaky calls somewhat like those of the Fish Hawk but much weaker and more nearly resembling the sound of a slate pencil scratching a slate, they were given in sets of four or eight—*kip, kip, kip, kip—kip, kip, kip, kip.*

As the Eagle flies close by one is struck by the similarity in the time of the wingbeats to those of the Crow as opposed to the much more labored efforts of the Turkey Buzzard, or perhaps this is really due to our closeness to the bird. The intense blackness of the under parts, too, is characteristic, the counter shading on a naturally dark color intensifying it to a noticeable degree. In the case of the birds seen in the early morning mists, when the white head and tail are not distinguishable, the dark color and the character of the wingbeats made them look remarkably like big Crows. On alighting on a tree top the feet are thrown forward to grasp the perch, displaying the flank feathers to good advantage and in the sailing bird the tips of the primaries are seen to be separated from one another but not to the same extent as in the Turkey Buzzard. One of the pair before us disappeared almost immediately and the other also eventually flew away. At neither nest did they show any inclination to attack the intruder as he climbed to the eyrie. The year previous Fletcher Street tells me that the birds belonging to the fallen nest continued to sail anxiously about while he made the ascent and the female several times swooped down past the tree making a distinct roar with her wings so rapid was her speed, but she made no attempt to attack him.

As the birds fly past the head looks very large and white, but at close quarters it always appears soiled, and is not impressive—not nearly so much as the rich dark hue of the body. The hooked bill is clearly seen but the yellow color is not striking and it seems to differ little from the color of the head. As the Eagle flies from perch to perch the head is lowered and the body bent over at right angles to the legs, just as it is when the bird walks on the ground, and often the wings are held half open awaiting the moment to make the leap into the air.

In building their nest the Eagles undoubtedly pick up fallen sticks from the ground and George Stuart and Fletcher Street have seen these Bombay Hook Eagles sailing down close over the ground apparently in search of material though they did not actually see them gather any. At Cold Spring Otway Brown and I have several times seen Eagles flying overhead carrying sticks or other material to nests back in the woodland to the north, but Henry Fowler and Walker Hand made the most interesting observation along these lines. On April 5, 1912, they were seated on the edge of the marsh near South Dennis, Cape May County, where there were a number of muskrat houses scattered about. Presently a mere speck in the sky attracted their attention and as it got larger they recognized it as an Eagle. On it came lower and lower and then, when still at a considerable height, it raised its wings and shot at terrific speed straight down toward them and alighted on the top of the nearest muskrat house. Folding its wings it gazed this way and that and then began to work its feet down into the sods of the house. It gave several flaps with its wings but with no result and resumed its foot action. The next effort was more successful and with powerful strokes it arose in the air and with it went the whole top of the muskrats' domicile. As the bird rose higher and higher showers of dead grass, dirt etc., sifted down but the mass that it took away with it was considerably larger in bulk than its body. It disappeared in the direction of the Great Cedar Swamp back of Ludlam's Bay where it doubtless had a nest under construction. Eagles must gather a large amount of sods judging by the proportion of them that go to make up the nest mass and it seems likely that this robbing of the muskrat houses may be a favorite method of procuring this desired material though I have never found any mention of it in works on the subject. On March 13, 1927, Frank Dickinson saw an Eagle, on his farm near Erma, which descended in the manner just described on a small haystack that he had raked up and succeeded in carrying off almost the entire mass to the nest that was being built that year in the vicinity.

I have several times seen Eagles in the Cape May district pursuing one another, doubtless in play or as part of the mating activities, but sometimes these pursuits may be of a hostile nature. Dr. William L. Abbott has described such a case where the birds were fighting in the air in the upper part of Cape May County and their talons became locked together so that they fell to the ground. A farmer upon running to the spot found that one of the contestants had freed himself and was just taking wing. The other however offered fight and was killed with a club. Norris DeHaven told me of an Eagle which had been shot in the act of carrying off one of his duck decoys. It fell into the water near to the boat and at once attempted to come aboard. Doubtless any wounded Eagle will show fight but my experience fails to show any such disposition in a free wild bird, even when its nest is in danger.

Soaring Eagles are, after all, the most satisfying to the bird-lover and to those who have been brought up on the conventional Eagle of the average artist, with wide spread wings and graceful pose, the Eagle close at hand must be a disappointment, for Eagles like Turkey Buzzards appear to better advantage at a distance. I shall never forget my first near view of a Bald Eagle in the Zoo when I utterly failed to harmonize with my ideal the thick-set, short-legged bird with great ill-fitting wings and dingy white head, perching listlessly on a rock or waddling clumsily from side to side as it walked on the ground.

To arrive at a just estimate of our Bald Eagle we must, I think, first dissociate him entirely from the conventional American Eagle of the orator and poet—remove him as it were from our national shield and let him stand on his own feet. Had Benjamin Franklin had his way the Bald Eagle would probably never have gained the prestige that he now enjoys. Franklin would seem to have been a better ornithologist than those who chose our national bird and he deplored the selection of a bird of such a cowardly disposition, though he humorously remarked that after all it made little difference as few would be able to distinguish an Eagle from a turkey, but whether this was a reflection on the artist or the general state of scientific knowledge history does not state. The committee evidently had in mind the Golden Eagle, a very different bird, both in appearance and habit, and did not realize their mistake. Thus the Bald Eagle, like many another character in our history has achieved high station through the machinations of ignorant politicians!

Viewed solely upon his merits as a bird the Bald Eagle is an interesting and unique species. A degenerate member of the eagle tribe whose whole life and character have been modified by a fish diet. In his case however the specialization has not advanced so far as in the Fish Hawk, which feeds exclusively upon the finny tribe, and curiously enough it is upon him that the Eagle has developed a practise of parasitism which is one of his most interesting and unique traits. Probably this habit is responsible for the fact that the Bald Eagle never became an expert fisherman on his own account and as a result, following the line of least resistance, when the Fish Hawk fails him he ekes out his existence by becoming a scavenger. Withal however he retains traces of his predaceous character and can, upon necessity, kill his own game. His great wing power, except in the pursuit of the Fish Hawk, would seem to have become to a great extent useless, unless to keep him out of harm's way! What, we wonder, has been the history of the Bald Eagle's evolution? Did he first acquire a taste for fish and then resort to the mean practise of enslaving the poor Fish Hawk or did the latter habit develop first and lead him to adopt a diet of fish? These are some of the problems connected with the history of this interesting bird. There are many others and among them one of especial interest is the cause of the peculiar plumage pattern almost unique among birds. But we are not likely to solve any of them.

Unfortunately Bald Eagles are steadily becoming scarcer year by year, Game Commissions, intent only on man's sole right to fish and game, see to it that the Eagle gets little or no protection; farmers relying more on theories and tradition than upon the facts before their eyes, wage relentless war upon the Eagle eyries whenever possible; and sportsmen and gunners desirous of adding to their fame take a shot at an Eagle whenever opportunity offers, while egg collectors steal the eggs in the name of science long after all the science connected with them has been made known.

However all this bids fair to be changed and the well directed efforts now being put forward will we hope give this much persecuted and unjustly maligned bird the protection that he deserves. The irony of the whole Eagle situation seems to me to lie in the fact that this bird in the first place a coward and a parasite, and emblematic of nothing in particular, and later the victim of persecution almost to the verge of extermination, is referred to as the Bird of Freedom! And yet to any lover of the outdoors and friend of wildlife there is always something peculiarly impressive about the appearance of a Bald Eagle within the range of our daily walks. Perhaps in spite of all drawbacks we recognize him as our national bird; perhaps it is the element of mystery which surrounds him—whence he came and whither he is bound, and the ease with which he can range at pleasure over wide areas of both land and sea. Whatever may be the cause there is always that involuntary thrill when we gaze aloft and recognize this great bird sailing in majestic circles against the deep blue vault of the sky. We forget his weak defence of his nest and his utterly inadequate voice; we forget his lazy plundering of the luckless and hard-working Fish Hawk; we forget his tendency to share in the Turkey Buzzards' feast of carrion; and we agree that the days upon which we can record: "I saw an Eagle today" are red letter days on our calendar.

MARSH HAWK

Circus hudsonius (Linnaeus)

The only hawk that is present at Cape May the year round, and with the exception of the Osprey and Bald Eagle, the only "bird of prey" that breeds here at present, is the Marsh Hawk. Adapted in every way for a life in the open he finds in the great coastal marshes a peculiarly congenial environment. Here it is that his chief article of diet, the meadow mouse, is to be found in abundance, and here on the higher spots, near the upland, his nest is built.

On almost any day one or more of these hawks may be seen about the town. If we sweep the meadows with the glass we shall find one sailing gracefully over the marsh, rising and falling in undulating flight as he clears the tussocks and low bushes among which he seeks his prey. He flaps his wings occasionally but apparently just enough to maintain his poise and direction, and if there be a little wind he will shear up to right and left like a toy kite straining at its cord, displaying at such times the telltale white rump, which stands out distinctly at a great distance—an infallible mark of identification. We see the brown female Marsh Hawk more frequently than the beautiful silvery gray male but in late summer and autumn nearly all that we see are the rusty reddish brown birds of the year.

In ordinary flight the Marsh Hawk gives about six rather heavy but powerful wingbeats and then an undulating sail, followed by another series of beats and another sail, the two alternating regularly. The tail is closed during flight and is held straight out behind like a rudder and only rarely, usually when the bird makes a sharp turn, is it even partly expanded. The head is bent over so that the face is directed downward and the ruff becomes very prominent as we look up at the bird as he passes overhead. Every now and then spying some prey in the grass below the hawk will pause and "harry," or hover in the air, on rapidly beating wings. When he alights or pounces on some luckless mouse the tail is flirted up and down in a most conspicuous manner, as the wings are closed and folded just above it. Sometimes a bird will alight in the same way on a low post on the meadows but I have never seen one perch in a tree.

The Marsh Hawk seems to prefer to hunt over the border of the marsh where bushes are scattered about, either because his food is more plentiful there or because it is there that his nest is located. The Fill above the town is also a favorite resort and he may be seen threading his way among the bayberry bushes, sometimes following this shelter close up to the backs of the nearby cottages. During the great autumn hawk flights at Cape May and less frequently at other times Marsh Hawks may be seen flying high overhead and we note then that the tips of the primary feathers are slightly separated, that the feet, as in all hawks, are stretched straight out behind and look surprisingly long. The neck is very short which together with the peculiar facial ruff gives the head a bullet-like appearance. In autumn, too, Marsh Hawks will join other soaring hawks and Turkey Buzzards that congregate in a veritable maze above Cape May Point and we find it difficult to realize that these circling birds high up among the clouds are the same as the lilting harriers of the marsh. They may readily be recognized during these evolutions by their long straight tails so different from the fan-shaped tails of the other large species, while their wingbeats are more ponderous and more frequent. I have seen them flying thus high overhead in late August, September and early October of nearly every year, and on October 1, 1923, there were upwards of one hundred seen in the air during the day.

Once in the autumn I saw a Marsh Hawk dash right through a piece of low woodland after the manner of a Sharp-shin, doubtless pursuing some small bird for he does not disdain to take such prey when the congested migration at the Point offers them in such abundance, but this was the only occasion on which I saw one enter the woods. On September 21, 1929, I startled a Marsh Hawk from behind a bayberry bush on the Fill, where he had been feasting on one of the thousands of Tree Swallows that thronged the air. The feathers had been almost entirely stripped off and lay scattered all about the spot, but every bit of the bird, with the exception of one leg, had been eaten. As an illustration of the apparent rarity of such a feeding habit I saw a Marsh Hawk following a ditch in the meadows, probably in search of mice, pay not the slightest attention to swallows that were swarming all about him, while another on November 8, 1930, flew through a throng of Tree Swallows which were feeding over a small pond near the Lighthouse without making any attempt to attack them. Possibly in the first instance the bird had been injured and was fluttering on the ground.

There is an increase in the number of Marsh Hawks in September when the resident birds and their broods are augmented by the arrival of migrants from the north. I have seen as many as twenty at once on September 4, 1930, over the meadows north of Cape May Point. Some were soaring high others resting on cut hay on the marshes and others, still, pursuing one another in the air and going through all sorts of evolutions. I have also seen young birds of the year chasing one another in this way and resting on pieces of sod on the meadows near South Cape May as early as August 9, 1922. From twenty to fifty are to be seen during the height of the flight about September 20 to October 10, but usually the daily count is less, and at other times from one to six are all that are seen, while on certain days I have searched in vain for a single bird. There seems to be an increase in numbers in April when migrants are presumably passing north. The Audubon Association's wardens have counted two hundred and sixty-four from September 15 to October 29, 1932; and two hundred and seventy-four from August 4 to November 6, 1935.

Association with other hawks, Turkey Buzzards or Eagles, in soaring evolutions is probably voluntary with the Marsh Hawk; but contact with other birds seems to be forced upon him. Many times Marsh Hawks have been viciously pursued by Red-winged Blackbirds as they crossed the Fill or the marshes where the latter had nests or young, and on one occasion they forced the hawk to mount in the air until he disappeared from sight. On another day a pair of Barn Swallows also forced a Marsh Hawk to mount high in the air in his efforts to throw them off, so closely did they press him and another individual that crossed the Harbor was immediately attacked by three Laughing Gulls as he attempted to continue his flight over the meadows to the north. One was pursued by a Fish Hawk whose nest he had inadvertently approached, while yet another sailed close to a soaring Bald Eagle and beat a hasty retreat when he recognized the identity of his associate, although the Eagle apparently paid no attention to him. Once, on October 16, 1927, I saw a female Marsh Hawk swoop several times at a nearby Turkey Buzzard but it is very unusual, in my experience, for them to take the offensive.

Three times I have been close to nesting places of the Marsh Hawk without being able to locate the nest. On Two-Mile Beach on May 22, 1922, a female flew out from the edge of the woods and disappeared while her place was at once taken by the male which was much excited and flew back and forth over my head constantly uttering a throaty or woody call: *kelk, kelk, kelk, kelk*, etc., the note being uttered once in two seconds. A thorough search of the vicinity was fruitless and next day there was no trace of the birds. Possibly they had young somewhere near and later led them elsewhere. On July 5, 1923, on the edge of the marsh at Brier Island a female Marsh Hawk circled over head as long as I remained in the neighborhood calling *whi hi hi hi; hi hi hi hi; hi hi hi hi; hi hi hi hi*—sixteen short notes in blocks of four and then a rest, followed by a similar series, recalling somewhat the continuous spring calling of the Flicker heard at a distance (the sound of the "*i*" was short). A male acted in precisely the same way on Pond Creek Meadows on July 12, 1928. Nests of the Marsh Hawk have been found near Cape May by others more fortunate than I. Richard Miller tells me of one near Cape May Court House found on May 21, 1922, which was situated on the ground among bushes and tall grass on the edge of the salt meadows and contained five eggs about half incubated and another discovered on May 13, 1923, on the meadows north of Cape May Point which stood in sedge fifteen inches high and contained two fresh eggs. A third nest found July 19, 1924, had full grown young birds still in it, and another, June 15, 1930, held four eggs.

The meadows east of Cape May Court House seem to be a favorite breeding place for Marsh Hawks and Turner McMullen has given me data for eighteen nests that he has examined there. Four were found in 1924, three on one day, May 25, three in 1925 and three in 1927 all on May 8; in other years only one or two were found. The dates run May 3, two nests; May 4 and 5, one each; May 8, three; May 9, two; May 10, 18, and 23, one each; May 25, three; June 8, one. Three nests contained four eggs, four contained five and seven contained six. Another nest found April 22, 1928, held two eggs; one on May 25, 1924, three eggs and a young bird; June 23, 1923, one egg and two young; another May 25, four young. He found these nests either on salt hay or in sedge grass, sometimes in grass over two feet in height, near bushes.

The Marsh Hawk seems to stand apart from our other hawks—an open ground as distinguished from a forest loving species, while his owl-like facial ruff gives him a distinctive appearance and his white rump patch an ever present mark of identification. Adapted for a life on the ground all of his important activities—feeding, nesting and resting are carried on there, while his search for food is conducted in the air usually at low elevations. About Cape May the broad green marshes bordering the sea are his home and on these he spends his life tirelessly sailing or harrying as he seeks his prey, and as this consists largely of meadow mice and other rodents he deserves the protection of every farmer.

OSPREY

Pandion haliaëtus carolinensis (Gmelin)

The Osprey or "Fish Hawk," as it is universally known on the New Jersey coast, is probably more closely associated with the seashore, in the popular mind, than any other bird. As we approach Cape May either by train or motor, and turn into the base of the long peninsula, we almost immediately begin to see the great nests of the Fish Hawk here and there in the tree tops and hear the peculiar whistling notes of the birds, so different from the cries of our inland birds of prey, and we realize that we are approaching the sea.

All summer long out over the ocean or across the broad expanse of salt meadows we constantly see the Fish Hawks flapping heavily on their way back and forth from their nests, easily distinguished from the Laughing Gulls, which are even more abundant, by their dark coloration and more labored action, and from the Turkey Buzzards by their downcurved instead of upturned wings when soaring. During July they have little time to waste and their line of flight lies straight as an arrow, from their fishing grounds beyond the breakers to the nest back on the mainland, where hungry young are ever ready for food. At a distance they appear uniform black and it is only when they are nearer at hand or flying directly overhead that one sees the white of the belly, the white patches on the under side of the wings and the white crown and face markings. The wingbeats of the Fish Hawk in direct flight seem to be continuous, heavy, and labored, as if it took all of the bird's energy to keep his body afloat, but if we watch carefully we shall see that there is a short sail every now and then between the recurrent series of beats. About their nests and at other times, especially later in the season when the young are awing, the Fish Hawks engage in much soaring, either in wide circles or straight away against a steady wind, and then, more than in the heavy flapping flight, do they demonstrate their mastery of the air. With long wings held rigid and somewhat arched and decurved at the tips, they circle with infinite grace mounting higher and higher in the heavens, or perhaps skim in a straight line overhead at astonishing speed. In May we find the female at the nest and the male often perched nearby and on May 7 and 8, 1932, driving along the shore roads of Cape May County I have counted as many as seventy Fish Hawks established for the summer. During the period of incubation we do not see many of the birds out in the open or along the strand but later their numbers increase steadily until the maximum is reached when all of the young are on the wing. In late summer family groups of four or five or several families together, may soar overhead mounting up and up until they are almost lost to view, and on such occasions they may associate with soaring Turkey Buzzards or with migrant hawks but after mid-September they begin to depart for the south and their numbers steadily decrease.

Out over the ocean the Fish Hawk flaps his way up or down the coast and upon sighting his prey in the water below, with wings partly closed and feet extended, he drops like a plummet, striking the waves with a splash and with a force that frequently carries him entirely beneath the surface. One diving bird (August 5, 1920) disappeared for five seconds but emerged with his fish safe in his talons and made off at once for the mainland. On other occasions the strike was not successful, indeed I think that the birds miss more often than they score. One that was seen to fail in his strike immediately mounted in the air and shaking his wings violently for a moment was once more on the hunt and within less than two minutes made another attempt. Another Fish Hawk (May 9, 1924) which had made a successful catch was seen to have great difficulty in bringing his prey to shore. He flew very close to the water and twice ruffled up his plumage and shook his wings vigorously and upon reaching shore alighted on the strand to rest. Another bird that had been completely submerged in his strike seemed to have great difficulty mounting in the air in the face of a strong wind doubtless due to his failure to shake the water from his plumage.

Once a fish is secured and grasped firmly in both feet the legs are moved so that one foot is directly in front of the other and the fish is held parallel to the body of the bird with the head always directed forward and as a bird passes overhead we can see the unfortunate fish wriggling and glistening in the sunlight. Occasionally a moderate sized fish will be held in one foot only. The Fish Hawks also seek their prey back on the Bay or at the "ripps" where the waters of the Bay and ocean meet and where fish seem to be more plentiful, and here on August 26, 1920, amongst a flapping maze of over five hundred gulls of several species and many terns I counted no less than twelve Fish Hawks. Their black bodies stood out prominently and through the glass their heavy wing action could be distinctly seen as well as their plummet-like dives as they passed vertically through the maze of flapping gulls. At Pierce's Point on the Bay side I saw several Fish Hawks perched on the stakes of the fish and crab pounds (July 11, 1920) and I saw them several times swoop down to the surface of the water and snatch up fish that had been discarded by nearby fishermen. I am told, also, that they will sometimes pick up mossbunkers that have been stranded on exposed sand bars, but they will not descend to the ground to recover a fish that has been dropped from the nest. On one occasion I saw a single Fish Hawk fly directly out to sea and disappear in the distance bound, doubtless, for some fishing bank that lay well offshore.

Toward the end of the summer more and more Fish Hawks are observed devouring their prey on stakes out on the meadows, on telegraph poles, or actually on the ground, while some of the birds sailing round and round high in the air carry untouched fish in their claws. It is probable that some of these are young birds but just when they begin to fish for themselves or whether the parents feed them at other places than the nest I have not been able to determine. It is probable, too, that the old birds also feed away from the nest after the young are on the wing although earlier in the season the nest seems to be their feeding stand, as birds that have been flushed from the nest carried pieces of fish in their talons as they flew around protesting at my intrusion. The birds that I have found feeding on the tops of stakes or on dead trees usually have released one foot from their prey so as to secure a hold on the perch, but they may not do this when the fish is not quite dead. One bird that flew directly from the ocean to a telegraph pole with a rather large fish grasped in both feet was unable to hold his position and flew from one pole to another, flapping his wings at each stop, in a vain effort to secure a perch without losing his prey. Another bird (September 4, 1924) alighted on a telegraph pole with a wriggling fish which he held firmly in one foot against the top of the pole. He would gaze intently down at it and apparently press heavily upon it as if to force his claws more deeply into its body and then gaze leisurely out over the landscape, returning later to a contemplation of his catch. Another bird that emerged after a successful dive, was seen to have only one foot on his fish and was compelled to alight on the beach to adjust it, in the usual way, before flying off to the nest.

As the season advances we see more and more Fish Hawks perched on stakes out on the meadows, especially along drainage ditches or about shallow ponds. On August 3, 1920, there were five birds and on August 30, seven, perched on stakes by a small pond east of the Lighthouse among a flock of white herons which had assembled there, and on July 29, 1931, there were twelve sitting on the meadows above the Harbor. While these were doubtless young birds I have seen adults resting on stakes and telegraph poles back of Two Mile Beach (May 22–25, 1922) while I was living in a little cabin on the edge of the sounds. On September 1, 1920, several Fish Hawks were sitting out on a sand flat above South Cape May and on August

9, 1922, and August 4, 1925, three were standing and feeding on a log or on the ground in an open spot on the Fill. Some birds seem to have regular feeding stakes to which they repair day after day leaving bones, opercula, entrails and scraps of meat strewn about over the ground which the waiting Turkey Vultures make way with at the first opportunity.

The head of the flying Fish Hawk is usually held straight forward although it may be turned slightly to the right or left and I observed one individual gazing straight down 'as it passed overhead, the position, I presume, that is taken when the bird is fishing. In alighting on a perch the Fish Hawk throws his feet forward as he approaches and flaps his wings forward to check his advance, poising them above his body as he settles. He then reaches his head forward and looks straight down to the right and left after which he assumes the characteristic position with head lowered slightly below the shoulders, feathers of the crown elevated, body plumage somewhat ruffled and wings hung rather loosely, like a shaggy overcoat. One bird that came to rest on a dead tree on Lake Lily (September 5, 1920) immediately began to plume himself and stretched his beak far over the back to reach the feathers of the rump.

A Fish Hawk that had been roosting for a long time on a stake well out on the meadows (May 22, 1921) flew to a flooded sand bar where it stood in' water nearly up to its body and later returned to its perch. Other birds have been seen to alight in shallow pools on the Pond Creek Meadows and bend the head down to the water as if drinking, while another was very busy bathing, fluffing up the plumage and churning the wings up and down in the water. While this may be the Fish Hawk's usual method of bathing some individuals seem to prefer deeper water. On August 22, 1927, I saw two birds diving continually into Lake Lily where the water was at least four feet deep. They came down through the air head first and turning slightly entered the water at a low angle, the head and shoulders going under at each plunge. They would then rest on the surface with wings partly spread and finally flapped heavily out again, rising with some difficulty. They shook themselves vigorously in the air with a wriggling action and promptly dived again. Each bird went in fifteen or twenty times during the few minutes that I watched them, sometimes in quick succession without pausing to shake themselves. On July 23 and again on August 13, 1936, I saw Fish Hawks plunge into Price's Pond above New England Creek, in the same way, and flap about in the water before arising. None of these birds were fishing and their object seemed obviously to bathe, and doubtless to try to rid the plumage of vermin. On September 3, 1925, two Fish Hawks which had been flying about their nest during a heavy thunder shower finally came to rest in the nest tree. Their plumage was evidently thoroughly wet and for sometime after the storm had ceased they remained on their perches with tails and wings spread wide in a ludicrous attitude, the wings held out away from the body and slightly drooped.

The Fish Hawk's nest is an immense structure normally placed in the top of some tree, either alive or dead, but often in past years on a cart wheel supported on top of a pole close to a farmer's house. In recent times when the automobile has largely replaced horse-drawn vehicles, wheels have become scarcer and we see fewer of this type of Fish Hawk nest but in some instances a skeleton framework has been erected by the farmer in place of the wheel. Rubber tires do not lend themselves as supports for Osprey nests! I have seen nests built on the chimneys of unoccupied or deserted houses and one on an old chimney from around which the house had entirely disappeared. Some birds, too, have selected chimneys of occupied houses and the owners have been forced to construct wire netting covers to keep them away. Another pair built on the top of a water tank supported on a high iron framework and when the entire structure blew down the birds next year resorted to the chimney of the adjacent house which was at the time unoccupied. For many years a pair of Fish Hawks nested on top of a small belfry cupola on the Fishing Creek schoolhouse. The nest was repeatedly torn down in the autumn and as regularly rebuilt in the spring. Later they resorted to the chimney of the building but upon its being transformed into a dwelling they departed elsewhere. The little fishermen's shacks which stand on stilt-like legs out on the meadows have always been favorite nesting sites but the birds have trouble in making the first branches and sticks remain in position on the ridge pole, most of the material sliding down the sloping roof to the ground, and in one case, back of Seven Mile Beach, the occupants have constructed a little platform in order to give the birds a substantial foundation for their domicile. One year a pair of birds endeavored to start a nest in the fork of a tree on the edge of the meadows near Cape May Point and when I visited the spot there was more material on the ground than in the crotch. Curiously enough this nest was under construction during July and August (1920) at which time young birds had been hatched in all of the nearby nests. The structure was not completed that summer. Other birds were seen collecting building material on July 7 and August 8, 1921; July 27, 1922; and July 17, 1926; which included sticks of varied sizes, seaweed and masses of trash from truck patches. Similar nest building has been noticed in progress on April 4, 1924; May 30, 1925; and May 1, 1926, which dates are of course the normal time for this occupation. Sticks and branches are carried by the birds in exactly the same manner as fish, one foot being placed before the other so that the stick is held parallel to the bird's body and in line with the direction of flight. On July 5, 1920, a single Fish Hawk was seen sailing over the meadows back of South Cape May gathering nesting material while its mate perched in a tree not far away. The first bird made a swoop at a dead limb sticking up from a nearby sand flat but failed to carry it off and was almost brought to earth when the branch remained firmly fixed in the sand and the hawk nearly turned a somersault. Upon examining the branch later I found that it was part of a large limb deeply imbedded in the sand and while the bird had been unable to loosen it he had succeeded in tearing up quite a good deal of ground in the effort. This bird finally did secure a single branch about three feet in length which, however, was dropped before he reached the tree where the nest had been begun. The bird then joined his mate and they paired. Just what this late nest building means I do not know but so far as I am aware no eggs are laid. Possibly the original nest in such cases was blown down after the eggs were deposited.

Richard Miller has given me the following data on nests that he has examined on Seven Mile Beach where Fish Hawks have always been frequent breeders:

May 10, 1931, three eggs.	May 30, 1912, three eggs.
May 12, 1917, three eggs.	May 30, 1919, four eggs.
May 12, 1917, three eggs.	May 31, 1920, three eggs.
May 13, 1927, three eggs.	June 16, 1918, three eggs.
May 13, 1927, three eggs.	June 17, 1923, three young just hatched.
May 15, 1926, three eggs.	June 24, 1928, one egg and one young.
May 13, 1927, three eggs.	June 24, 1928, two eggs.
May 18, 1927, three eggs.	July 4, 1931, one young.
May 26, 1929, three eggs.	July 7, 1928, one young, one egg.
May 30, 1911, three eggs.	

At Wildwood Junction he found the following:

May 16, 1915, four eggs.	May 21, 1916, three eggs.
May 21, 1916, three eggs.	May 21, 1921, three eggs.

These nests and others that he has examined in Cape May County ranged from ten to forty-five feet from the ground; fourteen were in dead trees and thirty-nine in live ones, while three were on cart wheel poles. All kinds of trees were selected as nest sites; thirteen in black gums, nine in wild cherries, six in red cedars, and others in oaks, black walnuts, hickories and sassafras.

J. Parker Norris has recorded (Ornith. and Oöl. 1891, p. 162) thirteen nests apparently on Seven Mile Beach from which eggs were collected in a single day, May 29, 1886, all sets of three with the exception of two of two. Charles S. Shick, who was well acquainted with the bird life of Seven Mile states that in 1884 fully one hundred pairs of Ospreys nested on the island but since then they have gradually become scarcer every year and in 1890 not more than twenty-five pairs were to be found (Auk, 1890, p. 328). Turner McMullen has given me data on nests that he has examined on the mainland of the Cape May Peninsula as follows:

May 14, 1927, three eggs.	May 20, 1922, four, four, three, three, and two eggs.
May 15, 1915, three, three, four, and two eggs.	May 20, 1923, three eggs.
May 16, 1926, three and four eggs.	May 23, 1920, three eggs.
May 18, 1918, three eggs.	May 25, 1934, three eggs.
May 20, 1916, three eggs.	June 23, 1923, three young several days old (Miller).

On June 27, 1926, John Gillespie tells me that on Seven Mile Beach twelve nests were examined one of which contained a young bird, another a single egg while the remainder were empty. In 1927, Turner McMullen, states that thirty were occupied. On April 27, 1930, Edward Weyl counted seventeen nests from the shore road in Cape May County, all apparently occupied, while on July 28, 1935, Gillespie examined eighteen nests in the county of which ten were empty while eight held young, there being fifteen in all. Oölogists were undoubtedly responsible for many of the empty nests. In the summer of 1936, in the southern portion of the peninsula below the Town Bank Road, Ernest Choate and I located twenty occupied nests. During his examination of nests on June 27, John Gillespie found the following articles of decoration(?) placed loosely on the top but not entering into the composition of the structures: a woman's belt; shells of conch, oyster and mussel; a folded newspaper; dried wings of a duck and a gull; a portion of an umbrella and a shingle, all obviously picked up by the birds!

I have seen young birds in the nest being fed by the parents from July 3 to August 26 but I have also seen the young leave the nest as early as August 2 and 15 and Walker Hand tells me that they regularly return to the nest to roost, and apparently to be fed, long after they first take wing, which is in accordance with my own observations. Frank Dickinson, who for years had a nest close to his house at Erma, states that the young birds left on August 2 and returned every evening up to August 23, being absent during the mornings. On August 13, 1921, I watched four birds soaring about a nest above Mill Lane two of which, evidently young, returned to the nest to roost at dusk, while six that were together on the wing at Bennett on August 21, 1930, disappeared about sundown and were located in two nest trees nearby. Two full grown young were seen on a nest at North Cape May on July 23, 1934, which were flapping their wings vigorously as if trying them out and both eventually arose in the air, whether this was their first attempt at flight or whether they had done the same thing before I, of course, could not determine.

The actions of the Fish Hawks about their nest are interesting. One nest which formerly occupied a dead tree on the edge of the meadows just north of the town, and in plain view from the golf links, still contained young on July 27, 1920, although nothing could be seen of them from the ground. The female was on the nest as I approached it while the male perched on a dead limb close by. The former bird raised her head and watched me intently and finally, as I reached what she regarded as the danger zone, she took wing and circled about over my head, drawing in and expanding her wings alternately in order to maintain her desired position. At first she called, at intervals, a rapidly repeated *few, few, few, few, few, few,* and then as I came nearer there was added a much louder, more penetrating, *keuh, keuh,* both notes with a peculiar whistling quality and easily imitated by whistling with the lips. Finally she varied the performance with a peculiar throaty *keg-keg-keg-keg-keg* much like the grating of a rusty pulley wheel. The male bird did not move from his perch until nearly ten minutes after the female had taken wing, by which time I had reached the base of the tree and the excitement was at its height, but even then he made no sound whatever simply sailing about with the female. Not all males however are so silent. When the female left the nest she carried in one foot a piece of dried fish or wood which she eventually dropped. On another occasion an adult bird circling with her young uttered a clear, high pitched sibilant cry, *seee leee*, rising on the first and falling on the second syllable. At a nest on the Rutherford farm on the edge of the Pond Creek Meadows the female called in the same way as the one above described *few, few, few, few, few, few*, but her loud notes sounded more like the syllables *weet, weet*. This bird had also a shrill whistle like the *seee leee* just described as well as the guttural *keg-keg-keg-keg-keg*, which, however, was only uttered occasionally. Other birds have had eight repetitions of the *few* note instead of six and in one instance the two loud notes sounded more like *shreek, shreek*.

On Frank Dickinson's farm at Erma a cart wheel had been placed on the top of a dead pine tree which proved a satisfactory nesting place for many years but in 1920 it became somewhat tilted and the birds decided to build a new nest on the top of another pine trunk standing close to the original one. The nest was but a few yards from the house and the birds paid no attention whatever to members of the family who were constantly passing, but whenever a stranger appeared the female Fish Hawk would raise herself in the nest where she was setting, utter the regular cry *few, few, few, few, few, few*, and then sail off and circle about, still calling. On June 7, 1923, this nest was blown to the ground where it lay partly on one side with the three young still in it. Next morning it had settled in an almost horizontal position and the old birds were bringing food to the young. By August 2, only one young was left in the nest, the others having resorted to nearby trees, and he, too, took wing when I approached. A live mossbunker was flapping about in the nest having just been brought in by one of the parents, and the young bird returned later and ate it. The schoolhouse nest contained three young in 1929 which began to fly off by August 10 but returned to roost until September 1 when only a single one was present. A young Fish Hawk, apparently taking its first prolonged flight, was seen on July 29, 1921, over the meadows back of South Cape May. Its actions were very peculiar. It flapped its wings very rapidly and continuously all the while calling, *pilly-ate, pilly-ate, pilly-ate*, and holding its legs pendant below the body, its toes curved in so that at a distance it looked as if it were carrying some prey, but it was not. Other young after similar flights were seen to perch in peculiar places, one on a church spire at the Point and another on the top of a standpipe, having evidently been forced to rest before reaching a more desirable perch.

A peculiar performance, doubtless part of the mating activities, was noted on two occasions. On April 30, 1923, a Fish Hawk, presumably a male, hovered on rapidly beating wings high over the marsh at South Cape

May with legs dangling below while he drifted very slowly backward with the wind. Every minute or two he would set his wings and drop some twenty feet through the air and then, flapping vigorously, he would rise again all the while calling loudly *killee, killee, killee, killee.* Meanwhile another bird, probably the female, was sailing about close to the ground picking up bunches of trash for the nest. When she alighted on the ground the male came down and joined her and when she again took wing he mounted high in air and repeated his performance. A similar action took place on May 2, 1926, when a single Fish Hawk was seen mounting rapidly in the air with vigorous flapping of the wings and then swooping down again to its former level. This bird, however, was silent and I noticed no other on the ground.

While very solicitous for their nest the Fish Hawks do not, as a rule, do more in its defence than make vigorous vocal protests. I saw a pair on one occasion, however, pursue a passing Bald Eagle and drive him away, and Richard Miller tells me that in climbing to a nest on May 21, 1916, the parent struck him several times.

I have seen Ospreys standing on their nests as early as April 1, probably males, who take possession of nests as soon as they arrive from the south, and await the coming of the female. The date of first arrival is, however, much earlier than that. Records of first arrival at Cape May for some thirty years are appended compiled from the notes of Walker Hand and other members of the Delaware Valley Club:

1903.	March 14.	1914.	March 27.	1926.	March 15.
1904.	March 9.	1915.	March 23.	1928.	March 20.
1905.	March 14.	1916.	March 21.	1929.	March 18.
1907.	March 21.	1917.	March 16.	1930.	March 30.
1908.	March 14.	1918.	March 27.	1932.	March 25.
1909.	March 21.	1919.	March 23.	1933.	March 21.
1910.	March 12.	1920.	March 18.	1934.	March 25.
1911.	March 20.	1922.	March 25.	1935.	March 23.
1912.	March 18.	1923.	March 23.	1936.	March 25.
1913.	March 21.	1924.	March 29.		

At an old nest at Buckshutem, Salem County, Mrs. Alice K. Prince has seen a Fish Hawk on February 18, 1921, and on March 5, 1922, but the average date of arrival at Cape May would seem to be from March 21 to 25.

A number of last records for Fish Hawks in the autumn are as follows:

1918.	October 20.	1926.	November 6.	1932.	October 30.
1921.	October 16.	1927.	October 23.	1933.	October 29.
1923.	October 7.	1928.	October 15.	1933.	October 29.
1925.	October 18.	1929.	October 25.	1935.	November 1.
		1930.	October 11.		

Walker Hand had two exceptional records, November 16, 1909, and November 17, 1917.

While we know but little about the actual winter quarters of Cape May Fish Hawks John Gillespie's bird banding work has thrown some light upon this interesting problem.   One bird banded as a nestling on Seven Mile Beach on August 15, 1926, was killed at Dorothy, W. Va., on September 25 following, while another young bird banded at the same place on June 26, 1927, was shot at Upper Tract, W. Va., in September of that year.   The similarity of their lines of flight is remarkable.

I always think of the Fish Hawk as the connecting link between the life of the sea and of the upland and the first intimation of a change of environment as we travel through the pinelands and scattered farms of South Jersey on our way to the coast.   The immense nest topping some old sour gum tree along the fence row, the great birds circling overhead bearing shimmering fish in their talons, their fearless solicitude for nest and young, their apparent confidence in the security of their exposed eyrie and the querulous whistling of both old and young—all make up a never to be forgotten picture of that delectable borderland where the elements of land and sea meet and intermingle.

DUCK HAWK

Falco peregrinus anatum Bonaparte

This splendid falcon, the famed Peregrine, is frequently seen about Cape May during September and October. My earliest record is August 29, 1926, when a single individual was seen at Stone Harbor and presumably the same bird was seen at the same spot on September 5. William Rusling, Audubon Association warden in 1935, saw one at Cape May Point on August 22 and 23 of that year and, being present every day during the hawk flight, he was able to obtain a far better idea of the abundance of these birds than could be gotten on my casual visits to the shore at this season. He saw another single early bird on September 7 and his total count, August 22 to November 3, reached fifty-six. They occurred on twenty-eight days, usually single birds but on five days two were seen, on three days three, on September 12, four, and on September 15, seven. There may have been some duplication.

In the autumn of 1931, George Saunders, Audubon warden for that year, identified twenty Duck Hawks between September 15 and October 15, while Robert Allen, in 1932, counted forty-two between September 15 and October 29. We have records of two seen on September 22, 1925; two on

October 2, 1927; two on October 2, 1932; five on September 28, 1924; all by Julian Potter. There seem to be but few remaining to winter at the Cape; our only records after October are single birds on November 3, 1935; November 12, 1922; November 16, 1929; December 30, 1920; December 23, 1934; two on December 26, 1932; December 22, 1935; the December records being on the Christmas census of the Delaware Valley Club.

Of the fifty-six seen in 1935 only eight were old birds.

On several occasions Julian Potter has seen Duck Hawks in pursuit of other birds. On September 20, 1925, one seen about the Point caused the assembled Laughing Gulls and shore birds to scatter widely every time that he put in an appearance, while on November 12, 1922, one pursued a Bonaparte's Gull which, however, eluded him repeatedly. Another on September 17, 1922, struck at a Sharp-shin. A Sparrow Hawk that flew at a passing Duck Hawk was entirely ignored by the larger bird. Julian Potter saw a Duck Hawk (September 26, 1926) at Cape May Point swoop at a Marsh Hawk from above with great speed. "It looked as if the latter was doomed. At the very instant, however, that the Duck Hawk would have struck him he twisted in the air and presented his extended talons to the enemy who turned sharply up and prepared for another onslaught. Repeated dashes were always met at just the right instant with extended claws and in a short time the Duck Hawk flew off."

Twice I have been able to study a Duck Hawk at close quarters. On September 28, 1924, I found one roosting on a pile of dead grass on the edge of the meadows just above the Harbor and by using a thicket of marsh elder as a screen I was able to approach to within twenty-five feet of it. The general tone of the plumage of the back was purplish brown while at rest but when the feathers were ruffled up and in a different light it took on a bluer tint. All of the feathers were banded and the ground color of the head was buffy white, even the back of the head was light with darker lines, while the face was beautifully marked and the cere leaden blue. The bird was pluming itself and swung its tail far around so as to reach the feathers of the rump.

The other bird was seen at Cape May Point on October 24, 1920. As it flew overhead, comparatively low down, the sharp-pointed wings were characteristic as well as the wonderfully fine mottling on their under surface and the whitish tip to the tail. It alighted on a dead treetop on the bank of Lake Lily, a favorite perch for hawks at that time, and was in excellent view. Its dark ear-coverts were clearly seen as well as the longitudinal streaks on the under parts, marking it as a bird of the year.

Duck Hawks occasionally swoop too impetuously upon their prey as one was found floundering about in the mud and water on the edge of the Lighthouse Pond (October 30, 1921), too wet and helpless to get away. Later it was seen by Julian Potter on top of a nearby post with drooping wings in a very bedraggled condition but after drying off it managed, with some difficulty, to fly away.

PIGEON HAWK

Falco columbarius columbarius Linnaeus

The Pigeon Hawk is a common transient in September and October at Cape May Point and less commonly elsewhere in the peninsula. My earliest observations for a number of years are as follows:

1921.	September 5.	1925.	September 3.	1929.	August 29.
1922.	September 1.	1926.	September 12.	1930.	August 30.
1923.	September 16.	1927.	September 25.	1932.	August 22.
1924.	September 10.	1928.	August 27.	1935.	August 31.

The latest dates of observation are:

1920.	October 17.	1927.	October 30.	1929.	October 25.
1925.	October 18.	1928.	October 28.	1935.	November 8.

As I have not been present regularly after the middle of September these latter dates probably do not represent the last stragglers and it is interesting to consider William Rusling's record for 1935. He was present every day from early August until November 15 and in that time counted no less than 402 Pigeon Hawks passing in migration. The first one was identified on

August 31 and the largest number seen on a single day was 115 on October 15. That seemed to mark the peak of the flight as they were seen on only seven days thereafter and only one to three per day. I am inclined to think that they are always rare in November and I have so frequently been deceived by the dark appearance of young female Sparrow Hawks, under poor light conditions, that I am skeptical of the alleged sight records of Pigeon Hawks in December which appear on several of the Christmas censuses. There is no positive record of Pigeon Hawks in winter, I believe, north of southern Florida.

In 1932 Robert Allen counted no less than 1707 of these handsome hawks between September 15 and October 29. As in most of the migrant hawks there is no evidence whatever of a northward flight in spring Nearly all that we see in the autumn are birds of the year and an old gray-backed adult is rare indeed.

The Pigeon Hawk seeks its prey in the open as does its relative the Sparrow Hawk and not in the woods, and flies with great velocity over the marshes and fields. The brown birds of the year, which are almost the only ones that we see, appear very dark, practically black above, with a narrow white tip to the tail and narrow white bars which are only seen when the tail is spread, as the bird wheels or endeavors to check its flight. As it flies directly toward us the buffy front edge to the wing stands out very clearly against the general dark color, and the breast, buffy with heavy longitudinal black stripes, is very conspicuous.

The flight of the Pigeon Hawk is a steady rapid beating of the long pointed wings with very full strokes followed by a short sail. Some individuals, skimming over the marshes, have been seen to flap continuously for long distances. The tail in flight is closed and seems to act as a rudder. At rest the Pigeon Hawks perch on small dead trees more frequently than anywhere else. The body is generally held bolt upright and the bird peers out in front or leans far to one side as it looks after some passing quarry. At other times they lean forward and flirt the tail up and down like a Sparrow Hawk.

One that I watched for some time, perched on a dead pine on the shore of Lake Lily, suddenly dropped to the edge of the water and skirted the whole west bank flying close to the surface and examining the bushes evidently seeking some sort of prey. Twice I have seen these hawks alight on logs on the ground or at the edge of a pond, only a few inches from the water. Once I saw one strike at a Spotted Sandpiper which was winging its way across the Lighthouse Pond, but the latter promptly dived and escaped the hawk. Another in pursuit of a Song Sparrow checked itself suddenly with a great display of spread tail to prevent colliding with a bush into which the sparrow had plunged headlong. Out on the great Fill a Pigeon Hawk swooped down on a Flicker feeding on the ground and flushed it. Both birds flew away together although the hawk soon veered off, having evidently found too powerful a quarry. I have also seen these hawks attack a Sparrow

Hawk and a Sharp-shin and in the former instance the Pigeon Hawk clearly showed its superiority in speed as well as in aggressiveness.

While Pigeon Hawks on migration do catch small birds, as does the Sharp-shin, they do not comprise the bulk of its diet, which is composed of meadow mice, grasshoppers and dragonflies. I have seen them catch the latter insects on the wing and the hawk would fly off firmly grasping its prey in its claws, leaning over now and then to pick at it with its bill. Another caught a large grasshopper on the wing and fed on it as he flew, while a third hawk carried a grasshopper to a dead treetop and devoured it there.

The Pigeon Hawk seems to be particularly characteristic of the seashore as I have seen many more there than in other parts of the state. It is in all respects a typical falcon almost as swift on the wing as its larger counterpart, the Duck Hawk, and may be distinguished from our other small hawks by its much darker coloration at all times.

Next to the Sharp-shinned and Cooper's Hawks it was in the past a favorite target for the hawk gunners of the Point but its insectivorous habits have won immunity for it.

J. F. S.

SPARROW HAWK

Falco sparverius sparverius Linnaeus

This little falcon, the smallest of our birds of prey and the most attractive, is abundant about Cape May only during the autumnal migration. A few individuals, left over from the great flights of September and October, may be seen throughout the winter perched on telegraph wires along the railroads and highways, or on the small trees of the fence rows, and in the Christmas census of the Delaware Valley Ornithological Club it is recorded as follows:

December 23, 1928, nine.	December 26, 1932, ten.
December 22, 1929, seven.	December 24, 1933, nineteen.
December 28, 1930, fifteen.	December 23, 1934, thirty.
December 27, 1931, nine.	December 22, 1935, twenty-five.

but these are few indeed as compared with the numbers seen during the southward migration. Robert Allen counted 322 passing Cape May Point from September 15 to October 27, 1932; and William Rusling from August 23 to November 15, 1935, counted no less than 777, his highest daily records being eighty-nine on September 10, 117 on September 16, and 217 on September 20. There may, however, have been some duplication in these counts.

While no Sparrow Hawks have nested immediately about Cape May during the past ten years, they do breed occasionally a little farther north on the peninsula. Richard Miller tells me that on July 1, 1923, he found three young and their parents on telegraph wires on the upper part of Seven Mile Beach. The young were not yet able to fly and had apparently been hatched in an old Flicker's hole in one of the poles nearby as there was no other possible nesting site in the vicinity. Turner McMullen in his long experience tells me that he has never found a Sparrow Hawk's nest in either Cape May, Cumberland or Salem Counties. W. B. Crispin, however, has recorded two nests with four eggs—April 22 and 28 and four with five eggs—May 2, 3, 9, and 15 in the last county.

I have several times seen Sparrow Hawks in July at the Cape. One I encountered on the meadows just back of the town on July 27, 1920, following one of the numerous ditches at that time overgrown with tall grass and as I stood perfectly still it nearly struck me on the head. It alighted on a wild cherry tree just behind me and after nervously flirting its tail in characteristic manner it was off again, several rapid wingbeats alternating with short sails. The other bird was seen near the same spot on July 29, 1922, and it is just possible that a pair bred about some of the farm buildings not far away in both of these years. A pair was seen at Bennett on July 30, 1928.

In earlier years Walker Hand several times found Sparrow Hawks nesting close to the town. For two years at least a pair occupied an old Flicker's hole in a railroad tie left standing upright out on the Fill and another year he saw young ones which had obviously been hatched nearby, perched along the ridge pole of an unoccupied cottage north of Madison Avenue.

The regular migration begins about the middle of August:

1920. August 25.	1924. August 29.	1929. August 16.
1921. August 12.	1925. August 20.	1930. August 20.
1922. August 21.	1926. August 27.	1932. August 14.
1923. August 21.	1927. August 20.	1934. August 19.

As we approach Cape May by train at this season we shall notice Sparrow Hawks flying from the telegraph wires all along the railroad from Dennisville south and soon after that we begin to see them regularly on fence posts about the old fields and on dead tree tops on the dunes at Cape May Point. In 1920 I first saw two in an old orchard at Rutherford's and the next day there were six, obviously a family group. Some of them sat on the posts and others right down on the clods of earth in a plowed field where they stood very erect; apparently they were feeding on grasshoppers. In September, however, they become really abundant. On the eighth of the month in 1921, I counted forty Sparrow Hawks flying from the wires as the evening train approached the town and the next morning there were many on the telephone wires back of South Cape May from which they flew out over the open marsh. As they took wing they would sail a short distance and then there would be a wheel, more very rapid strokes and an upward swoop to the wire a little farther on. Now and then one would pause and hover over one spot with continuous wingbeats and tail partly spread as it located prey of some sort in the grass below.

On the morning of September 14 of the same year Sparrow Hawks were in evidence everywhere west of the town. They shot back and forth over the low dunes below Broadway and out over the marsh nearby, with rapid beats of their pointed wings, seven to twelve strokes and then a short sail followed by another series of beats. As they come to perch on the wires or poles they spread the tail to check their speed and instantly close it again, swinging it up and down several times as if using it as a balancer. They sit erect with the legs straight down and the body clear above the feet. The head is held down on the shoulders and the tail just a little above the slope of the back, sometimes nearly horizontal, while the breast appears somewhat puffed out. Swinging the glasses across the meadows I counted sixty-five Sparrow Hawks perched on the wires besides many in the air, while on the broad sand flat back of South Cape May there were thirty-five resting on dead branches partly buried in the sand by the last high tide or on other supports only a few inches from the ground. These birds were constantly skimming the sand like swallows, wheeling and chasing one another and coming back again to their low perches. If they were after food I was unable to detect its character, as cicindela beetles and small grasshoppers seemed to be all that the place offered and I was forced to the conclusion that they were merely resting. On August 25, 1927, there were a dozen Sparrow Hawks at Cape May Point hovering over the grassy meadows very evidently hunting grasshoppers which abounded there and one hawk was seen perched on a stake busily engaged in devouring one of these insects.

The migrant Sparrow Hawks seem always to prefer the open and during the flights there are practically none about the wooded regions of the Point where the Sharp-shins abound, and conversely there are hardly any Sharp-shins on the meadows or sand flats.

The winter birds gradually disappear as spring approaches, some are killed and others retire northward, and by late April or early May all have gone except the possible breeders. The late stragglers as well as the winter individuals are solitary birds with none of the flocking characteristics of the autumn migrants.

While easily identified in good light and at a reasonably short distance I have more than once been misled by the suffused coloration of female birds of the year which on misty days appeared so dark that I took them for Pigeon Hawks until changing light conditions enabled me to realize my mistake and I consider the winter occurrences of the latter species at Cape May Point, which have sometimes been reported, to be explainable in this way. As we see the Sparrow Hawk about Cape May it is usually a silent bird but Julian Potter tells me that he heard one on November 14, 1929, uttering the characteristic call of the breeding season—*killy-killy-killy*.

RUFFED GROUSE

Bonasa umbellus umbellus (Linnaeus)

While the Ruffed Grouse, or "Pheasant" as it was called in southern New Jersey, before the introduction of the Ring-necked Pheasant complicated our nomenclature, is still found in lessening numbers throughout the Pine Barrens and the northern portion of Cape May County, it is exceedingly rare immediately about the Cape. The last one I saw there was in January, 1892, when one was flushed several times between the 26th and 29th on the bushy sand dunes north of the turnpike at Cape May Point, when Samuel Rhoads and I were making a study of the midwinter bird life of the region. My only other record is a bird flushed by Earl Poole, who is very familiar with the Grouse in Pennsylvania, from the woods on the south side of the turnpike on August 11, 1921. Richard Miller saw one near South Dennis on June 6, 1921, and in the most northern part of the county Walker Hand reported several shot in the winter of 1922–1923 where doubtless some may be found every year. There have been rumors of Grouse seen in recent years near Cape May Court House and at Erma but they have not been substantiated. At Manumuskin in Cumberland County, William Baily found a nest with nine eggs on May 30, 1898, and Turner McMullen found another near Dividing Creek on April 26, 1936, which held three eggs.

*HEATH HEN

Tympanuchus cupido cupido (Linnaeus)

While this bird, which is only a very slight variant of the Prairie Chicken of the Middle West, never occurred in Cape May County so far as we know, it may have done so in earlier times and a few words on its occurrence in the state may not be out of place. Its known habitat, from which it was exterminated certainly by 1870, was the so called "Plains" of Ocean and Burlington Counties, where for miles the vegetation of pines and oaks are so stunted that they reach barely to a man's knees. (Stone, Flora of Southern New Jersey, 1911, pp. 70–72.)

J. Doughty writing in 1832 says: "In former years they were in great abundance on these barren grounds which were then visited by old and scientific sportsmen, who regarded the laws of shooting. But lately, through great persecution by those who have no claims to the principles which constitute sportsmen and who visit these grounds months before the season commences by law, and while the birds are in an unfledged state, the Grouse are driven from this favorite abode. Year after year has this unhallowed persecution of the Grouse been carried on, until the species has been almost exterminated from the state." (Cabinet of Natural History, Vol. II, pp. 15–16.) John Krider states that he shot them there in 1840 and William P. Turnbull, in 1869, says that they have been found there "within the last year or two." (Birds of East Pennsylvania and New Jersey, p. 35.)

George H. Van Note writing in 1903 from personal recollection, says, "They were larger than a Guinea Hen. They would have several places to collect; generally a clear place on the Plains. When together, the male bird would start around with his wings on the ground, like a Turkey gobbler, giving a sort of whistle. When flying they would rise ten to twelve feet high and go straight as a line. They bred on the Plains, and were always found on them. There were lots of them forty years ago [i. e. 1863]. The way they killed them at that time was to dig a hole in the ground and remove all sand, so as to make the ground level, then hide in this hole until they came to you. If you killed one the others would stay and fight it, and you could keep on shooting until you killed as many as you liked. If you missed the first shot they would fly away. This kind of gunning went on until about thirty years ago. They gunned in all seasons, and soon killed them off. Since that time there has not been a Grouse killed on our Plains." (Bird-Lore, 1903, pp. 50–51.) Mr. Van Note also refers to the noise that they made and "which could be heard for two or three miles" and sounded like "a man blowing in a conch-shell." This was the drumming of the wings which he erroneously considered was made by blowing out air from a pouch under the throat. It seems likely that the bird was exterminated a few years earlier than he estimates.

The late Thomas A. Robinson has told me of shooting the Grouse on the Plains in the forties and fifties on his way to and from Sammy Perrine's, on Long Beach, where he went for duck and beach-bird shooting. They started from Burlington in wagons equipped with broad tires for travel over the deep sand roads. The grouse were so plentiful that the gunners could shoot all they needed in a short time. Mr. Cranmer of Warren Grove has shown me the little bare hills near his home where his father had told him he used to go at daybreak and await the birds which came there to conduct their courtship performances and fights.

That the shooting of grouse was a well recognized pastime as early as 1823 is shown by an advertisement in "Poulson's American Daily Advertiser" of July 23 of that year, wherein Seth Crane "respectfully informs the public that he has commenced running a stage between Mount Holly and Mannahawkin for the accommodation of persons disposed to visit the Grouse Plains, Mannahawkin or Tuckerton. The stage will leave Mannahawkin every Monday and Thursday mornings at 6 o'clock and arrive at Griffith Owens' Tavern in Mount Holly, same afternoon at 4 o'clock. From whence passengers will be conveyed to Burlington on the following morning in time to meet the Steam Boat for Philadelphia and Trenton. Returning will leave Mount Holly every Wednesday and Saturday morning at 6 and arrive at Mannahawkin same afternoon at 4 o'clock. Where Ladies and Gentlemen can be accommodated with genteel Boarding and Lodging at the moderate rate of $3. per week; and conveyed at any time across the bay to James Cranmer's, Hazleton Cranmer's or Stephen Inman's. Fare through

$1.75. A conveyance will be in readiness at Mannahawkin for Tuckerton." Thus did one reach Beach Haven and the Plains in those days.

Alexander Wilson, curiously enough, has nothing to say about the habits of these birds in New Jersey and probably never visited the Plains. Audubon is equally silent upon the subject although he writes in 1834 that they have become "so rare in the markets of Philadelphia, New York and Boston that they sell at from five to ten dollars the pair" and that a friend in New York told him that "he refused one hundred dollars for ten brace which he had shot on the Pocano [sic] mountains of Pennsylvania." This was another of the last strongholds of the bird; the New York markets being supplied also from Long Island, and Boston from Martha's Vineyard, where the last surviving Heath Hen was seen in 1932. Audubon adds in 1835 that it had been "nearly extirpated in New Jersey."

Charles S. Westcott ("Homo") writing in "Forest and Stream" (August 9, 1884) states that he had met an old resident of Barnegat who remembered the last "grouse" [Heath Hen] that was killed in New Jersey which "was about twenty-three years ago" [i. e. 1861]. He recollected when the East Plains held hundreds of them and told how he used to make a train of grain in a section which they frequented and from the shelter of a nearby bush had once killed a dozen at a shot while they fed. This was the favorite way of hunting them.

I noticed also in "Forest and Stream" that in a synopsis of the New Jersey game laws for the year 1884 the Pinnated Grouse was still listed with a definite open season, October 15 to December 1, although it had apparently been exterminated for close to twenty years!

It is interesting to learn from Westcott (Forest and Stream, November 1880) that a half-grown Prairie Chicken was brought to Krider's gun store, which had been shot in lower Delaware or Maryland "evidently an offspring of the several pairs introduced some years ago by Dr. Purnell of Berlin or Snow Hill, Md. Nests of these birds were located and young seen. It is said that a similar attempt was made to introduce the western Prairie Chicken in New Jersey."

BOB-WHITE

Colinus virginianus virginianus (Linnaeus)

The cheerful call of the Bob-white, or "Quail" of the Cape May residents, may be heard throughout the spring and summer immediately about the town, from the fields lying north of the turnpike leading to the Point and from the Point itself even down to the open ground and thickets by the Lighthouse. Farther north in the peninsula it is common on farmlands and wood edges and when approaching Cape May by train one sees coveys of Quail, of various sizes, according to season, flushing from the sides of the track and seeking shelter in the nearby woods. They like to come out to feed along the right-of-way where the vegetation is mowed down as a fire prevention.

In the autumn coveys of full grown birds may be flushed from old briery fields about the Cape and Walker Hand told me that if the acorn crop is good they frequent the woodland at this season to feed upon them. Quail can prosper during the ordinary mild winters of the Cape May region but the severe weather of the past few years must have been very hard on them when the temperature fell to the unprecedented low of ten below zero. While the gunning season takes a toll from them some coveys always survive and I have seen a flock of sixteen near the old Race Track Pond on January 3, 1929, and six at Cold Spring on February 19, 1933, while the Delaware Valley Club members have found a few present on their Christmas census nearly every year:

December 26, 1932, eight.	December 23, 1934, one.
December 24, 1933, thirty-three.	December 22, 1935, twenty.

Walker Hand tells me that he has seen Quail feeding on eight inches of snow by jumping up and knocking down the weed seeds from the old dead stalks and then picking them up from the surface.

While we regard the Quail as a resident there is often a marked migration in autumn and doubtless some of the local birds pass farther south at this time. In the fall of 1902 Walker Hand told me that the town was full of migrating Quail and many were killed by flying against the houses and telegraph wires. There was a similar flight on September 28, 1924, when no less than eight were seen in a single yard on Washington Street. On another occasion (October 21, 1902) Hand saw a flock on the Bay shore at Cape May Point evidently on migration which took wing and started across the water flying due south and at a rather high elevation for Quail. Charles York, moreover, tells me that he had had a similar experience when one autumn he saw two coveys run out under the boardwalk to the edge of the water where they took wing and continued southward in the direction of the Delaware shore.

Quail are more often heard than seen unless one is hunting them with a dog and in the summer months most of our records are made by ear rather than by sight. I have heard them calling as late as July 6, 1925, and July 27, 1929, and once on the exceptional date of August 13, in 1922. It is probable that some are calling regularly during the former month especially as these summer calls are generally attributed to unmated males. The "Bob-white" call sounds very different when heard close at hand as compared with the whistle that floats to us from far over the fields. I remember that on one occasion (July 16, 1920) while I was sitting on an old snake fence a Quail alighted on the top rail not fifteen feet from where I sat and while I remained rigid he called several times. The volume of the sound when heard so close at hand was astonishing and I never before realized the power of the call and the carrying power that it must possess. I was not surprised, therefore, a year later when living for a few days in a small shack out on the meadows back of Two Mile Beach to hear a Quail calling on the mainland more than a mile away and it was a pleasant contrast to recognize among the harsh and raucous cries of the terns and gulls, the clear whistled "*Bob-white, Bob-white.*"

Our most frequent sight of Quail is when we flush a single bird or a covey from a grass field or a briery pasture; there is a whir of wings and they sail off to cover a little farther on. Now and then we are lucky enough to see one run across a road with short legs moving rapidly, head erect and neck stretched to the utmost so that the bird's body seems almost perpendicular. The feathers of the crown, too, are usually somewhat elevated especially if the bird is excited. Once or twice I have had a Bob-white approach me while I stood perfectly still. Once while standing in the edge of a wild crab apple thicket a bird came in on down-curved wings like a bullet and dropped immediately in front of me but on my slightest movement he was off again. Quail seem always to see you before you can see them and I have never been able to approach a covey lying in the grass without being detected. Occasionally I have come upon a Quail perched in a tree and once I heard one calling from such a position. It may be that they seek such an elevated perch for the purpose of giving their call though they also call from the ground.

Quail breed in the immediate vicinity of the town and a nest with twelve eggs was found on May 27, 1923, on a bushy ridge near the old Race Track Pond, about five feet above the surrounding level, situated among grass about ten inches high. Another nest, also with twelve eggs, was found on May 19 of the same year, one with nine eggs on May 21, 1922, and another with eleven eggs at Rio Grande on May 24, 1925. The normal time for eggs would therefore seem to be the latter part of May but I have records of exceptional nests at much later dates. A man cutting grass with a scythe in a churchyard near Erma on August 11, 1923, came upon two nests with eggs while another containing eight eggs was found in the same way on September 9, 1928. These may have been deserted but the finders thought not.

Coveys of young in the down are often found with their parents during June and I have seen them still in this condition on July 11, 1935, and July 15, 1929, the female in the latter instance being accompanied by no less than twenty little ones. In early July 1916, I frequently came upon a family that fed close to the edge of the meadow in West Cape May and whenever I approached, the female would run about in great anxiety with drooping wings and crest erect until the young had had time to scatter in the short grass. Another family was raised on the Fill and fed close to the edge of the Harbor. A favorite spot for Quail is along the Harbor branch of the railroad where it leaves the meadows and enters the upland fields and there is scarcely a summer that I do not find a family feeding there. On July 19, 1926, I saw the female cross the track ahead of me and upon approaching I saw several half-grown young in the scant grass; as soon as she saw me and realized that her family was still on the other side of the rails she returned to them and immediately took wing followed by two of the brood, the others apparently preferring to seek shelter in the cover. They were about the size of a Towhee. On July 11, 1935, I flushed another family from almost the same spot but they were more advanced than the former brood and all, to the number of twenty, took wing with the mother bird. On August 13, 1928, I found a solitary little downy Quail running along the roadbed of the freight line to Cape May Point. He was quite unable to climb over the rails and the ballast was packed so closely under them that he could not escape in that way. How he came to be deserted is hard to explain as the female Quail is very solicitous for her young. Probably the covey was frightened and this bird failed to follow the others where escape was possible and so became entrapped. My latest covey was observed at Cape May Point on September 28, 1930, and consisted of twelve birds not over half-grown. The largest covey of which I have a record was seen on Frank Dickinson's farm at Erma on August 17, 1928, and consisted of thirty-three birds nearly full grown. Thirty of them were seen again on August 31; probably this covey was formed by the union of two families.

There is no bird dearer to the hearts of the countryfolk than the Quail and justly so since quite apart from his charm of form and voice he stands in the front rank of destroyers of noxious insects. Such being the case it is regrettable that New Jersey cannot follow certain other states in giving the Quail absolute protection at all times. As it is the increase in gunners has resulted in shooting out the Quail in many sections and birds have to be imported or artificially bred and liberated in order to keep up the stock which today must be of very mixed blood and not the pure New Jersey Bobwhite of years gone by. As early as 1883 the West Jersey Fish and Game Protective Association liberated four hundred Quail from Indian Territory and in 1885 five thousand were imported from the West. Of course we have no means of knowing how many of these birds survived the winter or whether they actually bred.

RING-NECKED PHEASANT

Phasianus colchicus torquatus (Gmelin)

This old world game bird was introduced in most parts of New Jersey in 1897 and doubtless in later years as well. It bred and increased only to be shot off and reduced in numbers again.

I have no early records for Cape May but since I began an intensive study of the bird life in 1920 I have seen a few every year. A pair was present about the Point during August, 1921, and on July 14, 1927, I traced a fine male, by his footprints, along a road through the fields east of Bennett, and eventually came up with him near the edge of the meadows. On August 31, 1928, a flock of six nearly full grown birds in the brown plumage was seen in farmland back of the town and several single young birds feeding in truck gardens near Erma. When flushed they will fly for long distances and one kept on the wing for about one hundred yards as he crossed Pond Creek Meadows on July 14, 1928, while three full-plumaged adults flew across Lake Lily on October 17, 1920; keeping low over the water with their long tails streaming behind they made a rather unusual spectacle. The brilliant metallic plumage of the Pheasant makes a great show in strong sunlight and the snow white collar stands out conspicuously.

While I have no records of nests in Cape May County one found near Buckshutem across the line in Cumberland Co., contained fourteen eggs on May 21, 1920. Turner McMullen examined several nests in Salem County, one on April 30, 1932, containing ten eggs; another on May 30, 1933, with nine eggs; two in Cumberland County on April 23 and May 4, 1933, the former with fifteen eggs the latter with two, both situated in tall grass on the edge of a marsh.

(327)

WILD TURKEY

Meleagris gallopavo silvestris Vieillot

That Wild Turkeys at one time inhabited Cape May County, as well as other parts of New Jersey, we have abundant evidence in the accounts of the early settlers and explorers but they came to be regarded so much as a matter of course that we have no detailed account of them and no evidence as to when they became extinct.

Captain Thomas Young in his "Voyage to Virginia and Delaware Bay and River" (1634) records "an infinite number of Turkeys" in the latter region which may well have included Cape May, while Master Evelin in a letter from "New Albion," dated 1648, describes a visit to Cape May and states that he "saw there a store of Turkeys of which I tried one to weigh forty and six pounds." Beauchamp Plantagenet (1648) also writing of New Albion mentions a flock of five hundred Turkeys "got by nets" but this was in "the uplands" (cf. A. H. Wright, Auk, 1915, p. 71). Samuel Smith's "History of the Colony of Nova Caesarea or New Jersey" (1765) states that "there are a great plenty of Wild Turkeys." While this may refer to the northern counties it is certain that if Turkeys were still plentiful there they must have been equally abundant in the less settled southern parts of the state. Charles Bonaparte in his "American Ornithology" (1825) says that he is "credibly informed that Wild Turkeys are yet to be found in the mountainous districts of Sussex County, New Jersey. In Pennsylvania he states that they are found as far east as Lancaster County and adds that "those occasionally brought to the New York and Philadelphia markets are chiefly obtained in Pennsylvania and New Jersey."

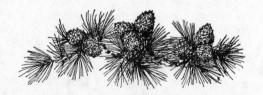

SANDHILL CRANE

Grus canadensis tabida (Peters)

We have had no record of cranes of any kind in New Jersey for almost a century. Alexander Wilson (about 1812) states that "A few [Whooping Cranes] sometimes make their appearance in the marshes of Cape May, in December, particularly on and near Egg Island [at the head of Delaware Bay], where they are known by the name of *Storks*." How much more of his account relates to New Jersey it is impossible to say, nor whether he had any personal experience with the birds in the state.

Turnbull writing in 1869, says of this bird: "Now very rare. While at Beesley's Point in 1857, I saw three off the inlet; they were very wary and could not be approached. In Wilson's time it bred at Cape May." The last statement is obviously incorrect and is the result of misreading Wilson's account, which is rather puzzling as to the locality of his observations.

Peter Kalm who travelled in New Jersey in 1748–49 and spent some time at Swedesboro, mentions cranes passing there on their northward migration in February. They usually alighted but remained for a short time only and in comparatively limited numbers. Kalm was assured by an old colonist, then over ninety years of age, that in his youth (about 1670) cranes came in numbers. Turnbull in commenting upon this refers these cranes to the "Brown or Sandhill" Crane.

To which species these early records apply we shall never be able to determine as no specimen is extant.

KING RAIL

Rallus elegans elegans Audubon

King Rails breed sparingly in the fresh water bogs about Cape May where there is a thick cover of cattails or rushes. Usually their call or a momentary view of a bird disappearing in the vegetation is all the evidence of their presence that we obtain. On several occasions, however, I have been more fortunate.

In the summer of 1920, a pair of King Rails occupied the Race Track Pond, then covered with a few inches of water over a muddy bottom, with a dense growth of cattails at the western end. Bushes about the bank offered convenient shelter for an observer and while the birds usually took alarm at my approach, I several times had an opportunity to study them for a few minutes before they recognized my presence. When first seen, on July 12, they were in an open area at the edge of the cattails. One of them immediately ran to cover, but the other remained for some minutes probing for food in the mud, finally following its mate into the seclusion of the vegetation. He had apparently taken alarm at the sudden cry of a Killdeer feeding close by. The rails were always exceedingly wary and ready for instant flight, the immediate cause being not always apparent. They ran with head lowered and neck stretched out in front and with the short tail held vertically. On July 26, while I was standing motionless on the edge of the pond, one of the rails came out of the cattails and ran nervously a few steps, when it paused a moment and then ran a short distance through the edge of the reeds, with neck hunched up and tail held vertically. Then suddenly it extended its neck to its full length and with head held low and wings flapping, it ran along the border of the cattails out in full view for some twenty feet. This was repeated several times, without apparent object, when it finally withdrew into the dense vegetation. On July 27, I flushed one of these rails from the grass and brambles twenty feet from the pond. It uttered a sharp double call *kip, kip,* as it took wing and flew with rather deliberate wing strokes and with neck fully extended, until directly over the cattails, when it allowed itself to drop vertically out of sight. On August 3 one rail was feeding on a grassy mud flat in the middle of the pond, and it immediately rushed for shelter, with neck stretched low over the ground and tail erect.

On May 23, 1921, another pair was found established in a bog near Bennett, a wet sphagnum bog covered with rushes and sedges with a dense growth of taller plants in the middle, where the water was deeper. As I stood unobserved on the edge of the bog a King Rail was seen walking leisurely about among the grass and sedge which partly concealed it. It was a brightly colored bird with a particularly rich rusty red breast, I presumed that it was the male. His movements were deliberate, almost like those of a feeding chicken. He turned his head as he advanced, drawing the neck back and extending it again with each step and picking now and then at something on the ground. The tail was held vertically, or nearly so, and was slightly flirted now and then. Finally, securing a good sized insect of some sort, the rail turned and ran rapidly with neck stretched out in front to the thickly covered part of the marsh. He soon reappeared apparently on a regular path or runway and the whole performance was repeated in exactly the same way. I stepped into the bog and approached the patch of heavy vegetation and when a few feet distant the female rail, a much paler bird, with a notably lighter breast, walked out into the open with feathers all ruffled up, those of the neck standing out straight, and the wings partly spread. She called at intervals of five seconds *kick—ick—ick* and walked about some ten feet from me. Suddenly the male came charging in from where he had been hunting food, with wings spread like an old hen in defence of her chicks, passing so close to me that he brushed my ankles. He wheeled about and charged me again, all the time calling like his mate. These birds apparently had young in the dense cover but I could not find either them or the nest and a week later there was no sign of rails in the marsh.

On August 13, 1924, I came upon a King Rail in a small pond one hundred yards back of South Cape May which was bathing deliberately as it stood close to the overhanging sedge and then spent a long time pluming itself, turning its head over sideways onto the back and poking its bill down among the feathers of the breast. All the time it was shaking its wings from the shoulders and every now and then spreading and erecting the short tail. When these operations were finished it stood motionless gazing down into the water and remained so up to the time I left it. The white throat of this bird was conspicuous even at a considerable distance. On July 25, 1925, I saw another in nearly the same spot, very early in the morning, and I was impressed by the extreme ruddiness of its breast as the slanting rays of the rising sun struck it. It was running about on the black mud bottom of the pond at low tide but never ventured far from the sheltering sedge. Quite different was the action of one seen on an adjoining pond on September 5, 1928, which walked right out in the open poking its bill into the mud and water and sometimes nearly submerging its head. Usually however they are seen running to cover and when they pause for a moment there is conspicuous, nervous, flirting of the tail.

On July 23, 1924, I was told of a King Rail crossing the railroad at a swampy spot near Bennett, followed by a whole brood of black downy young, and Otway Brown saw the same thing on the Bay Shore Road in the summer of 1926 and upon subsequent occasions. On July 11, 1928, and in June, 1929, several persons saw an old King Rail and a flock of downy young feeding in corn and bean fields in West Cape May, walking about like chickens, the old bird in one instance coming within three feet of the observer.

Richard Miller tells me of a nest with ten highly incubated eggs which he found on the Pond Creek Meadows on May 30, 1921. It was situated in a fresh water cattail swamp and stood nearly a foot above the water which was about four inches deep. The female was very reluctant to leave. He found another nest on May 23, 1915, containing fourteen eggs.

At Marshalltown, in Salem County, Turner McMullen found three nests on May 15, 1921, containing seven, ten and four eggs respectively, and on the 29th two more with eight and eleven eggs, while he discovered one on June 3, 1922, which also held eleven.

While the King Rail is mainly migratory Walker Hand reports it as wintering at Teal's Beach in 1915–16 and one was caught in a muskrat trap on November 29, 1926, and another killed on December 5, 1877.

A few are killed by gunners in October and we see them more frequently in autumn than in spring; our earliest record is May 5, 1934, although they undoubtedly arrive long before that. The autumn birds are often very highly colored the breasts rich ruddy and the black and white barring on the flanks very conspicuous. Two seen August 28, 1929, differed materially in color, one, doubtless the male, was slightly smaller and exceptionally dark.

Like all its tribe the King Rail is a victim of the mosquito draining operations and as its haunts are drained or poisoned with oil it disappears and in a short time this splendid bird, except for a few migrants pausing on their way, will be a thing of the past at Cape May.

Charles S. Westcott ("Homo") writing in "Forest and Stream," in the early eighties, tells us that many of them were brought to Philadelphia for sale and that the epicures recognized them as far superior to the Clapper Rails. Many of the latter, however, carefully plucked were palmed off for King Rails on those less expert in identifying them.

CLAPPER RAIL

Rallus longirostris crepitans Gmelin

The most characteristic bird of the great salt meadows which line the New Jersey coast for a width of from one to five miles, is the Clapper Rail—the "Mud Hen" of the local gunners, and yet so secretive is it, and so completely does the growth of rank grass and reeds among which it lives conceal its activities, that we know comparatively little of its life history, and all that we ordinarily see of it is a glimpse of a rather large grayish brown bird rising above the grass tops for a moment in rather labored flight and dropping out of sight again a few yards beyond. On most of our trips across the meadows we do not see it at all and are aware of its presence only by its harsh cackling notes coming to us from the bed of some narrow tidewater creek.

At the time of Alexander Wilson's visits to the New Jersey meadows, at the beginning of the last century, the Mud Hen was considered to outnumber all the other marsh birds put together, but in recent years its numbers have been sadly depleted and immediately about Cape May it has become rare or actually extinct in many a spot where it formerly abounded, while throughout the coastal meadows but a fraction of the former hoards of these interesting birds remain. Wasteful gunning in the past probably had much to do with this and in the early nineties, when the exceptionally high tides of September flooded the meadows and the poor birds were forced to swim about and take shelter behind any tuft of grass that remained above the water, I have seen gunners, back of Atlantic City, shoot them until their guns became too hot to hold and dead birds were left to rot in piles on the boat landings and adjacent mud flats. Norris DeHaven describes one of these Mud Hen tides on August 28, 1882, on the Atlantic City meadows

(Forest and Stream, September 28, 1882). He says that although many of the birds were too young to fly they were slaughtered indiscriminately by boys and men armed with guns and clubs. He saw one female followed by twenty-three half-grown young all swimming and although his boat was grounded within ten feet of them the old bird made no effort to fly or desert her offspring. DeHaven had no trouble in shooting thirty old birds in an hour and a half, all that he had any use for. In the stomach of one of these he found a meadow mouse (*Microtus*) two and a half inches long. He described a similar slaughter in late September, 1896, before a meeting of the Delaware Valley Club and estimated that ten thousand rails had been killed. Egging, persistently carried on from the earliest times as a source of food, began to have its effect when the birds were reduced in numbers, and the washing away of nests and eggs in spring tides and storms was also a factor, as was the constant draining of favorite nesting grounds. Potent as these causes undoubtedly have been there would seem to be some other reason for the remarkable decrease in the number of Mud Hens in recent years.

About the immediate environs of Cape May draining has been the most serious factor in the disappearance of the Clapper Rails. The filling in of Cape Island Sound in 1904, destroyed a favorite nesting ground and the ditching of Pond Creek Meadows did away with another, while the more recent "development" of the marsh between Cape May and the Point completed their extermination south of the Harbor except for an occasional migrant that drops down to pass the night.

So, while twenty or even ten years ago gunners could obtain good bags of Mud Hens close to Cape May, it is now not worth while to look for them below Schellenger's Landing and on the meadows to the north the game is hardly plentiful enough to warrant the effort involved in its pursuit. In September, 1925, numerous bags of twenty were reported and the Mud Hen shooting was considered better than for ten years past. Today Raymond Otter tells me that while the breeding birds seem to have gone elsewhere owing, he thinks, to the throngs of people which visit the meadows fishing and crabbing, there are still many of them to be seen during migration when the high tides drive them from their retreats. They are not, however, nearly so plentiful as in earlier times.

Walker Hand who kept migration records at Cape May for over twenty years showed that March 15 was the average time of arrival of the Mud Hen in spring, his dates running from March 2 to 27 but with twenty-eight birds shot on November 28 and eight on December 1, 1929, it is evident that some individuals do not go south. We have also several midwinter records for the meadows back of Atlantic City where Norris DeHaven shot one on January 19, 1892, while Dr. J. F. Prendergast found two dead back of Seven Mile Beach after the great blizzard of February 9–11, 1899.

The main south-bound flight seems to pass along the New Jersey coast in September and October. My earliest autumn date is August 28, 1922, when several were found near South Cape May where none had been seen before and others on dry ground just behind the sand dunes, evidently birds that had dropped down there from their night flight.

The nest of the Mud Hen consists of matted grass and sedge and is built up to a height of from six to fifteen inches above the floor of the meadow while it is sometimes arched over above by the growing grass amongst which it is built. The favorite location is along the tidewater creeks which thread the meadows and along which the grasses grow to a much greater height offering better protection to both nest and birds. Full sets of eggs are deposited by May 30 and Walker Hand records one in which the first egg was deposited on April 14, 1903. The number ranges from eleven to thirteen but he found one nest on May 18, 1919, which contained no less than twenty eggs, possibly the product of two birds using the same nest. Two nests found on June 10, 1928, back of Ludlam's Beach, contained eleven eggs while three on June 24, held a like number and I have a record of one with nine eggs found on July 13, 1924, and another with the same number on August 3, 1930. These latter were obviously second sets due no doubt to the destruction of the first nests by high tides. In nests that I have found where the bird was apparently incubating she did not take wing until I had almost trodden upon her, and in one case she spread her wings like a hen and ruffling up her feathers strutted about for a moment before running off through the grass.

The following list of nests examined on Seven Mile Beach with the number of eggs in each, is furnished by Richard Miller with a few additional records from Turner McMullen:

May 21, 1927, eleven and nine eggs.(McM.). June 18, 1922, twelve eggs (McM.).
May 30, 1911, seven, nine, nine, and   June 19, 1927, nine eggs and three young.
  eleven eggs.   June 19, 1921, nine eggs.
May 30, 1912, nine, twelve, twelve eggs.   June 24, 1928, twelve and thirteen eggs.
May 30, 1913, ten and ten eggs.   June 26, 1921, eleven eggs.
May 30, 1915, seven, seven, and ten eggs.   June 27, 1920, six eggs (McM.).
May 31, 1920, twelve eggs.   July 3, 1921, four eggs.
June 4, 1916, six, ten, ten, twelve and   July 4, 1920, eleven and eight eggs.
  thirteen eggs.   July 10, 1921, ten eggs.
June 12, 1921, eight and ten eggs.   July 20, 1932, two with eggs (McM.).
June 17, 1923, twelve eggs.

At Corson's Inlet Miller reports the following:

June 12, 1927, nine, ten, eleven eggs.   June 21, 1925, ten eggs and two young;
June 14, 1931, thirteen eggs.   five eggs.
June 15, 1933, ten and twelve eggs.   June 24, 1928, thirteen eggs.
June 18, 1927, ten eggs.   July 2, 1927, nine apparently fresh eggs.

One nest contained a runt egg another a single egg five-eighths of an inch longer than the others in the set. Most of the nests were in sedge fifteen to twenty inches tall, along tidewater creeks in the meadows, but a few were on dry spots and one on a dune among bayberry bushes one hundred yards from water; still another ten inches above the water which was six inches deep. One was situated within three feet of a nest of the Laughing Gull.

The downy black young may sometimes be seen on the meadows with their parents and on one occasion Walker Hand came upon a whole brood swimming a thoroughfare with the parent in the lead, while on July 4, 1925, I saw several partly grown birds, still largely black and downy, feeding on the edge of a narrow thoroughfare at ebb tide and as my skiff drifted into sight they scrambled up the bank and into the overhanging vegetation. In similar situations, where the vertical banks honeycombed with the burrows of the fiddler crabs rise from three to four feet above the slippery black ooze of the creek bottom and are capped above with green meadow grass, I have frequently seen an old Mud Hen running, now rapidly now with hesitating steps, along the narrow shore, picking up food among the scuttling fiddlers. Suddenly he takes to the water and swims across to the opposite bank. Head and neck are stretched well forward with the long bill conspicuous. The fore part of the body is held well out of the water while the bird strikes out vigorously with his feet. Although Mud Hens make fair progress in the water they always seem to me to be straining to the utmost, an appearance due perhaps to the outstretched neck.

While living for a week in one of those little stilted gunners' shacks on the meadows on May 20, 1922, and May 16, 1924, the Mud Hens were our constant companions and as soon as the sun began to sink in the late afternoon their rattling chorus started, gaining in strength and volubility as the evening advanced and all through the night the *kek-kek-kek-kek-kek-kek-kek* was the only sound to break the silence of the lonely marsh. One evening a bird came out on a raft of dry trash resting on the meadow nearby and I could see his mandibles move as he called. Suddenly he ducked his head and ran through an opening under the trash and emerged some distance farther on. He disappeared and reappeared in this manner several times. Then another bird came in sight which seemed slightly paler and less active, perhaps a female, then a third appeared out of the grass and was vigorously pursued by the first. Finally both took wing and flew off across the creek renewing the chase in the grass on the farther side. At sunrise too the Mud Hens were active and calling vigorously. One came out of the grass at the water's edge and swam the ditch, selecting the only place where there was sufficient water for the purpose, since at low tide the ditch was almost dry. The bird could have walked over anywhere and why it preferred to swim was a mystery. The next evening one walked out along the muddy bank of the creek, in and out of the grass, flirting its stubby tail up and down with a nervous twitch. It looked browner and less gray than those just mentioned, with breast more ruddy but this was due in part at least to the light. This bird swam the ditch three times, always at the same spot; with head well up and neck stretched out and upward and not low down over the water. Again a bird came completely out of the sedge as we remained perfectly still and ran from the far side of the little creek crossing it on some planks which had been placed there as a boat landing, thence he passed right under the shack, which was raised on stilts a couple of feet above the meadow, and down the muddy bank of the thoroughfare into the rank grass which bordered it. When he returned a little later he followed the same course but swam the creek to his starting point.

On the evening of May 17, 1924, there occurred one of those high tides which occasionally flood the meadows and cause such havoc with the nests of Laughing Gulls, Common Terns and Mud Hens. At the height of the flood we saw several of our neighbor Mud Hens driven into a smaller and smaller area of higher meadow until they were compelled to swim and to seek refuge in a bunch of tall sedge the tips of which still projected above the water and here they clung until the flood subsided with the turning of the tide. It is a weird sight to see the waters gradually creeping up until the creeks and thoroughfares are full and then overflowing the banks and spreading over the flat meadows until one is surrounded by an apparently endless sea with small waves running under the house. It is a curious instinct that leads the Mud Hens and other marsh breeding birds to stick to their ancestral nesting grounds in spite of the repeated calamities, in the form of floods, which overwhelm them and destroy their eggs and young so that in some seasons it is questionable whether they raise any broods at all. Walker Hand tells me that when crowded together at the times of high tides the Mud Hens sometimes fight one another viciously.

On August 10 and 13, 1928, I saw at least two families of Mud Hens back of Five Mile Beach where the Wildwood Road crosses the thoroughfare. A cluster of boat houses has been built there on a point of meadow land about twenty-five yards square which is under water at high tide. Looking down from the board walks connecting the boathouses, which are elevated about six feet above the level of the marsh, I saw the birds at low tide walking about and feeding directly below me. The first day there were three adults and several downy young walking about like chickens and paying no attention whatever to persons crossing on the planks above them. The old birds held their wings slightly lowered so that the tips of the primaries were a little below the tail, while the latter was erect and kept twitching nervously up and down. They walked out on the open part of the mud flat and fed in the little puddles left by the tide, plunging the head completely under the water. Now and then one of them would lower the body to a strictly horizontal position and would run rat-like through the reeds with neck stretched out in front close to the ground, twisting right and left as it steered its course among the coarser reed stems. On the second day one old bird was closely followed by a partly grown young one which kept close to her legs but she finally ran away and left it to its own devices and it proved perfectly able to find food for itself, as did others of the same age which were there shifting for themselves. Another female and six downy young were seen on this occasion. The young walked ahead feeding slowly while the parent brought up the rear. It is possible that two of the adults seen the first day were really birds of the year of a much earlier brood. Birds in this plumage are darker and browner than breeding birds.

In 1930, a pair of Mud Hens bred on the meadows directly back of the golf links along Cape Island Creek and on July 8, I saw a female and four downy young feeding on the bed of the creek at low tide; another, probably the male, was seen running along a path in the short meadow grass with tail up and neck stretched out in front. On September 4 of the same year a full grown bird was seen in the same spot feeding on the mud. It not only kept flirting the tail up and down but every now and then shook it violently from side to side.

Now and then I have been able to approach migrant Mud Hens on the marsh formerly existing back of South Cape May when there was sufficient water there for their needs, and to study their actions. On August 28, 1922, I flushed seven at this spot, where I had seen none during the breeding season. They arose one at a time with legs dangling awkwardly and neck stretched out forward and upward, as if straining to make progress. In spite of the fact that the rounded wings were beating steadily, the bird's progress was not rapid and though the legs eventually were directed backward it soon dropped into the grass again. On September 3, 1921, I saw five in the same vicinity, on the borders of a shallow pool surrounded by a dense growth of Spartina grass. They waded from side to side or stood resting under the arching tufts of herbage. When standing thus, undisturbed, the body was held at an angle of about 45° with the head drawn pretty well down on the shoulders. When one of them stepped out it moved slowly and daintily, the head gliding back and forward with each step and the short erect tail making nervous jerks at similar intervals. Suddenly, either in pursuit of prey or from fear, it started to run, lowering the neck and holding the body nearly horizontal, neck straight forward, and tail erect. Sometimes when standing at rest it shook the tail violently from side to side; and when feeding it pecked at the surface of the water or mud but did not probe. Another individual stood in water nearly up to its body and with neck outstretched, flirted the tail nervously at quarter minute intervals. Once in early autumn when standing on the meadows a Mud Hen alarmed at something came running rapidly through the grass and crouched suddenly directly between my feet doubtless thinking that he had found safety between two convenient lumps of mud and I did not abuse his confidence.

As we gaze over the broad expanse of green salt meadow stretching away northward to the horizon, we do not suspect the presence of the scores of Mud Hens that skulk through the grass and sedge but if at sundown we push our skiff through one of the narrow tidewater creeks we shall soon become aware of them;—*kek-kek-kek-kek-kek-kek* goes the rapidly repeated cry of an early bird and soon it is answered, now on this side now on that, until the chorus extends far and wide over the meadow. If we rattle the rowlock on our boat we produce a fairly good imitation of the call and immediately one or more slender necks appear above the short sedge anxious to locate the intruder. Once seen in these, his native haunts, the Mud Hen is forever associated in our mind with the great salt meadows, especially at the close of day when the light is beginning to fade. The rather harsh faded gray green of his plumage seems to harmonize with the salt- or mud-incrusted grass stems among which he is so much at home. If we stand stock still we may perhaps see him run past us unaware, twisting and turning among the grass stems like an elongated rat. So gracefully and so rapidly does he thread his way among the stems, and so perfectly does the marsh grass shelter him, that one wonders why he ever attempts passage by either air or water. The full history of how the Mud Hens live under the green canopy of sheltering grass we shall probably never learn. We know them only from the chance glimpses that we get of them as, unaware of our presence, they come out for a moment from their sanctuary and like other rails, the best sheltered perhaps of all water birds, they will keep their secrets to themselves.

J. F. S.

NIGHT HERON

VIRGINIA RAIL

Rallus limicola limicola Vieillot

In habits, haunts and coloration this bird is a miniature of the King Rail and, like it, is usually seen only on chance occasions when it is come upon unawares away from the shelter of the marsh, or is flushed from the thick growth of sedge or cattails among which it spends its life. When thus it takes wing it is seen only for a few moments as with neck outstretched and elevated, and wings beating rapidly, it flutters for a dozen yards in heavy flight over the top of the vegetation and drops back again to safety. The Virginia Rail seems to like to come out on the edge of the reeds when sure that the coast is clear and to stand in the sun or feed along the borders of small ponds or streams, ever ready to dart back into the maze of reed and cattail stalks through which it threads its way like a rat. It also occurs along the edge of the salt meadows and I have seen individuals on the inner side of Two Mile and Seven Mile Beaches; doubtless they occur on the other coastal islands as well.

As I stood by the roadside at Green Creek one day (July 11, 1920) a Virginia Rail came suddenly out of the sedge of a nearby marsh and seeing me rushed back precipitately to safety. I remained perfectly still and in a few minutes it reappeared within a few feet of me and stood for some time apparently sunning itself. It made no movement except to stretch one wing to its fullest extent downward and backward, until its tip touched the ground behind the bird's feet. Two others seen in the summer of 1891 acted in much the same way. One of them fluffed up its feathers and strutted about as if it might have had young in the vicinity. From July 15 to August 1, 1931, on a little pond on Seven Mile Beach, which up to that time had escaped the oil distributors of the Mosquito Commission, we repeatedly saw a Virginia Rail which came out of the sedge in search of food and followed exactly the same route in every instance so that we knew precisely where to look for it. On one occasion it brought several young out with it. On April 30, 1932, and subsequently, another was observed on the edge of the Lighthouse Pond which had a similar line of travel. On July 2–4, 1928, Julian Potter saw Virginia Rails on three different ponds between Cape May and the Point, two with young, but now both ponds and rails have gone. Our records for these birds in the Cape May district run from April 8 (in 1922) to October 31 (in 1920) with scattering occurrences in November and during the winter, indicating that in open seasons, at least, some Virginia Rails do not go south. One was shot at Newport, Cumberland County, by C. H. Newcomb, February 15, 1922, and Walker Hand saw two on March 10, 1929, running on the ice on a pond that formed on the Fill. He also reported birds present on January 7, 1908, and December 12 and 26, 1910, while Julian Potter saw one on November 9, 1930, and two on April 8, 1928.

The Virginia Rail formerly bred commonly in semibrackish marshes all about Cape May but like all the rails it is now becoming very scarce near the town. A nest that I found on May 22, 1892, was raised a little from the mud at the edge of a cattail swamp and was made of cattail and rush stems matted down into a bowl-like platform, in which were nine eggs. No matter how carefully I approached this nest the bird always slipped off before I could see her. Other nests were found on May 22, 1921, which were raised four and twelve inches from the ground respectively, and contained eight and six eggs, while one on Ludlam's Beach, June 24, 1928, contained only broken egg shells and in another June 29, 1919, the eggs had just hatched. Nests examined by Richard Miller and Turner McMullen are as follows: On Seven Mile Beach, May 20, 1922, six eggs; June 29, 1924, six eggs and one more a week later. On Ludlam's Beach, June 28, 1925, six eggs; June 12, 1927, three nests containing nine, eight and two eggs respectively; June 26, nine eggs; July 10, six eggs; July 22, 1928, six eggs.

Some Virginia Rails in autumn are largely melanistic a tendency apparently peculiar to this species. Two seen below Broadway on the edge of a small pool on July 16, 1927, were largely black below with a white line down the middle of the belly, and I collected a similar one near the same spot on August 29, 1891. This Rail, although never comparing with the Sora or Mud Hen as a game bird, is shot regularly by gunners to whom it is known as the "Red Rail."

SORA RAIL
Porzana carolina (Linnaeus)

The Sora is distinctly a fresh water bird, abounding in autumn on the marshes of the Delaware River down as far as the tidewater streams in the upper part of the Cape May Peninsula, while it occurs also in fresh water marshes on the ocean side of the country.

About Cape May I know the Sora only as an autumnal transient. Like other rails it is not often seen except when one searches the marshes for it and then we have but a fleeting glance as it flushes and drops back into the cattails and sedges. On October 5, 1930, however, Clifford Marburger was fortunate enough to find six feeding on an exposed mud flat on Pond Creek Meadows. That there must be a very heavy migration is evident from the number of dead Soras that are to be seen along the roads during the early autumn, killed by striking the telegraph wires. On the nights of September 7 and 16, 1921, and September 24, 1932, a great many were killed in this manner and I was informed by farmers driving down the turnpike that on the first date they saw them every few yards all the way from Erma to Cape May. On August 26, 1925, one was struck by an automobile on the edge of the town. Dead Soras have been seen every autumn, the dates running from

August 26 to September 26, while live ones have been seen as late as October 11, 1930. We have no records for either winter or spring.

I have flushed Soras from wet ditches on the Pond Creek Meadows on September 1, 1926, and crept up on one on a small pond below Broadway on August 28, 1929, which stood on the edge of an exposed mud flat and now and then retreated into the sedge for shelter when alarmed. Two on the Lighthouse Pond on October 11, 1930, acted in the same way, running rapidly along the edge of the overhanging sedge usually with the tail held straight out but occasionally elevated. They were picking up some sort of food from the mud. These were apparently birds of the year.

While the breeding grounds of the Sora are normally much farther to the north they do nest occasionally on the marshes of the Delaware River. When the Tinicum Meadows below Philadelphia were flooded in 1928, Victor Debes found a Sora with a brood of downy young on July 7, which gives some idea of their time of nesting here. Nests and eggs were found in the same locality in June 1900 and on several other occasions by Dr. William E. Hughes. Fletcher Street saw an old bird leading young along a narrow run near Beverly, New Jersey, a little farther up the river on June 19, 1932.

The Sora is *the* rail of the marshes of the Delaware and is usually designated by sportsmen simply as "Rail" or "Rail-bird." Thousands used to be killed every autumn in September and October from Philadelphia almost to the Cape May County line. Gunners were pushed in flat-bottomed boats among the rank growth of wild rice (*Zizania aquatica*) and as the birds flushed in their heavy, labored flight they formed an easy mark. High tides were a necessity for good rail shooting as otherwise it was not possible to reach their feeding grounds in the shallower parts of the marsh. Sporting journals published a schedule of tides for the autumn to aid the gunners in planning their trips but even so, exceptional tides, such as we often have at this season, were necessary, and when there were none it was regarded as a poor rail year. Charles S. Westcott ("Homo") wrote regularly in "Forest and Stream" about the occurrence of Soras on the Delaware and states that on September 29, 1882, single boats got bags of seventy, eighty and ninety birds a day, and on the Tinicum Meadows 3720 were reported shot in 1884, to October 11, which was by no means the end of the season. On the Maurice River he states that one man got seventy as late as October 1, in 1884. Great numbers were heard passing over Philadelphia on September 15, 1883, and September 19, 1885, and the flight continued for hours. He makes no mention of their return in spring. The reeds and wild rice are of course not sufficiently grown to give them shelter at this season and they may have passed without stopping but it seems hardly possible that they ever have occurred abundantly in spring or the gunners of those days, with little or no limitation to shooting, would certainly have been out after them, and their bags would have been recorded. There seems to be the same lack of spring data today but the Sora is still shot in the autumn all the way down to Cape May Point, although in somewhat smaller numbers than in the past.

YELLOW RAIL

Coturnicops noveboracensis (Gmelin)

The little Yellow Rail seems to be a fresh water bird and an associate of the Sora in migration; all of our records have been in October when a few are seen or shot by local sportsmen who are hunting Mud Hens or Soras. We have no evidence of a northward flight in spring. One was picked up dead on the Bay side on October 15, 1924, and given to Walker Hand while Raymond Otter shot three on the Pond Creek Meadows on October 7, 1932, and has seen others. We have record also of one shot by B. H. Koons on the Maurice River on October 7, 1929, and another at Rehoboth, Delaware, across the Bay secured by Charles Pennock on September 26, 1908. All of the above specimens are in the collection of the Academy of Natural Sciences of Philadelphia. My own personal record is one seen on the Fill on October 11, 1927. I was crossing a low area from which the black grass had been cut and which is usually more or less covered with water in spring when the Snipe are migrating, though quite dry in the autumn. The little rail took wing from almost under my feet and then flushed several times, flying only short distances and finally refused again to take wing. It held its head up with legs dangling and wings beating rapidly in the typical clumsy flight characteristic of the rails. The general impression of the bird was buff, darker and streaked above, but the conspicuous feature was the broad white band across the back of each wing formed by the tips of the secondaries and giving the impression of a piece of white muslin attached to the ends of the feathers.

In view of the scarcity of observations on this bird we quote Charles Urner's comments on its song as heard farther north in the state: "Early on the morning of April 27, 1930, I was 'railing' over the broad expanse of fresh water marsh along the Whippany River, not far from Boonton, N. J., known as Troy Meadows. Upon a dry path I stood listening to a chorus of bird notes—the rolling whinny and frog-like *ker-wee* of the Soras; the sequenced grunts and paired *kaks* of the Virginia Rail a-wooing, and a mixture of whines and wails establishing the presence of a number of Florida Gallinules.

"Suddenly close to me—barely fifteen feet away—came a note which I had never heard. It was repeated several times. I wrote it *kik, kik, kik, kik, kik-kee er* with the last notes slightly rasping but full and of a decidedly musical quality. It was in fact a song. As I looked toward the focus of these new sound waves a small bird crossed and recrossed, quickly but in plain sight, over five feet of open marsh between tussocks. It was very close and very black—much darker than the Virginia Rail still standing in the open, and I saw below the cocked tail on a dark-feathered background, narrow white bars.

"I had never heard these notes from the Black Rail inhabiting the Barnegat region where I am fairly familiar with the species—but since the notes came from the spot where I had seen the bird I assumed that it was the Black Rail calling, and I got some added confidence after reading William Brewster's comments on the notes of a supposed Black Rail heard in the Cambridge, Massachusetts region (Auk, XVIII, 1901, p. 321). However, upon consulting Forbush's "Birds of Massachusetts" I found that Mr. Brewster's rail notes, identical with those I had heard, were in turn identical with the notes uttered by a Yellow Rail kept alive by Mr. J. H. Ames. Mr. Ames described the call as a series of *kiks*, ending in a *ki-queah*. What I saw at Troy Meadows my eye says was a Black Rail. What I heard from the same spot Mr. Ames says were the notes of a Yellow Rail!" (Auk, 1930, p. 560.)

J. F. S.

BITTERN

BLACK RAIL

Creciscus jamaicensis stoddardi Coale

This is the smallest and most secretive of our rails and it is rarely possible to get a view of one. Their habitat is the heavy growth of fine grasses, *Distichlys* and *Spartina*, which often cover considerable areas on the edge of the salt meadows. These attain a height of eighteen inches or two feet and then are bent over by the wind and rain forming dense recumbent mats more or less supported by the stems of the grass beneath. Under this concealing thatch the Little Black Rails spend their lives running about like mice between the roots of the grasses and here too they build their nests, deep pocket-like affairs, eight inches across and a foot deep, lined with dry wisps of grass and covered completely by the live green thatch so that one sees no trace of them from above.

All my efforts to find this bird about Cape May have been without result but several have been found by others. One individual, probably forced out of the marsh by high tides, took refuge in Walker Hand's garden close by, and was caught by him and examined on September 25, 1909. Another was picked up dead on a Cape May street by Raymond Otter on May 7, 1933, who recognized it as an unusual bird and correctly identified it. A third one was found dead, entangled in a wire backstop on the tennis court at Cape May Point, by Robert Allen on September 5, 1934, who kindly sent me the wing, and still another by William Rusling on the meadows back of South Cape May on September 11, 1935.

The Little Black Rail was found nesting years ago by Thomas Beesley on the brackish marshes at Beesley's Point across the bay from where Ocean

City now stands. Three nests were uncovered when the black grass was being cut in 1844 and 1845 (Auk, 1900, p. 172). Many years later Richard Harlow discovered a number of nests near Tuckerton and elsewhere in the coastal portions of Ocean, Atlantic and Burlington Counties. He examined twenty-four nests in all, the average number of eggs being seven and ranging from six to nine with one set of thirteen. The dates for full sets ranged from June 6 to July 23, average, June 14. His first nest discovered back of Brigantine on June 22, 1912, was "skillfully concealed in a thick mass of green and dead grass, so that it was completely hidden from above. It was little larger than the average nest of the Robin but deeper cupped and built entirely of the dry yellow stalks of the sedges. Another nest, June 29, was similar and was quite invisible until the grass was parted from above, it was interwoven on all sides with the surrounding stalks. Still another was found by George Stuart on Little Beach Island on July 4, 1919, containing eight eggs upon which the female sat so closely that she was easily caught in the hand. (Auk, 1920, p. 292.) On June 12, 1921, Richard Miller found a nest with seven eggs on Seven Mile Beach and four others in subsequent years; June 14, 1925, six eggs; July 4, 1928, five eggs; June 21, 1925, seven eggs; July 4, 1929, nine eggs.

In June 1926 Turner McMullen found quite a colony of these little rails nesting in the usual sort of grassy patches along the inner edge of Ludlam's Beach and I examined a nest there on June 24, 1928, in company with McMullen, Arthur C. Bent and others. When found the bird ran quickly over the matted grass as the thatch was raised. Nests examined at this locality by Turner McMullen and a few by Richard Miller are as follows, with the number of eggs in each:

June 12, 1932, eight eggs.
June 13, 1926, two with eight and nine eggs.
June 18, 1927, eight eggs.
June 20, 1931, eight eggs.
June 23, 1928, seven eggs.
June 27, 1926, six eggs.

July 2, 1927, three with five, six and seven eggs.
July 3, 1932, eight eggs.
July 10, 1927, six eggs.
July 10, 1934, eight eggs.

CORN CRAKE

Crex crex (Linnaeus)

A specimen of this European bird was shot by Walker Hand while gunning for Quail in corn stubble near Dennisville on November 11, 1905, and presented to the Academy of Natural Sciences of Philadelphia (Cassinia, 1905, p. 75). Another specimen, also in the Academy's collection, was shot by William Patterson at Salem, in the fall of 1854 (Proc. Acad. Nat. Sci., Phila., 1855, p. 265). These, so far as I am aware, are the only records for the state.

PURPLE GALLINULE

Ionornis martinica (Linnaeus)

This beautiful southern gallinule has several times been blown north by tropical storms and stranded on our coast. In May, 1928, Mrs. E. H. Fisher found one sprawled out exhausted in a privet hedge, along the beach drive facing the ocean, and while unacquainted with the bird at the time, she noted its "brilliant blue head, shading to steel blue on the back and tail, and the vermilion shield at the base of the bill." Upon being touched it managed to fly around the house and could not be found. Another individual was caught on the Anglesea beach by Harry Callahan on May 8, 1932, and kept for some time in captivity. I saw it a few days after its capture. A third specimen was caught on the Cape May ocean front on May 10, 1907, and was identified by Walker Hand, and still another caught there by C. F. Gardner, May, 1892, was mounted by Charles A. Voelker who reported it to me. On May 28, 1934, Otway Brown almost stepped on a Purple Gallinule while crossing a bog near Bennett and, as the bird flew away in front of him, he saw the general blue color and the great pea green feet stretched out behind. On August 31, 1935, he and I were exploring a swampy thicket near Cold Spring when we flushed one of these birds which flew past Brown in good view and an hour later I saw it return to the swamp and run through the low vegetation covering the water.

Purple Gallinules have been found at points farther up the coast: several near Tuckerton by Jillson Brothers, prior to 1894; two by William H. Werner at Longport, May 23, 1898, and Ventnor, May, 1902, and one at Beach Haven by C. W. Beck, May, 10, 1907, the last now in the collection of the Academy of Natural Sciences of Philadelphia. A notable occurrence, although quite out of our range, was a dead bird in very fresh condition picked up on the Tinicum marshes below Philadelphia by Brooke Worth on June 15, 1934. This was at the time that these marshes were flooded, when not a few rare water birds were found there and Pied-billed Grebes, Coots, and Soras remained to breed. No other Purple Gallinules were found, however, and no indication that a pair had nested there (Auk, 1934, p. 519).

FLORIDA GALLINULE

Gallinula chloropus cachinnans Bangs

Whatever its former status may have been in the immediate vicinity of Cape May the Florida Gallinule seems at present to be a regular autumn transient with occasional occurrences in spring. The constant draining of the swamps and marshes have destroyed what may have been regular nesting places in years gone by. It seems to be most frequent from September 20 to October 25, when one is likely to find single birds, pairs or groups of four to six, leisurely swimming on the deep pond at the Lighthouse. On the nearby Lake Lily I have never seen it but this is undoubtedly due to the absence of cattail thickets on this body of water, such shelters being apparently necessary to the bird's existence.

Gallinules progress slowly on the water, the head and neck moving forward and back in unison with the alternate strokes of the feet. The wing tips are elevated slightly above the base of the tail and the white edgings of the flank feathers appear against the sides of the body outside the folded wings, while the white spots on the under tail coverts are conspicuous, especially when the tail is nervously flirted. Now and then the head is lowered to the surface of the water to pick up some morsel of food. The olive brown of the back is deeper in some individuals than in others, doubtless birds of the year, and is always more or less in contrast with the slaty gray of the lower parts.

When they sight danger the Gallinules on the Lighthouse Pond almost always take wing and seek safety in the dense cover of cattails which surrounds the open water. Usually they fly close to the surface, splashing or skittering through the water for some distance, at other times, especially when crossing the pond from one side to the other, they fly higher with rather rapid steady wing strokes, their yellow green legs dangling below the body. On one occasion a Florida Gallinule was accompanied by a Pied-billed

Grebe and on another two were associated with a female Baldpate and remained with her after my appearance on the edge of the pond, although four other Gallinules which were nearby immediately took wing for the cattails. On only one occasion have I heard them cackle while on the pond, on September 25, 1927, when the familiar call came from the middle of the cattail growth.

Our earliest record for the southward flight is August 6 and 7, 1931, when Julian Potter found a pair present, and August 31, 1923, when one was caught by the coast guards which had apparently struck a wire during its night migration. Our latest date is October 27, 1928. In spring there are but two records, a single individual on the pond on May 7, 1932, and another which flew onto the Five Fathom Bank Light Ship, some twenty miles off Cape May, on May 18, 1924. It was caught by one of the crew and brought to land, when their boat came in for the mail, and liberated. Florida Gallinules are common summer residents at Delaware City and on June 4, 1927, Julian Potter saw five downy black red-billed young there, following a clucking mother like so many domestic chicks.

I do not doubt but that the Florida Gallinules breed in some of the extensive cattail swamps bordering the Bay in the vicinity of Dennisville and they certainly do along Riggin's Ditch in Cumberland County, where Julian Potter and I saw several families swimming about in the shallow water on August 18, 1933, and where he had seen them on several previous occasions.

From Salem north to the Bridesburg marshes in the northern part of Philadelphia they nest in any favorable spot. Richard Miller who has found many nests in the Philadelphia area tells me that they are constructed of the dry flat leaves of the cattails and lined with narrower cattail blades, sedges and grass stems, they are well cupped and about two feet thick, always over water. The eggs number from nine to fourteen, usually ten to twelve, while he once found a set of eighteen. His extreme dates for full sets are from May 22 to June 14. Turner McMullen tells me that he examined a nest near Marshalltown, Salem County, on June 10, 1922, which contained eight eggs while three found near Camden contained on May 25, 1935, eight, nine and eleven. A few Florida Gallinules may winter in Cape May County as Richard Miller found one in the Bridesburg marshes, Philadelphia, on February 12, 1913.

COOT

Fulica americana americana Gmelin

Among the Cape May gunners the name "Coot" is universally applied to the Scoter ducks while the present bird, the true Coot, is known as "Blue Peter." It is not very common about the Cape, occurring in October and November and again in April, while an occasional individual may be seen in December and possibly remains through the winter.

While here it seems to prefer sedgy marshes where it can swim slowly about among the flooded tussocks of grass, and all that I have come across, have been on the ponds at the Lighthouse. One that I watched for some time on the deep pond, October 18, 1925, worked its way in and out along the edge of the thick vegetation, now completely lost to sight, now reappearing. Its most conspicuous feature was the peculiar short deep, white or greenish white bill so strongly contrasted with the black head. The contrast between the latter and the leaden blue gray of the back was also striking, while from the rear the white spots on either side, below the tail, stood out conspicuously. As the Coot flapped its wings, while still resting on the water, their white posterior edging was clearly seen, and when swimming the head moved slowly forward and back as the bird struck out with its feet.

There were two present on the shallow pond on October 23, 1927, where the border of tall grass was especially to their liking, and on the deep pond, on November 6 of the same year, there were eleven. These birds were swimming leisurely and constantly dipping the head under the water sometimes turning tail-up like a duck for a quarter of a minute at a time as they sought some sort of food nearer the bottom of the pond.

There was a heavy gale blowing and the Coots were gradually blown across the water and were apparently unable to swim back so that one after another they took wing and flew over the pond to the far side where they again began to feed. While in the air the bill was pointed downward and the legs trailed behind drooping a little from the horizontal. Now and then, when there was a clear sky or white cloud for a background to the flying bird, the white bill was lost entirely and the impression was of a bird with a peculiar truncated black head and no bill at all.

My earliest record for Coots at Cape May is September 30, 1928, when I found four on the Lighthouse Pond and I also saw one there on November 9, 1930, but most of our records are for October. Julian Potter saw three at the Lighthouse on November 11, and two on December 26, 1927, and John Gillespie one on the Water Works Pond near Rio Grande on December 27, 1931, which indicates that they occasionally winter. They do not seem to be so common in spring or else they pass through much more rapidly or possibly keep out to sea. We have a record of one seen on the Lighthouse Pond, on April 18 and 24. 1926, by Julian Potter, and five found by him on a pond at Ocean View a few miles farther up the coast. One was also seen at Heislerville, on the lower Maurice River, in Cumberland County by Edward Weyl on May 1, 1932.

Coots, like ducks, require larger areas of water for wintering than are afforded about Cape May and it is quite natural, therefore, to find them much more abundant on Barnegat Bay than farther south in the state. Charles Urner has recorded seven hundred there on November 13 and four thousand on December 11, 1927, while at Delaware City, Delaware, Norman McDonald has found thirty-nine on November 13 and seventeen on December 13, 1927.

When unusual conditions prevail Coots may be induced to nest farther south than usual and a nest was found on the Newark marshes on May 30, 1907, by Clinton G. Abbott (Auk, 1907, p. 436). Another nesting was clearly indicated when the Tinicum marshes below Philadelphia were flooded and Victor Debes, on July 29, 1933, found an adult with a brood of downy young feeding there, which had been reported previously by Delos Culver. On June 3, of the following year, John Gillespie saw three adults there and Benjamin Hiatt two hundred on April 19, 1935, and April 5, 1936. Very late birds at this locality, possible breeders, were seen on May 30, 1934, and May 3, 1936 (Hiatt).

CAPE MAY SHORE BIRDS

The shore birds constitute the dominant element in the bird life of the coastal "meadows" and so closely do they adhere to the peculiar environment that it is a matter of surprise to find them elsewhere. There seems to be an invisible barrier between the upland and meadow across which they do not trespass. We make this statement, however, with some reservations. There are, for example, the Woodcock and the Snipe, the Solitary and Spotted Sandpipers, and the Killdeer, all of them shore birds so far as structure and systematic position are concerned, but in habits and distribution they are as much birds of the upland as of the marsh—the Woodcock not a salt meadow bird at all. These we have therefore excluded in our general discussion of the shore birds which follows.

While we might infer, from our constant association of shore birds with the coastal marshes, that their flight lines covered these areas exclusively but such is apparently not the case. We know that many of them travel regularly along the Mississippi Valley and it would appear that flights of waders must be continually passing high overhead in our uplands during their periods of migration. Should heavy rains flood some of the farmer's low lying fields, or a new reservoir be established near a town or city, or an extensive dredging operation be undertaken, shore birds, sometimes in great numbers, will promptly appear and descend to feed where under normal conditions they had been unknown. On Lake Ontelaune, the recently established reservoir of the city of Reading, Pennsylvania, nearly one hundred miles from the sea, Earl Poole has observed no less than twenty-four kinds of shore birds including such typically maritime species as the Sanderling, Dowitcher, Stilt and Red-backed Sandpipers, Hudsonian Curlew etc. (Cassinia, XXIX, 1931–32); on the Springfield Reservoir in Delaware County, Pa., on July 29, 1930, John Gillespie found seven species with three more in the next ten days (Bird-Lore, 1930, p. 358); on a small mud-rimmed pond in the city of Camden, New Jersey, Julian Potter has identified thirteen species, while on low grass-covered fields a mile of more from the shore back of Cape May I have found ten or a dozen species feeding immediately after heavy rains which have converted the fields into shallow ponds.

Nevertheless it is the great salt meadows which lie behind the barrier islands of our coast, all the way from Cape May to Barnegat, and the marshes

about Newark Bay, farther north, that constitute the real home of the
host of migrant shore birds that passes in spring and late summer, and of the
few which tarry with us during the winter months. These meadows form
a natural ecological unit and what may be said of one section applies equally
well to another. At the same time it is dangerous to draw general conclusions from observations over a relatively restricted area, for, as Charles Urner,
our leading authority on New Jersey shore birds, says: "several factors may
influence the number of individuals passing or stopping to feed, giving an
appearance of increase or decrease which may not be in accordance with the
facts. Weather conditions during the flight undoubtedly influence the
numbers of several species and feeding conditions influence the numbers
which tarry upon the marshes and flats in any locality. It is possible also
that the route of migration of some shore birds is subject to considerable
variation from year to year."

"In many species" he adds "it was noted that the season's flight occurred
in definite waves, sometimes widely separated. There frequently was a
large main movement and either earlier smaller flights or later recurrences
on a smaller scale, due possibly to differences in time when the birds left the
same breeding grounds, or the filtering south from favorite feeding places
along the line of flight, as the urge to continue the migration was recreated
by changing weather conditions. In a number of species observed from the
seashore the main flights coincided with a steady south breeze, but they
were not always so associated, and considerable movements of several
species were observed when the wind was very light from west or south."

It is obvious, therefore, that it is impracticable to divide the coastal meadows into northern and southern areas since it is an easy matter for a flock
of migrating shore birds to cover long distances in a surprisingly short time,
and the first migrants may be as likely to come to rest at some South Jersey
locality as at Barnegat Bay on the more northern coast. The fact that one
spot is better adapted to the birds' needs has more to do with their coming
to rest than relative latitudinal position, and in late years man, with his endless ditching and draining, has played a major part in ruining many a shore
bird resort.

It is the lure of the shore birds that draws us to the "meadows" in May
when the congested northward flight is in progress and in July, August and
early September, when the more protracted south-bound migration is passing. The flocks of Sanderling, Semipalmated Sandpipers and Semipalmated
Plover, and the bunches of Knot and Willet that frequent the beaches have
their charm but the great masses of Curlew, Yellow-legs, Black-bellied Plover,
Turnstones and Dowitchers that arise from the grass of the meadows or the
nearby mud and sand flats are even more impressive, and then there are the
great shoals of little "peeps" (Least and Semipalmated Sandpipers) and the
ever present possibility of detecting among them one of the several rarer
species that associate with them—Baird's, White-rumped, Western and Buff-
breasted Sandpipers.

Charles Urner has spent many years in the study of the shore birds of the Newark, Barnegat Bay and Brigantine areas and has published most interesting résumés of the flights. As his conclusions coincide with mine and are based upon more extended observations we present extracts from his published reports to which he has generously furnished additional data. As already explained these studies are usually as applicable to the Cape May district as to that in which they were made—the two areas are but a bird's flight apart! (Auk, 1929, p. 314; 1930, p. 424; 1931, p. 418; 1932, p. 470, Proc. Linn. Soc., New York, No. 47, 1935, p. 77.)

In order to arrive at some idea of the relative abundance of the several species Mr. Urner has made careful counts of the numbers of each seen on every field trip through his region and, by combining his data for all trips taken during a season, he is able to ascertain (1) the number of trips upon which a given species was seen, (2) the largest number of individuals seen on one trip, and (3) the total number seen on all trips (this last figure probably involves some duplication as the same flock of birds might be counted on successive trips to the same spot but, while perhaps not accurate as to actual number of individuals present, it is perfectly satisfactory for purposes of comparison). The rank of each species in each of these three categories is then ascertained and the average of the three gives the bird's rating for the season. The Semipalmated Sandpiper, for example, was first in each list in the autumn flight of 1928, seen on every one of the sixty-seven trips, with 6000 as the maximum number for a single trip, and 30,774 for all trips combined. The Sanderling fourth in number of times seen, second in maximum per trip, third in season total, was by average in second place, etc.

While I cannot speak too highly of the great value of Mr. Urner's painstaking records it is only fair to quote his own appraisement: "Duplication (in counts of total numbers present) is most marked in the fall in such species as Semipalmated Sandpiper, Lesser Yellow-legs, Dowitcher, Least Sandpiper, and Semipalmated Plover, and improves the showing of these in comparison with birds which are seen merely passing on migration. Moreover trips were not made every day and the main flight of some species may have been witnessed one year and missed the next.

"Because of the variable repetitive factor and other factors mentioned, no single year's counts can be considered certainly comparable either between different species the same year or between the same species in different years. However the data for a series of years will offer indications, in a general way, of any marked change in the status of the various shore bird species on the New Jersey coast." (Proc. Linn. Soc. N. Y. No. 47, p. 87.)

The ranking of the species for the past seven years, as compiled from Mr. Urner's reports, is as follows, the numbers at the left being the average of the seven years (no counts were made for the spring flight of 1928). I have omitted the Woodcock and other species previously mentioned which occur also on the upland; as well as the phalaropes and a few others of rare or irregular occurrence.

Autumn Migration—Southward

	1928	1929	1930	1931	1932	1933	1934
1. Semipalmated Sandpiper	1	1	1	1	1	1	1
2. Lesser Yellow-legs	3	3	3	5	2	2	3
3. Semipalmated Plover	4	2	2	4	6	3	4
4. Dowitcher	7	4	5	2	3	4	2
5. Sanderling	2	5	4	3	4	6	5
6. Least Sandpiper	5	7	8	6	5	5	6
7. Black-bellied Plover	6	13	6	12	9	8	8
8. Greater Yellow-legs	8	8	14	10	10	7	11
9. Hudsonian Curlew	10	6	7	7	12	14	12
10. Pectoral Sandpiper	13	14	10	9	8	10	7
11. Knot	12	9	9	8	7	17	10
12. Red-backed Sandpiper	9	10	13	11	13	11	9
13. Turnstone	11	11	12	13	14	15	14
14. Western Sandpiper	18	15	16	14	16	9	13
15. Golden Plover	14	17	11	17	11	16	17
16. Stilt Sandpiper	17	18	15	15	15	13	16
17. White-rumped Sandpiper	16	12	17	16	18	12	18
18. Willet	15	16	18	18	17	18	15
19. Purple Sandpiper	19	19	—	19	19	19	19

Spring Migration—Northward

	1929	1930	1931	1932	1933	1934
1. Semipalmated Sandpiper	1	1	1	1	1	2
2. Black-bellied Plover	7	4	3	3	2	3
3. Dowitcher	2	6	8	4	3	1
4. Semipalmated Plover	6	2	4	2	5	7
5. Greater Yellow-legs	3	3	5	6	9	5
6. Least Sandpiper	9	5	7	5	4	4
7. Turnstone	5	7	2	7	7	8
8. Red-backed Sandpiper	4	8	11	11	6	6
9. Sanderling	11	9	6	8	8	9
10. Knot	8	10	10	10	10	11
11. Hudsonian Curlew	12	11	9	9	11	10
12. White-rumped Sandpiper	10	12	12	13	12	13
13. Lesser Yellow-legs	14	13	—	12	14	12
14. Western Sandpiper	10	12	12	13	12	13
15. Willet	—	—	14	—	14	—
16. Pectoral Sandpiper	—	14	—	—	15	14
17. Purple Sandpiper	—	—	—	—	—	16
18. Golden Plover	—	—	—	—	16	—
19. Stilt Sandpiper	—	—	—	—	—	17

In comparing the tables species for species it will be noticed that the
Lesser Yellow-legs, which ranks second in abundance in the southward flight,
is placed thirteenth in spring. This is because the northward flight of this

species is up the Mississippi Valley and it is represented on the Atlantic seaboard at this season by only a few straggling individuals. The same is true of the Pectoral and Stilt Sandpipers and the Golden Plover which were seen in only one or two of the spring flights and then usually scattered individuals. The records of the Purple Sandpiper were at Cape May in most cases.

Considering all of the shore birds in Charles Urner's lists—breeding birds and rarities as well as regular transients—thirty species have been recorded on the northward flight.

I. Of these sixteen have been observed every year:

Piping Plover	Hudsonian Curlew	Red-backed Sandpiper
Semipalmated Plover	Spotted Sandpiper	Dowitcher
Killdeer	Greater Yellow-legs	Semipalmated Sandpiper
Black-bellied Plover	Knot	Western Sandpiper
Turnstone	White-rumped Sandpiper	Sanderling
	Least Sandpiper	

II. Four have been recorded in five years:

Woodcock	Solitary Sandpiper	Lesser Yellow-legs
Wilson's Snipe		

III. Eleven in four years or less:

· Red Phalarope (four years)	Oyster-catcher (one year)
Upland Plover (three years)	Golden Plover (one year)
Pectoral Sandpiper (three years)	Purple Sandpiper (one year)
Northern Phalarope (three years)	Stilt Sandpiper (one year)
Willet (two years)	Hudsonian Godwit (one year)
· Wilson's Phalarope (two years)	

The species in section III are all rare and the records are often based upon single individuals. The Woodcock is mainly restricted to inland localities while the Golden Plover, Lesser Yellow-legs, Pectoral and Stilt Sandpipers, except for a few stragglers, travel north by way of the Mississippi Valley and are absent or very rare on the Atlantic coast in spring, although on the southward flight the second and third are abundant on our coast. The Willet is common in southwestern New Jersey where it breeds.

In the southward migration thirty-eight species have been recorded:

I. Of these twenty-five have been observed in each of the seven years:

Piping Plover	Spotted Sandpiper	White-rumped Sandpiper
Semipalmated Plover	Solitary Sandpiper	Least Sandpiper
Killdeer	Willet	Red-backed Sandpiper
Golden Plover	Greater Yellow-legs	Dowitcher
Black-bellied Plover	Lesser Yellow-legs	Stilt Sandpiper
Turnstone	Knot	Semipalmated Sandpiper
Wilson's Snipe	Purple Sandpiper	Western Sandpiper
Hudsonian Curlew	Pectoral Sandpiper	Sanderling
Upland Plover		

II. Three have been observed in six years:

Baird's Sandpiper   • Marbled Godwit   Northern Phalarope

III. Three in five years:

Woodcock   Buff-breasted Sandpiper • Wilson's Phalarope

IV. Seven in four years or less:

Long-billed Dowitcher (four years)   • Curlew Sandpiper (one year)
Hudsonian Godwit (four years)   ♪ Ruff (one year)
• Red Phalarope (three years)   , Avocet (one year)
• Wilson's Plover (one year)

All species, except in the first section, are rare and records are often of single individuals while the same is the case with the Golden Plover and Purple Sandpiper and, of late years, with the Upland Plover. The Woodcock is of course common in upland localities.

Quoting Urner's own arrangement (Proc. Linnaean Soc. New York, No. 47, p. 88) as to rarity, we have, numbered in the order of abundance:

Abundant or very common:

1. Semipalmated Sandpiper   4. Dowitcher
2. Lesser Yellow-legs   5. Sanderling
3. Semipalmated Plover   6. Least Sandpiper

Common:

7. Killdeer   12. Knot
8. Black-bellied Plover   13. Red-backed Sandpiper
9. Greater Yellow-legs   14. Spotted Sandpiper
10. Hudsonian Curlew   15. Turnstone
11. Pectoral Sandpiper   16. Western Sandpiper

Irregularly and locally tolerably common:

17. Golden Plover   21. Willet
18. Piping Plover   22. Upland Plover
19. White-rumped Sandpiper   23. Wilson's Snipe
20. Stilt Sandpiper   24. Solitary Sandpiper

Rare:

25. Purple Sandpiper   29. Woodcock
, 26. Marbled Godwit   30. Northern Phalarope
. 27. Wilson's Phalarope   31. Buff-breasted Sandpiper
28. Baird's Sandpiper   32. Hudsonian Godwit

Very Rare:

33. Long-billed Dowitcher   • 37. Ruff
• 34. Red Phalarope   , 38. Wilson's Plover
, 35. Avocet   • 39. Curlew Sandpiper
36. Oyster-catcher

Charles Urner has also recorded the extreme dates between which each species has been seen on each flight; the period over which the bulk of the migration took place; and the maximum number of individuals seen on a single day. These dates and figures will be found in the detailed accounts of the several species which follow.

Considering observations along the entire coast, but more especially the vicinity of Newark Meadows, Barnegat and Brigantine, he finds that twenty-one species have been recorded in the winter months (November to February),—the Piping Plover mostly in small numbers or as individuals. These birds have occurred as follows:

Throughout the winter: Killdeer, Black-bellied Plover, Wilson's Snipe, Purple Sandpiper, Red-backed Sandpiper, Sanderling.

To January: Knot, Semipalmated Sandpiper.

To December: Semipalmated Plover, Golden Plover, Turnstone, Greater Yellow-legs, Least Sandpiper, Western Sandpiper.

To November: Piping Plover, Woodcock, Lesser Yellow-legs, Pectoral Sandpiper, White-rumped Sandpiper, Common Dowitcher, Long-billed Dowitcher.

The Christmas census of the Delaware Valley Ornithological Club from 1931 to 1936, in the Cape May area, shows fourteen species which tallies exactly with Mr. Urner's total for the same period, although the species differ slightly. The Woodcock and Lesser Yellow-legs are added for December and the Golden Plover and Turnstone were missed, although the latter was seen on another occasion in December.

The Purple Sandpiper is the only shore bird that occurs on our coast exclusively in winter, and our records are in most cases on the artificial stone jetties at the mouth of Cape May Harbor. The fluctuations in abundance of the several species of shore birds during migration make it difficult to predict what one may see on any given day. Furthermore a few species are restricted almost entirely to the beach while others are distinctly inland birds, so that these areas must be covered as well as the salt meadows if a complete list is to be secured.

At the height of the northward migration in May it is safe to say that the following species will be present about Cape May, all on the meadows unless otherwise indicated:

Piping Plover (beach only)	Greater Yellow-legs
Killdeer (mainly inland)	Knot (also on the beach)
Black-bellied Plover	Least Sandpiper
Turnstone	Red-backed Sandpiper
Woodcock (inland)	Dowitcher
Hudsonian Curlew	Semipalmated Sandpiper (also on
Spotted Sandpiper (mainly inland)	the beach)
Willet	Sanderling (beach only)

Wilson's Snipe occurs regularly on inland meadows and those close to the salt marsh but it passes through earlier in the season and has gone before the other migratory species have arrived.

At the height of the southward flight, from July to September, we are likely to see all of the above except the Red-backed Sandpiper which does not arrive in numbers until after most of the other species have passed south and when it does come will be found mainly on the beaches, Wilson's Snipe is likewise a late arrival. We are also likely to see four other species at this season which are rare or accidental in spring:

Solitary Sandpiper (rare on the coast in spring and seen mainly about fresh water ponds on the southward flight).

Lesser Yellow-legs (very rare in spring, on the Atlantic coast; it goes north via the Mississippi Valley).

Pectoral Sandpiper (migration similar; when here on the southward flight it is found mainly on the inner edge of the meadows).

Stilt Sandpiper (migration similar, but somewhat variable in occurrence, rarely or never seen in spring in the East).

Seventeen additional species of shore birds have been recorded in recent years on the New Jersey coast all of which are rare or very rare at Cape May. According to Charles Urner's records the Golden Plover has been more plentiful on the Newark Meadows then anywhere else; it is decidedly rare at Cape May. The White-rumped and Western Sandpipers also seem to be more frequent on the northern coast than they are in the Cape May district, while on the other hand the Hudsonian Curlew certainly pauses more regularly on the southern meadows and in greater numbers than about Barnegat. There may be some difference in the food that is available on the different sections of meadows which accounts for the preference of the migrant birds for one as opposed to another, and it may be that the prominence of the coast line at Brigantine and its southwestward trend from there to Cape May may have something to do with their direction of flight. Certain it is that a number of species of both shore birds and gulls are rarer south of Barnegat, or do not occur there at all, while others prefer the southern coast and meadows.

As the lists just presented are to some extent hypothetical it may be of interest to quote some actual lists. The first six are for the northward migration:

May 21, 1922. Two Mile Beach. Stone and Hand. Fourteen species.

Piping Plover	Hudsonian Curlew	Red-backed Sandpiper
Semipalmated Plover	Spotted Sandpiper	Dowitcher
Killdeer	Greater Yellow-legs	Semipalmated Sandpiper
Black-bellied Plover	Knot	Sanderling
Turnstone	Least Sandpiper	

May 17, 1929. Newark to Brigantine. Urner and others. Eighteen species.

Piping Plover, two.
Semipalmated Plover, one hundred.
Killdeer, four.
Black-bellied Plover, one thousand.
Turnstone, two thousand.
Hudsonian Curlew.
Spotted Sandpiper, six.
Solitary Sandpiper, four.
Greater Yellow-legs, forty-eight.

Lesser Yellow-legs, one.
Knot, five hundred.
White-rumped Sandpiper, twelve.
Least Sandpiper, twelve.
Red-backed Sandpiper, five hundred.
Dowitcher, five hundred.
Semipalmated Sandpiper, one thousand.
Western Sandpiper, one.
Sanderling, two.

May 20, 1934, Brigantine. Julian Potter. Fourteen species.

Piping Plover, ten.
Semipalmated Plover, two hundred.
Killdeer, ten.
Black-bellied Plover, eight hundred.
Turnstone, 150.
Hudsonian Curlew, two.
Greater Yellow-legs, four.

Knot, twenty-five.
White-rumped Sandpiper, five.
Least Sandpiper, one.
Red-backed Sandpiper, fifty.
Dowitcher, one thousand.
Semipalmated Sandpiper, five hundred.
Sanderling, forty.

May 30, 1925, Cape May. Julian Potter. Eleven species.

Piping Plover, one.
Semipalmated Plover, twenty.
Killdeer, ten.
Turnstone, five.
Spotted Sandpiper, five.
Greater Yellow-legs, one.

Knot, six.
White-rumped Sandpiper, two.
Dowitcher, one.
Semipalmated Sandpiper, twenty.
Sanderling, thirty.

June 10, 1928, Ludlam's Beach. Julian Potter. Six species. This and the next show how the last northbound migrants linger into June.

Semipalmated Plover, six.
Turnstone, two.
Knot, seven.

White-rumped Sandpiper, one.
Semipalmated Sandpiper, forty.
Sanderling, two.

June 20, 1923, Grassy Sound. Julian Potter. Six species.

Black-bellied Plover, two.
Turnstone, four.
Knot, five.

Least Sandpiper, three.
Red-backed Sandpiper, two.
Sanderling, one.

The largest daily lists of shore birds seem to have been obtained on the southward flight, probably because the birds apparently linger on good feeding grounds at this season, and it is possible for a number of different species to accumulate; while in the spring they seem to be more concentrated and do not linger so long at any one spot. Some daily records for the southward flight are appended:

July 26, 1930, Brigantine. Julian Potter. Eighteen species.

Piping Plover, thirty.
Semipalmated Plover, twenty.
Killdeer, ten.
Turnstone, thirteen.
Hudsonian Curlew, one.
Upland Plover, one.
Spotted Sandpiper, fifteen.
Willet, two.
Greater Yellow-legs, two.

Lesser Yellow-Legs, ten.
Knot, 170.
White-rumped Sandpiper, five.
Least Sandpiper, five.
Dowitcher, six hundred.
Stilt Sandpiper, forty.
Semipalmated Sandpiper, thirty.
Western Sandpiper, two.
Sanderling, one hundred.

July 29, 1931, Brigantine. Julian Potter. Fifteen species.

Piping Plover, twenty.
Semipalmated Plover, seventy-five.
Black-bellied Plover, fifteen.
Turnstone, six.
Hudsonian Curlew, thirty.
Spotted Sandpiper, four.
Lesser Yellow-legs, fifteen.
Knot, 525.

Pectoral Sandpiper, one.
Least Sandpiper, three.
Dowitcher, 107.
Stilt Sandpiper, four.
Semipalmated Sandpiper, two hundred.
Western Sandpiper, twenty.
Sanderling, 250.

September 9, 1928, Brigantine. Julian Potter. Nineteen species.

Piping Plover, ten.
Semipalmated Plover, twenty.
Killdeer, two.
Black-bellied Plover, forty.
Turnstone, fifteen.
Hudsonian Curlew, six.
Spotted Sandpiper, one.
Willet, thirty.
Greater Yellow-legs, two.
Lesser Yellow-legs, twelve.

Knot, six.
Baird's Sandpiper, one.
Pectoral Sandpiper, ten.
White-rumped Sandpiper, six.
Least Sandpiper, 4.
Dowitcher, twenty-five.
Semipalmated Sandpiper, one hundred.
Sanderling, one hundred.
Buff-breasted Sandpiper, six.

September 11, 1932, Brigantine. Julian Potter and others. Twenty-one species.

Piping Plover, six.
Semipalmated Plover, ten.
Killdeer, ten.
Golden Plover, six.
Black-bellied Plover, twenty.
Turnstone, one.
Hudsonian Curlew, one.
Willet, two.
Greater Yellow-legs, two.
Lesser Yellow-legs, thirty.
Knot, one.

Pectoral Sandpiper, one.
White-rumped Sandpiper, one.
Least Sandpiper, one.
Dowitcher, two.
Stilt Sandpiper, two.
Semipalmated Sandpiper, fifty.
Western Sandpiper, twenty.
Buff-breasted Sandpiper, one.
Sanderling, forty.
Wilson's Phalarope, one.

September 15, 1929, Brigantine. Julian Potter and others. Twenty-one species.

Piping Plover, one.
Semipalmated Plover, forty.
Killdeer, fourteen.
Black-bellied Plover, one.
Turnstone, four.
Hudsonian Curlew, eight.
Spotted Sandpiper, one.
Willet, thirty-six.
Greater Yellow-legs, two.
Lesser Yellow-legs, five.
Knot, five.

Pectoral Sandpiper, six.
White-rumped Sandpiper, two.
Least Sandpiper, two.
Red-backed Sandpiper, one.
Dowitcher, sixty-five.
Stilt Sandpiper, two.
Semipalmated Sandpiper, sixty.
Western Sandpiper, five.
Marbled Godwit, two.
Sanderling, 125.

September 6, 1935, Brigantine. F. W. Loetscher, Jr. Twenty-three species.

Piping Plover, one.
Semipalmated Plover, 133
Killdeer, sixteen.
Golden Plover, two.
Black-bellied Plover, one hundred.
Turnstone, six.
Hudsonian Curlew, fourteen.
Upland Plover, one.
Spotted Sandpiper, three.
Solitary Sandpiper, one.
Willet, one.
Greater Yellow-legs, eighteen.

Lesser Yellow-legs, forty-five.
Pectoral Sandpiper, eighty.
White-rumped Sandpiper, one.
Least Sandpiper, five.
Dowitcher, five.
Long-billed Dowitcher, one.
Stilt Sandpiper, fifteen.
Semipalmated Sandpiper, ninety-four.
Western Sandpiper, four.
Sanderling, sixty-five.
Red Phalarope, one.

In addition to the above five other species were seen by Loetscher a day or two previously—Knot, Baird's Sandpiper, Wilson's Plover, Red-backed Sandpiper and Buff-breasted Sandpiper. Counts were made on every day from August 23 to September 6; on four days they reached nineteen and on six others from thirteen to eighteen. The daily total of individuals reached from five hundred to upwards of a thousand; the Sanderling, Semipalmated Sandpiper and Semipalmated Plover leading with daily averages of 302, 192, and 131 respectively.

This last list shows a good record for mid-winter.

January 13, 1929, Brigantine. Charles Urner. Six species.

Black-bellied Plover, four.
Turnstone, one.
Knot, three.

Red-backed Sandpiper, one hundred.
Semipalmated Sandpiper, two.
Sanderling, twenty-five.

It is astonishing to present day students of the shore birds to look upon conditions in years gone by. At the time of my first visit to the Cape, in 1890, these birds were apparently given no protection at all, unless to some extent in spring, but the laws were mainly dead letters and, in late August and September, every one who was inclined to do so, shot Sanderling and other beach birds with impunity. The birds were far less numerous than at present and infinitely more wary. To realize how this condition came about one has but to read the letters sent to "Forest and Stream" from 1880 to 1886 by Charles S. Westcott ("Homo"). This journal, like all other sporting magazines of the time, published information as to the best spots for "bay bird" shooting and the names and addresses of competent men who would take sportsmen out on the sounds, and their terms. Passing flocks were decoyed by whistling and the guide who was familiar with the calls of the various species was in great demand. "Dory" Shute of Townsend's Inlet, at the Curlew Bay Club House, was recommended as a man "who could erect a good blind and had great prowess as a whistler." The time of the tides was published every week and little booklets could be had for fifteen cents giving the habits and calls of the bay birds and the best methods of shooting them and this publication was described in an editorial review as "instructive, entertaining and timely"! In fact everything possible was done to encourage the sport and help the gunners. Westcott's voice was apparently the only one raised even against spring shooting.

By 1884 there were many complaints that the bay bird shooting was not what it used to be on the northern beaches and the explanation was that they were being made into summer resorts and the birds were compelled to pass farther south to the "unfrequented regions about Townsend's and Corson's Inlets." Westcott writes that "at any spot along Long Beach one can see flocks of Robin Snipe, Curlew, Marlin [Godwits], Willet, Brownbacks [Dowitchers] etc. making their way south outside the breakers" and he urged gunners who have to shoot bay birds to go farther south where the meadows are "less molested and I dare say the mosquitos more plentiful!"

He also writes (1883) that "For summer shore bird shooting, when no other sport with the gun can be had, I am told that Ocean City, Maryland, after August 1, is a capital place. All the varieties that pass down our New Jersey coast during July and August and do not stop nowadays, make a halt in this section and shooting over stools for Curlew and Willet is good all through August." He also mentions on May 12, 1884, that in the Philadelphia markets "are many Golden [probably Black-bellied] Plover and Grass Plover" and again, October 17, 1885, "our markets are full of Yellowlegs." Once more he says, June 7, 1883, "a large flight of shore birds reached the New Jersey coast during the past week and many of the late migrating varieties can be seen in our markets. How much better if they are to be killed at all to let them pass on to their nesting grounds unmolested in the spring to return in August when there are more young birds in the flocks and then they would be more palatable, if one can abide them in any condition!" Westcott was evidently not a shore bird gunner at heart.

In 1882 he deplores the great increase in market shooters who are on the grounds when the first flights make their appearance and frighten them so that none remain for the "sportsmen," and says that there is no good shooting "this side of Cape May," yet as early as 1843 Baird had bewailed the scarcity of birds and the abundance of gunners at the Cape as compared with still earlier times! Quoting again from Westcott's letters we read that a new danger threatened the birds. "In the summer of 1883, two trained taxidermists settled at Long Beach, one at Beach Haven and the other at Barnegat Inlet, and were busy all summer and part of the autumn killing and preparing birds for the millinery market; everything was bought; fifteen cents for terns and ten cents for sand-snipe." In the next year he writes, May 31, 1884, "the market gunners have put away their guns and are getting the yachts and boats ready for the summer season unless the demand for water birds for hat ornaments should offer a more lucrative occupation. If the rage continues another year the millinery establishments will send taxidermists who will remain at the hotels all summer and purchase everything in the shape of feathers from a strand-snipe to a Black-headed Gull and better wages can be earned by bay-men than by sailing seaside visitors at three to four dollars a day. It is to be hoped that the fashion has spent itself and that gulls and beach birds may be spared to ornament our seaside and not the hats of city belles."

Sometime in the nineties there must have been some restriction on shooting shore birds but when the autumn season opened on September 1 any sort of wader seemed to be fair game at Cape May. In the late eighties the Cape May hotels were in the market for "sand-snipe" and men of my acquaintance have told me of lying in wait for the flocks of Least and Semipalmated Sandpipers at the edge of favorite sand flats and mowing them down by the hundred.

Charles Urner has written (Bird-Lore, 1935, p. 265) that when he first made the acquaintance of the Newark Meadows as a lad, in 1895, market gunning was actively practised there and of the flocks that stooled to the decoys few ever departed. Pectoral Sandpipers and Lesser Yellow-legs outnumbered all of the medium-sized and larger species in the locality and while he can remember reports of fairly heavy flights he recalls no great numbers actually seen. The hunting became steadily poorer and by 1905 such species as the Dowitcher, Curlew and Stilt Sandpiper were never seen. From then until 1927 the yearly kill, although variable, showed an average decrease. He quotes the testimony of H. Walter Sapp, a veteran native hunter, before the state legislature, in 1909, to the effect that shore birds had decreased during the preceding ten years and were not present in the swarms that he remembered as a young man. While most species had decreased he saw no change in the numbers of the Hudsonian Curlew and Cape May gunners have told me the same thing.

In 1913 the Federal migratory Bird Law went into effect and the season was closed on all shore birds except the Woodcock, Black-bellied Plover, Golden Plover, Wilson's Snipe and the two species of Yellow-legs. In 1926 the two Plover were placed on the protected list and in 1927 the Yellow-legs followed them. The season on Woodcock and Wilson's Snipe, the only shore birds remaining on the game list, is now reduced to one month with a bag limit of four and fifteen respectively.

Charles Urner continues to record the recent decrease in shore birds on the Newark meadows and in the Barnegat region and says that in 1923 Dowitchers and Knot were hard to find and the only birds left of game size were the Yellow-legs and Black-bellied Plover. With the stopping of all shooting of these birds the increase at once set in and the 1935 spring flight was the largest he ever saw while in 1934 thirty-nine species were recorded from Brigantine northward. The same increase has taken place in the Cape May area, as many of us can testify, but it is a matter of regret that we have no detailed accurate record of the occurrence of shore birds from 1890 to 1920 to compare with present day counts. The substitution of shore bird watchers for shore bird gunners did not take place quickly enough!

Our present day students of the shore birds are aware of the tameness that the birds have attained through this protection and the ease with which one may study them at close quarters. While we realize that we, at best, can see but a mere fragment of the hordes of these beautiful birds that thronged our meadows and beaches in the time of Wilson, we should be thankful for what protection has done for them and do our utmost to check the craze for draining the marshes which seems to have obsessed those in authority and bids fair to destroy all the wildlife of our coasts that we have taken so much pains to preserve. As one menace is disposed of another seems inevitably to develop!

H. B.

OYSTERCATCHER

Haematopus palliatus palliatus Temminck

These splendid shore birds were of regular occurrence on the Cape May beaches in the time of Alexander Wilson when, according to George Ord, they frequented sandy sea beaches in small parties of two or three pairs and nests were found between May 15 and 25. Beesley (1857) and Turnbull (1869) regarded them as rare summer residents.

While they probably continued to nest in diminishing numbers much later than this, they have been for many years past mere straggling visitors and they have become today purely accidental in occurrence. For the last seven years none have been recorded this far north.

Our records for the New Jersey coast are as follows:

Autumn 1880, Samuel Rhoads saw three on Long Beach where W. E. D. Scott failed to find any in 1877.

May, 1894, one was shot by Jillson on Barnegat Bay near Tuckerton; August 9 and 12, 1896, William Baily saw three on Five Mile Beach; October 14, 1903, two were seen on the sounds north of Cape May by gunners whose accuracy was vouched for by Walker Hand. July 31, 1907, one was shot by C. K. Drinker, on Anchoring Island, Little Egg Harbor, in whose collection it was seen by Reynold Spaeth. June 18, 1922, Julian Potter saw one at the southern extremity of Seven Mile Beach where the Black Skimmers were nesting, "it manifested no apprehension such as nesting Oyster-catchers at Cobb's Island, Va., had been observed to do, and a very good view of the bird was obtained. True to its wild nature, however, it did not permit a very close approach but as it flew the striking wing pattern was plainly seen" (Auk, 1922, p. 564). Turner McMullen who also saw the bird says that this was the third successive year that he had seen one at this spot, and Richard Miller gives the date of the 1921 observation (three birds) as July 3. He saw another on June 22, 1924, (Auk, 1926, p. 93), while John Emlen saw one there on June 3, 1930. In August, 1928, two were seen by Oscar T. Eayre near Barnegat Inlet "in late summer." He knew them as "Oyster-crackers."

PIPING PLOVER

Charadrius melodus Ord

Of all the birds that frequent the New Jersey coast the Least Tern and Piping Plover have suffered most from the spread of summer resorts along the shore. A few years ago, indeed, it was feared that they had utterly disappeared but fortunately some small remote areas have offered them sanctuary and they have been saved from extinction. Whether the Federal protection of shore birds and the aroused public interest in their welfare will prove their salvation remains to be seen.

While no longer breeding on Cape Island Piping Plovers have, since 1914, occurred regularly every season from late July to September even on the ocean front of the town; usually only two or three at a time or perhaps only a single bird, while a few are to be seen also on the return flight in spring.

The reason for their rapid decrease in numbers lay in the fact that they, like the Least Terns, were nesting birds of the strand, and the constant annoyance of a throng of summer visitors, apart from any actual molestation, has been enough to drive them from most of their former haunts. Added to this the steady conversion of wild beaches into cottage colonies makes their nesting grounds no longer tenable, while the habits of visitors to the shore of running their dogs on the strand results in the active pursuit of any birds that may be found there. Of late years, owing to the establishment of nesting colonies north of Cape May, they have become more plentiful.

The Piping Plover is almost exclusively a bird of the upper dry sands of the beach and so exactly does its coloration match this background that it is very difficult to detect it—quite impossible, in fact, so long as it remains still. In size, habits and actions it is the counterpart of the Semipalmated Plover, or "Ring-neck," but although the pattern of coloration is the same in both, the tints are utterly different. Both are white below and both have more or less of a black collar across the chest, but while this is complete in the Ring-neck, in the Piping there is usually only a blotch on either side of the breast. Above, the Piping Plover is as pale as the dry sands on which it lives, while the dark brown of the Ring-neck's back seems equally designed to match the wet strand or the black mud flats upon which it is at home. When running on the sand the legs of the Piping Plover move so fast that they cannot be clearly distinguished and the little bird seems to fairly glide over the beach like one of the pale, hairy-legged ghost crabs that live close by, or a scudding bit of spume blown by the wind. On a broad dry sand flat just back of the beach I found one of these plovers on August 5, 1920, and as it sped along before me, it alternately faded away, as it crossed the areas of white sand, and sprang suddenly into view again when it passed over patches of exposed black mud against which it stood out in beautiful relief. It was possible then to see the bright orange yellow of the rapidly moving legs and the yellow orange of its bill so different from the orange red or vermilion of the Ring-neck's legs, and its bill of slightly paler tint. The black half collar, too, stood out sharply defined against the prevailing pale sandy gray and white of the plumage. Curiously enough the first bird seen that year had the breast band nearly complete though narrow, a rare thing in Atlantic coast individuals, while later, on August 21, I studied one that showed scarcely a trace of black on the breast—doubtless a female. Viewed from the side the narrow black frontlet which is visible over each eye bears a striking resemblance to the "horn" of the Horned Lark, and when seen from directly in front, when the black on the breast is also in evidence, the resemblance is striking. In flight there are black markings on the wings which stand out prominently and the edge of the tail appears as a narrow black line, even when closed.

Besides the striking color differences between this plover and the Ring-neck, it seems to be of chunkier build with shorter legs and with the neck and head, when at rest, held straight up and not bent over in the listening, hunchbacked attitude so characteristic of the Ring-neck.

When running it sometimes has the habit of the Killdeer of looking at a pursuer over the shoulder, first on one side, then on the other, but when not alarmed it holds the head straight out and the shoulders slightly humped, as it speeds away like a shadow across the sand. In a group of six observed in the spring (April 3) the tails appeared more prominent than in the Ring-neck, slightly elevated at the tip and very slender. These birds were feeding on a little patch of high dry sand that had not been washed over by the waves and kept stepping round and round in this limited area pecking vigorously with heads continually bent down.

The call of the Piping Plover is a plaintive, almost bell-like, note repeated at intervals: *peenk!*; *peenk!*; *peenk!*; which it utters as it speeds past on rapid wings. I have heard this note on the breeding grounds as well as on Two Mile Beach, May 22, 1922, and at Cape May Point, on July 21 and August 5, 1921, so that it does not seem to be restricted to the nesting season. Another call heard only when breeding is a continual repetition of a single note, reminding one of the prolonged re-iteration of the call of the Flicker or of the Mud Hen although not quite so rapid.

I have found the Piping Plover nesting on Brigantine Beach farther up the coast, while an egg washed up at Anglesea and young birds seen on Seven Mile and Ludlam's Beaches show that it breeds there as well. On July 18, 1921, at Brigantine Beach, we found as many as twenty of these little birds, seven being in sight at once, and they glided here and there over the great stretches of dry white sand, now in clear view, now vanishing— swallowed up in the background which they so perfectly matched. At this remarkably late date we found one nest, a mere hollow in the loose sand, in a sort of basin surrounded by several little dune-like knolls. The four eggs were tilted slightly over on their points and fitted closely in the little cup that was the nest, the pointed ends all directed inward. The old birds did not come near the nest while we were there but we found them running about behind some dunes not far away. As they took wing they uttered their plaintive *peenk!*; *peenk!*; like the stroke of some silver bell, clear cut and delicate as the little bird itself, and suited to the solitude of the spot, with its broad stretches of white sand flanked by green-topped dunes, and with the distant pounding of the surf on the outer bar. It seemed to represent the antithesis of the raucous bedlam of the tern colony on the beach not far away. Many more Piping Plovers have spent the summer on Brigantine in later years. Turner McMullen found five pairs nesting there in June, 1922, and Richard Miller estimated that at least ten pairs bred in 1923. Sets of four eggs were found in both years on June 24. Julian Potter counted twenty birds present on June 27, 1925, and Richard Pough found a nest with two eggs and a recently hatched young as early as May 28, 1934.

Others have seen young plovers here earlier in the season which had but recently emerged from the eggs, minute balls of pale down, veritable sprites of the sand. Of these Charles Rogers writes: "On June 18, 1921, I found a pair of Piping Plover wearing an anxious mien, I retired to the top of a nearby dune and lay down in the long grass, and after a few minutes, I noticed that running about with the old plovers were three fuzzy bumble-bees on stilts. When I walked toward these downy chicks they hid, but I caught one eventually and no pin feathers were visible. In scurrying over the beach before me, it would outspread its white, half inch wings, like a miniature running Ostrich. One of its parents would run about with tail spread to the utmost and wings partly spread and quivering strongly but if this was an attempt to play wounded and lure me away it was not well done, for the bird kept at such a distance from me that I had to use my glass to observe clearly its attitude!" (Auk, 1921, p. 600). Julian Potter found a downy young on Seven Mile Beach as late as July 30, 1922.

On June 5, 1925, I found two Piping Plovers on the narrow sand spit at Corson's Inlet where the Least Terns bred and it was evident that they had a nest there although I failed to locate it. The birds kept widely separated and when I followed one of them it ran very rapidly, fairly gliding over the sand like a mouse, its head drawn in and slightly bowed over in front. Every few yards it would gradually sink to the sand as a Killdeer will frequently do and, when I failed to approach, it would sometimes stand erect with neck elevated and give a spasmodic jerk to the body. Now and then it called *peenk!*; *peenk!*; and once it stood still and uttered its rapidly repeated call for half a minute without pause—*peent, peent, peent, peent, peent, peent,* etc.

This bird appeared beautifully snow white as it ran down on the wet beach to join a Ring-neck that was feeding there but upon returning to the dry white sand it completely disappeared from view until a careful survey of the ground with the glass located it again. The yellow orange, black-tipped bill, black cap and orange yellow legs were more conspicuous against the white sand than was the body of the bird. Returning to this spot on July 16 I found six adult birds and several young on the sand. The latter were half-fledged and ran rapidly about. Sometimes, when pursued, they went down almost to the edge of the waves and at others retreated into the grass at the back of the beach.

On June 10, 1928, Fletcher Street found a nest with four eggs and also a young bird on a recently made sand flat near Sea Isle City to which the Least Terns from Corson's Inlet, driven out by building operations, had largely removed, and on June 24 we found the parent plover very solicitous for the young which were doubtless concealed among the scattered oyster shells and bunches of dune grass. A few pairs apparently bred there also in 1929, 1930, 1931, and later. Julian Potter found young just able to fly on Peck's Beach on July 17, 1927. Turner McMullen found Piping Plover breeding on the lower end of Seven Mile Beach—three nests on June 29, 1924, and one on May 30, 1926, all with eggs. At Corson's Inlet he examined seven nests with eggs on May 30, 1926; on Ludlam's Beach near Sea Isle City, two nests with eggs on June 10, 1934, and one with an egg and two young on June 16, 1929; on Peck's Beach, May 26, 1935, four nests with eggs.

While typically birds of the dry beach the Piping Plovers occasionally stray onto the wet sand and on August 18, 1920, when the sky was overcast with constant threat of rain, I saw two of them run out on slightly submerged sand bars and even through the shallow water of the receding waves; the same thing occurred on another stormy day, August 3, 1921. Twice also I saw single birds fly from the sand flat at South Cape May down onto the wet beach where other shore birds were feeding. Such cases, however, are exceptional and the plover do not long remain away from their natural environment.

Frequently there will be a group of Ring-necks feeding on the wet sand and a Piping Plover on the dry beach just opposite to them and as they advance along the shore they maintain exactly the same relative positions. On July 21, 1921, moreover, a pair of them formed the land end of a string of thirteen Semipalmated Sandpipers feeding on the wet strand, and I have occasionally seen them join mixed flocks of Ring-necks, Turnstones and Sanderling feeding at the water's edge. A very unusual occurrence was the presence of three Piping Plovers on a mud flat above Atlantic City in company with Ring-necks and Least Sandpipers for they seem to have a marked repugnance to flying over the meadows or resorting to them for food.

The Piping Plover is an early migrant in spring. Julian Potter saw one at Cape May on March 11, 1923, while I saw six on the beach on April 3, 1922. Farther up the coast Potter found twenty-five on Ludlam's Beach on April 9, 1922, and one at Corson's Inlet on April 8, 1928, while Dr. W. L. Abbott obtained specimens on Five Mile Beach on April 11 and 13, 1879. At Ventnor, on Absecon Beach, Charles Urner has seen them as early as March 19, 1922, and at Barnegat on March 14, 1925, while Julian Potter saw one on Brigantine Beach on March 25, 1932, and nineteen on March 30, 1934. On May 22, 1921, I found one on the sand bar at Turtle Gut Inlet feeding with Ring-necks and another on the same day in 1922, on Two Mile Beach. These may have been breeding birds but more likely marked the end of the northward migration as they were associated with unquestioned migrants.

From the middle of July throughout August and well into September single birds or bunches of two or three may be seen at intervals on all of the beaches of Cape May County. My dates of arrival at Cape May run:

1919.	July 27.	1923.	July 21.	1925.	July 17.
1920.	August 1.	1924.	July 28.	1926.	July 24.
1921.	July 21.	1925.	July 17.	1929.	July 20.
1922.	July 25.	1926.	July 24.	1930.	July 24.

Our latest autumn dates are:

1914.	September 13.	1924.	September 9.	1925.	September 23
1921.	September 13.				

There is also a record of three observed on Brigantine Beach on October 4, 1931.

On July 23, 1922, I saw my greatest number together—sixteen feeding on Two Mile Beach. These may have been migrants from farther north but it would seem more likely that they were breeding birds and their offspring, especially since they apparently bred on this very beach.

Charles Urner's records for the region from Brigantine to Newark Meadows, showing extreme dates, period of regular or common occurrence, and maximum count for one day at a single locailty, are as follows:

1928. —— to September 15 (August 18–19). Twenty on July 15.
1929. April 13–September 15 (July 13–27). Thirty-five on July 20.
1930. April 13–September 14 (July 16–August 9). Thirty on July 16.
1931. —— to October 4 (July 11–29). Twenty-three on July 29.
1932. April 17–November 6 (July 16–30). Twenty on July 16.
1933. March 26–October 8 (July 2–August 21). Ten on July 2.
1934. March 18–September 23 (July 17–August 26). Fifty on July 17.

On the basis of his studies he regards the Piping Plover as "irregularly and locally tolerably common" today. The presence of this little bird with its bell-like whistle adds a peculiar charm to the sea beaches and if it must be crowded off of these, its true habitat, let us hope that the sand flats, formed by dredging of the inland channels through the meadows, will furnish a satisfactory substitute for them during the nesting season, as they seem to be doing at the present time.

Alexander Wilson discovered the Piping Plover on "Somers's [Peck's] Beach at the mouth of Great Egg Harbor," probably about 1810. He at fiest regarded it as the Semipalmated Plover, or Ring-neck, in summer plumage, but later corrected his error and pointed out the differences between the two and their different times of occurrence. He did not, however, give the paler bird a name fearful, doubtless, that someone else had already done so, and his friend George Ord later bestowed upon it the name that it now bears. Wilson had studied the bird carefully and has presented an accurate picture of it. He writes: "The voice of these little birds, as they move along the sand, is soft and musical, consisting of a single plaintive note occasionally repeated. As you approach near their nests, they seem to court your attention, and the moment they think you observe them, they spread out their wings and tail, dragging themselves along, and imitating the squeaking of young birds; if you turn from them they immediately resume their proper posture until they have again caught your eye, when they display the same attempts to deception as before. A flat dry sandy beach, just beyond the reach of the summer tides, is their favorite place for breeding." (American Ornithology, Vol. V, p. 30; VII, p. 65; Ord reprint, Vol. VII, p. 71.)

SEMIPALMATED PLOVER

Charadrius semipalmatus Bonaparte

The little "Ring-neck" is not only the most common plover of our coast but is also of regular occurrence on the ocean beaches where the ever abundant Sanderling and Semipalmated Sandpiper are almost the only other shore birds to be found. It is, however, by no means restricted to the strand but occurs in even greater numbers back on the exposed flats of the sounds and about the salt ponds on the meadows. The sandy stretches of the Fill, before it became overgrown with bayberry bushes, were another of its favorite resorts when the northeasters of August flooded them with water. On the sea beach the Ring-necks associate most intimately with the Semi-palmated Sandpipers and Sanderlings. But while they all take wing to-gether and fly out over the surf to seek new feeding grounds farther along the beach, the Ring-necks always alight farther back from the waves though still on the wet part of the strand. Occasionally a wave threatens them and they run rapidly away, indeed they always run away from the water, or parallel to it, and never follow the receding waves as the little sandpipers delight in doing.

They run rapidly for a short distance throwing their feet forward in such a way as to give one the impression that the legs are trying to run away from the body, the head all the while being held close down on the shoulders with no neck in evidence. When feeding the Ring-neck takes a few quick steps and then pauses and stands motionless with feet together and head bowed over as if listening intently. The resemblance of this attitude to that of the Robin seeking earthworms on the lawn is striking. Detecting some minute prey in the wet sand there is a quick dab, or a succession of dabs, with the bill, but the whole action is over in a moment and there is another rapid dash forward and again the sudden pause and the rigid pose.

The long runs and bewilderingly rapid foot action are always to escape the waves when an individual has ventured too near the water, or an unusually high wave pushes up the strand and floods the feeding grounds. The birds are always standing still longer than they are in action so that, as one looks down the beach, their statuesque pose and bowed heads contrast strikingly with the constant activity of the sandpipers and even in silhouette, against the light, when no color can be seen, they can be identified as far as they can be distinguished. When viewed from this position, even when comparatively near at hand, they seem very dark and the contrasted black and white of the breast is concealed by the counter shading. When we get them in proper light, however, with the sun behind us, what a contrast! The snowy under parts and the velvety black breast band, the narrow white frontlet and reddish orange bill and the orange red legs all stand out in brilliant contrast against the dark background of wet beach.

When running away from the waves the leg action of the Ring-neck as compared with that of the Sanderling seems to be just as rapid but the bird does not attain quite so great a speed nor does it seem able to keep up the pace so continuously as does the tireless Sanderling. In feeding, when the head goes down for a probe in the sand, the rear end of the body goes up in a seesaw fashion, but there is no independent tilt of the tail as in the Spotted Sandpiper, nor convulsive jerk of the whole body as in the Yellow-legs. Viewed from behind the Ring-neck appears somewhat bowlegged as do other plump-bodied shore birds, since its feet are closer together than its hips.

Single stray Turnstones or Piping Plovers show a marked preference for the Ring-neck's company and I have found them associated on the beach on many occasions; the Piping Plover, will almost always hold to the dry white sand of the upper beach keeping opposite to the Ring-necks which prefer the dark wet strand nearer to the water with the Sanderlings still farther out, foraging on the very edge of the waves.

On the mussel beds of Jarvis Sound, May 20–24, 1922, where shore birds of all kinds gathered to feed at low tide, there were about one hundred Ring-necks although the height of their migration had apparently passed. As we lay there in our skiff they came in small bunches of six to twelve, about fifty in the first half hour of shoal water. They remained on the mussel beds and did not venture out on the mud flats as did some of the other species. Their actions while feeding were again strikingly suggestive of those of the Robin, now pausing rigid with head bent, now running forward to pause again and look for food. Sometimes one would seize the end of a marine worm and drag it from its burrow among the shells, tugging at it just as the Robin struggles with the luckless earthworm. Often, too, another individual would pursue the fortunate captor and endeavor to take his prize away from him, adding not a little to the excitement and action displayed in these interesting assemblages. When a large part of the flock took wing it was easy to distinguish the Ring-necks from the Least and Semipalmated Sandpipers—the "peeps" of the baymen, by their longer wings and bulkier body.

During their migrations Ring-necks like other shore birds, vary greatly, in numbers and on April 30 and May 1, 1932, there were hundreds of them scattered all over the meadows back of Seven Mile Beach while on May 7, there were scarcely any to be seen. Conditions of tide, wind and food supply all contribute to this variation. On the beach, too, their numbers vary from day to day; on May 9 they were very abundant on the meadows and the next day on the strand on Ludlam's Beach they were exceptionally numerous, seventy-five to one hundred in every flock of little sandpipers while some of the largest shore bird flocks were made up entirely of Ring-necks.

Walker Hand's records of arrival, 1902 to 1918, together with my own for subsequent years are as follows:

1902.	May 1.	1916.	April 27.	1926.	May 2.
1905.	May 4.	1918.	April 26.	1927.	May 2.
1909.	May 2.	1922.	May 5.	1931.	May 9.
1914.	April 24.	1924.	May 9.	1932.	April 30.
1915.	May 9.	1925.	May 8.	1933.	May 6.

It is probable, however, that Ring-necks arrive during the latter days of April in every year; the dates given as our spring records are based on scattered trips to the shore and not on continuous observation. Dr. William L. Abbott obtained specimens as early as April 13, 1879, and April 11, 1883, on Five Mile Beach. There have been frequent visits to the Cape May region on Memorial Day, May 30, and on nearly every one a few Ring-necks were found while we have several records for June, possibly non-breeding birds in some cases at least: June 18, 1922; June 17, 1923, thirteen at Stone Harbor; June 5, 1925, one at Corson's Inlet; June 19, 1926, four at Stone Harbor; June 12, 1927; June 10, 1928, two at the same place. Undoubted migrants from the northern breeding grounds begin to arrive about the middle of July and my records for Cape May run as follows:

1920.	July 25.	1925.	July 13.	1930.	July 19.
1921.	July 22.	1926.	July 19.	1931.	July 13.
1922.	July 23.	1927.	July 12.	1932.	July 10.
1923.	July 21.	1928.	July 18.	1933.	July 26.
1924.	July 19.	1929.	July 20.	1934.	July 14.

A single bird on Grassy Sound, July 8, 1923, and one seen by Walker Hand on the Harbor, July 5, 1924, may have been belated spring migrants which did not go farther north and may have remained here all summer. Ring-necks seem to arrive on the great meadows north of Cape May Harbor some days before they appear on the ocean beaches but the above dates cover both localities.

The bulk of the Ring-necks pass during August but they are to be seen all through September while we have not a few October records: Last seen dates are as follows: October 18, 1919; September 29, 1925; September 27, 1926; October 14, 1928; October 13, 1929; September 28, 1930; October 22,

1932; October 18, 1931; October 29, 1933. Exceptional observations were: November 5 and 6, 1926, seven at Sea Isle City by Fletcher Street; November 28, 1926, twelve back of Seven Mile Beach; and several November 22, 1931, at Corson's Inlet by Wharton Huber. The presence of ninety Ringnecks on the meadows west of Avalon on October 9, 1932, was the latest large assemblage of autumn birds of which we have knowledge.

Charles Urner's records for the region from Brigantine to Newark Meadows showing extreme dates, period of regular or common occurrence, and maximum count for one day at a single locality, are as follows:

Northward migration.

1929. May 5–June 22 (May 17–30).   500 on May 19.
1930. May 11–June 22 (May 11–31).   1500 on May 14 and 31.
1931. —— to May 31 (May 14–31).   1500 on May 17.
1932. May 1–June 4 (May 8–June 4).   4000 on May 18.
1933. April 29–June 4 (May 5–June 4).   550 on May 7.
1934. April 4–June 18 (May 6–27).   200 on May 20.

Southward migration.

1928. July 1–October 28 (July 28–September 8).   300 on September 8.
1929. July 14–November 10 (July 28–September 8).   2000 on August 17.
1930. July 12–October 12 (July 26–September 6).   1200 on August 17.
1931. July 7–October 29 (July 29–August 30).   200 on August 2.
1932. July 8–November 6 (July 26–September 11).   300 on August 6.
1933. July 13–November 6 (July 30–September 30).   500 on September 3.
1934. June 27–December 24 (July 29–September 23).   350 on August 12.

On the basis of his comparative studies he regards the Ring-neck as one of the six "abundant" shore birds of our coast today (see p. 358).

WILSON'S PLOVER

Pagolla wilsonia wilsonia (Ord)

The first specimen of this plover was shot on Cape Island, the present Cape May City, by Alexander Wilson, on May 13, 1813. He drew it for his "Ornithology" but his death ensued before he had prepared his account of it and we are indebted for our knowledge of the discovery to George Ord, who was with Wilson at the time and who brought out the last volumes of his great work. Ord tells us that there were two males and a female all of which were secured. He adds that Titian R. Peale and himself on an excursion along the New Jersey coast in May (probably about 1820) found these plovers "pretty common in the vicinity of Brigantine Beach and also observed them in various places between Great Egg Harbor and Long Beach." (American Ornithology, Vol ix, p. 77.)

William M. Baird on a collecting trip to Cape May in 1843 (see page 29) secured two specimens on July 15 and 17.

While there have been references to the presence of Wilson's Plover on the coast in later years, in several publications, they are not convincing and we are led to believe that they were not based upon personal experience or were misidentifications of the Semipalmated Plover. I have examined many hundred Semipalmated Plovers on the Cape May beaches in the hope of finding a Wilson's but without success and the experience of my ornithological friends who have visited the Jersey coast has been, for the most part, the same. We have, however, three unquestioned records.

On September 15, 1933, Fletcher Street is convinced that he saw this long-sought species on Ludlam's Beach. He says "During the period of a severe northeast storm, I noticed two waders seeking shelter from the high winds beside a cement coping of a flower bed, I watched them from a distance of twenty feet. One was a Semipalmated Plover but the fact that the other was considerably larger aroused my suspicion. A closer inspection revealed a larger and heavier bill, which was entirely black, a lighter mantle and a neck ring which did not show black all around and I have no doubt that the bird was a Wilson's Plover. My observations were corroborated by a friend who was unaware of the significance of the points of distinction that I was endeavoring to determine." (Cassinia, XXIX, p. 5.)

On exactly the same date in the following year Julian Potter detected a Wilson's Plover on Brigantine Beach and F. W. Loetscher found one there on September 4 and 5, 1935. Potter writes: "A direct comparison with the Piping and Semipalmated Plovers was obtained but the bird under observation was so unlike them that it could hardly be mistaken by any experienced bird student. It looked somewhat like a dull colored Semipalmated Plover with a large black bill, a distinct light line over the eye and a dark Shrike-like line passing through it. The brownish back of the bird was somewhat mottled. The legs were pink compared with the yellow or orange colored legs of the others" (Auk, 1935, p. 80).

Charles Urner justly regards Wilson's Plover, a straggler from the south, as one of the seven "very rare" shore birds of the New Jersey coast today (see p. 358).

KILLDEER

Oxyechus vociferus vociferus (Linnaeus)

About Cape May, as elsewhere in its range, the Killdeer is primarily a bird of the upland, frequenting old pastures when feeding, and the farmer's plowed fields or truck patches at nesting time, while in winter we often find it in the cowyards close to the barn seeking shelter during cold or stormy weather. It is one of the familiar and characteristic birds of the Cape and not infrequently we see one or more crossing over the town itself with strong rapid wingbeats in dove-like flight, or hear their wild calls even in the dead of night. As soon as the young are well-grown the Killdeer families seek the borders of some inland pool such as the old Race Track Pond or some more remote brackish pond like the shallow one east of the Lighthouse. Sometimes these family parties may be found feeding along the shores of the Harbor, while I have come upon single birds on the sea beach itself although these occurrences have been usually close to the mouth of Delaware Bay. In autumn we find Killdeers about the shallow pools on the edge of the salt meadows but usually not far from the mainland and all through the winter a few scattered individuals may be looked for there.

Before the great Fill, east of the town, became grown up with bayberry bushes it was a favorite resort for Killdeers at all times of year, the varied conditions which it presented—open mud flats, broad gravelly stretches and low grassy spaces, furnished both feeding and nesting grounds to the birds' liking. The golf links with the smooth close-cropped turf also offer them an attractive, if artificial, feeding ground and in the early morning a number may often be seen scattered about over the greens.

Here we have an excellent opportunity to study the appearance of the birds and the effect of light and shadow upon their coloration. When one of them happens to pause face-on the satin white breast with its double band of black stands out conspicuously in the sunlight, but when it turns, the dull gray brown color of the back blends immediately with the background and it all but disappears. I have noticed this also in wintertime when the plumage is perhaps a trifle less contrasted than in summer and birds that were on the ground were entirely overlooked until they moved; while on another occasion a flock of fifteen pitched down into a plowed field and seemed to disappear as soon as they touched the ground. They were standing perfectly still and I was able to detect only eight although I knew that the others were right there before me.

The Killdeer when feeding exhibits all of the typical behavior of the plover family—a short rapid run, a sudden stop and rigid pose, a dab or two of the bill to the ground to pick up a morsel of food and then again the short run. The bird's chief characteristic seems to be wariness; in the open fields he is constantly alert and usually sees us before we can see him and our first knowledge of his presence is the cry of alarm as he takes wing.

While we know that the Killdeers are back on their upland breeding grounds by mid-February there is some question as to whether these are merely local birds that have moved inland from the edge of the marshes and from other retreats where they have passed the winter, or are regular migrants from farther south coming back along with the Robins, Grackles and Bluebirds which show migratory activities at this time. At all events Killdeers seem to be pretty well scattered about the Cape May region by the first opening of spring and from then until the young are hatched we rarely see more than two or three together. Walker Hand has reported a regular migration of Killdeers through Cape May from March 1 to 19 which may have been merely passing individuals heard calling during the night, at any rate none of my observations indicates any concentration at this time.

Nests with full sets of eggs have been found in the Cape May region on:

April 17, 1921.	May 27, 1923.	June 18, 1923.
April 20, 1924.	May 27, 1929.	June 23, 1923.
May 26, 1923.	June 3, 1925.	July 4, 1927.

Turner McMullen has furnished me with the following additional data:

Seven Mile Beach

June 23, 1924, four eggs.
June 29, 1924, two eggs.
July 4, 1924, two eggs.

Peck's Beach

May 26, 1935, two with three eggs each.
June 10, 1933, four with one, four, four, and four eggs.

Ludlam's Beach

May 22, 1922, two with one egg each.   June 23, 1928, four eggs.
June 10, 1933, four eggs.

McMullen's records for Salem County seem to indicate that the bird nest earlier there viz:

Salem County:

April 1, 1923, two eggs.
April 13, 1925, four eggs.
April 14, 1923, three eggs.

April 16, 1922, four eggs.
April 16, 1927, four eggs.
April 17, 1927, four young.

And the same is true of Cumberland County where Mrs. Alice K. Prince found a nest with eggs at Buckshutem on the exceptionally early date of March 31, 1922, and another at the same place on April 7, 1923.

It will be seen from these records that the nesting activities of the birds are spread over a long period and this is borne out by observations on the young. Walker Hand has seen the little downy birds, a few days out of the eggs, running about as early as May 4, 1918, and April 27, 1922, while he found an adult with a brood of young about the size of sparrows as late as July 9, 1922. On June 28, 1930, I saw a group of four young and two adults feeding on the edge of the Harbor, the former full-plumaged but with excessively long tails, due no doubt to the adherence of the shafts of the early tail feathers to the tips of the later ones, as frequently occurs in the Spotted Sandpiper and doubtless in other shore birds.

The favorite nesting site of the Killdeer is in plowed or cultivated ground, or in sandy or gravelly spots in upland pastures, or on the edge of ponds. Such localities are the nearest approach to the ocean beaches which we must suppose to have been the original home of the Killdeer ancestors, since all plovers must have evolved from the same stock. Another similar habitat in the uplands of which the Killdeer also has taken advantage is the roadbed of the railroads. Walker Hand found a nest and eggs in such a location many years ago, just outside the rails but between the ends of the ties, where fast moving express trains passed within a foot of the nest several times a day. On July 4, 1927, Julian Potter found another nest similarly placed within six feet of the rail, and on May 27, 1929, I located a nest on the freight line to Cape May Point which contained five eggs. It was placed in the center of the track equidistant from the two rails and between two of the ties; the upper surface of the eggs was only an inch and a half below the top of the rails. Over this road at least two trains passed each day. Still another nest was placed on the cinders of a siding just north of the town and within three feet of the nearest rail. The nest hollows in each case were scooped out of the black cinders and one of them was lined with flat pieces of white cinder; the lining of a nest in a truck patch consisted of small pieces of sticks and coarse grass stems. The eggs in all the nests were arranged with their pointed ends directed inward and downward at an angle of about 45° thus presenting the smallest possible surface for the parent to cover during incubation.

Julian Potter has given me some data on two nests found farther north in the state. In one the four eggs hatched on June 12 at 4 p. m. and the young had left the nest by 5.30 the following morning; in the other the first egg was slightly pipped on July 1, this process continued on the three succeeding days but the young were not hatched until the 5th when they were found in the nest still damp from the egg contents. In a third nest found April 20, 1919, only three of the five eggs hatched.

The little downy birds are very hard to discover in the grassy cover which they frequent and their mottled coloration blends perfectly with their background so long as they remain still. I have repeatedly sought them near the nest from which they had but recently departed and although the frantic actions of the parents indicated that they were close at hand they completely eluded me.

In early July, 1920, some adults and full-plumaged young sought the borders of the old Race Track Pond, obedient to the lure of water which seems to date back to the early ancestry of the Killdeer and which its upland nesting habits have in no way destroyed. Here I made a study of them day after day. The pond at that time had not been drained and its shallow water was covered with an abundance of water lilies while there were little exposed sandy spits and flat muddy borders where from two to four Killdeers could always be found. Later in the season their numbers increased to eight or ten. There were old briery fields on all sides and some cultivated ground as well, where the birds doubtless had bred and from which they were still flushed occasionally. There were probably downy young in some of this cover when my observations began but I never found them, unless some of these apparently adult birds, even at this time, were really young of an early hatching and the distress of the parents at my intrusion had to do with them. That this was the case later on there is no question.

The birds fed in the regular plover fashion; the head was held well down on the shoulders and the body bent forward; there would be a number of rapid steps followed by a pause and rigid pose, a dab of the bill and then on again. One individual, perhaps a young one, ran more rapidly and recklessly with head bent over most of the time lunging with his bill to the right and left. He was not so wary as the others and failed to assume the watchful rigid posture so regularly. He also had the habit of flirting the tail from side to side which was not observed in the others. All of them, however, occasionally shook themselves and ruffled up their plumage for a moment or two and on one occasion a bird, with plumage all on end, charged another and drove him across the pond.

So long as I remained hidden the Killdeers made no noise whatever, running and pausing in their usual manner in the pursuit of food. Painted turtles paddled over the shallow water close to them and a great snapper plowed through the mud without occasioning any alarm, while Green Herons stalked past and caught frogs directly in front of them without attracting their notice. The moment I arose, however, there was an uproar and usually all of the Killdeers would take wing together, sometimes settling again, sometimes beating across the open fields in full cry. The same actions obtained in other places too. So long as no intruder was in evidence the birds went on feeding or resting in silence, but the minute one appeared they arose with their heartrending cries of alarm usually startling into flight all other birds which happened to be near; the Killdeers being the first to detect danger appeared to act unconsciously as warners for the others.

The first alarm cry of the Killdeer as it takes wing is a long drawn *tweeeeee* almost like a scream in its suddenness, then, with short intervals between, there follows a series of notes, *dee, dee, dee,* changing to *deee, diddy; deee, diddy,* or a still more prolonged, *deee diddity.* These are repeated in varying order often, for a few moments, from the ground after the bird has alighted. The familiar long drawn, *kill déer, kill déer, kill déer,* I have heard only from the bird on the wing and in full flight with no marked feeling of alarm. The pitch and accent of the cries vary and it is sometimes necessary to employ other syllables to represent the calls of different individuals. One, for instance, heard on July 27, 1920, had a harsher cry and seemed to say, *keu, keu, keu, keee kikky.* These alarm calls of one sort or another are kept up as long as the intruder is present, the long drawn notes being particularly plaintive and almost human in their appeal. By the last days of July much of the excitement has abated and usually the only call heard at the Race Track Pond was, *skeep, skeep, skeep,* corresponding in a way to the *dee, dee, dee,* of the early summer, and uttered only when the bird is taking wing or alighting. One individual on August 31, called *kée-dee, kée-dee, kée-dikky,* a rather remarkable performance for so late in the season. Usually the late summer and autumn birds, if they call at all, utter only one or two notes, either the familiar *kill déer, kill déer,* (or *kill dée, kill dée,* as some hear it) or a rather prolonged, *deer, deer, deer,* a mere suggestion of the wild alarm cries of early summer. October birds sometimes further vary the *kill déer* cry into *dickadee, dickadee, dickadee,* as they take wing.

Spring birds have a varied repertoire but are not so vociferous as they become just after the young are hatched. There is the familiar *kill déer, kill déer,* of the bird on the wing and on the Lighthouse Pond, on April 24, one individual called, *skeek, skeek, skeek, tee diddy, tee dee, tee dee, tee diddity,* and another, perhaps with young, which may have been the case with the preceding one as well, flew round and round in much anxiety calling *seep, seep, seep, seep, see diddy,* and when he took wing for the second time varied it to *tee dicky-dicky-dicky.* A January bird which flushed suddenly called *teeee dididity* a very unusual performance as at this season they are usually silent. The vocal variation in Killdeer calls is no doubt largely individual as is also, to some extent, the seasonal difference but there is a regular and gradual diminution in the length and variety of the calls after the breeding season is over, reaching the point of almost absolute silence in midwinter and increasing in spring until the young are hatched when it is at its maximum. There are two other notes to be mentioned, one a low *seeeeeep* when a bird is feeding and unaware of the presence of an intruder, and a peculiar low liquid trill, *tul-ul-ul-ul-ul,* difficult to express in syllables but somewhat recalling the song of the Pine Warbler. This I have heard at various times from mid-April to mid-August and several times, at least, it was from birds performing the peculiar squatting action referred to beyond.

One of the interesting features of the Killdeer's nesting behavior is the practice of the so-called broken wing ruse, a habit shared by a number of other species but particularly conspicuous in the Killdeer. While usually attributed to the desire of the bird to lure intruders from its nest or young this habit has lately been interpreted by Dr. Herbert Friedmann as the result of conflict between an urge to escape danger and a call to remain at the nest, which renders the bird more or less incapable of action, a paralysis which disappears as the bird succeeds in getting farther and farther away. I have watched several nestings of Killdeers at Cape May and have come to the conclusion that whatever its cause this "ruse" does not develop until the eggs are nearly ready to hatch. A nest was found on May 23, 1923, in a strawberry patch at Erma, and was placed on top of one of the little hills upon which the plants were grown. The bird which was incubating, pre-sumably the female, left the nest before I got near it but I caught sight of her twenty-five yards away running between the rows of strawberry plants like a mouse, crouching low and taking advantage of every concealment that offered. I took a position in a nearby thicket and soon got a glimpse of her returning. She would run a short distance and then pause to pick up some morsel of food and every now and then would turn slowly around look-ing intently in every direction. Finally she approached the nest and settled herself slowly upon the eggs. At my first movement she slipped off, slinking away in a crouching position, and I almost immediately lost sight of her. I must almost have stepped upon her as I walked down the furrow along which she had disappeared, for upon looking back after I had reached the nest I found that she was behind me. As I turned she must have been aware that I had seen her as she ran directly away from the nest and settled down in a hollow, and then, as I approached, ran on again and settled in another hollow, sometimes gaining complete concealment behind an intervening clod of earth. Even when I went rapidly after her she did not take wing nor attempt the broken wing ruse but ran steadily uttering once or twice toward the end, the single call *deer, deer*. The male bird (or at least the mate, as the sex of the incubating bird in this and other shore birds is always open to question) could not be detected anywhere in the vicinity. The eggs in this nest ap-peared to be only in the early stage of incubation.

In the case of a nest found on the railroad track on May 27, 1929, the setting bird went off the eggs in full sight as I approached without trying to sneak away, and immediately practised the "ruse." She rested on her breast with tail fully expanded and wings half opened and pointed upward and away from the body. She keeled over first on one side then on the other waving or flapping the wing on the opposite side, and all the time kept her head up and watched me intently. Later, after I had followed her for some distance down the track, she flew over to some plowed ground nearby and ran about rapidly with head crouched down between the shoulders and uttered no cries whatever. Upon coming back some time later she went through much the same performance; both wings were thrown forward and the plumage all ruffled up and suddenly turning she ran a short distance toward me like an enraged hen defending her chicks; then turned back and falling upon her breast went through the ruse as before. When I pursued her rapidly she flew and ran ahead of me and several times called *sheep, sheep, sheee-peep*. Then again she took to the plowed ground. The eggs in this nest appeared to be far gone in incubation. In another case where I knew that the eggs were soon to hatch the incubating bird on every occasion flew at least twenty-five feet from a railroad bank out onto the adjoining marsh, spread her tail and resting on her breast flapped the half-closed wing on one side or the other. The mate, which appeared on the scene whenever I approached, went through the same performance but not so perfectly.

On several occasions I have seen birds go through a partial broken wing performance often at seasons when direct association with nest or young was not apparent. On July 6, 1922, a flock of ten flushed with much clamor from the shallow pond at the Lighthouse and upon settling again one of them fluffed up its rump feathers, partly spread its tail and turning on its side waved the opposite wing slightly and then repeated the action on the other side. On July 11, following, with a flock of twenty-five Killdeers present, one individual ran along before me with tail fully spread and finally squatted down with wings about half open and wagged first one and then the other, the action seeming to involve only the primaries. On July 1, 1931, one of three on the golf links squatted, spread its tail and rolled over on one side in the usual way, while still another on July 6, 1921, performed in the same manner although it seemed to be entirely alone. These may of course have been remains of the broken wing performance left over from the height of the nesting season. Still another case of which I have more detailed observation was of a bird at the Race Track Pond, June 30, 1920, which, with its back toward me, proceeded to settle slowly bending the legs under the body but without actually touching it to the ground. The slightly spread wings were then moved up and down alternately while the long tertials fluttered slightly as if blown by a breeze, although no air was stirring. Again this bird would run away from me with tail elevated posteriorly, holding it over to the left in a peculiarly lopsided position while it looked back at me over the right shoulder. Then shifting the tail to the right it looked over the left shoulder.

Still more difficult to explain are performances of this sort early in the season. On April 29, 1923, at a shallow pool on the Fill I came upon a bird feeding; it ran across an open stretch of mud to a grassy spot beyond, where it settled slowly down in the manner already described, holding the wings slightly away from the body, but not spread, so that the patch of orange buff on the rump was fully displayed while the tail was spread and occasionally vibrated with a tremulous motion. The bird was perfectly silent and looked over its shoulder for some time. I did not move and if finally arose and went on feeding as before, flirting its tail from side to side at every few steps. A still earlier performance, March 19, 1924, was by one of a pair on the golf links which settled down and elevated its spread tail while it churned its wings alternately. In neither instance could I find any evidence of nesting. The Killdeers of the Race Track Pond showed less anxiety after the middle of July and by the end of that month paid comparatively little attention to my approach, often passing in flight without calling, although they continued to sound the alarm when alighting or taking wing. On August 4 they remained silent without attempting to fly when I walked along the edge of the pond, something that never occurred earlier in the summer. A severe storm on August 5 and 6, with much rain, flooded the pond and the Killdeers disappeared from these haunts.

While Killdeers remain on the uplands they keep to themselves either as pairs or families, or later the union of several families, but when they begin to gather on the brackish coastal ponds they become associated with various migrant shore birds such as Ring-necked Plover, Yellow-legs and Semipalmated Sandpipers, but when the assemblage takes wing on the approach of danger the Killdeers not only do not fly with the other species but do not form a definite flock of their own, usually taking wing as individuals and not with the simultaneous action characteristic of truly gregarious birds. When feeding on the ponds they seem to associate most intimately with the Spotted Sandpiper which, like the Killdeer, usually frequents the dryer spots and does not habitually walk in the water like the Yellow-legs. On one occasion, however, I saw a Killdeer run deliberately through water nearly up to its body. In autumn and winter Killdeers are more nearly gregarious and we often see small compact flocks silently flying from one point to another. In winter, too, in upland retreats, they will associate with Meadowlarks, Starlings and blackbirds of several kinds which often form large gatherings at this season, but here again the association is not intimate.

I have never seen Killdeers on Jarvis Sound, where migrant shore birds congregate in such numbers, and their occurrence on the sea beaches is unusual and generally near to some harbor or inlet. A single bird was seen at Turtle Gut Inlet, May 22, 1921, and I once saw two accompanying some Ring-neck Plovers down to the edge of the waves while on August 17, 1920, a lone Killdeer was joined on the beach by a Turnstone and the two fed there for some minutes. A few days later at the same spot, a Killdeer accompanied a flock of Semipalmated Sandpipers and, when flushed, flew after them out over the surf and came in to the beach again farther on.

In flight the Killdeer is an exceedingly graceful bird with relatively longer wingbeats than the Black-bellied Plover but not so speedy. Both of them recall the flying dove or pigeon. I have on one occasion seen a flying Killdeer bend down its head and scratch the back of it with its foot as I have several times seen gulls do.

After the young are on the wing we find gatherings of Killdeers on various available ponds and even temporary rain water puddles will attract them. These parties number from half a dozen, members of a single family, to twenty-five or even fifty, and through August and September flocks of the same size are of frequent occurrence. On September 2, 1924, Julian Potter saw sixty together, and on October 19, 1924, I counted eighty on the shallow pond at the Lighthouse. There are, however, days during these months when no Killdeers can be found in their usual haunts. Such occasions usually follow northeasters and it would seem that the local Killdeers must pass on south with the storm and that their places are taken later by migrants from farther north. This however is merely surmise and it should not be forgotten that special food conditions may temporarily attract most of the birds of a general region to one spot. On August 27, 1921, for instance, Otway Brown saw over one hundred Killdeers in a field from which a crop of bean vines had been removed and on August 2, 1934, there was a large number in a field where an unharvested wheat crop had been plowed under.

From November to early February we rarely see more than five to ten Killdeers together although there are exceptions, such as a flock of thirty in a field near Cape May Courthouse, November 7, 1931; twenty-five at Cold Spring, January 10, 1932; forty in a day's walk, on November 26, 1922, and again, December 4, 1923; while on November 16, 1924, Potter saw flocks of twenty, thirty and forty. The Christmas census of the Delaware Valley Club always shows some of these birds present but on the years of abundant wintering there may have been some duplication:

December 22, 1929, three.	December 24, 1933, forty-one.
December 28, 1930, fifteen.	December 23, 1934, eighteen.
December 27, 1931, eight.	December 22, 1935, eighty-four.
December 26, 1932, eleven.	December 27, 1936, twenty-seven.

The favorite resorts of the Killdeers in winter are cowyards close to the farm buildings where they gather in small groups and run about close to the cattle; or the inner borders of the salt meadows where we may come upon single individuals foraging along some depression which is wet by the rising tide. At this season the birds are, for the most part, quiet and it is very difficult to distinguish them as they stand motionless against brown turf or black mud flats. Their colors in winter are especially soft and blended and match the background even more perfectly than at other times so that doubtless many a small group is missed entirely by the observers.

The silent Killdeer of winter, seeking shelter and concealment along the banks of the Harbor, and trusting to a rigid pose and protective coloration rather than to the use of its wings or legs, is but a sad contrast to the active bird of early summer, consumed with anxiety for its young. The latter fearlessly challenged every intruder and coursing over field and marsh in its graceful flight dominates all bird voices with its appealing, heartrending cries. Its clear call *killdéer, killdéer, déee diddy* is so identified in my mind with old wind-swept farm lands, lying just back of the sea, that it seems to be part of the environment and when we cross these same fields in winter, when all is bare and brown and there is a hint of snow in the air, and do not hear the well remembered cry there is a distinct feeling of something lacking.

GOLDEN PLOVER

Pluvialis dominica dominica (Müller)

The Golden Plover is an extremely rare bird on the southern New Jersey coast and it is evident that its normal line of migration follows some other course. We have but three records of its occurrence at Cape May in recent years, all on the southward migration nor can I find that gunners were ever familiar with it in times past.

On August 29, 1925, while scanning the shallow pond east of the Lighthouse I detected a bird standing in the grass on the far side, perfectly still with its head held close down on its shoulders. It appeared conspicuously speckled and showed some black on the belly. When a bunch of Killdeers feeding nearby took wing it also flew, coming to rest on the edge of a small pond back of the Coast Guard Station. It flew rather slowly, low over the ground, and its flight in this instance recalled that of a Whip-poor-will. In color it appeared dark—a mixture of black and brown. Under the shelter of some bayberry bushes I was able to approach to within fifty feet of it and at once confirmed my suspicion as to its identity as a Golden Plover. It took the same pose as before and I discovered that one leg was broken which accounted for its sluggishness.

The black on the under parts was restricted to the middle of the belly, the rest being grayish white. The sides of the face were gray with a dusky patch about the eye and a light line over it, while the dark color of the crown gave the bird the appearance of having a dark eyebrow when viewed from the side. The entire upper parts were blackish brown covered with round spots or speckles of white or gray, becoming bright golden yellow on the shoulders and especially brilliant when the sun's rays struck them. The bill and feet were dusky. The bird was of distinctly lighter build than the Black-bellied Plover and appeared somewhat smaller.

Julian Potter saw an immature Golden Plover at this spot on September 28 of the same year, also feeding with Killdeers. He noticed that it was smaller than the Black-belly and more slender, and was light colored above. As it walked and flew it uttered a rather high pitched and harsh *tu—ee, tu—ee*, very different from the melancholy call of the Black-belly. Fletcher Street has recorded the note of this bird as *quee-e-ip* while John T. Nichols describes it as "a ringing *que-e-e-a*, less clear and whistled than that of the Black-belly, with a suggestion of the Killdeer in it." (Auk, October, 1920.)

On October 2, 1927, on the same pond I was again fortunate enough to see one of these birds, also an immature bird of the year. Again it was associated with Killdeers and seemed to be of about the same size. At a distance it appeared uniform buffy brown all speckled with yellow spots on the back and wings. As I approached closer the belly appeared pale ash gray and the forehead, as it stood face-on, appeared white in contrast to the brown crown. There was a dusky line from the eye and the head looked large and broad like that of other plovers. Its actions were typically plover-like, dabbing the bill to the ground, while between each thrust it raised its head and paused a moment before advancing. Fletcher Street found a single individual with a flock of Black-bellied Plover at Crandall's Point on Ludlam's Bay on September 30, 1933, which he recognized by the white axillary feathers, uniform with the under surface of the body, the bird being in winter plumage; and by its peculiar call. We have two other records for Cape May County, both birds shot by Wharton Huber, one on the meadows back of Ludlam's Beach September 28, 1900, and the other at Corson's Inlet, November 23, 1922.

Charles Urner's records of Golden Plover for the region from Brigantine to Newark Meadows, showing extreme dates, period of regular or common occurrence, and maximum count for one day at a single locality, are presented below. He has but one spring record, a single bird seen on the Newark Meadows on May 30, 1933. Nearly all of his autumn records came from this locality which he says "still stands pre-eminent as a resting place and feeding ground for the species in this region. Its regular occurrence there and rarity elsewhere indicate not only favorable food conditions but the probability that this stretch of marsh has been an established stopping point with the species for centuries. Hunters still live in Elizabeth who recall the days when 'Golden-backs' were gathered here by the bushel basket for market." It will be seen from the records that there was an increase of these birds in 1928 and especially in 1932 after which a sharp decline has set in which Mr. Urner informs me continued through 1935 and 1936.

1928. August 31–October 28 (September 23–30). Ninety-three on September 30.
1929. August 28–November 9 (September 18–October 6). Thirty on October 6.
1930. August 17–October 5 (September 21–28). One hundred and fifty on September 27.
1931. August 27–October 3 (September 19–21). Eleven on September 19.
1932. September 1–December 4 (September 8–October 1). Three hundred on September 15.
1933. August 20–October 22 (September 9–20). Thirty-two on September 20.
1934. August 30–November 11 (September 3–8). Forty on September 8.

At the present time Charles Urner, as a result of his intensive studies of the shore birds, regards the Golden Plover as one of the "irregularly and locally tolerably common" species but, as explained above, Cape May is not one of its favorite localities.

In 1810 Alexander Wilson says that while the Golden Plover occurs occasionally along our Atlantic coast in September and October it is nowhere numerous and the same is true today. As a matter of fact the bird seems to fly directly south from Nova Scotia to South America in the autumn and, while more plentiful on the coast of New England, and especially on the island of Nantucket, its presence anywhere on the mainland seems to be due to northeast storms which drive the flights inland. It would seem likely that the birds which gather on the Newark Meadows again put out to sea either there or at Barnegat and so, for the most part, miss the south Jersey coast. The spring flight is up the Mississippi Valley as in the case of the Lesser Yellow-legs and several other species.

The Golder Plover was a favorite game bird in the past and was ruthlessly slaughtered all along its migratory route and in its winter home on the pampas of Argentina. Had it not been for the Migratory Bird Treaty it would probably have been exterminated as was the Eskimo Curlew, a bird of similar distribution and line of flight. Charles S. Westcott (Forest and Stream, May 12, 1884) states that "there are many Golden Plover in the Philadelphia markets" but as the bird was practically unknown on our coasts in spring this statement must have referred to the Black-bellied Plover, unless it were possible at that time to bring birds here from the Mississippi Valley without spoiling. That they were shipped in barrels to cities in the middle west is well known.

Brigantine records are:
September 11, 1932, six (Urner); August 30, 1933, two flocks (Warren Eaton); September 15, 1933, one (Pough); September 17, 1933, one (Urner and others); October 6, 1935, one (Tatum). Julian Potter found a Golden Plover at Camden on October 1, 1932, and a dead individual was picked up by Brooke Worth on the Tinicum Meadows below Philadelphia, on November 3, 1934. These birds were of course stragglers blown inland from the ocean or up Delaware Bay.

BLACK-BELLIED PLOVER

Squatarola squatarola (Linnaeus)

This bird, the largest of the plovers and known to sportsmen as the "Bullhead," is an inhabitant of the sounds and salt meadows where it associates especially with the Turnstone and Curlew, gathering in numbers on the broad mud flats at low tide and resting about the meadow ponds when the water rises. When several are together they seem to hold aloof from the other species and even if they take wing with them the Black-bellies keep to themselves in a close bunch and do not necessarily alight when the others do. Single individuals or small parties occur at intervals upon sand bars near the inlets and rarely on the ocean beach itself, although there, too, they are usually loath to mingle with other shore birds but may join such Turnstones or Knots as are feeding nearby.

The color pattern of the Black-belly in full spring plumage is unique. Generally speaking it is white above and solid black below and at a considerable distance the effect is of a white bird with a black belly. Closer at hand the white, except on the rump and tail, is found to be more or less mottled with dusky producing a grayish effect, while the black is found to be absent from the under tail-coverts although it extends upward on the sides of the neck involving the entire face to above the eyes while it also reaches under the wings. The light upper parts render the bird conspicuous on the black mud flats as far as the eye can distinguish it, while the black belly disappears, merging completely with the background, so that when seen in full side view the effect is that of a small white bird in silhouette against a black field.

Black-bellies are most conspicuous when standing diagonally with the head directed away from the observer, in which position the greatest amount of white is visible; and when resting on the meadows with scattered Curlew, Turnstones and Yellow-legs they are noticeable at long distances especially when their white backs are turned toward the sun. They are easily the most conspicuous of the four species. On the sandy ocean beach the peculiar coloration of the Bullhead produces an entirely different effect, for now it is the black under parts that stand out in bold relief while the whitish back blends with the beach sand and when it is seen running directly away, so that the black is completely concealed, the gray mottled back of the bird merges perfectly into the background of the beach with its scattered bits of white and black shells and pebbles. One individual which I followed for some time along the high tide line, where much trash and broken shells had been washed up, disappeared from view every time I looked for it after taking my eyes from it even for a moment, and more than once I thought it had flown away. The young birds of the year lack the black bellies and are browner above than the adults, while the latter by wintertime have also acquired white bellies. When they arrive on the meadows in August they are usually in full nuptial plumage but later on and in the early autumn individuals in all stages of molt may be seen, some with the bellies still quite black others with only a few black feathers remaining. Most of the earliest migrants seem to be adults.

I have studied the Black-bellies in spring very satisfactorily from a skiff stranded on the edge of the mud flats where they come to feed in mid-May, and Walker Hand and I have several times spent hours watching them on the mussel beds of Jarvis Sound. No matter how they may come in, whether in small or large flocks, they soon scatter widely over the flats taking up positions at nearly equal distances from one another. They stand perfectly still with neck erect, then suddenly a bird will take several quick steps and give a peck at the ground, then there is another pause and another short rush and dab with the bill, the whole action being exactly like that of their smaller relative the Ring-neck.

As the tide rises on the feeding flats the birds gather in a smaller and smaller area and crowd more closely together. When associated with other shore birds the Black-bellies seem very loath to leave and are the last to take wing, standing for hours until they are literally washed off by the water. They go in small detachments following a regular line of flight to their favorite salt meadow where they await the next ebbing of the tide. In the flocks of shore birds which gather in May back of Little Beach Island, farther up the coast, where I studied them with Dr. Harry C. Oberholser, I estimated that the Black-bellies numbered about one-tenth to one-twelfth of the entire assemblage and in one flock of feeding birds that totalled 2300 individuals I counted 250 of these plovers. As they passed our boat *en route* for the resting meadow we counted flocks of twenty-four, thirty-six, sixty-five, eleven, fifteen etc. Meadows with short grass and patches of salicornia, and with many shallow ponds, seem to be their favorites yet there are often other meadows close by, apparently just as satisfactory, which will not attract a single bird. The small detachments that leave the flats seem to unite again on the meadows and, when flushed repeatedly, the meadow flocks seem to grow larger every time they take wing, small scattered groups continually joining their ranks.

A single bird feeding on the beach at Cape May on August 30, 1921, seemed to prefer searching the line of trash that marked the limit of the last high tide, though occasionally it ran down onto the wet sand. It was completely at ease and not alarmed at my presence. It gave one the impression of being rather long-legged especially when compared with some nearby Killdeers. The neck was relatively short and was held erect and when the bird picked up food from the sand it tilted the whole body with the hips as a fulcrum, leaning over until the bill just touched the sand, with an immediate recovery to the upright position. The pecks or dabs were always made rapidly in a nervous fashion as if in apprehension of approaching danger. This bird would take several rather hesitating steps, between a walk and a run, pausing sometimes with one foot raised until only the tips of the toes touched the ground. Then it would run rapidly for ten or twenty feet and once more begin to feed, with the short dashes intervening. Its legs did not move so fast as those of the smaller Ring-necks which were feeding close by. Once or twice while running it kept its head partly turned, looking back over its shoulder, and once it paused and shook its tail violently from side to side. I was unable to see what it picked up from the sand except on one occasion when it raised a small fish several times in its bill only to drop it again. In all its actions it was exceedingly dainty. When it flew, as it did once or twice for short distances, it raised its wings deliberately and seemed to take a running start, like a gull, and on alighting again held them aloft for a moment, at which time one could see the black axillary feathers. The pure white rump tapering to the tail tip behind was very conspicuous as contrasted with the gray of the back and wings. The bird looked chunky in build and the body was about the size of that of a young tern when it first takes wing.

One who sees the Bullheads on the wing for the first time, when some distance away, will be impressed with their general resemblance to doves or pigeons, with their narrow wings and long, rapid beats. The neck is so short that the head appears to rest immediately on the shoulders and the short bill is quite inconspicuous. If any color can be distinguished in the flying Bullheads they will appear as white birds with a broad black belly band and a white stripe running through the rapidly moving wings but against the varying brightness of the sky and with the light striking them at different angles they look like gray or black silhouettes. As compared with other species on the wing they most closely resemble the Turnstones, though decidedly larger.

The call of the Black-belly, which is frequently uttered in flight is a plaintive *pee-a-wee, pee-a-wee,* which can readily be imitated and the birds decoyed with little trouble, though they are often wary, coming only near enough to satisfy their curiosity, and will not alight if anyone is in plain view within a reasonable distance.

Walker Hand's records for arrival of the Black-bellied Plover in spring are as follows:

1902.	May 1.	1911.	April 29.	1919.	May 4.
1903.	April 29.	1912.	May 8.	1925.	May 9.
1905.	May 1.	1914.	May 6.	1929.	May 11.
1906.	April 29.	1915.	May 3.	1931.	May 10.
1908.	May 5.	1916.	May 7.	1932.	April 30.
1909.	April 21.	1917.	April 23.	1934.	May 5.
		1918.	April 26.		

They remain throughout the month of May in large flocks up to at least the 25th, and we have records for the first weeks of June—June 5, 1925, a flock of thirty at Corson's Inlet; June 20, 1923, two on Five Mile Beach; June 12 and 26, 1927, two on Seven Mile Beach; two on the latter beach on July 9, 1927. Some of these were doubtless non-breeding birds that did not complete their northward migration.

An immature bird seen on Seven Mile Beach on July 29, 1936, was probably a very early south-bound transient and a flock of fifty at Little Beach Island, farther up the coast on July 21, 1921, may also have been on migration. Henry Hazelhurst's gunning record shows that he shot single birds at Cape May on July 27, 1901; July 26, 1903; July 24, 1905, but his other records are mainly after the middle of August. My records of first arrivals are all in August as follows:

1921.	August 30.	1927.	August 11.	1931.	August 1.
1922.	August 8.	1928.	August 19.	1932.	August 13.
1924.	August 17.	1929.	August 12.	1933.	August 12.
1926.	August 8.	1930.	August 10.	1934.	August 15.

They remain through September and October in varying numbers and we have records for November and December as follows: November 8, 1930, two single birds and a flock of twenty; November 7, 1931, eighteen; November 12 and 13, 1932, flocks of seven and twelve; all on the meadows back of Seven Mile Beach. The records of the Delaware Valley Club's Christmas census follow:

December 26, 1927, two.	December 24, 1933, twenty-four.
December 27, 1931, thirteen.	December 22, 1935, eleven.
December 26, 1932, one.	December 27, 1936, twenty-six.

On Ludlam's Beach Julian Potter saw ten on December 4, 1921, and six on January 13, 1929, and eight at Corson's Inlet on November 28, 1926. At Brigantine there are other winter records: six on January 28, 1934, and sixteen on January 4, 1931 (Marburger); sixty on February 12, 1933 (Urner).

Charles Urner is summarizing his painstaking observations on the Black-bellies in the region from Brigantine to Newark Meadows says "present most of the year in some numbers; sometimes all year." His detailed records follow giving extreme dates, time of regular or common occurrence, and maximum count for one day at a single locality, for the past six years.

Northward migration:

1929. —— to June 22 (May 11–30). 1000 on May 17.
1930. —— to June 22 (May 18–23). 300 on May 23.
1931. May 3–May 31 (May 14–26). 800 on May 17.
1932. Wintered to May 25 (May 15–25). 2000 on May 18.
1933. Wintered to June 24 (April 29–June 4). 600 on May 28.
1934. April 8–June 18 (April 15–May 27). 800 on May 20.

Southward migration:

1928. June 27–January 13 (August 26–October 14). 200 on October 4.
1929. July 14–November 10 (August 10–October 13). 100 on October 13
1930. July 12–November 15 (August 10–October 12). 200 on August 17.
1931. July 18–November 15 (August 21–September 13). 40 on September 13.
1932. July 16–wintered (August 1–October 16). 75 on October 2.
1933. June 27–February 18 (August 13–October 22). 350 on September 17.
1934. July 15–January 13 (August 5–November 4). 150 on November 4.

In his summary of relative abundance the Black-belly is included among the ten "common" species today, ranking below the six that are termed "abundant" (see p. 358).

The flocks of Bullheads are larger in spring and for that reason, probably, they seem more abundant at that season. The birds are much fatter on the southward flight and Walker Hand estimates that they are at least fifteen percent heavier.

The Bullhead always impresses one as a wild suspicious bird, a bird of the open stretches of marsh and water, or of remote sandbars, seeking no communion with mankind or with shore birds of other species. Standing like a sentinel he is ever alert ready for instant flight, and winging his way rapidly from point to point, over long intervals, he signals his presence by that weird and far reaching whistle, *pee-a-wee, pee-a-wee*, and perhaps before we can locate him in the sky he is but a flicker of wings in the distance.

TURNSTONE
Arenaria interpres morinella (Linnaeus)

The Turnstone occupies a unique position among the shore birds; neither plover nor sandpiper, it partakes somewhat of the characters of both while it maintains a marked individuality both in habits and make-up. Its curiously mottled color pattern with mingled patches of white, black and maroon chestnut, has earned for it the name of "Calico-back" among the fishermen and gunners, which is often shortened into "Calico," while its unique habit of turning over shells with its beak, in search of food that may lie hidden beneath them, is responsible for its more widely known sobriquet.

The full-plumaged Calico-back is by all odds the most brilliant of our shore birds and in bright sunlight, against the black mud flats, there is no gayer sight than a flock of these birds busily feeding. The closed wings are largely maroon chestnut, the back mottled with chestnut and black, the breast, belly and head mostly white with broad black breast bands joined to a sort of black bridle. Viewed from the side the white collar seems to separate the black into two narrow bars and there is a conspicuous white patch on the wing. In flight there is much white on both wings and tail and a prominent black and white U-shaped patch covers the lower back and rump. Then there are the bright red feet, forming altogether a color combination that easily distinguishes the Calico-back from any of the other shore birds. One highly colored individual seen on May 17, 1924, had the usual white markings of the head and face replaced by a rich buff.

When the Calicos return from the north in midsummer both old and young are duller in color and the maroon chestnut patches are for the most part lacking. Several early arrivals, however, seen on Jarvis Sound, July 26, 1924, were almost as brilliantly colored as spring birds and had evidently not yet begun to molt. Similar birds were seen on August 15, 1931; August 10, 1926; August 13 and 15, 1927; August 31, 1928.

The only species with which the Turnstone need be compared are the Black-bellied Plover and the Killdeer. In flight it strongly resembles the former. There is the same short rather thick neck,—"no neck at all" as I have sometimes recorded it, the bill so short that it is indistinguishable as the bird passes rapidly by, the same long narrow wings with their full beats, though the motion is rather quicker than in the plover. The white and black color combination is also similar in a general way though the black of the under parts terminates much farther forward, while the prominent U-shaped mark on the rump is always characteristic of the Turnstone, as are the chestnut patches if the bird is on the ground where one can see the upper surface. At rest the two are easily distinguished. The Turnstone not infrequently associates with the Killdeer and on August 17, 1920, two birds, one of each species, were standing side by side on the upper beach near the Lighthouse offering an excellent opportunity for comparison. The size and general coloration at this season were much the same, but the Turnstone was slightly longer and heavier while its legs and neck were shorter. There was more black on the back of the Turnstone and the black breast band was continued as black stripes on either side of the head and back over the shoulder, making a U-shaped marking while in the Killdeer there were simply two breast bands. When they flew the Turnstone had the greater spread of wings, but the strikingly different rump and tail and the different calls, of course, rendered confusion impossible and yet the general similarity of dull-colored, later summer individuals, when at rest, is rather striking.

The favorite feeding grounds of the Turnstone are the mussel beds and mud flats of the sounds, where at low tide they gather in large numbers during their migrations along with Least and Semipalmated Sandpipers, Dowitchers, Ring-necks, Black-bellied Plover, Curlew etc. Next to the Semipalmated Sandpiper I have found the Turnstone the most numerous species in the spring assemblages on Jarvis Sound. Here on May 21, 1922, we counted two hundred arriving on one mussel bed in forty minutes after the water began to shoal and the next day 172 came in in the same time with 150 in sight on the mud flat a little to the north. They were early in arriving and occasionally came in before the shells had begun to show above the water, flying over the exact spot where the first shoal would appear, then indicated only by a few short tips of grass. The birds bent off on another course to return later when landing conditions were more favorable, but no matter how early they came, they were always preceded by the Dowitchers.

Once the bar began to show, the Calicos came in rapidly in bunches of six to twenty and scattered about among the mussels. They seemed to be the most active of all the shore birds that assembled there and seldom rested as did many of the others, but started at once turning over shells with their bills in the manner peculiar to them, and poking into the tufts of sea cabbage in search of small crustaceans, worms and other prey. One bird that I watched carefully went systematically over the flat turning over oyster shells and apparently did not miss one that lay in his way. Later he poked his head and shoulders under a mass of kelp (*Fucus*) which he attempted to overturn and finally succeeded, the operation reminding one of turning over a cock of hay. Another Calico-back feeding on the Bay shore (September 13, 1924) in company with a Sanderling, was devouring blow flies that had gathered about some dead crabs and once or twice he tried to turn over one of the crabs. Those feeding on the mussel beds were constantly bathing and fighting among themselves, pecking at each other and pursuing one another through the shallows. When bathing they duck the head under water sideways and churn the wings against the body in the usual way. When pluming themselves they raise the neck straight up and probe the bill down among the feathers of the breast. Their cry is a grating, fluttering, $k'r'r'r'r'r'r'r'k$, and uttered simultaneously by many birds constitutes much of the constant murmur that arises from the great crowd of birds busy feeding on the flats.

I have two records of Turnstones alighting on boat landings and pilings which is unusual in a maritime shore bird. When the tide rises the Turnstones resort to the high grassy meadows where they await the next turning of the tide.

While studying the shore birds with Dr. Harry C. Oberholser on the meadows between Tuckerton and Little Beach Island, on May 16–19, 1927, we found that about one quarter of the birds feeding on the exposed flats were Turnstones and they came in flocks of twenty-five, seventy-five, three hundred and five hundred. They mingled constantly with the Black-bellies in flight but in the feeding assemblage stood a little farther out, between the latter and the Dowitchers which always fed in the deeper water, while in taking wing as the water rose they were always the last to leave.

The Calicos are social birds and if an individual becomes separated from his own kind he will join a flock of some other species. This perhaps accounts for the presence of one or two Calicos on the sea beach with flocks of Sanderling, Semipalmated Plover, or even roosting terns. The association is most frequently with the Sanderling, and as these nimble birds of the strand follow the edge of the waves back and forth, the ever active but short-legged Turnstone waddles about farther back on the beach, busily turning over shells and pebbles until something frightens the flock and away they go, the Turnstone prominent with his longer wings and conspicuous rump mark.

One Turnstone watched on August 30, 1920, followed the beach birds back onto a large sand flat behind the low dunes. Here he stood at rest with Semipalmated Sandpipers and Ring-necked Plovers and later fed with the latter, running forward much as they did, though his short legs did not move so rapidly as theirs. He was very wary as I followed slowly behind and although loath to leave his companions he kept well to the front of the procession. Finally, upon my remaining perfectly still, he began to feed, running forward a few steps and then pausing to turn over shells and pebbles and explore the sand beneath. He took the shells in regular order as he came to them, picking up whatever food they might conceal, and once pried up a fairly large oyster shell which remained standing upright in the sand as he passed on to fresh fields. Julian Potter tells me of one that he watched on Ludlam's Beach which waddled up to a piece of soft coal stranded there, which weighed more than the bird, and after eying it in a contemplative manner for a few moments inserted his bill under it and with a single twist of the head turned it completely over, apparently with the greatest ease. Again I have found Turnstones at Turtle Gut Inlet (September 3, 1921) at low tide, associated with Semipalmated Sandpipers, Sanderlings or Knots, and busily engaged in turning over bunches of black mussels left there by the tide. Once more (August 3, 1922) three Turnstones had joined company with half a dozen Ring-necks and a Least Tern.

Walker Hand tells me that both Turnstones and Black-bellies between sundown and dark will leave the sounds and seek roosting places on the higher beaches, always following the northern shore of the inlets, and he quotes old tradition of a regular flight back and forth across the Cape May Peninsula to feed on king crabs which are washed up in numbers on the Bay shore. There seems to be no question that they do feed on the dead crabs as I have several times seen them so engaged, unless, which did not seem possible at the time, they were eating insects or small crustaceans which were themselves attracted by the crabs.

The Turnstones arrive in the Cape May area early in May and in some years during the last days of April. Walker Hand's records prior to 1920 together with some later observations by myself are as follows:

1905.	May 1.	1916.	May 7.	1925.	May 9.
1908.	May 5.	1917.	May 2.	1931.	May 9.
1909.	May 2.	1918.	May 4.	1932.	April 30.
1911.	April 29.	1920.	May 1.	1935.	May 5.
1914.	May 6.	1922.	May 5.		

They occur in flocks throughout May and some north-bound birds remain on the meadows until early June—June 3, 1879 (Dr. W. L. Abbott); a flock, June 21, 1909 (Hand); flock of twenty-five at Corson's Inlet, June 5, 1925 (Stone); Seven Mile Beach, June 12 and 26, 1927 (Potter); June 20, 1923 (Potter). Some of these may well have been non-breeding birds which did not travel farther north.

On its return flight the Turnstone reaches Cape May by the last of July and we have the following arrival records:

1888.	July 29.	1923.	July 22.	1930.	August 4.
1902.	July 16.	1924.	July 26.	1931.	July 27.
1909.	July 9.	1926.	July 25.	1932.	July 26.
1922.	August 3.	1928.	July 28.		

The bulk of the southern flight takes place during August but they remain until the middle of September in smaller numbers and we have several later dates: September 23, 1925, a single bird; two back of Avalon October 12, 1930; two on the stone jetty at the mouth of the inlet on October 22, 1922. These last were observed by Henry Gaede and William Yoder who tell me that the birds were very tame as they fed among the huge rocks and came within ten feet of them.

Charles Urner's records for the region from Brigantine to Newark Meadows showing extreme dates, period of regular or common occurrence, and maximum count for one day at a single locality, are as follows:

Northward migration:

1929. —— to June 22 (May 17–30). 2000 on May 17.
1930. —— to June 22 (May 18–23). 150 on May 23.
1931. —— to May 31 (May 17–24). 2500 on May 17.
1932. May 8–May 25 (May 15–25). 500 on May 18.
1933. May 7–June 4 (May 7–June 4). 700 on May 21 and 28.
1934. May 13–May 27 (May 16–27). 150 on May 20.

Southward migration:

1928. June 29–October 24 (August 1–September 16). 100 on August 11.
1929. July 14–October 13 (August 1–September 8). 125 on August 1.
1930. July 14–September 14 (July 26–August 31). Thirty on August 9.
1931. August 1–September 20 (August 1–September 6). 100 on August 9.
1932. July 23–November 6 (August 1–September 11). Seventy-five on August 20.
1933. July 30–November 18 (August 12–September 10). Thirty-four on August 25.
1934. July 23–December 2 (July 29–September 16). Fifty on August 12.

In point of relative abundance he places the Turnstone in the group of ten "common" species just below the six "abundant" kinds (see p. 358).

On rare occasions a Turnstone has been found on the coast in winter— one on Brigantine on January 13, 1929 (Urner), and one at Cape May December 27, 1936 (Street).

There is something very attractive about the Calico-back entirely apart from his gay holiday attire. He seems to stand out as the shore bird personification of diligence and sociability. Ever on the move, carefully and systematically turning over his shells and pebbles, and gathering up such lurking prey as the more impetuous hunters have passed by. And then there is his dislike for solitude which impels him to strike up a friendship with any possible associates that may come his way, and when foregathering with those of his own kind on his favorite mussel shoals, in the sounds, there is that genial flow of conversation which it would seem must cheer up all who are within hearing.

1936
CR.

AMERICAN WOODCOCK

Philohela minor (Gmelin)

The Woodcock is notable as our nocturnal shore bird—solitary, sluggish, almost a recluse as it were, averse to light and to extended flight except on migration, and with the true flocking instinct, so characteristic of most of the family, all but lost.

About Cape May it is present every month in the year—a regular summer resident but now in reduced numbers; a regular transient in varying numbers sometimes, especially in late autumn, in great congested flights; and a rather rare but possibly regular winter resident. Whether any individual birds are present continuously is open to question and it would seem more likely that the breeding birds go south in the autumn and that the wintering individuals are relicts from the congested flights from farther north.

One's experiences with summer Woodcock about the Cape are likely to be few and brief. Walking leisurely along some grown-up pathway through low wet woods or huckleberry thickets, there will be a sudden explosion from the ground before us and we catch a momentary flicker of wings, as a heavy-bodied bird whirls over the bushes and is immediately lost to sight; or perhaps it may be on the edge of some wood-girt bog when we get a glimpse of a large head, short, rounded wings and a long bill pointed downward like that of some giant mosquito. Search may result in flushing the bird into another short flight but always its sense of sight or hearing is better than ours and it is off before we can spot it on the ground. A striking concealing coloration—a combination of browns and buffy tints like dead leaves and dry grass—aids it in no small degree in escaping notice and doubtless we unconsciously pass many a crouching Woodcock which does not happen to be directly in our path and which, confident in its security, does not take the trouble to move. Certain it is that when a dog, whose senses more nearly match those of the bird, points to a resting Woodcock which he has detected, it may be a long time before our eyes are able to distinguish the bird among the dry leaves and the light and shadow of the woods floor. This trait of the Woodcock, more pronounced, perhaps, in the autumnal flights than in the case of summer birds, is well appreciated by the gunners who usually use trained dogs when hunting them.

However, it is by no means impossible to approach a Woodcock without forcing it to flight and even to see it first, if we know its haunts and are ever keenly alert. The narrow strip of woodland skirting Anthony's Branch, near Cold Spring, is a favorite resort for summer Woodcock. A clear stream of cold water runs through a shallow ravine, tall trees—ash, red maple and sour gum tower above a dense growth of holly, wild cherry, spicewood and viburnum, intertwined with greenbrier and Japanese honeysuckle. In some spots it is quite dark and the growth is so dense and so low that one has to crawl on hands and knees to force a passage. Penetrating this tangle on July 18, 1924, in the hope of finding a Woodcock I came face to face with one in the midst of the dark cover. I doubt if I should have seen him had he not moved, in spite of the fact that I was fully expecting to run into him and was examining every foot of the ground. I stopped instantly and saw him walk a few feet with a somewhat waddling gait, his head moving forward and back with every step, his bill sloping toward the ground at an angle of 45°, while his eyes, high on the sides of his head, gave him a ridiculously important mien. At my first move he took wing fluttering away under the low vegetation to come to rest a little farther on, while on the next attempt he broke through and disappeared. I have flushed Woodcock in July and August from various other spots about Cape May: from the low woods north of the turnpike at Cape May Point; from the overgrown shrubby shores of Lake Lily; from the edges of the Bennett bogs and from Two Mile Beach where there are swampy spots with very dense cover.

During long periods of summer drought, when the bogs and swampy haunts of the Woodcock are largely dry, the birds not infrequently come into the town where the lawns are kept fresh by sprinkling and remain for days at a time. On June 27, 1925, while looking into a secluded garden on Lafayette Street, I noticed a movement under one of the bushes and discovered a Woodcock busily boring in the soft earth. It walked slowly about, its legs appearing ridiculously short in proportion to its heavy body, and its long bill sloping diagonally forward so that the tip just touched the ground. With each step the bird apparently poked the bill into the soil as if feeling for prey, sometimes directly forward, sometimes slightly to the right or left.

Every now and then it paused as if it felt some motion in the earth. If it was mistaken it resumed its walk, if not the bill was lunged into the ground for half its length or more, the head working slightly from side to side as it was forced in, while a tremor seemed to pass through the entire body to the tail. As the bill was withdrawn the mandibles could be seen working up and down as if food was being devoured, the head nodding slightly back and forward. After a few minutes the bird walked from under the bush and came directly toward me across the lawn. Face-on its appearance was ludicrous, the eyes high up on either side of the head resembled somewhat those of a stalk-eyed crustacean, while the transverse bars on the head, peculiar to the Woodcock, I believe, among all North American birds, formed a striking marking. Startled at something it ran rapidly to the shelter of another bush, the shortness of its legs being emphasized by their rapid action. Here it came to rest not four feet from me, as I stood on the other side of the fence, and squatted down on its belly. At all times whether feeding, running or resting, the bill was held at an angle of 45° to the ground. This bird had been reported in this or adjoining gardens for at least four days previously and doubtless found an abundance of food in the loose earth of the flower beds or on the moistened lawns.

On August 28, 1930, I saw Woodcock just at dusk feeding on the lawn of the Dougherty place on Washington Street in the heart of the town, where a sprinkler had been spraying the grass. At first sight I took it for a Flicker so closely did it crouch on the ground. It moved forward, apparently by several rapid steps, although the effect was as if the body advanced by jerks from one pause to the next. At each stand it probed rapidly to right and left with the bill held nearly vertically the tip just touching the ground; every minute or so it would be forced deep into the earth while the head moved slightly from side to side as if to aid in forcing it down. The tail was held down at all times so that the back formed a regular arch. Although automobiles were passing constantly and people walking past just outside the fence, the bird, not over twenty-five feet away, paid not the slightest attention to them. Several Woodcock have been seen in Walker Hand's garden on Washington St. in different years during July and August and one found on July 19, 1929, had a broken wing probably from contact with a telegraph wire. When I saw it it was resting on the ground under a bush where it had been all day. As I approached it arose, elevated and spread its tail so that the whole pattern was displayed, and walked about with both wings drooping to the ground like a strutting chicken. Another was found resting in the same way but made no effort to escape and was picked up without trouble. It almost immediately died in my hand from no apparent cause; possibly it had been poisoned by some insecticide sprayed in the garden; its stomach was empty.

It would seem that a summer Woodcock which finds a satisfactory feeding place will occupy it for some time, since one was flushed from the edge of a little thicket near the head of the Lighthouse Pond on July 2, 1926, and what may have been the same bird was started from the identical spot several times later on. On July 27, 1929, Julian Potter found two at the same place one of which took wing at once the other standing motionless for a minute before flying. Next morning at 9:00 a. m. there was no trace of them but an hour later the edge of the mud was found to be dotted with their borings and with patches of excrement showing that they had been feeding in bright daylight. On the Physick property on Washingtion St., they frequently occur singly at all times of year and during August, 1929, up to early September a Woodcock constantly frequented the manure pile where worms were plentiful.

Walker Hand, whose long and intimate experience with Woodcock at Cape May covered periods of the year when my visits have been limited to a few days, has furnished me with much valuable information on the species. He tells me that in addition to their favorite woodland haunts Woodcock sometimes spend the day in corn fields when the corn has grown high enough to afford them satisfactory shelter, and when once a bird has selected a cover it is likely to remain there all through the summer, coming out at dusk to feed in the adjoining cultivated ground, boring where the ground is soft, especially in spring, or turning over leaves and bits of twigs etc., in search of insects and worms which constitute its food. Hand has seen a Woodcock boring at dusk and noticed that it poked its long bill straight down into the ground and then seemed to work it from side to side, which corresponds with my observations. Old orchards, and especially peach orchards, are favorite feeding grounds for Woodcock and it is surprising to see how many birds will come out of the nearby woods into such spots on the approach of darkness. Sometimes, Hand tells me, a gunner may have searched an old orchard or field so thoroughly that he is satisfied that there are no Woodcock there and returning at dusk will find it swarming with them. They seem to feed during most of the night and are often flushed along the edge of a public road by passing vehicles. They are much more active and alert after dark and will usually take wing at such times long before you come up to them. Woodcock usually make but short flights, just far enough to find a satisfactory cover in which to drop, but when forced to make a more protracted journey they fly rather low, flapping and sailing alternately.

The breeding season of the Woodcock is marked by the peculiar aërial performance of the male bird at dusk. He mounts into the air in successive spirals uttering his characteristic call and then drops to earth near where his mate is located. I have heard this performance as late as April 30, 1904, which is unusual, and on this occasion it was too dark to distinguish the bird. Walker Hand, who has frequently watched the performance, tells me that the male after plunging to earth spreads his tail, squats on the ground and opening his bill gives a single squawk. He has seen the aërial performance as early as February 16, 1925; February 17, 1926; February 19 and 20, 1929.

The Woodcock's nest is a mere depression among the dead leaves and the eggs, as in the case of most shore birds, are four in number. Walker Hand records one set on April 5, 1909, which hatched on the 8th, and others on February 20, 1931; March 23, 1919; and April 25, 1920, while Richard Harlow found two nests in the vicinity of the Cape on April 20, 1914, and May 9, 1908, the eggs in the latter being pipped and on the point of hatching. William Baily found a set in the same condition at Tuckahoe on April 12, 1910. This nest was visited several times and on each occasion the female flushed from the eggs while the male was found standing perfectly still about one hundred feet away always in the same spot. On March 20, 1931, Walker Hand reports the finding of three dead downy Woodcock just hatched, immediately after a heavy and unusual fall of snow. He also found young just able to fly on May 1, 1918, and a female with three young as late as June 18, 1922. Woodcock nest in some numbers in the vicinity of Pennsville and Alloway in Salem County, along the Delaware River, a region which seems quite as acceptable as Cape May for breeding purposes. I quote dates for several nests examined in this region by Turner McMullen and Richard Miller for comparison with Cape May records: March 30, 1924, (three), April 6, 1924; April 4, 1926 (two); April 2, 1927; April 3, 1933. All of these contained four eggs while in another, April 3, 1927, the eggs were just hatching.

Although not at Cape May regularly during the spring I have on several occasions found young Woodcock elsewhere that had just hatched, little balls of buff and brown down; once on April 18, 1919, just across the Bay in Delaware, and once on May 6, 1911, at West Creek, a little farther up the coast. In each case the parent flushed from the little birds which she had evidently been brooding and flew low over the ground alighting a few yards away where she spread her tail and fluttered her wings in an apparent effort to lure me away (but cf. Killdeer p. 385). Two of the young in the first brood had been turned completely over on their backs when she took wing but all four remained rigid without attempting to move or even to right themselves. Marking the spot I withdrew a few feet and then returned in a minute or two to find no trace of the little birds. Doubtless they were close at hand but were rendered invisible by their admirable concealing coloration. Walker Hand found an adult with young at Price's Beach on the Bay shore on May 1, 1910, and other families on April 28, 1910, and May 5, 1930. He tells me that full grown young in summer may be told by their heavier bodies and more labored wing action.

Interesting as is the study of the breeding Woodcock and of the scattered individuals of summer, it is but a small item in the history of the bird at Cape May. It is during the migrations, especially in autumn, that the species becomes a notable feature of our bird life. Ordinarily the south-bound migrants pass along during October and November about as they do elsewhere, the birds being more numerous than in summer but not notably abundant. At some time, however, during the period of its flight, there comes in nearly every year, a day of strong cold wind out of the northwest when the skies are beautifully clear and the air crisp and tingling with frost. On such days we are sure to have what is termed a Woodcock flight. The birds following the coast line southward are apparently blown out to sea by the strong winds or are bewildered by the broad expanse of water presented by Delaware Bay after passing the point of the Cape, and hesitating to cross until the gale moderates, they beat their way back into the wooded area bordering the Bay shore. Coast Guards patrolling their beats on the ocean front between the town and Cape May Point have frequently seen them coming in in the night and winging their way across the open stretches of meadow and marsh headed for the shelter of the woods, and have notified the gunners of the morrow's sport. The birds drop down to rest all along the Bay traveling as far north at least as Fishing Creek and next morning as soon as it is light, or even before, gunners are covering the ground with dogs and Woodcock are put up at almost every step. One sportsman sometimes has as many as seven birds in the air before him at one time so that it is confusing to decide which one to shoot at. The birds squat close on the ground awaiting the abatement of the wind when they will resume their flight. Walker Hand tells me that at such times they flush with difficulty and without a dog a man would see few indeed.

Once the wind dies down they are gone and the next morning one may search for them in vain. One of the greatest flights ever known at the Cape occurred on October 29 and 30, 1920, and for two days the country about Cape May Point and along the Bay shore fairly swarmed with Woodcock but the wind subsiding on the afternoon of the 30th, they passed on. On the morning of the 31st, I tramped all over the Woodcock area and not a bird could be found. Walker Hand tells me that they start to fly about dusk and he has been fortunate enough to see them starting across the Bay for the Delaware coast. One after another they came over the low treetops, skirting the shore, and descending to the lower beach kept on over the water at a height of not more than three feet above the surface, flying, as they always do, when a protracted flight is necessary, with alternating wingbeats and sails.

The autumnal flights of the Woodcock are exactly like those of the Flickers, hawks and small passerine birds all of which occur at times of high northwest winds and should there be no such winds during the period of the migration there are no congested flights. Notable flights of Woodcock have been those of November 4–7, 1908; November 25–29, 1909 (three hundred shot); November 8 and 14, 1911, when many came to rest in sheltered spots in the town itself and those shot were very thin and apparently exhausted; November 5–7, 1913; November 3–6, 1917 (eight hundred shot in two days); November 21–23, 1918; November 6 (one thousand killed); December 3, 1919; October 29–30, 1920, the greatest flight known; October 25–26, 1922; November 18 and 27, 1923; November 17, 1924 (one thousand killed); November 10–11, and 18, 1926; November 7–8, 1927 (three hundred shot); October 29, 1928 (one man shot twenty-eight); November 9 and 21, 1928, the greatest "kill" on record, individual bags of sixteen, twenty-seven, twelve and thirty-two were counted while two men got forty-two between them and two others a half bushel of birds which were sold in the town; a boy picked up twelve dead birds in the woods next day which the gunners had failed to recover; November 3, 1930, one man got forty-two and two men together sixty; November 27–28, 1930, another large flight, two "guns" getting one hundred and twenty; October 26–27, 1931. These facts and estimates were furnished to me by resident gunners who participated in the sport.

In December, 1886, Walker Hand tells me that there was an entirely exceptional flight during a severe snow storm and dead birds were picked up all over the town. Whether this was a migratory flight or merely the birds of nearby swamps seeking shelter it is impossible to say. Some Woodcock remain about Cape May nearly, if not quite, every winter and we have records of single birds for January 12, 1922; December 23, 1924 (two); December 22, 1925; December 9, 1928; January 7, 1929; January 22, 1929; December 27, 1931. Many of these birds were flushed from the protection of privet hedges in gardens in the town.

In spring the northward migration occurs from February 25 to March 15, but does not seem to be influenced in any way by the wind nor is it usually as concentrated or noticeable as the autumnal movements. It progresses in a normal manner and spreads over a wider area. The most notable spring flight was in February 1906, when as many as two hundred were flushed by Walker Hand with a dog in a single afternoon's walk. What was the cause of such a concentration was not evident. Dates of spring arrival from Hand's records are as follows:

1905.	March 15.	1914.	March 21.	1924.	February 20.
1906.	February 28.	1915.	March 16.	1925.	March 19.
1907.	March 14.	1919.	March 16.	1927.	February 25.
1908.	February 17.	1921.	March 13.	1928.	March 1.
1909.	March 9.	1922.	March 12.	1929.	March 10.
1911.	March 1.	1923.	March 5.	1930.	March 5.
1913.	February 27.				

Concentrations which might be called flights occurred on March 12, 1922; February 15, 1927; March 1, 1928; February 16 and March 5, 1930.

One of the crimes of the past was the maintenance of an open Woodcock season in July when the young birds were hardly able to fly and old ones more or less in the molt, yet this practice continued in New Jersey through the eighties if not later. Charles S. Westcott, who wrote to "Forest and Stream" under the name "Homo," uttered many a protest against summer Woodcock shooting but to no avail, and we find him on August 30, 1883, commenting on the growing scarcity of the birds and attributing it to this cause although, as he says, "sportsmen will not see it," and "not even content to await the legal season the gunners shoot the birds in June!" The flagrant abuse of the bag limit, well known to all who shoot about the Cape, will eventually exterminate the bird if not checked. It is not a local question since the Woodcock which mass at Cape May in the so-called autumnal flights represent the Woodcock of all the states to the north, not merely New Jersey. It may be, however, that the bulk of the migrating Woodcock pass by the Cape without stopping, when the winds are from the south and congested flights do not occur. This we do not know but if true then greater damage to the bird is done where there is continuous shooting over a longer period than where one or two congested flights occur. In any case there is no possible excuse for disregarding the bag limit or other legal restrictions.

In 1886, Westcott again deplores the scarcity of Woodcock and says: "Cannot we go fishing, yachting etc., and enjoy a hundred and one things in the summer months and let these poor gasping birds grow and fatten so that when October comes we may go out for a day's shooting—not butchering." Yet in the same article he protests against giving any legal protection to herons, bitterns, hawks, owls, blackbirds, butcherbirds etc. It would seem that the average sportsman cannot appreciate conservation except so far as it benefits his own immediate interests! Some idea of the slaughter of Woodcock in the past may be gathered from an account in Doughty's quaint "Cabinet of Natural History" which states that in 1825, on the Cohansey River, three men in two hours killed upwards of forty on a very small tract of ground and, of course, with muzzle-loading guns!

The Woodcock is a bird of mystery, his activities shrouded in the dusk of evening or the darkness of night. One wonders what may have been his history and origin. Does he represent the primitive shore bird stock from which, through the influence of a maritime habitat, the active birds of beach and tundra have developed, or is he a comparatively recent offshoot, degenerate if you like, or perhaps highly specialized, which has changed entirely his manner of life and has become, to all intents and purposes, a "land bird" of the woodland rather than a shore bird of the beaches.

*EUROPEAN WOODCOCK

Scolopax rusticola rusticola Linnaeus

Several specimens of this Old World bird have been shot in the eastern United States of which two have been attributed to New Jersey. One of these was obtained in Washington Market, New York City, on December 6, 1859, which was said to have been shot at Shrewsbury, N. J., while the other was obtained from a game dealer in Philadelphia and said to have been obtained in New Jersey. There is no suggestion that either of these came from Cape May county and mention of them is made simply to make the list of New Jersey water birds complete.

WILSON'S SNIPE

Capella delicata (Ord)

Equally removed from the other shore birds in structure and habits the Snipe has much in common with the Woodcock—the relatively short legs and heavy body, the long straight bill and the crepuscular or nocturnal activity. It is not however a bird of wood edges and thickets like the Woodcock but chooses as its haunts the open grassy meadows or marshes bordering on fresh water streams or larger bodies of water. As might be supposed it is therefore not as abundant about Cape May as in some other portions of the state. The varied environment of the Cape however offers many acceptable spots to migrant Snipe especially along the Bay shore and we find them also not averse to spreading onto the brackish marshes where they adjoin the upland, or even straggling to bits of salt marsh immediately behind the coastal sand dunes, although this is exceptional.

We find Wilson's Snipe a regular spring and fall transient all the way from Dennisville to Cape May Point, varying in abundance from year to year, while occasional individuals remain through the winter. Walker Hand's arrival dates prior to 1920 with my own for later years are as follows:

1903. March 3.	1912. March 17.	1919. March 30.
1904. March 9.	1913. March 12.	1920. March 24.
1905. March 10.	1914. March 14.	1921. March 17.
1907. March 14.	1915. March 20.	1922. March 26.
1909. March 20.	1916. March 23.	1923. March 18.
1910. March 10.	1917. March 16.	1925. March 21.
1911. March 14.	1918. March 9.	1929. March 10.

They usually remain throughout April and our latest dates are May 6, 1928; May 7, 1932; and May 8, 1912. Thus for about six weeks in spring we may count on finding Snipe about Cape May. In a morning's walk we may flush only a few or there may come days when a notable migration is in progress. On May 8, 1912, Walker Hand reported forty seen on a three mile walk between Bennett and Cape May while on April 14, 1914, forty-three were seen and other notable flights occurred in the vicinity of South Dennis, April 8–10,

1907, and at Cape May, April 2, 1924. On all occasions they are much more abundant on the Bay side of the peninsula than on the ocean side.

The autumn Snipe flight at Cape May is not nearly so regular as the spring flight and when one occurs it is more congested but the birds seem to pass on immediately. Walker Hand considers that they are composed of birds that have been driven from some temporary stopping place by changing weather conditions. Such flights have been recorded on December 5, 1916; December 3, 1919; October 25, 1923. Of scattering autumn occurrences our earliest was a bird flushed several times in the Pond Creek Meadow on September 16, 1923, but this was quite unusual, most of our autumn records falling between October 2 and November 25. A still more remarkable record is a single bird seen by Julian Potter at Camden on August 28, 1923. We have records of wintering birds on January 12, 1919; December 26, 1927; January 27, and February 8, 1929; December 24, 1933; December 22, 1935. Also one at Bridgeport on the Delaware River, in Salem County, on December 7, 1930 (Weyl), and another at Barnegat on January 8, 1928 (Kuerzi). There are also winter records elsewhere in the state— Haddonfield, about 1875 (Samuel Rhoads); Moorestown, December 25, 1902 (W. B. Evans); Princeton, January 4, 1900 (W. A. Babson).

Wilson's Snipe has always been a favorite game bird along the Delaware River marshes and on damp low grounds at many points inland and still is shot but in a much curtailed season and with a small bag limit. It and the Woodcock are today the only shore birds which are allowed to be shot in the United States. While found mainly on the Bay side of southern New Jersey Snipe seem to have been abundant on fresh water marshes on the ocean side of the Cape May Peninsula in certain favorite spots. A writer in "Forest and Stream" in April, 1884, states that the Snipe shooting is very good on the Littleworth Meadows on Cedar Swamp Creek some miles inland from the Curlew Club at Ocean View, which fronted Ludlam's Bay and was cared for by Capt. T. F. ("Dory") Shute. On a recent trip he stated that four Snipe were sometimes flushed at once; one man got forty-one in twenty minutes while two others got thirty-four and thirty-three respectively.

The early days of April are typical Snipe days at the Cape. Days of blue sky, with water still lying on the ground in low spots, with sunshine enough to bring the green grass tips up through the brown earth, but with still a touch of March in the winds that blow from the north. We tramp over the elevated pasture land surrounding the Race Track Pond tearing away with every step the old wiry stems of the dewberry vines which cover the ground like a network. Suddenly from a moist depression just ahead there arises with the suddenness of an explosion a long-billed yellowish brown bird which goes on twisting erratic flight low over the dead grass and, topping a slight rise in the ground, is down again before we can see just where it landed. As we approach it again takes wing always in the same explosive, unexpected manner and seldom from the spot from which we look for it to arise. Its rapid wingbeats recall somewhat the flight of a dove but it gives a peculiar twist to its body, now turning up the breast and now the back, as it curves to the right or left. The long bill held straight out in front, the plump body, white on the belly and beautifully streaked above with longitudinal lines of buff on dark brown—all are clearly seen with the glass as we follow the bird in its rapid flight. Then there is the characteristic call, *scaaap, scaaaap*, as it leaves the ground.

On a late afternoon walk on April 3, 1922, I flushed five Snipe. They took wing individually, at some distance from one another, for Snipe will very rarely rise together, being, like Woodcock individual in action and with no trace of the flock movement so characteristic of the true shore birds. Try as I would I was quite unable to approach any one of these birds near enough to see it on the ground; they always took wing before I was prepared for them. By easy stages they circled around so that each of them eventually returned to almost the spot from which it first arose, never leaving the moist low ground. On April 2 and 3, 1924, a number were flushed from low ground just back of the town and from marshy spots close to the brackish water, some from the stubble from which a crop of black grass had recently been cut. One bird I followed carefully and, although I finally located the exact spot where it dropped down in rather scant stubble, I could not see a trace of it with the glass and yet when I had approached to within fifty feet it arose from the very spot that I had had under observation. This time following it with the glass I saw it alight and run a few steps, moving its head back and forward as it went, before settling behind a tuft of dead grass. Here it seemed to squat flat on its belly and this is perhaps the habitual posture in open ground during the daytime. At any rate when hugging the ground closely the wonderful concealing coloration of the bird would render it practically invisible, while its keen senses would enable it to detect the approach of an intruder when still some distance away, so that it would have plenty of time to spring into flight and escape danger. Walker Hand tells me that on one or two occasions he has almost trodden on a Snipe concealed in a bunch of indian grass before it took wing, but this is probably very unusual although at the same time we have no means of knowing how many birds may lie close in this way and allow us to pass nearby without flushing. On the Fill, to the north of the town, there were often considerable accumulations of water in the spring time and the Snipe took to frequenting these wet depressions. I have also on several occasions flushed Snipe from the edge of the deep pond at the Lighthouse where there was only a small open area between the large growths of cattails.

Only once have I had the opportunity of studying the activities of Snipe that were entirely unaware of my presence. On April 24, 1927, Richard Erskine, with whom I was driving, stopped his automobile at the roadside bordering the Maurice River marshes at the head of the Cape May Peninsula to watch some Greater Yellow-legs. They were feeding in a considerable area of shallow water with bunches of sedge and various water plants growing on all sides but not forming a dense mass. Among these the birds were moving about. Presently we realized that there were other birds there besides the Yellow-legs and we soon identified a half dozen Wilson's Snipe. They stood in the water, sometimes up to their bodies, with their bills thrust down into the soft mud. They would probe many times without raising the bill above the level of the water, poking it in a sort of semicircle and then there would be a slow step or two and the probing would begin as before. Now and then one of the birds would rise above the vegetation with several rapid strokes of the wings and drop back again a few feet farther on. They fed almost entirely in spots in which they were more or less sheltered by the sedge which accounts for the fact that we did not discover them immediately, but later they came out into clear view for a few moments at a time. The buff lines down the back were very conspicuous and produced the beautiful striped appearance so characteristic of the Snipe, while the vertical bars on the sides of the body were prominent whenever a bird came out in good light. They kept on feeding as long as we remained and apparently did not associate the car with danger. Doubtless in well sheltered spots like this Snipe may feed at all hours of the day but it is as the crouching bird of the open meadow and black grass marsh, exploding into the air at our approach, that we know them best at Old Cape May.

LONG-BILLED CURLEW

Numenius americanus americanus Bechstein

This great wader with its immense bill has long been gone from the New Jersey meadows where it once occurred regularly and is now found mainly in central Canada and our extreme North Central States. Writing in 1812 Alexander Wilson says that "they appear in the salt marshes of New Jersey about the middle of May, on their way to the north; and in September on their return from their breeding places. Their food consists chiefly of small crabs, which they are very dexterous at probing for, and pulling out of the holes with their long bills; they also feed on the small sea snails so abundant on the marshes, and on various worms and insects. They are likewise fond of bramble berries, frequenting the fields and uplands in search of this fruit, on which they get very fat, and are then tender and good eating, altogether free from the sedgy taste with which their flesh is usually tainted while they feed in the salt marshes.

"The Long-billed Curlews fly high, generally in a wedge-like form, somewhat resembling certain Ducks, occasionally uttering their loud whistling note, by a dexterous imitation of which the whole flock may sometimes be enticed within gunshot, while the cries of the wounded are sure to detain them until the gunner has made repeated shots and great havoc among them.

"A few instances have been known of one or two pair remaining in the salt marshes of Cape May all summer. A person of respectability informed me that he once started a Curlew from her nest, which was composed of a little dry grass, and contained four eggs, very much resembling in size and color those of the Mud Hen, or Clapper Rail. This was in the month of July." If this be correct it was, as Wilson himself says, a very unusual occurrence. But the chances are that the eggs not only resembled those of the rail but actually were rail's eggs!

Giraud in 1844 states that the Long-bills are regular visitors at Egg Harbor and Long Island in the spring and summer, and have been seen in the latter place as late as the middle of November," while Turnbull, in 1869, says that on the New Jersey coast they are "frequent, arriving early in May and again in September." By 1877 W. E. D. Scott found it rare on Long Beach, N. J., "very shy; seen middle April." Subsequently we have only a specimen in the collection of The Academy of Natural Sciences of Philadelphia shot by Dr. W. L. Abbott, on Five Mile Beach, on September 14, 1880; and one reported to me by William Baily, shot on the same island on September 8, 1898. On Long Island, N. Y., specimens were taken August 20, 1873, and August 26, 1885, (cf. Braislin, Birds of Long Island) while Dr. W. Todd Helmuth claims to have seen this species there on five occasions between 1910 and 1930 (Auk, 1924 and 1930). A few have been recorded in recent years from South Carolina and Georgia but whether these birds came down the Atlantic coast or flew southeast from their Canadian breeding grounds, as other birds of the same region have done, cannot be determined. A few alleged recent observations of this bird on the New Jersey marshes have not been corroborated.

The bird was known to Cape May gunners of former years as the "Sickle-bill."

J. F. S.

YELLOW-LEGS

HUDSONIAN CURLEW

Phaeopus hudsonicus (Latham)

The Hudsonian Curlew, the only Curlew now found on the New Jersey coast, since the apparent extermination of the Eskimo Curlew and the practical disappearance of the Long-billed Curlew from the Atlantic seaboard, may be considered our largest shore bird in the sense that it is much larger bodied than any other species, although its relatively shorter legs may give it a stature apparently little if any greater than that of the Willet or Greater Yellow-legs. The Oyster-catcher, of course, is decidedly larger but is a mere straggler this far north.

The Curlew is distinctly a bird of the sounds delighting in the broad expanse of open meadow and the many waterways with their bars and mud flats exposed at low tide, which stretch away for miles to the north of Cape May Harbor. It is a wary bird well able to take care of itself and in its favorite environment can readily detect the approach of danger from afar and be on its guard. Thus in spite of the fact that it was a favorite game bird in former days, when the shooting of all shore birds was permitted, it has not materially decreased in numbers, as have many of the smaller species, and may still be seen in flocks of hundreds and probably thousands at the height of the spring migration. On the return flight it does not usually occur in such large numbers, or at least the flocks are more scattered, and its occurrence on our meadows is much more extended. On the vast marshes surrounding Great Sound, some miles north of Cape May, the largest flocks of Curlew are to be found. Here on some favorable location on the inner edge of the coastal islands they may be seen through the glass rising and settling in enormous flocks at the height of the May migration. At a distance of several miles through the haze that hangs over the horizon the closely massed birds resemble whisps of smoke or bits of cloud floating over the meadows. As they turn in one direction their dark bodies are exposed to the light and the "cloud" seems to darken but the next turn brings the light at a different angle and it disappears completely for a moment only to loom up again out of the haze. Then as the flock comes end on with greater apparent congestion of individuals it seems like a great cyclonic cloud settling point foremost on the meadows. Small detachments may be seen to leave the main flock for a time and as they drift about by themselves they look exactly like airplanes seen at a distance. Other flocks that arise somewhat nearer to the observer, so that the individual birds can be distinguished as dark specks, resemble swarms of mosquitoes or gnats. It is difficult to estimate the number of birds present in these assemblages but there must be several thousand.

As all rules have their exceptions so on rare occasions we come upon Curlew away from their usual haunts, but they are rare indeed. On August 27, 1925, a single bird, evidently lost, was found on the shallow pond near the Lighthouse. It took wing several times and flew about calling in a bewildered manner and on July 25, 1923, after a heavy rain a flock estimated at a thousand birds alighted on a shallow pool on the Fill, but as this lay right on the line of their southward flight it was not so remarkable. Again on July 9, 1916, Walker Hand and I found a single Curlew on the strand of Two Mile Beach associated with some Semipalmated Sandpipers.

The distribution and habits of the Curlew on the New Jersey coast are rather peculiar. It is abundant on all of the salt meadows and sounds of Cape May County as far north as Corson's Inlet, resting and feeding here constantly during the migration period, but varying in numbers from day to day according to the direction of the wind or condition of the weather. At flood tide the birds resort to the higher grassy meadows and at low water take possession of the exposed bars and mud flats. On the meadows of the northern half of the coast, however, for some reason or other, they seldom stop to feed and are seen as "birds of passage" only. They do not seem to be so abundant there as they are in Cape May County and it is possible that their line of flight is, to some extent, offshore. While covering the meadows from Barnegat Bay to Cape May with Dr. Harry C. Oberholser on a Biological Survey boat, May 16–19, 1927, we saw not a single Curlew until we reached the meadows back of Ludlam's Beach when they arose in clouds in the usual fashion.

The dates of spring arrival of the Curlew at Cape May as recorded by Walker Hand and myself from 1902 to 1932 are as follows:

1902.	April 30.	1910.	April 19.	1918.	April 26.
1903.	April 27.	1911.	April 17.	1920.	April 25.
1904.	April 21.	1912.	April 14.	1921.	April 16.
1905.	April 25.	1913.	April 18.	1922.	April 23.
1906.	April 29.	1914.	April 23.	1925.	April 8.
1907.	April 12.	1915.	April 19.	1926.	April 18.
1908.	April 18.	1916.	April 14.	1931.	April 25.
1909.	April 16.	1917.	April 23.	1932.	April 30.

Some of these dates are doubtless late but it may be that the birds are present in considerable numbers on the first day of their occurrence. Certain it is that on April 25, 1931, I have recorded them as rising in a great cloud far north of the Harbor while on April 30, 1932, Otway Brown and I saw one flock of eighty and another of fifty feeding on the meadows back of Seven Mile Beach, very conspicuous among the short grass which had only just begun its spring growth. Curlew are present in numbers until at least May 25 in most years and doubtless stay on in small detachments throughout that month.

On their return flight we find Curlew passing regularly by the first week of July. We see them crossing the Fill from the Harbor to the bathing beach near the pier and thence out to sea across the mouth of Delaware Bay, bound for Cape Henlopen or the meadows farther down the coast. Their line of flight apparently follows the edge of the old Cape Island Sound, now entirely filled in, and they are seen most frequently in the evening or late afternoon, especially between six and seven o'clock, and only when the wind is from the southeast. The passing flocks consist usually of from ten to thirty individuals. My arrival records for the southward flight follow:

1916.	July 9.	1926.	July 6.	1932.	July 8.
1921.	July 2.	1927.	July 4.	1933.	July 1.
1922.	July 12.	1928.	July 4.	1934.	July 12.
1923.	July 4.	1929.	July 11.	1935.	June 30.
1924.	July 14.	1930.	July 19.	1936.	July 15.
1925.	July 12.	1931.	July 5.		

Some of these dates may be late as the early flights are often small and could easily be missed, while I was not regularly on the meadows in some years. On July 18, 1926, there was an unusual number of Curlew passing, several small bunches being observed as early as 2:00 p. m., and from then until 3:30 flocks were seen at intervals of ten minutes numbering eight, four, two, two, one, five and twenty-nine; while between 6:00 and 6:30 p. m., when we watched for them again, they passed at about the same intervals but in larger bunches, usually fourteen to twenty together. On July 21 of the same year several much larger flocks were seen between 6:00 and 7:00 p. m., while on August 9, 1924, they continued to pass until dark and probably on into the night. There was a strong southeast wind blowing on all of these days and the birds were constantly shifting and changing position causing their line to bulge and waver. Their wingbeats were continuous with no sailing and the strokes distinctly faster than those of a Laughing Gull. Their heavy bodies and long curved bills could be distinctly seen against the evening sky and the horizontal rays of the setting sun gave a bright ruddy glow to their breasts as they came toward us. When the flocks numbered a dozen or more the birds flew in a V or in an irregular line but in either case they were somewhat congested in front. Now and then a flock would contain a few Lesser Yellow-legs, bunched together and not scattered promiscuously among the Curlew, and sometimes there would be a few Pectoral Sandpipers, some Dowitchers or a Knot. Charles Urner has seen Hudsonian Godwits in flocks of Curlew at Barnegat Bay, on July 17, 1927, and I saw two of these birds in a large spring flight of Curlew near the Avalon Road on May 9, 1931. On July 26, 1930, when we watched for passing Curlew for an hour about 7:00 p. m., six flocks passed consisting of five, eleven, one, four, two, fourteen; and on the following evening—eight, sixteen, sixteen, six, seven, five, eight, passed in the same time. On July 13, 1927, a flock of forty-five passed over the town at 7:15 p. m., and at Stone Harbor on July 24, of the same year flocks of eighteen, fourteen, twelve, thirteen, passed within an hour all headed in a southeasterly direction.

Charles Urner's records for the region from Barnegat to Newark Meadows showing extreme dates, period of regular or common occurrence, and maximum count for one day at a single locality, are as follows:

Northward migration:

1929. —— to May 30 (?). Sixty on May 20.
1930. May 10–June 22 (May 11–23). Fifty on May 18.
1931. —— to May 24 (May 17–24). Four hundred on May 17.
1932. May 1–May 22 (May 15–22). Four hundred on May 18.
1933. April 29–May 21 (May 7–21). Fifteen on May 14.
1934. April 29–May 27 (April 29–May 16). Two hundred on May 16.

Southward migration:

1928. July 4–September 16 (July 18–August 11). 196 on July 18.
1929. July 7–September 15 (July 7–August 11). 940 July 23.
1930. June 22–September 14 (July 26–August 10). 500 on August 2.
1931. July 9–September 20 (July 9–August 2). 450 on July 11.
1932. July 10–September 11 (July 16–August 13). 350 on July 23.
1933. July 2–September 17 (July 30–September 3). 250 on July 30.
1934. July 8–September 16 (July 21–August 5). 960 on July 21.

He considers the Curlew as one of the ten "common" species standing below the six that are regarded as "abundant" (see p. 358).

While our early July Curlew, immediately about Cape May, are almost all seen on the wing they apparently settle to feed on the broader meadows north of the Harbor as soon as they arrive and gather in considerable flocks even this early in the season. Fletcher Street reported a flock of two hundred on the meadows behind Ludlam's Beach on July 8, 1932, and on July 9 and 13, of the same year, I saw a flock of twenty-six back of Five Mile

Beach while we see them back of Seven Mile Beach on almost every visit
during July and August. Curlew linger on the meadows of Cape May
County until the end of August and we have several September records:
fifty on September 1, 1924, and several on September 12, 1926, September 12,
1931, and September 1, 1932, while Dr. William L. Abbott shot one as late
as September 14, 1880. Doubtless some will be found every year during the
first half of the month if looked for carefully, especially now that gunning
on the meadows has practically ceased, except during the ducking season.

It seems probable, from our observations, that the flocks of Curlew which
feed on the Cape May meadows, especially on the southward flight, remain
there for some time awaiting a change in wind or weather to urge them on
their way, and that other small bunches then accumulate until another
exodus takes place. Certainly there are days when no south-bound flocks
are to be seen and there are days when but few if any are found on the meadows. The discovery of a roosting place on the meadows to which the birds
evidently resorted night after night also accords with such a theory. On
May 23, 1907, Walker Hand found such a roost. The birds came in about
8:00 p. m., with little or no whistling, though their wings, from where he lay
concealed, made a roar in the dark, so large were the flocks. Their tracks
and droppings were all about, the ground being white with the latter, showing
that it was not a one night stand.

The flight of the Curlew takes on an added interest when we realize that
scarcely a month has elapsed after the last flocks leave us, bound north, until
the first south-bound birds arrive and even if we compare the earliest spring
record with the earliest return there is barely ten weeks time for the birds
to reach the Artic Circle, raise their broods, and return again.

The Curlew with its heavy body and relatively short legs seems low in
stature and a flock feeding on the meadows reminds one somewhat of a lot
of chickens. Their actions are deliberate and slow; they walk leisurely a few
steps at a time and then give a quick probe with the bill into the mud or
down among the grass, and immediately look up nervously as if in fear of
approaching danger. The body is carried horizontally, with neck bent over
and head directed forward, the long downwardly-curved bill being a striking
and unique character. On several occasions I have watched flocks which
acted rather differently apparently feeding on some more active prey.
These birds walked more quickly, sometimes almost running, and the head
moved backward and forward in unison with their steps, every now and then
one of them would slow up for a step or two and stab quickly at something
on the ground. The long bill, curving out from the head, seemed entirely un-
suited for such action and looked to be more of a hindrance than a help in
feeding, while it gave to the bird a top heavy appearance. Sometimes late
in the summer when the marsh grass has attained its full growth the birds
are almost entirely concealed by it and would likely remain undetected did
not some individual every now and then raise its head and display its characteristic curved bill and when all are alarmed and preparing for flight the
"sickle" bills appear above the grass in every direction.

Frequently the individuals of the flock walk about in all directions each one for himself with no concerted action. At other times one bird after another will leap into the air and settle again a little in advance of the flock, or perhaps will continue on the wing for a hundred yards or so with the result that the whole flock will gradually advance, much as do the flocks of blackbirds in early spring which seem to "roll" over the fields. When resting the Curlew usually stand at various angles not all facing the wind as is usual with shore birds, gulls and terns.

In color the Curlew appears uniform dark brown but in bright light the wings are seen to be strongly ruddy while the lower abdomen is paler rusty with a pinkish tone, the brilliancy of the ruddy tints being dependent upon the sunlight and the position of the birds with respect to it. At close quarters one can also distinguish a light eyebrow and stripe down the middle of the crown.

At low tide the Curlew resort to the sand bars and mud flats and I have had exceptionally good opportunities to study them while watching the smaller shore birds gathering at the mussel beds of Jarvis Sound. On May 21 and 23, 1922, there were upwards of five hundred in sight at one time within easy range of our skiff as it lay stranded on the mud. They came in from the great meadows to the north in long wavering lines, company front, or in irregular V formation, like ducks, and flew with rapid wingbeats, the wings pulsing through a small arc so that they made short strokes and seemed to merely pat the air. This and the heavy body with its solid dark color gave them a decided resemblance to Black Ducks while on the wing. They sail down to the mud flats on set decurved wings and then beat them rapidly as they allow their bodies to settle gradually until the outstretched feet touch the ground. The wings are then held aloft for a few seconds and folded against the body as the bird begins to walk about. The long curved bill is conspicuous in flight but the neck appears short and thick. The birds always came to rest at some distance from our boat although later on they approached nearer, but were more or less uneasy, ever ready to take wing at the slightest disturbance. They were more wary and suspicious than any of the other shore birds that gathered to feed in the vicinity.

When pluming themselves the Curlew would frequently stretch the neck straight up and bending the bill down against the breast would poke it up and down among the feathers; others would raise the body and flap the wings vigorously in the air. When bathing they stooped over in the water and churned the wings up and down, scattering the spray over their backs. One feeding bird waded into the water up to its body and then submerged the head entirely, apparently in search for food, but in all other cases they fed entirely from the mud or in very shallow water. There must have been upwards of a thousand Curlew on the sound that morning. Victor Debes has estimated a thousand back of Seven Mile Beach on May 11, 1935, while Julian Potter counted five hundred at the same place on April 27, 1930.

While some birds came each day to the mussel beds the large flocks sought the stretches of black mud which were not exposed till later. It was a beautiful sight when one hundred and fifty Curlew in a long wavering line came in on arched wings, settled on the far side of one of these flats and began walking about like so many chickens while I lay concealed in the skiff only a few yards away. They were much less conspicuous here than one would suppose for such big-bodied birds, their crouching attitude and uniform brown plumage concealing them admirably against the background of mud. The black and white Bullheads on the contrary, with their contrasting colors and upright carriage, stood out prominently as far over the flats as the eye could reach.

As Curlew come in to alight, or as they fly past the boat, they have a long-drawn, tremulous, interrogatory whistle, but when suddenly alarmed they utter a harsh double note *kek-kek*, *kek-kek*, as they take wing.

While other smaller shore birds associate with the Curlew in flight, when on migration or when on the grassy meadows, the latter do not seem to seek cummunion with any other species, keeping to themselves so far as possible. On several occasions I have seen them in rather intimate association with Laughing Gulls when resting on the meadows, but the gulls were nesting nearby and the association was doubtless accidental.

The Curlew is the characteristic shore bird of the meadows just as the Sanderling is the dominant species of the ocean beaches. But one unacquainted with the habits of the birds may miss them entirely as they feed or rest among the grass. We may gaze for miles over the stretches of green sedge and open sounds without a suspicion of their presence, when far off on the horizon there arises what seems to be a long wisp of smoke trailing away across the meadows. Now it is entirely free from the ground and forms a great V as it advances toward us and we realize that it is made up of scores, yes hundreds, of birds. On they come and now as they pass we can see the duck-like wing action, the heavy bodies, black against the sky or with a bright ruddy glow on their breasts as the late rays of the sun strike them. There are the long decurved bills and short thick necks which mark them as Curlew and perhaps we catch the querulous call trilling down to us as their line wavers and bulges forward, now at one point now at another, and passes on the long flight to their tundra breeding grounds far to the north, which is their real home though, as a matter of fact, they spend more of their life with us on the New Jersey marshes than on their nesting grounds.

* ESKIMO CURLEW

Phaeopus borealis (Forster)

The Eskimo Curlew, now regarded as extinct, was confused by Wilson with the Hudsonian and the bird described by him under this name was the latter species. Charles Bonaparte (1833), in his continuation of Wilson's "American Ornithology," says "we have occasionally met with it in the markets of New York and Philadelphia: in the Middle States, however, it is by no means common, having escaped the industrious Wilson." Audubon knew it, but as he makes no mention of New Jersey in his account, he was apparently unaware of its occurrence on our coast and it is Giraud to whom we are indebted for the first mention of it in this state. He writes, in 1844: "In New Jersey, New York, Massachusetts and Rhode Island, this species is seen every season. It frequents the open grounds in the vicinity of the seacoast, feeding on grasshoppers, insects, seeds, worms and berries. It arrives among us in the latter part of August, and remains until the first of November, when it assembles in flocks, and moves off to its winter quarters. I have shot a few stragglers in this vicinity as late as the twentieth of November. It occasionally associates with the Golden Plover. In the autumn it is generally in fine condition, and unlike the former two [Long-billed and Hudsonian Curlew], its flesh is well flavored."

Turnbull, in 1869, regards it as "rather rare, appearing in May and again in September" on the New Jersey coast.

Great flights of the Eskimo Curlew, came across from the northwest to Labrador, in Audubon's time, and continued down the Atlantic coast. As stated by Dr. Braislin (Birds of Long Island) the migrations probably were accomplished well off the coast and heavy easterly storms occasionally drove individuals or small flocks ashore, when they resorted to the dryer parts of the salt meadows to feed. "In earlier days," he states, "and then at very long intervals, 'flights' of many thousands have been seen on Long Island." They went north by way of the Mississippi Valley.

Of its last occurrence on Long Island Newbold T. Lawrence states: "During a period of about twelve years' Bay Snipe shooting at Far Rockaway and vicinity, I have only four records of this bird,—September 12, 1875, September 10, 1876, and two September 26, 1884." Ludlow Griscom gives a record of one shot at Good Ground, Long Island, on August 3, 1893. There have been several "sight records" since then and also corrections of earlier records of birds shot and which proved to have been Hudsonian Curlew.

Curiously enough we do not seem to have a single definite record of a specimen secured on the New Jersey coast.

The extermination of this bird was brought about by shooting for food on its spring flight up the Mississippi Valley and on its return along the New England coast as well as throughout its winter home in the Argentine where its associates the Golden Plover and Upland Plover suffer similar persecution today and may also disappear.

UPLAND PLOVER

Bartramia longicauda (Bechstein)

The Upland Plover is distinctly a bird of the grass lands, breeding in old fields and frequenting the hayfields after the haymaking when a second crop of grass and weeds is growing up. It is therefore anything but a maritime species and would seem out of place at Cape May were it not for the intimate mingling of upland and marsh which characterizes the region and forms one of the interesting features of the Cape.

It so happens that within sight of the sea there are old hayfields admirably adapted to the needs of the Upland Plover while the Fill, once a waste of dredged sand, for several years grew a thick covering of grass just to the liking of the bird, although this later gave way to a jungle of bayberry bushes. During August, and sometimes earlier, we used to hear the unmistakable call of this fine shore bird as it passed overhead on its southward flight and nearly every year a few of them stopped for a few days, at least, and

(424)

we flushed them from the grassland back along the railroad or on the Fill. Another favorite resort of the Upland Plover was the field back of the Lighthouse Pond which is relatively high ground, and the fields bordering the meadows east of Erma, which one reaches by way of the old wood roads which cut through the narrow strip of forest that runs parallel to the marsh. Here Walker Hand saw five as late as August 24, 1924, and three August 2, 1925. Of late years these birds have become so rare that we hear but few passing over and in the last six years only two have actually been seen.

In days gone by—some twenty-five years ago—Walker Hand tells me that they were common throughout July and August on all the fields about the town, even where the golf links now are, and were shot regularly by the gunners who approached them with an open wagon and were thus able to get within easy range. He considered, in 1903, that ninety-nine per cent of the former number had disappeared. Back in the eighties the Upland Plover was still more abundant and Charles S. Westcott ("Homo") writing in "Forest and Stream" says that "good Grass Plover shooting could be had near Swedesboro in southwestern New Jersey by using a wagon and leaving the fence rails down so as to drive from one field to another." Still earlier, as a boy [probably about 1860] he states that he shot hundreds in the fields at Long Branch, "between Howland's and Deal, in potato fields and pastures right up to the edge of the bluff which rose fifteen to twenty feet above the surf," but he never saw a single bird on the strand.

While there is no definite record of their nesting about the town, or even in Cape May County, they did nest in Salem County where W. B. Crispin found young birds and where on June 21, 1910, near Yorktown, I found a pair evidently with nest or young in the immediate vicinity. Julian Potter also saw them near Salem on April 28, 1918. In 1922 several individuals were found on the Fill as early as July 6 and were reported to me as present in June but we have no further information that would indicate nesting.

In 1920, the last year in which the birds were at all common, there were several seen on various dates from July 20 to September 4. I studied two of them on the Fill on several occasions. They were very wary and always managed to see me before I could locate them and I was thus never able to approach them when they were still at ease and unalarmed. This was due to the excellent concealment offered by the grass and to the general buffy brown coloration of the birds which is very characteristic and matches the scattered tufts of dry Indian grass so perfectly that, when the plover stands still, it is very difficult to distinguish it. By noting the spot where the flushed bird alighted, and creeping up under cover of the scattered bayberry bushes, I was several times able to see it on the ground but it seemed always to be aware of my presence and ready for instant flight.

The exceptionally long tail is a noticeable character of the Upland Plover as well as the slender neck and bullet-like head and the very short bill, but the most striking thing about it on the ground is its erect carriage. When it runs its neck is held vertically, stretched up at full length and the whole attitude is more like that of a quail than a shore bird. When standing still the head is moved back and forward suddenly, at intervals, an action analogous to the convulsive tilt of the Yellow-legs. When on the wing it seems larger than it really is and the spread of the wings is surprisingly broad. The body seems heavy in proportion and tapers away equally each way, the shortness of the bill and the length of the tail seeming to balance one another as it were.

The flight of the bird consists of long rapid wing beats which continue until it has attained a considerable altitude, in the gaining of which it describes quite a wide circle. The wings are then set in a decurved bow and it sails gracefully downward continuing on its circular course. Eventually it makes a direct and precipitous pitch to the ground, when still at a considerable height, and frequently lands not very far from the point of departure.

One individual that I had flushed flew only twenty feet but usually they make long circuits high in air and seem invariably to alight in grassy places, although with the increase in bushes on the Fill this became more and more difficult. Usually two or three birds were found together but often they were flushed singly. Quite often, too, they associated with Meadowlarks which bred commonly in the grassy spots that the plover frequented. On August 10, 1924, Julian Potter saw two feeding on a mud flat with other shore birds on the Lighthouse Pond, a rather unusual occurrence.

On August 6, 1925, I saw an Upland Plover running along the edge of a broad rainwater pool on the Fill and was struck with the yellowness of its plumage, somewhat recalling that of the Snipe but still lighter and when it spread its tail the buffy yellow on the sides was noticeable.

One bird (July 30, 1920) called once as it took wing *quit—it, quit—it* and another on August 15 called as it passed over head *quip-it-ip, quip-it-ip* while a third uttered a whistling call just once, as it flushed, *willa-willa-willa*, apparently part of the long drawn tremulous call of the species at the height of the nesting season, the same call that is heard from passing birds at night.

Our records for the arrival of the Upland Plover are:

1920. July 20.	1924. July 19.	1930. July 25.
1921. August 8.	1925. July 29.	1931. August 7.
1922. July 6.	1927. August 2.	1932. July 30.
1923. July 8.	1928. July 24.	

Our latest dates are: September 24, 1924; September 4, 1920; August 31, 1931. On the northward flight we have Walker Hand's records: April 24, 1911; April 12, 1915; but the bird does not seem to have been nearly so frequent at this season.

A very recent record is a family of two adults and a like number of young seen at Camden by Julian Potter on June 25, 1933. The Upland Plover is not so generally known to the baymen as are the more strictly maritime species. As in most other localities it is popularly called the "Field

Plover" and, incidentally, it is not a plover at all but a sandpiper allied to the Spotted Sandpiper. Like that species it has the habit of alighting on fences, stakes or any convenient perch, but this practice is probably limited to the nesting grounds as I have never seen one resort to it in the Cape May region. It is probable that the bullet-like head, short bill and general plover-like build have been responsible for its widespread misnomer.

On the drier parts of the Newark Meadows the Upland Plover was found breeding by Charles Urner in 1929, 1932 and 1933, and ninety per cent of his New Jersey records came from there. He has also heard it passing down the coast at night. His spring records are: two on May 14, 1929; five on May 5, 1932; three on May 14, 1933; one on May 16, 1934. On the southward flight he has found it much more frequent though he rates it only as "irregularly and locally tolerably common." It has certainly not reached the low ebb that it has attained at Cape May and one wonders by what route the Newark birds pass south; probably offshore. His records of extreme dates, time of regular or common occurrence, and maximum count on one day at a single locality, follow:

 1928. July 11–September 2 (August 8–22). Thirteen on August 22.
 1929. July 4–September 8 (July 28–August 22). Eight on July 28.
 1930. July 25–September 13 (August 1–17). Twenty-one on August 16.
 1931. July 19–August 29 (August 7–29). Seventeen on August 13.
 1932. July 13–September 8 (July 28–September 1). Twenty-five on Aug. 14.
 1933. (Bred)–September 9 (July 29–August 26). Thirty on August 15.
 1934. (Bred)–September 5 (July 29–August 30). Fifty-five on August 11.

This is one of our shore birds that no amount of legislation or restraint on the part of hunters in North America can save from extermination so long as it is ruthlessly killed and eaten in its winter home in the Argentine and it will soon go the way of the Eskimo Curlew which owes its extermination largely to the same cause.

SPOTTED SANDPIPER

Actitis macularia (Linnaeus)

With the exception of the Killdeer the Spotted Sandpiper is the most generally distributed breeding shore bird of Cape May—and there are but five. Though really a fresh water bird and always to be looked for on Lake Lily and the old Race Track Pond, it is equally common about pools and creeks on the edge of the salt meadows and even, in small numbers, along the larger thoroughfares out near the sounds.

On the sea beach it is very rare and its occurrence there is accidental due in some cases, at least, to its having followed a flock of Semipalmated Sandpipers out from some brackish pond on the meadows. I have seen Spotted Sandpipers on the beach between South Cape May and the Point where the shore begins to curve in to the Bay and when approached these beach feeding individuals have usually run or flown back to the marsh or have disappeared in the bayberry bushes behind the dunes. On August 8, 1920, one was busy feeding along the line of beach trash at the high water mark and on the 17th several were down on the lower beach near the water. On August 5, 1921, there were two in the same place associated with a Semipalmated Sandpiper, while on July 15 two were feeding there with a Killdeer and some roosting terns. On the last occasion the Spotted Sandpipers ran right out into the shallow water of the incoming waves but flew to the meadows as soon as they were flushed. Where buried sod banks are exposed along the beach these birds are very likely to resort to them for food. On August 19, 1920, I was surprised to find a Spotted Sandpiper on the true ocean beach at East Cape May but closer investigation showed that a drainpipe at this point emptied a stream of rain water from the Fill out onto the

(428)

upper beach and my sandpiper was walking up and down this little channel of fresh water never leaving it for the ocean. Another seen later in the same vicinity and still another on the Stone Harbor beach apparently had no such attraction. Only once or twice have I seen a Spotted Sandpiper fly out over the surf and in each case the bird made a short circuit and landed back on the meadows.

From all of our other sandpipers the Spotted may be distinguished by the constant up and down motion of the tail. This is quite different from the convulsive heave of the whole body so familiar in the Yellow-legs and certain other species, and involves the tail only which acts as if it were hinged at the base. It is wagged up and down in rapid succession followed by a pause and then the seesaw is repeated. This action occurs not only when the bird is walking but also at certain times when it is standing still and during the breeding season, at least, there is an occasional bob of the head, a movement analagous to the hiccough jerk of the Yellow-legs.

The action of the Spotted Sandpiper on the wing is equally distinctive. Instead of the continued and uniform rapid wingbeats of most of the shore birds, it has several very quick beats almost like a flutter and then a short sail on arched wings and again the flutter. The progress of the bird as one follows it along a stream or the border of a pond is a succession of loops as it flies out over the water and curves back to the shore a little farther on. The flicker of white on the wings in flight is conspicuous as is also the white edging of the outer tail feathers. As soon as the bird alights the tail action begins and even a soaking wet individual fresh from a bath in Lake Lily on August 26, 1920, was continually tilting its tail while busily engaged in preening its plumage.

So distinctive are the actions of the Spotted Sandpiper that we do not think much about its proportions or color characters in identifying it. It is somewhat larger than the Semipalmated Sandpiper, more olive above and the plumage slightly glossy. The adult in summer is marked with some irregular black transverse marks on the upper surface and with large round black spots below, covering the belly and flanks as well as the breast, and standing boldly out on a silky white background. There is also a diagonal white mark just in front of the bend of the wing. The tail is rather long for a shore bird and the bill longer than that of the Semipalmated and dull reddish brown in color with a black tip. The young birds and adults in winter are readily recognized by the lack of spots below, but their actions still proclaim them Spotted Sandpipers even with plain white unspotted breasts.

The downy young, mottled with brown and black, have been seen as late as July 22 in 1923, but usually they have developed the juvenal plumage by the middle or early part of the month and often by the latter part of June. The downy tail feathers frequently remain for some time firmly attached to the ends of the regular feathers which have pushed them out and one individual observed on July 27, 1923, had at least an inch of these appendages which were rendered all the more conspicuous by the constant tail action. On June 22, 1924, on a sand flat below South Cape May, I found an adult accompanied by a young bird which was nearly as large as its parent but clothed with down, with tips of the true feathers just beginning to appear. The tail seemed longer in proportion than in the adult and the wagging action was more exaggerated. It seemed to be continually in motion and was raised above the level of the back and flapped down so far that it often touched the ground or struck the back of the bird's legs. There were three or four tilts of the tail and then a convulsive backward jerk of the head and neck. This youngster was much lighter in color than the old bird and, compared with the grace of the latter, was distinctly ungainly. A second young one came out of the salt grass in response to the excited calls of the parent. It took several running steps and then a pause and between each set of steps there was tail action as just described.

When feeding the Spotted Sandpiper walks deliberately about with its neck stretched a little forward, the head moving slowly back and forth in unison with its steps. Every now and then the head is lowered and the bird makes a dab at something in the water or mud while the tail tilting goes on independent of gait or other actions. One excited bird, apparently with young, ran rapidly, tilting all the time, but the usual gait when feeding is a walk.

This sandpiper has a habit not usually shared by the truly maritime shore birds, as we know them at Cape May, of alighting upon unusual objects that happen to come in its way and will thus seek perches that other species avoid. Old partly submerged logs or pieces of lumber or boxes floating in pools are favorite perches, as well as rafts and boats, old boards cast up on the marsh, and even the top of a standing bill board. Out on the sounds at low tide they will perch on the oyster stakes and on July 26, 1924, I saw three at one time holding such positions back of Seven Mile Beach, while on another occasion one perched repeatedly on the telephone wire leading to the Coast Guard station. On August 18 and 21, 1924, and many subsequent occasions I saw them walking over the floating leaves of the water lilies on Lake Lily. Where the leaves were thickest the birds often ran rapidly over them and seemed hardly to depress them at all, but where they were somewhat scattered they leaped from one to another and sometimes gave a slight flap of the wings where the distance was greatest.

The usual flight—the flutter and sail, is never very long-sustained but during nesting and mating activities it is somewhat varied. A bird studied on the Fill, May 24, 1923, evidently with a nest nearby, ran rapidly and then took flight and glided slowly through the air on constantly quivering wings calling continually as long as the fluttering was maintained: *weet—weet— weet—weet*—etc. The same individual also took swift sailing flights, remarkably long-sustained and apparently more definitely directed than the fluttering flight, and during their duration it was silent. It is possible that this performance may have had to do with the courtship behavior. On May 22, 1922, two were observed which were evidently mating. One bird strutted about like a diminutive turkey cock with head and neck extended upward and raised itself as high as possible on its legs. Its tail was widespread and deflexed, wings loosely spread and held away from the sides of the body. The second bird paid not the slightest attention to the display so far as I could see.

Another individual seen on the Fill north of the town on July 30, 1920, was greatly excited by my presence and doubtless had young nearby although it was late in the season. There was no water anywhere about and the sand was covered by a growth of various kinds of grasses and weeds. No other adult appeared while I was there and a careful search failed to disclose either nest or young. The single adult flew about nervously, alighting on the tops of the dead mullein stalks and uttering a metallic call: *spink; spink; spink; spink;* occasionally varied by a double note, *spee-dink*, rising in intensity on the first note and falling on the second. This was similar to the usual call of the running or perching bird which I have written on various occasions: *peet-weet; peet-weet;* or *spee-peep; spee-peep*, other renderings are given below. The peculiar perching of this bird on the mullein stalks was not accidental as it resorted to them again and again and rarely alighted on the ground. Its legs were bent as it balanced itself on the tip of the stalk and its neck was stretched far forward and constantly moved back and forth in an effort to retain its equilibrium in the strong wind that was blowing. Other individuals observed on July 10, 1924, in the same locality, ran on the ground and fluttered through the air on decurved wings and occasionally perched on a low stake and a piece of timber lying on the ground. Their cry was the same metallic *pink; pink;* which seemed to increase in intensity as I approached into *spink; spink;* one bird gave the double call when on the wing; *see-peep; see-peep;* but on alighting relapsed into the single note again. These birds were feeding as they walked about, in and out of the low sedge, and deftly picked small insects from the leaves. One of them raised and fluttered its wings as it walked away from me, which may have been an attempt at the broken wing ruse. Another individual found in the same neighborhood on June 26, 1928, which had downy young running about among the vegetation, acted in exactly the same way as those described, uttering the double call *spee-pit, spee-pit,* and frequently alighted on the tops of the bayberry bushes. It sometimes ran very rapidly over the open sandy areas with neck stretched out in front, close to the ground, giving it a crouching appearance.

On August 18, 1925, several Spotted Sandpipers were feeding on the beach near Sewell's Point when two of them, an adult and a bird of the year, suddenly began fighting, breasting one another with heads held high, and jumping like game chickens for several seconds, until one turned tail and the other pursued it for a short distance. On July 11, 1931, one was seen to repeatedly drive off a Least Sandpiper which approached a feeding group of Spotteds, pursuing it for twenty feet across the sand and mud.

An individual observed bathing in a rainwater pool on the Fill on August 1, 1921, first sat down in the water and then stood up and ducked its head under, churning the wings and tilting the tail; finally it sat down again with the water half way up to its back. On May 22, 1922, I saw one swim across the thoroughfare back of Two Mile Beach and then climb up a three foot vertical mud bank which was exposed at low tide. It took a zigzag course using the irregular projections of mud for foot holds and disappeared into the grass of the meadows above. I cound see no reason whatever for its swimming. A second case of a bird taking to water was on September 5, 1921, when one seen crossing the Lighthouse Pond with its usual fluttering and sailing flight, was suddenly struck at by a Pigeon Hawk which appeared overhead. It promptly dived out of sight in the water and the hawk swooped up and away without its prey. The sandpiper bobbed up again almost instantly, like a cork, and remained resting on the water for nearly a minute before resuming its flight.

Spotted Sandpipers of the uplands, away from the coast, often nest quite a distance from water and Chreswell Hunt has found nests in strawberry patches, while others have been in gardens or cultivated fields. I have found young on the sandy stretches of the Fill where they were undoubtedly raised but the birds that live closest to the sea nest on the sand hills and flats immediately back of the dunes, where a scant growth of dune grass offers them sufficient cover. On June 22, 1924, I flushed a bird from a low sand hill a hundred yards from the beach at South Cape May. It flew but a few feet and then alighting walked slowly away in a crouching attitude. At the point from which it started I found a nest containing the usual four eggs. It was a mere hollow in the sand an inch and a half deep and was located under a tuft of beach grass, a few strands of the grass forming a sort of wreath around the rim. Three eggs had their pointed ends directed inward, the fourth being slightly disarranged. The bird uttered no call and flew off to some distance, eventually coming back around the other side of the sand hill. A second nest, a few yards beyond, was also on a low sand mound under a tuft of grass but contained only two eggs. The surrounding grass wreath was heavier and there were some bayberry leaves in the bottom of the hollow which was rather deeper than in the first nest. This bird called once as it took wing. Only a single bird was seen at each nest although I kept a sharp lookout for the mate, and in neither, apparently, had incubation begun. Chreswell Hunt tells me that the sitting bird may be approached quite closely and when it darts off the tail is spread and the feathers of the top of the head raised into a distinct crest. Richard Miller has found nests with full sets of four eggs on Seven Mile Beach on May 21, and June 22, 1922, which would indicate that in some instances, at least, two broods are raised in a season and similar nests seen by Turner McMullen at Brigantine, farther up the coast, on May 20, 1934, and June 14, 1931, seem to constitute further evidence in the same direction. Miller and McMullen have examined several nests at Corson's Inlet in the sand dunes not far from the beach all of which contained full sets of eggs as follows: June 13, 1925 (two); June 14,

1925; June 28, 1925; June 13, 1926 (two); and on Ludlam's Beach Fletcher Street found a parent and brood of four young only a day or so old, on June 23, 1928.

The Spotted Sandpiper is not gregarious as is the Semipalmated and most other small sandpipers. There may be six or even more feeding in proximity on some pond or marsh edge but they are scattered and act as individuals and not as a flock, taking wing usually one at a time and not simultaneously. While often close to other shore birds when feeding and perhaps attracted by a gathering of related species, they do not associate intimately with them and usually hold aloof taking their own course when they happen, through some sudden alarm, to take wing all together.

This sandpiper is most plentiful about Cape May during July and early August but by the middle of the latter month it often becomes scarce and on some days, at this season, I have hunted for them in vain. They seem to pass on southward in the wake of some favorable wind or change in the weather conditions, and their place is taken by other individuals from farther north. They become still scarcer by early September although individuals remain throughout that month and I have records for September 23, 1928; September 27, 1926; September 29 and 30, 1923. Unusual records were one seen on the edge of the Harbor on October 2, 1924, and another on Lake Lily on October 18, 1925. First arrival dates in spring are as follows:

1909.	May 2.	1922. April 23.	1927. April 30.
1917.	April 19.	1923. April 30.	1928. May 5.
1912.	April 14.	1924. May 9.	1929. May 10.
1921.	April 15.	1925. May 8.	1932. May 7.
		1926. May 1.	

Doubtless some of these dates are late due to the impossibility of my being regularly present during the spring migration.

Charles Urner's records for the region from Brigantine to Newark Meadows showing extreme dates, period of regular or common occurrence, and maximum count for one day at a single locality, are as follows:

1928. — to October 4 (July 15–September 15). Thirteen on July 8.

1929. April 21–September 15 (July 7–August 17). Six on May 26; thirteen on July 8.

1930. May 4–September 21 (July 6–August 20). Ten on May 18; twenty-seven on August 17.

1931. May 12–September 20 (July 5–August 30). Ten on May 17; forty on July 19.

1932. May 5–October 1 (July 10–August 27). Ten on May 25; twenty-five on July 16.

1933. May 7–October 1 (July 23–August 23). Ten on May 28; twenty-five on August 19.

1934. May 12–September 23 (July 8–August 19). Twenty on May 13; thirty on Aug. 18.

W. A. Babson has recorded one at Princeton on November 1 (Birds of Princeton, p. 44).

The usual call of the Spotted Sandpiper as it takes wing is *tweet-weet*, *tweet-weet-weet-weet-weet*, *tweet-weet-weet-weet-weet*, the shorter more metallic cries previously described being characteristic of the nesting season and more in the nature of alarm notes.

The Spotted Sandpiper is the only sandpiper that breeds about Cape May and is thus the only one that we can study, as it were, at home. It is confiding and dainty like all of its kind, with more characteristic actions, perhaps, than any of our other shore birds. We recognize it as far as we can see it by its tail tilting and its fluttering, sailing flight, and there is something cheering about its modest call *tweet-weet*, *tweet-weet*, which it utters as it goes from one feeding spot to another, and has been interpreted as *water-here*, *water-here*. And this rings true in the majority of cases no matter where the bird may be, whether on the sandy shores of Delaware Bay, the border of some shallow pond on the edge of the salt marshes, some narrow brook gurgling through a grassy meadow, or merely a mud puddle left by a recent squall. It is only at nesting time that it seeks drier situations.

E. L. P.

SPOTTED SANDPIPER IN AUTUMN AND WINTER

SOLITARY SANDPIPER

Tringa solitaria solitaria Wilson

The Solitary Sandpiper, although associated in our minds with inland localities and fresh water, occurs regularly in the vicinity of Cape May within sound of the ocean surf, and its distribution in this respect recalls that of the Spotted Sandpiper. Its favorite haunts here have been the old Race Track Pond, the Lighthouse Pond and Lake Lily at Cape May Point. The first of these, before it was drained, was always choked with aquatic vegetation—myriophyllums, utricularias etc., and into this the birds would wade until the water reached their bodies, turning their heads from side to side and deftly picking up minute insects or other forms of life from the leaves or the surface of the pond. At the Lighthouse Pond, a similar but much deeper body of water surrounded by dense growths of cattail and coarse sedges, the Solitaries choose the outlet of a drainage ditch where one may be found every season feeding in the running water. On Lake Lily they run along the sandy edges or walk about on the lily pads which in midsummer rest almost on the bottom, so low does the water become.

Farther back in the country the Solitary Sandpiper frequents open pasture ponds and woodland pools and everywhere, after a storm, they may be looked for on the temporary rainwater pools that are formed in low grassy spots. Occasionally, too, one may be found on shallow brackish ponds bordering the salt meadows. In the summer of 1920 I saw the first migrant individuals on August 13, running about a little pool of rainwater on a vacant lot on Windsor Ave. This was not more than four feet in diameter and was surrounded by a dense growth of tall weeds which completely screened the birds from the eyes of the many passers-by. In this retreat for several days there could be found from one to three Solitaries usually accompanied by a Northern Water-Thrush but as the pool gradually dried up the birds departed one by one, until September 3, when the last one was seen standing in the shrunken puddle now scarcely a foot across.

As its name implies the Solitary Sandpiper is more solitary than any of the other sandpipers, although two or three may often be seen together and on September 4 and 5, 1921, I saw five and four individuals respectively, on the Lighthouse Pond. Even so, however, they acted individually and not as a flock. Whether purposely or accidentally the Solitary Sandpiper often associates with individuals of other species or perhaps the initiative in feeding together may have come from the latter. On August 15, 1920, one was found feeding with two Least Sandpipers, a Killdeer and a Lesser Yellow-legs while on several occasions individuals of the last two species or a Spotted Sandpiper were found associated with Solitaries. In the large flock of Lesser Yellow-legs, Stilt Sandpipers and Pectorals which gathered on a piece of wet stubble just back of the Beach Drive on August 25, 1926, there were several Solitaries which had undoubtedly joined the assemblage individually and they remained as long as any of the others were left.

Though the general style of coloration of the Solitary Sandpiper is that of the Yellow-legs it is with the Spotted Sandpiper that we are accustomed to compare it both on account of the greater similarity in size and because these are the only two of our sandpipers which regularly frequent fresh water. The Solitary is distinctly larger than the Spotted, it is true, though nearer to it than to the Lesser Yellow-legs and its dark legs are also distinctly longer. The small whitish dots scattered regularly over its upper parts contrast with the plain or black marked back of the Spotted. It is also more dainty in all of its actions than this species and tilts with its whole body like a Yellow-legs instead of wagging the tail alone. At close quarters, however, there is sometimes discernible a slight undulation of the tail independant of the body tilt. When the Solitary takes wing it appears to be a much larger bird than when at rest and we are usually surprised at the wing spread, while the white, black-barred feathers at the sides of the tail are strikingly conspicuous. When seen side by side the contrast in the color of the back of the Stilt and Solitary Sandpipers is striking; the former very light and the latter very dark; with the Yellow-legs between the two.

Usually when feeding the head is drawn down on the shoulders although when wading in deeper water the neck is much more in evidence. The only case of conflict among them that I have noticed was in a group of three feeding on the Lighthouse Pond one of which attacked another and both flew off over the water with feet dangling and tails widespread.

The Solitary seems to be the least wary of our sandpipers and with caution we can approach to within a few feet of it before it will take wing. If we then remain perfectly still, or seek nearby shelter, the bird, if it has found a good feeding place, will very likely circle about and come to rest again at the very spot from which it was flushed. In any case it may confidently be looked for at the same stand in the course of half an hour. When on the ground the Solitary is silent but it has a double note that is almost always uttered as it takes wing—*peet-peet,* or *peent-peent,* as I have written it down on different occasions, while one individual with a very soft rendition seemed to say, *pip-pip.*

My arrival records for Cape May on the southward migration are:

1920.	August 13.	1924.	July 19.	1929.	August 5.
1921.	August 12.	1925.	August 18.	1930.	August 10.
1922,	July 15.	1926.	July 21.	1931.	August 13.
1923.	July 18.	1927.	July 19.	1932.	August 14.
				1937.	July 10.

The remarkably early bird of 1937, was feeding with a single Killdeer in a small boggy spot close to the turnpike leading to Cape May Point and remained for some days until the little patch of water in the middle had entirely dried up.

But it is probable that a few are present each year during the last two weeks of July. They stay regularly until the middle of September and a few linger for another month. Julian Potter saw from two to six daily from September 21 to 29, 1925, and we have records for October 3, 1920; October 8, 1932; October 10, 1927; and latest of all a single bird seen by Julian Potter on October 23, of the same year.

In spring we have no evidence of the occurrence of the Solitary immediately about the Cape but Richard Miller saw one at Ocean View on June 9, 1907, and Elliot Underdown another at the same place on May 20, 1928, while I saw one on a pond at Dennisville on May 12, 1929. It would seem to pass through rapidly on its northern flight or else its route is farther inland which is more likely.

The rather unexpected occurrence of this fine sandpiper about Cape May is another result of the mingling of upland and seaside environment so characteristic of the locality.

J. F. S.

FOOT PRINT RECORD ON THE SAND

EASTERN WILLET

Catoptrophorus semipalmatus semipalmatus (Gmelin)

This splendid big wader I know in the immediate vicinity of Cape May only during the migrations from the last days of July to the end of August or early September and occasionally on the return flight in May. They sometimes occur singly but more often in twos or threes and once or twice a flock of six has been seen. As the marshes about Cape May have been drained and the beaches more frequented by pedestrians most of our records of Willets have come from the great meadows back of Five Mile and Seven Mile Beaches where they were probably always more plentiful. Walker Hand has reported a flock of between forty and fifty there on August, 1907, and Julian Potter found twenty-six on August 20, 1916, while Joseph Tatum saw flocks of thirty-two and twenty-one on August 4, 1935.

My records of arrival on the southward flight are as follows:

1920. July 17.	1926. July 18.	1927. August 21.
1921. July 3.	1929. July 11.	1933. August 5.
1922. August 14.	1930. August 25.	1934. August 1.
1923. July 22.	1931. July 23.	
1924. July 14.	1932. August 20.	

William Baily saw a number on Five Mile Beach from July 23 to August 17, 1896. These dates show much variation and the irregularity of the bird's occurrence is beyond question, but my inability to be on the meadows every day may easily account for failure to record some early arrivals. I have only five September dates:

1917. September 29.   1918. September 6.   1928. September 2.
1898. September 8.   1923. September 2.

In 1936, throughout the month of August and in early September the Willet was of frequent occurrence on the beaches all the way from Cape May to Corson's Inlet, running about close to the bathers with no evidence of fear and attracting general attention from persons who had never seen it before. There was a very definite increase in the numbers of the bird.

On Brigantine Beach flocks of considerable size have been reported; thirty on September 19, 1928 (Potter); twenty-seven on September 2, 1927, (Urner); twenty-four on September 9, 1924 (McDonald).

Charles Urner has but three records for the Willet in spring on the northern coast district—one on May 25, 1931; two on May 20, 1934; one on May 7–21. His records of south-bound birds from the area from Brigantine to Newark Meadows, showing extreme dates, period of regular or common occurrence, and maximum count on one day at a single locality, are as follows:

1928. August 5–October 4 (September 2–15). Thirty-seven on September 2.
1929. July 16–September 29 (September 1–15). Thirty-six on September 15.
1930. July 6–September 7 (?). Three on September 7.
1931. August 1–October 4 (August 22–30). Five on August 22.
1932. July 10–September 13 (August 18–September 3). Thirty-two on September 3.
1933. July 30–September 17 (August 13–September 10). Thirty on August 15.
1934. July 15–October 14 (August 4–September 16). Thirty-five on August 25.

As a coastal transient the Willet falls in Urner's group of "irregularly and locally tolerably common" species. This rating, however, does not consider the breeding birds of southwestern New Jersey referred to beyond.

On the beach near the Lighthouse where Willet used to occur in small numbers every year, up to 1928, I studied one at leisure on August 11, 1921. As it stood on the sand I was struck with the length of its legs, for the Willet is one of our tallest waders, appearing nearly as tall as the Greater Yellow-legs but relatively more bulky with thicker legs and bill. Neither of course equals the Curlew in bulk, though the legs of the latter are relatively shorter than those of either the Yellow-legs or Willet, and its height therefore not much greater. Viewed from the side there is visible an angular bend in the Willet's neck such as is seen in the herons and the bill, while perfectly straight, seems as thick as that of a Clapper Rail. When alarmed and standing erect the Willet every now and then gives a spasmodic jerk, like a hiccough, raising the fore part of the body in the manner of a Yellow-legs but less vigorously. In color the Willet is unique. At first sight it appears uniform pale sandy gray with black bill and dark legs, conspicuous enough against the black mud of the marsh but nearly invisible against the beach sand which it almost exactly matches. A more careful examination with the glass shows that the under parts are paler, while the feathers of the upper parts are beautifully margined with narrow pale buff or dull white edgings.

At a greater distance it is merely a plain gray bird. But when it takes flight, what a transformation! The spread wings disclose hitherto concealed patches of black and white in brilliant contrast. A white band runs the whole breadth of the wing bordered deeply behind with black and with a black triangular area on the fore middle portion, the rump too is pure white. The obscure gray bird of the sand has now become the most conspicuous object on the beach. None of our shore birds but the Turnstone can compare in beauty with the Willet and a flock of them passing down the coast, with their splendid black and white wings flashing in the sunlight, against the background of blue waves and snow white surf, is a picture long to be remembered.

The Willet is extremely graceful on the beach, wading into the water until the waves reach its belly, and plunging the head under as it darts its bill into the sand after the crustacea that seem to form the greater part of its food. It walks and runs quickly and turns rapidly, wheeling about with the legs somewhat widely spread. Now it will pause and, leaning over, scratch its head with its foot, or run leisurely along in the face of an on-coming wave or directly toward it, glancing right and left as it goes and lunging viciously on either side with its long bill, leaning well over as it does so. It shows no apparent fear of the waves and turns and runs only just in time to escape them. Frequently the water rises above the bird's tarsi and sometimes up to the body although it is apparently never allowed to touch the plumage. The individual that I was watching was busy catching Hippa crabs and when it secured a big one it carried it out, gripped by one leg and hanging from the tip of the bill, and deposited it on the beach where it stabbed it with the bill in an effort to crush it and secure some especially desirable portion. When dropped on soft sand, however, the crab usually buried itself instantly according to its habit and the bird lost it. Small crabs I think were swallowed at once, as frequently the bill was give several vigorous lunges when well buried in the sand without withdrawing it. Another bird was seen to pick up a small fish about three inches in length and still another was observed to eject something from its mouth, apparently a pellet of indigestible material—crab shell and the like.

The association of this long-legged bird with his lesser brethren of the beach is interesting. One seen on September 29, 1917, was close to a group of Sanderlings and took wing with them, or rather just after them, and following their flight parallel to the beach he passed from the rear, with his stronger wingbeats, completely through the flock, until he had gained a comfortable lead and, coming to rest again, took up a position as before a little removed from the Sanderlings, as if to say "I am with you but not of you." On August 14, 1922, I saw precisely the same thing. Four Willet feeding near a flock of roosting terns invariably took wing just a few moments later then they did, and outflew them, returning each time to alight near them. These birds were preparing to rest and when they alighted held their wings aloft for a moment before closing them, displaying to great advantage the black and white pattern. Eventually they squatted down on the sand with their long legs bent under them.

Another Willet which Dr. A. K. Fisher and I studied on July 30, 1923, was feeding on the narrow beach bordering a small cove on the Harbor. It was very tame and allowed a close approach until one could see the pinkish gray color of the upper parts, with the very fine paler mottlings, the whitish breast, black eye, and olive tinge to the legs, as well as a narrow line of interrupted black and white on the edge of the closed wing, a mere suggestion of the contrasting color display concealed by the gray coverts. This bird would take two or three rapid steps, then one or two in a slow hesitating manner, and then another short dash. Now and then it darted its bill into the mud and water snatching up some morsel of food, while it often gave a lateral twist to the neck, throwing aside small scraps of seaweed etc., as if separating them from food that it held between its mandibles.

On the meadows Willets usually are found associating with the Curlew, generally only a few in a great flock of the latter, but their flash of black and white as they take wing makes them conspicuous among their dull-colored companions even at great distances. At other times I have found them with flocks of Yellow-legs or Black-bellied Plover.

On August 25, 1930, I found six Willets feeding about the shallow meadow ponds just below the Avalon Road. Their paleness at once attracted attention and made them much more conspicuous than they appear on the sands of the sea beach. They stood still most of the time and when they fed they walked slowly and bent the body over in a continuous curve from head to tail, with the bill pointing vertically down. When they took wing the double white band, the black dividing stripe, and black triangle at the bend (wrist) were strikingly conspicuous. As usual they held the wings aloft for a moment after alighting.

In my experience, during the past sixteen years, Willets have been much rarer in spring than in the southward flight in August. I found three on the exposed mud flats on Jarvis Sound on May 19, 1924, they seemed darker than the summer individuals and were hard to distinguish against the sandy areas upon which they stood. They seemed to be of about the same bulk as the Black-bellied Plover, with which they associated, and as they bent over did not stand any higher; they occasionally raised their wings without taking flight and thus attracted immediate attention. Another seen farther up the coast on the Barnegat Bay marshes opposite Tuckerton, on May 16, 1927, was feeding in three inches of water on a piece of sandy beach; it was beautifully barred on the flanks and was mottled above apparently not quite in full breeding plumage. When it raised its wings the black under wing-coverts were clearly in evidence. On June 24, 1928, I found a single Willet on the meadows back of Ludlam's Beach which looked very dark probably in full nuptial dress with dark gray and brownish upper parts although I could not get close enough for a clear view. We have additional May records: May 21, 1922, fifteen at Stone Harbor—one flock of six, the others in pairs or singly (Richard Miller); May 26, 1929, six on Jarvis Sound (Walker Hand); May 14, 1927, one on Seven Mile Beach; Five Mile Beach, May 21, 1898 (Baily); also some June records: June 25, 1922, June 17 and 23, 1923, Stone Harbor (Miller); June 24, 1928 (McMullen).

The identity of our New Jersey coast Willets in recent years is somewhat in doubt. It had been inferred that all Willet records from north of Virginia, since the eighties, were of the western race which shapes its return migration, at least in part, down the Atlantic coast, certainly such few specimens as we have examined belong to that form, the last example of the Eastern Willet being one (doubtless a breeding bird) taken by Dr. William L. Abbott back of Five Mile Beach on May 15, 1877. The discovery, in 1930, that Willets were breeding on the Bay shore at the head of the Cape May Peninsula complicated the matter not a little, and it is probable that some, at least, of the birds seen on the ocean side late in summer may have come from this colony. Those coming south past Barnegat Bay are however almost certainly the Western Willet.

As to its former breeding on the New Jersey coast Alexander Wilson says "This is one of the most noisy and noted birds that inhabit our salt marshes in summer. . . . It breeds in great numbers and has eggs May 20th." On the marshes bordering Barnegat Bay, at Long Beach, W. E. D. Scott states that in 1879 while "said to have been formerly one of the most abundant breeding species it is fast becoming rare by the inroads of gunners and egg hunters." C. E. Bellows found a nest near the Warner House on Delaware Bay, on May 19, 1884 (O. and O., 1892, p. 53) and C. S. Shick records two nests found on "Gull Island" back of Seven Mile Beach in 1889. Since then we have no nesting records for the ocean side of the state.

The situation on the Delaware Bay side of south Jersey is somewhat different; in the '80s and early '90s Willets were known to Walker Hand as breeders on the marshes back of Egg Island at the head of the Bay but no ornithologist had recently visited this rather isolated locality. Whether they became absolutely extinct as breeders in this neighborhood or merely decreased in numbers we cannot determine but an old resident stated, in 1931, that they had been there as long as he could remember. The region was difficult of access before the advent of the automobile and the swarms of green-headed and strawberry flies serve as a deterrent to any who would visit the spot. In the spring of 1929 David Leas reported seeing several Willets in the vicinity of Fortesque at the head of the Bay and John Emlen found four there on June 4, 1930. On June 29 of the same year Julian Potter and Edward Weyl visited the locality to find if possible a breeding pair. The former has furnished me with an account of their experiences:

"For fifty miles along the Delaware Bay shore of New Jersey there stretches a great marsh. Between Pierce's Point, Cape May County, and Alloway Creek, Salem County, only four roads cut through to the edge of the Bay and in southern Cumberland County the broad reaches sweep from Beaver Dam to Egg Island Point, a distance of eight miles. Rivers, creeks, channels, ponds and ditches divide and subdivide the whole expanse into many sections difficult to reach and explore either by boat or on foot. Near Beaver Dam a veritable network of ponds and channels aptly named "The Glades" presents to the explorer a mystic maze.

"Still largely undrained and natural this area remains a paradise for marsh-loving birds. Egg Island, where in former years water birds nested and were systematically robbed and finally exterminated, may be Egg Island in name only and Turkey Point may likewise be destitute of turkeys, but other forms of bird life remain. Along the edge of the marsh the "*zewick*" of the Henslow's Sparrow and the curious clicking trills of the Short-billed Marsh Wren can be heard. Out in the marsh the *chuck-wheeze* of numberless Seaside Sparrows mingle with the *clack-clack-clack* of the Clapper Rails. Night and Green Herons breed in the bordering thickets and feed in the shallow pools and in the depths of the Bear Swamp the Bald Eagle is said to nest—one was observed on March 9, 1930, and two immature birds on July 17. Ospreys' nests may be seen in the woods of the various necks and points—and on these marshes we found the Willets breeding.

"I saw the first Willet flying about the marsh as we neared Fortesque and getting out of our car we had scarcely entered the marsh when one began its *quip, quip, quip*, and presently advanced toward us sounding the alarm. It circled over our heads calling incessantly, sometimes uttering the characteristic *pill-will-willet* note. Another bird voiced its disapproval from the top of a signpost some distance away, a position which he held throughout our investigations. The actions of the birds showed clearly that eggs or young were nearby and due to the lateness of the date we suspected the latter and forthwith searched the short salt grass which was here covered with about an inch of water from the recent rains.

"Our efforts produced no result and we retired to the car to watch the birds with the glasses from a distance. They at once ceased their frantic cries and more subdued notes followed. Soon we saw a movement in the grass near one of the old birds and detected a white chick. The downy young of the Willet should not be white but nevertheless this one was white although in a few minutes two more appeared which were normal brown downy young Willets. We watched the group for some time. The old bird appeared to act as a guard only, while the young picked up their own living from the marsh. Once the adult caught a fiddler crab, crushed it and dropped it apparently for the young. Desiring to obtain a closer view of the young I once more entered the marsh my companion directing me from the car and at once the old bird took to the air and started the everlasting *quip, quip, quip*. The young squatted in their tracks, according to my companion, and guided by his suggestions I soon discovered the albino chick. It was creamy white with here and there a grayish wash; the eyes and soft parts normal, and the bill and legs slaty blue as in the adult. The other chicks could not be found so closely did their little brown bodies merge into the background of mud. The young Willet is slow and awkward on its feet compared with young Spotted Sandpipers or Killdeer and apparently depends more on freezing than on running to escape notice from intruders.

"We later located two more pairs of the birds apparently with young but the hordes of vicious flies literally drove us out of the marsh. On July 16, with Clifford Marburger, we found two young Willets just able to fly with down still noticeable about the head and with tails only partly grown. They took wing at the frantic urge of their parents and flew about fifty yards. They made no attempt to conceal themselves by freezing as did the downy young. We counted fourteen adult Willets on this trip, doubtless all local birds.

"A native of Beaver Dam, Phineas H. Lupton, informed us that he had first noticed Willets during the breeding season about five years ago and that they had become fairly common during the past two years."

On June 25 and 26, 1931, Charles Urner with several companions made a systematic search of these marshes for eleven miles from Dividing Creek to Back Creek, counted 223 Willets and found several nests. Turner McMullen examined six nests at Fortesque on June 6, 1931; four on May 21, 1932; two on May 29, 1932; and one near Turkey Point on May 21, 1932, all of which contained four eggs. I visited the Fortesque Marsh with Julian Potter in June, 1934, and saw birds evidently with young as well as an adult perched on a dead tree stub, a habit shared only by the Spotted Sandpiper and Upland Plover in this area. On June 19, 1935, Robert Allen and Roger Peterson made a trip there and saw an albino bird, doubtless Potter's chick grown to maturity, and found two nests.

The call of the bird which has been referred to by Julian Potter is as characteristic as its plumage and while the south-bound migrants that we see in August are not usually vociferous a single Willet that flew into the shallow pond at the Lighthouse on August 11, 1921, was in full voice *pill-will*, he cried, *pill-will*, *pill-will-will*, but never did he reach the full phrase of the breeding bird *pill-will-willet* the last syllables of which have given the bird its name.

To one who has seen the Willet on its nesting ground this call brings back the widespread marsh with the nest of dead grass and sedge resting in a hollow and the four large pyriform speckled eggs, or the mottled downy youngsters hiding in the green carpet while the parents in the air above, frantic with fear, flash their beautiful black and white wings and make the air ring with their cries of protest *pill-will-willet*.

While it is a matter of deep satisfaction to know that we still can claim the Willet as a breeding bird of New Jersey it is a pity that this picture once so characteristic of the Cape May meadows is denied to us of the present generation through the avarice and thoughtlessness of those who have gone before.

Julian Potter writing in the June, 1934, "Bird-Lore" says: "The Cumberland County Willet colony continues to flourish under the natural protection of undrained marshes and an abundance of biting flies. The birds are more active late in the day and at this time it is an easy matter to see from fifteen to twenty in a comparatively small area of marsh." And now, in 1936, when one would have supposed that the needs of wildlife had been made sufficiently patent to our Federal Government, we find C. C. C. camps established close to the Willets' last stand and hundreds of men, under the plea of finding work for the unemployed, ordered to ditch and drain all the Bay side meadows from Dennisville to Fortesque and unless some way is found to check the outrage, to destroy forever one of our last natural sanctuaries for wintering ducks, migrant shore birds, and nesting Willets!

* WESTERN WILLET

Catoptrophorus semipalmatus inornatus (Brewster)

Until it was ascertained that the Willet bred regularly on the marshes at the head of Delaware Bay it was supposed that the migrant birds seen every year along the ocean beaches and meadows were the Western Willet which had travelled in a southeasterly direction to the coast, although no specimens had been collected. Now, however, the presence of this race is somewhat open to doubt although skins from New England, Long Island and Virginia prove to be the western form, and were taken in mid-August.

It is impossible to distinguish the two races in life; this form is, however, somewhat larger with longer, more slender bill, and paler throughout in summer plumage. Wing 193–220 as against 180–195; exposed culmen, 58–65 as against 53–59 mm. (Ridgway).

It may well be that all of the ocean front birds really are Western Willets and that the breeding birds of Fortesque do not visit the ocean nor migrate farther north. The south-bound Willets that are seen to the north of Cape May County are almost certainly of the former race.

GREATER YELLOW-LEGS

Totanus melanoleucus (Gmelin)

The Greater Yellow-legs is to all appearances a larger counterpart of the Lesser, their relationship being similar to that between the Hairy and Downy Woodpeckers. Its haunts and habits are much the same as those of the Lesser Yellow-legs during the southward migration but unlike the latter it is equally abundant in spring. Immediately about Cape May it is not so plentiful as is the smaller species but, before the draining of the meadows between the town and Cape May Point, it was frequently to be seen resting or feeding on the shallow ponds. It is much the more wary of the two and takes wing at the slightest alarm. The calls of the two species are also quite different the Greater Yellow-legs uttering from one to four notes in succession instead of the couplet call of the Lesser. One bird that I watched on the Race Track Pond on May 25, 1922, gave a single call as it stood in the shallow water, which was repeated at half second intervals for some time before it took wing, and then, as it flew off, it called *pheu-pheu-pheu-pheu, pheu-pheu, pheu-pheu-pheu, pheu-pheu-pheu*, which was a fair example of the vocal performance of the species.

Two Greater Yellow-legs were observed on the Race Track Pond on September 1, 1920, one in perfectly fresh winter plumage beautifully mottled above and pure white below, the other with remains of the old breeding plumage, showing scattered worn brownish feathers above, and traces of the black streaks below, a difference often seen in flocks of this species which, like many other shore birds molts while in migration. The former bird stood with it head well down on its shoulders raising it now and then to perform the convulsive "tilt" so characteristic of both the Yellow-legs. The other individual kept its head elevated all the time but "tilted" with the same convulsive effort. Both gave a triple call before taking wing. On the Pond Creek Meadows on August 2 of the same year three Greater Yellow-legs were associated with the Lesser Yellow-legs, Pectoral, Least and Semi-palmated Sandpipers. Two of them took wing as soon as they saw me and, while the other smaller birds went on feeding, the remaining Greater Yellow-legs stood stock still with head erect performing his hiccough-like "tilt" with monotonous regularity and never for a moment relaxing his vigilance until the flock took wing together. On July 30, 1921, a flock of seventeen was seen on a shallow pond east of the Lighthouse some of them having waded into the water nearly up to their bodies. They flew as soon as I appeared with triple cries—*pheu-pheu-pheu*, and a great display of white rumps and yellow legs. On September 14, following, a single bird with several Lesser Yellow-legs was feeding at the inlet of a small ditch on the Lighthouse Pond. When I approached he took wing but circled about as if loath to leave and, as I remained motionless, he finally came to rest again within twenty feet of me. He stood in water nearly up to his body but waded out where it was somewhat shallower and began to rush madly about in pursuit of small fishes which were coming in with the current. I saw him catch one and swallow it but it may be that there were other more minute forms of life which were the main object of his chase. He held his neck sloping downward at an angle, with head and bill horizontal. His "tilt," when he performed it, was much more convulsive than that of the smaller species which was close at hand for comparison.

The flight of the Greater Yellow-legs consists of easy rapid wingbeats as they pass overhead on a protracted journey but if they fly low and contemplate alighting the wings are set with the tips arching downward, and they sail for some distance. With alternate wing beating and sailing they either come to rest or make off again for some distant point. Several times small flocks have circled the Race Track Pond in this manner only to go off to some other feeding ground. Doubtless they had caught sight of me in spite of my efforts to remain motionless and had suspected possible danger. One flying over had answered the call of an individual already feeding on the pond and in doing so raised his head and neck above the level of his back and deflecting his wings sailed through the air at a surprising speed.

The legs of the Greater Yellow-legs seem to be of a deeper more orange tint while those of the Lesser are paler and more lemon yellow. As they are stretched out behind the flying bird they are quite conspicuous and the white rump, especially as they fly away from one, is also prominent. An individual feeding on the Lighthouse Pond on May 9, 1924, took a few rapid steps immediately upon alighting and then turned completely around and ran a few steps back again. After several repetitions of these rapid dashes he began to probe in the usual way and I wondered if the preliminary maneuvers might not be intended to stir up possible prey. A second bird which accompanied him almost turned a somersault upon alighting, suddenly throwing one wing up in the air and the other downward until its tip almost touched the water, with orange legs widespread. Next day four of the birds were feeding at the same spot one of which, as soon as he saw me, broke into a peculiar cackling, the repetition of a single note, somewhat like the cackle of a Mud Hen or the spring call of the Flicker.

On May 8, 1925, another individual gave the repeated call as I approached a little salt pond on the meadows just west of the town. It was kept up for a couple of minutes without interruption and strongly resembled the Flicker call but was not so rapid, running three notes to the second while that of the Flicker is four. This bird walked as he called and his head moved slightly back and forward with each step, while I could distinctly see his mandibles open and shut with each call. When he made a short flight his legs dangled loosely below and he plumped down into the water nearly up to his body when he again came to rest. Flushing once more he uttered a loud *keu-keu-keu-keu; keu-keu;* and then lapsed into a more liquid, mellow, note *toor-loo; toor-loo; toor-loo;* etc. As he again alighted he reverted gradually to the original call and finally to the prolonged cackle viz: *toor-loo; toor-loo; loo; loo; loo; loo; keu-keu-keu-keu-keu-keu* etc. These elaborate callings I have not heard during the fall migration.

On May 1, 1926, there were ten Greater Yellow-legs at their favorite resort on the shallow pond at the Lighthouse, feeding and bathing in water almost up to their bellies through which they ran rapidly, wheeling and turning and occasionally flapping their wings. Some of them held the head high up and the "heel" of the foot kicked out of the water behind as they ran; others stood stock still with head drawn close down on the shoulders, making a strikingly different appearance and looking at first sight like some different kind of bird.

On the great meadows north of the Harbor the Greater Yellow-legs are abundant, showing a distinct increase since the shooting of these and other shore birds was abolished. There they resort to the mud flats and mussel beds at low tide to feed with the hordes of other waders. They always seem to form the outermost line of such assemblages, beyond the Dowitchers which in turn feed farther out than the plover and little sandpipers. So deep is the water where the Yellow-legs feed that they frequently have to submerge their heads to secure their prey, but always in quick lunges with a prompt withdrawal. When watched from a distance, where the presence of the observer has no effect upon their actions, they are still the first of the assemblage to take wing, being simply washed off their feet by the rising tide. Some of them are always loath to leave and one bird "drowned out" on the bar mounted a lump of black mud and held that position for another half hour.

When they finally leave the bars or flats they resort to the shallow ponds on the grassy meadows and there await the next turning of the tide. If food be present they may also feed there but usually they stand still for hours at a time; some of them hunched down with apparently no neck at all, head down on shoulders and slender bill directed forward; others with neck erect and head entirely clear of the grass and sedge which conceals the body—a totally different pose.

The slenderness of the Yellow-legs both in flight and at rest is characteristic. In flight they appear uniform gray except for the white rump and tail unless one is near enough to distinguish the white of the underparts. The wings are narrow and pointed and as the bird comes to rest they are set in a bow, curving backward from the tips. In this position the bird takes a long sail, now and then undulating slightly like a sailing Barn Swallow. The slender bill and legs add to the contrast between the Yellow-legs and the heavier built Black-breasted Plovers and Turnstones which are his close associates on the meadows.

Sometimes we see migrant Yellow-legs winging their way south with the Curlew or in flocks of their own passing high over the town. One flock of twenty-eight which on September 4, 1931, had been flushed by a passing Marsh Hawk from the shallow pond at the Lighthouse was seen later very high up passing over our cottage and looking like swallows against a white cloud. They seem to associate in larger flocks as the season advances.

Walker Hand's arrival records together with my own covering years between 1902 and 1930 follow but it is possible that some of the early dates refer to birds that may have wintered in the neighborhood:

1902.	April 2.	1913.	March 28.	1921.	March 20.
1903.	March 2.	1914.	March 18.	1922.	March 28.
1904.	March 18.	1915.	April 15.	1923.	April 8.
1905.	March 19.	1916.	March 21.	1925.	April 8.
1907.	March 14.	1917.	March 16.	1926.	April 11.
1909.	March 21.	1918.	March 19.	1928.	April 8.
1910.	March 20.	1920.	April 5.	1929.	March 29.
1911.	March 14.	1921.	March 20.	1930.	April 1.

The main flight seems to be in late April and May and, although the bulk has departed by May 21, some linger to the end of the month and we have several records for May 30, while Richard Miller saw one bird on June 8, 1919, and Julian Potter one on June 19, 1929, and another on June 21, 1923. After leaving us in May or early June it is but a short time before we see these fine shore birds again about Cape May coming south on their return flight from their breeding places in the far north. We have dates of arrival as follows:

1920. July 23.	1925. August 1.	1932. July 24.
1921. July 17.	1926. July 18.	1933. July 17.
1921. July 17.	1926. July 18.	1933. July 17.
1922. July 25.	1927. July 30.	1934. August 1.
1923. July 21.	1928. July 25.	1936. July 16.
1924. July 22.	1931. July 11.	

They become more plentiful in August and September, decrease during October, and linger on into November, our latest records being November 7, 1877, when one was shot by Dr. W. L. Abbott; November 5, 1913; November 15, 1928; November 8, 1930; December 27, 1931; November 13, 1932. In other years the latest dates of observation average October 25.

Charles Urner's records for the region between Brigantine and Newark Meadows showing the extreme dates, period of regular or common occurrence, and maximum count for one day at a single locality, are as follows:

Northward migration:

1929. April 13–? (May 11–17). Three hundred and fifty on May 11.
1930. April 13–June 22 (May 11–31). Two hundred on May 11.
1931. April 15–May 31 (April 24–May 26). Seventy-five on May 12.
1932. March 25–May 25 (April 15–May 25). One hundred on May 7.
1933. March 31–June 24 (April 23–June 4). Seventy-nine on April 29.
1934. March 25–June 18 (April 8–May 16). One hundred on May 5.

Southward migration:

1928. July 12–December 6 (July 22–October 12). 175 on August 29.
1929. July 4–November 28 (July 14–October 12). 100 on September 18.
1930. July 19–October 12 (August 1–September 6). 93 on September 6.
1931. July 3–December 27 (July 12–September 20). 85 on August 16.
1932. July 14–November 12 (July 16–September 24). 75 on August 13.
1933. June 27–November 30 (July 27–October 22). 150 on September 2.
1934. July 2–December 2 (August 4–October 20). 60 on September 29.

He places the Greater Yellow-legs in the group of ten "common" species as opposed to the six that are regarded as "abundant or very common" to which the Lesser Yellow-legs belongs (see p. 358).

The Greater and Lesser Yellow-legs constituted for many years the most important shore birds from the gunner's viewpoint. While they distinguished the two species, they did not trouble to do so in such records as remain of the early days, so that it is difficult to form estimates of their relative abundance for comparison with more recent years. Both species were shot in numbers and they were fast decreasing when saved by the Migratory Bird Treaty.

The Greater Yellow-legs are the most vociferous of our shore birds with the single exception of the Killdeer. To all other shore birds the note of the Yellow-legs is recognized as the alarm call and earned for it, years ago, the nickname of the "Tell-tale" from the disgruntled gunner who has approached a promising flock of lesser game only to have them suddenly depart at the sound of the Yellow-legs' warning.

LESSER YELLOW-LEGS

Totanus flavipes (Gmelin)

One of the best known of the Cape May shore birds is the Lesser or Summer Yellow-legs. It is distinctly a bird of the shallow ponds of the salt or brackish marshes though it sometimes frequents fresh ponds near the beach or just back of the meadows and after storms may occur in considerable numbers on grassy areas of the great Fill, north of the town, where temporary shallow rain water pools have formed. On the ocean strand it is never seen. The individual groups of Lesser Yellow-legs usually consist of from two to twelve birds, though sometimes flocks of twenty or even forty may be found and the greatest number that I have seen in close association was seventy. On the meadows, where large numbers of various species of shore birds congregate at high tide, many individual groups may for the time be merged in the assemblage and the number of Yellow-legs present may be even greater.

While very scarce on the Atlantic coast during spring the Lesser Yellow-legs is to be found about Cape May from early July to mid-September in more or less abundance and on favorable evenings bunches are seen passing overhead on their southward migration, usually from six to twenty together while during July it is not unusual to find them flying with the south-bound Curlew. In spring they go north by way of the Mississippi Valley.

When feeding they are exceedingly graceful birds, walking daintily about often in water up to their bodies. They never lunge the bill several times from one stand, as do some of the waders, but keep always on the move and seem to pick up their food rather than probe for it in the mud or sand. When they come to rest the feet are placed side by side while the head is held erect but with the neck and body seeming to lean a little forward. Then every few seconds there is a convulsive "tilt" of the whole body, the head and breast going up and the tail down with an immediate recovery of the former position, the whole action being a sort of seesaw with the hips as a fulcrum. The impression is of a severe hiccough or as if the bird had swallowed some object which it was trying to cast up again. This tilting action, however, really seems to be one of nervous alarm or indecision. When I have come suddenly upon several of the birds feeding within ten or fifteen feet of me and all of us have stopped simultaneously, they at once assumed the tilting posture, evidently in doubt as to their next move, and I have seen them suddenly assume it on the approach of another intruder, when I, myself, was completely concealed. On August 4, 1920, I saw a bird stand on one leg and repeatedly make the spasmodic tilt as easily as if both feet were on the ground. On rare occasions, too, I have noticed the birds perform the tilt at intervals in their feeding, while walking about in a shallow pond.

On September 13, 1921, I approached to within ten feet of a Lesser Yellow-legs feeding on the Lighthouse Pond. His neck was always extended and sloping toward the ground and every now and then he wiped his bill on the surface of the mud as if to cleanse it of something, first on one side then on the other. He was accompanied by a Solitary Sandpiper and it was interesting to compare the two. The tilt of the Yellow-legs was much more convulsive than the modest dip of the sandpiper, while in flight his rump showed pure white strikingly different from the barred pattern of the Solitary. The next day I found three of the Lesser accompanied by a Greater Yellow-legs feeding where a small ditch enters the Lighthouse Pond. They ran rapidly through the water in pursuit of small fishes but were apparently not so expert in catching them as was the larger bird. Now and then they would hop up and down in one spot as if to avoid sinking in the soft mud.

I have several times watched them preening their plumage while they stood at rest after feeding. All of the feathers were ruffled up, the head held high and the neck bent over at the throat so that the bill could be poked vertically downward among the feathers of the breast. It was then thrust back over the shoulder while the plumage of the back and rump received attention, with a further ruffling of the feathers, and finally the wings were elevated slightly and the bill and head poked beneath them to reach the under coverts and sides of the body. Another individual dipped the point of its bill into the shallow water in which it stood, shook it, and then proceeded preening as above described. It also leaned the head over sideways and, raising one foot leisurely, scratched the top of its head. This was repeated several times. Another bird, engaged in bathing, squatted deliberately in the water until it reached the middle line of the body, fluffed up all its feathers and worked its wings from the shoulders up and down, apparently alternately. It churned up the water, while at the same time the head was several times ducked completely under and suddenly withdrawn, causing a shower of spray.

When alighting in shallow water or chasing one another, as they frequently do, when a passing bunch joins a flock that is busy feeding, the wings are elevated over the back and the birds patter along the surface of the water making a fine display, and bringing the barred axillaries prominently into view while the yellow legs are bent sharply as they are raised in making each forward step.

The call of the Lesser Yellow-legs, as it takes wing, is a double whistle *pheu-pheu, pheu-pheu.* Sometimes it may be a single note but never the more numerous syllables uttered by the larger species. When I was very close to a small flock of birds on July 26, 1920, their calls sounded like *séa-pink, séa-pink,* possibly the closeness accounted for the apparent metallic quality of the second syllable, or perhaps it was another note with a different significance, as the birds were on the ground when they called. Other individuals feeding on the shallow pond at the Lighthouse during July, 1923, were continually calling as they ran about feeding, their notes sounding like the syllables *kip-kip, kip-kip,* and combined with the constant chatter of the Least and Semipalmated Sandpipers, which were there in abundance, made quite a medley of shore bird music.

Besides associating with the little "peeps" and with the Solitary Sandpiper the Lesser Yellow-legs seeks the company of various other shore birds when feeding, notably the Greater Yellow-legs, Dowitcher and the Pectoral and Stilt Sandpipers. On one occasion several passing birds decoyed to a couple of Laughing Gulls resting on the shallow pond by the Lighthouse and again twenty came to rest with a large flock of Common Terns on a sand flat below South Cape May, while after a heavy rain storm on July 25, 1925, I found, on a flooded piece of grass stubble on the Fill, six Lesser Yellow-legs, a number of Pectorals, several Meadowlarks and about one hundred Red-winged Blackbirds—surely a strange assemblage. The largest flock of Lesser Yellow-legs that I ever saw immediately about Cape May gathered on a small piece of rain-flooded meadow east of Broadway and within fifty yards of the Beach Drive but screened from it by a dense growth of bayberry bushes. The flock was first noticed on August 23, 1926, and resorted there until the 28th, although the Yellow-legs decreased in that time from seventy to twelve, associated with them there were a few Stilt Sandpipers, some Solitaries and Pectorals and a number of the two little sandpipers called "peeps." They fed all around in the grass which almost concealed them from view and every now and then would flush and return again to their feeding. On one occasion when I visited the spot they were absent but returned later in the day in reduced numbers.

As they take wing the Lesser Yellow-legs are conspicuous from their pure white rump and tail and their long yellow legs trailing behind, while the rather long needle-like bill is characteristic. On the ground the distinctly dark slate or blackish tone of the back is in strong contrast to the warm brown tones of the little sandpipers and the adult Dowitchers. Old birds, on their arrival from the north, have coarse black streaks on the under side of the body, while the upper parts are mottled with old and new feathers the latter much paler gray. The young of the year are easily distinguished by the smoothness of their fresh new plumage and the suffusion of gray on the breast. Old birds, too, have heavier barring on the sides of the tail. In a good light, at close quarters, one can distinguish the small white dots on the feathers of the back of the young birds which shine like gold in the sun.

Compared with the Greater Yellow-legs there is the difference in size but the Lesser has a much more slender bill and legs, while the color of the upper parts is paler and the yellow of the legs not so orange—at least at certain seasons and in certain individuals.

Of spring records we have but few: Walker Hand saw them twice at this season, on April 29, 1914, and May 3, 1915, while I saw two on the Light-house Pond on April 30, 1932, in company with a Greater.

On the southward flight my records of first arrival are:

1916.	July 8.	1924.	July 12.	1931.	June 28.
1917.	July 14.	1925.	July 3.	1932.	July 9.
1920.	July 13.	1927.	July 4.	1933.	July 7.
1922.	July 7.	1928.	July 2.	1934.	July 13.
1923.	July 5.	1929.	July 11.	1935.	July 13.
		1930.	July 12.		

The birds remain regularly to September 25 and 30 in most years and on September 10, 1913, there was a big flight and many birds were shot. I have also seen Lesser Yellow-legs on October 10, 1925; October 7, 1928; October 12, 1930; October 18, 1931; October 8, 1932; while Dr. W. L. Abbott shot one on October 11, 1880, and William Rusling saw several on October 30, 1935. Doubtless a few remain every year to the middle of the month and we have a few later records viz: one on Ludlam's Beach, November 13, 1927 (Potter) and three at Cape May on December 22, 1935, on the Christmas census of the Delaware Valley Club.

On the northern part of the shore Charles Urner has been more fortunate in finding Lesser Yellow-legs in spring and has scattered records for five years:

1929.	May 5–19. Two on May 5.	1933.	May 14. One.
1930.	May 4–18. Four on May 14.	1934.	May 5–16. Tewnty-four
1932.	May 1–15. Ten on May 7.		on May 12.

On the southward flight his records for the area between Brigantine and Newark Meadows showing extreme dates, period of regular or common occurrence, and maximum count for one day at a single locality, are as follows:

1928. June 29–October 14 (July 11–September 23). 450 on August 1.
1929. June 30–October 12 (July 10–October 6). 600 on September 18.
1930. July 6–October 5 (July 16–October 5). 350 on August 16.
1931. July 3–September 20 (July 9–September 20). 175 on July 30.
1932. June 26–November 12 (July 3–October 12). 800 on July 28.
1933. June 27–October 29 (July 4–October 7). 700 on July 19.
1934. June 30–November 4 (July 7–October 12). 400 on August 30.

The Lesser Yellow-legs is considered one of the six "abundant shore birds" today as a result of Urner's careful comparative studies (see p. 385).

As an index of their abundance at Cape May the late Henry Hazelhurst's shooting record gives the following information and, although he did not separate the two species in his notes, he killed but few of the larger birds during the time he shot which did not extend beyond the first few days of September.

    1898, shooting on twenty-one days, he killed........169 birds
    1899, shooting on thirteen days, he killed............108 birds
    1900, shooting on nineteen days, he killed............140 birds
    1901, shooting on seventeen days, he killed..........154 birds
    1902, shooting on twenty-six days, he killed.........343 birds
    1903, shooting on twenty-eight days, he killed.......321 birds
    1904, shooting on twenty-nine days, he killed.......320 birds
    1905, shooting on twenty-seven days, he killed......337 birds
    1906, shooting on twenty-five days, he killed........450 birds
    1907, shooting on thirteen days, he killed............202 birds
    1908, shooting on twenty days, he killed.............233 birds
    1909, shooting on twenty-five days, he killed........323 birds
    1910, shooting on twenty-three days, he killed.......276 birds

Total............3376

An average of twelve for 286 shooting days—and he was but one of a large number of shore bird gunners and covered only the meadows close to the town.

The filling in of Cape Island Sound and the draining of most of the marsh land south of the Harbor, in recent years, greatly affected the abundance of these birds immediately about Cape May, and Walker Hand was of the opinion that not more than 250–500 per year were killed here during the last years that Yellow-legs shooting was legal.

Let us hope that the change in the law has saved this attractive member of the Cape May avifauna for there is no more beautiful sight than a group of these dainty, active birds feeding in one of the shallow ponds that dot the broad expanse of green meadows, and their cheerful double whistle is the dominant bird voice of the coastal marshes during the period of their southward migration.

KNOT

Calidris canutus rufus (Wilson)

Before the shooting of shore birds was abolished the Knot, or as the baymen know it the "Robin-breasted Snipe" or "Robin Snipe," was, along with the Dowitcher, the most desirable species from the gunners' standpoint and as they both decoyed easily they most nearly approached extermination. The Robin Snipe is, and apparently always was, primarily a beach bird, frequenting especially the entrances to the inlets although it occurs in smaller numbers out on the meadow ponds and on the exposed flats of the sounds at low tide. It is not surprising, therefore, that Henry Hazelhurst's shooting record at Cape May for July and August, 1898–1910, which we have quoted as an index of relative abundance of the various species, shows only eleven Knots while in eight of these years he got none at all. His shooting was done on the meadows and consequently he did not encounter the passing flocks of these birds on the beaches and his figures possibly do not reflect the real abundance of the species. I am inclined to think that, while its numbers were sadly depleted, the Knot never reached such a low ebb on our coast as some have supposed, and since the abolishing of the shooting of shore birds it has steadily increased in abundance.

While not generally distributed along our Cape May beaches, and avoiding those most frequented by summer visitors, a flock of Robin Snipe may almost always be found on the lower end of Seven Mile Beach near to the entrance to Hereford Inlet during the month of August; my earliest dates being August 5, 1926; August 10, 1930; August 4, 1932; but as I have been unable to visit this spot regularly it is safe to assume that the birds were present at least a few days earlier. William Baily has recorded a flock of ten at this spot on July 26, 1899, and Julian Potter saw them there on July 24, 1927. Farther north on the coast Potter found flocks aggregating 525 at Brigantine on July 29, 1931, and 750 on August 1, 1932, while Clifford Marburger saw 150 on September 21, 1925.

On June 4 and 5, 1925, just at dusk, I saw a flock of seventy-one Robin Snipe on the beach below Corson's Inlet and two other flocks far down the strand to the south. The birds near to me looked very large and dark in the gathering gloom while the massing of the flock was very striking, recalling a company of soldiers marching in close ranks, and they never scattered promiscuously like other shore birds that were present. When they flew the uniform coloration was very manifest with the lack of any wing bars or other conspicuous character.

Robin Snipe on the mud flats of Jarvis Sound in May had the same feeding habits as those of the beach. Here they associated with the Dow-itchers but never probed into the mud as they did, simply picking up morsels of food from the surface of the flat, while their short bills contrasted strongly with the very long Woodcock-like bills of the Dowitchers.

The dark color of Robin Snipe on the beach is always noticeable due apparently to the ruddy color of the under surface which combines with the natural shadow to produce this effect. On the black mud however they exhibit their natural colors and appear as very light birds especially in contrast to the Dowitchers. As we lay in our skiff, one day, stranded on the flat with shore birds gathering to feed close to us on all sides, we had an excellent opportunity to compare the several species. As two Robin Snipe flew in to join the assemblage they seemed very light, the upper surface being pale pinkish gray finely speckled with white; this contrasted strongly with the rich ruddy color of the lower parts which extended over the face and cheeks much as the black of the breast of the Black-bellied Plover encroaches on the white of its head. The effect is such that when they face us they appear entirely rusty red, while turning sideways again, they fade suddenly to pale gray birds with reddish breasts and bellies, a curious transformation. The flight feathers were dull brownish and the wings showed no white bands or stripes either in flight or at rest; the rump was pale gray appearing almost white in bright sunlight, though not so conspicuous as in the Dowitcher because of lack of contrast. Two other individuals a little farther off seemed nearly uniform grayish brown and were doubtless females with their paler under parts.

As one looks down the long stretch of wild strand and dunes covering some two miles of the southern tip of Seven Mile Beach, Knots are usually present in August and can be distinguished from the multitude of small Sandpipers, Sanderling and Ring-neck Plover, which gather there, by their large size, their dark color and especially by their massed formation. They are in worn breeding plumage when they arrive and during most of their sojourn on our coasts, and this, with their ruddy breasts, make them appear dark in the strong light of the beach as compared with the pale plumage and white bellies of the Sanderlings. As we approach closer to them we can see that they are gathered close together, shoulder to shoulder as it were, and that the whole mass of birds moves forward and back as a unit, slowly walking or trotting all together and just fast enough to avoid the on-coming wavelets; very different from the quicker movements of the Sanderlings, where every individual is acting for himself, and although the latter do advance and retreat in unison it is in a long line with plenty of room for individual action. When feeding undisturbed the Robin Snipe will spread out a little but the crowding is always evident and when they are forced to form a single line they still stand close to one another. Their manner of feeding is more like that of a plover than of the Sanderlings, which are their usual associates, as they do not probe into the sand but peck at objects on the surface, the action being straight in front of them not to right and left as in many beach birds. They feed, too, on the wet beach but do not venture to step into the water, and keep farther back from the waves than do the Sanderlings. When the mass moves it seems to stream, or flow, along the beach and we lose the individual action. When approached too closely they move ahead of the intruder, first at a walk then a trot and finally a run, still holding their close formation; then they take wing and in a long stream fly out over the surf to alight farther up the strand, displaying their dull rather pale gray rumps.

On August 8, 1926, Alexander Wetmore and I found a flock of one hundred and eight Robin Snipe at this same locality on lower Seven Mile Beach. They were gathered in a long line, the individuals very close together, their heads all directed outward toward the water and downward, while their tails all pointed diagonally upward. They were actively pecking at the edge of the narrow strand left by the tide where it cuts away the sand as it enters the inlet. A few birds were feeding immediately behind the main line failing to find an opening in the front rank but crowding in as closely as possible. Now and then the entire flock, for no apparent reason, would turn from the water and advance along the beach in solid ranks shoulder to shoulder, the huddled group reminding one of the crowding of a flock of sheep. All of these birds were ruddy-breasted, so far as I could determine, so that the Robin Snipe, like the Sanderling, and other shore birds, must leave its breeding grounds before the molt begins and assume the winter plumage during its protracted migration.

Similar flocks were seen on the same beach on other occasions: seventy-three on August 23, 1930; one hundred and fifty on August 4, 1932; etc. Sometimes the flock will divide into several smaller sections and when flushed will unite again while at other times single individuals are seen or parties of two to six, the earliest arrival records being often of this sort. One first arrival flew over the Fill with a flock of twenty-nine Curlew on July 18, 1926; another on Five Mile Beach, on the south side of Hereford Inlet, came with Turnstones and Sanderling on July 24, of the same year, and another on Seven Mile Beach on July 24, 1927.

The Robin Snipe are rather heavy-bodied birds with relatively short legs and plover-like bills. Sometimes, when there is a mirage on the beach and all things are distorted and heights exaggerated, these birds seem to be standing on long slender legs and, were it not for the characteristic massing, they might at a distance be mistaken for some long-legged species.

In the spring the character of occurrence of the Robin Snipe is practically the same as on the southward flight. They are found in considerable flocks on the beaches as well as in small detachments and likewise occur, perhaps a little more frequently, on the meadows and sounds. Walker Hand has recorded them as early as May 1, 1905, and May 9, 1920. I have found them present daily on the flats of Jarvis Sound during the weeks of May 17, 1923, and May 22, 1922, while William Baily records flocks aggregating about one hundred-and-fifty on Five Mile Beach, May 21, 1898, and Walker Hand tells me of twenty-five shot at Cape May on May 29, 1907. They remain regularly through the early part of June and we have records for Seven Mile Beach on June 12, 1927 (four); June 18, 1922 (two); June 26, 1927; and on Five Mile Beach, June 20, 1923; while William Baily saw a flock of thirty at Ocean View on the lower side of Hereford Inlet on June 30, 1900. Just below Corson's Inlet I saw flocks on June 5, 1925, and seven on June 10, 1928. The latest records may be south-bound birds.

The Knots are still to be seen in the month of September and I have records for September 10, 1898; September 1, 1929; September 2, 1895; September 3, 1921; September 6, 1918; September 8, 1928; October 1, 1922.

Charles Urner's records covering the more northern beaches from Brigantine to Barnegat showing extreme dates, period of regular or common occurrence, and maximum count for one day at a single locality, are as follows:

Northward migration:

1929. May 14–May 30 (May 17–26). 500 on May 26.
1930. May 11–June 22 (May 23–27). 12 on May 23.
1931. May 17–May 31 (May 23–31). 150 on May 24.
1932. May 15–May 25 (May 15–25). 100 on May 18.
1933. May 7–June 4 (May 7–28). 250 on May 28.
1934. May 13–May 27 (May 16–27). 50 on May 21.

Southward migration:

1928. June 29–October 28 (July 14–September 16). 150 on July 14.
1929. July 13–September 27 (July 22–September 15). 160 on August 10.
1930. July 6–September 14 (July 13–September 7). 170 on July 20.
1931. July 12–September 13 (July 12–September 7). 525 on July 29.
1932. July 16–October 4 (July 16–September 4). 790 on August 1.
1933. July 24–November 2 (July 30–October 7). 25 on August 6.
1934. July 23–January 13 (July 23–November 11). 1000 on July 28.

As a result of his careful studies the Knot is placed in the group of ten "common" species on our coast today following the six that are regarded as "abundant" (see p. 385).

By September the Robin Snipe have all changed to the winter plumage, pale gray above and white below, and become the "Gray-backs" of the gunners. The young birds of the year are in an almost identical plumage from the time of their arrival. They apparently come south considerably later than the adults as all the early flights are composed entirely of old ruddy-breasted birds.

At Turtle Gut Inlet, before it was filled up and Two Mile Beach joined to Five Mile, I found a flock of twenty-one young Robin Snipe feeding on the sand flat at low tide. They kept close together pecking at the bunches of young mussels and other objects left by the receding water, but never probing; sometimes they fluffed up their plumage and stood at rest. They were quite tame and when not startled one could approach to within twenty feet of them. Their coloration was very beautiful and delicate. At first they appeared uniform pale gray with a slight pinkish tinge; but when closer they were seen to be finely mottled with a darker shade of gray, the feathers of the upper side being lighter on the margins but all the coloring was soft and blended, even the tail was soft gray faintly barred or mottled with darker; breast grayish and belly white with pale gray barring on the flanks. The birds occasionally stretched one wing at full length over the side of the body and leg, its tip touching the sand, and also stretched the leg out in the rear. These birds kept to themselves and it is probable that the association with other species, that has been noticed, has its initiative with the latter rather than with the Robin Snipe. Their note was a short *quick-quick, quick-quick*.

Julian Potter saw an adult in full gray plumage with a tinge of salmon on the breast on the Cape May beach on October 1, 1922. It was very tame.

Our Robin Snipe is distinguished more by the character of his activities than by any outstanding peculiarities of form or color, unless one is close enough to him for detailed study. With the bulk of the Dowitcher and almost twice the size of the Sanderling, plover-like in bill and method of feeding, unique in his mass formation, lacking the spasmodic nervous tilting of the Yellow-legs and their kind, he is always trim in form and deliberate in action—one of the aristocrats of the Limicoline world.

PURPLE SANDPIPER

Arquatella maritima (Brünnich)

Until the winter of 1924–25 there was, so far as I am aware, only two definite records of the Purple Sandpiper on the New Jersey coast, a specimen shot by Archiclaus Willets at Beach Haven, on October 31, 1896, and presented to the local collection at the Academy of Natural Sciences of Philadelphia; and another obtained on Barnegat Bay by J. A. Weber on November 19, 1921. With the construction of the stone jetty at the mouth of the Cape May Harbor, however, a few of these birds are to be found, probably every winter, feeding among the rocks at the extreme end of this stone pile. It would seem that they must have come down the coast to Cape May in former years but, as the low sand beaches were not attractive to them, they did not remain and the scarcity of observers in winter prevented any records being made of possible straggling individuals.

The great rocks at the ocean end of the jetty, nearly a mile at sea, are loosely piled together leaving many passageways and crevices below and between them, and these the birds frequent in their search for food. (Plate 3.)

The discovery of the Purple Sandpipers at this cold and desolate spot is due to Henry Gaede and William Yoder who chose the jetty as their post of observation for the Christmas census of the Delaware Valley Ornithological Club on December 28, 1924, hoping to pick up some unusual sea birds. While scanning the sea for bird life they suddenly noticed the head of a small shore bird protruding from behind one of the rocks. It had been out of sight for the half hour or more that they had been on the watch. A moment later it came out into full view and was immediately identified as a Purple Sandpiper. Shortly afterward two others appeared. They were not over twenty-five feet distant and with the glass every detail of their plumage could be seen—the dark back and orange toes and tarsi.

The birds continued to climb about the rocks and to fly from one to another, one of them approaching to within eight feet of the observers. They seemed to be feeding on some minute form of animal life inhabiting the clumps of sea weed that covered the lower parts of the rocks, and were very sure-footed as they ran down the sloping surfaces after the receding waves and dashed back again before the on-coming water could overtake them. Two were seen by Gaede on January 28, 1925, and on December 26, 1927, William Yoder again saw two Purple Sandpipers on the jetty and heard them down under the rocks, uttering a note like the squeeking of a rusty hinge, for some time before they came out into view. Indeed they seem to spend most of their time in these invisible passageways.

Subsequent visits to the jetty by several members of the Club have usually discovered a few of the birds—December 23, 1928, twenty; December 9 to 11 five; December 22, 1929, one; December 27, 1931, two; December 26, 1932, two; December 24, 1933, one; December 23, 1934, ten; December 22, 1935, one. On some of these occasions one or more of the birds were seen as much as two hundred yards from the end of the jetty. None were seen on October 25, 1929, and we have not ascertained just when they arrive from the north. It is a long and rough trip out to the end of the "rock pile," as it is termed, and not many visits are paid to the spot which renders it impossible to gain much idea of the time and duration of the birds' occurrence at this, their most southern regular resort on the Atlantic coast. Charles Rogers writes me that on December 24, 1929, he saw fourteen Purple Sandpipers on a small breakwater in front of the Hotel Traymore in the heart of the beach front of Atlantic City. A northeast storm had been in progress immediately before and the birds came in from the north. They were quite tame and allowed a close approach so that it was possible to see the orange feet, streaked flanks and light base to the bill without the use of the glass. After a few minutes the flock passed on to the south. A jetty at Manasquan has also proved attractive to these birds and Charles Urner reports them present there from January to March, 1934, while J. L. Edwards saw one as late as April 7.

Additional recent records for the Barnegat Bay region by Charles Urner and others are as follows: January 18, 1933 (Cruikshank); December 24, 1933, two (Vogt); December 22, 1935; December 26, 1936, three. On the Tuckerton Meadows Charles Urner saw one or two from November 24 to December 28, 1935. In his table of comparative abundance he regards the Purple Sandpiper as rare (see p. 385).

1936
CR.

PECTORAL SANDPIPER

Pisobia melanotos (Vieillot)

None of the shore birds is more particular in its feeding grounds than the Pectoral Sandpiper. It is not a bird of the great open meadows and tidal flats where most of the species congregate, nor is it ever seen on the sea beaches, but it is sure to be found in season on the drier parts of the marshes next to the upland or on isolated patches of salt meadow at the head of some inlet or bay. In such places grows the so called black grass (*Juncus gerardi*) so frequently cut for hay by farmers whose lands run down to the salt water, and as soon as the hay is cut the Pectoral Sandpipers will be found feeding amongst the stubble, a habit which has resulted in the local name of "Hay-bird." They occur, too, in other wet grassy spots especially on Pond Creek Meadows and on rain water pools in grassy bottoms on the Fill and elsewhere. Their most frequent associates are the Lesser Yellow-legs although they may be found feeding with Killdeers, stray Dowitchers or Least and Semipalmated Sandpipers. While so frequently seen with the Yellow-legs each species keeps more or less to itself, the Yellow-legs in the shallow pools and the Pectorals in the drier spots among the grass and stubble.

In general pattern and tone the Pectoral resembles the small sandpipers and is in almost every detail a larger edition of the Least. There is the same uniform buff overspreading the breast and sharply separated from the white belly, while this buff background is finely streaked with black, presenting a beautiful appearance when close at hand. The dark buffy breast, like the rusty under parts of the Dowitcher, makes the bird look especially dark, particularly when viewed from the front or when flying head-on. The coloration of the upper parts, while not conspicuous, presents in a good light such a beautiful combination of buff edgings and spots on a rich dark brown background that it never fails to arouse one's admiration. The legs are olive yellow lighter than those of the Least Sandpiper.

The Pectoral is more than twice the size of the Least and nearly the bulk of a Killdeer while the legs and bill seem relatively short for a bird of its size. When standing at rest the head is held well up but the short thick neck seems to merge rapidly into the body. In action it is decidedly sluggish compared with the little sandpipers or the Yellow-legs, and when feeding it moves about in a leisurely manner somewhat like the Dowitcher, making many probes of the bill from one position and advancing slowly. Its method of progression when not frightened is a walk and apparently never a run although it sometimes quickens its pace to a sort of trot. The Pectorals will often gather to rest on the edge of some grassy pond just as do other species when awaiting the ebb of the tide on the Jarvis Sound flats. A flock of eleven that flew into the old Race Track Pond to join some Lesser Yellow-legs and Killdeers came to rest almost at once, turning the head back over the shoulder and burying the bill among the scapular feathers. Some of them fluffed up their entire plumage, jerking the tail sharply from side to side during the process.

On August 3, 1920, a flock of two dozen came into a grassy stubble on Pond Creek Meadow and alighted all facing the same way but at once began feeding in their usual deliberate fashion. They scattered about over the ground much as do the Least Sandpipers, but every one ready for instant flight, and all took wing simultaneously in the same mysterious way as they wheel and turn, in perfect unison. The movements of the Pectorals from one spot to another are frequent for they are restless birds, flying without any apparent cause, doubtless trying one spot after another until a satisfactory feeding ground is found.

On July 21 and 22, 1923, there were six Pectorals on the shallow pond at the Lighthouse which fed much like Dowitchers constantly probing but in no hurry about it. At intervals one of them would stand rigid with head and neck held straight up, presenting a totally different appearance from that of the feeding bird with body in a horizontal position. On July 26, of the same year I found two associated with Yellow-legs and Red-winged Blackbirds in a depression on the Fill where rain water had settled on a sedge stubble.

On September 16, 1923, four Pectorals were feeding with five Lesser Yellow-legs and some Least and Semipalmated Sandpipers and the afternoon sunshine, coming across the Pond Creek Meadows, illuminated their plumage in a remarkable manner so that every feather seemed to stand out clearly and never did the brilliant buff and rich dark brown stripes on the backs of the Pectorals show to better advantage. These birds fed more rapidly than usual, wading into the water until it nearly reached their bellies and dabbing here and there with their bills. One of them was distinctly larger than the others, evidently an adult male, as there is much more disparity in size between the sexes in this species than in any other sandpipers. The whole group of fourteen birds flushed several times but in spite of my presence seemed loath to leave and always returned to the same spot, so that it was possible to approach to within thirty feet of them and to slowly withdraw again without causing alarm.

On August 25, 1931, after a heavy rain a low-lying bit of fill covered with grass, immediately back of South Cape May, was flooded to a depth of several inches and for two days, or as long as the water remained, it was thronged with shore birds—fifty Lesser Yellow-legs, twenty-five Pectorals, five Stilt Sandpipers, two White-rumped, one Solitary and a host of little Leasts and Semipalmateds. The Pectorals were conspicuous from their dark coloration, appearing blacker than any of the other species present. These birds were, however, studied against the light while two Pectorals on a small grassy pond on Seven Mile Beach with the light playing on them were noted as "very light with all markings clearly defined"—another illustration of the light factor in the appearance of shore birds.

In the more open places near to the town we usually see Pectorals in small parties of from two to six but in more remote and sheltered spots, especially on the grassy meadows adjacent to Dennis Creek and other streams emptying into the Bay, they occur in larger flocks and several times I have seen flocks of one hundred or more, and when they concentrate on temporary rain water pools on the Fill we often see as many as twenty-five together.

When shore bird shooting was legal the "Hay-birds" were much esteemed and were shot in large numbers along with the Yellow-legs with which they so regularly associated. Henry Hazelhurst's shooting record shows that from 1891 to 1910, during a shooting season of twenty-two days in July and August, he averaged seventy Hay-birds a year, a toll which did not apparently affect the abundance of the species, but were shore bird shooting permitted today, the automobile would make it possible to visit meadow after meadow and to follow the birds no matter where they went for shelter, something quite impossible when the gunner went entirely on foot.

The Pectoral Sandpiper is in my experience seen only on the southward flight and I have never found one in New Jersey in spring. Charles Urner has recorded single birds at this season on three occasions: May 16–18, 1930; May 7, 1933; May 12–16, 1934, all on the upper New Jersey coast, but the bird is certainly very rare on the Atlantic seaboard at this season, the flight going north by way of the Mississippi Valley.

Our dates for the arrival of south-bound Pectorals are:

1903.	July 8.	1924.	July 26.	1932.	July 10.
1905.	July 9.	1926.	July 18.	1933.	July 26.
1910.	July 9.	1929.	July 12.	1935.	July 26.
1920.	July 11.	1931.	July 13.	1936.	July 26.
1923.	July 21.				

Our latest dates for several years are:

1879.	October 4.	1926.	September 25.	1929.	September 21.
1921.	October 9.	1927.	October 2.	1933.	October 9.
1925.	October 11.	1928.	September 30.	1935.	September 29.

An extraordinary observation was a single bird at Camden on the Delaware River seen by Julian Potter on November 10, 1929.

Urner's records for the region from Brigantine to Newark Meadows showing extreme dates, period of regular or common occurrence, and maximum count for one day at a single locality, are as follows:

1928.	July 15–October 28 (August 11–October 13). 75 on September 12.
1929.	July 26–October 12 (July 31–October 12). 50 on September 7.
1930.	July 19–October 12 (August 10–September 13). 60 on August 15.
1931.	July 15–September 27 (July 18–September 20). 200 on July 28.
1932.	July 10–November 12 (July 23–October 12). 125 on September 15.
1933.	July 12–November 11 (July 23–October 29). 300 on September 17.
1934.	July 14–November 10 (August 5–September 30). 350 on August 18.

In point of abundance he groups it as one of the ten "common" transient shore birds on our coast today, being outranked only by the six which he regards as "abundant" (see p. 385).

J. F. S.

KILLDEER

WHITE-RUMPED SANDPIPER
Pisobia fuscicollis (Vieillot)

My opportunities for observing this bird have been limited, except during the last few years, to the immediate vicinity of Cape May and the Point, and my records do not satisfactorily cover the great meadows lying to the north, but, even so, I do not think that the White-rump is nearly as plentiful here as it is on the marshes from Brigantine to Newark in the upper part of the coast where Charles Urner has studied it. Its close resemblance to the Semipalmated Sandpiper, with which I have usually found it associated, might account in part for its apparent scarcity, but when I consider the many hundred flocks of the latter species that have been studied carefully without detecting a single White-rump, it would seem to be anything but a common species about Cape May.

Our records are usually of single birds though sometimes two or three are seen together while Julian Potter saw six at Corson's Inlet on June 10, 1928. They have been noted from May 20, to June 10, in spring, and from August 4 to October 25, in autumn.

One or more White-rumps are seen at Cape May on the Memorial Day trips of the Delaware Valley Ornithological Club whenever these are directed to the shore, which indicates that the bird is a late spring migrant, while the October dates indicate a tendency to linger on the southward flight.

Charles Urner's records for the northern coast region from Brigantine to Newark Meadows, showing extreme dates, period of regular or common occurrence, and maximum count for one day at a single locality, are as follows:

Northward migration:

1929. May 11–May 30 (May 17–19). 75 on May 19.
1930. May 11–May 31 (May 17–31). 5 on one day.

1931. May 26–May 31 (May 26–May 31). 6 on May 31.
1932. May 8–19 (?). One seen on May 8 and 19.
1933. May 7–May 30 (May 7–28). 15 on May 21.
1934. May 6–May 27 (May 20–27). 10 on May 27.

Southward migration:

1928. August 11–October 14 (August 29–September 23). 14 on August 29.
1929. August 31–October 25 (September 1–October 6). 400 on October 6.
1930. July 16–October 12 (August 24–September 21). 12 on September 21.
1931. July 5–September 13 (August 30–September 13). 25 on August 30.
1932. July 31–November 12 (August 20–October 15). 8 on September 28.
1933. July 2–November 11 (August 25–October 1). 200 on September 24.
1934. July 14–November 11 (September 2–16). 10 on September 2.

On the spring flight the White-rump was seen on seven days in 1929; five days in 1930; two days in 1931 etc. On the autumn flight, on sixteen days in 1928; thirteen days in 1929; sixteen days in 1930; nine days in 1931. It is rated as a "tolerably common species irregularly and locally" by Urner.

The largest numbers quoted above were due to unusual flocks feeding on a garbage dump on the Newark Meadows.

In earlier years W. E. D. Scott has recorded it as common on Long Beach in 1877 but Philip Laurent found it rare on Five Mile Beach in the eighties. William Baily saw none there in 1896, but George Morris shot one at Cape May Point on August 5, 1886.

Two individuals studied on the shallow pond at the Lighthouse on May 30, 1925, appeared longer and relatively more slender than the Semipalmated Sandpipers which were nearby, the actual thickness of the body being about the same. The streaking of the under side was distinctly black on white with no buff tint and extended farther down on the sides of the chest than in the other species. There was a fairly well defined eyebrow and prominent buff streaks on the back, which was also slightly marked with rusty. The birds were near a flock of Semipalmateds but seemed to keep more or less aloof, and flew at one of the latter on several occasions when it approached too close. Their actions seemed slower and reminded one more of the Dowitcher as they often probed with the bill several times in one spot instead of dabbing quickly to the right and left and moving constantly, as do the Semipalmated Sandpipers. They waded out in the water up to their bodies, too, and often completely submerged the head in feeding.

The distinctive character—the white rump, could only be seen when they took wing or spread the wings for a moment as their feet sank into the soft mud. Then a narrow band of white was visible, reaching clear across the rump close to the tail, but it was not wide enough to be very conspicuous. The method of feeding and the movements of the birds were sufficient to distinguish them once one became familiar with these actions.

Julian Potter saw a White-rump on the ocean beach at the Point on September 29, 1929, associated with Semipalmated Sandpipers and another or the same individual, on October 13 at the same spot, while on October 25, 1929, the American Ornithologists' Union field party saw three individuals there, so that its occurrence on the strand seems to be more than a casual matter.

On August 28, 1933, I watched one of these birds on the meadows back of the upper part of Seven Mile Beach which ran along the edges of the small salt pools close to the road that crosses the meadows at this point. It was with several Semipalmateds and was seen to be distinctly larger and showed its white rump to perfection whenever it spread its wings.

The rain water pools that form in hollows in grassy fields or meadows following severe storms, are favorite resorts for sandpipers of this and other species. On August 25, 1930, a long narrow lake of this sort formed on a piece of filled-in meadow back of South Cape May and an active group of sandpipers gathered there for several days, busily engaged in seeking food among the flooded grass tufts. They were entirely unmindful of our close approach so that we could watch them at a few feet distance. There were Semipalmated, Least, Pectoral and Stilt Sandpipers present, and feeding shoulder to shoulder with the Semipalmateds were two White-rumps. We judged them to be three-quarters of an inch longer than the former and they continually displayed their narrow white rump patch as they held their wings aloft to retain their balance, as one foot or the other sank into the soft bottom.

There is a great fascination about the study of the various little sandpipers and the facility with which one can approach them today is in strong contrast to their wildness when gunning of shore birds in both spring and fall was legal, as was the case when my studies began. Our first task, when the general use of binoculars became customary, was the differentiation of the Least and Semipalmated, a distinction which today is given scarcely a thought, if the birds are reasonably close to the observer and in good light, then came the separation of the White-rump—and the Baird's, if one was fortunate enough to find it, and finally the separation of the Western which is a refinement only certainly possible in extreme individuals, in breeding-ing plumage, but there is always the possibility of the three rare species among the hordes of the two common ones and hence the lure of the "peeps" as they are collectively termed.

BAIRD'S SANDPIPER

Pisobia bairdi (Coues)

This little sandpiper, first described from the Far North in 1861, and for the next ten years supposed to be exclusively restricted to the interior regions of North America, was obtained in 1870 in the vicinity of Boston, Mass., and since then has been known as a rare transient along the Atlantic coast during the southward migration.

The only specimen of which we have record from the shores of Cape May County was shot on Seven Mile Beach, on September 5, 1898, by David McCadden who did not distinguish it from the Semipalmated Sandpipers with which it was associated. It later came into the collection of the Academy of Natural Sciences of Philadelphia and its true identity was determined.

Since modern binoculars have made intimate studies of shore birds possible, and added attention has been given to critical field characters, several Baird's Sandpipers have been identified by specialists in this group. Julian Potter found one in a mixed assemblage of shore birds at Brigantine on September 9, 1928, while John Gillespie found another at the same place on August 16, 1931. Charles Urner saw this species on four occasions between September 8 and 18, 1929, in the Barnegat-Newark area, three at one time and one on each of the other days, He also detected one in the spring flight, May 30, 1929, a most unusual record. Urner had two records of this sandpiper on the Newark Marshes on September 10 and 15, 1932; and five in 1933, August 30, September 24, 30 (two), October 1, (two) and 7; while F. W. Loetscher found one on Brigantine Island on August 18 of the same year. In 1934, there were three observations on the Newark Marshes, September 8 (Edwards), September 15 (Urner), September 22 (Rebell); while in 1935, Loetscher saw three on Brigantine on August 30 and one each on August 31 and September 3. It always seems possible that in such cases, when birds are seen on several successive days, that some at least are the same individuals and that the group had spent several days feeding at the same spot.

Julian Potter has also found Baird's Sandpiper at shore bird habitats on the Delaware River near Camden, N. J.—one, probably the same individual on August 24, 25 and 26, 1923; three on September 17, 1932; two, doubtless part of the same group, on September 21 following.

Baird's Sandpiper resembles the White-rumped Sandpiper in size and general color of the upper surface although there is more of a buffy tint and the pattern is less streaked and more shell-like; the absence of the white rump however is diagnostic. Below it is pale buffy and the streaks may be quite indistinct, the White-rump has a white breast with distinct black streaks, while the larger Pectoral and smaller Least have a brownish buff breast streaked with black and a white abdomen instead of a uniform buffy under surface.

Charles Urner in his list of comparative abundance of our shore birds (see p. 385) regards the Baird's Sandpiper as one of the "rare" species.

LEAST SANDPIPER

Pisobia minutilla (Vieillot)

The habits of this, the smallest of our shore birds, are almost identical with those of the Semipalmated Sandpiper, but while that species occurs both on the sea beach and the salt meadows, the Least Sandpiper is almost entirely restricted to the latter and only on a very few occasions have I seen individuals on the beach, each time with flocks of the Semipalmated.

On the muddy bottoms of shallow meadow ponds, which the ebbing tide has left bare, both species intermingle freely but often when a mixed flock settles on a mud flat surrounded by a growth of low salt grass, the Least. Sandpipers will seek the shelter of the grass and work their way in and out among the tufts leaving the open spaces to the Semipalmateds. On the great meadows north of the Harbor single individuals or groups of two or three Leasts will often be flushed from areas uniformally covered with grass, situations not frequented by the other. On the meadows at the height of the northward migration, as on May 16, 1927, the Least Sandpiper is perhaps the most abundant bird. Shoals of them frequent the moist flats of the meadows and later fly to some dryer spot which they have selected as their resting place until feeding time again comes round, skimming over the ground in great flocks, they wheel back and forth and finally disappear into the low grass where they scatter and go scurrying about like the little fiddler crabs which swarm in the same localities. Once again they rise, alarmed at we know not what, and again we see the wheeling flock of little birds, and again the settling on the very spot, perhaps, which they had occupied before. Now and then the mouth of one of the little tidewater creeks which empty into the larger thoroughfares attracts their attention and several hundred of them may cluster about the spot hovering over it like the swarms of yellow butterflies that one sees about some roadside puddle. Next we find them settling and alighting all over the vertical mud banks at the mouth of the creek like gulls or murres on the ledges of the northern cliffs. Always, too, there are scattered individuals in the air, like lone swallows, flying here and there seeking the flock from which they have become separated.

Least Sandpipers also associate with bunches of Dowitchers and Lesser Yellow-legs or with Pectoral Sandpipers, of which they are veritable miniatures in coloration. All of these birds find suitable feeding grounds about some little shallow pool on the marshes at the Lighthouse, or on the Pond Creek Meadows, which are favorite haunts of the Least as well. Out on the great meadows almost every gathering of larger shore birds will have its attendant flock of Least and Semipalmated Sandpipers. On the Pond Creek Meadows on September 16, 1923, a characteristic gathering contained five Lesser Yellow-legs, four Pectorals and two Leasts which were joined later by three Semipalmateds, while another group found near the Lighthouse, July 12, 1924, consisted of eight Dowitchers, a dozen Leasts and one Lesser Yellow-legs whose loud cry of warning put the whole flock to flight. Unlike the Semipalmated, the Least Sandpiper rarely occurs on the bare flats and shell beds of Jarvis Sound which are exposed at low tide and are favorite feeding grounds for a countless horde of shore birds; another evidence of their avoidance of open spots and their preference for grass-covered meadows. I have also seen a flock of fifteen of these little birds busily feeding on some kind of small insect on the short cropped grass of the golf links on perfectly dry ground.

At a little distance, or in poor light, it is by no means easy to distinguish the two little sandpipers from one another and it is often necessary to record an apparently mixed flock as "peeps" a name given by gunners to all small sandpipers of this general type, regardless of species. While the style of coloration in the Least and Semipalmated is nearly the same and there is but little difference in size, nevertheless, under favorable conditions, one can distinguish the paler legs of the Least—olive instead of black as in the Semipalmated; also the darker, richer and browner colors of the Least as contrasted with the lighter, grayer tone of the other. The breast of the Least is darker with a buff background and numerous narrow black streaks as in the larger Pectoral, while in the Semipalmated it is white with the streaks larger and more widely separated. In the younger birds of the year the spots are almost round and form a band across the breast.

Least Sandpipers are, I think, always present at Cape May by the end of the first week of July although my actual records do not bear this out in every year; they run:

1920. June 29.	1925. July 13.	1931. July 12.
1921. July 3.	1926. July 19.	1932. July 9.
1922. July 6.	1927. July 16.	1933. July 9.
1923. June 20.	1928. July 2.	1936. July 15.
1924. July 12.	1929. July 12.	

My latest dates are October 18, 1925; October 7, 1928; October 18, 1931; but the majority have usually departed by the middle of September. In the spring my dates are not entirely reliable, as it was impossible to be on the meadows regularly, but apparently the first Leasts arrive during the last week of April. I have dates of April 17, 1921; April 25, 1926; April 30, 1932; and other dates for the first week in May. My last dates of observation on the northern flight are May 31, 1926; June 9, 1907; June 5, 1910; June 16, 1918; the latter probably non-migrating birds.

Charles Urner's records for the region between Brigantine and Newark Meadows, on the more northern seaboard, are given below to show extreme dates, period of regular or common occurrence, and maximum count on one day at a single locality.

Northward migration:

1929. May 11–May 30 (May 10–30). 300 on May 30.
1930. May 10–June 22 (May 11–23). 100 on May 11.
1931. May 14–May 31 (May 17–24). 250 on May 25.
1932. May 1–25 (May 8–25). 1500 on May 18.
1933. April 29–June 4 (April 29–May 28). 700 on May 7.
1934. April 28–May 27 (April 28–May 16). 250 on May 12.

Southward migration:

1928. June 27–October 14 (July 11–September 15). 500 on July 18.
1929. June 30–October 20 (July 10–September 29). 200 on July 14.
1930. June 22–October 12 (July 13–September 7). 100 on July 23.
1931. July 3–October 4 (July 5–September 19). 300 on July 5.
1932. July 3–October 12 (July 4–September 18). 400 on July 13.
1933. June 27–November 4 (July 2–October 8). 400 on July 19.
1934. June 23–October 14 (July 7–October 12). 250 on July 14.

Urner lists the Least Sandpiper as one of the six "abundant" shore birds on our coast today (see p. 385).

CURLEW SANDPIPER

Erolia testacea (Pallas)

This Old World bird, which has occurred as a rare straggler in various parts of North America, has been recorded several times from Long Island. Giraud mentions ten from there obtained in Fulton Market and Ludlow Griscom adds two more—May 24, 1883, and June 9, 1891, while Helmuth and Fuertes secured two on September 7, 1923, and December 19, 1923, respectively. (Auk, 1924, p. 341.)

Audubon tells us that two were shot at Great Egg Harbor in the spring of 1829 and Turnbull (1869) says that it is occasionally shot there while he adds that Wilson must have known it "as in his portfolio of drawings I found a figure of this bird in Autumn plumage." Turnbull had obtained from one of Wilson's executors the remainder of his drawings, not used in completing his work, and some other drawings etc. (see Cassinia, 1913, pp. 4–5). Dr. Charles C. Abbott, in his list of New Jersey birds (1868) says that specimens have been taken at Tuckerton and at Cape May, but there are no details whatever.

There seem to be but two records for the New Jersey coast in recent years. One a specimen in the collection of John Lewis Childs of which he writes as follows: "On July 29, 1904, a friend shot at Long Beach, Barnegat Bay, N. J., a strange sandpiper. It was forwarded to me, but unfortunately the weather being exceedingly warm, the bird was spoiled beyond the possibility of skinning when I received it. I recognized it at once as *Erolia ferruginea*, evidently an adult male in full plumage. The rufous color of the breast and throat was very deep and rich. I have never seen any sandpiper, not even of this species, so highly and beautifully colored. I have the specimen preserved in alcohol." (Auk, 1904, p. 485.) The other bird was seen at Barnegat Inlet by Charles Urner on October 28, 1934. It was on the wing and "swung close several times and passed at varying levels with mixed flocks. Bird very light below, whiter than Red-backs, fairly dark above with distinct wing stripe but not as prominent as Red-back's; clear white rump and long evenly curved bill. Note definitely different from either White-rump or Red-back. Size comparison in conformity." (Abst. Proc. Linnaean Soc., No. 47, p. 84.) In his summary of shore birds with regard to abundance Urner regards the Curlew Sandpiper as one of the "very rare" group (see p. 385).

J. F. S.

BULLHEAD

1936
cR

RED-BACKED SANDPIPER

Pelidna alpina sakhalina (Vieillot)

The Red-backed Sandpiper is a regular transient in the Cape May district. Its spring flight occurs in May at the height of the northward migration of shore birds but, except for a few stragglers, it does not pass south in autumn until October after most of the other species have gone. Unlike our other shore birds the Red-back remains here in large flocks throughout the winter in favorite localities, the Sanderling being the only other species to be found regularly at this season. At Corson's Inlet we have records on November 10–12, 1919 (three hundred); December 4, 1921 (eighty); February 5 and March 12, 1922 (fifty); January 1, 1926 (sixty); November 28, 1926 (one hundred); November 13, 1927 (two hundred and fifty); March 11, 1928 (thirty-five); January 13, 1929; January 28, 1934 (two hundred) and on similar dates in other years. Also on Ludlam's Beach on November 13, 1927; January 13, 1929; January 28, 1934 (two hundred); and at Hereford Inlet on February 23, 1931 (five hundred). Farther north along the coast we have records of five hundred at Brigantine Beach on February 12, 1934, and one hundred and fifty at Barnegat on December 9, 1928. The Red-back undoubtedly winters at all these points but as we have visited them only occasionally during the winter, continuous observations have been impossible.

We have a number of April records which may very likely be of birds that have spent the winter nearby, especially as they had gone from Corson's Inlet by April 9 in 1922. W. E. D. Scott has recorded them at Beach Haven on April 17, 1877, and Dr. William L. Abbott obtained specimens on Five Mile Beach on April 18, 1878, while Julian Potter saw fifty on Ludlam's Beach on April 14, 1929, and a flock on Brigantine on April 15, 1934. Walker

Hand's earliest dates for Cape May are April 29, 1911, and May 9, 1920, the latter doubtless the first arrivals from farther south; I have found them still in abundance on Jarvis Sound on May 20, 1921; May 21–24, 1922; May 17, 1924; although a trip by boat through the sounds back of Tuckerton on May 16–19, 1927, showed only three in the great assemblage of shore birds present there. It is doubtful if they linger after the last days of May on the Cape May meadows although Julian Potter found two on the bar in Hereford Inlet on June 20, 1923. These may well have been migrants that failed to reach their nesting grounds.

I have seen a single bird back on Seven Mile Beach on July 19, 1931, two at the same place on August 28, 1929, and another on September 1, of the same year, all of them in full breeding plumage, which may have belonged in the same category. Wharton Huber secured one at Corson's Inlet on September 25, 1924, which seems to mark the beginning of the regular flight and they were there in force on October 4, 1931, 125 in one flock, with ninety on Ludlam's Beach on October 14, 1928.

Charles Urner's records for the region from Brigantine to Newark showing extreme dates, period of regular or common occurrence, and maximum count on one day at a single locality, follow:

Northward migration; "arrival dates difficult to ascertain as the species winters:"

1929.   April 13–June 22 (May 17–26).   500 on May 19.
1930.   to June 22 (?).   50 on May 11.
1931.   May 3–May 26 (May 3–26).   15 on May 3.
1932.   May 8–May 25 (May 15–25).   25 on May 15.
1933.   to June 4 (May 14–June 4).   75 on May 20.   500 on February 12.
1934.   to May 27 (April 8–May 27).   100 on April 29.

Southward migration:

1928.   July 8–January 13 (September 30–December 23).   155 on Dec. 9.
1929.   July 13–December 26 (October 6–December 26).   300 on December 26.
1930.   August 24–November 16 (October 5–November 16).   100 on November 2.
1931.   July 18–December 27 (October 4–December 27).   500 on December 27.
1932.   September 11–January 23 (October 1–December 26).   100 on Oct. 16.
1933.   July 9—wintered (September 30–February 18).   324 on December 24.
1934.   July 15—wintered (September 2–January 13).   500 on November 25.

In comparison with our other shore birds Urner regards the Red-back as one of the ten "common" species standing below the six designated as "abundant" (see p. 385).

The Red-backed Sandpiper, or "Black-breast" as it is also called here, is most frequent on the meadow pools at high tide and on the exposed flats of the sounds when the water is down, but it also is to be found on the beaches or bars at the entrance to the inlets, especially where long points project out on either side offering more or less isolated feeding grounds. In spring I have studied them on Jarvis Sound where Walker Hand and I used to allow our skiff to become stranded on the mussel flats as the tide ebbed and the various species of shore birds gathered all about us to feed. On May 21–24, 1925, about one hundred Red-backs assembled there each day, their bright nuptial dress conspicuous among the flocks of more somber plumaged Semipalmated Sandpipers with which they fed. They were half again as large as these little "peeps" and their legs seemed relatively shorter, while their longer slightly down-curved bill was their most notable structural peculiarity, looking absurdly long for the size of the bird's body. The black patch on the middle of the belly stood out boldly against the prevailing grayish white of the under parts and the rusty, almost coppery, red of the back shown brilliantly in the sunlight. They walked about rather rapidly turning now to this side now to that and probing here and there, sometimes burying the bill so deeply that the head was submerged in the shallow water. When walking where the water was deeper, nearly touching the belly, they held the head high and the body nearly vertical. They came into the exposed flat rather late in the procession and usually fed in the outer ranks. Sometimes they would squat on the mud and fluff up their feathers, while they plumed the breast with the bill; sometimes they bathed like the Dowitchers. On May 22, 1921, there were fifteen feeding on the sand flat at Turtle Gut Inlet with the Semipalmated Sandpipers which they resembled in their habits. They would flutter their wings in the air after bathing and occasionally jump clear of the flat, apparently startled at something, and were several times seen to scratch the head with the foot in the usual sandpiper manner.

In winter they are, perhaps, more likely to be seen at the mouth of the inlets and the long sandy point on the south side of Corson's Inlet is a favorite haunt. Here a flock of fifty to one hundred may be seen at almost any time from November to March. On January 1, 1926, a day of bitter cold gales from the north, I found a flock of sixty. They huddled together in a dense mass close to the water's edge, some of them standing on blocks of ice washed up by the tide. Their plumage was all fluffed up making them seem very fat and chunky. Their heads were drawn down closely on their shoulders and their curved bills, as always, conspicuous. Most of them stood on one leg, hopping along for short distances, and even pluming themselves, without touching the other foot to the ground. They showed the same flock action that characterizes the little "peeps," taking wing simultaneously and circling, wheeling and alighting as a unit as though following definite commands.

The winter birds are very different from the brilliant birds of spring being gray brown above with slightly lighter edgings to the feathers when in good light; the throat and breast are delicate gray with fine gray streaking visible on the latter and down the sides of the body. In flight there is a conspicuous light, longitudinal band down the middle of the wing.

While Red-backs are largely restricted to the inlets and adjoining beaches in winter, they also occur now and then on the meadow ponds back of Seven

Mile Beach where I have found flocks on November 28, 1926, and November 16, 1929. The Delaware Valley Club's Christmas census lists show the following records; for the Cape May district:

December 23, 1928, sixty-five	December 24, 1933, one thousand
December 22, 1929, three hundred	December 23, 1934, five hundred.
December 27, 1931, five hundred	December 22, 1935, one thousand
December 26, 1932, eight hundred	December 27, 1936, sixteen hundred

On the flats and sand bars of Barnegat bay back of Tuckerton, where they winter regularly, I have seen them on December 28, 1928, flying in small flocks from feeding grounds to roosting meadows and back again with every turn of the tide. There were upwards of two hundred there.

EASTERN DOWITCHER

Limnodromus griseus griseus (Gmelin)

The Dowitcher or "Sea Pigeon" of the local gunners, always one of the most desirable game birds during the years when shore bird shooting was permissible, suffered serious depletion in its numbers but has in the last decade recovered much of its former abundance and is now a common and regular transient about Cape May. So greatly has the Dowitcher increased in numbers of late years that it is now regarded by Charles Urner as one of the six abundant species of shore birds. It is distinctly a bird of the meadows and I have never seen one on the sea beach. Immediately about the town it occurs usually in parties of from two to six on little ponds about Cape May Point or on the Pond Creek Meadows, while larger assemblages will gather for a day or two on temporary rain water pools that form in low lying spots. The center of abundance of the Dowitcher, however, is on the great meadows that stretch away north of the Harbor and here flocks of fifty, one hundred, or more may be found resting about the shallow salt ponds or feeding actively on the exposed mud flats of the sounds at low tide.

My dates of arrival of the Dowitchers in the Cape May region on their southward flight run from July 6 to 25, although they are probably always present by the 20th, as I have been unable to visit the meadows daily.

My records are as follows:

1921.	July 17.	1926.	July 23.	1932.	July 10.
1922.	July 23.	1928.	July 25.	1933.	July 1.
1923.	July 8.	1929.	July 7.	1934.	July 13.
1924.	July 12.	1930.	July 25.	1935.	July 13.
1925.	July 17.	1931.	July 6.	1936.	July 15.

Latest records are:

| 1921. | September 9. | 1926. | August 29. | 1930. | August 30. |
| 1924. | August 27. | 1928. | August 19. | 1934. | September 5. |

The Dowitchers undoubtedly stay later than here indicated, at least in small numbers, as our opportunities to observe them in autumn have been limited. In spring we have the following arrival dates:

1905.	May 1.	1926.	April 25.	1932.	April 30.
1916.	May 7.	1929.	May 11.	1933.	May 6.
1925.	May 9.	1931.	May 9.	1934.	May 5.

I have found them usually present until the end of May—regularly, I think, until at least May 25, while Dr. William L. Abbott secured a specimen back of Five Mile Beach on June 10, 1879.

Charles Urner's north Jersey records for the region from Brigantine to Newark Meadows, showing extreme dates, period of regular or common occurrence, and maximum count for one day at a single locality, are as follows:

Northward migration:

1929. May 11–May 30 (May 17–19). 500 on May 17.
1930. May 10–June 22 (May 18–31). 100 on May 31.
1931. May 12–May 31 (May 17–26). 100 on May 17.
1932. May 1–May 25 (May 1–18). 1500 on May 18.
1933. April 29–June 4 (April 29–June 4). 5000 on May 7.
1934. April 28–June 18 (May 6–27). 1000 on May 20.

Southward migration:

1928. June 29–October 29 (July 14–September 23). 325 on July 29.
1929. July 7–September 22 (July 13–September 15). 600 on July 14.
1930. July 6–September 14 (July 16–August 31). 600 on July 16.
1931. July 3–September 20 (July 11–August 30). 815 on July 11.
1932. July 3–October 2 (July 7–September 11). 2450 on July 10.
1933. July 2–October 29 (July 9–September 17). 500 on July 16.
1934. June 30–November 4 (July 8–September 2). 1200 on July 28.

Single Dowitchers, or small bunches, usually will be found associated with other species and the large flocks on the meadows, while keeping to themselves in flight, will settle down with Curlew, Black-bellied Plover and other species, while hosts of little "peeps" seem to seek the company of the larger species. The first Dowitchers to arrive in the summer of 1920, were in a characteristic assemblage gathered around a shallow pool on the Pond Creek Meadows. There were a dozen Lesser Yellow-legs, an equal number of Semipalmated Sandpipers, a few Killdeer, a Pectoral and two Dowitchers. One of the Dowitchers was standing in a clump of dead reed stubs by the edge of the water. His general brown tone and the rich rusty of his breast showed brightly, and so closely did they match the colors of the dead sedge that it was some time before I was aware of his presence. When the Yellow-legs came close to him the contrast in color was striking, they being blackish slate, gray and white, with none of his brown and rust red tints. The light eyebrow of the Dowitcher was also very noticeable as was the very long bill which strongly recalled the Woodcock. The second bird was so well concealed by his color tones that I did not see him at all until the first one walked out into the water and began to feed when suddenly I realized that there were two of them close together. They remained side by side and did not attempt to follow the dainty, active Yellow-legs, although they did not avoid them as they passed again and again close by. The Dowitchers were distinctly more sluggish in their actions, probing slowly along in the mud without raising the bill from the water and with none of those quick stabs and instantaneous recoveries so characteristic of the Yellow-legs. When viewed from the side the Yellow-legs' body seems heaviest at the belly and tapers up gradually to the neck which is long and slender forming a somewhat triangular outline, while the Dowitcher's body is held horizontally and is oval in outline with a much shorter neck which is usually drawn down close onto the shoulders—the whole pose more duck-like, an impression which is emphasized by the relative shortness of the legs.

In another assemblage on the old Race Track Pond on August 3 of the same year there were three Dowitchers, seven Yellow-legs, some Killdeer, Semipalmated and Spotted Sandpipers. Again the Dowitchers appeared strikingly brown and their breasts very ruddy, especially in the case of one highly plumaged individual. They appeared chunky and short-legged and held the head well back on the shoulders when resting, with the long bill always conspicuous and seeming quite out of proportion to the size of the bird. They walked slowly for a few steps and then paused and probed many times from one position, plunging the bill vigorously into the mud without raising it from the water. Another individual feeding on the edge of the Lighthouse Pond walked deliberately, often pausing with one foot slightly elevated and the toes bent over so that only their tips touched the ground. The head was drawn down when walking but when startled the neck was stretched up.

A flock of twenty feeding on an exposed mud bar on July 17, 1921. poked the bill straight down until it, and most of the head as well, were completely buried in the water and soft ooze and, when only partly submerged, it was possible to see the mandibles open and shut as the bird apparently seized something far down in the mud. These birds bent over so that the back was somewhat humped and the tail elevated at an angle of 45° and at a distance they resembled a row of small stocky herons standing rigid on the bar, the elevated tail representing the heron's head.

On Jarvis Sound in the springtime the Dowitchers come in considerable numbers to feed on the black mussel beds at low water. They are the first of the various shore birds to arrive after the tide begins to ebb, small bunches circling about the spot where the shoal will appear fully half an hour before the first shell is exposed, and flying off again to return later. When the shoal begins to bare they alight, sometimes with their feet in the water, and at once assume an attitude of rest with the head bent back over the shoulders and the long bill largely buried among the feathers of the back and rump. Later on, when most of the other species have arrived and become active, the Dowitchers wake up and begin to feed and bathe. They go out on the edge of the shoal where the water has not entirely receded and form, as it were, the outer zone of feeding birds, except for such Yellow-legs as may be present. Here, as on other occasions, they move deliberately and constantly submerge the entire head as they probe for food. When bathing they squat down in the water and churn the wings up and down, apparently alternately, holding them slightly away from the body but not spreading them. The long bill is then poked repeatedly through the feathers of the sides and under the wings as they plume themselves. On May 21, 1922, when Walker Hand and I studied them carefully on this shoal, we counted thirty-seven arriving in the first forty minutes and thirty-eight in the next period of equal length, while on the 22d, fifty came in during the first thirty-nine minutes after the beds began to bare. Over one hundred came in each day singly or in parties of from two to ten.

When studying shore birds on the bars and flats of lower Barnegat Bay, opposite Tuckerton, on May 16–19, 1927, with Dr. H. C. Oberholser, Dowitchers were abundant and passed regularly, as the tide turned, from certain chosen meadows where they rested when the tide was up, to the great flats where they fed when the water receded. The roosting meadows were selected apparently for some specific reason as other meadows close by were entirely bare of shore birds at all times and were deliberately passed by. In a flock of eight hundred shore birds feeding on a flat there were no less than one hundred and fifty Dowitchers and on the average they constituted one-tenth of every flock, with sometimes three hundred in sight at once. They fed in the water, when possible, standing next in line to the Greater Yellow-legs which formed the outer rim of the flock. They walked about slowly probing the long bill many times in one spot in their characteristic way and frequently submerged the head. When occasionally the neck was raised high up the bird looked potbellied, somewhat like the Yellow-legs, which is usually in a similar position but whose neck is much longer. When these birds plumed themselves I noticed that the bill was usually dipped into the water before each lunge among the feathers. The generally dark coloration of the Dowitcher was especially noticeable in these flocks on the flats and contrasted strikingly with the whitish backs of the Turnstones and Black-bellied Plovers which thronged the bars with them.

The dark appearance is due not only to the general dark brown coloration but also to the fact that their lower parts are almost or quite as dark as the upper, or appear so in strong light, which accentuates the natural shadow of the lower surface and makes the bird appear darker than it really is. In flight they also look very dark as we see them passing by but when going directly away from us the grayish white rump and tail are conspicuous, this color extending up the middle of the back for some distance. The flying Dowitcher seems to have the body somewhat curved, or bent over, both head and tail pointing diagonally downward, and, with wings in rapid motion and the long bills conspicuous, a passing flock reminds one of a lot of gigantic mosquitos.

In spring they are dark brown above except for the gray longitudinal band down the back, and rusty red below, with transverse black barring on the flanks, if one be close enough to see it. The females are paler than the males. The birds that arrive in July and most of those that are present in August are also in this dark plumage, the molt to the gray dress of winter taking place *en route*. Birds of the year are gray when they arrive and make up most of the late flocks but I have seen gray birds with white under parts as early as July 31 and August 27. Some August adults present a mottled appearance above as the molt begins.

The habits of the Dowitchers on the meadows, during their return flight, in July and August, are the same as in spring. They are the earliest of the several species to arrive on the feeding grounds and rest, often on one leg, until the tide ebbs sufficiently for them to begin their feast.

The call of the Dowitcher is a series of triplet notes *pheu-pheu-pheu*, *pheu-pheu-pheu*, and there is another mellow fluttering call, somewhat like the gurgling note of the Red-winged Blackbird settling in the grass.

While there is much similarity in the habits of the shore birds there is also individuality, as we become better acquainted with them. The Dowitcher thus becomes fixed in memory as one of the middle-sized waders of body bulk comparable with that of the Killdeer, the Lesser Yellow-legs and the Knot. A dark bird, short-legged and deliberate in action, with a plain gray white tail and back stripe and a bill half as long as himself which he pushes deep down in the mud and keeps there while he probes, preferring to walk, the while, in the shallow water or soft ooze.

I can also picture in mind long lines of these "Sea Pigeons" standing Woodcock-like on the rapidly flooding sand bars of the sound loath to leave and waiting patiently for the rising tide literally to wash them off their feet, and force them to seek the meadow ponds until the next ebb again uncovers their feeding grounds. Meanwhile the horizontal rays of the fast sinking sun line the rippling waves with edges of crimson and illuminate the ruddy breasts of the birds until they fairly glow with color far brighter than one can imagine who knows them only as dark birds of the marsh.

LONG-BILLED DOWITCHER

Limnodromus griseus scolopaceus (Say)

This bird differs from the common Eastern Dowitcher very much as the Western Sandpiper differs from the Semipalmated, but the characters are even less clear and it would seem to be impossible to distinguish them positively in life. It is the Dowitcher of central and western Canada, and the Mississippi Valley in migration, and would seem to be very rare on the Atlantic coast. Very long-billed birds might be referred to this race but the bill lengths of the two overlap when both sexes are considered and some individuals of the western form would thus be left with the eastern. Charles Urner considers it to be slightly larger than the Eastern Dowitcher and darker with the under parts more deeply tinted and more heavily marked. This is discernible in birds just arrived from the north in which the under tail coverts are widely barred. He also finds a difference in the common call note of the two forms. He "feels" that most of the November Dowitchers, which are frequently seen at Tuckerton, are referable to this race. His records of birds that he would refer to it are as follows:

1932.   August 20, Brigantine.   Charles Urner.
1933.   August 27, Brigantine.   Urner and Edwards.
        September 9, Newark Meadows.   Urner and Walsh.
        September 20 and 30, Newark Meadows.   Urner and Kuerzi.
1934.   August 18, Tuckerton.   Urner.
        August 25, Tuckerton.   Urner.
        September 2, Tuckerton.   Urner.
        November 4, Tuckerton.   Urner.

Fletcher Street records a Long-Billed Dowitcher on Ludlam's Beach on August 20, 1933, which had a longer bill and fed apart from the Eastern Dowitchers with which it was associated. It also seemed to be slightly larger and darker. Urner places the Long-billed Dowitcher in the "very rare" group in his comparative abundance list (see p. 385) but adds that this rating is doubtless due to the failure to positively identify the bird and thinks that it is probably more frequent than we suppose. Ludlow Griscom (Birds of New York region) regards it as a rare fall transient on Long Island with a number of records in past years when it seems to have been more plentiful but this was doubtless because in the days of shore bird shooting many more specimens were available for careful examination. We know of no actual specimens collected on the New Jersey coast.

STILT SANDPIPER

Micropalama himantopus (Bonaparte)

The Stilt Sandpiper seems to be of irregular occurrence on the New Jersey coast, present in some years and rare or absent in others. It is moreover an autumnal migrant like the Lesser Yellow-legs and Pectoral Sandpiper, the main flight passing up the Mississippi Valley in spring; in fact we have no spring record whatever for the Cape May region.

Our first Cape May records were in 1897 when William Baily secured specimens on August 11 and 20 from little flocks of three and four and we have no records from then until 1924. But this is not remarkable when we consider the lack of observers and of familiarity with the bird as well as the irregularity of its occurrence. On August 3, 1924, Julian Potter identified a specimen on the shallow pond at the Lighthouse and others were found there until August 24. In 1926 they were present from August 23 to 28; in 1927, July 31 to August 21; in 1930 from July 12 to August 23; on August 25, 1931; and from August 5 to September 5 in 1934. A few were seen on August 7, 1935, and August 23, 1936. Julian Potter saw one at Camden on the Delaware River on October 20, 1929, which would indicate that more continuous observations at Cape May would show the birds present there much later in the autumn. On Brigantine Beach the Stilt Sandpiper would seem to be much more common in years when it is of general occurrence, as Potter found no less than forty there on July 26, 1930.

Charles Urner's records for the more northern section of the coast from Brigantine to Newark Meadows, where the Stilt Sandpiper seems to occur much more frequently than at Cape May, are presented below. They show extreme dates, period of regular or common occurrence, and maximum count for one day at a single locality.

1928. July 3–September 30 (August 29–September 30). Fourteen on Sept. 9.
1929. July 14–September 15 (August 30–September 8). Four on Sept. 8.
1930. July 6–September 21 (July 16–August 10). Forty on July 26.
1931. July 9–September 27 (August 8–30). Thirty-five on August 30.
1932. July 7–October 12 (July 16–September 18). Thirty on August 31.
1933. July 12–October 7 (July 19–September 30). Sixty on September 2.
1934. July 7–October 12 (July 14–September 23). Thirty on August 5.

Urner also saw a single bird on May 12, 1934, the only spring record for New Jersey so far as I know. In his grouping of the shore birds as to relative abundance he regards the Stilt Sandpiper as "irregularly and locally tolerably common."

Alexander Wilson was not acquainted with this bird and we owe our first account of it to Charles Lucien Bonaparte who tells us that "he met with it in July, 1826, near a fresh water pond at Long Branch being in company with Mr. Cooper. We observed a flock flying about at which I fired and killed the one here represented." (American Ornithology, IV, p. 89.) Turnbull (1869) states that it occurs in May and August and a specimen in his collection is labeled Brigantine. Dr. Jonathan Dwight secured ten specimens at Squan Beach between July 15 and September 15, 1879, which seem to be the only other early records for the New Jersey shore.

Julian Potter's bird of August 3, 1924, was associated with some Lesser Yellow-legs and there were from one to five present there throughout the month. In general form they resemble the Yellow-legs but are distinctly smaller, a difference easily recognized when they are feeding side by side or when flying together; the general color is somewhat paler, too, the prominent eye strip is distinctive and the legs are black instead of yellow. One individual showed traces of the barring on the flanks which is characteristic of the breeding season when it covers the entire under parts. Others in this plumage were seen by John Gillespie on July 31, 1927, on Two Mile Beach and by Charles Urner on July 18, 1928, at Barnegat.

In feeding the Stilt Sandpiper pokes its bill into the mud or sand several times in one spot without withdrawing it and often submerges its head in the operation resembling the Dowitcher in both particulars.

Julian Potter saw Stilts at Camden on the Delaware River on October 28, 1929, and found a flock of fifty feeding on a recent dredging at the same spot on September 17, 1932. On August 13, 1934, there were twelve at Cape May Point.

On August 23, 1926, I found a single Stilt with a large assemblage of Lesser Yellow-legs and other smaller species on a shallow rain water pool which had formed on a bit of meadow a little north of the beach drive above Broadway where they were sheltered from the nearby traffic by a dense growth of bayberry bushes. In this seclusion the flock remained for several days and many excellent opportunities for comparison of the Stilt with the other birds were offered. Its body was slightly smaller than that of the Yellow-legs and the bird was not quite so tall; legs and bill black and the latter seemed actually a trifle longer than that of the Yellow-legs. This was determined a number of times when the birds were in a position for accurate comparison. The back of the Stilt was much paler than that of any of the other birds with which it was associated and a Solitary Sandpiper looked black in comparison. When the Stilt Sandpiper flew its expanded tail was dull white. Another Stilt Sandpiper was seen that same day which had sustained a broken leg and was feeding in another rain water pool at South Cape May along with the Semipalmated Sandpipers, Killdeers and a Pectoral Sandpiper. It showed very clearly the dark banding on the flanks.

On August 25, 1931, a rain water pool formed on a patch of grass stubble back of South Cape May and for several days was well populated with a host of shore birds. There were five Stilt Sandpipers present which waded into the water until it was up to their bodies and in feeding repeatedly plunged the head under, probing into the bottom of the pool. Their actions were very different from those of the delicate stalking Lesser Yellow-legs which held their heads up and their necks constantly moving, or from the sluggish Pectorals which seemed more plover-like in action. Some of these Stilts showed the barring below while others, probably birds of the year, were plain gray above and white below.

On August 7, 1927, a group of shore birds was studied on a meadow pool back of Seven Mile Beach which contained several Stilt Sandpipers; they not only submerged their heads but tipped up like ducks in order to reach the bottom, the water being rather deep. The Dowitchers that were present also tipped up and submerged but the Lesser Yellow-legs fed entirely from the surface.

SEMIPALMATED SANDPIPER

Ereunetes pusillus (Linnaeus)

The most abundant shore bird about Cape May is undoubtedly the little Semipalmated Sandpiper, the smallest of the tribe with the exception of the Least Sandpiper, its frequent associate, and so like it in general coloration and habits, that, unless close at hand, we often find it difficult to distinguish them. Apart from the slightly larger size, however, the black instead of olive legs, lack of buffy tints on the breast (the ground color white throughout) and less blended colors above, will enable one to recognize the present bird although the two latter characters are most conspicuous in springtime. The partial webbing between two of its toes, to which its name is due, is not visible in life except under unusual conditions.

From early July throughout the summer the Semipalmateds occur in ever increasing numbers usually dwindling away again during September although small bunches may remain until November and we have records for December. In the spring flight they are with us from April to June and it is quite possible that if one were on the spot continuously they could be found in every month of the year.

They first appear on the mud flats of the sounds or on the muddy banks of salt ponds on the meadows and may be seen about the shallow pond at the Lighthouse or on the bottoms of smaller ponds which become entirely dry with the ebbing of the tide. The fresh pond formerly existing at the old race track, with its broad muddy margins, and the shores of Lake Lily at Cape May Point, also furnished satisfactory feeding grounds for these active little birds. About the middle of July or even earlier they occur in numbers on the sea beaches where they later associate with the Sanderlings and Ring-necked Plover, the three constituting the bulk, of our "beach birds."

The Semipalmated Sandpipers are typically gregarious, feeding together in rather close assemblages and taking wing simultaneously in that curious and puzzling manner of flocking birds, the individuals being able in some way to signal to each other so that the flock moves as a unit, instantaneously and with a definite objective. As one of these flocks circles about over the meadows, or wheels out over the surf, its actions are marked by the same unanimity, the individuals all turning, rising, or falling, together. The turning of their bodies in the air while in full flight produces a curious effect; at one moment one sees a series of brown backs exposed to view with rapidly vibrating wings and a slight flicker of white and then, presto, there is a sudden flash of silvery white as they turn their white bellies to the light. For a moment or two they sparkle and then with another turn they are gone and we have trouble in finding the flock at all. At a distance one may lose them completely so closely do their brown backs merge with the dark color of the marsh or water. In a fog, or in dull light with gray skies, passing flocks are practically invisible until they turn their bodies over and we catch the magic flash of white.

When they alight all are facing the same way, and when flying down the beach with the wind, they turn and face it as they come to rest. While they always occur in flocks there are some individuals that have become detached from their fellows and scattered birds may be seen all through the summer passing overhead, and what would, at a casual glance, be put down as a swallow in springtime, will now doubtless prove to be one of these little sandpipers.

On the exposed flats or on the grassy meadows, when feeding, they are all action from the moment their feet touch the ground, moving about in every direction like a colony of fiddler crabs, although even now the individuals somehow keep in close touch with one another and the flock keeps fairly well together. They walk deliberately with head lowered, probing with the bill every few seconds, now to the right and now to the left. Now and then two individuals will engage in an encounter fluffing up the plumage and jumping in the air like miniature gamecocks, the victor often pursuing his antagonist for short distances, their tails elevated throughout the performance.

On the beach they seem even more active than on the meadows, spurred on perhaps by the constantly recurring waves from which they must retreat, and also by the greater activity of their prey. They keep well down on the wet sand and run rapidly, sometimes through the shallow water of a receding wave, although this seems to be accidental and not habitual as it is in the case of the Sanderling. The latter, however, has always a closer association with the waves and water than has the present species.

At high water, when the coastal mud flats are flooded and the feeding on the beach is poor, the Semipalmated Sandpipers resort to the dry sand flats behind the dunes to rest until the turn of the tide. Here on August 30, 1920, I came upon several large flocks and by advancing slowly and silently I was able to get within fifteen feet of them and, so long as I stood rigid, they sensed no danger and went right on with their activities. It is only at such close quarters that one realizes the beauty and trimness of these little birds. They were now all in fully molted fresh plumage. Many were at rest squatting on the sand often in little depressions, some with heads turned back and bills buried in the scapular feathers, apparently asleep. A whole flock resting in this position looked precisely like little lumps of dark wet sand. Other birds were standing but with heads turned back as just described and still other with bills directed forward. Many of them were on one foot only and went hopping about preferring to advance in this way rather than to put the other foot down, although it would come into action when I started to follow them and kept them on the move. Occasionally one of the standing birds would lower the leg that had been drawn up and give it a rapid shake, as if in this cramped position it had "gone to sleep." When a bird squatted down on its belly it always placed its feet together and bending the legs at the heel (tibio-tarsal joint) would allow the body to settle down upon them. Other individuals would raise their wings high over their backs as terns and Yellow-legs so often do, and bring the axillary feathers into view.

There were little pools of water here and there and in these some of the birds were bathing. A few stood in the water and dipped their heads into it possibly seeking food. Others sat down and repeatedly ducked the whole head under, at the same time churning up the water by slightly spreading the wings at the shoulders and working them up and down. On another occasion bathing birds fluttered rapidly in the water and then raised up and did the same in the air, sometimes leaping six inches or more above the surface, possibly with the object of shaking off the water and drying the plumage. On August 28, 1933, just after the great storm and hurricane which flooded much of the Cape May district and drove salt spray far inland killing much of the vegetation, I watched a number of Semipalmated Sandpipers feeding in a flooded field north of Cold Spring, some of which submerged their heads completely in probing for food.

At a distance a large flock of these sandpipers at rest often completely escapes notice so perfectly do the birds resemble their background, and when they suddenly take wing it seems as if hundreds of the clam and oyster shells, that we supposed we had been looking at, had come to life and developed the power of flight.

On the large shallow pond to the east of the Lighthouse flocks of Semipalmated and Least Sandpipers are often seen feeding late in July or in early August when the water has become stagnant through lack of rain and is covered with a thick scum of vegetation which offers only partial support to the many shore birds that attempt to walk upon it. At every few steps the Semipalmated Sandpipers would sink in and immediately raise their wings over their backs and flap them to aid in regaining their footing. When one hundred or more were so engaged their wings made a continual flicker. On a small pond near South Cape May, under similar conditions, the sandpipers gathered on a submerged log to rest, crowding onto it in such numbers that there was not foot room for another individual.

During the northward migration the Semipalmated Sandpipers seem to be even more abundant on the sounds and meadows than they are in summer, on the return flight, or possibly their time is shorter and they are more concentrated. From one of the little gunning shacks standing on stilts back of Two Mile Beach, from which many of my studies were made, they are almost constantly in view, flying past singly or in small bunches like swallows, or gathering on the exposed flats in flocks of from twenty-five to several hundred, feeding, rising and settling again in restless activity. On May 20–24, 1922, Walker Hand and I watched them coming in to feed on the mussel beds of Jarvis Sound, as the water ebbed away and left acres of exposed mud and sand. Upwards of fifteen hundred Semipalmated Sandpipers gathered here every day exceeding the combined total of all other species of shore birds which frequented the spot. On May 22, no less than 276 came in during the first forty-five minutes of our count as against 245 of all other kinds and the same proportion was maintained, so far as we could judge, when the main body arrived. The Semipalmateds preferred the mussel beds and but few were to be seen on the smooth black mud, which the Black-bellied Plovers and certain other species frequented. They came in groups of six to thirty herding close together on the small area that first showed above the water, standing shoulder to shoulder, the outer ones with feet submerged, waiting patiently until the tide would subside sufficiently to make feeding practicable; then they would scatter in every direction and resume their characteristic activity. When the in-coming tide again began to flood the flats they reluctantly left in small flocks to seek the little salt ponds in the grass, or to rest on the dry sand stretches behind the beaches and on the higher parts of the meadows. On May 18, 1924, during a strong wind these resting birds sought shelter behind every tuft of grass or sedge on the dry sand that was large enough to act as a windbreak, or behind clods of mud or large shells which stood above the surface of the ground. It was comical to see them standing, mostly on one foot, one behind another in close packed lines, three or four feet long to leeward of the grass tuft, the first in line sheltered by the grass itself the others by the birds immediately in front.

In the spring they are more conspicuously colored than during the southward flight and are all definitely streaked on the breast with black, the elimination of the plain-breasted young birds moreover, makes identification more easy.

The Semipalmated Sandpiper arrives at Cape May during the last week of April or the first of May, the fact that some of the birds winter here, or only a short distance to the south, makes it difficult to decide just when the true transients arrive. Our records are as follows:

1903. April 29.	1912. May 8.	1925. May 8.	
1904. April 30.	1913. April 18.	1931. May 9.	
1905. May 2.	1916. April 22.	1932. April 30.	
1907. April 25.	1917. April 19.	1933. May 6.	
1908. April 28.	1920. May 1.	1934. May 5.	
1910. May 1.	1924. May 9.		

These dates may not be entirely accurate as I have never been present continuously at the Cape during the spring, but I have found birds plentiful in several years by April 30. Their height of abundance is in May and they begin to decrease by the 25th of that month. We have however many records for June: June 7, 1914; June 4, 1916; June 12, 1921; June 21, 1923; June 5, 1925; June 19, 1926; June 12 and 26, 1927; June 10 and 24, 1928; June 9, 1929; June 25, 1932; June 17, 1933; June 30, 1935. Just how these dates are to be interpreted it is difficult to say. They may represent very late north-bound migrants, or, as seems more likely, non-breeding birds which never reached their summer homes. Within a week or two from the last June record we begin to get the van of the south-bound birds my arrival dates being:

1916.	July 8.	1926.	July 16.	1932.	July 10.
1921.	July 2.	1927.	July 4.	1933.	July 8.
1923.	July 3.	1928.	July 7.	1934.	July 13.
1924.	July 12.	1929.	July 15.	1935.	July 13.
1925.	July 4.	1931.	July 6.	1936.	July 4.

Our latest autumn dates are:

1930.	November 9.	1932.	November 13.	1934.	November 3.
1931.	November 7.	1933.	November 19.		

The only December records were obtained on the Christmas census of the Delaware Valley Ornithological Club on December 24, 1933—nine seen by Richard Miller on Seven Mile Beach and two by Fletcher Street from a boat on the sounds, and three on December 22, 1935.

I saw several Semipalmated Sandpipers on the meadows west of Seven Mile Beach on March 18, 1933, and fifty on March 28, 1936, our only records for that month which leaves only two months in the year—January and February, in which these little sandpipers have not been recorded from Cape May.

Charles Urner's records of the migration of the Semipalmated Sandpiper for the region between Brigantine and Newark Meadows on the northern seaboard of the state are given below and show the extreme dates and period of regular or common occurrence, as well as the maximum count for one day at a single locality.

Northward migration:

1929. to June 22 (May 11–30). 4000 on May 19.
1930. to June 22 (May 11–31). 3500 on May 31.
1931. May 3–May 31 (May 17–31). 2000 on May 17.
1932. May 1–June 4 (May 5–25). 4000 on May 18.
1933. April 29–June 4 (May 5–June 4). 2500 on May 21.
1934. May 5–June 18 (May 12–27). 800 on May 16.

Southward migration:

1928. July 1–October 14 (July 1–October 14). 6000 on August 1.
1929. July 4–October 20 (July 13–October 6). 4000 on July 31.
1930. July 7–October 12 (July 7–September 21). 3000 on August 3.
1931. July 4–October 4 (July 11–September 27). 1400 on August 2.
1932. July 4–November 12 (July 9–October 15). 2700 on July 31.
1933. July 2–December 24 (July 2–October 8). 4000 on July 30.
1934. July 7–December 23 (July 11–September 30). 4000 on August 26.

In point of numbers the Semipalmated Sandpiper stands at the head of the six species grouped by Urner as "abundant" on our coast (see page 385).

WESTERN SANDPIPER

Ereunetes mauri Cabanis

While there is no question as to the distinctness of this Western race from the Semipalmated Sandpiper it seems probable that, with our changing conception of the differences between species and sub-species, the two forms may eventually be regarded as differing only subspecifically, as has already been done in the case of the Eastern and Western Dowitchers, birds which have exactly the same relation to one another.

The older writers up to the time of Baird, Cassin and Lawrence's classic "Birds of North America" did not recognize two forms although the latter authors did comment upon the extraordinary variation in the length of the bill. The two birds were apparently first separated by Lawrence in 1864 and while recognized by Ridgway in 1887, Coues in the same year regarded the western form: as "an alleged variety, probably untenable." Ridgway moreover stated that this race was "chiefly restricted" to the West, thereby implying that it also occurred in the East. C. W. Beckam (Auk 1885, p. 110) recorded one specimen taken in Virginia in 1884 saying that eastern records were few, and Hugh M. Smith mentioned fourteen collected at Piney Pt., Md., in 1885. Other records followed as collectors learned to separate skins of the two forms. The species was apparently first definitely recorded from New Jersey in my "Birds of Eastern Pennsylvania and New Jersey" (1894) on the basis of one taken and identified by Norris DeHaven at Atlantic City, May 17, 1892, and one found in Dr. W. L. Abbott's collection, Cape May Co., September 14, 1880. Other specimens previously regarded as Semipalmated Sandpipers were found in various collections, one obtained as early as 1885, by George Morris, at Beach Haven; one on Seven Mile Beach by David McCadden on September 5, 1898, and another by Dr. William E. Hughes on July 23, 1899, while William Baily secured a series of thirty-five little "peeps" on Two Mile Beach, Cape May Co., on September 1, 1895, of which he identified twenty as Westerns. (Auk, 1896, p. 174). All of the New Jersey specimens with one exception were taken in the autumn and Ludlow Griscom states (Birds of N. Y. Region) that "no specimens have been *collected* there in spring."

With the advent of "sight records" on such a large scale as we have to-day, we are confronted with numerous occurrences of Western Sandpipers on every hand. Charles Urner records them in numbers on the Newark Meadows, Barnegat and Brigantine in his reports on the shore bird migration on the New Jersey coast. In connection with his first year's record he adds "difficulty of satisfactory field identification with many individuals reduced the number of records," but the number increased in the later years and he had apparently found identification characters satisfactory to himself.

As Charles Urner is one of our most experienced students of the living birds we feel that his figures probably present as near an approximation to the relative abundance and time of occurrence of these two birds on the New Jersey coast as it is possible to reach. For ordinary observers, however, only a small proportion of the Semipalmated Sandpipers that one sees can be identified as Westerns with any pretense to scientific accuracy. And we agree with Griscom that "only the merest handful of ornithologists who have given special study to these birds are competent to make a sight record worthy of serious consideration."

In order to better appreciate the problem presented by these birds we may quote from two of the leading authorities on field identification. Roger T. Peterson says: the Western Sandpiper is "one of the most difficult species to identify in the field. When mixed with Semipalmateds it appears a little larger and more coarsely marked. The bill is very noticeably longer and thicker at the base. In the breeding plumage it is much rustier on the back and crown than the Semipalmated. A trace of this rusty is often evident in the fall. In late summer or early fall the more complete band of breast streaking is a good point. The breast band in the Semipalmated is reduced at this season to a dusky smudge at each side of the breast. The black feet distinguish it from the Least which is more or less rusty on the upper parts."

Ludlow Griscom in his "Birds of the New York Region" says: "An even more difficult proposition is to determine the Western Sandpiper. The greater average length of bill is not a field character. There remains simply color. In summer the Western differs from the Semipalmated exactly as does the Least but the difference is slightly more intensified (i.e. rustier on the back and more streaked below). I have yet to be convinced that it is possible to distinguish the two species satisfactorily in winter plumage."

The bill measurements given by Ridgway are, for the Semipalmated males .68–.75 in., females .80–.90 in. and for the Western, males .68–.90, females .85–1.15 in. As we manifestly cannot determine the sex in the field we have a range of .68–.90 in. against .85–1.15 in. The largest Semipalmateds therefore exceed the smallest Westerns and there must be a considerable number of individuals on either side of the 85–90 series that the human eye could not possibly determine. I am unable to say just how many hundredths of an inch difference in length can be so recognized but it is easy to see the meaning of Griscom's statement regarding bill length as a field character. All that can be done with regard to the length of bill is to refer the manifestly long bills in any flock to the Western and leave the others as Semipalmated, even though we know that some of them belong, almost certainly, to the former. Thickness of bill at base is a better character. Fortunately many of our late summer migrants are still in more or less complete breeding plumage and the rusty tints of these are diagnostic.

With regard to identifications by our Cape May observers: Fletcher Street finds the Western Sandpipers on Ludlam's Beach present every summer after the middle of July in association with the Semipalmated and has a few spring records; Julian Potter has recorded a number on the southward flight but none in spring and one on Seven Mile Beach on December 26, 1932. He has seen a flock of twenty as late as October 25, 1933, on the beach, where they seem more common than on the meadows after mid-September.

Personally I have every summer during July and August picked out a certain number of individuals with decidedly long bills and very rusty backs which I have felt sure were Westerns. One of these studied on July 28, 1929, with John T. Nichols, on the meadows back of Five Mile Beach possessed these characters while it was also very fully streaked below. It also kept to itself out on the end of the flock and did not mingle with the other birds. Ernest Choate and I found another with Semipalmateds on an overflowed field at Cold Spring on August 29, 1933, after the great storm of that month, which seemed slightly longer and more slender than the other species. It seemed to prefer to feed in the deeper water, a peculiarity mentioned by Urner.

Charles Urner's records for the northern beaches and meadows from Brigantine to Newark indicate that this species, as he identifies it, is very much more plentiful there than in the Cape May region. It is also, as we have found it in the southern areas, rare in spring, at which season his records are meager. He saw five on May 12, 1929; and a few others May 11 and 26; two on June 22, 1930; six on May 24, 1931; one on May 25, 1932; five on May 7, 1933 with a few more until the 21st and one on May 12, 1934.

On the southward migration his extreme dates, period of regular or common occurrence and maximum count for one day at a single locality are:

1928. July 22–October 12 (August 29–September 15). 10 on September 15.
1929. July 7–November 2 (July 13–September 15). 50 on September 13.
1930. July 16–October 12 (July 16–September 28). 30 on August 10.
1931. July 12–October 4 (July 12–September 20). 66 on August 30.
1932. July 4–November 8 (July 16–September 11). 25 on August 28.
1933. July 19–November 5 (July 19–October 1). 500 on September 10.
1934. July 14–November 11 (August 4–September 30). 100 on September 16.

Urner regards the Western Sandpiper as one of the ten "common" species standing just below the group of six designated as "abundant" but relegates it to last place in the group (see p. 385). In southern New Jersey we should not give it nearly so high a rating.

BUFF-BREASTED SANDPIPER

Tryngites subruficollis (Vieillot)

My personal experience with this rare little sandpiper is limited to one individual found on September 25, 1926, on a shallow rain water pool back of the low dunes at South Cape May. I came upon a bunch of six Pectorals and some Semipalmateds and with them was a bird that I had never before seen in life but I at once recognized it as a Buff-breasted Sandpiper.

The general buffy tone of coloration was very characteristic. The under parts were entirely uniform buff, with strong buff tints above, while the dark markings of the upper parts formed a peculiar shell pattern. The bill was very short and slender, shorter in proportion than in any other Sandpiper. All things considered it recalled the Upland Plover more than any other species and it is, in fact, a miniature of it in many respects. In size it seemed intermediate between the Pectoral and the Semipalmated Sandpipers with which it was associated.

The bird had one leg broken and hanging useless and it stood perfectly still on the other leg or occasionally hopped a few steps. It flew a few yards when I tried to approach still closer and then took wing and disappeared in the direction of the Delaware coast. It was not to be found in the afternoon nor on the following day although a diligent search was made.

On September 9, 1928, Julian Potter was fortunate enough to find six of these birds on Brigantine Beach. He first saw two and then two more.

"Their small heads, thin necks, brown and buff plumage and yellowish brown legs told me at once that they were Buff-breasted Sandpipers. Scarcely had I secured a good look at them when they flew and settled among a mixed flock of shore birds some forty yards distant. The striking white, dark edged under wing pattern noticed in the first birds was also shown by these. There were nineteen species of waders on the edges of the pool, among them thirty Willets, six White-rumped Sandpipers and a Baird's Sandpiper but not race of my Buff-breasted birds.

"Returning to where I had first seen them I found that the first two had returned and brought two more with them. Upon alighting they had immediately frozen but soon relaxed and while four started to feed in a rather indifferent manner, the others bathed. This was not a vigorous process as is the rule with most waders. One dipped the rear portion of its body in the water by teetering exactly like a Spotted Sandpiper; the other just wet the under side of its body by a series of squats. The wings were fluttered without touching the water, and both birds seemed in fear of disarranging their immaculate plumage. They moved about in a very deliberate manner, the folded wings extending just beyond the tail. Suddenly one uttered a short throaty err, err. Immediately alert, they all took flight, settling on the Brigantine Golf Course. Here two of them indulged in the curious performance of stretching one wing straight up over the back, apparently they were the bathers pluming themselves." (Auk, 1929, p. 109.)

This sandpiper was quite unknown to Wilson and Bonaparte and Audubon never saw a live one. It was discovered in Louisiana by the French ornithologist Vieillot, when visiting America about 1818, and Nuttall tells us that in 1834, a few were found in some seasons in the Boston markets in August and September. In 1844 Giraud had procured it on Long Island but I find no mention of its occurrence in New Jersey until Turnbull's brief statement in 1869 and little or nothing since then.

The first New Jersey specimens that I ever saw were two shot by W. M. Swain and presented to the Academy of Natural Sciences of Philadelphia, one secured on Barnegat Bay below Cedar Creek between the 7th and 21st of September, 1898, and the other at Toms River on September 9, 1899.

Charles Urner saw one on Brigantine Island on September 24, 1926, the day before I found mine at Cape May, another at the same spot on September 4, 1927, and a third on September 16, 1928; Potter's birds already mentioned were seen on September 9, of the same year.

In 1931, one was found on the Newark Marshes on September 6 (Kassoy and Herbert) while there are seven additional records in 1932 from the same locality on September 8, 18, 19, 22, 28, October 1 and 8 (Urner, Kuerzi and Edwards) and one for Brigantine golf course on September 11 (Henry Collins). In 1933 Clifford Marburger found a dead specimen on Brigantine on September 10 and one was seen there on August 24, possibly the same bird. In 1934 Charles Urner has three records for the Newark Marshes, August 30, September 15 and 23 and on August 30, 1935, F. W. Loetscher saw one on Brigantine.

In his list showing the comparative abundance of New Jersey shore birds Charles Urner places the Buff-breasted Sandpiper in the "rare" group (cf, p. 385).

MARBLED GODWIT

Limosa fedoa (Linnaeus)

Wilson tells us that in his time, about 1810, this splendid bird was found on the salt marshes of Cape May "in May and for some time in June, and also on their return in October and November; at which last season they are usually fat, and in high esteem for the table." He found them cautious and watchful and only to be approached by imitating their call or whistle. They were "much less numerous than the Short-billed Curlews [i. e. Hudsonian Curlew], with whom, however, they not unfrequently associate."

To the Cape May gunners the Marbled Godwit was known as the "Marlin."

Examining our scanty sources of subsequent information we find that on Long Island Giraud (1844) regarded it as not abundant but of regular occurrence in spring and autumn. Turnbull (1869) calls it "not uncommon" and W. E. D. Scott, in 1877, secured an adult in May and two young birds late in July on Long Beach. Dr. William L. Abbott on his many trips to Five Mile Beach, obtained but two specimens both taken on September 14, 1880. While old gunners, when I first visited Cape May in the "nineties," told me of shooting "Marlin," as these birds were known along the New Jersey coast, and did not seem to realize that they had become scarce, I was unable, however, to locate any specimens or secure any definite records. On Long Island, two had been shot on August 10, 1910, and a single indi-

(498)

vidual seen by a reliable observer on August 20, 1909. In view of the apparent disappearance of the species on our coast I was astonished to find one near Cape May Point on August 9, 1920. As I was walking down the beach below South Cape May I noticed a large brown shore bird with a very long straight bill coming across the meadows on set wings with the apparent intention of alighting on the strand but some persons walking there caused it to change its mind and it continued on beyond the Lighthouse and disappeared. I had a good look at it as it passed me and was in no doubt as to its identity. About half an hour later when I was standing at almost the same spot the bird reappeared sailing in on set wings as before, from the northwest, directly toward me, and alighted on the meadow in a patch of short dead grass. I had a splendid view of it, if only of short duration, and could see the slight up-curve of the bill at the tip, the very black narrow border to the front edge of the wing and the dark brown and buffy brown mottling on the back. The rump was plain buffy brown. Although I sank at once to the ground the bird apparently saw me and after a few moments was off again in the direction of the Race Track Pond where on set wings it again alighted. It was gone again, however, before I could reach the spot. On August 14, following, John T. Nichols and Charles Rogers saw a single individual on Long Island, perhaps the same bird or an associate!

For several years following these observations we find no record of this species in New Jersey but on August 22, 1928, J. S. Edwards saw two on the Newark Marshes where the same birds or others were found on August 29, 31, and September 1, of the same year. On September 2, 1928, Charles Urner and J. M. Johnson saw three on a fresh water pond on the Brigantine golf course, they wrote: "They were feeding within a few feet of the car in which we sat, running about on the partly submerged sidewalks, and we had opportunity to observe their method of drinking, curving the lowered neck apparently enough to allow the water to run into the throat. We could also observe their ability to flex the tips of their long pink-based bills." (Auk, 1929, p. 321.) Julian Potter saw two on Brigantine and Charles Urner one on Newark marshes, on September 15, 1929, while in 1931 Urner saw one at Tuckerton on September 6 and one on Brigantine, on the 20th.

In 1932, Charles Urner has recorded the Marbled Godwit on Brigantine on August 20, 27 and September 3 (two) and on September 4, a single bird at Barnegat Inlet. In 1933 single birds were seen on Brigantine August 23 (Lester Walsh) and August 27 (Urner) while Joseph Tatum found three on the Absecon meadow on September 12. In 1934 Julian Potter found one on Brigantine July 28 and the same individual or another was seen by various observers on August 5, 21 and 28. One was also present there in 1935 on August 15 (Lehrman), September 1 and 29 (Tatum), and November 10 (Potter) the last being our latest date for the species.

So while we have no further Cape May records the numerous observations on Brigantine Island and the recent conversion of that spot into a bird sanctuary bid fair to restore this fine bird as a regular autumnal migrant along our coast.

HUDSONIAN GODWIT

Limosa heamastica (Linnaeus)

While this species, the Ring-tailed Marlin of the gunners at Cape May, must have been present in the time of Wilson and Audubon neither mention a New Jersey specimen and the former was entirely unacquainted with the bird. The first mention of its occurrence in the state is by J. Doughty in his quaint "Cabinet of Natural History" (1832) in which he describes and figures an adult specimen in full nuptial plumage which "was shot in May 1828, by Mr. Titian R. Peale at Cape May and preserved in the Phila-delphia Museum [Peale's Museum]." He was aware of its occurrence oc-casionally in autumn in a different plumage but whether in New Jersey he does not say.

Turnbull (1869) knew the bird and regarded it as "rather scarce arriving late in September."

There are several old specimens in the collection of the Philadelphia Academy—obtained in the Philadelphia market by A. Galbraith in 1855, and another from the Delaware River, from Dr. S. W. Woodhouse, while there is a record of one shot by C. D. Wood, the noted taxidermist, on the Schuylkill River below the city in September, 1878.

Norris DeHaven told me that when he gunned on the Atlantic City mead-ows in the eighties this Marlin occurred with the Willet usually two or three in a flock of the latter species. Henry Hazelhurst showed me a speci-men that he had secured at Cape May in September 1900 and had had mounted but he had never seen another. H. W. Wenzel shot two on the upper end of Five Mile Beach on August 26, 1901, one of which is in the Philadelphia Academy collection.

On Long Island between 1881 and 1893 Ludlow Griscom considers that there were about twenty-five definite records and since then they have occurred on five occasions with quite a flight on August 28, 1922. In 1929, Charles Urner has recorded four observations of this bird on the Newark Marshes—August 31, September 29 and 30 and October 13, two on each occasion, probably six in all. On Barnegat Bay, he was informed by Capt. Chadwick that he had seen three of these birds in the fall of 1924 and at Elizabeth he himself saw an individual in full breeding plumage flying north over the Newark marshes on July 3, 1925. At Barnegat, again, on July 17, 1927, he was told by the guides that "there were a few 'Marlin' in the big flight of Curlew" that was passing at the time and upon locating one of the lines of flight on August 31 he saw "two birds somewhat smaller than the Curlew, with apparently straight bills, darker, not brown, upper parts, a wing pattern, and dark tails with a conspicuous white band at the base. Under the wing the feathers seemed dark but the breasts relatively lighter." Later five others passed at a greater distance but seemed to be the same. I saw two flying with Curlew in the same manner back of Seven Mile Beach on May 9, 1931. One was seen by John Kuerzi on August 31, 1931, at Beach Haven, one by Charles Urner at Tuckerton on May 15, 1932, and another by Lester Walsh on August 23, 1933, on Brigantine Island. Additional records for 1933 were Brigantine, September 10 (Tatum); Newark Meadows, September 17, 18, 20, 23, 24, 27 and 30 (Urner or Kuerzi). In 1934, Urner saw one at Absecon on May 16. He regards both of the Godwits as "rare" species in his list of comparative abundance of our shore birds (see p. 385).

All of the specimens of this Godwit that I have seen have been taken in summer or early autumn and were molting, presenting a very mottled appearance; a mixture of the nuptial and winter plumages.

*RUFF

Philomachus pugnax (Linnaeus)

There have been casual records of this European bird all along the Atlantic coast and more, apparently, on the island of Barbados than at any other single spot. Just how or why it crosses the ocean we do not know but doubtless storms have something to do with it.

On the coast of New Jersey we have a record of a specimen in the American Museum of Natural History labelled Barnegat (Chapman, Birds within Fifty Miles of New York City) and a statement by Turnbull which may well refer to the same specimen, while there are three records for Long Island.

No others were observed on our coast until October 2, 1932, when Charles Urner saw two near Tuckerton, N. J. which he identified first by sight and later by a careful examination of skins. He states that "the birds were seen first at a distance of fully one hundred yards, walking about on the salt meadows in search of food. They appeared at that distance, with the rising sun striking them, quite light colored, the under parts of one being especially light, the other definitely tinted with and finely streaked with buffy across the breast. They were not as long-legged as the Greater Yellow-legs of which many were present, but their bodies were fully equal in size. They were decidedly larger than the Stilt Sandpipers which were close by. The bodies were rather chunky and when the birds stood at attention they reminded one very much, in general proportions and profile, of overgrown Buff-breasted Sandpipers.

"The length of bill in relation to head conformed to that of the Ruff skins examined. The upper parts, while considerably darker than the breasts seemed, when the birds took flight (they circled twice and lit again), a bit lighter and buffier in color tone than the Greater Yellow-legs. The span in flight was somewhat less than in that species. The outstretched wings showed a narrow white line contrasting, when the birds were near at hand, with the darker wing. The rump and tail showed two conspicuous white areas on the sides, divided by a darker medial line, broadening out at the tip of the tail. I got very satisfactory views at forty to fifty yards." (Auk, 1933, p. 101.)

SANDERLING
Crocethia alba (Pallas)

The Sanderling is preëminently the sandpiper of the sea beach and I have almost never seen it elsewhere. It is unknown on the sounds or on the grassy edged ponds, on the borders of the upland, where Yellow-legs often congregate. On the beach its favorite place is down on the edge of the water as close to the waves as it can get with safety and when it retires to rest it seldom goes farther than the high beach where the sand is dry, or perhaps back on some extensive sand flat that the storms have formed by breaking through the dunes and spreading them out over the salt marsh.

In general appearance, as well as in habits, the Sanderling closely resembles the Semipalmated Sandpiper of which, indeed, it is in many respects a larger and paler, edition. It is half again as big and the head seems larger in proportion to the body while the tail tapers off so rapidly that the whole bird seems to have "run to head" or become top heavy. While its style of coloration is the same as that of the little sandpipers it is very much whiter, with little brown in the plumage, so that the general effect is distinctly lighter. In spring, however, some of the Sanderlings have assumed the rusty red-brown breeding plumage, with bright ruddy breasts, and the first arrivals from the north in July are still in this plumage, while most of the adults of July and August are in various stages of molt, to the gray and white dress of winter. While the adult winter plumage is plain gray above and white below, the young of the year are white above heavily streaked with dusky on crown and back. The Sanderling's frequent associate is the Semipalmated Sandpiper and flocks of the two often intermingle on the beach. The former occupy the outer zone closest to the water, next the Semipalmateds while beyond them may be some scattered Ring-necks which take wing with the others when disturbed, though the three species usually hold together in separate bunches within the flock.

It can hardly be said that the Sanderlings seek the company of any other species. They usually occur in large or small flocks, rarely singly, and are sufficient unto themselves. Scattered individuals of other species however will join the Sanderlings and we thus frequently find one or two Turnstones or a few of the two species above mentioned intimately associated with them.

There are no more nervously active shore birds than these Sanderlings of the strand. Almost any day in late summer or early fall we find them running rapidly, close to the edge of a receding wave, or actually through the shallow water, probing here and there as they go. All the while they manage to keep an eye on the incoming wave and just in time they will turn and scamper back, going only far enough to avoid a wetting. Then they once again face the sea and rush back with the outflowing water picking up such food as is being carried along with it, or such small animals as are burying themselves in the soft wet sand. As a bird slows up before turning for the rush, its humped up attitude gives one the impression of shrugging the shoulders. The sallies through the receding water and the dashes back again to safety are usually made on diagonal lines so that the birds are always making progress along the beach, while if anyone follows them they run directly parallel to the shore line, temporarily abandoning their feeding. When convinced that no harm is intended they begin again to follow the waves, glancing once in a while over the shoulder to see that their confidence is not abused. If pressed too closely they take wing and wheel out over the surf coming in to the beach a little farther on. As they fly there is a great show of white, the bands on the wings being especially conspicuous. The body seems heavy and the head and neck are bowed over as if straining every muscle to increase their speed, when they alight one might suppose that the bill touched the sand as soon as the feet so quickly is it in action again, pushing along through water and wet sand for several short steps without being withdrawn. Then the head is raised, the bird runs rapidly a few steps, probes once, and is off again, or else it slows down and pushes the bill along for a few steps as at first. Coming up to spots where the birds have been feeding thus, in wet places on the beach, I have found strings of little double holes as if the mandibles had been separated when the probing was done and in some spots the whole surface of the sand was fairly plowed up for an area of ten feet square. Sometimes, too, I have seen them at close range digging small Hippa crabs from the sand, plunging the bill again and again into the same hole until the entire face was covered.

Frequently one feeding bird will fluff up its feathers until he appears very much humpbacked and like a chicken will charge another one, pursuing him until he takes wing. Sometimes a third individual assumes the offensive and drives the first aggressor to flight, the latter relaxing his plumage the moment he is attacked. Sometimes, the attacked bird resents the onslaught and they jump at one another in true game cock style.

The Sanderlings bathe in pools along the beach which have formed around old pilings, squatting down and ducking their heads under with a great fluffing of the plumage.

When resting, usually at high tide, they retire to the higher sands of the back beach and collect closely together. A roosting flock observed on Two Mile Beach on August 2, 1920, contained over a hundred birds which were huddled closely together, looking at a distance like a mass of sand and shell fragments. Some were squatting on their bellies, others standing with their heads turned over on their backs and their bills buried in the feathers, and some of the latter resting on one leg with the other drawn up under the body. When approached, instead of taking wing, they scuttled away towards the water like a host of fiddler crabs retreating from an intruder, and their bodies glide so smoothly over the sand that it gives one the curious impression that they are impelled by some other force and are not connected with the rapidly moving legs. Many of the birds on one leg refused for some time to put the other one to the ground, stumping along like cripples over the sand, and I have seen feeding birds do the same thing until I was almost convinced that they actually had but one leg.

When standing in groups on the higher beach or the dry sand flats, they seem always to select the highest spots and blend so perfectly with the sand that one often fails to detect them until they begin to scatter and scuttle away.

Once or twice I have found two or three Sanderlings feeding along the line of trash marking the last high tide, fully forty feet from the water and once on October 17, 1920, two of them came to rest on a sand spit in the shallow pond above the Lighthouse just back of the beach, but this was a very exceptional occurrence and even in flight they prefer to follow the beach line or the surf. As in the case of the gulls, light materially affects the appearance of shore birds under certain circumstances. Once early in the morning, September 1, 1921, when the sun's rays were still not far from horizontal, a group of Sanderlings far down the beach were illuminated so that their white breasts shown like spots of glistening silver and again, September 15, when the sun was shining brilliantly on the beach I waded out into the water and approached a large flock of feeding Sanderlings from the ocean side, and in this direct light they appeared to splendid advantage, their breasts gleaming like snow.

Rarely Sanderlings are blown from the beach by storms and I so regard several seen on the shallow pond east of the Lighthouse and one observed flying over the sounds back of Two Mile Beach. At the time of the famous hurricane of August 23–25, 1933, Sanderlings were found along the turnpike leading to Cape May Point, but this at the time was practically the beach line as everything between it and the sea, with the exception of a few of the highest sand dunes, was under water.

My records for the arrival of the Sanderlings at Cape May on their southward flight are:

1916. August 7.	1920. August 1.	1923. July 21.
1917. July 28.	1921. July 22	1924. July 22.
1918. August 8.	1922. July 27.	1925. July 22.

1926. July 24.	1929. July 27.	1933. July 26.
1927. July 24.	1930. July 24.	1934. August 1.
1928. July 18.	1931. July 17.	1935. July 21.
	1932. July 19.	

The August dates may be rejected as late, since it has not been possible to watch the beach continuously, and small bunches of early migrants may easily have escaped notice. The flocks of summer number usually from six to thirty but in September and October they become much larger, often from one to five hundred. The constant annoyance occasioned by people strolling on the beach in summer doubtless tends to break up the flocks, although the birds are really not afraid and continue their search for food close up to where persons are bathing. The tendency, moreover, is to congregate at the entrances to the inlets as the season advances and our largest flocks have been seen at the ends of Seven and Five Mile Beaches, four hundred on the former, October 18, 1931; one hundred and fifty, October 14, 1928; five hundred, September 14, 1916. On the first occasion they lined the shore for a long distance, running along the water's edge and now and then turning head-on to the tide to feed. Sometimes a portion of the assemblage would take wing and the flock would be broken into several units and later united again.

While always to be seen at such places, and less frequently on the sea beaches, in October, the Sanderlings seem to decrease in numbers in November, although a few are to be seen on almost every trip to the Cape during this month. The Delaware Valley Club always finds Sanderlings at the time of its Christmas census as the record will show:

December 26, 1927, one (Potter)	December 26, 1932, sixty
December 23, 1928, twenty	December 24, 1933, 195
December 22, 1929, twenty-five	December 23, 1934, eighty-six
December 28, 1930, twenty-eight	December 22, 1935, 136
December 27, 1931, fifty	December 27, 1936, 211

Doubtless some of the large flocks remain at one or other of the inlets all winter, or smaller scattered groups somewhere along the shore, since we have records of eight at Stone Harbor January 23, 1927; four at the same place, February 22, 1928; four hundred on Gull Bar, March 26, 1932; and one at Corson's Inlet, April 8, 1928. Absence of observations through late April and May leave the actual arrival dates for the northward flight somewhat in doubt but the following will give some idea of their first appearance:

1909. May 4	1925. May 10.	1928. May 10.
1924. May 10.	1927. May 16.	1929. May 10.
		1931. May 9.

They remain usually until May 30, and there are several June records: Five Mile Beach, June 13, 1879, June 20, 1923; Seven Mile, June 16, 1918, June 18, 1922, June 12, 1927, and June 19, 1926; Ludlam's Beach, June 10, 1928.

These like other June shore bird records may probably have been non-breeding birds which never completed the northward migration, two seen at Anglesea on July 8, 1922, were certainly of this nature as they were in the gray plumage of winter, while the June birds were more or less ruddy. Charles Urner's records for the upper beaches from Brigantine north showing extreme dates, period of regular or common occurrence, and maximum count for one day at a single locality, are:
Northward migration:

1929.	March 10–May 30 (May 11–30).	50 on May 11.
1930.	April 13–June 22 (April 13–May 23).	50 on May 23.
1931.	to May 31 (May 12–31).	300 on May 31.
1932.	wintered to May 25 (May 8–25).	160 on May 8.
1933.	wintered to June 4 (May 7–28).	400 on May 14.
1934.	wintered to May 27 (April 8–May 27).	50 on April 8.

Southward migration:

1928.	July 1–January 13 (July 15–December 9).	500 on July 29 and Sept. 8.
1929.	July 4–December 8 (July 19–November 10).	600 on August 31.
1930.	July 6 through winter (July 16–December 20).	600 on July 26.
1931.	July 11–January 17 (July 18–December 27).	1100 on July 25.
1932.	July 10 through winter (July 17–Dec. 26).	1000 on July 29 and Oct. 30.
1933.	July 2 through winter (July 22–December 24).	500 on July 30.
1934.	July 8 through winter (July 15–October 14).	500 on August 25.

He regards the Sanderling as one of the six "abundant" species of shore birds on our coast today, in his comparative list (see p. 358).

The note of the Sanderling is a continuous twitter, a sociable conversational sound, fitting for birds of such eminently gregarious habits. Just as each bird brings to mind some scene, some particular environment, with which it is inseparably associated, so to me the thought of the Sanderling brings up recollections of a day late in summer, a glittering wind-swept ocean with bright sunlight reflected from every wave cap. The white surf booming and breaking on the beach and the sheets of shallow frothy water rushing up the strand, slower and slower until the turning point is reached, when they slip back to meet another wave advancing. And in their wake go the white and gray Sanderlings. A whole line of them push forward as the water retreats down, down, dabbing right and left as they run, on to the last moment of safety and then scampering back out of harm's way, the roaring of the breakers and the rush of water ever in their ears. Then as something disturbs them they take wing and away they go with rapid mutterings, wheeling out over the surf. They gleam white in the sun as their snowy breasts are turned toward the light, away in mad flight to new feeding grounds farther down the beach. Wild, hardy birds of the sea!

AVOCET

Recurvirostra americana Gmelin

Alexander Wilson found Avocets and Black-necked Stilts associated on the salt marshes of Cape May County on the 20th of May about 1810. "They were then breeding. Individuals of the present species were few in respect to the other. They flew around the shallow pools, exactly in the manner of the Long-legs [Stilts], uttering the like sharp note of *click click click*, alighting on the marsh, or in the water, indiscriminately, fluttering their loose wings, and shaking their half bent legs, as if ready to tumble over, keeping up a continual yelping note. They were, however, rather more shy, and kept at a greater distance. One which I wounded, attempted repeatedly to dive; but the water was too shallow to permit him to do this with facility. The nest was built among the thick tufts of grass, at a small distance from one of these pools. It was composed of small twigs, of a seaside shrub, dry grass, sea weed, &c. raised to the height of several inches. The eggs were four, of a dull olive color, marked with large irregular blotches of black, and with others of a fainter tint.

"The species arrives on the coast of Cape May late in April; rears its young, and departs again to the south early in October. While here it almost constantly frequents the shallow pools in the salt marshes; wading about, often to the belly, in search of food, viz., marine worms, snails and various insects that abound among the soft muddy bottoms of the pools." (American Ornithology, Vol. VII.)

Audubon visited Great Egg Harbor in May, 1829, and saw but three Avocets and found no nests. Perhaps they had, even then, all but disappeared as breeding birds, while they seem never to have been plentiful.

The older ornithologists have left us no further records, although Turnbull (1869) states that it is "rather rare." As his dates of occurrence are evidently taken from Wilson he may have had no personal experience with the bird. There is also an old specimen in the Philadelphia Academy collection received from Samuel Ashmead of Beasley's Point, which is opposite Peck's Beach, where Wilson found his Avocets.

Later records follow.

Long Beach, May 20, 1877 one seen by W. E. D. Scott (Bull. Nutt. Orn. Club, 1879, p. 224). Barnegat, May 31, 1880, one shot by John Fonda (Auk, 1905, p. 78). Tuckerton, one shot by I. Norris DeHaven, last of of August, 1886 (Stone, Birds of E. Penn. and N. J., p. 70). Seven Mile Beach (Avalon), middle of September, 1908, one seen by I .W. Griscom. Krider (Forty Years' Notes) states that both the Avocet and Stilt nested on Egg Island in Delaware Bay but many of his statements must be taken with reservations.

Then after a lapse of twenty-four years Charles Urner had the good fortune to find three Avocets "on September 15, 1932, on what was formerly a salt marsh near the Newark, N. J., airport, at a point where the high tide still partly floods the fill with a mixture of salt water and sewage. They were among flocks of hundreds of assorted shore birds. There were two black and white and one dark brown and white bird. The heads and necks showed no color tint. The legs appeared clay-colored. One old bird called occasionally when disturbed though they were very tame. Its notes reminded one of the Lesser Yellow-legs, but louder and fuller, at times with a shade of hoarseness. The individual notes, when uttered in sequence, were often spaced with an appreciable interval between. The birds fed both by side-swiping the surface of the water, as does the Lesser Yellow-legs, and by probing.

"They remained until October and were seen to swim on October 1; when pursued by a Duck Hawk they took to deep water where their profile afloat with stern carried high, seemed distinctive and a good field mark at a distance." (Auk, 1933, p. 100.)

In publishing his account of the lone Avocet seen by him on the beach near Avalon (Seven Mile Beach) I. W. Griscom writes "It seems that these singular birds with their long, slender, up-turned bills were at one time quite abundant on this part of the Atlantic Coast but are now very rare east of the Alleghanies. It came across from the meadows to the beach and flew so close by me that identification was certain and its discordant clamor indicated that the bird had lost none of its lawyer-like qualities. It opened up its boisterous address upon seeing me in the distance and did not change its course in the least to avoid me" (Forest and Stream, January 31, 1909)

BLACK-NECKED STILT

Himantopus mexicanus (Müller)

Alexander Wilson presents an interesting account of this species which has long departed from the Cape May meadows where he found it of regular occurrence, in 1810. He writes:

"This species arrives on the sea coast of New Jersey about the twenty-fifth of April, in small detached flocks, of twenty or thirty together. These sometimes again subdivide into lesser parties; but it rarely happens that a pair is found solitary, as during the breeding season they usually associate in small companies. On their first arrival, and indeed during the whole of their residence, they inhabit those particular parts of the salt marshes pretty high up towards the land, that are broken into numerous shallow pools, but are not usually overflowed by the tides during the summer. These pools, or ponds, are generally so shallow, that with their long legs the Avocets [i. e. Stilts—Wilson called them Long-legged as opposed to the Red-necked Avocet] can easily wade them in every direction. . . . In the vicinity of these bald places as they are called by the country people, and at the distance of forty or fifty yards off, among the thick tufts of grass, one of these small associations, consisting perhaps of six or eight pair, takes up its residence during the breeding season. About the first week in May they begin to construct their nests, which are at first slightly formed of a small quantity of old grass, scarcely sufficient to keep the eggs from the wet marsh. As they lay and sit, however, either dreading the rise of the tides, or from some other purpose, the nest is increased in height, with dry twigs of a shrub very common in the marshes, roots of the salt grass, seaweed, and various other substances, the whole weighing between two and three pounds. This habit of adding materials to the nest, after the female begins sitting, is common to almost all other birds that breed in the marshes. The eggs are four in number, of dark yellowish clay color, thickly marked with large blotches of black. These nests are often placed within fifteen or twenty yards of each other, but the greatest harmony seems to prevail among the proprietors.

"While the females are sitting, the males are either wading through the ponds, or roaming over the adjoining marshes; but should a person make his appearance, the whole collect together in the air, flying with their long legs extended behind them, keeping up a continual yelping note of *click click click*. Their flight is steady, and not in short sudden jerks like that of the plover. As they frequently alight on the bare marsh, they drop their wings, stand with their legs half bent, and tremble as if unable to sustain the burden of their bodies. In this ridiculous posture they will sometimes stand for several minutes, uttering a purring sound, while from the corresponding quivering of their wings and long legs, they seem to balance themselves with great difficulty. This singular manoeuvre is, no doubt, intended to induce a belief that they may be easily caught, and so turn the attention of the person from the pursuit of their nests and young to themselves.

"The Red-necked Avocet, practises the very same deception, in the same ludicrous manner, and both alight indiscriminately on the ground, or in the water. Both will also occasionally swim for a few feet, when they chance in wading to lose their depth, as I have had several times an opportunity of observing.

"The name by which this bird is known on the seacoast is the Stilt, or Tilt, or Longshanks. They are but sparingly dispersed over the marshes, having, as has already been observed, their particular favorite spots; while in large intermediate tracts, there are few or none to be found.

"They occasionally visit the shore, wading about in the water, and in the mud, in search of food, which they scoop up very dexterously with their delicately formed bills. On being wounded while in the water, they attempt to escape by diving, at which they are by no means expert. In autumn, their flesh is tender, and well tasted. They seldom raise more than one brood in a season, and depart for the south early in September." (American Ornithology, Vol. VII.)

Audubon has nothing to say of their occurrence in New Jersey except that he never noticed them raising the height of their nests as described by Wilson.

William M. Baird wrote to his brother Spencer from Cape May, July 16, 1843, that there were still some of them present and later, on July 21, that he had secured a specimen, at Cape May Court House. Wilson seems to have gotten the local names of this and the preceding bird confused as he calls the Avocet the "Lawyer" whereas that name belonged to the Stilt, the Avocet being known as the "Blue Stocking." Turnbull (1869) says that the Stilt is "rather scarce, I have found its nest on Egg Island, Delaware Bay." Doubtless the extensive marshes where the Willet has recently been found breeding abundantly, were the last stand of the Stilt.

From Turnbull's time to date I have been unable to find a single reference to this bird in New Jersey except the specimen mounted by Charles A. Voelker which was said to have been shot on Seven Mile Beach on April 27, 1894, and concerning which I have no further information.

NORTHERN PHALAROPE
Lobipes lobatus (Linnaeus)

Phalaropes are essentially seagoing sandpipers, and so far as New Jersey is concerned, are distinctly birds of the ocean like the Kittiwake, the jaegers and the petrels, and only on exceptional occasions are they found along the coast or farther inland, driven by severe storms from their true habitat on the open sea. During their spring and autumn migrations between their winter home in the southern hemisphere and their breeding grounds in the Far North, both the Northern and Red Phalaropes are seen by passengers on ocean steamships floating in great rafts on the waves. We have several records of such observations off the coast of New Jersey. Dr. Frank M. Chapman has recorded both this species and the Red Phalarope off the Delaware coast on May 9, 1897, and Dr. James P. Chapin and others saw phalaropes, probably the Red, off New Jersey, November 27, 1928. Occasional specimens of the Northern Phalarope have been shot, captured or picked up dead on the beaches of the New Jersey coast but usually by persons who have kept no record of the weather or of the behavior of the birds. On May 6, 1933, in company with Otway Brown, I was fortunate enough to be on the strand of Seven Mile Beach during one of the most notable occurrences of these birds of which we have record. There was a strong northeast wind and overcast skies with a fine misty rain and heavy surf. As soon as we came out on the beach from the dunes we noticed a number of small birds flying just beyond and among the breakers. They kept close to the water constantly alighting upon it and rising again to avoid the breaking waves. They appeared in the wind much as a flock of Tree Swallows under similar conditions, their bodies short and compact with neither head nor tail projecting as they beat into the gale. But they differed from swallows in showing a distinct flicker of white in the wings such as most shore birds do in flight. Approaching closer I could see them swimming with head more or less erect and turning to this side or that, ever alert for an approachng wave. Others that we found farther along the shore were swimming in shallow beach pools just beyond the reach of the tide, where they were not annoyed by the waves. They held the neck erect and apparently stretched upward to the limit, while the delicate head with its needle-like bill was turning constantly this way and that. When progressing straight ahead they swam rapidly but for the most part they kept twisting to right and left sometimes whirling entirely around in a circle. They floated buoyantly, like corks, and were evidently feeding as they would dab the bill to the surface of the water continually sometimes ducking the head completely under.

A number of those first seen came in and alighted on the wet sand where they also fed. They walked in a somewhat bowlegged fashion and progressed rather slowly, with many short steps. Some of them squatted on the beach holding the head stiffly erect and turning it as did those that were swimming in the pools. Others plumed themselves rapidly poking the bill over the shoulders and down into the breast feathers all the time seeming to rub the side of the head against the plumage.

There was a Semipalmated Sandpiper and several Ring-necks nearby on the strand and by comparison the Phalaropes seemed about the size of the latter but shaped more like the former, having the sandpiper build rather than the plumpness of the plover, but more delicately formed than either. In the dull light and the rainy mist they seemed quite black above, many showing clearly the longitudinal bands of brownish chestnut down each side of the back and a white stripe on the wing, some also showed a white tuft at the shoulder. Below they appeared black across the breast and down the sides of the body mixed with white toward the belly which was pure white. There was also a white throat spot spreading on each side so as to form more or less of a collar and below it a band of chestnut.

They were very tame, or perhaps exhausted, and allowed us to approach to within fifteen feet without taking wing. There were twenty on the sand at one time and about one hundred over the surf at this point with other similar flocks farther along. At one spot there was a large band of Barn Swallows sitting on the beach and they appeared slightly larger than the Phalaropes which were resting close to them. Some of the Northern Phalaropes that were feeding on the beach ran right into an oncoming wave and floated back on it but took wing when threatened by a curling crest.

The next day Joseph Tatum found a similar flock of these birds at Ship Bottom, on Long Beach, some distance up the coast and I quote from his account of his experience. He says: "I saw a bird out on the water just inside the first line of breakers which looked like a Semipalmated Sandpiper. A few yards away was another and still others to the number of perhaps a dozen. They were quite active swinging from side to side or turning completely around with incredible rapidity. They were constantly picking something from the surface of the water with their needle-like bills making twenty-five dabs per minute. None were seen to dive but they would frequently arise from the water with no apparent effort and fly a few yards dropping back again onto the water where they at once resumed their feeding.

"In flight they looked like sandpipers but with a little more sweep and grace to the wing stroke. The head seemed rather large in proportion and distinctly round, while the neck was held very erect, at right angles to the plane of the body. As they bobbed around in their feeding the head would swing back and forth, like a pigeon's head when walking. In fact the whole impression of the bird's progress in the water is of walking rather than swimming. They are the most buoyant birds I have ever seen, riding the crest of a breaker with the easy nonchalance of an empty bottle. I could hear them utter no sound except an occasional peep as they flew. At no time did I see any of them on the beach. They were found at different points along the coast during the day, probably more than one hundred in all.

"On the jetty at 4:00 p. m. I had a dozen of them within fifty feet of me. I could look almost vertically down upon them and noticed a flange of feathers projecting from the body onto the water like the overhanging decks of a ferryboat, which apparently could be extended or withdrawn at will thus keeping the bird always in perfect balance. This may account for their remarkable buoyancy. As they float they seem to use the feet for wheeling and for maintaining equilibrium, the action being a down stroke like treading water rather than a backward pushing stroke.

"Their outstanding markings were the white chin and upper throat; white eye ring and rich chestnut red patches on the sides of the neck."

Storms occasionally carry phalaropes far inland and Earl Poole has four records of this species at Reading, Pa.—September 14, 1909; October 3, 1923 (two); August 28, 1932; August 24, 1933. Julian Potter also found a pair on an overflowed meadow near Camden, N. J., on August 13, 1916. He writes me that they seemed to be feeding on small insects on the surface of the water and swam very swiftly and easily against the strong wind, while they twirled in the water at times turning completely around. The larger bird continually sought the company of the smaller; they uttered a sharp *peet, peet*, when on the wing. He watched them for half an hour at distance of from fifteen to forty feet and they were still dashing about when he left. There had been a northeast storm the day before.

Additional records for the New Jersey coast are:

1894. May 23, Peck's Beach, F. L. Burns.
1895. September 13, Barnegat Bay, A. P. Willets.
1903. September 4, Seven Mile Beach, David McCadden.
1909. May 4, Cape May, Walker Hand.
1911. May 16, Cape May, Walker Hand.
1928. Barnegat Bay, August 12, September 15, Charles Urner.
1931. August 30, Brigantine Island, Charles Urner.
1932. May 12, Beach Haven, Charles Urner.
1932. September 8, Newark Meadows, Charles Urner.
1933. May 7, Long Beach, Joseph Tatum.
1933. Newark Meadows, August 23 (2), September 9, 10, 13 (Charles Urner).
1934. August 4, Cape May, J. L. Edwards.
1934. September 18, Newark Meadows, J. L. Edwards.

RED PHALAROPE

Phalaropus fulicarius (Linnaeus)

This larger and heavier species has precisely the same status and habits as the Northern Phalarope although it appears to be less frequently blown inshore, or perhaps through force of circumstance, has been less frequently observed or recorded. Dr. Frank Chapman has seen it off the Delaware coast as he returned by sea from a sojourn in Mexico and others aboard ship have noticed these little birds disporting themselves on the ocean.

Capt. Horace O. Hillman of the Cape Cod mackerel fleet, which puts in to Cape May every spring, reported immense numbers of them resting in great rafts out on the open ocean where the fishermen were casting their nets sixty-five miles south-southeast of the Cape and at the request of Walker Hand secured eleven specimens for the Academy of Natural Sciences of Philadelphia, on April 28, 1929. These were all in mottled plumage in various stages of the molt from the gray and white dress of winter to the rusty red of summer. The fishermen of the fleet were all familiar with the phalaropes and called them "Bull-birds," "Sea Geese" and "Sea Plover," names which they apparently applied as well to the small species.

While the occurrence of this bird offshore was well known from early times I had but two definite records for Cape May County up to 1910—one obtained by William H. Werner on Peck's Beach, May 6, 1907, and another secured by Walker Hand at Cape May, May 3, 1909, and so far as I am aware there were no records from the northern beaches of the state. While living in a little fisherman's cabin on the meadows back of Two Mile Beach on May 18, 1924, I found a Red Phalarope's foot close to a stake which had been driven into the ground and which served as a feeding perch for visiting hawks and Ospreys. Although the scattered remains of fish and bird skeletons were carefully examined no other phalarope bones could be found. This is my only personal record of the species at the Cape!

In December, 1918, there seems to have been a flight of Red Phalaropes extending inland to Pennsylvania where two were picked up dead, or dying, one at George School, Bucks Co., on December 15 (Cocks), and the other near Lenape, Chester Co. in mid-December (Ehinger). Both of these birds were in the gray plumage of winter. While these captures were entirely outside the limits of our study they are of interest as evidence that the species must winter to some extent at least on the north Atlantic. Further interest attaches to these records because the specimen figured by Alexander Wilson was of similar occurrence, one of three obtained near Philadelphia in May 1812, evidently blown in by a storm. On May 26, 1931, Laidlaw Williams and others saw a female in full summer plumage on Barnegat Bay (Auk, 1931, p. 597), and Charles Urner has a record for that year in the same region.

On May 12, 1932, during a northeast storm, occurred the greatest invasion of Red Phalaropes that has ever been recorded on the New Jersey coast, and fortunately it was witnessed by several observers who realized the importance of what they saw and made a satisfactory record. Charles Urner and J. L. Edwards have published the most detailed account of the flight (Auk, 1932, p. 475). They saw a few of the birds on the Tuckerton meadows fighting against a strong wind and the first one observed appeared, in the poor light, absolutely black like a Black Tern, with a white stripe in the extended wing. "As we crossed Barnegat Bay" writes Mr. Urner "over the Manahawkin Bridge to Long Beach another individual was seen. Driving south toward Beach Haven we found others in the bay and one swimming in a puddle by the roadside. Soon we became conscious of the fact that all the shore birds that were passing over the dunes making slow headway against a heavy wind and rain, were of this species.

"When we reached Beach Haven Inlet, a rare sight greeted us. The place fairly teemed with Red Phalaropes. We stood on a small spit of sand, while in a sheltered bit of water, literally right at out feet, a large flock of these striking and agile birds fed over a mass of seaweed and garbage. We collected two and picked up another dead on the road.

"The birds were in every degree of plumage change, about forty percent fully colored. A few were in almost complete winter plumage except that the forehead, white in the winter birds, was dark. The darkening of the forehead is probably one of the first noticeable changes toward the summer attire. A good many birds were fully red but showed little or no definite white area on the side of the head (not even as much as the male bird shows in summer). The white face is thus probably the last feature of the breeding plumage to be acquired. We saw fully three hundred birds and probably more."

Mr. Urner further comments on the resemblance of this species to the Northern Phalarope saying that "those still in winter dress could easily have been mistaken for Northerns but for the yellow on the base of the bill, visible only at short distances. Seen alone, without contrast, the bills of the

Reds did not seem particularly heavy." In the case of a single Northern Phalarope riding the waves with the Reds its thinner bill and the absence of yellow could be noticed.

Stuart Cramer who was at Brigantine Beach three days later saw two Red Phalaropes swimming on one of the channels about one hundred yards apart in fairly deep water. They were picking up food of some sort from the surface of the water holding the bill vertically, point down, and when a morsel would float past they would spin about to secure it. The white bar on the wing was very evident. Natives informed him that the water was full of these birds a few days before, especially farther back on the meadows. Rainy, stormy weather prevailed during the week prior to his visit which doubtless drove the birds inshore.

On May 14 members of the West Chester, Pa., Bird Club found a flock of twenty-five Red Phalaropes on the point of the island below Beach Haven and picked up another exhausted individual (Isaac G. Roberts *in lit.*).

On May 7, 1933, on Long Beach, Joseph Tatum observed a few of these birds in the flight of Northern Phalaropes which he has described as quoted above. Their habits were identical but they could easily be distinguished from the latter species by their reddish breasts, being in nearly full summer plumage, and by the white patches on the sides of the head and the lighter bills. The amount of red on the breasts, however, varied in different individuals. Oscar Eayre also saw several in June of that year at sea sixteen miles off the same beach and on September 6, 1935, F. W. Loetscher saw a single Red Phalarope on Brigantine Island. Charles Urner has given me three additional records: May 7, 1933, and August 18, 1934, both at Beach Haven and May 12, 1934, Absecon.

WILSON'S PHALAROPE

Steganopus tricolor Vieillot

Unlike the other two phalaropes this species instead of summering in the far north, breeds in the middle western United States and southern Canada and its occurrence on the North Atlantic, or on our eastern sea coast, would seem to be entirely abnormal. Nevertheless I am convinced that such occasional individuals as have been recorded from New Jersey have come in from the sea as in the case of the other species. They may very likely have accompanied flocks of shore birds with which they always seem to associate, and have travelled with them in a southeasterly direction from their breeding grounds in the interior.

Our earliest records are very indefinite. George Ord records a specimen shot near Philadelphia on May 7, 1818, and prepared for Peale's Museum; Audubon was shown two that his informant told him were shot in July near Cape May "close to their nest which contained four eggs," a statement that tends to discredit the entire record, while Dr. C. C. Abbott records two captures without dates or authority—one taken at Deal Beach and the other at Atlantic City—but they may have been referable to one of the other species.

The first Wilson's Phalarope to come to my personal attention on the New Jersey Coast was one shot by Gilbert H. Moore on Peck's Beach, on May 19, 1898, and now in the collection of the Academy of Natural Sciences. On May 4, 1909, two were shot at Cape May one of which was procured by Walker Hand and presented to the Academy. I have no information as to the circumstances of the capture of either of these specimens nor as to their habits or actions.

On September 2, 1929, a single individual in the dull plumage of winter, or of immaturity, was found by Julian Potter on a tidal mud flat on Newton Creek in West Collingswood, Camden Co., N. J. It was feeding with a group of Lesser Yellow-legs and its thin neck, small head and light coloration were in marked contrast to its associates. It was very active, swimming about in a small pool, darting its slender dark bill from side to side and sometimes turning its body half way around as it fed. Once it came out on the bare mud to plume and arrange its feathers, showing its pale yellow legs, white rump and pale wings. During the half hour that the bird was under observation it confined its activities to a space of a few square yards.

On September 15, what was apparently the same bird was present in almost the same spot and was studied by four members of the Delaware Valley Ornithological Club and on the 18th it was found again. On each occasion it was feeding with the Yellow-legs (Auk, 1930, p. 76). In 1930, two Wilson's Phalaropes were seen on Brigantine Island by John Gillespie on August 10, and two on September 1, by Charles Urner. One was also observed at Secaucus by J. L. Edwards and others, on September 13. On August 28, 1932, Julian Potter was fortunate enough to find another of these rare visitors on Brigantine, running about with a large flight of shore birds on a mud flat. It presented a comical sight with tail held up at an angle and neck stretched out in front while it held the body in more or less of a crouching position. On September 11 he saw another individual at the same place while Charles Urner saw single birds (or the same individual) on the Newark Meadows on September 18, October 1, 6 and 8 and two at Tuckerton on September 3. In 1934, F. W. Loetscher found a Wilson's Phalarope on Brigantine Island on August 18 and again on the 25th, while Warren Eaton saw probably the same bird on the 20th, and Richard Pough saw one on Ludlam's Beach, Cape May County, on September 10. There were a large number of observations of this phalarope on the Newark Meadows during this year, single birds on September 4, 8, 9, 10, 13, 16, 17, 18, 20, 23 (Urner, Edwards, Herbert, Kuerzi and Loetscher) while Urner saw three on the 24th, two on the 27th six on the 30th and four on October 1; a remarkable influx of these birds due undoubtedly to severe storms at sea. In 1934, J. L. Edwards saw a Wilson's Phalarope on Peck's Beach, Cape May County, August 4, in company with a Northern Phalarope while Charles Urner found one at Tuckerton on August 5 and four on the Newark Meadows on August 8. He has also records of one on the northern beaches on May 12 of the same year.

In 1936, Julian Potter found one of these birds on the Killcohook Wild Life Refuge on the Delaware River in Salem County, on October 11.

PARASITIC JAEGER

Stercorarius parasiticus (Linnaeus)

Jaegers, like petrels, are birds of the open ocean so far as the New Jersey coast is concerned and their occurrence inshore is exceptional and usually due to storms. They are the "hawks of the sea," gulls modified into birds of prey, as it were, and their parasitic habit of forcing other sea birds to disgorge food for their benefit is unique. At Cape May nearly all of our records of jaegers have been during late September and October when on their southward migration from Arctic breeding grounds.

Julian Potter saw one on October 1, 1925, pursuing a Laughing Gull at the Point. It appeared quite dark in contrast with the gull and showed a distinct flicker of white in the wings and a rounded tail although the usually extended central tail-feathers were not noted. Next day he saw two Parasitic Jaegers harrassing the gulls and when one of the latter was hard-pressed "it spread the contents of its stomach on the waves and immediately both jaegers and gulls settled on the water to devour their ill-gotten gains."

On September 30, 1928, Potter saw another Parasitic Jaeger come flying in from the ocean and begin to hunt over the marsh back of South Cape May like a Marsh Hawk. It hovered and dived into the grass several times before a passing Pigeon Hawk gave chase, when it immediately put out to sea and rapidly disappeared from view. Philip Livingston saw one in the same place on October 5, 1930.

(521)

On September 27, 1931, one of these birds was repeatedly flushed from the ocean beach and seemed exhausted and unable to remain on the wing, possibly a victim of oil poisoning; and on August 24, 1933, following the great hurricane, I saw one fly in from the sea and cross the Fill to the Harbor, the white flashes in the wings showing conspicuously.

While jaegers seem to be very rare in spring and early summer, Walker Hand and Frank Dickinson evidently saw one on Delaware Bay off Town Bank on June 6, 1922. They were fishing from a skiff and their attention was attracted to the bird by the calls of a Laughing Gull which was dodging this way and that in its efforts to escape the persecutions of the strange dark bird which followed it relentlessly. Finally as it was passing down the beach the gull disgorged a mass of food which its pursuer caught up and swallowed and then put off down the Bay for the ocean. It appeared black above and white below, was not quite so large as the gull and flew differently with more rapid wing strokes. Frank Dickinson reports another instance when he saw a bird, similar to this one, pursue a Red-breasted Merganser on the sounds and force it to disgorge its food which was immediately eaten. He was closer to this individual and it appeared grayish brown above.

One or two jaegers, probably of this species, were seen off the coast below Cape May on October 25, 1929, when members of the American Ornithologists' Union visited the Cape. Audubon was apparently the first to record a jaeger on the New Jersey coast as he says, in his account of the Laughing Gull at Egg Harbor, in May, 1829, just one hundred years before our last mentioned record, "Like other Gulls the *Larus atricilla* disgorges its food when attacked by a Lestris." As the Parasitic Jaeger is the only one that he mentions as ranging this far south he must have referred to it, although he says that he never saw it here except in winter! We have additional records of three of these birds seen off Peck's Beach by William Baily on November 9, 1895; one collected by Capt. John Taylor on Five Mile Beach, on October 23, 1891 (in the Reading, Pa., Museum); and one seen off Seven Mile Beach by Charles Voelker, on May 27, 1901. Farnum Brown also examined one shot by a fisherman off Atlantic City, in March, 1892.

Farther north on the coast jaegers seem to be more frequent and occur on earlier dates, as Charles Urner and others have recorded them from Brigantine, Barnegat and Point Pleasant on July 6, 1927 (six); July 23, 1926; July 23–24, 1927 (six); August 22, 1925 (eleven); August 23, 1933; August 26, 1926; August 28, 1927 (eight); and on numerous occasions in the autumn. During September 1926, there was a great flight of these birds and at Point Pleasant they were recorded as follows: September 6 (seven); September 15 (fifteen); September 18 (forty); September 19 (seventy), and sometimes as many as six were seen pursuing a single gull. It would appear from observations elsewhere along the coast that jaegers remain on the North Atlantic in small numbers still later in the autumn, or early winter, although most of them pass on to the more southern seas. T. Donald Carter and Ralph Friedman have recorded eight at Barnegat on October 31, 1926, and Charles Urner one on November 29, 1924.

POMERINE JAEGER

Stercorarius pomarinus (Temminck)

The Pomerine Jaeger is a larger and heavier bird than the Parasitic and is much rarer on the coast although well known to occur offshore during the autumn migration. We have no record for Cape May but Dr. James P. Chapin has recorded them at sea not far from our coast (Auk, 1929, p. 102), as he saw twenty following the wake of a vessel upon which he was a passenger somewhere between New Jersey and Virginia.

On the North Jersey coast Charles Urner and others have recorded Pomerine Jaegers as follows: Barnegat, August 26, 1926 (two) (Kuerzi); August 28, 1926 (Urner); August 28, 1927 (Urner); October 11, 1925 (two) (Weber); September 6, 1925 (Weber); Pt. Pleasant, September 18 and 19, 1926 (Urner). W. E. D. Scott has an earlier record for Long Beach of two shot on Barnegat Bay in December, 1876 (Bull. Nuttall Orn. Club, 1879, p. 227), and I have seen a specimen mounted by Charles Voelker, which was said to have been shot on the Delaware River in October 1898.

Ludlow Griscom states that the Pomerine Jaeger is of regular occurrence from three to five miles offshore from Long Island, August 2 to October 30, and that adults may be distinguished from the Parasitic Jaeger by their square-ended not pointed, central tail feathers, although immature individuals are difficult to identify. (Birds of the New York Region.)

* LONG-TAILED JAEGER

Stercorarius longicaudus Vieillot

While this jaeger undoubtedly occurs at sea off the New Jersey coast the only positive evidence that we have is Dr. Frank M. Chapman's statement that he saw one eighty miles off Barnegat, from an ocean steamer, on May 6, 1894. Griscom records three occurrences on the coast of Long Island. It would seem to be much rarer than the other two species.

* GLAUCOUS GULL
Larus hyperboreus Gunnerus

This visitor from the Far North has not yet been found at Cape May but
is confidently looked for and considering the infrequent visits of ornithologists to the Cape during the severe snow and ice storms, of recent winters, it
may easily have been present and gone unrecorded. A single individual was
seen on the Delaware River at Philadelphia by Richard Erskine on January
1 and 4, 1918, and was reported by others in the interim, all observers commenting on the absence of black on the wings and the large size of the bird
compared with the numerous Herring Gulls which frequent the river. This
bird in all probability had come up from Delaware Bay and passed Cape
May on its course.

This gull is stated by Griscom to be an uncommon, but regular, winter
visitant to the coast of Long Island and New York Harbor, rarely seen much
before Christmas but lingering into May. Charles Urner has a number of
records for Barnegat Bay and on January 23, 1927, he and Ludlow Griscom
counted five individuals. At Seaside Park John Emlen saw a single bird on
February 26 of the same year accompanied by an Iceland Gull. On March
24, 1934, William Baily found a Glaucous Gull on Brigantine Beach feeding with a few Herring Gulls on the strand. "It was easily distinguished from
the latter by the absence of black tips to the primaries, and was practically
white all over but for a faint buffy edging to the feathers of the back. It was
of noticeably heavier build than the Herring Gulls with a greater wing spread,
and was easily 'cock of the walk' defending itself against all who would
claim its freshly opened clams." It was still present on March 30, when it
had been joined by an Iceland Gull. (Auk, 1934, p. 374.)

At Bridgeport in Gloucester County not far from the Delaware River Julian Potter found a single Glaucous Gull on March 2, 1935, with some five thousand Herring Gulls, feeding on frozen fish that had been exposed by a thaw. Other records of the Glaucous Gull have been published as follows:

Twenty miles off Long Branch, December 31, 1904, (Stackpole and Wiegman).
Barnegat Bay, May 22, 1926, (two), (Urner).
Barnegat Bay, April 30, 1927, (Urner).
Barnegat Bay, January 8, 1928, (Kuerzi).
Shark River, April 15, 1934, (Urner).

* ICELAND GULL

Larus leucopterus Vieillot

This smaller edition of the last is also a visitor from the Far North and like it has not yet been detected on Cape May waters, although there are records of straggling individuals much farther south on the coast and on the Delaware River, which doubtless passed the Cape.

In the New York region it is said by Griscom to be less common than the Glaucous Gull but probably of annual occurrence. Charles Urner records it on Newark Bay from January to April 1, 1922, and April 10, 1927.

At Seaside Park John Emlen found a single bird on February 26, 1927, in company with a Glaucous Gull, while on Brigantine William Baily found one of each species on March 30, 1934. He noticed that the Iceland Gull was four or five inches shorter than the other and was white with a little more of the buff tint on the back and upper tail coverts, while its bill was shorter and its legs a darker shade of flesh. It stood most of the time with head drawn down on the shoulders, plover-like. Both birds were quite tame and were easily approached (Auk, 1934, p. 375).

On May 12, 1934, Stuart Cramer found a dead Iceland Gull, possibly this same individual, on Brigantine Beach, it was badly decomposed but he saved the skull and from the measurements thought that is must be the so-called Kumlien's Gull. (Auk, 1934, p. 375.) The skull was later sent to Dr. Alexander Wetmore who pronounced it an Iceland Gull. (Auk, 1935, p. 186.)

On the Delaware at Philadelphia Julian Potter and Joseph Tatum saw an Iceland Gull on January 15, 1934, and again on April 10, following. Additional observations of the Iceland Gull in New Jersey are:

Barnegat Bay, December 10, 1926 (Griscom).
Barnegat Bay, February 22, 1925 (Griscom and Urner).
Seaside Park, March 28, 1926 (Yoder).
Shark River, April 15, 1934, in company with a Glaucous (Urner).

GREAT BLACK-BACKED GULL

Larus marinus Linnaeus

The Great Black-backed Gull, largest of our gulls, occurs regularly in
small numbers as far south on the coast as Hereford Inlet, during the winter
months. It appears to be somewhat more common at Townsend's and Corson's Inlets and still more so on Brigantine Beach and along the coast to
the north. Like the Herring Gull and certain other water birds it seems to
be particularly partial to the bars at the mouths of the inlets or to the
points of beach on either side; indeed the Black-back is rarely seen at rest
elsewhere. Doubtless it may have occurred in the past at Turtle Gut Inlet
and the closing of this channel and the lack of suitable bars at the entrance
to Delaware Bay may account for the absence of the bird nearer to Cape
May, while passing individuals, if such there be, doubtless travel too far
out to sea in crossing to the Delaware coast to be seen from the shore. Apparently the Black-backs never enter the Bay nor the sounds along the Cape
May coast.

On January 1, 1926, a day of bitter cold wind from the north, the long
sandy point which juts out on the southern side of Corson's Inlet was capped

and bordered with white ice while great floating blocks filled the thoroughfare and rushed out to sea on the swiftly ebbing tide. The whole upper
portion of the spit was covered with gulls. Some two hundred Herring Gulls,
in varied plumages, stood or squatted on the sand or hovered over the water.
There were a few of the smaller Ring-billed Gulls and, scattered through the
flock, four of the great Black-backs, distinctly taller and bulkier than the
others and distinguishable by their colors as far as they could be seen. The
entire back and most of the wings appear jet black in strong contrast to the
snowy white of the head, rump, tail and under parts. There is no suggestion
of gray in the plumage, although the "black" is found to be very dark slate
when we have a specimen in hand, and as we view the bird through the glass
it looks as if a vest of black velvet had been drawn over an otherwise white
gull.

On the wing these great gulls spread wider than the Herring Gulls and
their wingstrokes seem more powerful. In flight they display a narrow
white border along the front of the expanded wing and another along the
posterior margin; the under side of the flight feathers appears dusky gray
but the under wing-coverts, as well as the ground color of the bill, are white.

These Corson's Inlet birds at intervals flew out over the surf at the
mouth of the inlet, circling about among the Herring Gulls which were there,
and later returned to the strand to roost. They seemed to be more wary
than the Herring Gulls and ever on the alert for possible danger.

This inlet as I am informed by Wharton Huber has always been a winter
resort for these birds and both adults and immature individuals were to be
seen there on almost every visit.

Throughout the winter, on the sand bars that are exposed at low tide in
Townsend's and Herford Inlets, I have seldom failed to find, in the long
lines of Herring Gulls that resort there, from one to three Black-backs which
may easily be picked out even with the naked eye from the northern tips of
Seven and Five Mile Beaches respectively.

On February 6, 1932, I saw one on the ocean beach at Avalon, near to
Townsend's Inlet, which remained apart from the Herring Gulls that rested
at the same spot and I have occasionally seen individuals on the beach some
distance from Hereford Inlet, while on several occasions Black-backs have
been seen going from one inlet to the other along Seven Mile Beach out beyond the surf.

Usually one sees but one to three Black-backs together but we have
several records of four, five and six, while the number seen during a day has
reached much higher figures. The Delaware Valley Club's counts on the
Christmas census are as follows:

December 23, 1928, six	December 24, 1933. seven
December 22, 1929, six	December 23, 1934, seven
December 28, 1930, five	December 22, 1935, eighteen
December 27, 1931, five	December 27, 1936, eleven
December 26, 1932, eight	

John Emlen saw thirteen on Seven Mile Beach on January 23, 1927, while Charles Urner at Barnegat, on December 22, 1925, saw no less than twenty. While the Black-backs do not occur regularly until November on the Cape May inlets we have earlier records—September 12 to 17, 1914, one in Townsend's Inlet (Mrs. Prince); September 20, 1931, at the same place (Brooke Worth); September 20, 1936, (John Hess); September 28, 1935, Hereford Inlet (four) (Stone); September 28, 1935, Peck's Beach (Clarence Cottam, Auk, 1936, p. 81). Our latest observations in spring are: March 12, 1922, and March 14, 1931, Townsend's Inlet; March 26, 1932, Hereford Inlet; April 9, 1922, Corson's Inlet. There are several late April records for the Brigantine and Barnegat areas farther up the coast and Charles Urner saw one on Barnegat during the period from June 21, to July 2, 1926, when making a survey of the bird life of Ocean County.

* LESSER BLACK-BACKED GULL

Larus fuscus graellsi Brehm

On September 9, 1934, James L. Edwards and Charles Urner saw a gull on the ocean strand of Long Beach, just south of Beach Haven, which they identified as this European species. It was associated with a single Great Black-backed Gull and a number of Herring Gulls and while distinctly smaller than the former did not seem to exceed the latter. It was of the same color above as the Great Black-back but its legs were yellow instead of pink as is the case with the other species. The opportunity for comparison with the other species, as Mr. Edwards says, makes this "sight record" worthy of consideration. (Auk, 1935, p. 85.)

The species had never been recorded from North America before, but on December 15, 1934, another very similar bird was seen by the Messrs. Kuerzi in the east Bronx, New York. (Auk, 1935, p. 185.)

While there is of course no Cape May record of this gull it may possibly occur there in the future and the records are presented in order to complete the list of gulls seen on the New Jersey coast.

HERRING GULL
Larus argentatus smithsonianus Coues

One associates the big silent Herring Gulls with wintertime, when they constitute the most noticeable feature of the bird life of the shore, just as we identify the smaller black-headed Laughing Gulls with midsummer. But while most of these large winter gulls go north to the coast of Maine or to the Great Lakes, to breed, not a few remain along our New Jersey shores throughout the summer—mainly dusky gray brown immatures or birds in which, for some reason, the migratory impulse has failed to develop. Thus there is not a month in the year—not a day indeed—that Herring Gulls cannot be seen about Cape May if one knows where to look for them.

They may be found at all seasons on the sounds and meadows back of the island beaches, all the way from Two Mile Beach to Barnegat Bay, resting on the higher sand flats or scattered over the meadows where their white and gray plumage makes them very conspicuous, especially against the brown winter grass, and groups of them huddled together glisten in the sunshine like patches of snow. At this season they gather immediately back of the town where Cape Island Creek winds its way through the meadows and on its muddy banks they find an abundance of food. We see them rising above the level of the meadow for a moment and settling again and finally resting in close ranks back on the edge of the upland where a border of trees and bushes shields them from the northwest gale. No less than one hundred and fifty were counted here on February 12, 1921, while for a day or two in January 1932, after a high tide, they covered the meadows and the adjoining golf links right up to Lafayette St., which forms the edge of the town on this side. Their number was estimated as more than a thousand. Again on December 30, 1935, five hundred gathered there feeding along the edge of the flood.

On the great stretches of meadow lying north of the Harbor and back of Five Mile and Seven Mile Beaches they occur all through the winter but vary in abundance at any one spot, from day to day, as they are constantly shifting their position. Now they will gather in numbers on some sand flat left by the dredges apparently to rest or to await the ebbing of the tide. Again we see them scattered widely over the meadows after a spell of heavy rain or a high tide blown in by a winter gale, feeding like chickens upon stranded scraps of food as they walk about amongst the tufts of dead grass and sedge. On March 6, 1932, just after sunset, they were advancing across the meadows in a dense column in the face of a terrific gale. It was a weird sight as, with darkness fast settling down upon the marsh, thousands of the great birds passed us, now settling for a moment and then pushing on, those in the rear often flying forward and dropping down at the head of the column as if the mass were rolling over the ground. Then as the water continued to rise and flood every resting place they all took wing and battling with the wind sought some other shelter probably in the lee of the outer beaches.

At all seasons Herring Gulls like to gather on the points of beach on either side of the inlets or on the bars or sandy islands which form at their mouths. I have counted as many as five hundred on a bar in Hereford Inlet on February 1, 1931, at sunset, with more coming in every minute—a veritable migration of the meadow gulls to this spot to roost. Whether they spent the night there I could not determine but I feel sure that they did on the larger extent of Gull Bar nearby, where hundreds can be found at almost any time. Another favorite assembling point for winter Herring Gulls is the extreme southern point of Seven Mile Beach on the other side of the inlet. When these stations are crowded with birds the meadows appear almost deserted, just as they do at ebb tide when all of the gulls are busy feeding down on the exposed bottoms of the creeks and thoroughfares, below the level of the marsh.

Gulls that have been feeding along the Bay shore often, if not always, resort to the meadows or inlets at night and we see them crossing the peninsula in the neighborhood of Cold Spring flying in close formation in regular V-shaped flocks of six to ten individuals. Other birds that have been feeding out at the "ripps" at the mouth of the Bay may be seen at evening winging their way back over the ocean in single file low over the waves. The "ripps" are a favorite feeding place for these gulls, just as they are for the Laughing Gulls, and they may be seen there at various times during the winter as well as in spring and autumn, mingling with the other species in the latter season. From the shore we see a maze of moving wings as they flap back and forth over the rough water constantly swooping to the surface and rising again as they pick up food of one sort or another. In the autumn besides the two species of gulls there will be Common Terns weaving in and out of the maze and diving precipitately into the sea and one or two dark colored Fish Hawks laboriously flapping about.

From the "ripps" Herring Gulls follow the Bay and the Delaware River to and beyond Philadelphia shifting up and down with the ebb and flow of the tide but the individuals that we see between that city and Camden probably remain all winter in that vicinity picking up their food from the refuse that floats on the river. They rest on the ice cakes that crunch against one another in midstream or roost on the river marshes below the cities.

Herring Gulls are common on the sea beach in winter usually scattered at intervals along the strand, busily feeding. Occasionally one of these birds will take a few running steps in pursuit of some floating object or will wade out into the water and breast the smaller waves as they come rolling in. When they take wing from the beach it is usually one after another so that the flying birds are stretched in a line along the shore and not in a compact flock. On such occasions they will usually come to rest on the water out beyond the surf and may remain there for some time riding the waves with heads drawn down on their shoulders.

Herring Gulls perch regularly on the tops of pilings on the edge of the Harbor or along the beach. They usually stand with the body held well up leaning slightly forward and with the head a trifle lowered. Sometimes their position is more erect and they plume the feathers of the breast and sides as they rest or they may crane the neck over forward and peck at their feet. Again they may settle down on the feet and breast and, turning the head backward over the shoulder, will bury the bill among the scapulars and assume an attitude of perfect rest. We realize the complete mastery of the air that a Herring Gull possesses when we watch one settle on a piling in the face of a strong wind. It hovers over the perch and arches its wings, closing them just enough to maintain a stationary position in the air. Then with feet dangling the wings are gradually drawn in so that the body slowly drops until the toes touch the perch. Occasionally a bird will hover over a post already occupied and descend directly upon the back of the perching gull forcing it to take flight and usurping its position, only to be itself displaced a few moments later by the former occupant or another. Now and then a gull will resent being crowded off its perch and will raise its head and open its bill and perhaps elevate the wings and utter its call *keé-ough, keé-ough, keé-ough, kaw, kaw, kaw.* Usually however, there is no protest and the bird slides gracefully off alighting on the water or circling in the air preparatory to taking another perch. A bird is often forced off even if there be vacant perches available and sometimes several gulls come in at once and there is a general shifting of positions.

In summer Herring Gulls occupy much the same localities as in winter but are everywhere less numerous. There are always some assembled at the inlets and at favorite roosting places on sand fills or they may be scattered about feeding on the meadows. On the beaches however they are scarce although one or two may be seen flying up or down the coast well offshore. In general they seem to prefer remote spots at this season where they will not be disturbed and keep farther away from the bathing beaches and the immediate vicinity of the town, except where there is a regular supply of food as about the fishing wharves on the Harbor. As the summer wanes and autumn approaches they increase in numbers and we see more adult birds in gray and white mingling with the dusky immature individuals which make up most of the summer population, and the proportion increases as the season advances.

One of the nearest summer assemblages of Herring Gulls has for years been on a sand flat on the meadows back of Two Mile Beach a little north of the Harbor and here on May 21–24, 1922, and May 16–19, 1924, I counted flocks of thirty-one and forty respectively. Since the fish dock was built near this spot the birds have increased greatly and there were over one hundred Herring Gulls assembled there on July 16, 1932, and about five hundred on the 29th. When I studied this gathering in 1922 and 1924 from a nearby fisherman's cabin we saw them collect in nearly the same numbers every evening. In the morning they would sit motionless most of the time, or walk slowly about, but when disturbed by a passing boat they would take to the air and flap and sail about making a great display of moving wings in the early sunlight. Then they would settle again or if the tide were ebbing they would scatter to feed on the exposed flats. On the meadows close to our cabin there were many dead fish thrown overboard by the fishing boats and washed up by the tides, quantities of sea-robins and several huge Lophius fishes with their great gaping mouths and curious tentacle standing up from the upper jaw. These were a great attraction to both Herring Gulls and Fish Crows, though the latter held aloof while the gulls feasted. There were eighteen gulls feeding here on the morning of May 19 and as we watched them we were reminded of Turkey Vultures as they rushed at one another with partly spread wings in their efforts to secure the choice bits from the feast, and now and then one bird would pursue another for some distance with wings raised high. Like the Vultures, too, they seemed to pick out the eyes of the large fishes before tearing them asunder.

The gulls when moving quietly about had a heavy waddling gait and now and then, for no apparent reason except perhaps to gain time, one of them would leap into the air and settle again a few yards away, dropping down slowly with wings held aloft for a moment after alighting. When one gull would settle near another the latter would often raise its head and with neck outstretched nearly vertically, and bill pointed upward, would utter its call *koo-lick, koo-lick, koo-lick, koo-lick*, followed by *keeer, keeer, keeer* the neck being gradually lowered as the call proceeds. Herring Gulls do not utter many sounds while here or else they become vocal only when far-removed from the haunts of man. I once heard one on the beach in mid-September call, with neck stretched out horizontally a little below the level of the shoulders, and once in March one passing overhead called *kee-law, kee-law*, followed by a goose-like "*goggle*" still another called from a piling on September 7, 1931. I have copied the calls as I wrote them at the time.

While lying out in a skiff on Jarvis Sound, on July 23, 1922, I had an excellent opportunity to watch the Herring Gulls feeding at low tide on the exposed flats. Two in immature plumage arrived when the water was still a few inches deep and the mussel bed was exposed for an area of only about sixty by ten feet, then came one in nearly adult plumage and another immature bird alighted on an adjoining shoal. They all stood for a while and then squatted on their bellies awaiting the slipping away of the water. Soon there were six on one flat and eleven on the other, and now all began to feed as the area of exposed mussel beds increased. They plucked the mussels from the shells and hastily devoured them and when passing from one bed to another they were often forced to swim. When a boat disturbed them they took to the water and later mounted in the air one flock after another until there were sixty-four of them sailing and flapping about overhead. One feeding bird would often pursue another which had secured some choice morsel, and now and then a bird would leap a foot or two off the ground for some reason that was not apparent, possibly the spouting of a buried clam nearby. Some individuals seemed much less aggressive than others and we noticed one that was feeding on a dead sea-robin desert his food, when approached by another apparently more powerful bird, and return to it only after the intruder had flown away.

On September 14, 1921, large numbers of Herring Gulls were scattered over the sound some of which were feeding on live blue crabs. They would lift the crab in the bill and cast it down again its legs and claws wriggling in the air. Another individual was eating dead conchs and would raise an entire shell in its beak in an effort to dislodge the animal. A flock of immature gulls feeding on the mud flats back of Five Mile Beach on July 21, 1930, were eating dead blue crabs which had been left there in numbers by the receding tide. As we looked at them from our skiff, almost on a level with them, they seemed distinctly bob-tailed and appeared very clumsy as they ran or walked quickly about with neck held erect. Occasionally they would run at one another with shoulders hunched and heads down. When the tern colony, back of Five Mile Beach was washed out, in 1931, the Herring Gulls resorted there and ate the broken eggs and the little downy tern chicks which had been drowned in the flood. From the way in which the terns pursue them, when they fly near a breeding colony, they may also appropriate fresh eggs and living young if opportunity offers, but I have never actually seen them do so.

At Corson's Inlet, on January 1, 1926, a day of bitter cold and howling north winds, a number of Herring Gulls were catching winter crabs (*Cancer irroratus*) in the shallow water of the channel at low tide. A gull would hover over the water and suddenly drop to the surface with head stretched down so that it was submerged in the act of seizing the crab, and failing in the attempt the bird would rise again for another plunge but never did it make a true dive or submerge more than the head and that for only an instant. Having secured a crab by the leg the gull would carry or drag it to an exposed sand bar near at hand and walk around it jabbing at it with the bill and deftly avoiding the widespread claws. Once these were broken off it was an easy matter to crush the shell and devour the soft parts of the animal. Sometimes another individual would drive off the successful fisherman, appropriate his catch, and proceed with the attack. I have once or twice seen Herring Gulls swimming on shallow pools on the meadows feeding on something on the bottom, plunging the head under the surface and keeping it there for several seconds at a time. When arising from mud flats after feeding they often shake the wings and wriggle in the air as if to dislodge sand or water from the plumage.

I saw an immature Herring Gull on December 27, 1931, plunging down to the water in an effort to secure food of some kind and once or twice it splashed pretty well in but did not make a regular dive and its action, such as it was, was quite exceptional. In my experience I have never seen these gulls make a regular dive and such action is very unusual in any of our gulls except the little Bonaparte's.

Herring Gulls occur regularly on the Harbor and about the fish wharves at all times of year but much more abundantly in the winter. Here they feed on all sorts of refuse that may be thrown out on the water and in late summer follow the fishing craft that are returning from the banks, snatching up the offal and rejected fish that the fishermen are throwing overboard.

One well known habit of the Herring Gull is to be observed frequently on our New Jersey coasts in winter—that of carrying up clams and dropping them on the beach in order to break them. Usually this is a solo performance, the gull carrying the clam up to a height of fifteen or twenty feet and following it in its fall to the beach to devour the contents. Where many clams are washed up on the strand, however, a number of gulls may gather and the one which carries the clam rarely gets back to the beach in time to share in the feast though he probably profits on the attempt of the next individual that mounts in the air. Frequently the beach is not hard enough to break the clams and several attempts result in failure. As a result, in some places, as at Beach Haven a few years ago, the birds left the beach for the deserted board walk with great success, and the walk was literally strewn with broken clam shell. I watched a lone gull carrying up clams on the Avalon beach on March 5, 1932, and upon recovering a large one that he had dropped found it to measure five and a half by three and three quarter inches and two inches in thickness. I have seen Herring Gulls on the meadows carrying up large mussels and dropping them on the ground with, of course, no result although they tried it over and over again, one individual swooping down and catching the shell in its bill on the rebound from the ground. Another one had secured a large natica snail and made repeated but vain efforts to break it on the muddy shore of the Harbor.

On March 4, 1934, there was a great deposit of clams on the Brigantine Beach and Herring Gulls seemed to come from considerable distances to partake of the feast. Julian Potter estimated that there must have been ten thousand present. I have seen similar gatherings on Seven Mile Beach when the beaches of the entire northern half of the island were thronged with the birds. Potter also tells me of the freezing of a lagoon at Bridgeport, Gloucester County, not far from the Delaware River during the exceedingly cold weather of January, 1935, which killed tons of catfish, carp etc., and that with the thawing of the ice the exposed mass of fish attracted some five thousand Herring Gulls along with several Ring-bills and a single Glaucous Gull. They were seen busily at work on this unexpected feast on March 2 and in one week's time had entirely disposed of it. The below zero weather of February 11, 1934, brought a similar assemblage of Herring Gulls to the Atlantic City garbage dump where they hovered overhead in an immense swarm (Potter).

The flight of the Herring Gull along the coast is a continuous easy flapping, once the bird gets under way, and the body slides ahead with no apparent effort. But there is much labor at the start as can readily be seen if we are close at hand. I came up with some gulls, in a motorboat, one day, just as they had arisen from the water and before they could develop their usual momentum. Side by side we progressed ; the birds with head and neck stretched well forward and slightly downward, and the wings apparently beating clear over them as they swept forward, seemed the very picture of strain, until the steady restful gait had been attained. Under certain conditions Herring Gulls will sail in short arcs and, when the time comes for the northward migration, they will mount high in the air flying in wide circles as do the Laughing Gulls in their aërial evolutions.

When launching forth in flight these big Herring Gulls leave a clear record of their actions on the sands of the beach. The ordinary walking footsteps will be seen to pass into running steps in which the forepart of the foot is most strongly impressed and finally a print of the toes only, deeply dug into the sand, shows where the bird gave its final shove as it launched itself into the air. If we follow back to the other end of the trail we shall find two footprints deeply indented posteriorly where the bird, when first it touched the sand, "dug in its heels" to check its forward progress. When taking wing from the water the action is just the same and there are numerous vigorous strokes of the feet before the bird clears the surface.

At all times the Herring Gulls seem to prefer to keep to themselves and when they are associated with other birds it is apparently the latter that make the advances. On the beaches they prefer to be solitary and when roosting they gather by themselves, although Ring-bills, Laughing and Bonaparte's Gulls may join them. Occasionally immature dark-plumaged Herring Gulls do seem to join the large assemblages of Laughing Gulls but that is usually before the great autumn and winter flocks of the big gulls have arrived. A curious instance of association on the part of a Herring Gull was a single individual in adult plumage which joined a nesting colony of Laughing Gulls, at Little Beach on July 18, 1921, and circled about in the midst of the raucous assemblage that was violently protesting against the invasion of their domain. He seemed as much concerned as they although he certainly had no nest or young to defend and uttered no vocal protest!

I have seen Herring Gulls on the Harbor drive away Laughing Gulls which attempted to join them on the water and have also noticed one pursuing a Common Tern, endeavoring, but without success, to make him drop a fish which he was carrying. On the other hand the terns often attack a luckless Herring Gull which ventures too near their nesting grounds. On May 1, 1932, a Bald Eagle flying over the meadows passed near a gathering of Herring Gulls which were roosting along the edge of a creek and about fifty of them arose and pursued him. He was seen to drop a fish or some sort of food that he was carrying upon which they gave up the chase.

The large size of the Herring Gull readily distinguishes it from any of our other Cape May gulls although its adult plumage is so like that of the Ring-bill that solitary birds, seen when there is no opportunity for size comparison, may occasionally be confusing. The Herring Gull is longer necked and less graceful than the smaller species and its bill is noticeably longer, appearing almost curved or hooked, while in color it is yellowish or flesh color, usually with a reddish or dusky spot near the end of the lower mandible but without the vertical black bar so characteristic of the Ring-bill. In less mature individuals the dusky spot is more pronounced and often the entire terminal portion of the bill is dark.

The plumage varies considerably according to age but not at all as to sex or season. The adult has the mantle and wings pearl gray but is otherwise pure white, while the bird in its first breeding plumage has dusky or dark gray splotches here and there, and dusky lines on the head and on the sides and there is a conspicuous dark terminal band on the tail. The young birds of the year are entirely different being dark gray brown or chocolate throughout with darker centers to the feathers and buffy edges to some of them, there is a dusky band along the posterior edge of the spread wing and a dark blackish tail band. The bill is smaller than that of the adult and wholly dusky. A number of birds in this plumage and some in the one described just above fail to molt in the spring and likewise fail to acquire the migratory impulse and these are the birds that make up our summer colonies of Herring Gulls at Cape May. With their constant increase in numbers and the spread of the breeding range of the species southward we may yet see this great gull nesting on our New Jersey coast.

The autumnal migration of Herring Gulls seems to take place in September and October as it is usually about the middle of the former month before they are to be seen on the Delaware River at Philadelphia, but there would seem to be an earlier influx of birds from the north on the sea coast, for certain it is that the proportion of those in adult or semi-adult plumage is much greater in August than in May and June. In April most of the Herring Gulls move north and it is not uncommon to see them at this time circling high in the air, or sailing across to the Bay and back giving every indication of restlessness. The last of the migrants seem to have left us by the middle of May, leaving only the dusky-plumaged band which will remain throughout the summer.

The life and habits of these great silent scavengers of our winter sea coast offer a great field for study. Their very silence seems to increase their dignity and add to the impression of independence, self reliance and contempt for their lesser associates which characterize the Herring Gulls. In spite of their presence here in summer, I always associate them with winter and picture them in my mind's eye as accompanied by the angry roar of the surf and the rush of waters on some wild beach by the harbor's mouth, as they gather together for the night. Many times I have seen them on a narrow bar at sunset where they stand out silvery white and gray against the deep blue black of the darkening ocean, while every moment a long line of snowy surf boils up behind them as if to wash them from their narrow strand. Farther out other lines of surf appear successively and are lost again in the flood of waters. Every moment the pale yellow moon is growing brighter as the last rays of the sun die out in the west, and the night shadows creep closer and closer, while the great gulls stand there like a row of sentinels silhouetted against the sky until they are slowly and gradually swallowed up in the night.

* THAYER'S GULL

Larus argentatus thayeri Brooks

This slightly different subspecies of Herring Gull comes down to us in winter from far northern shores mingling with the individuals that breed on the New England and Canadian coasts. A specimen of this form, which probably cannot be distinguished in life, was secured by Samuel Rhoads near the Delaware River at Mt. Ephraim, N. J. on March 9, 1888, the only one to be recorded from the state so far as I know. In all probability it occurs occasionally with the ordinary Herring Gull about Cape May. It is slightly paler and has a different proportion of white on the primaries.

RING-BILLED GULL

Larus delawarensis Ord

The Ring-billed Gulls arrive at Cape May each year during the first week of August, or earlier—August 3, 1921 and 1922; August 4, 1923; July 2, 1924; July 30, 1925; August 8, 1926; July 24, 1927; July 18, 1928. They take up their position on the strand, often close to the bathing beaches, and at intervals of about twenty-five yards, as if each bird had its own feeding station. This habit of feeding on the beaches during August and September is rather distinctive as the other gulls rarely do so, at this season at least, and do not exhibit the tameness of the Ring-bill. The latter when busy feeding will allow a close approach but if one comes too near will wheel out over the surf and land again a little farther down the strand.

The feeding Ring-bills walk about through the shallows, now and then running suddenly and rapidly after a receding wave to pick up some morsel from the sand or water, often one of the little Hippa crabs which come in by the thousand with the waves and scuttle down in the sand as the water retreats. At such times the gulls remind one, in their actions, of gigantic plovers. In the excitement of the chase the wings are often raised above the back and now and then a bird will run along with the wings half spread as if about to take flight and then change its mind and come to rest again. When they do launch into the air several strokes of the wings are given before the feet leave the ground and the bird gets fairly under way.

Sometimes when feeding before an incoming wave the water suddenly becomes deeper and the gull, in picking up its prey, completely submerges its head and, again, it will wade right into an oncoming wave and jump through it like a surf bather. Sometimes, too, they will be carried off their feet by the sudden rush of water and swim off gracefully with breast well down and wing tips projecting above the tail.

When feeding out over the ocean, as they often do, they fly with steady rather quick wingbeats but all of them of equal length and regular, like those of the Herring Gull, and with none of the short half-beats of the Bonaparte's Gull. Every now and then a bird will pause for a moment and then drop or "dive" usually striking the water with the feet and belly and, snatching up some floating scrap of food, will be on the wing again. Sometimes the head will be bent far over and will be submerged as the bird strikes the water but there is never a true head-on dive and the body never goes under.

Ring-billed Gulls will often remain on the beaches throughout the winter but many retire at times to the more elevated dredgings and sand fills on the meadows and all probably go to such spots to spend the night. In both places they associate more or less closely with the larger Herring Gulls and sometimes, in late summer or early fall, will be attracted to a gathering of terns on the sea beach but they usually stand off by themselves and do not mingle intimately with the terns as do the Laughing Gulls. Occasionally I have seen them back on the sounds at low tide perched against the vertical mud banks much as they might rest on some rocky cliff in their summer home far to the north.

Ring-bills are not so abundant as the Herring or Laughing Gulls but while they frequently occur singly they also associate in moderate sized flocks; Julian Potter saw a flock of forty at Corson's Inlet on August 19, 1928, and counted a hundred birds on Seven Mile Beach on December 23, 1929, while I found nineteen gathered closely together on the ice of the Lighthouse Pond on February 14, 1925. A flock of twenty was found on August 25, 1931, on Seven Mile Beach and on September 27 we found that it had increased in number to over fifty. This flock remained until late October and then decreased in numbers, though six were still present on December 25.

The first birds to arrive on the sea beach in August are usually adults in beautifully fresh plumage, with snowy white head and breast and pale gray wings and mantle—slightly darker than in the Herring Gull but not nearly so dark as in the Laughing Gull. When the wings are spread the gray is so pale that the white tips of the secondaries, which form such a conspicuous border to the wing of the Laughing Gull, are scarcely distinguishable, but the black terminal portion of the primaries and their white tips are strongly contrasted and conspicuous both in flight and at rest.

Generally speaking the Ring-bill has the plumage of the Herring Gull but approaches more nearly to the size of the Laughing Gull; unless one of these two is present, however, for comparison, it is by no means easy to judge its size or to distinguish it at a distance from the Herring Gull. As further means of identification there are the yellow, instead of pink, toes and tarsi and the much shorter yellow bill with its dark vertical bar near the middle. The Herring Gull usually shows a red spot near the end of the lower mandible which is often more or less dusky but this is quite different from the vertical bar covering both mandibles of the Ring-bill, while the long, almost hooked, bill of the Herring Gull gives to its head a very different outline. The very dark upper surface of the Laughing Gull renders it easily separable from the other two at any season. While not so slender as the latter the Ring-bill is a more graceful, trimmer bird than the Herring Gull, especially as it stands on the beach with neck contracted and head drawn down on its shoulders in an almost dove-like pose.

Light plays curious tricks with the appearance of all gulls and two Ring-bills observed on the Bay off Cape May Point on March 29, 1936, appeared as dark as any Laughing Gulls and I was convinced that they were this species and could not account for the white heads so late in the spring. Suddenly, however, at a different angle their backs faded to pale gray!

Young Ring-bills resemble Herring Gulls in plumage, intermediate between the dark bird of the year and the gray and white adult. They are never so dark as young Herring Gulls and have more white in the plumage, but have a dark terminal tail band and dusky bill and feet. We are therefore dependent almost entirely upon size as a means of identification.

A few Ring-bills, as is so frequent in the case of the Herring Gull, do not migrate northward in the spring, evidently undeveloped non-breeding birds and they may be seen in the immature plumage all summer. One such was found with immature Herring Gulls on the meadows back of Ludlam's Beach on June 24, 1928.

During their winter sojourn at the Cape the Ring-bills ascend the Delaware River to Philadelphia, especially in spring, when, by March 12, they seem to reach their maximum number and gradually supplant the Herring Gulls until by April 15 Julian Potter has found only immature birds of the latter species on the river while the Ring-bills present were all adults. In the autumn John Carter has found them arriving at Chester on the Delaware in late August or early September, in advance of the Herring Gulls.

In Salem County along the upper Bay Julian Potter tells me of a flock of one hundred and fifty Ring-bills on April 6, 1935, foraging in plowed fields and of a small flock following a farmer who was plowing. In the same vicinity John Gillespie saw at least twelve hundred of these gulls in flocks of two hundred or more in several different pastures feeding on scattered earthworms which had been drowned out by encroaching water.

As spring approaches the Ring-bills at Cape May seem to increase in numbers and then decrease again as they leave for farther north. Most of them have left by April 1 but we usually have records of individuals or small flocks up to the last week of that month and last records as late as May 8, 1925, and May 19, 1924, while I saw a flock of twenty-five on May 10, 1925, flying steadily along over the surf following the flight line of scattered Loons and long ranks of Double-crested Cormorants which on the same day were winging their way to their breeding grounds in the Far North.

LAUGHING GULL

Larus atricilla Linnaeus

The most striking feature of summer bird life at Cape May, to one familiar only with the birds of the upland, is the presence of the graceful Laughing Gulls which all summer long pass and repass high overhead or beat their way up and down the coast low over the breakers. Looking at them from below one sees only the snow white under surface of the body and wings which now and then glistens like silver in the sunlight, and the black hood which envelops the head as far as the upper neck and throat. When skimming the ocean or wheeling in midair, however, the dark slaty mantle which spreads over the back and wings is conspicuous, contrasting with the white tail and hind neck and the white border which fringes the wings. When we find a bird at rest on the meadows and are able to approach closer we note that the head is not black but a darker shade of slate, darker than the mantle, and we distinguish the curious white crescent-shaped mark behind and around the eye which gives such an uncanny expression to the bird's face while we admire the dark carmine of the bill and feet. The delicate rosy flush which suffuses the breast is, however, usually visible only when the bird is in hand.

The manner of flight of the Laughing Gulls varies. Usually they go singly or in loose flocks, each bird for himself, with no attempt at concerted action.

The wings move with short easy beats as if there existed endless reserve power and the bird were not half exerting itself. In the face of a strong wind they take a few strokes and then a short sail, tilting up now on one wing, now on the other, and producing a rather erratic, zigzag flight. In their efforts to escape the force of the gale, too, they will come close down to the ground in crossing the Fill or the meadows and will skim the house tops as they pass over the town.

The Laughing Gulls have a regular line of flight from their nesting grounds on the marshy islands back of Seven Mile Beach to the "ripps" at the mouth of Delaware Bay, where the waters of the river meet the ocean swells and make a favorite fishing ground for various sea birds. The morning flight is usually southward, while at evening most of the birds are travelling northward again to roost in the vicinity of their nests. These evening flights, especially later in the summer, are much more business-like and direct, a dozen birds often traveling in close formation, sometimes in a more or less definite V, with steady even strokes of their long wings and no deviation whatever from their course.

Over the ocean their direction of flight is not always so definite and they may be seen traveling along both up and down the coast at almost any time of day. They fly about twenty feet above the water and every now and then one will sail down close to the waves on set wings, sometimes disappearing entirely in the trough of the sea. As the bird turns the white fan-shaped tail now full spread, catches the sunlight and flashes out conspicuously against the dark water. At other times, especially on days when the sky is overcast and the birds on steady flight, their slaty upper parts merge so perfectly into the steel-gray of the sea that it is difficult to distinguish them, and in the attempt to follow their course, even with the glass, one will lose them entirely for considerable distances, catching them again farther on.

The number of Laughing Gulls passing over Cape May increases steadily as the summer advances, from the few scattering individuals that venture so far from the breeding grounds in June, to the hundreds that make up the flights of late August and early September, after the young are able to shift for themselves. The factors that influence the flights and roosting habits of birds are not easy to discover and just as one imagines that he has all their movements reduced to a schedule they defy all the laws that careful observation seems to have established. These gulls are no exception.

In August, 1921, they drifted as usual down to the feeding grounds at the "ripps," every morning, and between six and seven in the evening passed north again in force, sometimes directly over the town and again in the lee of the woods which border the salt meadows on the west. On the 27th, however, contrary to all precedent, the northward flight took place in the morning in long straggling columns with gaps between, like Crows going to roost. At 8:30 I counted 160 birds and later on detachments of 130, 109 and forty all in half an hour, but I had undoubtedly missed many which passed before I was aware of the movement. On the next day the flight was also at this time. On both days a strong east wind prevailed which doubtless induced the birds to spend the night on the Bay but why return north in the morning? Again on August 23, 1924, under similar conditions they flew north in the morning in squads of 220, 240, 190 and seventy and on September 17, 1921, a similar northward flight took place at 3:00 p. m. this time out over the ocean; 210 in the first division and 108 a little later, each flock stretching out about two hundred yards in varying density.

Laughing Gulls have a peculiar aërial performance which I have not been able positively to explain. They gather together in numbers at some point, usually over the meadows, and fly back and forth within a limited area forming a veritable maze of flapping, turning birds and reminding one of swallows feeding on a cloud of gnats. In the case of the gulls, however, I have never been able to detect the presence of insects of any kind and rarely any movement of the bills of the birds, even though I have held the glass on them for many minutes at a time. One of these assemblages on July 30, 1921, and others at the same spot on August 25, 1917, and August 29, 1922, consisted of upwards of one hundred birds which in the first instance kept low down over the meadows none of them rising more than twenty-five feet in the air while those of the later gatherings mounted higher, the last including both old and young birds in all stages of plumage. On August 30, 1921, a somewhat similar gathering of fifty birds occurred high in the air, immediately over the town, but in this case the birds were, for the most part, sailing regularly in circles and mounting higher and higher until they appeared like small hawks and later even like martins. A precisely similar soaring occurred at Schellenger's Landing on July 5, 1923, and another on August 3, 1921, over Cape May Point, the gulls in the latter instance mingling with several Fish Hawks which were engaged in the same sort of evolutions. On one occasion I saw a lone Laughing Gull begin to circle over the town and watched him mount higher and higher until he became a mere speck and finally vanished in the sky.

A habit apparently peculiar to the Laughing Gulls, since I have never observed the other species practicing it, is their resorting to fresh water ponds for bathing. When the water in Lake Lily is sufficiently deep and free from vegetation they gather there regularly during intervals in their feeding at the "ripps." On July 29, 1920, I watched a number of them so engaged. As they came sailing down to the water they threw the feet forward and drew the wings in gradually as they settled down. Others observed on August 24, 1924, fairly threw themselves into the water from a height of several feet and did it repeatedly. Once on the water they would duck the head under the surface and withdraw it again with a sudden jerk. The plumage was then ruffled up and the closed wings, held a little way out from the body, were worked up and down, churning up the water all around. Several birds usually bathed together getting as close to one another as possible during the operation. Eventually they flew off, one at a time, while others came in to take their places. Occasionally a newly arrived bird would dip its head to the water several times without submerging it, apparently drinking, and I have seen birds fly down and skim over the surface dipping the bill in to sip up water. Some individuals after bathing sat quietly on the water as if loath to depart and several times a bird about to alight would raise its foot and scratch its head while still in the air. In certain years Lake Lily has been almost entirely deserted by the gulls as a bathing spot, due apparently to the great accumulation of bladder-wort and other aquatic plants but the birds have other fresh water ponds to which they resort. One of these is an old sand hole pond surrounded by trees and bushes just east of Wildwood Junction and much nearer to their breeding grounds than Lake Lily. Here on July 8, 1923, gulls were continually coming and going in pairs or singly, flying from the meadows to the mainland. Another favorite pond is the Sea Isle Reservoir, a picturesque spot on the Shore Road, surrounded by white cedar woods against which background the gulls stand forth in beautiful relief. Still another resort is the Poor House Pond also close to the Shore Road and some miles farther south. Here on August 15, 1931, there were about one hundred birds crowded closely together, energetically churning the water and ducking their heads; some always springing up into the air and settling again, causing constant motion throughout the mass while they made much clamor with their calling. Why with such a wide expanse of water available they must crowd together in a compact mass covering only an area of a few square feet is hard to understand but such is their invariable habit. Temporary pools are quite as satisfactory as these regular resorts and when on July 16, 1926, and at other times, heavy rains had made extensive shallow pools on the fill back of South Cape May they were crowded with gulls so long as the water remained.

Laughing Gulls frequently rest on the ocean, on the Harbor, or on the thoroughfares, just as do the Herring Gulls, especially late in the season when they may be found in immense masses riding the waves off Cape May Point, but they are rarely seen to alight on pilings a habit which is very characteristic of the larger species. When they do so it is usually in company with terns and Bonaparte's Gulls and on some continuous perch rather than on single posts. On the ground they are much quicker and more active than the large Herring Gulls and their legs seem proportionately longer and more slender. While they often associate with the larger gulls I have on several occasions seen the Herring Gulls pursue them and drive them away. They occur regularly on the Delaware River ascending at least as far as Fish House above Camden. According to Julian Potter they are most abundant there in spring, arriving about April 12 and remaining until about the 20th, although some are to be seen until May.

The Laughing Gulls arrive at Cape May from the south by April 20 our earliest records being April 14, 1904; April 8, 1913; April 6, 1919; April 1, 1928; March 31, 1929; April 1, 1930; some of these were probably stragglers coming in advance of the main flight, the March record being of a single individual. On April 23, 1922, Julian Potter saw a flock of twenty, apparently just arrived, circling high in the air and calling.

Of late years the Gloucester mackerel fleet has been in the habit of putting in to Cape May Harbor in April to ship their catches and the boats attract large numbers of Laughing Gulls, the earliest arrivals being frequently seen there. On April 29, 1923, over a hundred of the fishing boats were gathered close together about Schellenger's Landing, their slender masts looking like the trunks of a burned cedar swamp against the evening sky and all about them was a whirling mass of Laughing Gulls. About two hundred were in the air at once with others resting in ranks on the meadows beyond. They were flying in every direction and swooping right and left from higher to lower positions, now flapping vigorously, now sailing on set wings and pitching downward to the surface of the water. They displayed the whole upper surface of the body when, for the moment, they poised in air with one wing extended upward and the other down. All the while they called continually as they contended for the scraps that were constantly being cast overboard by the fishermen. I have seen similar flocks in late August following the fishermen's excursion boats into the Harbor and on August 29, 1922, over fifty were flying in the wake of one of these vessels swooping down and picking up scraps of bread that were thrown to them, without settling on the water. Many of these birds were young in the dusky plumage. I think that the early spring flocks on the Harbor may very likely be migrant birds on their way farther north for a large number, at least, of the local breeding gulls resort at once to their nesting grounds.

While anchored off Little Beach Island on May 18, 1927, not far from a breeding place of the Laughing Gulls, we could hear the calls of the birds floating across from the meadows and later encountered fifty-four of the birds feeding out over the water. They would drop to the surface and plunge their heads under after their prey, checking further immersion by prompt wing action. Later sixty-two of them gathered about the stern of our boat to pick up scraps of bread, calling all the while in harsh chorus. They flew low and made a dab at the bread as it floated along on the tide and were often compelled to bend the head under the body in their efforts to secure the prize, so that it appeared as if they were making a backward lunge between their legs.

None of our gulls seem to dive in the true sense of the word as do the terns but are ever alert to seize upon any fish that the latter may drop and are constantly swooping to pick up morsels from the surface and now and then, in the excitement, plunge the head under.

Most of the feeding of the Laughing Gulls is apparently done from the water and they may be seen on the Harbor, ocean, sounds or thoroughfares collecting whatever food may be found or coursing along looking for such floating matter as may be to their liking. One of their favorite feeding stations is at the "ripps" at the mouth of Delaware Bay. Here at almost any time during the summer and autumn, large numbers of them may be seen associated with other gulls, terns and Fish Hawks, flapping back and forth in a veritable maze. Occasionally the birds come in close to the shore when the wind changes the location of the rough water but usually we must resort to the glass to study them. Sometimes there will be only a few dozen present and at others, especially late in the season, there will be thousands. Occasionally they feed on the beach under exceptional circumstances. On July 3, 1917, there were a number of fish stranded in the shallow water and a Laughing Gull coming winging its way along the shore spied them, hovered for a moment, and dropped with zigzag darts, its red feet widely spread and pendant. When close to the sand it swooped forward and upward picking up one of the fish in its beak. This was repeated several times and then alighting on the beach it ran quickly about greedily picking up the fish right and left and gulping them down. Back on the channels, too, I have seen them swoop down to the surface of the water to pick up morsels of food and once a gull that had secured something attached to a piece of sea weed, dropped it and recovered it again from the water in a vain effort to detach the objectionable appendage but finally flew off with the weed streaming from its beak.

On August 31, 1920, there was a great gathering of Laughing Gulls over a shallow pond and exposed mud flat that formerly existed near the Lighthouse, the birds diving and rising again continually behind the dense growth of Spartina grass which surrounded it. They appeared to be feeding and subsequent examination disclosed numbers of small fish stranded there or swimming in the shallow water. I again saw them feeding in this way along Mill Creek on September 10, 1924, and on May 18, 1927, I saw a number walking about like shore birds on the black mud of an exposed bar, all headed to the wind and busy picking right and left. On other occasions they may be seen out on the salt meadows walking about like chickens picking up food from amongst the roots of the grass and sedges. This sort of feeding is practiced especially after high tides have flooded the meadows and receding have left all sorts of crustaceans and other forms of animal life stranded there.

Laughing Gulls are always more interested in live food than in dead and are not scavengers like their larger cousins the Herring Gulls. Besides the life that swarms over the exposed flats at low tide I have found them catching fiddler crabs on the sloping banks and bottoms of the tide water creeks to which they resort at the proper time, running after the crustaceans, breaking off their large claws and crunching the bodies between their mandibles. Alexander Wilson, writing about 1810, says that at Great Egg Harbor the Laughing Gulls resort to the plowed ground about the farmers' houses to feed.

While living in a gunner's shack on the meadows back of Two Mile Beach, May 16–19, 1924, I had an excellent opportunity to study the habits of the Laughing Gulls prior to the nesting season. They put in an appearance soon after sunrise each morning and from then until dark were constantly in sight, flying over the meadows and waterways and uttering an interrogatory *tuk, tuk*, as they winged their way past our cabin. They usually went in pairs and now and then rested on the channel before our door, floating leisurely along with the tide, their dove-like beauty a delight to the eye. Once during our sojourn here we experienced one of those phenomenally high tides which prove so disastrous to the nesting sea birds. The water rose to the top of the channel banks and then spread gradually over the meadows until only the tips of the tallest grasses remained in sight and the waves, driven by stiff winds, rushed under our little cabin as if a miniature surf were breaking there. The gulls at such times became very active and were constantly on the wing coursing back and forth over the flood attracted doubtless by the variety of food that must inevitably be washed up. They flew about ten feet above the water, pausing every now and then to drop to the surface and snatch up some floating morsel. At such times they have a rapid hovering flight with head bent far over and feet dangling so that head and feet strike the water simultaneously and the former goes slightly under in grasping the prey. Often a bird will rest for a moment on the meadow standing in water up to its belly and then takes wing again. And so they feed until darkness shuts them from our view.

During May and also through the summer large numbers of Laughing Gulls frequent the extensive flats of Jarvis Sound and other shallows which are exposed at ebb tide. Against the broad expanse of black mud they appear like splotches of white scattered here and there as far as the eye can reach. They stand often in pairs at the beginning of the season, or singly, and now and then a dozen of the barren white-headed individuals will be found in a group by themselves. Other gulls passing overhead are constantly calling to those on the flats which answer them as they pause from their search for food.

On these mud flats Laughing Gulls have been seen to pair and one curious performance was witnessed by Walker Hand and me on May 23 and 24, 1922, which probably is part of the courtship behavior of the species. A Laughing Gull came to rest on a mud flat near us and was joined by another which seemed smaller and slightly paler. The first bird was very sluggish but the new arrival was all activity walking back and forth close in front of the other one and throwing its head upward and backward as it uttered a low *kah, kah, kah, kah*, at short intervals while the first bird replied occasionally with a low *krrooo*. Now and then the active bird attempted to force its bill between the mandibles of the other, which would gape slightly or bend over and apparently sip water, the former action causing redoubled activity on the part of the other individual. Presently about five minutes after the performance began the sluggish bird gave a mighty convulsion and ejected a yellowish mass of partly digested food about the size of a hen's egg, which the other bird, ever alert, seized as it appeared and with several gulps swallowed it without allowing it to touch the ground. Both birds now seemed relieved and walked about sipping water, or washing their bills and soon after flew away. Precisely the same performance was observed at the same spot on the following day.

The Laughing Gulls always nest back on the salt meadows usually on a section that has been converted into more or less an island by the innumerable narrow tidewater creeks which wind their way backward and forward and eventually join with the broader channels or thoroughfares. They are distinctly social birds and nest in colonies, all of the flock joining in their defense. The nests are placed reasonably close to one another and are usually located in the coarse growth of marsh grass that skirts the little creeks, sometimes built directly on the growing grass but more often on masses of trash—dried grass stems—that have been washed up on the meadows by the high tides of the previous autumn and winter. There seem to be only two colonies of the birds nesting on the New Jersey coast in recent years, at least. One is located back of Seven Mile Beach about twelve miles north of Cape May, which contained some fifteen hundred birds on July 20, 1919, and has fluctuated in numbers in later years. The other is in the neighborhood of Little Beach Island north of Brigantine, in Atlantic County.

A visit to the Cape May County colony in 1923 showed that the nests are often mere depressions in the dry "trash" which is supported and raised somewhat from the ground by the growing reeds and grass stems underneath. Other nests are more carefully constructed, largely of this same material. The trash is composed mainly of coarse dead stems of the tall Spartina grass, one eighth to a quarter of an inch in diameter with some finer grass stems and strips of dried eel grass (*Zostera*) intermingled. Some nests, however, are constructed wholly of the stems of the finer marsh grasses. The average external diameter of the nests is eighteen inches but some of the longer stems, forming the base, project at least a foot farther. Some nests are located out among the short grass away from the creeks but the material for these had evidently been carried from the trash which had accumulated in the latter localities. At the Little Beach colony, which I visited in 1921, the nests were all built along the course of a larger creek where the Spartina grass grew to a much greater height and had been bent and beaten down into a great loose bed into which one sank up to the waist.

The eggs usually number three and are olive-brown splotched with purple and dark brown. The earliest date we have for a full set is May 18, 1919. The later dates vary greatly since the gulls are seriously affected by the high tides which flood the meadows and often wash out all the nests. In such cases the birds lay again so that in certain years eggs may be found until late in the summer. In 1922, as early as May 30, great numbers of eggs were found washed up along the causeway leading across the meadows from near Cape May Court House, evidently the result of a storm and high tide, and second sets were found in most of the nests a little later. On June 18 there were twenty-seven nests with three eggs, five with two and three with one. In the next year a similar flooding occurred and the whole beach at Corson's Inlet was strewn with eggs of gulls, terns and Clapper Rails. In that year some of the Seven Mile Beach nests contained eggs as late as July 13. High tides in the springs of 1933 and 1934 were very hard on the gulls and August 1 of the latter year found scarcely any immature birds in evidence, indicating a total failure of the nesting.

Turner McMullen and Richard Miller have given me the following data on nests examined by them in the breeding colony back of Seven Mile Beach.

July 4, 1920, fourteen nests with eggs.

June 7, 1921, fifty-one nests with eggs.

June 19, 1922, twenty-five nests with three eggs; one with four; many with two or one.

May 30, 1922, colony washed out; fourteen eggs found in ditches.

June 21, 1925, one nest with four eggs; nine with three; two with two; two with one.

June 20, 1926, 236 nests with eggs or young.

June 19, 1927, fifty-one nests examined; ten with one egg; fourteen with two; five with three; and others with downy young.

June 9, 1928, 181 nests with eggs or young.

June 6, 1931, forty-one nests with eggs.

July 20, 1932, 680 nests with eggs or young.

The young seem to leave the nest as soon as they are hatched and seek shelter in the tangled mass of grass stems upon which the nests are built. Here their mottled brown and black down renders them inconspicuous and we may gaze directly at one of these youngsters without detecting it, so perfectly does it blend with the black mud, dead grass stems and the play of light and shadow through the vegetation. Usually the young remain near the nest, often half submerged in the shallow water, but often they force their way through the grass to the open creek and swim with ease, even at a very early age, sometimes completing the crossing of the water and disappearing in the grass on the opposite side. When caught and placed back in the nest they immediately jumped out and pushed on through the grass, invariably on the side where it was rankest and where the water was deepest. Some nests had a sloping platform of trash at the side, accidental in all probability, but down this the young always scrambled in making their escape.

It seems probable that the young birds return to the nests or to the adjoining patches of trash when danger has passed, and are there fed by the parents. At any rate the flattening of the nest into a mere platform as the summer advances, and the plentiful excrement and scattered feathers show that the adults, at least, use it as a roosting place while the young are being raised. If the little birds do climb back to the nests to be fed, it is certain that they must scatter to seek shelter as soon as the old ones give the alarm cry, for not a young bird will be found in the nests when we approach them after the down dries and they are able to use their legs. The cry of the downy young is a shrill call resembling that of a young Robin *chir-r-r-rup; chir-r-r-rup.*

The actions of the Laughing Gulls on the nesting grounds are interesting. At a distance one sees them scattered here and there about the meadow with many others in the air, some putting out for the feeding grounds; some just arriving with food; dropping down into the grass or hovering for a moment just above, evidently passing food to the young below them. As

Wharton Huber and I landed from our skiff on the morning of August 14, 1923, and started across the level meadows back of Seven Mile Beach, it was high tide and about six inches of water covered the entire island, only the tips of the short grass and glasswort (*Salicornia*) showing above the surface while the winding lines of taller Spartina grass marked the course of the creeks. Most of the adult birds were evidently roosting on the patches of trash or on the flattened nests and they arose in several instalments as we approached, coming beating across the meadows to meet us, company front and all in full cry.

The most frequent call was *kek-kek* or *kek-kuh*, repeated continuously and rapidly, different individuals having slightly different expressions. Some also called in triplets—*kek-kek-kek; kek-kek-kek*, with more of the woody quality of the clarinet. Then some birds would begin to break through the chorus with a *kek-kek-kek-kek-keeeer, keeeer, keeeer*, or *haaaar, haaaar, haaaar*, like mocking laughter—the call that has given the bird its name of Laughing Gull.

The circling, soaring crowd of birds overhead, now numbering some two hundred individuals, looked beautiful in the sunlight. As seen from directly below the pure white of the breast and tail of the adults contrasted sharply with the apparently jet black head and wing tips. The sunlight moreover shown through the translucent posterior margin of the wings and tail producing a silvery white effect in contrast with the opaque white of the breast and belly. There were some white-headed birds with dusky ear patches which appeared to be individuals already in the molt since some of the white tail feathers were missing. There were also a number of birds of the year, uniformly dusky except for the lower abdomen and base of the tail which were white. Now the moving maze of birds drifts away as we turn to one side and some of them settle on the meadow, but as soon as we return toward the nesting ground they rise in a cloud and once more come toward us low over the ground. It is a striking picture. The tide has ebbed and the meadows stretch away in brilliant green. The little fiddler crabs have come out of their burrows and go scuttling and sideling away while the periwinkles are slowly climbing the grass stems where they hang by hundreds like great excrescences or curious seed pods. Beyond our meadow island are the sparkling waters of the sound and on the distant horizon the low-lying woods and shining sand dunes of Seven Mile Beach while high overhead the snowy white, dusky-winged birds pass and repass in intricate circles under the great blue vault of the sky.

Once in a while an intruder enters the gulls' domain—a Turkey Vulture innocently sailing out over the marsh to inspect some possible carrion or a Marsh Hawk beating the grass for meadow mice—and he is at once set upon by the angry gulls and made to twist and turn until he is glad to beat a hasty retreat. Similarly on Jarvis Sound I have seen feeding gulls, apparently forgetting that they are no longer on the nesting grounds, rise to attack a Bald Eagle which had started to cross the meadows to the sea. On May 22,

1922, a single gull routed this royal enemy while on July 23 of the same year three gulls joined in the pursuit. So also on September 4, four gulls pursued a Fish Hawk out over the surf at Cape May Point and made him pitch and toss and finally fall through the air for some distance, for he is not accustomed to rapid wing action and is no match for the gulls at twisting and turning. On August 12, 1929, I saw an adult Laughing Gull pursue a Common Tern which had caught a small fish and was carrying it in his bill. The gull kept close after him turning and twisting and following every movement of the smaller bird until the tern dropped his fish and the gull caught it, gulped it down and flew away. I do not however regard this as a common habit.

The Little Beach colony, which I have known since 1892, was visited last on July 18, 1921. It was then located on either side of a large salt creek which wound its way through the meadows and emptied into a thoroughfare directly opposite the cabins on the island. Some three hundred Laughing Gulls could be seen from a distance perching on the masses of trash and constantly jumping up into the air and settling again. Upon closer examination we found that there were only a few nests that still contained eggs but there were many broken egg shells from which young had recently hatched, and splashes of excrement covered the dead meadow trash on all sides. Search as we would not a single young bird could we find. The depth of the masses of trash made it exceedingly difficult to locate them in any case, and as we were, at the time, unaware of their habit of leaving the nest and remaining perfectly still amongst the grass, sometimes until almost trodden upon, we undoubtedly passed over many without detecting them. All night long from our cabin across the channel we could hear a continual clamor from the colony, evidence that the young were still there and that their parents had returned to watch over them.

When we had approached the nests during the day the birds arose *en masse* several hundred in number uttering their harsh *kek, kek, kek, kek, haaar, haaar, haaaar, haaar,* about four of the long calls in three seconds. When quite close to them it seemed that some of the calls had a "*k*" sound at the beginning and end and that they consisted of two slurred syllables— *kee-ahhk, kee-ahhk, kee-ahhk, kee-ahhk.* Some birds also omitted the short preliminary notes, *kek, kek, kek, kek,* at least at certain times. The birds formed an intricate maze as they wheeled and circled overhead or, catching a favorable wind, a number of them would sail off a little way together and return again to the fray. As they arose in the air they appeared smaller and smaller until one could scarcely realize that the large birds close overhead were of the same species as the little ones higher up. All the while the bedlam of noise continued. There were in the flock several white-headed, non-breeding birds, that had never molted, and one great adult Herring Gull which seemed to be just as much concerned as his lesser brethren, though we could discover no vocal effort on his part.

While in August, 1921, the gulls passing north over Cape May in the evening had as their immediate objective the extensive sand flats left by the dredges back of Two Mile Beach, where Turtle Gut Inlet then emptied into the sea, I think that this may have been only a temporary stop or a roosting place for only a part of the colony. Certain it is that up until close to the time of their departure for the south the greater part of the colony remains in, or returns to, the meadows where they nested, or to the adjoining meadows back of Five Mile Beach. On numerous trips across from the mainland to the beach islands during August and September they were present in abundance. On August 12 and 13, 1930, they were scattered all over the meadows south of the Wildwood Road while on the 23d, there were close to one thousand on the higher meadows near the mainland north of the road to Stone Harbor. They were seeking safety and shelter during a very high tide and a strong wind, and probably one thousand more occupied elevated spots on the meadows back of Five Mile Beach a few miles farther south. As the water fell the birds scattered in every direction seeking food left stranded by the tide. On September 27 of the same year at low tide there were six large assemblages of Laughing Gulls on the mud flats in the thoroughfares or resting on the water back of Five Mile Beach, and another gathering on the upper meadows; about three thousand birds in all.

In 1931 there were large assemblages of adults and full-feathered young on both stretches of meadow, on August 15, while on September 12 and 26, at high tide, we estimated three thousand present north of the Wildwood and Stone Harbor Roads. They were gathered in dense masses either on the grass or on the salt ponds. Every little while an entire flock would shift from one spot to another, arising section after section, and settling with much confusion and flapping of wings. Some of the ponds seemed simply paved with birds. There were no Herring Gulls mingled with them at this time and all of the adults seemed to have acquired their white heads. On the former date a long straggling flight was coming in from the south shortly before sunset; doubtless birds that had spent the day at the "ripps."

From the time of their first arrival there will be found some Laughing Gulls with white or mottled heads, birds which have never acquired the full nuptial plumage and which seem to be non-breeders. They are often seen standing solitary along the beach or on the mud flats at low tide, where, however, they frequently flock together, and a bunch of twelve with no full-plumaged individuals among them was seen on May 23, 1922. Some of these barren birds however frequent the nesting grounds and join the colony in protesting the presence of intruders.

During the latter weeks of July the first dusky-plumaged young are seen on the wing and they become increasingly plentiful in August. These birds in their brown-drab plumage contrast strongly with the gray and white adults. They show white only on the rump and at the base of the tail, the latter having also a very distinct black terminal border, a character shared by the barren birds but absent in the adult at all seasons.

As early as July 23 and August 8, in different years, I have seen adults which had begun to lose the dark hood and by mid-August most of them show white mottlings on the head while others lack some of the tail feathers, or are in other phases of molt. On July 30, 1927, the shallow pools and the sand fill back of South Cape May were covered with scattered feathers, the result of this annual molt of the adult gulls. Some black-headed individuals, however, are still to be seen as late as August 30, but by early September the molt is practically over and a flock seen on the 16th, 1927, and another on September 12, 1931, were all white-headed. The brown birds of the year do not change their plumage until later and on September 27 and 28, 1924, I found them in full molt. There was a flock of about fifteen hundred Laughing Gulls gathered on the shallow pond north of the Lighthouse, and so closely were they crowded together that the pond seemed literally paved with them, while their varied plumages produced a remarkable color effect. Close study of the brown immature birds showed that each body feather had a darker dusky border while the wings were edged behind with buffy white and the tail, dull white at the base, was mottled with darker and had a distinct dark terminal band. The adults which had, of course, completed the molt were clear dark slate on the back and wings with white breast, tail and head, the latter with a clearly marked dusky ear patch. The primary feathers, as always, were black. These two plumages were to be seen all through the flock while many young birds showed all sorts of intermediate conditions. Some had white feathers on the head and large white patches on the breast while blue gray areas were appearing on the back and all over the pond could be seen the cast-off feathers.

The whole flock was busy bathing and preening, the birds rising a few feet in the air and dropping back again with dangling legs. On alighting they would often hold their wings aloft for a few seconds and would now and then stretch out one wing and leg horizontally as if for exercise. Finally all took wing with much clamor, their call at this season being a short *keek, keek*, somewhat like the short note of the Common Tern.

During October the Laughing Gulls rapidly decrease in numbers on the meadows but increase correspondingly at the "ripps" and about Cape May Point, where they repair to rest on the beach or formerly on the shallow pond which for many years existed just east of the Lighthouse. In 1921, I noticed the first of these gatherings on August 2 and on the 5th there were thirty gulls on the beach associated with Common Terns which also habitually gather there. They lined up just back of the terns and after August 8 separated from them entirely. When at rest on the beach the gulls stand at the very edge of the water with heads all pointed toward the wind as is customary with resting water birds. Some individuals now and then walk rapidly about in the shallow water, apparently picking up food, while others are busy preening themselves, with head bent down on the breast or twisted around over the shoulder in order to reach the feathers of the back. Occasionally, too, a bird will rise in the air, give a half dozen wingbeats and alight farther along the line. On August 11, 1922, over three hundred Laughing Gulls were gathered on a sandy spit on the pond near the Lighthouse and the next day were resting on the strand where they resorted several times during September. On August 23, 1923, six hundred assembled here with many Common Terns. On September 16, 1921, there was a flock of fifty-eight on a wet spot on the beach near South Cape May which I was able to approach and study, creeping up under shelter of the sand dunes and bay-berries. Many were standing on one leg with the head bent over on the back while others in similar position had both feet on the ground. After watching the birds for some time I suddenly arose and the attitude of the group changed instantly. Every bird was alert and ready to take wing in strong contrast to the easy positions of rest of but a moment before. Soon they spread their wings and heading out to sea settled on the water in a long ir-regular line, several birds deep, just beyond the surf.

This habit of resting on the ocean seems more characteristic of the end of the season and is doubtless preparatory to migration. A flock on the beach, on September 15, 1921, settled on the sea when flushed and then arose in a cloud on the approach of a chugging power boat. On September 17, 1923, the ocean just off the Point, was fairly strewn with Laughing Gulls, all adults in full winter plumage, and on October 1 of the same year and October 11, 1930, there were approximately five thousand feeding at the "ripps" and three hundred more resting on the beach.

On October 24, 1921, I found some three thousand on the water just off the Point. They rode the waves very prettily all facing to windward, their heads drawn well down on their shoulders with the tips of the wings stand-ing out clear above the rump. On November 7 there remained about one thousand, while in 1921 there were several hundred feeding offshore on October 31, some of which remained until November 10, and in 1923 five hundred were present at the same spot on October 21. The Laughing Gulls linger during the first few days of November—five hundred on November 7, 1922, four hundred on November 4, 1923, and one hundred and fifty on November 7, 1931, but by the middle of the month they have passed south, my latest dates being November 12, 1922; November 14, 1926; November 16, 1929; a few individuals each day. My only winter record is a single bird seen feeding on the bed of the channel near Stone Harbor at low tide, December 25, 1931, and another in a similar situation at Avalon on the same day.

They occur on the Delaware River in autumn as well as spring and Julian Potter has seen them as early as August 27, 1927, and September 30, 1924, when many were blown in by heavy storms on the coast. They often occur through October in flocks of ten or more and single birds have been seen as late as November 13, 1924; November 10, 1921; November 8, 1928.

On September 9, 1924, many Laughing Gulls were seen about Cape May which seemed peculiarly restless, coursing back from Two Mile Beach to the Point in numbers. Their flight was very different from that of the summer birds, lacking in directness and definiteness of purpose. Many flew over the golf links and another flight passed along the ocean front, all converging at the Point where at least two thousand birds were soon assembled on the beach facing a southwest gale. These were perhaps birds from farther north, the first southbound migrants of fall, or local individuals stimulated by the change in the weather.

The Laughing Gulls of Cape May seem to be the last of the summer sojourners to leave the coast and in our studies of them we have, year after year seen the airy light-winged Black-heads of spring time, birds of green meadows and azure skies, change to these white-headed, leaden-backed denizens of the steel-gray November ocean; ready at any moment to start on their southward journey, leaving our shores to the great gaunt Herring Gulls and the dainty dove-like Bonaparte's Gulls which constitute the winter gull population of the Cape.

BONAPARTE'S GULL

Larus philadelphia (Ord)

At the first approach of cold weather there arrive at Cape May flocks of the beautiful little Bonaparte's Gull and all through the winter it is a characteristic bird of the sea coast. It is distinctly smaller than our other gulls and in its plump build and general proportions calls to mind a white Domestic Pigeon, while its restless activity and its habit of frequently holding its bill point down as it flies reminds one of a tern. Indeed George Ord, who originally discovered it on the Delaware River at Philadelphia, described it as a tern.

While typically a winter bird we have sporadic occurrences in late summer and early autumn—August 5, 1920; August 3, 1921; September 12, 1926; September 27, 1928; all single individuals and all in immature plumage; also several immature birds at Corson's Inlet on June 27, 1925 (Miller). The regular time of arrival at Cape May seems to be late October—October 25, 1929; October 30, 1921; etc., though they do not become common until November and from then until mid-March we are likely to find them present on any visit to the coast. Our latest records are April 2, 1921; April 4, 1924; April 18, 1926; April 1, 1928; April 7, 1929; April 12, 1930; with several exceptional dates—May 9, 1925; May 10, 1931; May 27, 1931. This little gull has been reported in large numbers ten miles off Long Branch on the upper New Jersey coast, in Christmas week, and we wonder if it may not always be more common out at sea, and if the earliest and latest occurrences at Cape May may not be stragglers from offshore flocks. Farther up the coast Charles Urner has seen them on Barnegat Bay in June on several occasions and in 1925 immature birds remained there throughout the month, while at Perth Amboy thirty were present on July 25, 1923.

The occurrence of Bonaparte's Gull at Cape May is somewhat erratic. It usually associates in flocks of from half a dozen to fifty, and these drift about from place to place so that on some days we may miss them entirely on a winter's walk. At other times the sea may be covered with them as flocks of one and two thousand individuals suddenly arrive from somewhere to be gone again the following day. The shallow pond near the Lighthouse used to be a favorite resort for them and even when it was largely covered with ice the birds would gather over whatever water remained. On other days they may all fly over to the Bay or perhaps put out to sea beyond the breakers. They are often found, too, about Schellenger's Landing and at other points on the Harbor while flocks may be seen coursing over the meadows back of Five Mile and Seven Mile Beaches, or gathering about the shallow salt ponds.

The actions of Bonaparte's Gull in the air are varied. When gathered together at some feeding station they wheel and quarter over the water with remarkable grace and dexterity, darting to the surface with all the agility of a tern to pick up some morsel of food while a hundred or more observed on the Bay shore on December 30, 1920, were performing intricate evolutions while every moment one or another of them would wheel suddenly and swoop to the surface. Their wingbeats on a straight away flight are more rapid than those of the larger gulls and less regular, while sometimes the wings appear as if their tips were pointing backward; they frequently flutter, too, in true tern style as they pause momentarily in their flight.

On March 13, 1921, I found three flying over a salt pond on the meadows and every moment others were arriving until there were twenty-six in the flock. A careful analysis of their flight showed that they give several rapid wingbeats and then sail softly down, pausing suddenly over a likely spot with quick short beats—often a distinct flutter, and then, just as they seem about to alight, and just as the dangling feet touch the water, they are up in the air again bouncing along in buoyant flight. With individual birds in all stages of this performance the result is a veritable maze of white wings. They will go crisscrossing through the air, swooping down and flapping up again and then suddenly, stirred by a common impulse, away they go all together like a flight of shore birds, out to sea beyond the surf line where they come to rest on the water.

On March 24, 1923, a small flock was observed over the meadows flying with short, quick wing strokes followed by a short sail. Sometimes they halted instantly with backward strokes, almost turning somersaults, and then hovered, for a moment, the actions being very like those of the Black Terns on the same spot in August. On February 14, 1925, a flock was seen putting directly out to sea in bullet-like flight.

On November 12, 1922, Julian Potter saw a flock, possibly on migration, which he estimated to number 2000 birds and which was performing wonderful evolutions off Cape May Point. The birds would rise from the water in close formation and fly up the coast a short distance, where they would suddenly settle in one large assemblage, pitching and diving through the air in every direction, before coming to rest on the surface. Then, after a few minutes rest, they would rise again *en masse* and fly back down the coast to repeat the whole performance.

Six birds watched on January 7, 1922, were wheeling about one another in a small area over a patch of open water in the ice-covered pond by the Lighthouse. They flew about twenty feet in the air and one after another pitched downward, actually plunging into the water, an unusual action for any gull. They kept their wings half spread and never submerged entirely though the head went under at every plunge. They would rise again almost instantly, flapping their wings which had never been closed and dangling their feet, all the while uttering a tern-like cry. A flock of ten watched on the Harbor on December 27, 1929, dived repeatedly and went entirely under in the deeper water but only for a second or two. Several Bonaparte's Gulls studied at very close range from the bridge at Schellenger's Landing (February 11, 1921) were beating against a strong wind. Their flight was very light and buoyant and many times they recalled pigeons as they turned against the wind with head and neck outstretched and slightly raised. Their bodies bounced up and down in the air as if the wings were fairly pounding it and causing the rebound.

On the water they float lightly and often revolve, whirling around suddenly like a phalarope. When swimming vigorously the neck is stretched forward and slightly elevated and the wings held clear of the water and crossed above the tail. One bird in this position suddenly plunged its head completely under the surface while another bent its head over and scratched the side of its face with its foot. When at rest they ride the waves like ducks. With heads drawn well down on their shoulders and all pointing to the wind, they bob up and down like corks when the water is rough. Even the most restful assemblage, however, is liable to take sudden flight and a flock floating peacefully on the Lighthouse Pond took wildly to wing and was out over the Bay in a few moments. The advent of a Cooper's Hawk proved to be the cause of their alarm although he made no apparent effort to attack them. On another occasion a Duck Hawk managed to isolate a gull from the flock but in the dodging that ensued the hawk was outmastered at every turn and eventually flew off.

Sometimes we see Bonaparte's Gulls resting on the ground, most frequently on the strand, and twenty gathered on April 2, 1922, near the Coast Guard Station at the Point, close to some Herring Gulls which seemed like perfect giants in comparison. I have also seen them standing close together in a line on a narrow sand spit in a salt pond on the meadows. These birds later swam about on the pond tilting and plunging their heads under for a moment with the tails held erect like surface-feeding ducks.

On November 7, 1931, there were eight Bonaparte's Gulls along with twenty Common Terns and a few immature Laughing Gulls all perched on a row of close set pilings at Cape May Point, the only ones I have noticed in such a situation, as they seem to prefer to rest on the water or the beach.

In his study of the great flock that assembled on the Delaware in April and May, 1922, Julian Potter writes me that they fed sometimes on small minnows which they caught by diving and swallowed while skittering over the waves like petrels. Sometimes a bird would drop its fish and was unable to recover it and once one went completely under and emerged with a fish only to be attacked by many others with the result that the catch was lost to all. Again they seemed to be picking some sort of insect food from the surface of the water dipping down with the utmost grace and barely touching the water with bill and feet. On one day the flock was spread out in a long line three-quarters of a mile in length, all of the birds headed into the breeze, and as they neared the shore the leaders would swing back to the rear so that the flock remained in the same position. They floated high on the water with tail and wings elevated so that the body seemed tilted forward. When facing a heavy wind, or resting, the head was drawn close to the shoulders and at a distance the birds seemed to be headless. The great majority were immature, or at least white-headed, birds with only a sprinkling of black-headed ones. They had an explosive call somewhat like the *haaa* of the Laughing Gull and a call that might best be imitated by a whistled *cheer* while there was also a succession of rapid notes recalling the spring "song" of the hop toad. One morning a flock of forty sailed high overhead apparently catching insects in the air as the Laughing Gulls seem to do and continually rising disappeared entirely from sight. They were first seen on April 12, had increased to five hundred by April 27, and were last seen on May 6, a mere half dozen.

During the time that they are with us the Bonaparte's Gulls exhibit much variety of plumage. The winter adults are delicate pearly gray above with white head, breast and tail, and an always conspicuous character is the white outer border to the front of the wing caused by several pure white primary feathers which contrast with the rest of the pearl gray upper parts. There is also a black area on the posterior wing border formed by the tips of the other primaries. Young birds are quite different having a dusky stripe immediately over the bones of the wing and a dark terminal band on the tail. There are also dusky patches on the head and back which vary in individuals and produce a strongly mottled appearance. There is a dusky spot on the side of the head in both plumages while the bill is always black and the feet pink. The adult bird in breeding plumage has a dark gray hood like the Laughing Gull, which seems almost black at a distance, this is apparently rarely assumed before the birds leave Cape May on spring migration. One individual seen on the Harbor at Schellenger's Landing, on April 27 and again May 10, 1931, had the black head and several seen on the Delaware River some miles above the Bay, on April 15, 1927, had partly completed the molt. No others in this plumage were recorded at the Cape, and one seen at Stone Harbor, May 9, 1925, still had the head white. A similar specimen was secured by Norris DeHaven at Atlantic City on April 4, 1896, and another at Gloucester on the Delaware River on April 10, 1890. Another Gloucester bird on the same date had only a slight dusky tinge on the top of the head while a similar bird was seen on the river below Philadelphia on March 3, 1890, and another was picked up dead at Cape May Point on April 4, 1924. These are perhaps just beginning the molt.

Light conditions affect the appearance of these gulls as they do other species. When resting on the water their snowy breasts are often concealed or turned away from the observer and their backs seem darker, so that they appear as slaty gray birds with white heads, and when flying low over the waves they often seem unnaturally dark. Their full beauty is best displayed in the brilliant winter sunshine against dull brown meadows or blue sky.

All in all they are delightful birds with more vigor and action than most of our winter sojourners, and with an apparent enjoyment of flight for flight's sake not appreciable in the sluggish, business-like actions of the Herring Gulls.

* LITTLE GULL

Larus minutus Pallas

This European bird has never been seen on Cape May waters but it has occurred at the northern extremity of the state's coast line.

One was seen by Warren F. Eaton in company with James L. Edwards and John Thompson on the Newark meadows on May 12, 1929. He writes: "In a mixed flock of about 200 Bonaparte's Gulls, fifty Ring-bills and a dozen Herring Gulls we picked out two birds with black heads. One of these showed black under the wings and the other was an adult Bonaparte's Gull. The former fluttered its wings like a tern and was several times an object of attack by the many Bonaparte's Gulls. In size it was about two inches shorter than the latter and markedly smaller when they were seen side by side. The color of the under surface of the wings looked absolutely black and the upper side was apparently not marked at all, the body, tail and back were white; head and neck black" (Auk, 1929, p. 376). On May 6, previous, James P. Chapin saw what he felt sure was one of these little gulls under particularly favorable circumstances in upper New York Bay (Auk, 1929, p. 377) and on August 11, 1929, James L. Edwards, with Charles Urner and others, saw still another, in the white-headed plumage, but with the black under surface of the wings very evident, with a flock of Common Terns on the Manasquan River, at Point Pleasant, N. J. (Auk, 1929, p. 532). The upper side of the wings was seen to be pale gray with no white area as in the Bonaparte's Gull, except for a white posterior margin.

* KITTIWAKE

Rissa tridactyla tridactyla (Linnaeus)

This little gull is a pelagic species which is occasionally blown to our beaches by severe storms.

William H. Werner told me that he secured quite a number of specimens from fishermen who went out to the banks off Atlantic City, in the winter of 1894–95, and that they were reported as present there every winter up to 1900, and doubtless later. They were especially plentiful in 1898.

From the lists of those who went to sea for their Christmas bird lists for "Bird-Lore" we get some interesting data. On December 31, 1904, Robert E. Stackpole and William H. Wiegmann, counted seventy-four adult and fifty-seven immature Kittiwakes while cruising from ten to twenty-five miles off Long Branch, N. J., but did not begin to see them until they had "almost lost sight of land." On December 27, 1908, Robert E. Stackpole and Charles H. Rogers saw fifteen off Seabright, N. J.; December 18, 1910, Noemi Pernessin and Charles Rogers saw twenty off Seabright; December 28, 1912, Wiegmann, Rogers and Waldron Miller, saw fifteen in the same region; and on December 19, 1913, the same with John T. Nichols, saw forty. On all of these trips the Bonaparte's Gulls outnumbered the Kittiwakes ten, or sometimes fifty, to one. On December 27, 1925, five were seen at Barnegat Bay, and Charles Urner has other records from that region for October 11, 1925; October 31, 1926; November 12, 1921.

Two Christmas census trips on fishing boats out of Cold Spring Inlet, Cape May, to the fishing banks failed to yield any Kittiwakes although conditions of weather may have had something to do with this. The bird is more frequently seen from the shores of Long Island than farther down the coast and there are records from November 4 to February 27 (Braislin). In size the Kittiwake is comparable only with Bonaparte's Gull which is a little smaller; in plumage the adult is of almost the same pattern as the much larger Ring-bill but the black at the tip of the wing cuts straight across at right angles and the feet are black; the immature bird is distinguished from the winter adult and young Bonaparte's by the black feet, the absence of white on the primaries, and a black bar across the back of the neck, instead of a spot behind the eye; both have a narrow black tail band (cf. Peterson). Other Kittiwakes have been recorded off Barnegat by Charles Urner on October 11, 1925; November 29, 1926 (six); November 12, 1927 (twelve); and by Carter and Friedmann on October 31, 1926 (two).

GULL-BILLED TERN

Gelochelidon nilotica aranea (Wilson)

Alexander Wilson discovered this tern on one of his several trips to Cape May probably in 1811 and announced the fact as follows: "This new species I first met with on the shores of Cape May, particularly over the salt marshes, and darting down after a kind of large black spider, plenty in such places. This spider can travel under water as well as above, and, during summer at least, seems to constitute the principal food of the present tern. In several which I opened, the stomach was crammed with a mass of these spiders alone; these they frequently pick up from the pools as well as from the grass, dashing down on them in the manner of their tribe. Their voice is sharper and stronger than that of the Common Tern; the bill is differently formed, being shorter, more rounded above, and thicker; the tail is also much shorter, and less forked. They do not associate with the others; but keep in small parties by themselves." "This species breeds in the salt marshes, the female drops her eggs, generally three or four in number, on the dry drift grass, without the slightest appearance of a nest."

By 1869 Turnbull regarded it as rare. In 1886 Harry G. Parker reported it as nesting at the lower end of Seven Mile Beach, but this spot is a sandy beach frequented by the Common Tern and it seems unlikely that this species could have nested there. Charles S. Shick, however, speaks of it as still present in 1890 and nesting at this same point "on the meadows and sand flats." These men were primarily egg collectors and whether they could distinguish the bird from the Common Tern is open to question. Turner McMullen has reported a nest and two eggs found on June 20, 1926, in the Laughing Gull colony on the meadows back of Seven Mile Beach. But the fact that I have found nests of the Common Tern on the meadows back of Atlantic City built on "trash" would indicate that location alone is not conclusive and none of the expert students of our coastal birds has been able to find a single individual of this species. We seem to have no records of the occurrence of the Gull-billed Tern in New Jersey, since Wilson's time, not even as a straggler, that seem convincing. While it is a southern species there are three records for Long Island—July 4, 1882; July 8, 1884; July 1, 1885. (Griscom, Birds of the New York City Region.)

COMMON TERN

Sterna hirundo hirundo Linnaeus

While Common Terns do not nest on the beaches immediately about Cape May a colony was established some years ago on Gull Bar, in Hereford Inlet, and on the points of beach on either side of the entrance and since then the birds are to be found somewhere about the Cape from the time of their arrival in May until their departure in October. They were abundant all along the coast in early times down to the eighties when the slaughter for millinery purposes bid fair to exterminate them. Detailed history since that time is lacking but they have increased greatly since 1890 when my acquaintance with the bird life of the Cape began. Scattered individuals may now be seen at low tide on Jarvis Sound resting on the exposed mussel beds or perched on channel stakes along the thoroughfares while others will be flying back and forth over the waters of the Harbor.

More regularly we see small bunches of terns travelling up and down the coast flying low over the waves out beyond the surf line or wheeling in over the bathing beaches. In certain lights they appear dull gray and blend so perfectly with the color of the sea that it is difficult to distinguish them, while in others they show silvery white against the dark waters, the apparent change in color being remarkable. Occasionally too we see detachments of twelve to twenty-five travelling at a higher elevation, dropping to the waves and rising again in irregular undulating flight, the picture of grace and agility. These birds come in to the strand now and then, especially in stormy weather when the sea is rough and the sky leaden gray with a spit of rain in the air, and not infrequently give voice to their harsh cries.

These passing birds are traveling from their fishing grounds on the "ripps" at the mouth of the Bay, to their nesting grounds or vice versa and there is a spot on the beach near the Lighthouse where they used to congregate in considerable numbers to rest from their vigorous fishing or later in the season to pause on their migrations, While they still gather here there are fewer of them and the increasing numbers of pedestrians on the beach causes them constant alarm. From late July until September this spot was a regular roosting place for terns although as the summer advanced part of them, at least, seemed to shift to a sand flat at the upper end of Two Mile Beach, a spot to which roosting gulls resort. Large numbers of terns *en route* for this later roosting place pass over the town at dusk on August evenings in compact flocks of twelve to fifty individuals, flying steadily with much more regular wingbeats than usual.

In May and June, also, some terns assemble on the beach opposite the "ripps" but not in such numbers nor so regularly as later in the season. The earliest of these are, I think, transients bound for farther north while some are immature non-breeding birds, from colonies not far distant, which come here to feed. Of some two hundred birds assembled there on June 21–22, 1923, nearly half were in the immature plumage with white foreheads. On May 20, 1927, there were fifty-one adults busily engaged in fishing close to the beach and twenty-five more resting on pilings a little west of their usual gathering place, while on July 11, 1922, there were two hundred on the beach, on June 1, 1923, ninety-five, and ninety-seven on June 27, 1925.

It is August, however, when the southbound migrants collect on this strip of beach, that the Common Terns become a leading feature in the bird life of the Cape. The first large assemblages from 1920 to 1931 were on:

1920.	August 2.	1923.	July 28.	1927.	July 27.
1921.	July 29.	1924.	July 22.	1929.	July 30.
1922.	July 11.	1925.	July 27.	1931.	August 1.

There seem always to be some Black Terns mingled with the larger birds and since they do not nest south of the Great Lakes their presence suggests that the Common Terns in these early southbound flocks may also come from far north and not from the New Jersey nesting colonies. The fact that the New Jersey birds still had downy young and hatching eggs as late as July 18, 1821, and unhatched eggs on July 9, 1923, lends weight to this theory. The Black Terns that come with the others seem gradually to leave the flock and disappear, at least in most seasons, and the stray Laughing Gulls that associate with them at first leave later to roost elsewhere with their own kind. The gathering of the terns on the beach at the Point was for many years a characteristic sight to those who follow the shore line during August but of late years they have sadly dwindled. The number of birds fluctuated from day to day, ranging from fifty to one thousand, but the variation was perhaps more apparent than real since a large part of the flock was often out at the "ripps" feeding. The frequent pedestrians passing along the beach constantly disturbed the terns and finally forced them to seek a resting place back of the dunes, on sand flats which formed there as result of winter storms and high tides, or on the borders and mud bars of the shallow pond east of the Lighthouse. In still later years dogs ranging over the beach have well nigh driven the flocks away while the beach itself has suffered erosion until its width is sadly reduced.

Usually when disturbed the birds simply fly up the beach a little way and settle again or else, especially after being flushed several times in succession, they scatter over the ocean and begin feeding. There is usually much excitement and the angry birds protest vigorously at the intruders. The racket begins as soon as they take wing which they sometimes do *en masse* or sometimes rank after rank, in regular order down the beach. There is a great flapping of the long narrow wings but not in unison, as every bird beats the air independently. The first birds to take wing wheel, and come back, meet the next installment going forward, so that ere long birds are flying in every direction; an intricate maze of moving wings ensues and there seems to be wild confusion. Now as the sunlight strikes them the Common Terns look snowy white and the Black Terns that are mingled with them blacker than ever, then the angle of light changes and they appear in varied shades of gray and pearl. One flock of about one hundred flying with exceptional regularity came toward me head-on and the wings, with only the forward edge visible, appeared white with a dusky border above and below, while the jet black heads of the birds showed prominently against their snowy breasts, a truly beautiful sight.

The volume of sound produced by a flock of terns in full cry can hardly be appreciated by one who has not heard them and, although these beach flocks hardly equal a nesting colony in vocal ability, they are not far behind, and their calling is really but a continuance of the clamor of the breeding grounds, which they have maintained on occasion every since their arrival there in May.

Terns have several notes. The most characteristic is the long drawn, grating *keer-r'eer; keeee-te'arrr;* or *teeee-a'rrr* as I have written it at various times. It consists of two equally accented long drawn out syllables with much of the harsh "*rr*" sound especially at the end. This is repeated over and over; and with a hundred or more birds all crying simultaneously the bedlam of noise can be imagined and will continue to ring in one's ears long after the birds have been left behind. There is another quite different call, short and emphatic but lacking the volume of the last. It resembles closely the call of the Robin in defense of its nest or young but is not so loud and I have written it on several occasions *sheep'; chip'* or *quip'*. I have heard this cry from birds gathered at Cape May Point in spring (May 20) and it is frequent on the nesting grounds. The common cry is given both on the wing and while resting on the beach—at least in the spring, but the short cry is always uttered by a flying bird. A third cry is usually restricted to the breeding grounds but on rare occasions has been heard in the August gatherings on the beach. It is uttered by an enraged bird as it dashes down at the head of an intruder and is a rasping *kek-kek-kek-kek-kek-kek-ke'eeeer* the last syllable delivered just as the bird veers off without actually striking its enemy and sounds like the tearing of cloth. Then there is the cry of the young bird as it runs to meet its parent bringing food; a low tremulous call *kruuuuurr* or *chuuuuuurr* the note usually repeated four times. The clamor incident to the flushing of one of the beach gatherings continues until the birds begin to alight again or scatter widely over the ocean when silence is once more established.

When reassembling on the sand the terns come directly into the wind and settle in long lines with their heads all pointing one way. With the prevailing south wind they always face the sea but on several occasions when strong northwest winds were blowing they faced landward. On August 2, 1920, the first assemblage of the season, probably unacquainted with the spot and naturally nervous, arose at once upon being disturbed and circled around, higher and higher, until they had reached a considerable altitude and looked no larger than swallows, and then drifted with the wind out to sea and disappeared. Whether the same birds returned or whether this was a continuance of their migratory flight I could not determine. The same restlessness sometimes prevails in the spring and a flock observed on May 26, 1923, flushed and returned to the beach many times with no apparent disturbing cause. One day when the terns had shifted their roosting place to the sand flat they were completely enveloped in a dense fog which for some days spread over the coast for a quarter of a mile inland. The effect as they arose was weird. The air was at first full of the cries of birds that were invisible and then, as they advanced, their white tails and under surface of the wings would catch the light and gleam for a moment like silvery sheen from polished metal and then disappear again, swallowed up in the gray blanket of the fog. By the time the air had cleared somewhat the birds had settled on the sand flat, covered with its scattered shells and pebbles, and the resemblance of the resting terns to this background was remarkable. The white and gray of the adult plumage and the black caps being duplicated by the shells in various stages of bleaching and decay, while the mottled brown and gray of the young birds was matched by the various colors of the sand.

As the terns gather normally on the strand they remain standing for a time on their short red legs, many of them preening their feathers, bending over and arching the back so that the plumage of the under parts may be reached, or twisting the head over the shoulders to reach the feathers of the back. All the while the tail is partly erected and is spread out for a moment, first on one side, then on the other. Occasionally the preening bird will ruffle up its entire plumage and stretch out its wings like a man stretching his arms after a sleep. At other times they are raised vertically but only for a moment or two, as we often see them do when alighting on the beach. Some of the birds walk about in a rather clumsy manner, others bathe in the shallow pools and one individual scratched its head with its foot. Eventually, however, all settled down, squatting directly on the sand. Now and then one would raise its head and open its bill as if calling to another in the air above, and when a screaming tern does fly over head, or a Fish Hawk sails past on his way to the sea, there is much commotion in the flock and it is amusing to see the heads turned over on the side so that one eye looks straight up at the intruder.

There is a curious habit that seems to belong in part to the courtship behavior of the terns although it is to be seen also in these late summer gatherings. A bird will fly into the assemblage with a small fish in its bill and immediately another bird, sometimes an adult, sometimes a young of the year, will beg for it with open bill, doubtless a relic of the feeding period. Then another adult will strut past with neck elevated, bill pointed skyward and wings drooping, a most ludicrous pose. This took place on the beach on August 3, 1922, but I have witnessed precisely the same performance in a spring beach flock (May 28, 1923) when two individuals strutted past the one holding the fish, evidently a courtship performance. On the mussel beds of Jarvis Sound (May 21–22, 1922) I saw one of a pair—presumably the male—come in and feed the fish to the other but when apparently the same individual returned with another fish a second male was already there. The first bird paraded past the female in the attitude described above holding the fish aloft but being driven away he swallowed it himself, as he did on another occasion when he found no terns at all on the spot. This bird moreover ducked his head under the water several times and churned it up with his wings before taking flight. During August I have repeatedly seen terns flying along the shore with a small fish crossways between their mandibles and supposed that they had been so harrassed by other birds that they were unable to swallow their prey when caught, but later I saw one alight on the beach entirely free from molestation and instead of eating his fish he carried it around and eventually flew away without making any effort to swallow it. As late as July 30, 1923, I saw an adult come in with a fish and join a flock resting on the beach. A full plumaged bird of the year came running up to it and was fed the fish while it uttered the peculiar *kruuuurr* cry of the nesting grounds.

Terns feed on small fish that have been stranded on the beach, swooping down to pick them up, and they also dive into pools and ponds on the salt meadows as well as into the Harbor and the Bay. Their favorite feeding ground, however, is the "ripps" at the mouth of Delaware Bay, to which large flocks resort daily and viewed from the shore through the glass the assembled birds make a lively scene. I watched them thus for a long time on August 26, 1920. It was a day of approaching storm and the water looked gray and dull but the light was ever changing, there were bright flashes as the sun broke through the clouds, followed by bands of shadow. The wind had roughened the water and where the currents from the Bay and ocean met there was a broad stretch of whitecaps. Over these hovered the birds, flying in every direction, the terns pitching head foremost into the water, the gulls skimming the surface or swooping down and alighting on the waves, and the dark heavy-winged Fish Hawks passing and repassing higher in the air than the lighter birds. Now and then one of them dropped like a plummet into the water and emerged with a fish in his claws and was off to his nest back in the country. The terns appear gray at one angle and snow white at another and the white wings are constantly flashing as they catch the sunlight, now here, now there, in the misty distance, like gigantic fireflies in the dusk of evening. It is a wonderfully animated scene and food evidently abounds.

On August 26, 1931 I found two adult terns and three young practically full fledged and able to fly, resting on Seven Mile Beach. The old birds were constantly feeding the young and flying back and forth in search of prey. There was a flock of Sanderlings also feeding on the beach, catching small Hippa crabs or shrimps, and every time a Sanderling made a successful catch a tern would dart after him and pursue him relentlessly out over the surf and back over the beach but I did not in any instance see him relinquish his prey. While the Sanderlings were frequently taking wing or running up and down the beach the tern never pursued one of them unless he carried a shrimp. While I have seen terns pursue one another I never again saw them try to seize prey from other birds. Terns dive from the air, usually from a considerable height, plunging into the water with a splash, beak first, and usually go entirely under though they sometimes suddenly shear off and deftly pick up some scrap of food from the surface. When they do go in they seem to fairly throw the body so forcibly does it strike the water. A flock feeding close in along the Bay shore on August 26, 1921. was diving continually and every five seconds one or more birds would strike the water. They would pause suddenly in their flight and turning tail up and bill straight down would go under with a splash, the wings still partly spread, and in two or three seconds would emerge again usually with a fish held firmly between the mandibles. Sometimes they would pause and hover like a Marsh Hawk without making a plunge.

Terns flying offshore and traveling low over the water have also been seen to dive but in their case there was not sufficient altitude for a vertical plunge and they went splashing in at a low angle with the momentum of their flight. Several were seen August 29, 1920, diving in shallow pools on the meadows and one drove his bill well into the muddy bottom, clouding up the water. Back on the thoroughfares above the Harbor they seem to dive mainly near some point of land where two currents of water meet and where fish no doubt congregate. One bird watched for some time (May 26, 1929) at close quarters, as it hovered over a partly submerged islet in Jarvis Sound, was diving repeatedly. It held the same position in the air all the time, about four feet from the water with wings beating rapidly and tail fully spread, the long feather tips constantly fluttering so that it was only now and then that their dark outer webs could be distinguished. The body seemed to rise and fall with the motion of the wings as if floating buoyantly on the waves of air. The head was held a little below the horizontal and was turned constantly from side to side, sometimes twisted far around. The bird usually splashed in at a slight angle, often among the grass tips, but seldom made a catch. Occasionally it would rest on the water and plunge the head and neck under while it churned the wings, evidently bathing. So far as I am aware the terns do not resort to fresh water to bathe as do the Laughing Gulls.

Straggling terns are seen flying over the meadows or more frequently over the thoroughfares and often come to rest on the stakes which mark the location of oyster beds on the sounds, or on the channel markers on the thoroughfares, and one bird will sometimes alight on the back of another which is already in possession of a stake and force it from its perch, just as the Herring Gulls do on the Harbor pilings.

The beach flocks near the Lighthouse, which used to be a permanent feature during August, seemed to diminish in September and I have no record of more than fifty in that month, while most of them usually leave by October 1. My latest records are October 13, 1929, thirty, and October 12, 1930, fifty, with two exceptional occurrences—nine seen by Ludlow Griscom on November 11, 1921, and twenty that I saw roosting on pilings a little farther on, on the Bay shore, in company with some Laughing and Bonaparte's Gulls on November 7, 1931. Julian Potter saw a flock of some fifteen hundred Common Terns arise from the beach at the Point on October 1, 1925, and after attaining a great height disappear to the southward, doubtless on migration. Owing to my inability to be on the nesting grounds of the terns regularly during the spring my dates of arrival are not entirely satisfactory. The earliest occurrences at Cape May are May 9, 1925; May 10, 1924; May 14, 1928; May 9, 1931, while on May 11, 1929, they were present in numbers at a nesting colony established that year on Ephraim's Island, back of Five Mile Beach.

Julian Potter saw Common Terns on the Delaware River at Philadelphia on April 22, 1922, and Richard Miller on the same date in 1929. As an illustration of where our terns go to in winter a bird banded as a downy young in the colony back of Five Mile Beach by John Gillespie on August 23, 1925, was recovered on the island of Trinidad on May 16, 1926. The favorite nesting grounds of the Common Tern seem to be beaches at the entrances to inlets or harbors or the low sand bars formed at the mouths of the inlets. Most spots of this sort have in recent years been rendered unsatisfactory for the birds by "improvements" of various sorts while the development of summer resorts all along the coast has also tended to drive the birds away. On the other hand the dredging of the channels to form the so-called inland water way has resulted in extensive sand flats where the sand and shells from the dredges have been spread upon the meadows and these have proved very acceptable nesting grounds for the terns in lieu of the beaches. Various factors however have operated to cause the birds to shift their breeding places from one year to the next and many former nesting sites have been deserted.

A visit to a breeding place on Brigantine Island, while well beyond the limits of Cape May County, gave me a good idea of the activities of the bird at this season which were later found to be the same on nesting sites nearer home. Our visit was on July 17–18, 1921, and the scene was one of wild activity. About three hundred birds arose at our approach and as long as we remained the air was rent with their harsh chorus. The excited birds hovered and wheeled over our heads in layers, as it were, those highest up appearing no larger than swallows. The call of the individual tern was *keeer, keeer, teeeeé ark', teeeeé ark'*, but with dozens of them calling at once the din was indescribable. There were young birds already on the wing, too, full-fledged except for the bobbed tail which was their chief characteristic, but their more labored flight was noticeable as also their straightaway course, in contrast to the more erratic and wheeling flight of the adults. Their wings and bill were seen to be shorter and when they came to rest they somewhat resembled heavy-billed shore birds, standing high with practically no tail in evidence. The details of their plumage are discussed beyond.

On the ground on every side were other young birds, some just hatching with down still wet, others like mottled pieces of fur, the youngest browner and the older ones more gray, in each case so closely resembling the sand and bare ground that they were easily passed by, especially as they lay flat on their bellies, with the head and neck stretched forward to the fullest extent. Other young had pin feathers appearing among the down while others, still, had the wings nearly half-grown. These heavy-headed, top-heavy little fellows waddled away in a curious fashion and sought shelter in the marsh grass nearby or under the scattered tufts of sea rocket and goldenrod. The nests were usually mere hollows in the sand, sometimes with a few pieces of dry reeds or small sticks laid about the edge, but others were fairly well lined with the same material and some built on patches of trash that had been stranded on the meadow by the high tides. The eggs were never more than three in number and varied in color from pale green to deep olive brown with large and small splotches of brown and purple. We counted thirty-nine nests still occupied, containing eggs or young. All the time we were examining the nests the pandemonium overhead continued, there was a *chip, chip*, Robin-like call mingled with the usual cries and now and then some particularly aggressive bird would come swooping down directly at my head with glaring eyes and wide open red mouth from which came the rattling *kek-kek-kek-kek* cry ending with the tearing *keeeeeeer'* as the bird checked itself just at my hat brim and veered off without actually striking me—the picture of rage and defiance.

Hereford Inlet with its numerous sand bars, and the wild stretches of lower Seven Mile Beach were favorite breeding places for Common Terns for many years, the latter constituting the last beach colony in Cape May County.

I first visited the inlet colony on June 20, 1923, when it was divided into two or more sections. One of these on Gull Bar at the mouth of the inlet consisted of about three hundred birds and another of two hundred was located on the sand point on the northern extremity of Five Mile Beach. On the date of our visit we counted in the former forty-six nests with eggs and two with small downy young. Another visit on July 8 found the colony largely washed away by the high tides which had swept over the bar, but the terns, undaunted, had begun to nest again and we counted twenty-nine nests with a single egg each, sixty-seven with two eggs, and fourteen with three. There were also probably fifty downy young on the higher part of the bar which had survived the flood.

On July 11, 1926, one section of this Gull Bar community numbered seventy-two nests, eighteen with one egg, thirty-eight with two, and fifteen with three, while thirty-five downy young were counted. Later in this year a high tide again swept over the bar and on August 10 no nests were to be seen and we found only three young. These were as yet unable to fly and ran fearlessly down the strand when we attempted to catch them and even out into the surf ten feet from the shore, where they swam the waves while their parents hovered anxiously over our heads. There were several hundred adult terns on the bar which arose as we approached with all the clamor of a nesting colony and later arranged themselves in ranks on the beach.

It is surprising that these high tides which nearly every year, and sometimes twice in a season, sweep over the low bars and beaches which the terns select as their nesting grounds, do not discourage the birds and drive them elsewhere, but they return year after year and when they do desert an old breeding ground it is usually for some other reason.

The other portion of the colony located on the northern point of Five Mile Beach apparently escaped the tides of 1923 and on July 8, while there were but few nests with eggs, we found a large number of young birds in the short grass of the adjacent meadow where they had taken refuge and were being fed by their parents. In 1924 this section of the colony was much reduced and not more than fifteen pairs of birds were seen. As late as July 21 of that year there were four nests with eggs and one with downy young while many fledged young were on the wing. In 1925 and 1926 it was deserted but a few pairs returned in 1927 and 1928.

On the lower point of Seven Mile Beach a third section of the Hereford Inlet tern colony was located. This, according to Richard Miller, consisted, on June 26, 1921, of fifteen pairs of birds and twelve nests were found four with three eggs, four with two and four with one. On July 10 it was reported washed out but by July 17 six new nests had been made in which laying had just begun. On June 25, 1922, Miller reported sixty pairs of terns present with many full sets of eggs and fifty on July 17, but on June 17, 1923, only five pairs were in evidence. I visited the spot on May 9, 1925, and found some birds already there but while they showed some resentment at my intrusion there was no indication of nesting, nor was there in 1926.

In 1927 Gull Bar being entirely under water except at high tide the birds that had nested there, or at least a part of them, moved over to this spot and on July 9 there were about fifty pairs present with young running about and many nests with two eggs each. The colony had evidently been washed out earlier in the season as numbers of eggs were found buried in the wet sand and only the nests located on several isolated sand dunes had survived. On July 24 of the same year Julian Potter counted five hundred terns here, and they have continued to nest here in varying numbers ever since.

In 1928 while a few pairs still continued to nest here the majority moved over to a sand flat on Ephraim's Island, back of the southern part of Five Mile Beach, where the Black Skimmers that had nested on Gull Bar had already repaired. This is a low meadow island, which has been partly covered by dredgings of sand from the channel, and is protected from approach from the mainland by broad thoroughfares. I first discovered the presence of terns and Skimmers here on July 28 by noticing a great concourse of birds in the air as I crossed in a boat from Cape May to Wildwood. Visiting it on August 4, Walker Hand and I found from eight hundred to one thousand terns present and many Black Skimmers. There were young terns in all stages of development and many eggs. The nests frequently had little strips or stems of straw or grass laid around their edges but in many cases the shells of razor clams, which were strewn in great abundance over the sand, were used instead, making a remarkable appearance.

The young downy birds showed much variation in color apparently independent of age; some were buff with scarcely any spots while others ranged from vivid buff to chocolate and light gray with definite blackish spots or irregular blotches. The eggs likewise showed a wide range of color usually some shade of olive brown or chocolate with dark brown or purplish spots or blotches; once or twice we have found sets that were plain bluish green with practically no spotting and one found on this visit consisted of two eggs, one dark chocolate color and the other Robin's egg blue without spots.

The parent birds were brooding the downy young, sitting over them breast down on the sand with wing tips slightly elevated. They scratched in the sand as if to deepen the nest cavity and also revolved. Those that were out seeking food would bring in small fishes about an inch in length and the well-fledged young would run up to them to be fed. The downy young when offered a fish did not seem to respond but I did not see them fed in any other way.

On August 14, after a heavy northeast storm (August 11-12) this colony was visited again and I found that most of the island had been submerged and the higher spots subjected to a pelting rain. Most of the little terns were drowned and were lying all about attracting swarms of flies while the eggs that had not hatched were broken or buried in the wet sand. One young bird was discovered that had just hatched and it would seem that the parent must have weathered the storm and saved its young by sitting on the nest. There were many old birds in the air but not a quarter as many as on the previous visit.

In 1929 this colony was visited on May 11 and two hundred and fifty terns were found with twenty-five Black Skimmers. The former were sitting on the sand, some strutting about in the absurd courting poses and many in the air, but no sign of nests. There was much calling but quite irrespective of my presence, and the Robin-like note was frequently heard. The flying birds were usually in groups of three, sometimes four and occasionally two, one bird always carrying a small fish, the others in pursuit. So far as I could see the pursuers never secured the prey although they turned and twisted most adroitly. The whole performance was perhaps part of the mating behavior. The tail feathers of the flying bird were always held tightly together forming a single slender whisp. One solitary bird on the strand had a small fish which he swallowed and then seemed to gag, holding the mouth open and taking water several times before closing it. The birds flying overhead were at all levels and as one looked up through the maze of wings those in the uppermost strata seemed no larger than swallows while in the brilliant sunlight all seemed rose-tinted below and their wings curiously translucent.

The Crows and Turkey Vultures that occasionally passed over were promptly attacked and forced to beat a hasty retreat, with redoubled flapping of their wings.

On June 30 there were young of all sizes in the nests and about two hundred adults. Rats were discovered on the island on this occasion with burrows in the hard sand hills and they seemed to have made some inroads on the eggs and young. On July 17, no fledged young were seen among the adults in the air and there were no downy young on the ground and only one set of eggs. Possibly another high tide had destroyed the early broods.

On May 16, 1930, only fifty terns were present at the nesting site with no sign as yet of eggs or young. On July 4 it was found that a tide had again swept the island and only a few water-soaked eggs and four young birds could be found. Some of the birds persisted, however, and on July 21, there were twenty pairs present but only a few eggs or young and on July 25 several birds were still setting. They were much less timid than the Black Skimmers, some of which also remained, and when I erected a blind near a nest the parents returned the moment I disappeared inside. One parent, probably the male, stood nearby while the other was incubating the eggs and when a Skimmer approached too near it was promptly pursued. When a nest contained young the parent returned at once and stood over them to shelter them from the sun. On August 4 the tides had again covered the spot and no young birds were in evidence although two adults were seen carrying small fish.

On July 6, 1931, the colony had been again entirely washed out and the stranded and broken eggs were being devoured by Herring Gulls. Only four terns were seen and they were flying high in the air and showed no concern at our visit. We learned later that they had followed the Skimmers to a new location on a similar dredging on the meadows a mile farther north toward Grassy Sound. The move being attributed to the rats which had been increasing in numbers. In 1932 this new spot was occupied by the birds and on a visit to it on July 13, we found approximately two hundred terns and sixty Skimmers. There had evidently been a high tide some time previously as eggs lay scattered all about and in rift rows against stranded patches of meadow trash. The terns, however, had begun to nest again and there were over fifty nests nearly all with full sets of eggs. Only one downy young could be found, a possible survivor from the flood. Conrad Roland and I studied the birds for some time from our boat nearby in the channel thus avoiding the disturbance that our presence on the island would occasion. When forced to leave their nests by the passing of an occasional Turkey Vulture, or other cause, the incubating birds returned at once and settled immediately upon the eggs. Sometimes a male bird stood close to his mate but in other cases he seemed to spend most of his time fishing, returning now and then with a small fish which he fed to her; sometimes she sat on the nest while receiving it, at others she walked a few steps to meet him and then resumed her position. All the time we were there there was an assemblage of terns on the adjoining mud flats which were exposed at low tide, sometimes as many as seventy. These birds remained perfectly still or were engaged in pluming themselves and in bathing in the shallow water and took flight only when the colony was disturbed. Whether this assemblage consisted of males whose mates were incubating or of birds that had not begun to make new nests I could not determine.

The innocent trespass of an occasional Turkey Vulture had amusing results as the entire colony arose and attacked him, from six to twenty continuing the pursuit far over the meadows. The great bird flapped his wings laboriously and twisted and turned to avoid the annoying assaults of the agile terns and in nearly every instance he disgorged part of his food although the terns paid no attention to this tribute. Curiously enough a tern coming back from the chase would dart after any Skimmer that chanced to be awing apparently mistaking him for another enemy.

We found the nests on this island very attractive many of them built in the middle of a mass of reed stems which formed the meadow trash, or lined with the same material, while others had dry grass, reed stems and bleached razor shells in their make-up.

At various times the Common Terns nested on the southern extremity of Seven Mile Beach sometimes with Skimmers and Least Terns usually occupying several elevated sand hills. On May 16, 1936, Benjamin Hiatt found ten pairs present but no eggs had been deposited. On June 7 there were fifteen pairs with eggs and on the 20th fifty nests with well incubated eggs and a few young birds. A storm had swept over the point and washed out almost all of the Least Terns' nests as these birds built on the lowest stretches of the beach.

From the records of nests examined by Richard Miller and Turner Mc-
Mullen I have compiled the following data: Earliest date for eggs, one nest
containing a single egg at the southern end of Seven Mile Beach on May 27,
1928, all other egg dates are in June or July, as follows, all being at the above
locality unless otherwise stated:

June 26, 1921, four nests with two eggs and four with three.
July 3, 1921, three nests with one egg; two with two and three with three.
July 10, 1921, four nests with one egg; two with two.
July 17, 1921, two nests with one egg; one with two.
June 18, 1922, forty-six nests with three eggs; one with two.
June 25, 1922, fifty nests with two or three eggs.
June 17, 1923, only five pairs of birds present; two nests with two eggs.
July 8, 1923, several pairs but only one nest containing three eggs.
June 22, 1924, twenty nests with a single egg; ten with two and seven with
   three; some hatching; seventy-five pairs of birds present.
June 29, 1924, ten nests with a single egg; ten with two and thirty with three;
   four downy young.
July 4, 1924, forty-three nests with two or three eggs.
June 7, 1925, only a few pairs present; four nests with two eggs.
June 21, 1926, colony deserted, but thirty-one nests with two or three eggs on
   the northern point of Five Mile Beach.
June 19, 1927, eleven nests with one egg; nine with two; twenty-one with
   three; fifty pairs of birds present.
July 3, 1927, twenty-three nests with one egg; fifty-nine with two; twenty-one
   with three and a few downy young.
June 24, 1928, two nests with one egg; two with two; one with three and
   many young.
July 4, 1931, fifty birds present; eight nests with one egg; six with two; two
   with three and many young.
June 26, 1932, only four pairs present; one nest with three eggs.

The spring plumage of the Common Tern is pearly gray above while the
lower parts are washed with a paler tint of the same. The crown as far as
the back of the neck is glossy black and the bill and feet red, the former
dusky at the tip. Individuals in this plumage may be seen as late as the
last week of August and some perhaps leave for the south before undergoing any change. In the autumn and winter they are whiter below, and the
forepart of the head is white, and the bill and feet black, the molt usually
taking place in August or September. Some birds fail to change plumage
in the spring and, returning north in this dress, remain so throughout the
summer. These are non-breeding individuals in which neither the sexual
impulse nor the molt have developed. While they sometimes participate in
the excitement of the nesting colony they also may be found flocking by
themselves on the beach. In the flocks of late summer, besides the adults
in both summer and winter plumage, there will be found many young of the
year which have white heads bordered with dusky and with dusky patches
on the wings and back. They also lack the long outer tail feathers of the

adults, as do the latter when undergoing the molt, and appear somewhat bobtailed, while the bill is largely dusk.

Besides the Brigantine colony, already mentioned, there are several flourishing nesting colonies on islands in Barnegat Bay. On practically all of their nesting places they associate with the Black Skimmers.

The reëstablishment of nesting colonies, of Common Terns within the bounds of Cape May County, precarious and temporary as they seem to be, is interesting as illustrating conditions which formerly prevailed on the Cape May beaches themselves, and it is encouraging to those of us who have claimed that proper protection would bring about such results. Were it extended and maintained we might hope for still further restoration of former conditions. At present, however, the August gatherings of terns near the Lighthouse constitute the most striking feature of tern life immediately about Cape May and add greatly to the beauty and interest of the strand—even though sadly depleted by the inroads and activities of the human species.

No matter what the weather this assemblage of Common Terns seems always in place. On bright sunny days, when the sky is deep blue and the ocean pale green, the snowy breasts and glossy black caps of these graceful birds make a brilliant array against the shimmering sands as they rest in long rows facing the gentle south wind. Then there are days when the sea is roughened and the whole flock is out over the "ripps" in pursuit of their prey making a maze of white wings over the water. Again there are lowering days when the sky is dull and leaden with rain presaging in the east and the ocean steely gray with but little surf. The strand at ebb tide stretches away like a broad plain upon which are scattered bunches of Sanderlings, ever on the move and so pale that they almost escape notice, and solitary Ring-neck Plovers which appear jet black in the uncertain light. Still closer stands our long line of terns, rank upon rank, in varied plumage, awaiting the storm which they seem to know is coming. Every now and then they give voice with their grating, raucous calls so in keeping with the threatening squall and the only sound to reach our ears save the swish of the waves running up the beach.

FORSTER'S TERN

Sterna forsteri Nuttall

Forster's Tern is a bird of the interior of North America ranging south to the Gulf but like several other western birds it occurs as an irregular transient on the Atlantic coast and what is more remarkable it bred, for a time at least, on the Virginia shore and possibly in southern New Jersey. Turnbull (1869) states that he had found it breeding on Brigantine and Dr. William L. Abbott obtained specimens on Five Mile Beach on May 15, 1877, April 26–May 17, 1878, June 3, 1879, May 6, 1880, May 20, 1881 and May 22, 1882, all of which are in the collection of the Philadelphia Academy. Common Terns were taken at the same time and it looks as if they might have been nesting there together. W. E. D. Scott shot a pair on Long Beach on May 14, 1877, but regarded the species as rare. Charles S. Shick reports it as breeding on Seven Mile Beach in 1890, but I am not inclined to place much reliance upon this statement.

By 1923 Ludlow Griscom had relegated Forster's Tern to the extinct list, so far as the New York City Region was concerned, but John T. Nichols saw one in early September of that year and in 1925 there were many records extending from Long Island down the New Jersey coast, beginning with August 9, and no less than seventy-five were seen at Point Pleasant on August 30, sixty-two at Elizabeth on September 19 and fifty the next day with a final record on October 25.

In 1927 they were present on Barnegat Bay on various days between July 30 and September 10 with a single record for October 1, and in the same year Julian Potter found one on Ludlam's Beach on October 9.

In 1928 one was identified at Barnegat on September 2 and at Cape May on September 1 to 3.

1929 brought another influx and many were seen on Long Island from July 16 to October 16. In 1930 there were four records for the upper coast, one from Barnegat as late as November 9 (Urner). In 1931 two records for the Newark marshes, and in 1933, three records for Long Island with many records all along the coast in 1935, extending through September, with several at Cape May on September 7 (Eaton and Pough). In 1936 four immature birds were present on Barnegat Bay, June 21 to July 2 (Urner).

The "disappearance" of Forster's Tern may be due in part to the cessation of collecting which became prevalent about the time that the bird was supposed to be lost and its remarkable return to the better knowledge of field identification which developed later. In the autumn the adults in winter plumage may easily be distinguished from the Common Terns by the absence of the black band which extends around the back of the crown from eye to eye in that species, this being replaced by a black spot on either side of the white head.

In spring it is difficult to distinguish the two species except by voice. The color difference rests mainly in the fact that the outer web of the long tail feathers in Forster's Tern is lighter than the inner and vice versa in the Common Tern. I have tried in vain to ascertain this point in adult terns at Cape May in spring and summer and it may well be that Forster's Terns have occurred here in recent years at these seasons as they did in the years of Dr. Abbott's collecting and have gone unrecognized.

* ARCTIC TERN

Sterna paradisaea Brünnich

Bonaparte, Audubon and Turnbull all mention the Arctic Tern as rare on the New Jersey coast but without a definite record. Indeed it would seem that the two latter were simply quoting Bonaparte and as immature and winter tern plumages were very imperfectly known in early days they were probably all in error. There are no recent records and only one, I believe, for Long Island, and no new Jersey specimens are extant. As the bird breeds from the Arctic Regions south to Massachusetts and winters in the Antarctic seas it must pass our coast well off shore but like other pelagic birds it might occasionally be blown in by storms.

ROSEATE TERN

Sterna dougalli dougalli Montagu

This beautiful and graceful tern was stated to be a summer resident on the New Jersey coast by Turnbull (1869) while Harry G. Parker and Charles S. Shick have reported it as an abundant breeder on Seven Mile Beach up to 1885 but their identification, apparently based upon eggs, is not convincing.

We have no further record until a pair was detected by Julian Potter breeding in a colony of Common Terns on Gull Bar in Hereford Inlet on June 20, 1923. He described them as beautiful birds with black bills and red feet and longer outer tail feathers than the Common Tern. Their note was quite different too, resembling the "blink" of the Baltimore Oriole, although when charging they had a harsh tern-like cry. On July 8 when I visited this colony they were seen again, and on July 20, 1924, a brood of young about ready to fly was found by Julian Potter and banded by John Gillespie. In 1925 the breeding pair was not located.

On June 5, 1925, however, I came upon four terns resting on the beach at the entrance to Corson's Inlet, which seemed restless and particularly wary. They did not act exactly like Common Terns and when I obtained a good view of them I could see their slender black bills, with a slight reddish tinge at the base, the long outer tail feathers and the very white appearance of the rump and tail.

Once or twice I have seen other individuals in flocks of Common Terns on the Cape May beach that seemed to have the characteristics of this species but could not secure satisfactory studies of them.

Julian Potter saw two on Seven Mile Beach on June 12 and July 24, 1927, and William Yoder saw one feeding at sea off the end of the jetty on the mouth of the Harbor on July 18, 1926. On September 2, 1928, Julian Potter saw two at Cape May Point and another was seen resting on a piling there with Common Terns on May 30, 1931 (Parry).

On July 13, 1932, in company with Conrad Roland, I visited the tern colony which that year was located on a filled piece of meadow back of the northern part of Five Mile Beach and found a pair of Roseate Terns evidently nesting there. One of them rested for some time with a number of Common Terns on a mud flat close by and allowing our skiff to drift slowly down toward the flat we had an excellent view of the bird. We could distinguish its black bill with a slight red suffusion at the base, and note that it was slightly longer than that of the Common Terns which stood close by. We could not always pick it out when they were on the wing because they were above us, but when the upper parts came into view we could recognize the whiter appearance of the Roseate. As it sat upon the black mud the breast seemed slightly rosy but this was probably imagination, as this is usually only discernible when the bird is in hand.

There have been a number of records from farther north on the coast and one or more pairs have been reported breeding in the tern colony on an island in Barnegat Bay. One was seen there on August 9, 1925, and in June, 1928, there were twelve present (Potter) while John Gillespie found two pairs on July 15, 1934. In 1935 it was estimated that ten pairs bred on the New Jersey coast.

LEST TERN

Sterna antillarum antillarum (Lesson)

This, the smallest of our terns, was for many years very rare on the New Jersey coast. Time was when it was one of the most plentiful inhabitants of the strand, nesting in the sand just above high water mark, but it was sacrificed and all but exterminated to satisfy a passing whim of fashion which demanded its use as a millinery adornment. The birds that we see today are the descendents of the few that survived the slaughter or of others that pushed north from more southern shores and reoccupied the old breeding grounds.

Alexander Wilson says that when visiting the nesting places of the Least Tern about 1810, probably on Peck's Beach, "the birds flew in clouds around me, and often within a few yards of my head, squeaking like so many young pigs." W. E. D. Scott refers to them as "abundant" at Long Beach, in 1879, "breeding exclusively on the ocean beach." George Morris, writing to me in 1909, of conditions at the same spot in 1881, says: "It is difficult to give an estimate of numbers, but I can remember standing in one spot and seeing five or six nests within a radius of fifteen or twenty feet, but my recollections are that these conditions only pertained to an acre or so of the beach. In July 1884, I could no longer find Least Tern's eggs, and natives told me they no longer found eggs on the beach. During the period, 1881–1886, I saw a good deal of the slaughter of the birds in this region. I remember coming upon two professional millinery gunners, I think in the summer of 1885, who had two piles about knee high of Least and Common Terns, which they said they were sending to New York, my recollection being that they got twelve cents apiece for the birds."

Dr. B. H. Warren describes the same thing on Brigantine in the summer of 1883; he says: "The Least Terns were breeding in considerable numbers, laying their eggs in slight depressions in the dry sand and among the shells on the sand hills along the beach. I obtained the bodies of over 75 of these Terns from two taxidermists, who were collecting the skins for New York and Philadelphia dealers to be used for ladies' hats. These birds were all killed in one day."

On Seven Mile Beach, Charles S. Schick writes in 1890 (Auk, 1890, p. 326) that the Least Tern is a common breeder; "I must state, however, that all of the Terns are gradually forsaking their former breeding grounds on account of the new seaside resorts that are being started on all the islands. Formerly many hundred pairs occupied a small sand flat near Sea Isle City, but they are now all gone, not one pair breeding where a few years ago hundreds raised their young." Harry G. Parker estimated that there were thirty pairs nesting on Seven Mile Beach in 1888 and Philip Laurent said that there were still a few nesting in 1892. William Baily saw two birds on this beach in 1899 which he felt sure had nested and recorded individual birds on Five Mile Beach, June 1, 1893; Seven Mile Beach, August 28, 1896; Cape May, August 22, 1897.

On the beaches immediately around Cape May the Least Tern is today a transient visitor on its way to and from nesting places established a little farther up the coast and we see it most frequently during the month of August, always on the sea beach. The birds may be resting on the strand or flying out over the waves but seldom far from the shore. They are very light and active on the wing, bouncing up and down in the air in a distinctly erratic flight. The wings appear very narrow and seem to curve backward more than in the Common Tern, while the beats are about twice as frequent. The action is continuous and every now and then there is a distinct flutter as the bird pauses over some likely feeding spot, the body rising slightly the while. At other times a bird will suddenly dive, pitching headlong into the water. The bill is always pointed downward in flight as is customary with terns.

While the Least Terns associate with Common Terns as they rest on the beach, they do not mingle intimately with them but usually keep by themselves a few feet from the other assemblage. Indeed while seeking company itself this little tern seems to resent the approach of other species and when some passing Common Terns came in to join a group of Leasts the latter took wing. On one occasion a Least joined company with a single Ring-necked Plover, and several times I have found them with Sanderlings, but usually their association is with other species of terns.

The length of the Least Tern is about half that of the Common Tern and about equals that of the Sanderling although of course not nearly so bulky in body. They usually sit scattered along the beach and not in a close flock like the other terns. The head, held low on the shoulders, projects slightly forward, the wing and tail tips stand up above the line of the back. One bird busily engaged in pluming its feathers, shook both wings and tail vigorously, moving the wings back and forward so that the tips crossed and recrossed over the rump, while the tail was jerked from one side to the other.

On another occasion several birds that were engaged in pluming themselves bent the head over forwards in order to reach the breast, and back over each shoulder to the rump, while the long primary feathers were carefully drawn between the mandibles one at a time.

Two plumages are distinguishable among the birds that occur on our beaches in August. The breeding adult has a black cap and narrow white frontlet and line over the eye, black outer primaries, a yellow bill slightly tipped with dusky, and orange feet. The young of the year has much more white on the head, the black being restricted to the occipital region while the fore part of the spread wing is dusky and the bill black. The snowy white under parts and pale gray mantle are common to both.

I have several times seen Least Terns run on the beach for a few steps and they move more rapidly and easily than the Common Terns, although the extremely short legs of all terns make their movements on land somewhat grotesque.

The migrant birds have two calls to which they occasionally give voice one a high-pitched shrill, *pts-sek, pts-sek;* the other *chílik, chick, chick,* a double note followed by two short ones.

On one occasion two of the birds resting on the strand suddenly took up a fighting attitude like miniature game cocks bill to bill and heads stretched out close to the sand. They flew at one another once and then parted.

To illustrate the gradual increase in the number of migrant Least Terns on the beach at Cape May we saw single birds three times and two on one occasion during August, 1920. In 1921, none were seen although the strand was visited with about the same frequency. In 1922 there were two on July 25, fifteen on July 31, twenty on August 3, ten on August 4, one on the 7th, nine on the 9th, two on the 13th and four on the 19th.

In 1923, the first one came on July 21, another on the 24th and four on the 26th. On August 4 seven came in from the sea in single file and alighted on the beach, beginning at once to preen their plumage vigorously. Later another joined them making five adults and three birds of the year.

In 1924 only two were seen and in 1925 they were observed four times. Subsequently they have been more abundant and as many as twenty-six were seen together on August 27, 1926.

In the summer of 1924 a breeding colony of about twenty pairs of Least Terns was found established at Corson's Inlet, by Norman MacDonald and Horace Rolston. One or more pairs had nested on Brigantine Island farther up the coast in 1920, and probably had never been entirely exterminated from this or the nearby Little Beach Island, but this was our first knowledge of the return of the birds to Cape May County as breeders. On May 29, 1925, a visit to the Corson's Inlet colony showed several nests with two and three eggs each and on June 4 and 5 of the same year I found four nests with one, two, two, and three eggs respectively, none of them the same as those found previously, while from the actions of the birds there were many others that I failed to locate.

The nesting spot was on a narrow flat sand spit forming the northernmost point of Ludlam's Island and stretched up into Corson's Inlet. It is about two hundred yards wide and on the eastern side the ocean surf rolls in on a shelving beach while at low tide there is a series of short lines of rollers one behind another, marking the many bars and shoals with which the entrance to the inlet is filled. On the other side lie the smooth glassy waters of the thoroughfare. Several low sand hills covered with dune grass occupy the center of the flat while the sandy stretches all about are covered with shells of all sorts—clams of several kinds, scallops, naticas, young conchs, razors etc., some stained with iron and others dark blue gray from the blue mud banks off the shore, making a great mixture of colors as the sun shown down upon them. Here the terns had deposited their eggs in little hollows scooped in the sand, about four inches in diameter. They are cream colored often tinted with pale purple and spotted with purple and brown, a coloration that almost defied detection against the myriads of varicolored shell fragments with which the sand was strewn.

An examination of the nest hollows earlier in the season (May 19) indicated that they are deliberately scooped out by the birds as many were found all ready for the eggs which had not at that time been deposited, while the marks made by the birds' feet were evident. About half of the nests were in little oases of pure sand and not immediately among the shells, while telltale tracks leading to the nests from half a dozen directions, and extending for six to ten feet away, aided in locating them. Nests placed among the shells often had, apparently by chance, pieces of broken razor shells or clams as a flooring but others had none.

The tracks forming the approaches to the nests consisted of hundreds of little footprints made by the birds in going and coming; they were roughly diamond-shaped with a little scratch behind each, probably made by the toes as the foot was drawn forward with each step. When the birds were watched they were seen to run with a rather clumsy gait, waddling a little from side to side.

As one approaches the colony the terns may be seen in the air, even at a considerable distance. Their wings catch the light of the sun and flash out silvery white for a moment and disappear as quickly, like will-o-the-wisps, as the bird turns in its course. Approaching closer we see many of them flying back and forth over the nesting ground and with the first cry of alarm those that have been resting on the sand or incubating their eggs arise to join those already on the wing. The flying bird seems to have a very short neck and the wings have the appearance of springing directly from behind the head and to be always pointed diagonally backward, in graceful curves. The beats, however, are strong and the body seems at times to bound up and down in the air. When flying about undisturbed the birds have a rather clear call, though none of their vocal efforts are of much volume. *Tsíp; tsíp,* they cry or, as we get closer, *chísek, chísek.* Almost immediately upon the discovery of an intruder a harsher cry is substituted *zhweét, zhweét,* and for a few moments it may alternate with the other: *tsíp; zhweét; tsíp; zhweét* etc. Soon all are uttering the harsh cry, which recalls some of the notes of the Purple Martin and there is a perfect bedlam of harsh *zhweéts* all about, and as different individuals have slightly different accents the effect is much like a chorus of harsh voiced tree toads. Occasionally a bird comes diving headlong at me, his black crown, yellow bill and gleaming eye making a striking appearance and in wild rage he cries, still more harshly, *zkeeék* as he turns abruptly upward just before striking my face, or perhaps he may be hovering above my head and drops precipitately before making the upward swoop, but always with that harsh *zkeeék.* This note and the clear calls of other birds not yet alarmed vary the chorus.

Upon my retirement the excitement soon dies down and the nesting birds drop back again to the sand alighting a few yards from their nests and running to them by short stages, soon settling upon the eggs, with neck held erect and wing tips crossed over the base of the tail. They run rather rapidly and pause and then run on again much like a plover, holding the neck high all the while and ever alert.

At low tide the Least Terns of this colony resorted to the beach and perched on the wet strand where I counted thirty in sight at once. From here they went fishing among the riffles and shallows of the harbor mouth. Hovering in the air like a Fish Hawk, their wings beat so fast that they appear like a double pair, and effect often seen in rapid flying ducks. The bird will then turn suddenly tail up and, with wings partly closed and bill pointed straight down, will plunge into the water. It is gone for the moment and then emerges and is again on the wing. Sometimes it will have a small fish in its bill, and with a wriggle and shake of its plumage it is off to its mate with its plunder. Only one in several trials actually results in a plunge for more often than not the prospective prey disappears and the tern spreads its wings just before reaching the water and rights itself, swooping up again to the required height for diving. Near shore, in very shallow water, they often make a diagonal dive and swerve up again barely touching the water and apparently seizing something on the surface. One bird settled on the water as if to swim but instead began churning its wings violently up and down, splashing the spray over its back; it then arose from the surface and settled again repeating the process several times. It was apparently bathing.

When alighting on the beach the birds often hold their wings aloft, stretching them up to their fullest extent for a moment before folding them. The male, which has captured a fish, takes it to the female and struts with it in front of her in a grotesque manner and then offers it to her, evidently a part of the courtship behaviour. On May 20, 1927, before any eggs had been laid, I saw a male with a small fish approach a female on the sand and take up a position behind her with his body nearly upright, then, after mating, he presented the fish. On other occasions the male fed the female while she was incubating.

The yellow bill with slight black tip, the black cap with white frontlet, and the orange feet, are the most conspicuous color markings of the breeding bird, together with the black on the outer primaries which forms a dusky border to the wing. I saw one individual in the colony with a whitish head and somewhat mottled back, evidently a non-breeding bird which had not completed the molt.

Another visit to this colony on July 16 of the same year, found many young already on the wing, not such strong fliers as the adults and all with shorter and much less forked tails, the back browner and a brown band over the bones of the wing. They came to rest now and then on the wet sand near the inlet and were fed there by their parents. The bedlam of sound continued, as on my previous visit, whenever the nesting ground was approached but the harsh *zhweét* cry was now the only one used. Downy young, almost exactly the color of the sand, were found scattered about here and there on the upper beach or among the grass on the low sand hills. While some nests still held eggs, I found none with young. When the downy birds were picked up the parents became frantic and a new note was introduced, a rapidly repeated *chip-chip-chip-chip*. Wharton Huber noticed on the earlier visit that if we lay down flat on the sand the anxiety of the birds always lessened and the harsh cries gave way to the softer *tsíp*, or *chísek* notes or to shorter repetitions of the *chip-chip* note just described. When we stood upright the excitement began again.

In 1926, a visit to the Corson's Inlet colony on July 31 found only a half dozen birds present and these showed no excitement upon our visiting the nesting ground. Evidently all of the young were on the wing and had drifted away along the beaches to the southward.

On May 20, 1927, there were thirty-four birds sitting about on the sand and while freshly scooped nest hollows were found only one contained an egg. In 1928, most of the colony removed to a recent fill back of Sea Isle City several miles to the south where a dredge had spread fresh sand over the meadows. On May 20 nests had been scooped out here and on June 9 Fletcher Street counted thirty birds present. There were fewer birds to be seen on June 24 but several downy young were found which lay flat on their bellies, quite invisible at a little distance when on the white sand, but most conspicuous whenever they strayed onto the patches of black mud. Two nests were found with eggs—one and two respectively. Many Least Terns have continued to nest on this sand flat up to the present time, the Corson's Inlet locality having been rendered unavailable by building operations.

On July 17, 1925, four birds, adults and young were found at the upper extremity of Five Mile Beach where the Common Terns used to breed and they flew about calling as if they had nested somewhere in the vicinity.

Least terns have bred also in the Common Tern colony at the southern extremity of Seven Mile Beach since 1925, in some years, and on June 7, 1936, Benjamin Hiatt found twenty-five pairs present some with full sets of eggs others just beginning to build their "nests." By June 20, the point had been swept by a high tide and almost all of the Least Terns' nests washed over. Later in the summer I failed to find a single bird there.

In June, 1937, Otway Brown found a colony of Least Terns on the extensive sand flats back of the fish docks on what was formerly Two Mile Beach, and birds seen later in the season about Schellenger's Landing doubtless came from there. In July of this year I also found a number of the birds both old and young on the beach near Bidwell's Ditch on the Bay shore and while the latter were still being fed by their parents, they were perfectly able to fly and I do not think that they had been raised there, although they may have been.

Lists of nests examined by Turner McMullen, Richard Miller and Fletcher Street follow:

Corson's Inlet

June 14, 1925, sixteen with a single egg; fifteen with two.

June 28, 1925, four with one egg; eight with two; and one with three.

July 11, 1925, three nests with two eggs and several young running about (Street).

June 27, 1926, eleven with one egg; six with two; one with three; several young; thirty pairs present.

May 21, 1927, five with two eggs.

June 12, 1927, two with one; twenty-four with two; one with three; several young; fifty pairs present.

June 18, 1927, thirteen nests with two eggs not counted on the 12th.

July 3, 1927, one with two eggs and others just hatched (Street).

July 10, 1927, four nests with eggs.

May 26, 1928, twenty-five nests with eggs.

June 4, 1932, fifty-two nests with eggs.

Peck's Beach (Turner McMullen).

May 30, 1926, eleven nests with eggs.

June 7, 1931, seventy-one nests with eggs.

Lower end of Seven Mile Beach

July 5, 1925, one with two eggs; one with young (Miller).

June 16, 1935, nine nests with eggs (McMullen).

Brigantine Beach

June 7, 1921, twelve nests with eggs (McMullen).

June 25, 1922, eleven nests with eggs (Street).

June 24, 1926, fifteen nests with eggs (McMullen).

June 11, 1933, ninety nests with eggs (McMullen).

In June, 1925, Julian Potter had located five nesting colonies from Brigantine southward containing respectively thirty, ten, fifteen, four and two pairs. On July 17, 1927, the largest still maintained a population of thirty to forty and in 1928 the two Ludlam's Beach colonies contained one hundred birds, with twenty-five nests and eggs in the larger.

In June, 1933, Charles Urner found a few pairs beginning to nest on a gravel road across the meadows near Barnegat Bay and Potter estimated that there were two hundred Least Terns nesting on the coast in that year.

With the reëstablishment of this delightful little bird as a summer resident species in Cape May County it has increased as a transient on the Cape May beach. Thirty gathered at South Cape May on August 7, 1929 and a dozen were seen by Otway Brown at Cape May Point on June 12, 1932.

On the meadow ponds back of both Five and Seven Mile Beaches the Least Tern has also become of frequent occurrence throughout the summer and has been recorded there from July 9 to September 1 sometimes in flocks of twelve to forty-five, and on Price's Pond several were seen fishing during July and August, 1936. We have one record for Camden, on the Delaware River, August 26, 1933, (Potter) doubtless blown inland during the great hurricane a few days before.

Unfortunately the status of the Least Tern is rather precarious since the beaches which are its true home are almost entirely taken over by building operations and resort developments, while people and dogs constantly disturb the birds during the early summer, when they should be free from persecution. Were it not for the recent sand flats left by the dredges in deepening the inland waterway they would probably ere now have again taken their departure. Whether they will permanently establish themselves on these more or less artificial nesting grounds remains to be seen.

SOOTY TERN

Sterna fuscata fuscata Linnaeus

This southern tern has several times been blown by tropical storms to the New Jersey coast. The first record, so far as I am aware, is a specimen, since destroyed, which I have seen and identified, shot by Amos P. Brown, Sr., on Long Beach in the "seventies," doubtless 1876 or 1878, when several of these birds reached Long Island and New England. It was an adult in full plumage.

On September 7, 1916, Wharton Huber collected an adult female at Corson's Inlet. "It was resting in the long grass in the sand dunes, a very short distance back from the beach. It was very tame and allowed a close approach before flushing" (Auk, 1917, p. 206).

Another record is a dead and largely decomposed specimen found by William C. Doak and others on the salt meadows of Seven Mile Beach, about a quarter of a mile back of the Stone Harbor Coast Guard Station, on January 13, 1929 (Auk, 1929, p. 224). The bird had probably been blown north by the storm of September 19, 1928, which was responsible for seven birds of this species on Long Island, dead or alive, September 21–23.

The original specimen of Trudeau's Tern (*Sterna trudeaui* Audubon) which was supposed to have come from Great Egg Harbor seems to have been obtained in Chile.

ROYAL TERN

Thalasseus maximus maximus (Boddaert)

There seems to be no record of the occurrence of the Royal Tern on the coast of New Jersey until Turnbull listed it in his "Birds of East Pennsylvania and New Jersey," in 1869, as "very rare" and after that no records until quite recently. There are, it is true, several statements of its occurrence in the meantime but owing to the confusion on the part of most observers, between this and the Caspian Tern, it is not possible to say to which they refer and it seems more likely that they belong to the latter, since it has proven to be the more common on our coast in recent years.

The first "rediscovery" of this tern on the New Jersey coast was on September 17, 1933, when Julian Potter and Joseph Tatum found a flock of nine on Brigantine Beach following a heavy storm of September 14–16. All had "white foreheads extending back over the crown, bills more slender than those of the Caspian Terns and orange in color, the birds more buoyant in flight and more slender; cries less harsh and not so low." On June 30, 1935, Potter encountered four more of these birds at the lower extremity of Seven Mile Beach; they took wing as he approached and flew out to sea. "They were all adults with yellow bills but very little black on their heads, note not so low as that of the Caspian and less harsh." (Bird-Lore, 1933 and 1935—"The Season".)

On September 15, 1935, Joseph Tatum had the good fortune to find the two species together at the lower end of Brigantine Beach. There were two Caspians and one Royal. "The Royal was distinctly smaller with more

(589)

white on top of the head and a more slender bill, which was orange color in contrast to the strong red of the Caspian. There was an apparent difference, too, in the legs, those of the Caspian being longer and black, while the Royal's seemed to be greenish yellow or brownish. In flight the Royal was more graceful with darker wing mantle and with less black at the wing tips." (Auk, 1936, p. 95.)

CABOT'S TERN

Thalasseus sandvicensis acuflavidus (Cabot)

Turnbull states that "a specimen of this straggler from the Gulf States was shot on Grassy Bay (back of Five Mile Beach) in August, 1861." There is no other New Jersey record, nor has it been taken on Long Island where so many unusual birds have occurred. It was doubtless the victim of one of the tropical storms that have blown many southern birds to the lower New Jersey coast.

CASPIAN TERN

Hydroprogne caspia imperator (Coues)

The splendid Caspian Tern is but a casual visitor to Cape May and seems disinclined to pause in its passing flight along our coast. Doubtless if one were constantly on the watch, especially at some of the more remote inlets, more would be recorded but when we consider the short space of time that a bird flying along the beach is in sight, at any given point, it is not surprising that records of such stragglers as this are infrequent.

On August 30, 1930, I saw the only Caspian Terns that I have been fortunate enough to observe here. They passed me at South Cape May as I was walking down the beach and flew about thirty feet above the water just beyond the surf line. The atmosphere was somewhat hazy and as the two big birds approached I at first took them for gulls but was at once struck by the light color of the upper parts passing into pure white on the lower back and tail. The wings seemed relatively longer and narrower than gull's wings and the tail shorter. The great red bills and solid black caps, however, were their unmistakable and conspicuous characters. They kept calling as they pased along and I could see the bills open and shut but the calls were almost completely drowned in the noise of wind and surf. They pointed their bills downward in true tern manner and every now and then one of them would swoop down, turning tail up as if to dive but always caught itself a couple of feet from the water, sheared off and, rising again, continued on its way. The birds were in sight only about two minutes and were soon only specks in the distance as they headed across the mouth of the Bay for the Delaware coast.

On August 29, 1926, Julian Potter and William Yoder came upon three Caspian Terns on the strand at the southern extremity of Seven Mile Beach associated with a flock of Common Terns and Laughing Gulls; there were two adults and a bird of the year. The former were distinguished by their large red bills, dark feet and slightly forked tails and as they sat on the sand the wing tips seemed to extend slightly beyond the end of the tail. "Their call was a pitiful *kr-r-r-r-r*."

On Ludlam's Beach on August 21, 1927, William Yoder saw two Caspian Terns, and Richard Pough thirteen on Seven Mile Beach on September 10, 1933, while several were seen by the Audubon Society Wardens at Cape May Point in 1935 and 1936. Of these two were reported on August 28 and two on September 7, 1935 by William Rusling, the latter birds seen also by Warren Eaton and Richard Pough, and three were observed on September 19 and one on October 15, 1936, by James Tanner, one of the former also seen by John Hess on September 20.

There are two early records of large terns which, in all probability, refer to this species; one seen by William Baily at Avalon, on Seven Mile Beach, on August 26 and 27, 1896, which remained about the pier for several hours all told; and two recorded by W. E. D. Scott (Bull. Nutt. Orn. Club, 1897 p. 227) an adult and bird of the year seen on Long Beach on August 23, 1879.

Caspian Terns occur rarely on the Delaware River and three were seen by John Emlen and others near Palmyra on April 21, 1929, flying about with a flock of Herring Gulls during a drizzling rain. The whiter plumage of the terns was easily distinguished as well as the black cap, large red bill and the relatively short distinctly forked tail (Auk, 1929, p. 534). At Fish House near the same spot, on August 20, 1932, Julian Potter and John Gillespie saw two of these birds and following the great storm which swept the coast in 1933, John Gillespie saw two on the river opposite Camden on September 4, and Julian Potter five on September 13. Clifford Marburger saw four Caspian Terns at Delaware City farther down the river on May 3, 1931.

On the northern part of the New Jersey coast these birds seem more frequent than on the Cape May coast and many records have been published in Potter's "The Season" in "Bird-Lore" as follows:

1925, November 29, Barnegat Bay region, Charles Urner.
1926, August 18, Brigantine Beach, Herbert Beck and Clifford Marburger.
1926, September 12 (2), Seaside Park, Lester Walsh.
1926, September 19, Point Pleasant, Charles Urner.
1927, July 22, Barnegat Bay region, Charles Urner.
1927, September 18, Barnegat Bay region, Charles Urner.
1928, May 20, Manasquan, Charles Urner.
1928, July 1, (five) Point Pleasant, Charles Urner.
1929. May 1, (three), Manasquan, Charles Urner.
1929, September 7, Barnegat Bay region, Charles Urner.
1931. August 30, Brigantine, Julian Potter.
1932, May 1, Metedeconk River, Charles Urner.
1933, August 23, Brigantine, Lester Walsh.
1933, August 27, (two), Brigantine, Berkheimer.
1933, September 3, Brigantine, Julian Potter.
1933, September 10, (two), Joseph Tatum, Brigantine.
1933, September 19, (five), Brigantine, Charles Urner.
1934, September 9, (five) Brigantine, Brooke Worth.
1935, Charles Urner reported twenty records between July 22 and September 18.

It will be noticed that the greatest number of these big terns occurred after the famous hurricane of late August 1933.

I am of the opinion that, like many other species, they put out to sea after leaving Barnegat and therefore are not so often observed on the more southern beaches. Breeding as they do at many places in Canada and the Far North the Caspian Terns that we see are undoubtedly on regular migration, but the Royal Tern with which it may be confused is distinctly a southern species and the few individuals that have been recorded in New Jersey are accidental stragglers blown north by storms.

J. F. S.

BLACK TERN

Chlidonias nigra surinamensis (Gmelin)

The first large assemblage of migrant terns which formerly gathered on the beach during the last week of July usually included a few Black Terns and the same is true of the smaller flocks of today. There are often only one or two at first but their number increases later so that there used to be as many as one hundred in some years. Over the great meadows north of the Harbor during August there are often many times that number coursing about.

The dates of arrival at Cape May are as follows:

1921. July 29.	1925. July 31.	1931. August 1.
1922. July 23.	1926. August 1.	1932. July 13.
1923. July 28.	1927. July 24.	1933. August 5.
1924. July 21.	1928. July 14.	1934. July 19.
	1930. July 31.	

Late August is the time of their greatest abundance and by the middle of September they have usually all passed on; our latest records are September 20, 1925; September 19, 1926; September 30, 1928. They are however very irregular migrants; very abundant in some years and all but absent in others. Fletcher Street found them abundant far offshore when fishing on September 15, 1936, doubtless on migration.

The Black Tern visits us normally only in the autumn, its return flight in spring being confined to the Mississippi Valley. Nevertheless we have a few spring records, birds which probably have been wintering with Common Terns and have followed them in their northward flight rather than the main body of their own kind. These were all in full breeding plumage of black and gray and were always associated with Common Terns.

One was seen at Cape May Point, May 30, 1924, two at the same spot,

May 14, 1928, and another May 30 of the same year. Turner McMullen saw one at Corson's Inlet on May 21, 1927, and another on June 4, 1933. They occur also occasionally on the Delaware River in spring as well as autumn and Julian Potter saw four at Camden on May 10 and 11, 1919. On July 17, 1921, on a visit to one of the nesting colonies of Common Terns located on Brigantine Island, fifty miles north of the Cape May beaches, we found a lone Black Tern in perfect nuptial plumage which seemed just as much concerned over the welfare of the eggs and young of the other species as were the rightful owners. It followed me about from place to place poising in the air only a few feet above my head on beating wings and calling continually in great alarm, *sheep, sheep, sheep, sheep; sheep, sheep, sheep, sheep*, the notes always in blocks of four and uttered rapidly and emphatically in sharp contrast to the long drawn *keeé-árr* of the Common Terns. In spite of its apparent concern this bird was either an unmated and belated spring migrant, separated from its fellows, or possibly a very early southbound migrant. At all events, attracted by the assemblage of Common Terns, it was evidently influenced by the excitement of the nesting activities which recalled similar experiences of its own and renewed its parental anxiety. A precisely similar occurrence was observed by Julian Potter at the Common Tern colony on the northern end of Five Mile Beach, on July 7, 1929, when two of the Black Terns were present, and at the Ephraim's Island colony on August 10, 1926, I found two individuals greatly excited, and a single bird at the more northern colony on July 13, 1932.

When resting on the beach, where we see them most commonly at Cape May, they mingle intimately with the Common Terns but can always be distinguished by their smaller size and darker color. A single individual one day joined a flock of Semipalmated Sandpipers, in the absence of any tern companions, for they always seem to desire company. While they may outnumber the Common Terns in a composite flock I never have seen a beach flock composed wholly of Blacks. On August 22, 1923, a flock fishing off Cape May Point was composed of six hundred Black Terns, three hundred Common Terns and a few Laughing Gulls; all quite close to the shore and busily diving and hovering over the water.

Black Terns rest facing the wind as do all birds on the beach or water. They preen their plumage and often raise their wings high up above the body and hold them so for a moment or two. In flight their wings are at once seen to be distinctly shorter than those of the Common Terns, and relatively wider, while the strokes are not so full and powerful and cover a smaller arc. When a mixed flock of terns starts up the beach the Blacks are soon relegated to the rear, unable to keep pace with their larger stronger flying associates. They often swoop down, as do the other terns, to pick up small minnows stranded in the shallow beach pools and I have on one or two occasions seen them preen their breast feathers while on the wing, bending the head back under the body. The real feeding ground of the Black Terns, and their natural habitat when on the wing, is back on the green meadows where they secure insect food of one kind or another. They are widely scattered when thus feeding and course back and forth with easy turns quartering low down over the wide expanse of grass and sedge.

Theirs is an easy flapping flight, the body rising and falling without apparent effort, with now and then a short sail of a foot or two or a slight rise as the bird pauses and flutters for a moment when sighting some prey. There are usually several rapid wingbeats followed by one or two slow ones, producing a rather irregular flight, while their actual progress is slow on account of the frequent changes in direction. Every now and then a bird dives, bill foremost, into the grass or down to the surface of some shallow pool, but never, apparently, do the feet touch, so quickly is the recovery effected. The prey is snatched up and the tern, flapping rapidly, rises and is off again on its course.

In late August of 1922 there were many Black Terns busily engaged in feeding on the meadows back of South Cape May and mingled with them were numerous Barn Swallows; while the object and general activities of the two species were similar their method of feeding, and doubtless their food as well, differed materially. The swallows would glide or "stream" along close down over the tips of the grass at lightning speed, in marked contrast to the diving and the hovering, halting, and relatively slow flight of the terns. Over Lake Lily Black Terns act in the same way as on the meadows and doubtless there, too, they are feeding on insects of some kind. They swoop repeatedly down to the water and graze the surface without actually touching it.

On August 31 of this same year, a year of great abundance for these terns, one could see them with the glass scattered widely over the meadows from the Harbor away north beyond the Wildwood Road. There must have been hundreds of them in sight at once, and all through the first week of September they were to be seen quartering the meadows in their characteristic easy flight, rising and falling with little apparent effort.

The summer of 1928 was another great season for the Black Terns, like those of 1922 and 1926. There were many on the meadows back of Five Mile Beach on August 14 and hundreds back of Seven Mile Beach on August 31, while on September 5 they were hawking about on Pond Creek Meadows. On September 4 a flock was seen out over the ocean now flying in a close pack now scattering over the waves and once more gathering together. 1934 was another year of abundance and they occurred from August 26 to September 16.

On September 1, 1924, several of these terns apparently tired of the chase came to rest on oyster stakes on Jarvis Sound where they perched for some time and exactly the same thing occurred on September 7, back of Schellenger's Landing where twelve stakes each had an occupant.

During August, 1926, the Black Terns frequented the Lighthouse Pond, hawking about over the surface of the water, from three to fourteen gathering there almost every day. On August 27, while watching them, I was impressed by the similarity of their flight and general wing action to those of the Nighthawk, and to my astonishment I presently realized that one of the birds actually was a Nighthawk. These terns seem interested in almost any body of water and on August 25, 1931, a number were found feeding over a rain water pool which formed on the flat back of South Cape May and which also proved attractive to a variety of shore birds. While rarely actually diving, Black Terns do sometimes strike the water in swooping down to the surface for food. On August 26, 1926, I found a number of them on the water just below the bathing beach in company with Common and Least Terns. The latter were diving repeatedly with their usual abandon, the smaller birds fairly flinging themselves into the water. While the Black Terns did not dive so frequently they did strike the water on several occasions. Just what they were securing I could not determine.

As the season advances and cool northwest winds begin to blow one often sees compact groups of Black Terns come suddenly coursing straight over the meadows like a pack of hounds following the scent. They travel fast and keep close together and are soon past and disappearing in the distance. Flights of this sort seem to be for the sole pleasure of exercise, possibly preliminary to migration, as the birds certainly never pause to feed, nor do they appear to have any definite destination.

The first Black Terns to arrive in late July or early August are usually adults in full breeding plumage, dark slate gray above with black head, breast and belly. Later arrivals, from August 5 to September 12, are in all stages of molt tending to the white-crowned, white-breasted dress of winter. There come too, mostly in the later flocks, large numbers of immature birds of the year, much like winter adults but with brownish patches above. In 1926, however, young birds accompanied the first adults. In any plumage the very dark upper surface of the Black Tern will distinguish it from any other of our common New Jersey terns in which the gray of the back is so pale as often to appear actually white in certain lights.

Warren Eaton and Richard Pough saw a pure albino Black Tern at Cape May Point on September 5, 1935.

The only call note that I have heard from these late summer migrants is a weak *seep, seep*, but the birds seen in the summer nesting colonies of the Common Tern were more vocal.

BLACK SKIMMER

Rynchops nigra nigra Linnaeus

This grotesque bird forms one of the prominent features of the wild life of any coast that it favors with its presence. Although reported as common years ago on the beaches of Cape May County it is now a long time since it has been anything but a rare and casual visitor in the immediate vicinity of the town, if indeed it ever bred there. On the beach islands a few miles north, however, it has returned of late years to nest to the delight of all who are interested in wild bird life. In upwards of forty years of more or less intensive study of the Cape May avifauna we have but seven records of Skimmers on the Cape Island beach or between there and Cape May Point. Fletcher Street saw one skimming over the shallow pond near the Lighthouse in the summer of 1919 and Ludlow Griscom noted one off the Point on July 26, 1920, while a single bird came flying in over the bathing beach on July 10, 1921, about the middle of the afternoon, and another appeared near the same spot on September 25, 1927. William Yoder saw one flying past the end of the jetty at the entrance to the Harbor on September 27, 1925, and in 1924, Skimmers were found twice at the shallow pool near the Lighthouse. A single bird on August 27, 1924, associated with a flock of Common Terns, was obviously one of a flight of Skimmers which was in that year carried north by a severe storm and stranded all along the coast as far as Maine, while on September 27, of the same year, three were mingled with a great flock of Laughing Gulls which had gathered at the pond.

Alexander Wilson who, with George Ord, explored the beaches of Cape May County about 1810, says of the Skimmer: "Its favorite haunts are low sand bars raised above the reach of the summer tides, and also dry flat sands on the beach in front of the ocean. It lays in June. Half a bushel and more of its eggs has sometimes been collected from one sand bar within the compass of half an acre," but he is not explicit as to how many colonies he found. John Krider writing in 1879 states that they bred on all the beaches of the county, but his work, written from memory, is not very reliable. We know that they still bred on Seven Mile Beach (probably the southernmost extremity) as late as 1885 and 1886, when C. S. Shick counted seventy-five nests there, but by 1890 he tells us that while still present they had become very rare. On the other beaches we have no definite records.

Subsequent knowledge of the birds on our New Jersey coast shows them to be very local during the nesting period and seldom seen any distance from their breeding grounds, resting on the sand during most of the day and active mainly at night. These facts taken in connection with the bird's habit of selecting the more remote bars and sand flats for nesting purposes would make it seem very easy for small colonies to have been overlooked and we are therefore in some doubt whether the Skimmers really disappeared as breeding birds on the lower New Jersey coast, from 1895 to 1915, or were simply not observed by those who would realize the importance of making a record of their occurrence. Certain it is that there is a blank in their history on our coast during these years. While comparatively few ornithologists were interested in exploring the Cape May beaches during this time we know that William Baily failed to find Skimmers on Five Mile Beach during visits in 1896, 1897 and 1898 nor did Philip Laurent find them there in 1892, while Dr. William E. Hughes and David McCadden saw none during the summers of 1898–1900 on Seven Mile Beach. This is, however, not proof that small colonies did not exist.

The return or the increase of Skimmers on our coast began about 1920. Richard Miller found a single pair nesting at the southern extremity of Seven Mile Beach on June 26, 1921, while in 1922 there were three pairs present and two nests were found on June 25 not over six feet apart. In 1923 there was but a single pair and no nests were found. It was discovered later, however, that a colony of about forty birds had been established on Gull Bar at the entrance to Hereford Inlet not far away. Whether the birds had located there during the immediately preceding years I could not determine. This colony had doubled in size by 1926 but was apparently deserted in 1927 or 1928 as the bar decreased in size and was nearly washed away. In the latter year the birds moved to a sand fill on Ephraim's Island, a section of meadow back of the southern part of Five Mile Beach, and in 1931 they moved again to a sand flat about a mile farther north. In 1934, after being washed out several times the birds returned to Gull Bar but the same fate met them there and in 1936 they were back at the southern point of Seven Mile Beach. In this year they numbered about forty individuals. In 1927 when Gull Bar was deserted one pair of birds reared two young on a narrow spit of sand jutting out into the inlet from the south side, the northernmost point of Five Mile Beach, and several were seen at the southern extremity of Seven Mile Beach but apparently did not nest there, although they did so later, 1931 to 1936, and sporadically in earlier years.

It seems probable that these birds all belonged to a single colony which, as is usual with Skimmers, shifted its location from year to year just as the low bars and sand points which they select for their homes change in shape or disappear entirely at the mercy of the waves and storms. The population of the colony, or its several sections, changes too, one of them boasting the largest number in one year and another in the next. In every instance our New Jersey Skimmers have been associated with Common Terns on their nesting grounds and while the two species join in the defense of the colony they seem to nest, for the most part, in separate assemblages and do not intermingle promiscuously.

The history of the birds breeding on the more northern section of our coast is no more satisfactory than that of the Cape May Skimmers. W. E. D. Scott records them as abundant on Brigantine Island in 1877 and rare on Long Beach, a little farther north. We know that they still bred on Little Beach Island, between the two, in 1910 and immediately succeeding years when Richard Harlow found two pairs nesting, and in August 1915, I also found two pairs breeding there. We heard of their presence there from 1915 to 1920 and on July 17 and 18, 1921, in company with Fletcher Street and other members of the Delaware Valley Ornithological Club I visited this colony and one which had that year been established back of Brigantine, on a recently formed sand flat on the meadows, as a result of dredging out the channel. Charles Urner writes me that the first pair of Skimmers that he found "nesting north of Little Beach Island was at Brant Beach a few miles north of Beach Haven on June 30, 1925, on a high sand island created by the dredging of the channel. This colony grew rapidly and by July 25, 1926, there were six pairs. The following year there were thirty pairs and they have been present ever since with a maximum of about seventy-five pairs.

About 1930 Skimmers became established on another sand island created by dredging in the bay west of Beach Haven called Goose Bar Island and in 1931, there were Skimmer colonies on Brant Beach, Goose Bar, Little Island Beach, Brigantine, Shad Island in Little Bay, and Sandy Island west of lower Brigantine."

On our 1921 trip we approached by boat from Atlantic City but saw no Skimmers until within about a mile of the Brigantine breeding grounds, when one flew rapidly past us. A little farther on we found nine sitting all in a row facing us. Their legs were very short, which brought the body close down to the sand, and the wing tips were slightly elevated over the base of the tail. They looked jet black from a distance and very conspicuous on the little spit of sand and mud. From the side they appeared all black and their ample wings folded up against the body resembled a black overcoat. In front, the bulging white breast and the white lower part of the face were conspicuous, as was the remarkable red bill, the mandibles compressed like knife-blades set on edge one above the other, the lower one decidedly the longer. Farther on we encountered two birds wading or standing in the shallow water, poking their bills down in an apparent search for food. Another passed rapidly by, flying close in to the grass-capped mud banks of the channel, and not thirty-five feet from our boat. When just opposite he dropped the long lower mandible into the water, snapped up a fish about three inches in length and flew off with it. His red feet showed clearly, stretched back against the under side of his tail, as he passed directly over our heads.

In contrast to the "overcoated" appearance of the resting bird, as if its clothes were too large for it, nothing could be more graceful and pleasing to the eye than the flying Skimmer. Three passed us in tandem formation, their long wings beating in extended arcs apparently almost touching tips above the bird's neck which is stretched forward and downward, the remarkable bill suggesting a long nose or snout. Now they seem partly to sail and the wing beats are shorter, as if gently patting the air to keep their proper poise. Above, with the exception of a white posterior border to the wings, the plumage is jet black, contrasting with the snowy white lower parts, the line of demarkation passing along the side of the face just below the eye. Now dropping close to the water they proceed with the long lower mandible just touching the surface or immediately above it. The appearance is as if a string attached to the bill were being pulled rapidly forward by some unseen force beneath the water. In reality the bird flies in precisely the same position when some distance above the water as when "skimming," the apparent tilting of the body being more noticeable in the latter case. When skimming, the lower mandible is sometimes immersed, supposedly scooping up some sort of minute prey, but this theory has been questioned and in my experience it is always small fish that the bird catches in this way.

On July 28, 1929, I watched several Skimmers fishing on the thoroughfare just above the Wildwood Road flying back and forth over an area which evidently swarmed with small fish. Every now and then a bird would pause and elevating the tail slightly would stab its bill into the water; failing to catch a fish it would continue on its way but when the mandibles closed on one the bird made off instantly for the nesting grounds with its prey held crosswise. Three were seen on August 23, 1930, skimming over the shallow water off Seven Mile Beach and occasionally the lower mandible of one of them would strike the sand and the bird would almost turn a somersault. When we visited the Little Beach colony, in 1921, we found five pairs located there. As we landed and approached the nesting ground several birds came flying down the beach like a whirlwind, as if to carry us off our feet, but just as they reached us they sheared off to the right and left, turning up their snowy bellies and under wing surfaces which glistened in the sunlight with dazzling brilliancy. All the while the peculiar reedy calls echoed all about us—*aaar, aaar, aaar, aaar*, or as it sometimes seemed to our ears—*kaup, kaup, kaup*, with an angry *aaar*, at the end uttered as the bird passed close to my head, when his red eye was clearly seen and the brilliant red bill shading into black at the tip. They seemed unnatural—like painted birds, so striking were their colors and so sharply defined.

In the main colony on Brigantine there were twenty-five pairs nesting and we counted forty birds in sight at once. Thirteen nests were located containing in all thirty-three eggs while sixteen young were found. The eggs, usually four in a set, are cream color with large blotches of brown and purple. The nests are round depressions scooped out of the sand varying in diameter from eight to twelve inches and three inches deep with no lining of any kind. Some of them, where the parent had been incubating, showed radiating lines all around the edge apparently made by the long bill of the setting bird. The youngest birds were lying in the nests flat on their bellies with head and neck outstretched, the pale gray down exactly matching the color of the sand, and with similar black specks scattered through it. One could easily have stepped on them without seeing them. To one who has been on the western deserts they recalled the horned toads or similar flattened lizards. The older nestlings ran clumsily about calling vigorously to their parents. They had feathers sprouting out all over the body which carried on their tips little tufts of the down. The old birds arose in a great flock as we approached, and with them rank upon rank of silvery terns took wing, while shoals of little sandpipers skimmed the surface of the flat to settle farther on, early migrants from the north. The air was a moving maze of birds and their united cries produced a veritable bedlam; the harsh *keeé-arr'*, *keeé-arr'* of hundreds of terns and the woodwind *aaar'*, *aaar'*, of the Skimmers mingling in a confusion that is difficult to describe.

The parent Skimmers wheeled about over the sand, maddened at our presence and by the cries of their young as the latter scuttled away to take refuge in the coarse grass of the marsh edge, or even in shallow pools of water. The old birds traveled in parties of six to eight, or in pairs, and frequently charged us in company front, dashing past at apparently increased speed and uttering the climax note, *aaar'*, just as they turned aside to avoid striking us.

Just before sunset after we had withdrawn from the first colony and quiet reigned again, we saw from the water a pair of Skimmers performing evolutions near the nests and apparently about to mate. One pursued the other relentlessly; they wheeled and twisted right and left, now the snowy under parts gleaming in the horizontal rays of the sun, now the dead black of the back showing quite as conspicuously against the blue gray haze in the east. They rise one above the other alternately until they are some two hundred feet above the beach and then on set, curved wings they pitch to earth and settle on the bar. In a few minutes the pursuit begins anew but this time they go skimming over the smooth waters of a little inlet, between the bar and the beach, one close behind the other as if following a marked course. I saw this same action later in colonies on the Cape May beaches notably on July 8, 1923, on Gull Bar, and July 21, 1930, on Ephraim's Island, and on the adjacent colony on July 13, 1932.

I first visited the Gull Bar colony, in Hereford Inlet, on July 8, 1923, in company with Julian Potter who had been there on June 20. We found forty birds in the colony but egg laying had only just begun, earlier sets having apparently been washed away by the high tides. The birds flew in squads of from two to six and sometimes the entire flock took wing together charging us like a pack of hounds in full cry as they gave vent to their throaty barking calls. They did not however seem as much excited as did those of the other colony which had young to care for. Two birds arose high in the air attacking one another and fencing with their long bills. We also saw one individual flying about with a fish about three inches in length held crosswise in its bill, though there were no young to feed. Perhaps this is part of the mating behavior as in the case of the terns.

We were impressed with the way in which the birds remained close to their nesting grounds; we saw two in Grassy Sound about a mile away and several flew over to a narrow harbor at Anglesea, on the south side of the inlet, to fish. None seemed to stray more than two miles away, at least in the daytime, and they invariably followed the thoroughfares and other waterways, never crossing over land except on the nesting bar.

Viewing the bar from the point of beach at Anglesea on July 21, 1924, and July 17, 1925, we could see the birds sitting in rows on the strand and every now and then one would come over and fish in the shallow water at the end of the harbor. They skimmed back and forth with the lower mandible immersed and on each visit a bird caught a fish two or three inches in length and flew with it back to the bar. Sometimes they called as they passed me. Julian Potter visited the colony again on August 17, 1924, and found forty adults and twenty young.

In 1926 Horace McCann visited the Gull Bar colony on July 11 and found seventy-five birds present and all nests with eggs, but a subsequent visit on July 25, found the entire colony washed out by the tides. On August

10 when I visited it in company with Alexander Wetmore we found but six birds of the year in the flock of seventy-five assembled on the beach, and four others not yet able to fly, showing how completely had been the destruction of the first broods. Even at this late date some of the birds were nesting again and we found eight nests with from one to three eggs each. Five of these were seen again on the 17th but I was unable to learn whether any young were raised.

The young on the beach would run very rapidly and then stand stock still with legs rigid and head held forward in line with the body the heavy bill looking ridiculously out of proportion. Their action when running, and their pose when standing, reminded one strongly of the Killdeer. Their plumage was grayish lead color with buff edgings to the feathers. Those in the air were similar but showed more or less black on the under side of the wings their bills dusky with just a trace of red, while the mandibles were equal in length. They called occasionally as they flew about but not continously as did the adults and their note was shriller and more highly pitched. The adults had in many cases lost some of their flight feathers, apparently the beginning of the molt. They would settle themselves along the strand, facing the wind, and take wing when approached, yelping at a great rate. Sometimes one would "skim" over the sand its body almost dragging on the ground. They were still on the bar on August 28.

On August 13, 1927, we were surprised to see two Skimmers on a sand spit a little to the south of the lower end of Ephraim's Island back of Five Mile Beach. On September 3 David Baird saw a number in the same place, as he stood on the boat landing at Wildwood Crest, and on the 7th I counted one hundred and fifty gathered there, adults and young of the year, associated with many Laughing Gulls. I thought that these were birds making a temporary stop while on migration from the Little Beach colony or that on Gull Bar. Next year, however, passing Ephraim's Island on a boat I found a great throng of Common Terns and eighty Skimmers evidently nesting there and, having meanwhile learned of the desertion of Gull Bar, I realized that the colony had moved to this spot. Some birds in immature plumage were in the flock but whether young of the year or possibly birds that had not molted I could not determine although subsequent investigations would suggest the latter.

I visited this colony in company with Walker Hand on August 4 and again on the 10th. We counted eighty adult Skimmers and many downy young scattered about on the sand, the youngest were nearly white and the older ones pinkish gray. They lay prostrate the head and neck stretched out in front, the younger birds in the shallow saucer-like hollows which constitute the nests, the larger ones in depressions in the sand which they apparently make themselves in the process of digging in, for the marks of their feet are plainly evident in the rear. The sand seems to drift in all around them and makes their protective resemblance all the more perfect and the danger of treading on them more imminent!

Some of the larger young which are able to run about follow their parents when they arrive with a fish and are fed by them just as in the case of the terns. Some of them run ahead of the flying parent and take short experimental flights of a foot or two; the parent flies over and alights in advance waiting for them to catch up. Others ran down to the grass on the edge of the green meadow and even into the shallow water. The same thing was noticed on July 17, 1929, when they ran in droves down to the wet meadow and many holes in the sand were found which had evidently been occupied by young that had scampered off to take refuge in the grass at our approach.

Upon our first arrival on every visit the old birds came charging at us flying close to the sand and stretched out in a long line barking their alarm and almost striking us before they veered off. Later if we concealed ourselves among the marsh elder bushes, on a higher part of the island, they settled down and we could see through the glass that they lay prostrate just as the young do, with neck stretched out in front and throat appressed to the sand, the shoulders forming a sort of hump, and the bill slightly elevated at the tip. When studied from this shelter, on our first arrival, we found many standing in rows with their heads turned back and tucked under their wings. Possibly they were asleep in both positions. On August 14, 1928, the date of my third visit, I found the birds on the wing to number one hundred and fifty. There were some well-grown young in the grass but no downy birds and no eggs; the adults seemed just as solicitous as on the previous occasions.

On June 30, 1929, there were many eggs and downy young but on July 7 many of the latter seemed to have been destroyed and the presence of a number of rats on the island perhaps explained their disappearance, there were about fifty pairs of adults. On July 17, the young in mottled gray down were just beginning to develop pin feathers and were all active and running about, sometimes when pursued the tip of the bill would catch in the sand and the bird would be tilted over almost turning a somersault. There were no eggs or small downy young. On August 12 there was not a Skimmer in sight either on the sand or in the air; a high tide about a week before and a heavy rain storm the previous day, had apparently driven them elsewhere.

In 1930 there were about forty pairs present. On July 4 the colony was visited by Horace McCann who found that while the terns, which occupied the lower part of the flat, had been completely washed out, the Skimmers, whose nests were located on a slightly more elevated belt which supported some clumps of grass and sea rocket, had escaped the flood. He found forty-three young and twenty-five more nests with eggs. On the 21st and 25th when I visited the island there were eggs and young birds of all ages but most of the latter able to run about. On August 4 McCann reported it again submerged by high tides, in fact the sand was still wet. The young Skimmers were all on the wing and he estimated the flock to contain nearly two hundred birds.

On July 25, 1930, I spent some time in a shelter tent in the colony. After I was out of sight the birds settled down and most of them paid no attention to the tent but two pairs, with young buried in the sand close to me, flew by several times calling. There was no evidence of either feeding or seeking food while I was there and so far as I could detect the two young birds did not move or change their position a particle for three quarters of an hour, except for an occasional gape and a slight sideways movement of the head, but within five minutes after I left I found that they had disappeared. I noticed that the old birds, when apparently trying to lure you away from a nest or young, would "taxi" along over the sand getting closer and closer to it until the feet would touch and the bird would almost pitch over forward, seeming to be in distress. This action is perhaps a form of the "broken wing ruse." When really alighting on the ground they act quite differently fluttering down like a gull. The bridge tender on the Wildwood Road, who has paid considerable attention to these birds, tells me that when coming out to Grassy Sound to fish they always cross over the bridge never under, and that they are much more in evidence about dusk than during the middle of the day, while he hears them calling as they fly all through the night. Curiously enough, with this colony established within four miles of Cape May Harbor, we never saw a bird there nor did they occur on Jarvis Sound which is still nearer to their nesting place. They seemed always to go north to Grassy Sound to feed, but perhaps they journeyed farther afield at night.

On July 6, 1931, when I again visited the island I found that all the nests of both Common Terns and Black Skimmers had been washed out by high tides and the broken eggs were being eaten by Herring Gulls. There were only four Skimmers present and they several times flew across the sand and called feebly but showed no real interest in my presence. I ascertained later that the birds had moved to a new location about a mile north on another sand fill on the meadows. Horace McCann visited this spot on July 29 and found forty pairs of Skimmers and banded ninety young; there were about one hundred terns present.

In 1934 they nested on the southern extremity of Seven Mile Beach and Julian Potter saw sixty adults and three young there on August 1, and there were twelve pairs in 1936.

While the nesting date of the Black Skimmer may be seriously affected by high tides and wash-outs the following additional egg dates are presented to show the variation:

Lower end of Seven Mile Beach (Miller and McMullen):

> June 26, 1921, four eggs.
> July 10, 1921, three eggs.
> June 25, 1922, three and four eggs.
> July 4, 1931, two with one egg; one with two; two with three and two with four.
> June 6, 1931, eleven nests with eggs.
> June 16, 1935, seven nests with eggs.
> June 20, 1936, six nests with eggs.

Gull Bar, Hereford Inlet:

> June 21, 1925, eight nests with eggs (McMullen).

Little Beach Island:

> June 15, 1915, three and four eggs (R. C. Harlow).
> June 17, 1916, three eggs (R. C. Harlow).
> July 15, 1925, many nests with eggs (Street).
> June 16, 1931, one hundred and thirty-two nests with eggs (McMullen).

Brigantine Beach (Fletcher Street):

> June 18, 1921, three nests with one egg; two with four; six with three; two with four and twenty young birds.
> June 25, 1922, one nest with three eggs; four with two; six with four; six with five; and five not yet laid in.

I visited the Ephraim Island colony several times in May before nesting had begun. In 1929, on May 11, there were twenty-five birds present flying round in packs of five to nine close to the ground, barking like hounds. Now and then they would set their wings as if about to alight and then change their minds and start off again, there were several hollows in the sand apparently made by either Common Terns or Skimmers but no eggs had been deposited. The bridge tender said that the first Skimmers came on May 4. On May 16, 1930, there were thirty-two on the island; they were very restless and continually took wing and sped away to a bar north of the Wildwood Road where in each season the first arrivals seemed always to congregate. Then they would return in bunches of five, six, twenty etc. In 1931 there were six on this bar as early as May 9 and in 1927 they were present at Little Beach Island by May 17, but there were none at Seven Mile Beach on May 16, 1936, although nests had been scooped out by June 7 (Hiatt).

I have not been able to check the time of the Skimmers' departure in autumn but according to the bridge tender they go in early October. I have seen them present on September 2, 1930, and September 7, 1927, and individuals have been seen as late as September 25, 1927, and September 27, in 1924 and 1925, while there were thirty present on October 4, 1931, and a flock of 150 at Brigantine on the same date, apparently about to migrate.

The adult bird seen with terns at Cape May Point on August 27, 1924, rested with them on the beach bending its head back over the shoulder and burying its bill under the scapular feathers. A bird in immature plumage, at the same place, September 27, squatted in the water up to its breast and poked its head under repeatedly, first on one side then on the other, after which it sank forward submerging the breast, and, raising the wing tips a little, churned the water with the forward part of the wings. This bird was dull brownish above, very different from the adults, the feathers edged with white and the bill dusky but red at the base. This is apparently the full juvenal or bird of the year plumage and so far as my experience goes it is not changed to the adult plumage before the birds leave for the south.

The presence of the Black Skimmers furnishes an exotic touch to the more or less prosaic bird life of our Cape May coast and their breeding grounds constitute one of its most attractive ornithological features. On July 13, 1932, I was studying this Hereford Inlet colony and wrote: Picture a wide thoroughfare reaching north to a broad Sound with a gentle south wind ruffling the surface of its waters. It is ebb tide and broad black mud flats lie exposed on either side of the channel with tufts of sea cabbage and oyster shells scattered here and there. On the side opposite to our stranded boat are scores of great brown gray Herring Gulls searching for dead crabs or other food exposed by the receding water. Some heavy-winged Turkey Vultures perch on the edge of the meadow which stands three feet above the flat, its vertical banks honeycombed with the burrows of the fiddler crabs. Other Vultures sail high overhead against the deep blue of the sky and graceful black-headed Laughing Gulls go lilting past, while early flocks of Dowitchers, Yellow-legs and little "peeps" go wheeling away to feeding grounds in the sound. On our side of the channel stand rows of Common Terns resting on the mud or bathing in the shallow water and behind them rise the meadows covered here with white sand and bleached shells of clams, razors and oysters left by the dredges. There are tufts of glass-wort, salt-wort, sea rocket and coarse grasses with masses of trash stranded among them. On this, as well as out on the open sand, dainty terns are setting on their eggs with dozens of eggs of an earlier laying, washed out by the high tides, half buried in the sand. On a bare expanse of sand are the Skimmers, sixty of them in a single flock, standing close together some asleep others craning their heads up and strutting about among their fellows. Although all of their nests had been destroyed the undaunted birds are digging out new hollows in the sand and some already have eggs. An unsuspecting Turkey Vulture soars over the colony and the terns, to the last one, are awing and after him, their raucous cries filling the air as they drive the intruder from their domain. The Skimmers, too, responding to the alarm rise in a solid phalanx and circle a couple of times before alighting, adding their barking cries to the chorus and attracting the attention of a tern or two which for the moment take them for Vultures and give chase. Then once again silence settles on the community and before we realize it all the nesting birds are back on their precious eggs.

Only in the roosts of Grackles, Crows, Martins or herons are we of the Middle States privileged to see such examples of bird life *en masse* as are offered by these sea bird colonies of our Cape May coast and the fact that these are breeding communities where the life of both the individual and the species are at stake adds greatly to their interest.

RAZOR-BILLED AUK
Alca torda Linnaeus

Doubtless some Razor-bills are present off the coast of New Jersey every winter but our knowledge of them is mainly confined, as in the case of other pelagic birds, to individuals cast up on the beach in more or less decomposed condition, or to others blown into the sounds by storms. Those that have been shot have all been obtained by fishermen who have given us no details of the actions of the birds. The only detailed observation that has been recorded is of a bird seen by Ludlow Griscom at the mouth of the inlet near the lighthouse at Barnegat City in the midst of a cold wave on December 19, 1926, with the temperature at 5° F. "It flew in at moderate range the deep bill plainly noticeable; also the greater extension of white back of the eye and the clouded effect. Half an hour later another individual was seen at close range going out to sea."

Young birds in their first winter have the bill much less elevated and without the white band, so that they might easily be confused with Brunnich's Murre. The best distinguishing characters in this plumage are the lack of the white line along the base of the upper mandible and the elevated tail tip when swimming.

Cape May County records are as follows:

Grassy Sound, Anglesea, three observed on several occasions in February 1891, Philip Laurent.

Sea Isle City, one shot and others seen, January 23, 1909, Thomas Mitchell.

Five Mile Beach, one shot by coast guards, January 20, 1880, W. L. Abbott.

Ocean City, one shot, January 10, 1901.

Five Mile Beach, one found exhausted, on the beach, May, 1927, F. W. Cole. Gillespie, Auk 1928, p. 91.

Cape May Beach, one found dead, March 11, 1922, Walker Hand.

Cape May beach one found dead, April 27, 1932, Witmer Stone.

Seven Mile Beach, one found oiled, December 27, 1928, C. Brooke Worth.

Records for the upper part of the coast furnished by Charles Urner and others range from December 19 to March 13.

*BRÜNNICH'S MURRE

Uria lomvia lomvia (Linnaeus)

So closely does this bird resemble the Razor-bill that it is not easy to be sure to which records of occurrence belong without careful examination of specimens. The high white-banded bill of the adult Razor-bill is sufficiently diagnostic but the bill of the immature bird in its first winter is little larger than that of the Murre.

From Cape May County we have no definite observations of this species although there seems no doubt but that it occurs offshore perhaps every winter, though it is certainly not as abundant as the Razor-bill.

At Barnegat Light farther up the coast Ludlow Griscom states that he saw one flying into the bay in company with a Puffin, on December 19, 1926, during a spell of very cold weather. Although at long range he noted the slender bill and narrow white wing stripe (Auk, 1927, p. 535).

Records have been published as follows:

Brigantine Beach, January 7, 1933, R. F. Miller (found dead).
Barnegat, December 12, 1926, Watson (close approach on the beach).
Delaware Bay, December 24, 1896.
Delaware River at Byberry, January 11, 1901.
Middletown, Delaware, December 18, 1896, Charles Pennock.

Stuart Cramer has recorded an individual seen on January 23, 1932, at the end of the breakwater at the entrance to Cape May Harbor. It was studied at close range with a glass and the flesh colored stripe at the base of the mandible and a dusky band across the breast (oil stain?) were clearly seen. The bird was diving continually close at hand for several minutes and finally disappeared under water (Auk 1932, p. 219).

DOVEKIE

Alle alle (Linnaeus)

The Little Auk or Dovekie seems to be the most frequent of the Alcidae to visit the New Jersey coast and it probably occurs on the ocean off Cape May in greater or less abundance every winter. It is only when severe storms drive them into the bays or harbors, or when their dead bodies injured or oiled are washed up on the beaches that we see anything of these little sea birds. We have definite records of one or more individuals in eight out of the fourteen winters from 1921 to 1934 during which reasonably careful observations were made, and scattered occurrences prior to those years; back even to December, 1811, when the specimen figured by Alexander Wilson in his "Ornithology" was killed at Great Egg Harbor and sent to him "as a great curiosity."

I have never had the good fortune to see a live Dovekie nor even to find a dead one on the beach and those who have shot them have furnished no particulars as to their actions. One individual was blown in to Cape May on February 14, 1932, and was found dead in a garden on Washington Street and in the great invasion of Dovekies which occurred in November of that year along nearly the entire Atlantic coast a number of the birds were seen alive (cf. Murphy and Vogt, Auk 1933, p. 325, and Nichols, p. 448). Several were captured on the main streets of Cape May and others were seen on the Harbor where Otway Brown watched them swimming and diving in the deep water near the fish dock at Schellenger's Landing. They came within six feet of him and dived repeatedly, swimming under the water "like frogs," using their wings to propel themselves.

Oscar Eayre, cruising sixteen miles off Long Beach farther up the coast in January, 1933, counted eighty Dovekies from his boat.

A list of specimens obtained or observed in Cape May County is as follows:

December 17, 1878, Five Mile Beach. W. L. Abbott.
Early in 1898, Grassy Sound.
November, 1904, Grassy Sound. Walker Hand.
December 23, 1912, Peck's Beach. W. B. Davis.
December, 1921, Corson's Inlet. Wharton Huber.
December 31, 1922, Cape May.
January 7, 1924, Corson's Inlet. Wharton Huber.
November 21, 1930, Cape May. Walker Hand.
December 6, 1930. Wilbur McPherson.
December 8, 1930, Cape May. Walker Hand. (2).
January 25, 1931, Cape May. Schmidt.
January 31, 1931, Seven Mile Beach. Brooke Worth.
February 14, 1932, Cape May. Walker Hand.
November 20, 1932, Cape May. J. W. Mecray.
November 21, 1932, Cape May. Caught alive.
November 21, 1932, Five Mile Beach. Otway Brown.
November 24, 1932, East Cape May. Otway Brown. (Two dead.)
November 24, 1932, Schellenger's Landing. Otway Brown, several swimming.
November 27, 1932, Cape May. Charles C. Page. (Two dead).

There were other 1932 records from Seven Mile, Ludlam's and Peck's Beaches but we were unable to obtain details, and many from the Barnegat and Brigantine areas farther up the coast. Most of the birds recorded above were picked up dead but one was shot on Barnegat Bay by William McCall on January 9, 1930.

While the great flight of Little Auks in 1932 was unusual it is probable that there were somewhat similar occurrences in the past when winter storms drove numbers of these little birds in shore but absence of ornithologists was responsible for our lack of information. Lack of offshore records, where the birds are in their normal winter quarters, prevents the tabulation of dates of arrival from their summer home in Greenland in autumn or of their return in spring.

BLACK GUILLEMOT

Cephus grylle grylle (Linnaeus)

This rare winter visitor from the north has twice been recorded at Cape May, both times from the extreme end of the jetty, known as the "Stone pile" at the entrance to the harbor. The Guillemot like the Purple Sandpiper and the lobster prefers rocky coasts and like them, it seems to be attracted by the artificial rocky conditions offered by this rough stone jetty projecting out for a mile from an otherwise flat sandy shore.

Eliot Underdown reported the first Black Guillemot on December 10, 1929, an individual which flew from the northeast to comparative calm water on the lee side of the long jetty, which extends out from the entrance to Cold Spring Harbor, and remained within twenty feet of him for some fifteen minutes. The bird dived twice within this period and seemed to use its wings in submerging. It was in mottled winter plumage and seemed very light colored when in flight. The white wing markings were very conspicuous both in flight and when at rest, while the red feet at once attracted attention. (Auk, 1930, p. 242.)

On December 24, 1933, William Yoder saw two of these birds at the same spot, which was his post of observation during the Christmas census of the Delaware Valley Club. They flew across the channel and their light color and prominent white areas on the wings easily identified them.

A specimen mounted by Charles Voelker was said to have been taken on the Delaware River in December 1898, but we have no definite information concerning it. Charles Urner tells me of one picked up dead in the Barnegat region some years ago and Turnbull quotes John Krider as authority for two secured at Egg Harbor at some time prior to 1869 but we know of no other records for New Jersey.

PUFFIN
Fratercula arctica arctica (Linnaeus)

Of this very rare winter visitor to the New Jersey coast we have only Griscom's record of one flying into Barnegat Bay past the lighthouse on December 19, 1926, two reported by Oscar Eayre sixteen miles off of Long Beach Island in January 1933, in a flock of Dovekies, and one seen by Fletcher Street on Jarvis Sound during the Christmas census of the Delaware Valley Club on December 24, 1933. It has doubtless occurred in other years as far south as Cape May, but well out on the ocean.

Ludlow Griscom has given us an account of the occurrence of Puffins and other Auks in a remarkable flight at Barnegat Light on December 19, 1926: The bay was frozen and was crowded with grebes and gulls, as the tide was racing in from the sea. Brant, Scaup, Mergansers and Old-squaws were passing every moment. "Five minutes after my arrival, at dawn, a Brünnich's Murre followed by a Puffin flew in at long range, some Grebes just ahead, some Old-squaws just behind. The slender bill of the Murre and the narrow white wing stripe were plainly visible. The much smaller size of the Puffin and the absence of a wing stripe, the large head and buzzing flight were noticed. A little later a Razor-billed Auk flew in at a moderate range, the deep bill plainly noticeable, also the greater extension of white back of the eye, and the clouded effect. A lull of half an hour then ensued. Two Brünnich's Murres then flew out to sea; about five minutes later a Razor-billed Auk at close range came by tagging a flock of Old-squaws, also going out. This bird was near enough to be identified with the naked eye. Another five minutes passed and than a solitary Puffin came right past the light. This bird was also picked up and instantly recognized with the naked eye. It was buzzing in very rapidly, flying with the wind and was past me by the time I got my glasses on it, so that it was too late to make out the color of the bill-tip. The great depth was plainly seen, and of course all the other characters of the species." (Auk, 1927, p. 535.)

DOMESTIC PIGEON
Columba livia livia Gmelin

While the Domestic Pigeon, descendant of the Rock Dove of Europe, does not occur exactly as a wild bird in New Jersey, it would seem to merit mention as much as the English Pheasant, the Starling and the English Sparrow.

While Pigeons, so far as I know, nest entirely about houses and barns, so do the Sparrows, and in their feeding habits they are similar. The Pigeons gather on grain fields after the crops are cut, as well as on pastures, and frequently associate not only with Starlings and Sparrows but with Mourning Doves, blackbirds, and other native species. We also see Pigeons flying far away from man's habitations, over open ground, ocean beaches and the edge of the meadows. Occasionally, too, they may be seen alighting on trees.

MOURNING DOVE

Zenaidura macroura carolinensis (Linnaeus)

PLATE 86, p. 614

Doves occur about Cape May throughout the year but, as in the case of all resident species, it is a question whether the individuals that nest here are the same as those which pass the winter; certainly there is a considerable migratory movement, as the large flocks seen in late summer and autumn do not occur throughout the winter and can hardly be made up of the local breeding birds and their offspring, while in spring there are obvious arrivals from the south.

From November to March one may see single Doves or small groups, from one to ten on a day's walk, while the Christmas census of the Delaware Valley Ornithological Club has totalled as many as twenty. Late March and April bring additional birds from farther south and from now through May they are engaged in nesting and usually occur in pairs. While the breeding season continues throughout the summer we begin to find Doves collecting in small flocks as early as June, possibly largely young of the year, although flocks of ten and twelve flushed from the edge of the salt meadows on May 25 and 27, 1923, can hardly have been young, unless from very early nestings.

In millet fields, or in any fields where the crops have been cut, Doves gather in considerable numbers sometimes associated with scattered Flickers and Meadowlarks. On July 7, 1922, I counted thirty in a field just back of the town which took wing successively as I approached, while on July 5, 1921, and July 18, 1927, there were similar gatherings in exactly the same spot. When these feeding birds took wing they usually resolved themselves into pairs and often circled about and returned to the feast, quite unlike their usual straightaway flight. September and October seem to be the months of greatest abundance and flocks of one hundred or more are sometimes seen. These flocks are apparently not assembled for feeding but are gathered together for the winter and, for the most part, migrate farther south, although occasionally such large flocks remain in the Cape May district. One gathering of ninety birds was seen by Walker Hand on January 12, 1919, and another large flock on November 29, 1909.

One of the greatest Dove flights of which I have record was observed by Walker Hand passing over open truck fields opposite Bennett Station on September 6, 1924. For several hours they continued to pass in groups of from three to one hundred, stopping here and there to feed and then passing on again. They followed a comparatively narrow course and he estimated that at least 1200 Doves passed while he watched them. The next day at the same spot only one or two pairs could be found. The flight was to the northwest in the face of a strong wind, corresponding in this respect to the Flicker and Woodcock flights and the autumn migration of passerine birds. Such Dove flights apparently were of common occurrence in the past when Doves were game birds and I am told that the gunners took advantage of their definite courses and, hiding behind corn shocks on the line of flight, killed great numbers of the birds.

Sometimes we see Doves flying high up in a direct course probably on normal migration. Such were twenty seen crossing the Harbor on September 4, 1921, which kept on steadily southward over the town while a similar flock of nine flew over Cape May Point but when directly above the Bay shore turned backward, evidently in doubt about crossing the wide expanse of water.

We usually see Doves in pairs, or often in threes, passing overhead in rapid flight and in most cases they disappear in the distance without alighting. The group of three is of too frequent occurrence to be attributed to mere accident and may represent a pair and single young; the fact that I have seen groups of three more frequently in the late summer and autumn would support this theory.

The flight of the Mourning Dove is rapid and direct with strong short wingbeats, giving the impression of much reserved strength and with the body once in motion the wing action, though continuous, seems as if it were merely guiding a body hurtling through space of its own initiative. Occasionally, though rarely, Doves will sail for some distance when approaching a resting place, especially if they are compelled to descend from a higher level. One observed at Cape May Point on May 10, 1924, sailed on set wings, like a hawk, for at least two hundred yards in a gentle curve finally coming to rest in a tree. Another seen on March 22, 1935, flying in the usual manner suddenly set its wings and gave a remarkable exhibition of volplaning, jerking to the right and left in a zigzag glide. The tail was kept tightly closed during the glide but was opened suddenly with much display of white as the bird came to rest. This soaring performance is apparently a mating display as the male bird descends to the vicinity of the nest. I have seen it only in the spring.

When we flush Doves from the ground it is obvious that they have seen us first and they take wing, as a rule, while we are still at a safe distance and do not alight again within range of our vision. Occasionally, however, their action is different. On July 31, 1925, I came upon four on a shady wood road, which alighted on a nearby pine tree and peered out right and left poking the head forward and at intervals jerking it back and forth spasmodically like a shore bird. Possibly these birds were a pair of young and their parents. Again I had a Dove flush from almost under my feet as I walked along a wood road, an occurrence so unusual that I thought the bird might have been asleep.

I have frequently seen Doves arising from sandy spots on the Fill, north of the town, and by watching for their arrival in such places have been able to study them through the glass. Sometimes they would lie flat on their breasts at others stand with head bent back over the shoulder and in both cases would hold the position rigidly until disturbed by someone passing on a nearby path. They did not feed and seemed to be merely basking in the sun. Other individuals were flushed from sandy patches in open pine woods where from the appearance of the ground they may have been dusting themselves. There are many sand pits and hollows in the dunes on the Bay shore of Cape May Point where one may flush pairs of Doves at almost any time during the summer. I had supposed that the birds frequented these places for the purpose of obtaining gravel for their gizzards but they seem to occur there too regularly to be so engaged.

It is very difficult to stalk Doves so as to see just what they are doing in such places but when busy feeding they seem much tamer and more easily approached. I have come upon them in stubble fields where they were walking about like Domestic Pigeons, and several times in millet fields they were so busy picking up the seeds that they were apparently unaware of my approach. I have seen Doves close at hand on lawns, on Washington Street or in the Physick grounds, when they evidently had nests nearby and were naturally not so wary as are birds in the open.

Although it would seem that the natural habitat of the Dove when at rest is on the ground, except in the vicinity of the nest, they now and then alight in trees, temporarily at least, spreading the tail and thrusting it forward to check the momentum as the bird settles on its perch, and then swinging it up and down until a stable balance has been attained. The spreading of the tail displays the striking gray and white pattern which crosses the ends of the feathers and, with the rapid beating of the wings, makes the alighting Dove a conspicuous object. On several occasions I have found Doves resting on telegraph wires, quite a feat for so large a bird, and the descent from such a perch to the road involves another notable display of plumage.

A Dove is very wary when perching in a tree. Usually the neck is held high when the bird alights, as if to take its bearings; then it peers to right and left for possible danger and finally relaxes when all appears safe. One that came to rest in a pine tree directly over where I was standing looked down at me with head turned over sideways and immediately departed in precipitous flight. Another watching me from the top of a dead tree turned its head as I walked and kept me constantly in view, while two resting in a willow bush twisted their necks until they were looking at me over their backs. Both departed the moment they realized that they were detected. Doves nest in the trees of the town especially in the Physick grounds where they are to be found at almost any time both summer and winter and where familiarity with human beings seems to make them much less wary. They also build in the pines and small trees of Cape May Point and in orchards and other locations farther up in the peninsula.

I found a nest in a wild plum bush at the Point which contained two eggs as early as April 24, 1921, while two young were found in an old Robin's nest in a pine tree on May 30, 1924. The latest nesting of which I have record was in the Physick grounds on September 6, 1924, the nest being located on the horizontal limb of an oak tree twelve feet from the ground. It contained a single squab which flew out as I climbed the tree and one of the parents immediately flopped to the ground and went through the so-called broken wing ruse.

Turner McMullen examined a nest at Cape May Court House on June 20, 1920, which contained two eggs and in Salem County has found the following nests all with two eggs except one:

May 13, 1917.	April 15, 1923.	April 13, 1929, two nests.
April 15, 1922.	April 29, 1923, one egg.	April 13, 1930.
April 16, 1922.	April 20, 1924.	April 16, 1932.
April 22, 1922.	April 7, 1928, two nests.	April 23, 1932.
April 29, 1922.	April 14, 1928.	April 30, 1932.
	April 22, 1928.	

There are certain locations where Mourning Doves may always be looked for in winter when the stubble fields have lost their lure. Open sandy areas as already described are attractive at all seasons and thickets where the upland and salt meadow meet. A thicket of marsh elder just above Schellenger's Landing, in such a situation, seems to harbor from three to five Doves on every winter visit.

While a familiar species, inasmuch as it appears on almost every daily bird list that we make at Cape May, the Mourning Dove, as we know it here, is not a bird with which one feels on intimate terms. Its mournful call is not a pleasing sound and seems to speak of solitude and far off places and although it comes to our shade trees and orchards to nest it does so in a rather surreptitious manner. After long experience the only picture of the Dove that has made a lasting impression on my mind is a momentary glimpse of two dark-colored pointed-tailed birds hurtling across the sky, gone almost as soon as they are noted, leaving us in ignorance as to whence they came and whither they are going.

PASSENGER PIGEON

Ectopistes migratorius (Linnaeus)

While the Passenger Pigeon, now a thing of the past throughout its once extended range, formerly occurred in abundance in South Jersey it was regarded as rare by Thomas Beesley as early as 1856.

Our earliest record of the bird is by David Pieterson deVries who says in his journal that "an immense flight of wild pigeons in April obscured the sky" this was in 1633 apparently when he visited his colony near Cape Henlopen.

Then we have the account of the bird by Peter Kalm, who spent some time at the Swedish colony on the Delaware and probably his observations refer to the vicinity of Swedesboro. He speaks of the marked increase in the numbers of the birds in February and March and states that "in the spring of 1740, on the 11th, 12th, 15th, 16th, 17th, 18th, and 22d of March (old style), but more especially on the 11th, there came from the north [i. e. south] an incredible multitude of these pigeons to Pennsylvania and New Jersey. Their number, while in flight, extended three or four English miles in length, and more than one such mile in breadth, and they flew so closely together that the sky and the sun were obscured by them, the daylight becoming sensibly diminished by their shadow.

"In the beginning of February about the year 1729, according to the stories told by older men, an equally countless multitude of these pigeons as the one just mentioned, if not a still larger number, arrived in Pennsylvania and New Jersey. Even extremely aged men stated that on three, four, five, or several more occasions in their lifetime they had seen such overwhelming multitudes in these places." It was the opinion that eleven, twelve or more years elapsed between the unusual visitations. The birds roosted in such numbers that they broke the limbs of the trees. "The Swedes and others not only killed a great number with shot-guns, but they also slew a great quantity with sticks, without any particular difficulty." (Kongl. Vetenskaps-Akademiens Handlingar Vol. 20, 1759, translation by S. M. Grönberger, Auk, 1911.)

In Bergen County, John T. Waterhouse described the Pigeon flight in a letter to his parents in London dated March 23, 1838, as follows: "For the last fortnight the air has been almost black with Wild Pigeons emigrating from the Carolina swamps to more northerly latitudes. Within ten miles square during the first fortnight I suppose they have shot or netted at least twenty thousand. They fix up a kind of hut in a field made of limbs of trees and buckwheat stubble. They have one or two fliers which they throw out every time a flock passes; the fliers are of the Wild Pigeon breed usually wintered over or sometimes they take them directly from the flocks, tie their legs to a small piece of twine and throw them up. There is a floor cleared on the ground and buckwheat spread for a bait and they have a pigeon on the floor and also a stool pigeon which they move at pleasure by a rope fixed to it in the hut. There is then a net so fixed having a rope that fastens it to a stake in the ground at one end and soon as ever the Pigeons fly down the man in the hut pulls another rope fastened to the net and jerks it over them. They will sometimes net in this way at one haul three or four hundred. Whilst I am writing they are in the adjoining room picking seven Pigeons for our breakfast. They were shot this morning at one fire of the gun." (Condor, 1927, p. 273.)

Unfortunately there seem to be but very few later notes on the bird in New Jersey. Charles Westcott (Homo) mentions that he had shot Passenger Pigeons out on the beach but nothing further and beyond that we have only the records of last occurrences.

1878. Two recorded by F. M. Chapman, shot at Englewood, in September (Auk, 1889 p. 302).

1879. One shot by Dr. W. L. Abbott at Haddonfield, March 22 (Cassinia, 1907, p. 84).

1885. One seen by Thurber, shot at Morris Plains, September 16 (Morristown Democratic Banner, Nov. 1887).

1893. One shot by A. B. Frost from a flock of ten at Morristown, on October 7 (Cassinia, 1907, p. 84).

1896. One shot by C. Irving Wood, at Englewood, June 23, 1896 (Auk, 1896, p. 241).

GROUND DOVE

Columbigallina passerina passerina (Linnaeus)

This distinctly southern bird, which does not normally range north of South Carolina, has on one or two occasions been found in New Jersey doubtless carried by tropical storms which have been responsible for the occasional appearance of other stragglers from the south.

Turnbull (1869) states that John Krider, the noted Philadelphia taxidermist and gunsmith, shot one near Camden in 1858 and Krider himself in his "Field Notes" says: "I shot one specimen in the pine woods of New Jersey, which I suppose was a straggler. I had been out hunting quail in November, and on my return through a thick pine woods toward the ferry, this bird flew up from the ground."

Our only other record was a bird observed in the Physick garden by Otway Brown in October, 1935. He wrote me to ask whether there was more than one species of dove in this region stating that he had noticed a small dove feeding on the ground and thinking that it was a young bird and that the date was unusually late he approached it when it flew and he noticed that the tail was square and that it looked quite different from the ordinary Mourning Dove. This was the year of the severe August storm which may have been responsible for the presence of this waif from the south.

YELLOW-BILLED CUCKOO

Coccyzus americanus americanus (Linnaeus)

While I have not sufficient data to be sure, I am of the opinion that at Cape May the Yellow-bill is the more common cuckoo back in the farming country but that the Black-bill is equally abundant in the gardens and among the shade trees of the town itself. The habits of the two are essentially alike and their general appearance identical. They have a swooping flight from tree to tree in which the tail seems very long and somewhat curved upward by air pressure from below, especially when a bird descends from a considerable altitude as I have several times seen them do.

When moving about among the foliage of a tree a cuckoo makes clumsy jumps from limb to limb and often fluffs up its plumage or partly spreads its wings making quite a disturbance, but when it remains still it is exceedingly difficult to detect. I saw one fly into a tree at Cold Spring one day but failed to locate it and only after twenty minutes search with the binoculars was it discovered. Meanwhile I felt sure that it had flown away but my companion assured me that no bird had left the tree.

When perching the Yellow-billed Cuckoo has a curious way of craning its long neck to one side or the other often twisting the head at the same time, all of these actions being very slow and deliberate. It also has a way of raising the tail and letting it fall slowly to its former position and only occasionally do we see the tail feathers spread.

Like the Black-bill this cuckoo seeks caterpillars on the catalpa and buttonwood trees along the sidewalks and I have also flushed it from dense growths of grass and herbs in swampy spots where it must have been clinging to the stems close to the ground or resting on the ground itself, and one was seen to bathe in an open drain in a garden at Cold Spring on August 19, 1923.

The characteristic retarded call notes of the Yellow-bill—*kakakakakak-ka-ka-kow, kow, kow, kow, kow*—are often heard in the town and one bird was heard calling as late as August 18, 1931, while another, one evening, added his cries to the hubbub of the Martin roost on the Physick place.

Our earliest dates of arrival are May 8, 1912, and May 4, 1915, while cuckoos are present regularly until mid-September with exceptional records for October 5, 1930, and October 6, 1929. A nest found by Turner McMullen at Rio Grande on June 3, 1923, contained three eggs while another in Salem County on May 30, 1934, held two eggs and a young bird. A single egg was found on a lawn at Cold Spring on August 25, 1931.

Both of the cuckoos seem to have two broods in a season or else are very erratic in their nesting habits as young birds are found on exceptionally late dates. On September 5, 1914, Julian Potter found a nest full of well-feathered young birds near Camden which left the nest and were able to fly on the 9th. Another nest in the same vicinity on July 15, 1916, contained three eggs and one young just hatched, the others came out on the 16th and 17th. The skin of the young was black with a few hair-like down filaments.

BLACK-BILLED CUCKOO

Coccyzus erythropthalmus (Wilson)

When I began to study the birds of Cape May I was of the opinion that the Black-billed Cuckoo was a transient only. Strength was given to this view by the appearance of individuals about the middle of August along with the earliest south-bound migrants, which seemed to be lost and unfamiliar with their surroundings and obviously not local summer residents. These birds were found in the low dune thickets close to the ocean front from which one individual was seen to fly out over the surf and back again. They were also seen in the pine woods at the Point and in other places not usually frequented by cuckoos. Arrival dates for such birds were:

1921. August 15.	1923. August 30.	1929. August 5.
1922. August 9.	1927. August 25.	

Later I discovered that the Black-billed Cuckoo was quite as common as the Yellow-bill if not more so in the shade trees of the town throughout the summer, where it was a persistent hunter of caterpillars, while a young bird scarcely able to fly was found early in July; and Turner McMullen reported a nest with young at Rio Grande on June 8, 1924. These facts established the species as a summer resident although they did not affect the transient status of the dune thicket birds.

(622)

The caterpillar eating habits of the Black-bill are interesting. On July 19 and 20, 1934, the catalpa trees of the town were infested by a smooth yellow and black striped caterpillar (*Ceratomia catalpae*), about three inches in length, which threatened entirely to defoliate them. This cuckoo with some help from the other species, practically destroyed the pests on nearly all of the shade trees. Two birds were watched for some time at close range. One, evidently a bird of the year, had the white breast plumage rough and apparently in molt, while its eye ring was gray; the other, an adult, had the breast smooth and silky and the bare space around the eye red. They would seize the caterpillar just behind the head and switch it violently back and forth until it became perfectly limp and would wrap around the bill like a piece of string when the head of the bird was jerked sideways. The cuckoo would then raise its head and gulp the insect down sitting perfectly still for some minutes after. The adult bird sometimes jerked the caterpillar around for fifteen minutes before swallowing it but the young one would draw its victim through its bill from one end to the other and then back again, crushing it thoroughly between the mandibles and not wasting so much time switching it back and forth. Both birds would invariably wipe the bill on the twig upon which they stood after swallowing, but never beat the live caterpillar against it.

On September 7, 1927, the Black-billed Cuckoos were feeding upon a hairy caterpillar which infested the plane trees along the sidewalks. After jerking one of them about until dead they would deftly strip all of the hairs from the body before swallowing. Little tufts of whitish down were continually floating off from the feeding birds and helped to locate them. When a caterpillar was dropped the bird would fly to the ground and pick it up again, and Otway Brown tells me that he saw them repeatedly fly down to the ground and pick caterpillars from a fallen web-worm nest.

We have arrival dates for the Black-billed Cuckoo from May 10 to the middle of the month and while most of them have gone by the end of September there are several October records one as late as October 18, 1931.

Julian Potter has found two young birds apparently just out of the nest on September 14, 1919, which illustrate the late nesting of this species already alluded to in the case of the Yellow-bill.

The names given to our two cuckoos are unfortunate as they indicate the color of the bill as the best distinguishing mark whereas it is the color of the lower mandible only that is different while the rufous color of the inner portion of the wing feathers and conspicuous white tips to the tail feathers in the Yellow-bill, as contrasted with the uniform brown of the Black-bill's wings and its narrow grayish tail tips, are far better diagnostic characters.

BARN OWL

Tyto alba pratincola (Bonaparte)

The Barn Owl is a regular autumn transient in the pine woods at Cape May Point occurring most frequently during October; some remain through the winter and a few doubtless nest in the vicinity. Usually we see only a single bird but, as they are very secretive during the day, others may often be present that we do not discover. Our dates of occurrence run from October 11 to November 11 during the years 1923 to 1931. These were merely the result of scattered trips covering only single days, but in 1935, William Rusling, who was present throughout the autumn until early November watching the hawk flight, counted twenty-six Barn Owls, some of which of course may have been duplications and on some occasions he saw as many as five or six in the course of a day, which probably gives a fair idea of their maximum abundance. He saw none until September 16 and his last record was November 3. He tells me that they roosted quietly in the pine groves during the day and flew over the houses at night uttering their harsh whistling gasp *eeeeee séeek*.

One Barn Owl that I found in this vicinity on October 18, 1925, flushed repeatedly, flying from one pine tree to another as I tried to approach it. A number of these owls were found dead all about the Point during the hawk shooting season, victims of the "sportsmen" who were either ignorant of the wholly beneficial nature of these birds or bent on shooting anything that flew. Fortunately since the sanctuary has been established this indiscriminate slaughter has been largely abated. The old practice of shooting hawks and owls and nailing them on barns still persists through the farming country of South Jersey and Turner McMullen tells me of one barn near Pennsgrove which was ornamented by the carcasses of six Barn Owls, eleven Long-eared Owls, four Screech Owls, one Short-eared Owl, one Great Blue Heron, four Red-shouldered Hawks and one Cooper's Hawk!

In Cape May single Barn Owls have often been found in the safe retreat offered by the Physick place with its numerous shade trees and other vegetation, and with a thicket of red cedars flanking it in the rear. One was present here from August 1 to October 6, 1929, and probably earlier. Its pellets were found regularly and four of them examined on September 3 contained remains of thirteen meadow mice, and one long-tailed shrew, while another on September 5, contained three meadow mice and one shrew. Another Barn Owl seen on July 4, 1930, left the trees in great haste when fireworks were discharged in the street nearby, and as pellets were found later he was doubtless established there. Still another entered these grounds on July 19, 1923, at dusk causing great consternation among the Grackles that were roosting there and on March 19, 1932, one flew from a hollow tree that was being cut down and which may have harbored more than one of these interesting birds, at any rate, individuals had been seen there on December 2, 1923, and December 28, 1930.

One that I saw on August 20, 1921, could hardly have been a migrant and from his tameness I thought that he was probably a young bird hatched somewhere in the vicinity. He was perched in a clump of willow bushes on the Fill clasping the slender branches of a fork, one in each foot. His wings were drawn in closely against the tarsi and he stood very erect with head well up. As I walked around the bushes he turned his head in an effort to follow me but could not twist it for more than three quarters of the circle. He did not flush although I was within ten feet of him.

When the old Stockton Hotel, a great frame barn-like structure close to the ocean front, was closed, a pair of Barn Owls nested there and when it was later pulled down a brood of young was discovered in one of the lofts.

Farther north in the peninsula the Barn Owl is a more regular breeder and perhaps a permanent resident as more old hollow trees and old buildings are available, but like most other owls it seems more partial to the Delaware Bay side of New Jersey beyond the pine lands and Turner McMullen has found a number of nests there, in Salem County, as follows:

April 8, 1922, two eggs.	April 26, 1925, four eggs.
April 17, 1927, six eggs.	April 27, 1924, two eggs (four
April 23, 1932, three eggs.	by May 14).
	May 30, 1923, five eggs.

These six nests were mostly in cavities in old sour gum trees, one in a sweet gum in swampy woodland, and ranged from eighteen to thirty feet from the ground. The largest cavity was nearly three feet in diameter and in one of them both birds were present when the nest was examined and at once flew out and away.

The notebook of W. B. Crispin shows that he had found eight nests of this owl in Salem County from 1904 to 1913, April 10 and 11, four eggs; April 7, 9, 13 and 25 six eggs; April 13 and 27 seven eggs. In two instances the same pair had two sets of eggs in the season. A set of six was taken from a nest on April 9, 1910, and another set of five was found there on May 5. Similarly another pair losing a set of six eggs on April 7, 1913, had another set of six on May 3.

A nest found in a hollow tree at Mount Holly, Burlington County, held six young on May 30, 1932, but at Glenolden, Pa., across the Delaware, not many miles away, John Gillespie found young just able to fly on November 22, 1927, and another nest with young on December 7, 1929, two of which still remained in the nest on January 16, 1929.

He also found a family of five downy young Barn Owls in an old building on October 31, 1925, and shortly after John Emlen found three young at Riverton. Other families were found in Gloucester County on May 13, 1924 (seven young) and on June 21, 1925, while a female banded on the former date was found in the same place on November 15 indicating that this individual, at least, was resident. However another banded at the same time was shot at Somers Point fifty miles away, while one banded at Riverton on May 20, 1925, was shot at Wilmington, N. C., November 16, following, and another banded on November 18, 1925, was killed at Trappe, Md., on April 5, 1926, which are equally conclusive of an extensive migration (Gillespie)! Other nests have been found as far north in the state as Princeton, Plainfield and Summit and stragglers have occurred at Englewood, Chatham etc.

I have no spring Barn Owl records at the Cape that would indicate a northward migration but Julian Potter heard one flying overhead and uttering its characteristic screech on March 21, 1926, at Collingswood, Camden County, and at about the same time John Gillespie flushed one from near the ground at Glenolden, Pa., which evidently had no permanent retreat so that it appeared as if a migration might be under way.

The Barn Owl is a southern bird characteristic of the Carolinian Fauna and does not range much farther north than New Jersey. While usually of irregular occurrence in the northern parts of the state it has bred in Union and Essex Counties (Urner and Eaton). It is one of the species that give to the bird life of the Cape the southern character that renders it so interesting.

SCREECH OWL

Otus asio naevius (Gmelin)

Except for an occasional Barn Owl the little Screech Owl is the only owl to be found in Cape May in the summer and even it is by no means abundant or regular in its occurrence. While it is supposed to be resident almost all of my records are during the summer.

I have heard it hooting from shade trees along the sidewalks in the town in July 1918 and 1922 and again on August 28 of the latter year; also on July 20 and August 12, 1928, July 16, 1929, and August 29, 1931, while Otway Brown found one on June 17, 1932, in a bush on the Physick property on Washington Street with a pair of Cardinals and other nesting birds surrounding it and protesting in much excitement. A pair was reported to have wintered there in the season of 1907–1908.

At Cape May Point they do not seem to be any more plentiful and the only recent record that I have is of one in the red phase of plumage flushed by Warren Eaton on September 7, 1935, when watching the hawk flight.

Farther up in the peninsula in the farming district Screech Owls seem to be more numerous and Otway Brown has heard them hooting about his place at Cold Spring from May 7 to November 19 in different years.

In Salem County, west of the pine lands, they breed more plentifully and Turner McMullen tells me of nests that he examined on:

March 23, 1924, two eggs.	April 15, 1922, three eggs.
April 7, 1929, four eggs.	April 17, 1921, four eggs, and
April 8, 1923, three eggs.	another with three young.
April 10, 1927, one egg.	April 18, 1931, four eggs.
April 13, 1925, three eggs.	April 28, 1935, two eggs.
April 14, 1928, four eggs.	

These nests were in cavities in trees usually in apple trees but sometimes in red maples or oaks; often in old Flicker holes.

There seems to be no evidence of migration as practically no Screech Owls were noted by the hawk observers at the Point.

GREAT HORNED OWL

Bubo virginianus virginianus (Gmelin)

•

The Great Horned Owl is rarely seen today in the immediate vicinity of Cape May but wherever primeval woodland existed in the southern part of the county it was formerly to be found. Today, however, with the destruction of much of the old forest the Horned Owls have disappeared to a great extent although a few still remain. The big woods skirting the meadows opposite Rio Grande, from the Wildwood Road to Bennett, was one of their last strongholds and there Walker Hand and I came upon one of them on July 26, 1920. When flushed he flew to a nearby tree and stood there looking down at us and raising and lowering his ear-tufts. When he flew again a party of Crows gave chase and made a great disturbance. Before we saw him we heard him call once or twice which seemed unusual as it was only about midday but on September 12, 1920, Walker Hand heard another calling in the daytime in Rutherford's woods near Pond Creek. I have seen a number of mounted specimens that were shot in the Rio Grande woods and gunners shoot them on every occasion, deliberately hunting them out when their roosting places are known.

On August 17, 1891, I flushed a Horned Owl from the woods at Cape May Point north of the turnpike; the nearest approach of which I have record. John Gillespie flushed another from the grove where the Night Herons nest, back of Seven Mile Beach, and upon its taking wing all the Crows that were gathered in the neighborhood at once started in pursuit. This was the only one of which I have record on the island beaches except for a single one seen on Five Mile Beach by Philip Laurent when that island was covered with tall woodland.

Walker Hand tells me of one that used to come out in the early evening when it was still quite light and perch on a dead pine stub near Bennett Station where it was a very conspicuous object. In the same vicinity he once heard three Horned Owls calling to one another from trees and fence posts one of which swooped down close to a flock of Killdeers which had gathered there causing great consternation.

On November 21, 1934, and November 19, 1933, Otway Brown heard two calling at Cold Spring and on December 26, 1932, and December 24, 1933, Julian Potter heard one hooting near Cape May Court House.

Turner McMullen found a nest of the Great Horned Owl not far from Cape May on March 4, 1923; which was situated about twenty feet up in a walnut tree in an open field; it was an old Fish Hawk's nest which had been appropriated by the owls and contained the usual two eggs as in all other nests that he has seen. The same nest was found on the ground on March 2, 1924, having been blown down by a storm and held two broken eggs. In Salem County he has found a number of nests as listed below with several reported by William B. Crispin:

February 11, 1934.	March 23, 1924.
February 15, 1925.	April 8, 1922.
February 17, 1927.	April 13, 1935 (young birds).
February 17, 1935.	April 18, 1931 (young birds).
February 21, 1932 (two nests).	April 21, 1929 (young birds).
February 23, 1913 (Crispin).	April 23, 1933 (young birds).
February 27, 1910 (Crispin).	

The February 27 nests was robbed and held a second set on April 15. These nests, mostly old Fish Hawks' or Crows' nests, were in tall trees from twelve to eighty feet from the ground, usually forty to fifty, and situated either in swampy or dry woodland. The old bird was often found on the nest but generally flew away before a close approach was possible, sometimes coming back to within fifty yards of the nest but showing no inclination to attack the visitor. In one instance the old bird raised from the nest and spreading her wings plunged straight down to within ten feet of the ground when she checked herself and flew off. In another case both birds perched fifty feet away and remained there while the nest was being examined. Crispin has also recorded a pair of Great Horned Owls occupying an old nest of the Great Blue Heron and another pair took possession of a Bald Eagle's nest in the spring of 1928.

A nest described to me by Brooke Worth was found on April 7, 1928. It was in a grove of pitch pines on the mainland, opposite Avalon, between the meadows and the shore road. The pines were some fifty feet in height and bordered by dense growths of greenbrier and holly trees. The nest was in a tree taller than the others and had the appearance of being an old Fish Hawk's nest; it was thirty feet from the ground and was saddled on a large dead limb. There were two young and while they were being examined one of the parents remained in the vicinity but kept well hidden and only a few fluttering glimpses of it were obtained. It made no attack at any time. In the nest were the remains of a rabbit and the wing of a Black Duck.

On visiting the nest again on April 21, many pellets were found in the woods and one or both of the parents were hooting in low tones recalling the distant baying of hounds. About the base of the tree the ground was profusely littered with pellets and as the trunk was scaled one of the young flopped down to the ground, although the other remained crouched in the nest, its great yellow eyes glaring and its beak snapping with the rapidity of a machine gun. Every feather on the body was elevated to the fullest extent and the wings puffed out as the bird backed away. The one that had fallen went hopping and scrambling away with half-raised wings and when approached too closely would face about and go through the same performance as the one in the nest with the addition of a hissing sound made by the breath. The parents were some distance away in the wood, their position being indicated by the ravings of half a dozen taunting Crows.

Stirling Cole tells me of the capture of a young Horned Owl early in 1904, at a nest in a timber swamp a mile from his home near Seaville. It was kept in a large cage in the yard for over a month when it was liberated. One of the parents visited it every night bringing food—Doves, Flickers, Robins and sometimes mice. She came regularly at dusk hooting every few yards as she came and the young bird would answer with a most peculiar call. Her final flight was over an open field and she came along the ground apparently stopping every few yards. She spent the whole night with her offspring except when off for food. Finally she discovered the hen roost and thereafter took a young chicken every night although she had never molested them before. After a loss of about two dozen chickens it became necessary to shoot the old bird which proved a difficult matter as it was very hard to see her although she could be heard all night. The young owl resisted every attempt to tame him. A pet Crow seemed to be a natural enemy; it would stand in front of the owl's cage and seemed to taunt him while he hissed continually and endeavored to get at his tormentor.

The Great Horned Owl is resident throughout the state where woodland suitable for its nesting remains.

SNOWY OWL

Nyctea nyctea (Linnaeus)

The great Snowy Owl is apparently of very rare occurrence at the Cape as it seldom comes so far south on its periodic migrations from its home on the Arctic tundra. I have seen several mounted specimens which were shot in the vicinity of Cape May but never have seen a living individual.

Walker Hand tells me of one shot at Rio Grande, December 30, 1897, and two that arrived on the Fill during the late autumn of 1905 one of which was captured on November 26, the other remaining well into the winter.

Another was shot during the great flight of 1926–1927 and Prof. Gross in his report on this flight records twelve that reached the Jersey coast between Elizabeth and Atlantic City. (Auk, 1927, pp. 479–493.) During the flight of 1930–1931, while none was recorded from Cape May, one was seen on Long Beach by Earl Higgons and David Leas on December 22, and again on February 15, 1931, by T. Donald Carter. The former observers stated that the bird was sitting on the beach and allowed a close approach moving for a short distance rather reluctantly. After several such performances it flew out low over the water to an island in Barnegat Bay with very slow wingbeats and frequent soarings (Auk, 1931, p. 267). In the next flight, 1934–1935, one was found on Brigantine Island, on December 9, 1834, January 7, and February 12, 1935, by various observers. On one occasion it was feeding on the carcass of a Herring Gull which it carried with it when it was disturbed.

In earlier years W. E. D. Scott reports them "very abundant during the winter of 1876–77" on Long Beach (Bull. Nuttall Orn. Club, 1879, p. 223) while E. Carleton Thurber states that "four or five were shot near Morristown during the winter of 1886–87" (True Democratic Banner, Morristown, November 1887) and L. S. Foster writes of a number in northern New Jersey in November, 1889; one at Morriches on November 16 and another at Sea Isle City on November 20. (Forest and Stream, November 28, 1889.) On December 20, 1890, Philip Laurent saw one on Five Mile Beach (O. and O., April, 1892, p. 54). There are other northern records.

Albert Linton writes me of one seen near Riverton along the Delaware River above Camden on January 13, 1918, after a severe spell of weather when the river was completely frozen over. The bird was seen at a distance of 300 yards flying low over the marshy meadow. As he approached to get a better view the owl continued to fly over the frozen meadow remaining poised, now and then, in one spot on the lookout for prey. Suddenly he turned and came "directly toward me and his identity became evident. The owl face, the snowy plumage and great size marked him unmistakably as the Snowy Owl. I awaited his approach, not understanding the apparent lack of fear. About fifty feet away he poised and dropped down on the meadow with a distinct thud. As he dropped his back became visible and there were distinct bars of brown particularly on the wings. Apparently the pursuit had been successful, as the owl remained down for several seconds. He was completely hidden by the tall brown grass and I attempted to steal up on him. He was too alert, however, and in an instant was up and away, flying far out of sight up the river."

*HAWK OWL

Surnia ulula caparoch (Müller)

The Hawk Owl is a straggler from the Far North and is extremely rare in New Jersey. Dr. C. C. Abbott mentions two specimens shot in Mercer County in 1858 and Middlesex County in 1861, but with no authority or details. Eaton (Linn. Soc. Proc., 47) mentions one recorded from Essex Co. in 1904. These are the only records until December 19, 1926, when S. C. Brooks saw one near New Brunswick. It was "seen several times in the forenoon flying over open fields between scattered groups of trees, and later was observed perched in a small tree alongside a cattail-filled slough over which it made several short flights. There was ample opportunity for observation with 8 × glasses at about 150 feet; the long indistinctly barred tail, striking white spotting of the back, general light color of the top of the head, and the plain gray facial disc were noted" (Auk, 1927, p. 251). We have no record from Cape May County.

BARRED OWL

Strix varia varia Barton

The Barred Owl doubtless breeds in dense woodland of the middle or upper parts of Cape May County especially along the Bay shore but in the immediate vicinity of the Cape it is of irregular occurrence.

Walker Hand tells me that in the past the woods on either side of Pond Creek Meadows were always a favorite resort for them and that a number have been shot there. All of our recent records are from the same region.

On May 22, 1921, Richard Miller saw one above the Point pursued by a flock of some sixty Crows and in the pine woods on the edge of the New England Creek Meadows I saw one, also pursued by Crows, during a notable autumn flight of Flickers, on September 27, 1926. On December 28, 1924,

Julian Potter saw one in the woods a quarter of a mile below Higbee's Beach and at the same spot, on January 25, following, Henry Gaede saw probably the same bird. William Yoder saw one on October 20, 1929, and Dr. W. L. Abbott secured a specimen on January 29, 1878.

These records indicate that the species is a resident and we have no evidence of autumn migration. Farther to the west it seems to be a regular breeder and is doubtless resident. Turner McMullen has examined two nests in Salem and Cumberland Counties. One found April 11, 1926, in the former county, contained two eggs and was in a cavity in a gum tree twenty feet from the ground, two and a half feet deep and a foot in diameter; it contained leaves evidently the remains of an old squirrel's nest; one bird was on the nest and the other was seen in the vicinity. The second nest was found in Cumberland County March 17, 1933, and also contained two eggs it was in a cavity thirty-five feet from the ground in an old white oak in a marshy tract on the Maurice River and one bird was flushed from the nest. George Stuart found another nest near Salem on March 23, 1918, in a cavity in a swamp maple, twelve feet up, which contained two eggs.

William B. Crispin has left records of a number of nests in Salem County, February 28, March 5, 9, 12, 13, 17 with three eggs and March 27 with two eggs. Pairs that had been robbed had second sets as follows, March 12, 1911, three eggs and April 9, two eggs; March 17, 1912, three eggs and April 12, three eggs. Charles S. Schick has reported a pair of Barred Owls on Seven Mile Beach on May 10, 1890, which were probably nesting although neither nest nor young were found.

In woodland farther north along the coast the Barred Owl is of regular occurrence and comes close to the farm houses in winter in search of food. A male and female were caught in a steel trap on January 13 and 14, 1918, on the farm of J. W. Holman at West Creek.

The Barred Owl breeds in suitable woodland throughout the state and an interesting account of a nesting at Schraalenburgh in the extreme northeastern corner of Bergen County by William C. Clarke appears in "Bird-Lore" for 1908, pp. 99–102. Waldron Miller has also described a nest at Plainfield in which the female bird was imprisoned by a sheet of ice that formed over the aperture of the nest hole (Bird-Lore, 1907, p. 173).

*GREAT GRAY OWL

Scotiaptex nebulosa nebulosa (Forster)

While we have no record of this northern owl from Cape May there are two old records from the state. One shot near Mendham, Morris County, in 1887 (Thurber, True Democratic Banner, Morristown, November, 1887), and one killed in Sussex County, in December, 1859 (Abbott, Birds of New Jersey, 1868). Details are lacking in both cases.

LONG-EARED OWL

Asio wilsonianus (Lesson)

Immediately about Cape May the Long-eared Owl, like most of the other owls, is rather rare and most frequently seen in autumn and winter. Our records are few. On October 1, 1923, I came unexpectedly upon one of them sitting in the heart of a low thicket near Cape May Point through which passed a wood road. The bird was busily engaged in preening its feathers reaching its head far over to the base of the tail and drawing each tail feather carefully between its mandibles. Presently it saw me looking at it through the binoculars at a distance of about twenty feet and it immediately fluffed up its feathers, spread its wings out on each side and arched them over so as to make a fan-like circle with its head in the middle. It glared at me and slowly raised first one foot and then the other off of the perch like a cat working its toes when it is being stroked. Upon lowering my glasses the bird at once drew in its wings and flew to the other side of the thicket.

Julian Potter saw two Long-eared Owls in the nearby pine woods on November 11 of the same year and again on December 2, while Otway Brown saw a colony of several in the same vicinity on November 19, 1933. This winter roosting in considerable numbers has been observed in several localities farther north in the state; at Yardville, Miss Rachel Allinson had a colony in a Norway spruce tree in her yard from 1891 to 1906, every winter and the ground below their roost was strewn with pellets containing mouse hair and bones. Another colony was reported by W. A. Babson at Princeton, which persisted for a number of years (Birds of Princeton, p. 50, 1901). Philip Laurent found a few in winter in the cedars on Five Mile Beach when this island was wooded. Julian Potter found three Long-ears associated with four Barn Owls and a Saw-whet in a clump of pines in Gloucester County in February 1921. Only one has been recorded in the Delaware Valley Club's Christmas census at Cape May, i. e., on December 28, 1930.

Farther west, in Salem County near to the Delaware River, these owls seem to be much more common and doubtless they do not find the environment that they desire on the ocean side of the peninsula. Turner McMullen has examined the following nests in Salem County, April 10, 1921, and March 29, 1925, four eggs; April 17, 1921, April 10, 1927, and two on March 30, 1929, five eggs. One pair had made use of a squirrels' nest and in other cases doubtless old Crows' nests served as the basis for their own; they were from seven to twenty-two feet from the ground in oaks and pitch pines with one in a red cedar. W. B. Crispin has recorded two nests from the same region on March 21, 1898, and April, 1897, each with five eggs.

The Long-eared Owl is resident in suitable localities throughout the state but more common in winter.

SHORT-EARED OWL

Asio flammeus flammeus (Pontoppidan)

The Short-eared Owl is a winter visitant in the Cape May region and our records run from September 5 to January 25. Many of the birds fall victims to thoughtless gunners during the shooting season and this doubtless has something to do with the lack of records for late winter and spring. The Short-ear is distinctly an open ground bird being found exclusively on the meadows, the Fill and similar flat exposed situations.

I came upon one on October 1, 1923, on the marsh below South Cape May, which alighted successively on a pile of salt hay, a pile of dry mud left from ditching, and finally in a patch of dry grass. It stood with its body bent forward and the tail nearly touching the ground, giving it a crouching appearance. When it flew the wing strokes were long and somewhat irregular but so powerful that they gave the impression of bouncing the body up and down between them. The round bullet-like head projected immediately in front of the wings with no neck apparent.

Julian Potter saw one at about the same spot on November 14, 1920, which was pursued by a flock of Crows. When they abandoned the chase the owl alighted on a bare space from which the black-grass had been cut for hay and as Potter writes me: "by keeping the sun at my back and directly in the face of the owl I was able to approach to within three feet of him. He was really facing the other way but with his head twisted so that he gazed at me directly over his back. He was crouching somewhat like a Whip-poor-will but when I circled around him he took wing and alighted on the sand dunes where he stood erect for an instant and then relapsed into the crouching position. When flying he took heavy measured strokes, his wings almost meeting under the body."

Several times we have found Short-eared Owls resting on bare spots on the Fill or among short yellow grass and on December 28, 1921, Earl Poole saw one busily hunting there, flying close to the ground at about 1:00 p. m. on a perfectly clear day.

The Christmas census of the Delaware Valley Club showed these birds present as follows:

December 22, 1929, five.   December 23, 1933, two.
December 28, 1930, one.   December 22, 1935, one.
December 27, 1931, four.   December 27, 1936, one.

While the records of Short-eared Owls about Cape May are usually of single individuals, they have the same habit of roosting in winter communities that is characteristic of the Long-ear and Samuel Rhoads found such a roost on Barrel Island in Barnegat Bay in late February, 1894, the ground being covered with ejected pellets, scattered bones and a number of mouse skins which led him to think that the birds had, to some extent at least, skinned their prey before eating it (Proc. D. V. O. C., No. II, p. 12). W. E.

D. Scott describes a roost of one hundred and fifty to two hundred Short-eared Owls near Princeton, in the winter of 1878–79, in a field of some forty acres covered with a heavy growth of long dead grass. They arrived about November 1 and naturally many were shot as they flew about in the morning and evening hunting for mice. He later learned of two similar roosting places in the same vicinity (Bull. Nuttall Orn. Club, 1879, p. 83).

On May 9, 1905, William B. Crispin found a nest with six eggs on the marshes of Salem County at the head of Delaware Bay and several others were found later in the same vicinity; on May 6, 1923, Richard Miller and others saw a well-fledged young here on June 4, 1922, and found four young birds unable to fly but which had left the nest and were crouching at the bases of Juncus tussocks on May 6, 1923 (Oölogist, 1924, p. 109). Turner McMullen, who was with this party, later found two nests, one on May 21, 1932, in Cumberland County, containing seven eggs and another near Fortesque, on May 2, 1936, with five eggs. The former was located in short grass under a tuft of taller vegetation, on the edge of the meadows; the other in a tuft of dry yellow grass four feet in height and at least ten yards from the marsh.

Farther north along the coast W. E. D. Scott found a nest of Short-eared Owls containing seven eggs on the meadows adjoining Long Beach on June 28, 1878, and another was found in the same vicinity by Thomas D. Drown, while Richard Harlow found one on an island in Barnegat Bay, containing five half-fledged young, on June 17, 1915.

On the marshes near Elizabeth Charles Urner has found a number of nests as follows:

May 14, 1921, eight young of various sizes. (Auk, 1921, p. 602.)
May 22, 1922, six eggs. (Auk, 1923, p. 30.)
July 4, 1922, one egg and two young.
April 28, 1923, six eggs. (Auk, 1925, p. 32.)
May 5, 1923, six eggs.
June 8, 1923, young had left.
June 8, 1923, three young and seven eggs.
June 8, 1923, five young.

SAW-WHET OWL

Cryptoglaux acadica acadica (Gmelin)

This little owl is a rare winter visitant at the Cape and we have but three records of its occurrence here. Walker Hand found one dead in the town on December 1, 1904, and Otway Brown once found one at Cold Spring which had killed itself by flying into a wire netting, while Reynold Spaeth told me of finding one at Cape May Point about 1895.

The Saw-whet is doubtless more plentiful than these records would indicate as many have been reported in the central and northern parts of the state. While not roosting in close communities, like some of the other owls, the little Saw-whets seem to concentrate in limited areas, in some seasons at least, and W. E. D. Scott states that they were unusually plentiful in cedar groves near Princeton in the winter of 1878-79. He secured ten on December 10, 1878, and seven more the next day and many more through the winter. He states that "they roost close to the trunk and can frequently be taken alive in the hand. They seem to affect scattered groves where the trees do not grow too thickly. Most of the birds taken are females." (Bull. Nuttall Orn. Club, 1879, p. 85.)

It occurs as a casual winter resident throughout the state.

CHUCK-WILL'S-WIDOW

Antrostomus carolinensis (Gmelin)

Some years ago Otway Brown told me of a Whip-poor-will heard for several nights near his home at Cold Spring which had a call quite different from any that he had heard and his description tallied exactly with the call of the Chuck-will's-widow. Shortly afterward I learned from Dr. Harry Fox, who was quite familiar with both birds, that he had heard a Chuck-will's-widow calling in swampy woods at Cape May Point when he was out at night collecting orthoptera.

In June 22, 1930, Frank Dickinson heard a strange Whip-poor-will near his home at Erma and Walker Hand who was present identified it at once as a Chuck-will's-widow. I visited the spot on July 15 and was fortunate enough to hear it calling *chuck!—woo—woo; chuck!—woo—woo* repeated a number of times. The bird came from a swampy thicket into an oak clearing back of the Dickinson farm and although it was too dark to see it we could tell that it was moving about by the different directions from which the calls came to us. This bird had been heard many times during June and the early part of July but not after the middle of that month. It began to call just at dusk.

Early in this same summer a bird struck the headlight of an automobile and was identified by a local taxidermist as a Chuck-will's-widow and from the description that I received he was undoubtedly correct but I was unable to see the bird myself. Charles Page who had heard Chuck-will's-widows frequently in the South heard one repeatedly near his cabin in the woods at Cape May Point during June 1932 and 1934, but was unable to see the bird. The call contrasted definitely with the ever present Whip-poor-will chorus.

WHIP-POOR-WILL

Antrostomus vociferus vociferus (Wilson)

In the immediate vicinity of Cape May Whip-poor-wills are found only in the woods and thickets of the Point but farther up in the peninsula they are of regular occurrence in all swampy wooded localities, coming out into more open spots to call at dusk. They are particularly abundant at Higbee's Beach and other densely wooded localities along the Bay shore and take advantage of the cleared areas about houses, whether occupied or vacant, for calling grounds.

About Cape May Point on almost any evening from late April to early August their characteristic calls may be heard, sometimes from a single bird but more often from several and on the night of May 20, 1928, William Shryock was able to distinguish no less than eight performers and there were doubtless more.

Charles Page who for several seasons occupied a little cabin directly in the woods had an abundance of Whip-poor-wills immediately about his door and even on his roof. Their din seemed to reach its height on June 1 but they continued to call until August 22, 1932, and August 26, 1934, although the calls were not so long-continued and the repetitions less in number. His latest record of the call was September 13, 1932, but Conrad Roland heard a few calls as late as September 16, 1933, and September 20, 1934.

While heard more frequently than they are seen, about the Point, Whip-poor-wills can easily be flushed in the daytime from the huckleberry thickets if we search for them and are willing to brave the "chiggers" (red-bugs) which also abound in such places, and while gathering huckleberries I have frequently disturbed them. A bird will flush sometimes from almost under foot. There is a short, silent, bat-like flight of a few moments and it drops to the ground a few yards in advance shielded again by the bushes. Sometimes for a moment or two it will alight on a low branch of a tree in full

(637)

sight coming to rest cross-wise on the limb but wheeling at once to a position parallel to it. It will blink once of twice with half opened eyes and once again there is the silent flight as it vanishes in the thicket. One that I flushed in the pine woods on October 10, 1927, alighted on the sloping trunk of a large bayberry bush and perched there for several minutes with its great eyes wide open. Sometimes I have flushed three Whip-poor-wills in quick succession but have never tried to see how many could be roused from their midday siesta; doubtless many of them do not take wing unless almost stepped upon.

On May 9, 1924, I flushed two birds which were sitting close together in a dense thicket east of the pine woods which seemed loath to move far from the spot and I presumed that they had eggs in the immediate vicinity, and a pair was roused from the same spot on July 18 and 20, 1925. Curiously enough close to the May birds there grew a cluster of the lovely pink orchids (*Cypripedium acaule*) often called Whip-poor-will's shoes, so that there may have been some close association in folklore or legend of bird and flower which led to the name. Certainly their habitats are the same.

While I have not actually found a "nest" here Julian Potter did find a female Whip-poor-will and one egg on a bed of dead leaves in the same vicinity on June 9, 1929, with the other egg lying broken close by; Richard Miller found a pair of downy young about the size of Chipping Sparrows near Rio Grande on July 3, 1923, which were resting on a carpet of dead leaves on the woodland floor; and I found another pair of young at the Point, barely able to fly on July 20, 1929. Turner McMullen has found two sets of eggs of the Whip-poor-will, one at Whitesboro on May 20, 1922, and the other at Rio Grande on May 31, 1924, both placed on dead leaves in second growth woodland.

Walker Hand's dates of arrival with a few others run as follows:

1903.	April 26.	1913.	April 30.
1904.	May 1.	1914.	April 29.
1905.	April 26.	1915.	April 17.
1906.	April 21.	1916.	April 30.
1908.	April 13.	1920.	April 11.
1909.	April 23.	1927.	April 24 (Stone).
1911.	April 27.	1930.	May 1 (O. H. Brown).
1912.	April 19.	1933.	April 25 (O. H. Brown).

In the autumn various observers have seen them as late as:

1925.	September 29.	1932.	October 14.
1927.	October 10.	1935.	October 5.
1930.	September 28.		

Whip-poor-wills are summer residents throughout New Jersey in wilder sections of the country and are especially abundant in the Pine Barrens.

NIGHTHAWK

Chordeiles minor minor (Forster)

While it nests, casually at least, farther north in the peninsula, the Nighthawk is only a somewhat irregular late summer and autumn migrant about Cape May. Single birds have been seen on July 9, 1922; July 25, 1926; July 30, 1929; and at Fishing Creek, to the north, one was seen on July 5, 1923, and July 9, 1924. All of these may have nested in the vicinity but it is usually mid-August before Nighthawks are seen in any numbers or with any regularity at the Cape.

In the spring we have very few records of Nighthawks; one flying over the sand dunes at Cape May Point on May 24, 1892, and one over the farm land a few miles north of the town on May 20, 1927, and these birds could easily have been northbound migrants.

When Nighthawks occur in late August and September, we usually see them about dusk flying high over the town or at Cape May Point. Their flight is erratic, marked by a constant shifting this way and that. When progressing definitely forward there is a long sail followed by a series of rapid wingbeats, as the bird hovers or holds itself poised for a moment, and then one of the sudden dives downward or to the right or left which are so characteristic of the species. The wings of the bird impress one as being set very far forward, the neck being so short that the head is scarcely in evidence; the wings are also quite narrow, about the same width as the body and tail, so that the flying bird presents somewhat the outline of a tripod or a letter Y. But as the wings are bent or curved at the wrist two of the branches of the tripod are not quite straight. Sometimes the flying birds pass so high that they would not be noticed were it not for their erratic flight. On the evening of August 31, 1921, I was watching some gulls soaring high above the town and among them detected a Fish Hawk and later about a dozen Nighthawks. Except for their manner of flight it was really difficult to distinguish the several species as it was almost impossible to gage their relative altitudes. I have noticed this on many other occasions when swallows have been momentarily taken for hawks and even the numerous migrant dragon-flies, which occur so abundantly in late summer, have been projected until they were taken for birds soaring at a high altitude.

While some of these migrant Nighthawk flocks pass on immediately, others may be seen for several nights in succession in the same place, and from August 14 to 21, 1925, a number were seen every evening hawking about over the waters of the Bay. While usually numbering about a dozen some of the flocks are much larger and Conrad Roland saw at least 200 flying into a southeast wind at the Point on the evening of September 6, 1934.

Our dates of arrival of August Nighthawks run as follows:

1921.	August 16.	1926.	August 27.
1922.	August 20.	1930.	August 28.
1923.	August 20.	1931.	August 30.
1924.	August 27.	1932.	August 30.
1925.	August 14.		

The dates of last occurrence are:

1921.	September 18.	1929.	September 29.
1924.	October 2.	1932.	September 18.
1925.	September 28.	1935.	October 10.

William Rusling, Audubon Society representative during the autumn of 1935, made a careful record of Nighthawks at the Point. He saw the first one on August 20, five on the 23d, and one on the 29th. Then came a great flight of 117 on September 7 and then from twelve to fifteen daily until the 25th, and after that one or two, now and then, until October 10. It would appear that if one happened to be away on the day of the maximum flight he would probably be unaware that a notable migration had taken place and my inability to be in the field continuously after early September has no doubt been responsible for my scanty records of the movements of this interesting bird. Large flights, too, seem to be more frequent somewhat farther north and Julian Potter has recorded seventy-two passing Collingswood on September 8, 1929, ninety-two on the 9th and sixty-four on the 24th.

It is quite possible that the Nighthawk migration passes along the Bay shore rather than on the seacoast and Philip Laurent's experience on Five Mile Beach, where he saw but two in his many years' observation, would support this view.

Twice, on September 8, 1928, and September 7, 1931, I have seen one or two Nighthawks flying wildly over the pine woods at Cape May Point, diving precipitately into the trees and than sailing up again. They appeared as if intending to alight and then beat their way out again flapping wildly; the triangle appearance was well illustrated on these occasions. In the face of a strong wind I have seen a single bird hold almost the same position in the air for quite some time.

On August 27, 1928, while watching a number of Black Terns in their erratic flight over the deep pond at the Lighthouse I was struck with the resemblance of their actions to those of the Nighthawk when suddenly I realized that one of the nine birds before me actually was a Nighthawk; there was the same soft flapping, the same flutter and the same swoop to the surface of the water and it was by no means easy always to distinguish one species from the other. It was a very dull day, late in the afternoon, and the white patch usually so conspicuous in the wing of a Nighthawk was not easy to see.

I have several times detected Nighthawks resting during the daytime but so closely do they resemble the surroundings of their perches that there must be many more roosting about us than we realize. One was seen on the short dead branch of a pine tree, some thirty feet from the ground, which raised its head slightly displaying its white throat; another perched lengthwise on the limb, as is their custom, with its eyes nearly closed. One sat on a dead cedar tree on the Bay shore and still another on an open sandy spot at Cape May Point. On September 17, 1923, a Nighthawk sat for some time crosswise on the insulator of a telegraph wire in the heart of the town and at exactly the same spot and on the same date in 1932, one rested on the top of the pole!

In the northern counties it is usually a common transient but local or occasional as a breeder. Urner finds it a fairly regular but not common migrant in Union Co.

While I have no record of a "nest" in Cape May County Turner Mc-Mullen found two eggs on bare sandy ground at Malaga, Gloucester County, on June 16, 1928.

CHIMNEY SWIFT
Chaetura pelagica (Linnaeus)

The Chimney Swift is a common town bird at Cape May and during the late spring and early summer it is largely confined to the immediate vicinity of the houses, passing back and forth overhead with its narrow wings in rapid vibration or set in a bow-like curve as it sails in long arcs across the sky. When both sexes are on the wing a pair will pass in close pursuit with wings elevated in a V-like position, one bird almost touching the other, and while I have regarded such flights as mating flights I have never seen actual copulation in the air, as has been observed by others. Possibly some of these pursuit flights are of old and young. While I could never take seriously the claim that Swifts beat their wings alternately it is, nevertheless, no easy matter to prove by ordinary observation that they do not. The eye is not able to follow the detailed movements of the wings and the flickering flight is very confusing. By careful observation with binocular glasses, however, I have satisfied myself that the wing action is as in other birds.

Often from flying Swifts we catch the long drawn trilling chatter which is the characteristic note of the bird and from the chimney nest, when the young are being fed, there is an abundance of similar chattering and hissing sounds.

Swifts are always social birds and several pairs occupy the same chimney in amicable association and about every old farm house, in the open country north of the town, there will be a flock of Swifts coursing through the evening sky as well as the flock over the town itself and as the season advances several flocks seem to combine during their foraging flights.

So completely is the Swift a bird of the air that it has lost the power of perching, which is still retained by the Hummingbird and Swallow, nor does it rest on the ground or on large limbs as do the Nighthawk and Whip-poor-will—all birds of similar aërial ability. The Swift rests only in the interior of the chimneys in which it nests, clinging vertically to the soot-covered bricks. It has long since abandoned the hollow trees on the inside walls of which it originally glued its basket-like nest and although it occasionally places its nest in some deserted building or shed I have never come upon such a site at Cape May. Samuel Rhoads, however, did find a nest in an old barn at Haddonfield.

Exercise and the search for food, both for themselves and their young, are carried on in the air and even the gathering of small twigs with which to construct their nests. This latter activity may easily be witnessed if we can find a tree with dead and brittle branches at its top for here we are almost certain to see Swifts during the nesting season flying through the bare branches and endeavoring to break off small twigs. A line of old wild cherry trees growing along the Harbor railroad is a favorite source of such material and I have watched the birds time and again come sailing in and, fluttering opposite a likely branch, endeavor to break off a small piece by grasping it in the bill; occasionally the effort will be successful but more often the attempt fails and more often still, for some reason or other, the bird will change its mind and sail away over the marsh to return in a few minutes and repeat the performance. I have seen birds engaged in twig hunting on various days from May 28 to July 3. I am well aware that others have stated positively that Swifts grasp the twigs for their nests in their feet but it has always seemed to me that they use the mouth and how they could possibly alight on the wall of the chimney with their feet grasping a twig I cannot imagine.

It is in the early evening that the Chimney Swifts appear to be most active and vociferous and often at this time of day they will descend close to the ground and skim over the golf links or other level ground passing at arm's length in their precipitous flight. They skim in the same way back and forth over fields of new mown hay and I have seen them scouring the grass-covered sand dunes which used to lie back of the bathing beach. They are at such times evidently feeding on minute insects but so rapid is their flight that it is impossible to see the actual capture of the prey. I have several times tried to follow them with the glass and see the action but without success.

Swifts may be awing at any hour of the day, early morning being almost as congenial to them as evening, and I have seen as many as one hundred high in the air as late as ten oclock in the morning on August 21, though as a rule they are less active during the middle of the day.   They continue to feed until it is quite dark and I have seen them skimming close along the pavement of Washington Street until nearly 8:00 p. m. in early July, while on August 13, 1936, on Price's Pond I watched twenty-five Swifts feeding over the water and found that they left all together at 7:15 p. m.   The variation in their numbers, on different days and at different hours, is very noticeable. There are times when we may scan the sky in every direction without detecting a Swift and some entire days when we shall be unable to add the species to our list.   Doubtless this is to ·be explained by their local feeding and our failure to happen upon the spot where they are for the moment congregated.   After the young are on the wing they scatter and stray much farther afield, returning to the shelter of the chimneys where they nested, only on the approach of darkness.

In late July Chimney Swifts often frequent the open pine woods at Cape May Point coursing back and forth over the tree tops as well as threading their way in and out among the branches and coursing up and down a favorite overgrown wood road which passes close by.   In all such spots they evidently find insect food of various kinds.   Sometimes they travel only a foot from the ground along the narrow wood road or again they skim past shoulder high and barely miss my face as I stand perfectly still in the middle of the path.   In August, too, they will congregate over Lake Lily or other bodies of fresh water skimming back and forth just above the surface. Sometimes they apparently dip their bill for a drink while at others they strike the water with the body and not with the bill, probably taking a quick bath, which, like all of their activities, must perforce be done in flight.

Swifts are experts on the wing and their aërial evolutions are always interesting especially in the face of a strong wind.   On August 4, 1920, with a northeaster blowing, they would sail up against the wind until almost stationary, beating their wings for a moment or two to maintain their balance and then veering off to one side or the other, they came like lightning, spreading and closing their short tails as they guided themselves, and so held their equilibrium in the gale.   On July 22, 1917, six Swifts were seen poised absolutely stationary in the air over Congress Hall Hotel, on rigid wings, and evidently supported by the upward air currents deflected from the walls of the building.   Martins and Turkey Vultures do the same thing, an action which so far as I can see is instigated purely by the joy of flight.

Cloudy weather does not deter the Swifts from taking to the air and while I have not seen them actually flying in the rain they are awing almost immediately after a shower stops.

After August 1 of each year Swifts are noticeably more abundant and about the 20th of the month reach their maximum.   This is due, in part, to the appearance of the young birds on the wing but it would seem that there must also be a decided migratory movement with an influx of birds from farther north, or else a concentration of all the nearby colonies. My notes show such an increase on August 19, 1921; August 13, 1923; August 19, 1924; August 21, 1925, etc. On all of these dates, and on others later in the month, upwards of a hundred Swifts could be seen flying high over the town in an intricate maze. Then about August 30 they may disappear entirely to be replaced later by flocks apparently from the north. The same thing may be witnessed at Cape May Point, where on August 25, 1927, and August 30, 1932, I found the air literally full of Swifts while none were seen afterward up to the time of my departure about September 10. I presume that Swifts migrate in these comparatively large flocks, in daytime, and that these gatherings may actually have been on migration. On both occasions there was a strong northwest wind which always causes a congregation of migrant birds at the Point. Scattered Swifts seen following the beach line with south-bound Barn Swallows earlier in August may or may not have been migrating.

About the time of these late August gatherings the Swifts begin to roost together in large numbers making use of certain favorite chimneys in the town. On August 28, 1922, they selected an old house on Lafayette St., where they concentrated about 6:45 p. m. flying at first in wide circles and gradually contracting as darkness crept on until the south chimney became obviously the center of attraction. Now and then the flock would concentrate and pause for a moment just over the chimney on fluttering wings. Perhaps one would drop in, raising its wings high up over its back as it disappeared, or perhaps half a dozen would go down in quick succession. Then the whole mass would whirl away once more on their circuit only to congregate again in a few minutes. Occasionally one of those that had retired would emerge and once more join the flock influenced no doubt by flock attraction. The return of the circling flock to the chimney was more frequent as night approached and finally with rapidly narrowing orbits it formed a great funnel-shaped mass and poured steadily down into the chimney a few stragglers being crowded aside and forced to make another round before descending. Although I had not seen the beginning of the roosting flight, and may have missed some of the birds that went down first, I was able to count one hundred and sixty Swifts entering the chimney while I watched. The owner of the house told me that, while he had not noticed the Swifts previously that summer, they had used the chimney two years before. They continued to roost there in diminished numbers until September 2 and on the following year a flock of thirty went down on September 6.

In 1925 the Swifts made use of a chimney on Washington St., where on the evening of August 22 I saw upwards of 300 birds enter, the main body going down between 6:55 and 7:10 p. m. and they were still using this chimney up to August 31. The first time they were observed was an evening following a cold windy night and morning and it seemed that the aggregation may have consisted largely of migrants urged southward by the change in the temperature. How many chimneys may be in use in a single season or to what extent the birds change their roost from night to night I am unable to determine, but certainly the number of individuals using a given roosting chimney varies considerably. I have no reason, however, to infer that more than one or two chimneys are used for this mass roosting on a single night within the limits of the town.

My latest dates for Swifts at Cape May after the departure of the large flocks about August 25 to 30 are:

1914.	September 13.	1925.	September 20.
1916.	September 4.	1928.	September 8.
1921.	September 5.	1929.	September 2.
1923.	September 7.	1930.	September 4.

The earliest of these dates probably do not represent actual last appearances as I usually left Cape May early in September and subsequent data were not based on continuous observation. Doubtless September 18 or 20 is the usual time of final departure. Exceptional records are single birds seen on October 21, 1928, and October 25, 1929. Julian Potter made daily observations about Cape May and the Point from September 20 to October 2, 1925, and failed to find a single Swift but William Rusling saw thirty-one at the Point on September 10, 1935, and one each on September 20 and October 10.

Walker Hand's records of spring arrival for the Swift for a number of years supplemented by other observations after 1920 are:

1902.	April 27.	1911.	April 27.	1922.	April 23.
1903.	April 19.	1912.	April 25.	1923.	April 22.
1904.	April 20.	1914.	April 23.	1926.	April 25.
1905.	April 23.	1915.	April 20.	1927.	April 30.
1907.	April 30.	1916.	April 21.	1931.	April 25.
1908.	April 26.	1917.	April 19.	1932.	April 23.
1909.	April 29.	1918.	April 29.	1934.	April 23.
1910.	May 2.	1919.	April 27.		

Turner McMullen examined nests in a chimney at Ocean View on June 17, 1922, which contained one and two eggs respectively while nests at Lenola, Burlington County, June 20, 1915, and June 22, 1918, contained three and four eggs. Philip Laurant found them nesting on Five Mile Beach in the eighties.

Chimney Swifts are common summer residents throughout the state.

RUBY-THROATED HUMMINGBIRD

Archilochus colubris (Linnaeus)

While we not infrequently catch a momentary glimpse of a Hummingbird passing like a bumble bee in straightaway flight, most of our observations are of individuals feeding among the flowers in the old gardens of the town. On rapidly vibrating wings they pass from one blossom to another probing with needle-like bill for nectar or possibly for minute insects which may be concealed in the corolla. They force the head within the larger tube-like flowers or into the flaring cups of the nasturtiums, and back out again, with partly spread tail swinging back and forth beneath the body.

The earliest spring arrivals frequent the scarlet flowers of the *Pyrus japonica* while larkspur, Weigelia and Japanese honeysuckle are favorites as well as many other blooms, both native and cultivated, as the season advances. Above all, however, they delight in the clustered red tubes of the trumpet creeper (*Tecoma radicans*) which abounds in thickets and along fence rows at Cape May Point and elsewhere in the lower part of the peninsula. While flowers of all colors seem to attract the Hummingbirds red is apparently their favorite hue although I doubt very much if color has much to do with their search for food. They doubtless find what they are looking for without such flaring advertisements.

Where flowers are massed Hummingbirds will occur in numbers especially after the nesting season is over. I have seen seven feeding at once on a small dense growth of Bouncing Bet (*Saponaria officinalis*) in mid-July and Otway Brown found at least twelve in a bed of gladiolus on the last day of that month. About clusters of the trumpet creeper six to twelve are frequently seen, but in Hummingbird gatherings it is always everyone for himself, there is no flocking.

While both sexes frequent our garden early in the season the males are seen there more consistently than the females. Indeed most of our observations of male Hummers are in such places. They seem to desert their mates as soon as the nests are completed and eggs laid and are seldom seen in midsummer at Cape May Point where females and young abound during July and August.

On May 20, 1927, doubtless at the height of the mating season, I saw a number of males at the Point perching frequently on the top sprays of climbing vines or small trees with tails partly spread and flirted up and down at intervals; a female was roosting in the same thicket. On July 6, 1925, I saw two other males perched in the same spot which displayed the white tuft of downy feathers just behind each foot, while the distinct fork of their tails could be clearly seen. Another male, studied on May 12, 1929, was perched on a dead twig in a more or less humpbacked position, with bill slightly elevated while he turned his head right and left in rapid succession. One would suppose that the brilliant metallic ruby red throat of the male would be most conspicuous but this is by no means always the case. This bird was perched in such a position that the throat was never facing the sun nor fully directed toward me and it seemed dull black during all the time that I watched the bird. The sunlight on the back however produced a brilliant metallic green. On other occasions a bird would alight on a twig facing me and the sun but even then the throat was illuminated only when the head was turned to exactly the proper angle. When this was effected there flashed out a gorgeous spot of orange crimson only to be lost again as the head was turned farther around.

The aërial flight of the male is performed in the spring in the presence of the female and once I witnessed it as late as July 14, 1923, the bird swinging back and forth a dozen times or more over a concave arc of about thirty feet, keeping as truly to the same path as if it had been definitely traced for it through the air. The female was doubtless perched in the trumpet vine over which the male performed; it seems likely that her first nest had been destroyed and that this was a second mating.

Of the trumpet creeper thickets at Cape May Point one situated at a sewer outlet is an especial favorite for the Hummingbirds. Even before the blossoms open the birds may be seen in the vicinity perched on the vines or adjacent bushes, drawn there perhaps by previous association, since the spot seems to be quite as much a gathering place as a feeding station. As the summer advances a number of the little birds may be seen there busily chasing one another through thicket. One will pause on whirring wings for a moment and then make off after another, only to be itself pursued a moment later. Every now and then one will alight on a slender twig and tilt its tail up and down several times as if making sure of its balance or in mere nervous excitement. On July 18, 1923, a female and young were in constant pursuit, possibly a seeking for food by the young and an avoidance on the part of the parent. Another bird had its crest feathers slightly erect and constantly turned its head from side to side, alert for the approach of others. The white spot behind the eye was very prominent and the dusky lores less so. It stretched its neck forward and upward in a peculiar manner and then pointed the bill vertically upward as another Hummer passed overhead. It also protruded its long tongue several times drawing it slowly back again and immediately flirted the tail in the characteristic nervous manner. On July 22, 1931, an adult male was present in the gathering at this thicket, a very unusual occurrence. He constantly pursued a female diving down into the foliage as she tried to escape. There was much chasing and perching for some minutes.

I have seen Hummers that were feeding in the trumpet blossoms fly directly into the corolla and cling there for a moment with only the tip of the wings and tail visible but usually, when feeding at the mouth of the tube, the bird continues the rapid buzzing of the wings. On several occasions I have seen feeding birds alight on the outside of the flower and sip nectar from the slits which large bees habitually cut near the base in their search for honey or pollen. When probing far into the blossoms many Hummingbirds cover the top of the head with pollen so that the crown appears quite white or buff and some individuals, doubtless young of the year, exhibit a whitish frosting on the feathers of the back due to lighter colored tips. On at least two occasions I have seen nesting Hummingbirds scratch the side of the head, bringing the foot up between the wing and the body. The wings were slightly drooping.

On a number of occasions from mid-July throughout August, Hummingbirds in the open pine woods at Cape May Point systematically examined the small dead twigs or the lower branches of the trees. They would hover close to the twigs but never alighted on them nor could I see them pick anything from the bark. So intent was one of these birds that it came within a foot of my face as I stood rigid by one of the trees. Once I saw one of them peck at the twig with its bill but could not see what it secured, if anything. All of the birds so engaged were females or young and only one was in the woods at a time. I saw the same thing at Cold Spring where the bird was exploring the dead branches of a red cedar.

While we are accustomed to think of Hummingbirds on the wing they perch much more often than we imagine and by watching a feeding bird constantly we shall often see it resort to some nearby perch for a few moments rest. They are continually perching in the thickets at the Point but I have frequently seen them come to rest in shade trees in the gardens, on telegraph wires along the roads, or on clothes lines in our yard.

Sometimes we see Hummingbirds far from their usual haunts. On August 30, 1920, at the very extremity of Cape May Point a Hummer came swiftly past me, headed for the Delaware coast and disappeared flying low over the water. Others seen during August flying down the beach close to the water's edge and one seen out in the middle of the salt meadows on May 18, 1927, flying north, were all probably on migration.

Walker Hand's dates of arrival of Hummingbirds at Cape May are:

1902.	May 2.	1911.	April 30.	1926.	April 29.
1905.	May 4.	1914.	May 1.	1932.	April 27.
1907.	May 1.	1916.	April 30.	1933.	April 23
1909.	May 2.	1917.	May 6.		(O. H. Brown).

Our latest dates in the autumn are:

1907.	October 16.	1925.	September 24.	1928.	October 4.
1917.	October 1.	1924.	September 28.	1935.	September 29.
		1921.	October 9.		

A nest found May 30, 1924, on a branch of a pine tree four feet from the ground, at Cape May Point, offered exceptional opportunities for study. It contained two eggs and the female, which appeared as soon as we assembled about the nest, immediately settled upon the eggs right before our eyes and remained there even though I stroked her back. On June 8 she acted in exactly the same way, going onto the nest each time that I approached although she was always off when I arrived. By June 22 when I again visited the nest the young had hatched and completely filled the cup. They were entirely covered with pin feathers, the sheaths of those on the back being black, those of the breast white, the wings banded white and black. There were some very minute rusty feathers on the body just above the wings and some long narrow filaments on the lower back somewhat curled together. The bills were short and were pointed up over the side of the nest, both birds facing the same way. Their eyes were open and their bodies pulsated with each breath. As I watched them, one bird wriggled and raised its body and voided a fine spray over the side of the cup.

There was no sign of the parent although I remained for half an hour nearby. Upon returning later she appeared and took up a position on a twig some fifteen feet from the nest where she remained stationary except that she once flew directly at my face and veered up over my head. Doubtless on the previous occasion she may have been perched on a nearby branch but escaped observation. The young were still in the nest on June 28, but had flown by July 5 having probably gone a few days earlier.

A nest with eggs was found in a similar situation on May 30, 1928, and another on May 30, 1923, ten feet up in a red cedar, while a third held small young on May 31, 1926.

Hummingbirds are regular summer residents throughout the state.

While we usually associate them with bright sunshine I have seen them awing on very dark cloudy days and one in our garden fed at the flowers between showers on a day of continual rains, taking refuge in the nearby trees, or on sheltered wires, at other times.   They have also been noticed feeding at dusk as late as 7:30 p. m. when one would be more likely to see the sphinx moths which bear such a superficial resemblance to them. Indeed it is hard to convince many persons that the day-flying sphinx (*Haemorrhagia thysbe*) is not a hummingbird!

In spite of their small size Hummingbirds are very courageous and will drive intruders from their nests in a very determined manner.   I have seen a female pursue a Crow and drive him entirely out of her territory while another on August 13, 1923, attacked an Olive-sided Flycatcher, darting at him again and again.   The latter assault must have been shear pugnaciousness as the nesting season was well past by this time.

BELTED KINGFISHER

Ceryle alcyon alcyon (Linnaeus)

One or two Kingfishers may usually be found immediately about Cape May at any time of year, either in the vicinity of the Harbor, at Lake Lily or on the Lighthouse Pond, while in summer and autumn they extend their range to include any body of water that may contain small fishes or other prey. Before it was converted into a sewer they used to follow Cape Island Creek from South Cape May right into the town and they still come up the open creek from the Harbor to the edge of the golf links.

Kingfishers are most abundant in late summer and autumn after the young are on the wing, or when probable migrants come from farther north. In spring and early summer they resort to nesting places more or less remote and are of less regular occurrence while in winter their numbers reach the minimum. We can usually count on seeing one on a winter's walk and sometimes several, but there are times and indeed whole seasons when none can be found.

The Cape May Christmas census records are as follows:

December 24, 1922, one.	December 26, 1932, one.
December 23, 1923, one.	December 24, 1933, six.
December 27, 1926, one.	December 23, 1934, three.
December 28, 1930, two.	December 22, 1935, fifteen.
December 27, 1931, three.	December 27, 1936, six.

As a rule winter Kingfishers are solitary individuals which locate at some spot where open water is found and food may readily be obtained. They have been recorded in winter as far north as Plainfield (W. D. Miller) and Morristown (E. C. Thurber).

I have never seen a migration of Kingfishers and only on a few occasions have I come upon more than two birds on a day's walk. My records show six on July 6, and seven on August 25, 1927; six on September 23, 1928; five on April 20, 1924; four on September 17, 1922; September 11, 1926; September 30, 1923; while Julian Potter saw no less than twenty on September 30, 1928.

The flight of the Kingfisher is peculiar and easily recognized; there is a series of nervous flaps of the wings alternating with short sails which cause an undulatory movement somewhat like the flight of a Flicker, a resemblance which is emphasized by the bird's habit of occasionally turning the head from side to side while on the wing. Usually the head is held straight out which gives the bird the appearance of being unnaturally long. The flight is well sustained and direct and when passing from one point to another is often at a considerable altitude.

Strong wind may affect their flight and on one occasion a Kingfisher carried along with a northwest gale covered the entire length of Lake Lily on a continuous diagonal sail traveling at tremendous speed. A pair flying very low over the waters of the lake on April 2, 1922, and closely pursuing one another, held their heads well up while their wingbeats were short and spasmodic the effect being of a body bouncing up and down on an elastic surface. Another pair acted in the same way on June 28, 1928, and each bird in the course of the flight splashed once into the water, though evidently not in the pursuit of prey. Doubtless the whole performance was part of the mating behavior as was also the later mounting of one of the birds straight into the air to a height of a hundred feet or more.

When not passing rapidly on the wing we generally see Kingfishers resting or engaged in fishing. A favorite resting place for many years was an old wooden boat landing on the edge of the Lighthouse Pond where I could almost always count on seeing one of these birds apparently stretched out flat on its belly, the legs being so short that they do not raise the body perceptably from a flat surface. I noted an individual on this perch on August 19, 1921, busily engaged in preening his plumage, probing straight down among the breast feathers with his massive bill and stabbing viciously under the wings while he ruffled up the entire plumage, shook himself and partly spread his tail. When through with this performance he assumed again the squatting posture with the head drawn well down on the shoulders.

Any sort of post or piling standing in the water is an attractive perch for a Kingfisher and may be used either for resting or as a vantage point for fishing operations. Sometimes I have seen stakes projecting only a foot above the water occupied by resting Kingfishers and range poles in the salt water far out on the edge of Jarvis Sound served a similar purpose.

A dead tree top on the Bay shore dunes, quite a distance from water, was a favorite roosting place and a signboard near the pool at the Point was another. Quite frequently Kingfishers will be seen balancing themselves on telegraph wires and the supporting poles are even more to their liking. Twice I have flushed Kingfishers from the pine woods at Cape May Point a hundred yards from the lake and was unable to determine what brought them there, and on the evening of August 15, 1921, and on several subsequent occasions a lone Kingfisher came into the Martin roost on the Physick property in the heart of Cape May, and could be heard uttering his rattling cry, apparently trying to find a suitable perch among the horde of noisy birds gathered there.

Kingfishers procure their food by diving either from a perch or from the air. The first method was well shown by two individuals which I watched on the shores of Lake Lily on July 29, 1921. They were perched on pine limbs overhanging the water, resting on dense masses of cones and needles rather than upon the dead branches. Each bird kept a constant watch on the water below and every now and then would give his tail an upward flirt allowing it to drop again slowly and then, suddenly, he would dive head first perpendicularly into the water but generally without success. After each plunge he would resume his perch, shake off the water by ruffling the plumage, and again take up his vigil. On August 17, 1921, and September 11, 1927, I saw a bird perched on a piling which stood just beyond the surf line on the beach from which he repeatedly dived directly into the ocean.

Diving from the air seems to be the more common method about Cape May, probably because of the lack of suitable perches near the fishing grounds. One can detect fishing Kingfishers at a considerable distance as they hover in the air poising twenty to thirty feet above the water on constantly vibrating wings. After several minutes of this hovering the bird will dive, not vertically, but at an angle; I have seen them do this in various ponds both shallow and deep and on one occasion a bird struck a fish at least twelve feet beyond the point over which he had been hovering. They usually make a great splash when they go into the water and sometimes disappear entirely beneath the surface. One that dived into the Lighthouse Pond flapped several times upon emerging before he was able to take wing. He had apparently either become entangled in the abundant aquatic vegetation, or was loath to let go a fish too large for him to lift from the water. He repaired immediately to a stake where he rested for several minutes flirting his tail nervously as he sat there. His plumage seemed very wet and his crest feathers were plastered flat on his head. Kingfishers usually seek some stake or post after making a dive, whether successful or not, and even after hovering and failing to dive at all they will often drop to some stake to rest. Occasionally, however, they will hover at several points in succession, flying a few yards from one to the next without resting.

One Kingfisher which I found perched on a stump in Lake Lily had secured a rather large sunfish which he proceded to beat on the stump with vicious blows. Failing to break it or reduce it to a size suitable for swallowing he laid it down and pecked at it repeatedly with his bill but without result. Eventually he flew off carrying his prey with him.

The high bluff on the Bay shore of the Cape May Peninsula has long been a favorite nesting place for Kingfishers. For centuries the Bay has been cutting into the land and the vertical sandy wall is admirably suited for the tunneling of these birds, though of late years the development of small resorts all along this shore has sadly interfered with its bird life. A nest found here on May 24, 1925, contained five naked young. Elsewhere Kingfishers are forced to resort to artificial banks formed by digging away the soil or excavating sand. A nest found on May 23, 1920, near Rio Grande, about one foot from the top of a low sand bank, extended back for five feet with an enlarged chamber at the end which contained six eggs and was strewn with fish scales and bones from disintegrated pellets.

Another nest was located in a sand bank to the east of Lake Lily some fifty feet back from the water, and still another was found in a cut in a field south of Pond Creek Meadows where the surface soil had been cut away for filling, leaving a vertical bank on one side.

On the ocean side of the peninsula suitable nesting places are rare but on May 26, 1929, I found a nest hole in the side of a sand dune on Two Mile Beach about six feet above a shallow pool which had formed in a depression, The dune was a very old one and the sand had become hard enough to support the walls of the tunnel. On Seven Mile Beach Charles S. Shick found a pair of Kingfishers present all through the summer of 1886 and attributed to them a nest hole dug in a hollow stump (Auk, 1890, p. 328). Philip Laurent found them nesting on Five Mile Beach in the eighties.

Turner McMullen has furnished me with a list of Kingfisher nests examined by him in Camden and Burlington Counties which gives an idea of their nesting time here. They nest throughout the state.

May 18, 1909, seven eggs.	May 4, 1919, seven eggs.
May 10, 1914, six eggs.	May 3, 1921, seven eggs.
May 22, 1915, seven young.	May 25, 1922, six eggs.
May 7, 1916, seven eggs.	May 23, 1925, six young.
May 13, 1918, seven eggs.	May 16, 1930, seven eggs.

While young birds are on the wing about Cape May by June 1, I have seen a parent feeding a young one on a telegraph wire as late as July 5.

Kingfishers at Cape May are essentially solitary birds except at nesting time when pairs are seen and it would seem that the young must scatter soon after leaving the nest. This would be natural since, being lone fishermen, there would be nothing to gain and much to lose by close association which would inevitably involve trespass on a neighbor's preserve.

FLICKER

Colaptes auratus luteus Bangs

Flickers are to be seen about the wooded areas of Cape May Point and in the farming country north of the town throughout the year. Well distributed in summer and nesting commonly in orchard and shade trees; they are reduced in winter to scattered individuals perched in the top of some tall tree or passing in undulating flight across the sky.

Immediately about the town Fickers may be flushed in winter from the edge of the salt meadows, where the black grass has been cut off for hay, or from sandy spots on the Fill and now and then one may be found resting in the trees of the Physick place where there is often a nest in summer. In spring and autumn, too, they may be seen feeding on the lawns.

It is difficult to determine which are migrants and which are winter resident birds as spring advances, but there are always days when the number of Flickers that are present has obviously increased and this we interpret as the arrival of birds from the south. Such dates are:

1904.	March 20.	1911.	March 27.	1928.	March 25.
1905.	March 21.	1921.	March 31.	1929.	March 31.
1909.	March 30.	1923.	March 25.	1932.	March 26.
1910.	March 26.	1925.	March 21.	1933.	March 18.
		1927.	March 17.		

Even so it is usually early April before Flickers become common. On April 2, 1922, I counted twenty-five feeding on a burnt over area on the Fill and on April 4, 1924, the woods at the Point seemed full of Flickers. Most of these spring birds seem to pass on and in the nesting season Flickers are not much more abundant immediately about Cape May than in the winter. Back in the farming country, a few miles to the north, however, they are plentiful all summer.

In late summer and autumn Flickers become more frequent on the lawns along Washington St. and sometimes a portion of the great Flicker flights, mainly confined to the Bay shore, pass through the town. These flights usually occur from September 25 to October 8 and after they are over the Flicker population is reduced to its winter proportions. In an ordinary winter's walk one may see a single Flicker or perhaps two, while on some days none will be in evidence. On the Delaware Valley Club's Christmas census when the country from the Court House to the Point is systematically covered the observers' combined list shows Flickers present as follows:

December 23, 1923, one.	December 27, 1931, twelve.
December 26, 1926, five.	December 26, 1932, twenty.
December 26, 1927, eleven.	December 24, 1933, thirteen.
December 23, 1928, three.	December 23, 1934, eighteen.
December 22, 1929, seven.	December 22, 1935, thirty-five.
December 28, 1930, eight.	December 27, 1936, eleven.

In the open ground surrounding the town resting places for Flickers are rather scarce and I have frequently seen them on the tops of low bayberry bushes on the Fill while telegraph poles are favorite perches. A stake not over two feet in height placed in the middle of an open field was frequently occupied by a Flicker. He stood upright grasping the top of the stake and holding his tail close against the side, and remained rigid for many minutes.

In late summer or early autumn, at the Point or back in the farming country, they like to sit for long periods at the top of some dead tree or telegraph pole in this same frozen attitude, body hunched up, head low down on the shoulders, and bill pointed slightly upward as they gaze over the landscape turning the head occasionally to one side or the other.

The Flicker is our only ground-feeding woodpecker and this fact combined with the character of the country near the sea will explain why most of our Cape May Flickers are flushed from the ground.

When on the ground they can hop but they also progress for short distances at a surprisingly rapid gait by using the legs alternately in a sort of waddling shuffle and all the while the head is held aloft with the bill horizontal.

Flickers on the ground are usually feeding on ants and in one instance where I was able to locate the exact spot where the bird had been operating I found that he had dug out several small ant hills so that they were converted into funnels nearly two inches across and he was apparently picking up the ants as they appeared in them. This was along a shady wood road at the Point on August 9, 1920. Again, on August 1, 1921, I found a Flicker on the Fill, on a little sandy knoll covered with scant grass, where he had been probing into several ant nests, the entrances to which he had enlarged to about half an inch in diameter. A third bird on the edge of the meadows on May 1, 1926, was seen to bend far over forward apparently inserting his tongue into the burrow of an ant hill as there was no evidence of probing or digging into it with his bill as in the other cases.

A favorite nesting place in the town was a hole in a telegraph pole near Broadway, on the line of the abandoned trolley road, which was only three feet from the ground. Here on July 31, 1920, I flushed a male Flicker from a nearby sumac thicket and immediately a female appeared on the ground in front of me with tail spread out flat as she hopped along and looked back at me over her shoulder. Later I discovered two young birds clinging to a pole a little farther on and found the hole from which they had evidently just emerged.

On July 12, 1923, I found that the same hole contained a lively brood of pinfeather-covered young, which climbed excitedly up to the entrance when the pole was tapped. Those that gained the aperture called *choop, choop, choop-choop-choop-choop*, as they stretched out their necks while the others within made a peculiar buzzing sound like a swarm of bees. On July 17, the pinfeathers had burst open and the birds were fully clad in Flicker plumage but the swelling at the gape was still conspicuous. They did not now expose themselves as much as they did at first and by July 26 had all left the nest. The hole had become very offensive, due to the heat and the crowded condition of the brood. The parents always left the nest, if they happened to be within, while I was some distance away. This nest was not occupied in 1924, but was in use on May 8, 1925, and also in 1927. How many years the birds had used this hole I cannot say.

Another pair of Flickers dug a hole at the very top of a nearby pole and an old railroad tie, which had been stood upright when the Fill was being formed, offered a nest site for still another pair. A pair has bred nearly every year of late in a maple tree on the Physick place and had young still in the nest as late as June 30, 1931, while the ground was littered with chips for six feet around the tree. At Cape May Point Flickers nest in dead pines and on Seven Mile Beach Richard Miller found a nest in a dead holly, some fifteen feet from the ground, which contained five young on May 30, 1912. A nest at the Point contained seven eggs on May 25, 1919. Others are:

Seven Mile Beach, June 16, 1918, two nests containing seven and six young birds respectively, both in dead oak stubs (Turner McMullen).

Ludlam's Beach, June 9, 1933, five young, nest in a telegraph pole fifteen feet up (Fletcher Street).

During the mating season I have seen Flickers pursuing one another on telegraph poles and tree trunks and one pair rested in a bayberry bush the male raising and lowering his tail in a sort of display as he bent his body far over forward. A pair was observed copulating on a pole near the old nest site on May 25, 1922.

The alarm note of the Flicker as we hear it at Cape May is a *"kee-uke"* the same as is usually represented as *"yarrup"*; I have heared it from winter birds as they flushed from the ground and also in July and September. The familiar "spring song" *wick-wick-wick-wick-* etc., I have heard as late as July 15 and as late in the evening as 7:00 o'clock. I have several times counted the calls and have found a bird uttering as many as thirty *wicks* without a pause. The pairing birds utter a different call *whít-choo, whít-choo* as they dodge one another around a tree trunk, while a young bird of the year decoying to the excited calls of a Carolina Wren called *whée-o, whée-o.*

I have seen Flickers in the spring feeding on the ground with flocks of Purple Grackles while in the stubble fields, after the harvest, they associate with Red-wings and English Sparrows but these associations are always due simply to a common food attraction not to a social or flocking instinct. Indeed they do not form flocks of their own kind and even in the great fall migrations they travel in long scattered flights every bird for himself not in regular flocks as do the Grackles and Red-wings.

The autumnal migration at Cape May is one of the most striking ornithological sights of the region and one that I have not been able to explain to my entire satisfaction. It is exactly similar to the flights of Woodcock, Kingbirds and hawks, and the Flickers are usually accompanied by hosts of smaller birds, also on their way south.

The movement takes place mainly at night and what we see early in the morning is the closing of the flight as the birds settle down to feed and rest. While a few Flickers may be in evidence in the town at sunrise, drifting along to the southwest, we do not as a rule see many until after we cross the Bay Shore Road but from there to the Bay they abound, the thickest part of the flight being close to the shore itself. The curious thing about it is that the birds are all traveling north and the same is the case with the other migrant species which I have mentioned; moreover the flights always occur when there is a strong northwest wind and into this the birds fly.

It would seem that the normal flight is south along the sea coast, and that the birds are blown offshore by the wind and try to beat back to land or are in fear of such a catastrophe when they approach the Point, with the twenty mile stretch of Delaware Bay confronting them and the ocean on their left. Certain it is that I have seen hundreds of Robins and Bluebirds just above the Point beating their way in from the ocean in the early morning, appearing first as minute specks in the distance and looming larger until they make the shore, and flocks of warblers and scattered Woodcock have been observed coming in from the sea during the night. These birds do not stop but keep right on in the face of the wind following the long stretch of woodland that borders the Bay side of the peninsula. Probably the majority of the migrating column are not blown off of their course but slow up at the Point and cause the congestion there, while, by continuing to follow the shore line rather than launching out over the Bay, they are soon headed north as they round the point of the peninsula. The flights reach for twenty miles north of Cape May Point and so far as I have been able to ascertain they do not cross the Bay until the wind changes when they drift back to the Point and continue on their regular course. If there is no heavy northwest wind there is no flight; in other words everything progresses normally on nights when the wind is, as usual, from the south and hosts of birds may be heard passing south over the town and the Point. Philip Laurent states that when Five Mile Beach was covered with woodland Flickers were common in September but they did not nest; nor was there any concentrated flight.

To give a clearer picture of one of these Flicker flights we may select that of September 27, 1926. The 25th had been a mild autumn day with bird life at its lowest ebb. I had tramped nearly all day from 8:00 a. m. to 4:00 p. m. and besides the usual gulls and a few shore birds I was able to find only fifteen species—apparently the entire land bird population. It was cloudy with occasional showers and the next day was the same but by noon it had cleared and the temperature began to fall and by night the wind was coming strong from the northwest. At five o'clock the next morning there was a faint glimmer of light but sufficient to show an occasional Flicker passing overhead and it was evident that a flight was on. By sunrise Walker Hand and I were on the Bay Shore Road three miles above the Point. We had seen some thirty Flickers since leaving Cape May, scattered individuals passing steadily along overhead. As we skirted the northern edge of a dense woods stretching toward the Bay we saw Flickers constantly shooting out over the tree tops from the south and a hasty count showed twenty-five to thirty in sight at any moment within our range of vision. Occasionally one would alight on the top of some tree rising above the rest of the forest or perhaps standing alone in the open; others would decoy to him until there were six or more perched on the trunk or limbs then on they would go, leaving one at a time in close succession.

As we reached the head of New England Meadow we could see the Bay two-thirds of a mile away and the glistening white sand dunes with their dense black masses of prostrate cedars. Here we had a splendid view of the flight, Flickers were passing constantly, bursting out from the woods on the south, crossing the meadow and disappearing again over the northern woods barrier. Several counts showed that they were passing at the rate of thirty to fifty per minute.

Immediately over the dunes the flight was even more congested and, as we rested on the sand, the birds came close over head in a continuous stream, singly, then two together and now perhaps ten close together and then individual birds again, but there was no concerted action or semblance of flocking. The flight was strictly a stream, a follow-my-leader procession everyone for himself. In the bright sunlight which by now flooded both land and water we could see every detail of the Flicker's plumage; the pinkish gray of the under side with bold round spots of black; the black crescent on the breast; and the glorious golden lining of the wings and tail flashing at intervals as the birds spread and closed their wings. Now and then an individual would swing suddenly to the right or left, as he caught a glimpse of us, and would turn sufficiently to exhibit the pure white rump or more rarely the red crescent on the nape. The sexes could easily be distinguished by the black mustachial stripe of the male.

The tail was partly spread as they flew and there were rapid strokes of the wings—one, two or three, and then a glide through the air with wings folded closely to the body causing a slight loss of position and then the rapid beats as the bird regained its former level. This resulted in a somewhat undulatory flight but the birds were always going straight away in a definite direction as if they had no doubt as to where they were bound—head stretched out, and turning now to one side now to the other; the bill always slightly elevated; they appeared to be straining every muscle in some great effort. The impression was that some invisible but powerful force was impelling them on in their mad, uncanny flight.

As the sun came fairly up the flight slackened and the birds rested more frequently, flew in various directions, and began to feed. They were widely scattered over the country and most of them were resting quietly and all trace of the flight had disappeared. We estimated that at least five thousand Flickers had passed while we watched them, how many before sunrise we could not guess.

While the flight of the Flickers was the spectacle of the morning they were by no means alone. Thousands of smaller birds were passing continuously in a scattered stream or succession of waves. At first mere dusky spots against the sky but later, as the light grew stronger, identifiable as warblers, sparrows, flycatchers etc., though specific identification usually had to await their settling in the trees and shrubbery to rest and feed. We passed a little later down the Bay shore to the Point and the same woods and thickets which were practically bare of bird life two days before were now thronged with Catbirds, Thrashers, White-throated and Swamp Sparrows, Palm and Parula Warblers, Red-breasted Nuthatches, Red-eyed Vireos, Ruby-crowned Kinglets, etc., etc. Where on the 25th we had trouble in finding fifteen species of land birds we today listed fifty-five and some of them fairly swarmed. Impressive as this flight was to me Walker Hand considered that it fell far below some that he had seen in the past.

Not many years ago Flickers were lawful game in New Jersey and the Cape May gunners, well versed in the character and time of these flights, made regular preparation for them. Poles or fence rails were fastened to the tops of the low pines and cedars at the Point so that they projected upward above the topmost branches and formed resting places that the Flickers could not resist. The result was that the tired birds, stopping for a few moments on their wild flight, lined the poles so that at any moment six or more were clinging to the perches and a gunner concealed at the base of the tree "raked" them off with the greatest ease. Walker Hand informed me that two gunners of his acquaintance secured six peach baskets full of Flickers on a single morning and piles of dead birds as high as a man's knees were frequent sights. Another gunner got four hundred birds in an hour and a half and it was not unusual to get four at a shot. While this is now all a thing of the past it took constant effort in the face of the strongest kind of opposition before the State Legislature could be induced to make the practice illegal. Flickers are common summer residents throughout New Jersey and occur casually in winter in most localities.

Occasional Flickers have been found in New Jersey with a few red feathers in the mustachial stripe but I regard these as individual aberrations rather than indications of a strain of the western Red-shafted Flicker which has these stripes red.

PILEATED WOODPECKER

Ceophloeus pileatus abieticola Bangs

This splendid bird, the largest of our woodpeckers, was doubtless at one time common throughout the wooded parts of New Jersey but has long been extinct in most parts of the state. Today it is known as a resident of the wilder mountainous parts of Sussex and Passaic Counties in the extreme north—at Greenwood Lake, Newfoundland, Culver's Gap and perhaps elsewhere in the same region.

In Cape May County we have two records: birds shot by one of the Coast Guards at the Five Mile Beach station on November 8, 1878, and December 31, 1879, and now in the collection of the Academy of Natural Sciences of Philadelphia. They were presented to Dr. William L. Abbott, who was, at the age of eighteen, just beginning his collection of local birds and was accustomed to visit the station on what was then an almost uninhabited island with much virgin forest. Whether the birds were shot on the island or in the splendid tract of tall timber on the mainland opposite I have never been able to ascertain, but inasmuch as Schick mentions seeing one on Seven Mile Beach (Bay State Ornith., I, No. 2, p. 13) in June and Parker records one there in 1885 (Orn. & Oöl., 1886, p. 140), they may well have been shot on the island.

On June 4, 1983, Mark L. Wilde and J. Harris Reed saw one of these birds in Cumberland County just over the Cape May line on West Creek, and located its nest which contained five young birds (Bendire, Life Histories II, p. 107, and Atlantic Slope Naturalist, I, p. 27).

Our only subsequent record for southern New Jersey is a single bird seen by George Morris on the Egg Harbor River above May's Landing, on March 25, 1908.

The destruction of the primeval forest and the spread of civilization have been responsible for the disappearance of this fine bird not only in New Jersey but in all parts of the country. It is one of the species that cannot adapt itself to the presence of man and his activities.

RED-BELLIED WOODPECKER

Centurus carolinus (Linnaeus)

This southern woodpecker, which I have found, evidently resident and breeding, in southern Fulton County in Pennsylvania and along the Susquehanna River in Maryland, just below the Pennsylvania line, has been observed several times about Delaware City and has been recorded as breeding at Marydel, on the border between Delaware and Maryland, by Rhoads and Pennock (Auk, 1905, p. 200). In New Jersey it seems to be nowhere more than an accidental straggler. Charles Pennock shot a specimen at Cape May Point, near the head of Lake Lily, on April 11, 1903; Norman McDonald saw another on March 10, 1935, at Pennsville, Salem County; and Julian Potter saw one at Fish House on the Delaware, on October 11, 1908, which frequented the same grove for about two weeks; and another at Collingswood, Camden County, on November 27, 1927, while William B. Crispin shot one near Salem on December 20, 1912. There are also several records for the more northern counties.

Perhaps the fact that the Red-bellied Woodpecker is essentially resident in its breeding zone and not a migrant, is responsible for its not more frequently crossing Delaware Bay or River. Dr. C. C. Abbott records a nest and young in Ocean County, May 29, 1861, but without authority or details.

RED-HEADED WOODPECKER

Melanerpes erythrocephalus (Linnaeus)

The Red-headed Woodpecker is a rare autumn transient at Cape May with only one or two records each year and sometimes none at all; nearly all of the birds are young of the year.

Our autumn records arranged in monthly and daily sequence are as follows:

August 26, 1920.	October 2, 1924.
August 31, 1926.	October 5, 1929 (2)
September 2 and 3, 1920.	October 13, 1913.
September 7 and 10 (2), 1935.	October 17, 1920.
September 9, 1916.	October 18, 1925.
September 14, 1924.	October 21, 1923.
September 20, 1929 (2).	October 21, 1928.
September 29, 1925.	

In the winter of 1922–1923 one was resident at the Point frequenting an old frame church and storing acorns in holes in a large wooded cross on top of the building. It was recorded from December 24 to February 22 and doubtless had been there earlier. During the time that it was under observation the red feathers gradually appeared on the head.

I have but a few spring records of the bird at the Cape: one on May 13, 1923; another on May 10, 1928; a third near the junction of the Town

Bank road and the Bay Shore Road on May 29, 1921. This last bird flew from an orchard to a solitary tree standing in a field and it may have had a nest there. All were adults. Another single adult bird appeared at the Point on May 7, 1932, and remained until July 19, no mate appearing at any time, On May 20, it was busy digging a hole in an oak tree in the village while on June 23 it was watched for some time deliberately annoying a pair of Kingbirds which had a nest in the vicinity. It would fly close to the nest and both of the flycatchers would dart after it. After leading them away it would return and take up its position on a nearby house and call lustily after which the whole performance would be repeated. This occurred half a dozen times. I saw an adult north of Salem on May 7, 1896, and William Baily found another on Peck's Beach on May 5 of the same year.

While the Red-headed Woodpecker is, or was, a common summer resident through most of Pennsylvania it has always been rare and exceedingly erratic in its occurrence east of the Delaware and Hudson Rivers. I find but few breeding records for southern New Jersey and most of these are credited to Julian Potter and pertain to the vicinity of Camden. During Samuel Rhoads' many years residence at Haddonfield he never found the Red-headed Woodpecker nesting and had but few observations of the bird, but it seems to have increased of late years. Potter located one nest in a piece of woodland immediately on the Delaware at Fish House on May 19, 1910. It was located in a dead buttonwood tree forty feet from the ground and the birds were successful in raising their young which were seen out of the nest on July 9.

On November 27, 1910, a pair of the birds was located in another grove within the city limits of Camden and were seen continuously for the succeeding nine months. Their nest was located on May 21, 1911, in the dead top of a maple tree thirty feet from the ground and was found to contain young which left on June 25. On July 23, there was a second brood of young in the same nest which were able to climb up to the opening by August 6, while by the 16th they had left and a Starling had appropriated the nest. While the parents were attending to the second brood in the nest they persistently drove off the first brood and persecuted them to such an extent that they apparently left the locality by July 30. They fed the young in the nest at varying intervals sometimes every three or four minutes for half an hour and then failed to appear for twenty minutes or more. One parent spent the night in the nest and the other in an old hole in the same tree both entering about dusk. English Sparrows annoyed the woodpeckers to some extent but they were able to protect themselves and by the time the second brood was under way all the birds in the neighborhood seemed to have learned to leave them alone. (Bird-Lore, 1912, pp. 216–217.)

In May 29, 1916, a pair was located in a telegraph pole in Camden and two broods were raised there that year. This nest was an old Flicker cavity about twenty feet from the ground. Potter found the birds quite common at Camden in September 14 and saw individuals on October 6, 1912, and May

17, 1914, and one at South Vineland on June 19, 1914. On May 22, 1934, he located a nest in a public park in Collingswood which was apparently finished on June 2.

At Philadelphia there used to be regular flights of these birds in the autumn and the autumn records at Cape May are doubtless stragglers from such a movement. The Red-headed Woodpecker has always been a victim of the automobile, being apparently unable to adjust its movements so as to escape being struck. This may account for their scarcity in Pennsylvania in recent years and may have had some effect on their abundance in New Jersey as well.

YELLOW-BELLIED SAPSUCKER

Sphyrapicus varius varius (Linnaeus)

At Cape May, as in other parts of New Jersey, the Sapsucker is a regular autumn and rather rare spring transient. We seldom see more than one at a time usually in orchards or in rather open woodland. Unlike most other woodpeckers it is a real "sapsucker" puncturing the trunks of the trees for the sake of the sap which accumulates in the holes or for the cambium layer which underlies the outer bark. It doubtless feeds to some extent also upon small insects which are attracted to its borings. In spring I have several times found a Sapsucker in a little grove of hickory trees at the Point which it regularly girdled with holes. I watched one of these birds on April 4, 1924, which went from tree to tree probing into all of the holes which it had previously dug and sometimes waiting for them to fill up with the sap. At such times he hung against the trunk with belly and tail pressed close to the bark and legs widespread and directed a little upward. Now and then he would stab quickly at the bark on the side of the hole presumably at some visiting insect. A Myrtle Warbler followed the Sapsucker on his rounds seeking either sap or insects from the holes. The same trees were visited by the same of another Sapsucker on April 2 and 3, 1922, and once having made a chain of borings the bird will return for several days before continuing its migration.

In the autumn these birds exhibit quite a variety of plumage as the young start on their migration before the molt begins and change their dress *en route* as is the case with the Red-headed Woodpecker which often holds the juvenal plumage until well into the winter. William Rusling who was present at the Point throughout the autumn of 1935, watching the hawk flight, recorded his first Sapsucker on September 29, with eight on October 5, seven on the 7th, and one each on the 8th and 19th; this probably represents the normal autumn flight both in numbers and dates. Our records based upon scattered trips in autumn are as follows:

1913.	October 13.	1923.	October 1, 7.
1920.	October 3.	1924.	October 2.

1925.	September 29, October 11.	1931.	September 28, October 18.
1928.	October 21.	1932.	September 17, October 14.
1929.	October 5.	1934.	September 19.
1930.	October 5.		

Spring records are as follows:

1908.	April 6.	1924.	April 4.	1934.	May 7.
1922.	April 2.	1928.	April 1, May 16.		

Philip Laurent found it common on Five Mile Beach in the eighties in early October but saw none in spring. One Sapsucker was found by the Delaware Valley Club on the Christmas census trip of December 28, 1930, and there are several records of wintering individuals farther north in the state, but it does not nest in New Jersey. Babson in his "Birds of Princeton" (p. 53) states that a specimen was taken there by W. E. D. Scott on October 21, 1876, which "approached" the western race *S. v. nuchalis*, doubtless having some red feathers in the nuchal stripe, but this is probably an individual variation which may arise independently in eastern birds, similar to the Flickers with a few red feathers in the mustachial stripe which are occasionally come upon in the East.

HAIRY WOODPECKER

Dryobates villosus villosus (Linnaeus)

The Hairy Woodpecker is a larger counterpart of the Downy, resembling it closely both in plumage and in habits. It is, however, a bird of more remote and secluded spots and usually approaches gardens and orchards only in winter, so that we are likely to see it more frequently during the cold months. The census taken at Christmas time by the Delaware Valley Ornithological Club covering the region between Cape May Court House and the Point shows it present as follows:

December 23, 1923, one.	December 26, 1932, five.
December 26, 1927, one.	December 24, 1933, six.
December 23, 1928, two.	December 23, 1934, two.
December 22, 1929, two.	December 22, 1935, four.
December 28, 1930, five.	December 27, 1936, three.
December 27, 1931, one.	

I have seen a pair near Green Creek on July 3, 1921, and another in the Bear Swamp west of Rio Grande on September 4, 1924, while Richard Miller found a nest in an oak stub only five and a half feet from the ground at Burleigh on May 25, 1912, which contained three fledglings. Turner McMullen found another nest at Mays Landing in the Pine Barrens, farther north, on May 1, 1920, which contained four eggs and two at Haddonfield, on May 9, 1921, and May 23, 1923, both of which contained young birds.

It is a resident species throughout New Jersey.

DOWNY WOODPECKER

Dryobates pubescens medianus (Swainson)

Although not an abundant or conspicuous bird the Downy Woodpecker is to be found at Cape May in every month of the year. It is least common in June, when it probably seeks more remote spots for nesting, and most abundant in the late fall, winter and early spring, when it comes into the gardens and shade trees of the town and close about the farmhouses back in the country. A few were found in the woods of Five Mile Beach in the eighties by Philip Laurent.

Our observations are usually of single birds rarely two together, except at nesting time, and from one to four in the course of a day's walk although there are many days when none is seen. On the Christmas census of the Delaware Valley Club the record for the country from the Court House to the Point is:

December 26, 1927, three.	December 26, 1932, four.
December 23, 1928, three.	December 24, 1933, nine.
December 26, 1929, one.	December 23, 1932, twelve.
December 28, 1930, two.	December 22, 1935, fourteen.
December 27, 1931, seven.	December 27, 1936, twenty-four.

While they no doubt nest in the thicker woodland I have never found a nest, but a winter resident bird in an orchard at Cold Spring dug a hole in December in a broken tree stub perhaps for shelter but never occupied it.

Turner McMullen has examined nests in other counties as follows:

Salem County, April 15, 1921, six eggs.
Camden County, May 23, 1909, five eggs.
Camden County, May 2, 1915, five eggs.
Camden County, May 8, 1915, three eggs.
Camden County, May 16, 1931, five eggs.

During August a single bird is usually to be found in the pine woods at Cape May Point, possibly a young of the year, evidently finding satisfactory food in this spot. I have also watched another individual visiting the holes made by a Sapsucker in hickories at the Point although it dug none itself, preferring to explore a rotten branch of a nearby oak.

We often see Downies quite low down in alder swamps and on August 24, 1928, I found one climbing up mullein stalks in an old field probing into the seed pods for insects of some sort.

While not associating with its own kind to any extent the Downy is often found in one of the winter assemblages of chickadees, nuthatches, kinglets etc., which often scour the woods and thickets at this season. It is resident throughout the state.

*RED-COCKADED WOODPECKER

Dryobates borealis (Vieillot)

This woodpecker, a native of the pine forests of the Southern States, has been recorded as a New Jersey bird on the basis of a specimen in the collection of George N. Lawrence labeled as taken at Hoboken and published by Mr. Lawrence (Ann. N. Y. Lyceum, 1867, p. 201). There is also a specimen marked near Philadelphia in the collection of the Academy of Natural Sciences which was shot by C. D. Wood, the noted taxidermist, in 1861, and the record of another presented by William Wood on August 6, 1850. Whether these were secured in Pennsylvania or New Jersey is not certain but they were probably the basis of Turnbull's statement (1869) that the bird occurred rarely in Eastern Pennsylvania and New Jersey. We know of no record for either Delaware or Maryland and it seems strange that this bird so characteristic of the southern pine forests should have occurred so far north, without intermediate records.

*ARCTIC THREE-TOED WOODPECKER

Picoides arcticus (Swainson)

There are so far as I know but two records of this woodpecker for the state —a specimen now in the collection of the American Museum of Natural History discovered near Englewood on November 29, 1923, by S. V. LaDow and shot shortly afterward, on the same day, by J. A. Weber (cf. Griscom, Auk, 1924, p. 343), and one seen by R. H. Howland at Upper Montclair, February 10 and 11, 1926 (Eaton).

EASTERN KINGBIRD

Tyrannus tyrannus (Linnaeus)

From early May to early September Kingbirds are to be seen about Cape May perched upon tree tops or dead branches near to their nests, or on telegraph wires along the highways or railroads. Wherever found they are always in exposed positions and always conspicuous.

Nearly every summer a pair nests somewhere in the town and a pair or two at Cape May Point and along the turnpike connecting the two, while through the country to the north every farm has one or two breeding pairs. In 1920 a pair nested in the very top of a great silver poplar on the edge of the town and in 1921 an old pear tree on Lafayette Street harbored a family while in 1922 and 1923 a poplar along the sidewalk on Washington Street, in the very center of the town, furnished a building site. Kingbirds also nested on Five Mile Beach before the woods were cut down.

Sometimes the birds nest quite close to the ground and a nest in a wild cherry tree along the turnpike was less than six feet up while another in a scrub pine on the edge of Pond Creek was only three feet above the marsh. Other nests reported to me by Richard Miller were situated in oaks, cedars and apple trees at elevations ranging from eight to twenty-eight feet. Full sets of eggs have been found from June 1 to 23 and a set found on June 17 was just hatching. Full-fledged young were still in the nest on July 2, 1920, and July 10, 1921, but in the latter instance were beginning to climb out on the adjacent branches. Turner McMullen has given me data on the following nests which he had examined:

June 4, 1922, Cape May, three eggs.
June 8, 1919, Cape May Point, one egg.
June 15, 1924, Swain, four eggs.
June 18, 1922, Cape May Co., three eggs.
June 23, 1923, Seven Mile Beach, three eggs, and two eggs.
July 4, 1923, Seven Mile Beach, three eggs.
July 4, 1924, Seven Mile Beach, three eggs.
July 4, 1928, Seven Mile Beach, three young.

By July 7, 10, 11, and 12, in different years, the bobtailed young appeared on the telegraph wires where they were being fed by their parents and one youngster still unable to obtain food for himself was found on a bush on the Bay shore sand dunes as late as August 3.

The Kingbird is distinctly a bird of the air and high perches and most of his food consists of insects caught on the wing. He launches forth from his perch with head held high and wings rapidly beating and ascends at a steep diagonal toward his passing prey, seizes it, and returns in a long swoop to his perch to devour it. I have seen them catch monarch butterflies and many smaller insects and have noticed them fly straight as an arrow to meet some approaching victim when it was still a hundred feet away, and one Kingbird was seen in the middle of the Harbor with an insect in its bill which it had apparently pursued across the water. Sometimes an insect will turn and mount higher and higher in the air with the Kingbird in close pursuit on rapidly fluttering wings, while at other times the bird will descend to capture prey passing below the level of his perch and in one instance nearly turned a somersault in the operation and almost touched the ground.

I have seen Kingbirds dive down into hydrangea bushes apparently after some sort of insect and have also seen one hovering over fields of red-top grass, six or eight inches above the stalks, dart down at the heads of field garlic that were scattered through it, presumably after small insects which infested these blossoms. The bird would make eight or ten dives of this sort before coming to rest on a weed stalk and then would begin again after an interval of a minute or two. The wings moved continuously and very rapidly making a blur while the body was held in a partly upright position, the head held high and turning frequently from side to side. Other birds hovered in precisely the same way over a field of newly cut hay, apparently after grasshoppers, and I saw one dive repeatedly to the surface of a road finally flushing a "dusty-roads" grasshopper which he promptly seized and carried to his perch.

I have several times observed Kingbirds in the act of bathing. On July 14, 1920, one was seen hovering over the lily pond at the old race track. It was about a foot above the water and kept turning its head from side to side, as did the bird in the grass field, while it pecked at the surface of the water or at the lily pads. Finally it ducked its head into the water and retired to its perch. Almost immediately however it returned and plunged deliberately into the pond with quite a splash and was back again on its perch ruffling up its feathers and pluming itself. Another bird dived several times into the Lighthouse Pond retiring between dives to its perch nearby, and still another was seen to fly into the spray of a lawn sprinkler.

Kingbirds will occasionally alight on the ground which is rather unusual for such an aërial species. On May 27, 1929, I saw several of them in a plowed field with Robins, Grackles and Killdeers, and on July 3, 1922, one was sitting on the railroad track while another perched on a low weed nearby. On July 27, 1922, a Kingbird alighted on the beach of the Bay shore and on July 8, 1916, and July 21, 1917, one was seen sitting on the ocean beach picking up insects from the trash washed up at the high tide mark. Out on the Pond Creek Meadows I have seen them perching on reeds not two feet above shallow ponds where sandpipers were feeding and once I found one perched on a stake at least one hundred yards out on the salt meadows above Cold Spring. The early arrivals from the south seem to be more addicted to low perches and on May 18, 1928, I found no less than six resting on low weeds in an open field.

When on its perch the Kingbird's position is erect, the head is slightly bowed and every now and then is turned around sideways so that the bird seems to be looking over its shoulder—a peculiar and very characteristic attitude. When excited the crest is elevated which is always the case when one approaches the nest and it is then, or when he has driven away some intruder, that the Kingbird gives vent to his rapid, defiant call—*icky-icky-icky-icky-icky*. Kingbirds are valiant defenders of their nest and nearly every bird that passes near is likely to be attacked. I have seen one pursue a Bank Swallow and follow so closely upon a passing Chimney Swift that the latter bird was forced to the ground although apparently uninjured and able to take wing again shortly. Another Kingbird caused considerable annoyance to a Turkey Vulture which approached too close to its perch and a pair of the birds vigorously attacked two Meadowlarks which flushed from a grass field near to their nest tree. All four birds alighted on a telegraph wire each Kingbird close to his lark and when the latter again took wing the attack was renewed.

When pluming itself the Kingbird holds the tip of the wing away from the body with the bend, or wrist, close in to the shoulder just as the Swallows do, and probes below it.

The arrival dates for the Kingbird as recorded by Walker Hand are:

1902.	April 30.	1909.	May 2.	1916.	April 30.
1903.	April 27.	1910.	May 1.	1918.	May 4.
1904.	April 23.	1911.	April 24.	1919.	April 27.
1905.	May 4.	1912.	April 22.	1926.	April 25.
1906.	May 5.	1913.	April 18.	1927.	April 24.
1907.	May 1.	1914.	April 27.	1932.	May 2.
1908.	April 28.	1915.	April 28.		

The spring migration is not a conspicuous affair and it would seem that only the birds that nest in the vicinity or immediately to the north pass through Cape May, the main flight, perhaps, passing up the Delaware Valley as I think must be the case with a number of other species.

The return flight however is a very different matter and forms one of the ornithological features of the year and I am sure that anyone who sees it would be convinced that all the Kingbirds from a large part of the Eastern States must pass the Cape.

This migration first becomes noticeable in mid-August:

1921.	August 13.	1926.	August 23.	1931.	August 13.
1922.	August 19.	1927.	August 16.	1932.	August 9.
1923.	August 13.	1928.	August 13.	1933.	August 13.
1924.	August 24.	1929.	August 16.	1934.	August 13.
1925.	August 11.	1930.	August 12.		

First we see the Kingbird families of the neighborhood drifting about the country and perhaps uniting to a certain extent, but the August migrants from farther north are recognizably different. They usually come in high overhead at the Point in scattered bunches of from six to fifteeen and later in flocks of twenty-five to fifty or even more. One scattered flock follows another perhaps for several hours and fifty or more may be seen in the air at once. Their flight recalls that of a scattered flock of Robins and is different from the action during the breeding season. They stop to rest, as they arrive at the Point, on tree tops, wires and bayberry bushes, and a number will suddenly fly from a thicket as we approach. The birds seem lost and uncertain as to where to go next and when one individual takes wing others will follow, much like the Tree Swallows which a little later in the season will throng these same bushes. They will aimlessly chase one another about and when they take wing a number will go up to a considerable altitude in the diagonal lines that mark their pursuit of prey earlier in the summer. Even on the wires their attitude seems different from the birds that we knew in June and July. They perch stolidly with head bent over, somewhat in the pose of a Bluebird, with the tail slightly tilted up and the wings drooping a little below it. When first alighting, however, the head is held up and the wings tightly closed against the body.

While there may be several waves or flights of Kingbirds during the latter half of August the great migration occurs on or about the 30th of the month and after that only a few scattering individuals are to be seen and on many days none at all. The flights are mainly restricted to the vicinity of Cape May Point and northward along the Bay shore. We may have a veritable deluge of Kingbirds at the Point and practically none at Cape May only two miles to the east. It would seem that the birds come down the coast at a great height and when they sight the broad expanse of water where Delaware Bay joins the ocean they pause and descend forming the concentration that we find at Cape May Point. Or it may be that perhaps the birds which form the concentration may have been blown out to sea or in following the coast line lose it at the Point, and in either case, they beat in to land and follow the Bay shore into the wind as in the case of the Flickers. Kingbirds are supposed to migrate by day and my observations at the Cape bear out this theory. Certainly they are the most conspicuous birds that move in masses during the day with the possible exception of the Tree Swallows.

On August 27, 1920, I encountered my first notable migration of Kingbirds. Reaching the Point about nine o'clock in the morning I found the woods swarming with the birds; they were flying overhead in flocks, alighting and taking wing again continually. Many of them gathered about a sassafras tree on the edge of the pine woods and fed ravenously upon the berries. They would hover out at the ends of the slender branches or in the air and pick off the fruit, ten or twelve birds feeding at once, and then retire to perches in the nearby pines while others took their places.

As they came to rest there would be quite a tilt of the tail and then a bird would lean forward, lower the head and peer to this side and that in the manner of a cuckoo and then launch out once more into the sassafras tree, supporting himself on rapidly vibrating wings as he plucked off the berries. On September 11, 1927, a small belated flock was feeding in the same manner upon a cluster of pokeberry bushes, and on another occasion the migrants swooped down upon a patch of low huckleberry bushes and fed upon the fruit. This habit of an insectivorous bird feeding upon vegetable matter is characteristic of other species as well, when on migration, as I have noticed Tree Swallows and Myrtle Warblers devouring the berries of the bayberry and Red-eyed Vireos eating sassafras berries. On July 15, 1935, moreover I found a Kingbird, evidently not on migration, feeding on white mulberries in a tree on North Street.

When the great flight of 1920 took place, trees, bushes, and telegraph wires were literally teeming with Kingbirds all day. Flocks of fifty, forty or thirty-five would rise from trees on the Bay side and start out, widely scattered, across the water bound for the Delaware coast which, except for occasional views of the old Cape Henlopen lighthouse, at low tide, was entirely invisible. A few individuals which were kept in sight with the glass turned back and came to roost again but most of them kept on flapping their wings continually as they ascended at a very steep angle and attained a considerable height before advancing horizontally.

The birds that were thronging the tree-tops preparatory to making the crossing would pursue one another violently for short distances and then return to their perches. When I passed the Lighthouse on my return there were not more than twenty Kingbirds there and none at all at Cape May nor had there been all day. On August 30, 1930, there was the same mass movement in the face of a southeast wind.

On August 27, 1934, I happened to be at Cape May Point about 7:00 p. m. just before dusk and was surprised to find upwards of a thousand Kingbirds covering the bushes like Tree Swallows, their white breasts gleaming momentarily in the fading light as they arose in clouds and settled again in the same nervous manner characteristic of the swallows. They were still coming in, apparently from the north or down the ocean front, although it soon became impossible to distinguish them. They gave one the impression of settling for a night's rest and probably there are several distinct movements of Kingbirds during the day which are usually united at the Point where all pause, and that this was a delayed flock that had here reached the end of its day's journey.

When strong northwest winds are blowing at the time of the Kingbird migration the spectacle is somewhat different. There is the same congestion of birds at the Point but also a massed flight into the wind and north along the Bay. I was out on the sand dunes about a mile north of the Point on the morning of August 30, 1926, between 7:15 and 9:30 a. m. and witnessed the greatest Kingbird migration of my experience. The conditions were exactly those that would bring a Flicker flight later in the season, and the birds reacted precisely as do the Flickers.

There were always scattered Kingbirds in sight that morning flying along or perched on the tops of the gnarled and prostrate cedars which here dot the white sand hills. Then would come great straggling flocks of one hundred to five hundred birds spread out on a hundred yard front. They flapped their wings several times took a short sail, and then the wing beats again; some of them darted to right and left on short sallies and others passed low down close over my head. They seldom flew back but kept steadily on their way up the Bay. Every minute there were birds stopping to rest for a moment on the tops of the bushes or on leaning wind-blown trees, and others taking wing again. Sometimes a flock could be detected very high up coming in from the sea or from the ocean coast line I could not tell which, and apparently upon sighting the Bay they would drop down to the level of the tree tops. They seemed to come directly from the Lighthouse or a little farther east where I have seen ocean blown Robins beating their way back to land, and turn into the wind crossing the Pond Creek Meadows and on up the Bay shore. Some, however, seemed to come directly from Cape May Point having, perhaps, followed the coast line clear around the tip of the peninsula. On several counts I estimated that four hundred Kingbirds passed me within five minutes or about five thousand in an hour. This flight did not last more than two hours and there were considerable gaps between the individual flocks. Doubtless there were other flights during the day. It is a difficult matter to study these flights as it is impossible to be at several different spots at the same time to see what is going on at each, which is a necessary requisite to a proper solution of the activities of these migrants, but by combining a number of observations in different years and of different species of birds I have arrived at the conclusions stated above.

In 1935 William Rusling made a careful study of the August Kingbird flight and writes that "after several days of southerly winds there was a shift to the northwest during the night of the 22d. Immediately after dawn the next day there was a great flight of Kingbirds estimated at two thousand birds. They flew low just over the tree tops heading north into the wind and following the Bay shore rather closely, some of them directly over the beach. They appeared to be very tired and flew quite heavily. Occasional groups would alight on the tops of dead trees for a few minutes. The flight lasted for an hour with groups passing at intervals. There were five distinct flights; five hundred passed on September 29–30, 105 on September 6, 625 on September 7, and 265 on September 10. After that five were seen on the 13th, two on the 16th, four on the 21st, and a last individual on the 29th."

The dates of the massed flights of Kingbirds are as follows:

1923. August 27.	1929. August 31.	1932. August 29.
1924. August 29.	1930. August 30.	1934. August 27.
1926. August 30.	1931. September 1.	1935. August 23.

Small groups of Kingbirds are to be seen during the first ten days of September, usually not over ten to twenty, and after that only scattering individuals. My latest records are:

1921. September 13.	1925. September 22.	1928. September 23.
1924. October 2.	1926. September 12.	1929. October 5.
	1927. September 12.	

These were all single birds except on two or three occasions when as many as three were seen.

Throughout New Jersey the Kingbird is a common summer resident.

GRAY KINGBIRD

Tyrannus dominicensis dominicensis (Gmelin)

This southern bird was seen and studied at Cape May Point on May 30, 1923, by several members of the Delaware Valley Ornithological Club including Julian Potter, John Gillespie and David Baird, all of whom identified it positively and independently from specimens a day or two later. It was perched on a low cedar on the Pond Creek Meadows near the outlet, a short distance from Delaware Bay, and was studied for some time with binoculars. It darted out after insects and perched in exactly the manner of the common Kingbird while its attitude and actions were identical. The larger size, absence of white tail tip, and lighter gray back at once attracted attention and then the dark line through the eye and the large bill were clearly noted. The bird made no sound. (Auk, 1923, pp. 536 and 694.)

I searched the entire neighborhood two days later but could find no trace of it. This observation constitutes the only record of the species for New Jersey.

ARKANSAS KINGBIRD

Tyrannus verticalis Say

On September 17, 1923, I found one of these western birds perching on the posts of the fence surrounding the Coast Guard station at the Point. It flew from one post to another and also launched up into the air after passing insects, returning each time to the fence. As it perched it would turn the head, first to one side then to the other, looking diagonally upward. It was so tame that I was able to approach to within six feet of it where every detail of color could be studied.

I am thoroughly acquainted with the species in the West and even at quite a distance I recognized the pale gray color of the upper parts and the even lighter shade of the head, which appeared almost white, and I could detect the pale lemon tint of the breast. When close to it the light edgings of the tail feathers were plainly visible and the relatively small bill was noted. It made no sound.

There are several other records for New Jersey but no previous or subsequent observation of the species in Cape May County. Julian Potter saw one at Fort Mott on the Delaware, on October 6, 1929, and another on the same day at Oldman's Creek, Salem County; Charles Urner found one at Tuckerton on August 30, 1931; Joseph Tatum saw one at Barnegat on December 4, 1932; and another on Brigantine Island on October 13, 1935; Paff saw one at Barnegat Light on September 9, 1933. I have suggested (Auk, 1933, p. 221) that Tatum's Barnegat record was a Great Crested Flycatcher inasmuch as Marc C. Rich had seen one of these birds at the same spot nine days before but his statement sent to me later shows that I was in error, "the bird was always in the open, head and neck very gray and crest not conspicuous and tail short and square at end." A. H. Phillips secured an Arkansas Kingbird at Princeton on September 29, 1894, and there are I believe other records for the interior of the state. It is rather curious that so many individuals of this western species have found their way to the Atlantic coast.

*FORK-TAILED FLYCATCHER

Muscivora tyrannus (Linnaeus)

This tropical bird, which does not normally come farther north than southern Mexico and the Lesser Antilles, has, strangely enough been recorded three times from the state of New Jersey.

Bridgeton, December, 1820 (circa) (Bonaparte, Amer. Ornith. I. p. 1).
Camden, June, 1832 (Audubon, Ornith. Biog., II, p. 387).
Trenton, autumn, 1900 (Babson, Birds of Princeton, p. 56).

While all three birds were killed none seems to have been preserved. It would be idle to speculate as to how they found their way to New Jersey and whether man had anything to do with it.

*SCISSOR-TAILED FLYCATCHER

Muscivora forficata (Gmelin)

One specimen of this bird, which is a native of Texas and Mexico, was obtained by Dr. C. C. Abbott on the Delaware River at Crosswicks Meadow, five miles below Trenton on April 15, 1872, and is now in the museum at Salem, Mass. (Amer. Nat., VI, p. 267.)

GREAT CRESTED FLYCATCHER

Myiarchus crinitus boreus Bangs

Nearly every year one or two pairs of Great Crested Flycatchers nest in the town and a pair or more at Cape May Point while farther back in the country they are regularly distributed, a pair about almost every farm-house. In the pine woods at the Point we can nearly always count on seeing one of these birds on a morning walk in summer and in August an entire family is often encountered there. As soon as the young are on the wing, however, they drift about and there are days when none can be found in their usual haunts.

I have noted Great Crests at the Point on visits to the shore on May 8, 1925, and May 2, 1932, but probably they arrive a few days earlier, My latest observations in autumn are on September 16, 1923; September 27, 1926; September 17, 1932; September 29, 1935; but they probably remain later as in 1926 on the date mentioned there was a great migration and two dozen of the birds were counted among the many migrants. Their harsh grating call has been heard as late as July 16 and 31 in different years and on June 22, 1924, two Great Crests were pursuing one another through the woods calling continuously, doubtless a mating pair.

Like all flycatchers this species seems top-heavy with its large head and erectile crest, broad shoulders and tapering body. Its rusty tail, long and conspicuous, especially when partly spread, is always diagnostic. When perching the head is often thrust forward as the bird leans over and the crest feathers are elevated as it peers this way and that. In flying from one perch to another it seems a little sluggish or deliberate but upon alighting there is the same shuffling of the wings as in the Wood Pewee.

For several years a pair of these handsome flycatchers nested in a bird box 6 x 6 x 10 ins., fastened lengthwise to a pole in Walker Hand's yard in the heart of the town. The box was erected in 1923 with the hope of attracting a pair of House Wrens and to his astonishment a pair of Great Crests took possession about the middle of June and began to carry in building material. The female seemed to do most of the carrying and also did most of the nest building although the male accompanied her to and from the nest with great regularity. The sexes could be distinguished by the fact that the crest of the male was always elevated, while that of the female lay flat. The eggs hatched on July 3 and one of the young was seen to leave the box on July 28, flying directly from the hole to a tree some distance away with perfect confidence, and with none of the preliminary short flights of the usual fledgling. Its calls, too, were of exactly the same pitch as those of the parents. How many young were raised I am not sure.

During incubation the male fed the female and would lean into the hole but about dusk he would invariably leave the vicinity of the nest and go elsewhere to roost. After the young were hatched the female also apparently went away for the night. Both birds fed the young, the female often entering the box and remaining for some minutes, standing in the entrance hole and looking about in all directions. The male, however, departed at once after delivering the food. When he came to the nest he held onto the edge of the hole and leaned over inside meanwhile flattening his tail against the outside of the box. The parent birds caught many small moths and butterflies as well as other insects and an inspection of the nest after the young had left showed the hard shell of the abdomens of five cicadas. Another individual was seen to swallow a good-sized sphinx moth head first, without tearing it apart. Further inspection revealed the presence of ventral scutes and other fragments of the inevitable snake skins which the Great Crest places in its nest and which had been broken up evidently by the activities of the young. The nesting material consisted mainly of pieces of dry grass, a quantity of cow hair and some chicken feathers, the whole forming a dense mat two or three inches thick covering the bottom of the box.

The nest was about eight feet from the ground close to a garden fence and members of the family were constantly passing close to it. Hand was there most frequently and the birds showed perfect confidence in him, going to the nest to feed the young within a few feet of his head; they showed practically no alarm, however, at the presence of anyone in the garden.

In 1924 the male appeared about the box on May 15 and the female on May 29. By July 10 they were busy feeding young. The birds alighted, as the male had done in the previous year, leaning into the cavity and turning the flattened tail up so that the upper surface was close against the box above the opening. Sometimes one of them, doubtless the female, would enter for a moment and appear again with a ball of excrement which she carried away. I saw one of the parents catch a dragonfly and a grasshopper, each of which was beaten against the wire upon which the bird perched before it was fed to the young. Once a beetle failed to go into the mouth for which it was intended and the parent promptly hopped inside and recovered it.

The young birds left the nest on July 12, flying directly to a tree some distance away, as before. A pair of these birds used the box in 1925 but had considerable trouble with a pair of House Wrens which occupied another box not far away and apparently as a result of this they did not return the next year although in 1927 a male visited the box sometime in May.

A nest found by Richard Miller at Swain was located in a hollow limb of an apple tree and contained five eggs on June 6, 1920, while Turner McMullen tells me of one at Ocean View, June 17, 1922, which contained five young. Walker Hand tells me of another nest built in an old wooden pump propped against a house, the birds entering and leaving through the hole where the handle operated. Still another nest I found at Goshen built in a cylindrical tin mail tube fastened to a post on the roadside opposite an unoccupied house. It held three well grown young on June 28, 1933, and while we watched it one of the parents appeared with a yellow Colias butterfly.

Marc C. Rich has recorded a single Great Crest found at Barnegat Light on the remarkable date of November 25, 1932 (Auk, 1933, p. 221). The bird is a common summer resident in most parts of the state.

PHOEBE
Sayornis phoebe (Latham)

The Phobe seems to be only a transient in Cape May County as I have been unable to find any record of its nesting southeast of the lower Delaware Valley and central Pine Barrens, in both of which localities it is rare. It is more abundant in autumn than in spring but even then we usually see only scattered individuals, perhaps half a dozen in a day's tramp, although there are days of massed migration when there may be a considerable flight. I found them very abundant on September 27, 1926, during a Flicker flight, and again on October 2, 1924, while on September 30, 1923, Julian Potter counted thirty individuals. We usually see Phoebes at Cape May Point and along the Bay shore to the north but they also occur regularly about orchards and fence rows back in the open farming country.

Arrival dates for the Phoebe in autumn are:

1923.	September 23.	1928.	September 27.	1932.	September 25.
1925.	September 22.	1930.	September 28.	1935.	September 29.
1926.	September 27.	1931.	September 28.		

Latest dates are:

1923.	October 21.	1930.	November 9.	1935.	October 24.
1928.	October 27.	1933.	October 29.		

We also have some midwinter records which indicate that an occasional Phoebe remains as a winter resident; Norman McDonald found four scattered individuals on December 26, 1927; William Yoder and Henry Gaede found one on January 25, 1925; while on December 27, 1936, Otway Brown and I found one at Price's Pond and Norman McDonald another at the Point. Our spring records are as follows:

1924.	April 4–20.	1928.	April 1.
1925.	March 21.	1932.	March 25–26.
1926.	April 6–18.	1933.	April 1–17.
1927.	March 18–April 24.	1935.	March 23.

The migrant Phoebes of Cape May do not seem to have the assurance of the nesting birds that we are familiar with farther north and appear rather lost in the absence of their favorite surroundings of running brooks, stone springhouses and old bridges. They are usually seen at the Cape on low bushes or on the lowest branches of small trees. The behavior of one studied on April 3, 1924, at the Point was typical. It flew from one twig to another now and then darting down and alighting on a perch only a few inches from the ground but I never saw it actually rest on the ground. The tail is wagged constantly at intervals of one or two seconds, a quick elevation and a slower return, and sometimes there is a double "flirt" the second one less pronounced. When perching the wings are slightly drooping and the head well down on the shoulders and after a flight there is a rapid flutter and shuffling of the wings. Spring birds occasionally gave the familiar call but autumnal migrants were silent. The nearest breeding records to Cape May are furnished by Turner McMullen as follows:

Houck's Bridge, Salem Co., a nest on a ledge over a doorway in a deserted house: May 16, 1920, one egg and one young; May 15, 1921, five eggs; April 29, 1922, four eggs. Another nest on a beam in an old barn: April 29, 1922, four eggs; May 24, 1924, four eggs; April 26, 1925, five eggs; May 2, 1933, five eggs.

Pennsville, Salem Co., nest on a beam in deserted house: May 4, 1924, four eggs; April 15, 1928, two eggs.

Course's Landing, Salem Co., a nest on beam of a plank bridge: April 16, 1922, three eggs.

In the Pine Barrens we know of but a single nesting at Fourway Lodge some miles above Mays Landing on the Egg Harbor River, where a pair has bred for several years. This nest, placed on a ledge over a window, held young on May 16, 1936. The Phoebe is a common summer resident in the northern half of the state.

YELLOW-BELLIED FLYCATCHER

Empidonax flaviventris (Baird and Baird)

A regular autumn transient, always present in the first notable wave of late August or early September. Usually seen first in the pine woods at the Point, or in bayberry bushes immediately back of the sand dunes at South Cape May but occurs later in the Physick garden and along the Bay shore in thickets and open woodland. In the great flight of September 1, 1920, they were common everywhere.

Autumn arrival dates:

1920.	September 1.	1927.	August 25.	1932.	August 22.
1924.	August 27.	1928.	September 1.	1934.	August 27.
1925.	August 22.	1929.	August 29.	1935.	
1926.	August 28.	1930.	September 2.		

They have been seen as late as September 28, 1925; September 27, 1926; September 17, 1932; September 10, 1935. We have no spring records whatever and infer that the northward flight passes up the Delaware River Valley. It is not known to nest in the state.

ALDER FLYCATCHER

Empidonax trailli trailli (Audubon)

A specimen of this little Flycatcher was taken on Five Mile Beach on September 12, 1896, by William Baily. I also saw one in a thicket west of the Rutherford farm in West Cape May, on September 5, 1928, and Conrad Roland observed another at the Point on September 18, 1932.

In the northern part of the state they occur through the summer and nest locally. Samuel Rhoads found them at Alpine in June 1901, at Lake Hopatcong in late May, and at Greenwood and Wawyanda Lakes in June 1909 while Waldron Miller established their breeding at Plainfield in 1899. They are nowhere abundant and doubtless occur more frequently in migration in the Delaware Valley than along the coast.

ACADIAN FLYCATCHER

Empidonax virescens (Vieillot)

The Acadian Flycatcher is the summer resident little green flycatcher of the Delaware Valley especially on the Pennsylvania side and formerly or locally across central New Jersey north of the Pine Barrens. It seems to have decreased in the northern counties which were close to its northern breeding limit and this may have had something to do with its scarcity at Cape May where I have never been able to find it even as a migrant. Fletcher Street, however, located a pair, apparently nesting, at Ocean View on June 7, 1936.

Doubtless they migrate along the western side of the Delaware Valley.

LEAST FLYCATCHER

Empidonax minimus (Baird and Baird)

The Least Flycatcher is probably the commonest of the little green fly-catchers at Cape May. It occurs regularly as an autumnal transient at the same time as the Yellow-bellied Flycatcher and usually associated with it. I have records of occurrence as follows:

1920.	September 1.	1924.	August 18.	1931.	August 25.
1921.	August 19.	1925.	August 28.	1932.	August 29.
1922.	August 20.	1926.	August 28.	1934.	August 27.
1923.	August 31.	1927.	August 25.	1935.	August 23.
		1930.	September 4.		

They have been seen as late as September 14, 1921; September 17, 1932; September 5, 1928; September 7, 1935. We have no spring observations. The Least Flycatcher nests farther south in the state than does the Alder Flycatcher having been found in summer as far south as Haddonfield (Rhoads), Plainfield (Miller), Princeton (Babson), but its true summer home is in the northern counties, northward.

As we have only autumn records of any of the little flycatchers at the Cape they are of course not singing and one of the best means of identification is therefore lacking.

WOOD PEWEE

Myiochanes virens (Linnaeus)

One or two pairs of Wood Pewees are seen nearly every year at Cape May Point and in 1920 a pair bred in Cape May but this is unusual. To the north in the open farm land or along wood edges they are well distributed. They become more conspicuous in late summer when the young are on the wing and are then often quite abundant during the passing of the great flights of migrants from farther north.

Like all of the flycatchers the Wood Pewee spends most of its time on a perch on some dead limb or tree top from which it launches forth in pursuit of passing insects returning at once to its stand. At rest on its perch the Pewee looks peculiarly top-heavy and when the crest is raised the head seems out of all proportion to the slender body which tapers rapidly to the tail. A bird perching on a telegraph wire had great difficulty in keeping its equilibrium in the face of a strong wind and it was interesting to watch it swing its tightly closed tail far forward under the wire while the head was stretched over from above. On July 13, 1920, a pair nesting in the pine woods was studied for some time. The more active individual, presumably the female, held its crest erect and flew from perch to perch, shuffling its wings immediately after alighting and then settling them closely to the body. While I was near it uttered continuously the characteristic call of the species *tsu—wee* or *tur—eee*, as it has sounded to me at different times, or *pee—wee* as usually written, with an interval of four seconds between calls.

The other bird, watched from a distance through binoculars, seemed to have lost all apprehension and assumed a perch on a dead pine twig from which it sallied forth every few minutes after its prey, returning at once after each effort, and if successful wiped its bill on the branch after swallowing the insect. Sometimes it dived off the perch to quite a distance and turned half over sideways until one wing pointed straight down and the other up, and it seemed that the bird would turn a somersault; at other times in pitched right through the foliage of the tree and once vaulted high in the air and came down fluttering for some twenty feet in a long spiral apparently following some small insect in its devious flight. When returning to its perch it always flew around the trunk of the tree from behind and if it happened to come to rest facing the wrong way it immediately turned about into its usual position. Upon alighting there was always the same flutter of the wings and the shuffling of the tips as they were folded. Young birds, easily distinguished by the very pronounced buffy wing bands, would sometimes perch for many minutes without moving, on the top of an upright stub which they resembled so closely in color that they appeared to be part of it.

A nest found at the Point on June 23, 1932, was located on a horizontal branch of an oak about fifteen feet from the ground, and contained eggs. On August 28, 1926, a female and full grown young were busily engaged in devouring a large brown moth.

Last dates of occurrence in autumn: September 29, 1917; October 3, 1920; September 25, 1927; September 28, 1930; September 25, 1932. I have spring arrival dates: May 26, 1923; May 20, 1927; May 20, 1928; May 26, 1929, but as I was not present regularly through the spring these are probably late.

While not a characteristic species of the coastal area the Wood Pewee is a more abundant summer resident in the Delaware Valley and throughout the state north of the Pine Barrens, as well as in the cedar swamps of the latter region.

OLIVE-SIDED FLYCATCHER

Nuttallornis mesoleucus (Lichtenstein)

A rare but apparently regular autumn transient about Cape May Point where one or two have been seen every year. We have no spring records. I first saw one on the top of a dead cedar about fifty yards north of the turnpike, not far from the Bay, on August 13, 1923, and on September 4, 1924, collected one from the same perch. On August 25, 1927, this perch was again occupied by an Olive-sided Flycatcher and on August 14, 1932, one was located on the top of a tall dead tree in the sand dunes farther north on the Bay shore. This bird was quite restless and beside the usual sorties after passing insects it would make sustained flights to another dead tree top a quarter of a mile away and back again. The same, or another individual was found in the same place on August 29 and on August 9, 13, 26 and 30, 1934, one was perched on a dead pine at the head of Lake Lily.

The Olive-side is a glorified Wood Pewee with the head appearing still larger in proportion to the body; indeed with the crest erected the bird seems to be all head while the closed tail looks very short and inconspicuous. The blackish areas on the sides of the breast are very conspicuous, even at some distance, but the concealed white feathers at the sides of the rump when exposed are usually not visible from the ground. In the last individual mentioned, however, they were fluffed out and could easily be seen.

The habits of the Olive-sided Flycatcher are exactly like those of the Wood Pewee but there is a certain indescribable personality about it which, together with its size and habit of perching on the highest dead branches available, make it easy to identify. The first individual that I found was being harrassed by a Hummingbird, which may have had a nest or young in the vicinity, when it flew the Hummer was close after it and when the Olive-side rested the Hummingbird took its perch on a nearby twig.

The Olive-sided Flycatcher is a rather rare transient throughout New Jersey and so far as I know does not breed in the state.

*SKYLARK

Alauda arvensis arvensis Linnaeus

An article in the "Scientific American" for September 22, 1883, states that "eighty-four European Skylarks had been liberated on a farm in New Jersey and may now be seen apparently at home and quite happy." A subsequent note in the "Ornithologist and Oölogist" states that persons residing in the neighborhood of Winslow had seen Skylarks in a marshy scrub oak woods but there is no evidence that the observers knew the birds.

In the Patent Office Report for 1853 there is an article on the importation of Skylarks into Delaware by John Gorgas of Wilmington. He states that he imported from Liverpool two lots of these birds. The first twenty arrived on February 20, 1853, and were liberated on March 19. The second lot of twenty-two arrived on April 18 and were liberated next day. On July 24, following, the larks were ascending into the sky and singing "as cheerfully as they do in 'Merry England'."

In the vicinity of New York City Skylarks have been liberated from time to time and in 1887, according to Ludlow Griscom (Birds of the N. Y. City Region) a small colony became established near Flatbush, Long Island, but was destroyed by the advance of the city and none had been seen so far as he knew since 1913. So it appears that all of the introductions that might have established this bird in New Jersey have failed.

NORTHERN HORNED LARK

Otocoris alpestris alpestris (Linnaeus)

The "Shore Larks," as they are appropriately called along the coast, are of frequent occurrence throughout the winter on the upper beaches and sand flats of Cape May as well as in the old fields immediately behind the town, while they also are to be found on the golf links and on the baseball field. Their dull colors make them difficult to distinguish against the brown earth and the short dead grass and it may not be until they take wing right before us that we are aware of their presence.

While a flock may remain in the same neighborhood for some days at a time, or may return to it through the winter, they are usually restless birds and drift hither and thither, seeming to take particular delight in strong winds and blinding snow flurries. In a storm on February 5, 1933, when a gale was blowing and fine snow was drifting deep on fields and roads, I came on a flock of Shore Larks near Cold Spring which had just alighted in a small bare spot, temporarily clear of snow, where they were running about finding food of some sort in the clumps of dead grass that the wind had exposed. Farther up the coast, at Brigantine Island, Brooke Worth tells me of finding hundreds of Horned Larks in a violent snow storm which raged there on February 21, 1929.

The Shore Larks are probably present every winter in one place or another, not only in Cape May County but all along the New Jersey coast, while we may find them in old fields in the interior when least expecting them.

As we see a flock of the birds working their way along the sea beach unalarmed, they impress one as being very short-legged and they seem to waddle a little as they scurry along at a quick walk—almost a run, for they are one of the few species of small birds that walk instead of hop when on the ground. They hold the head well up and forward as if they were straining to get along as fast as possible and it bobs forward and back as the bird advances. As the flock proceeds the birds keep pecking at the sand to the right and left and sometimes turn completely around before making a dab, apparently sighting some choice morsel, and then resume their advance after a few steps backward. They never stop to feed at any one spot but are ever on the go, foraging as they advance, and although they may scatter widely and follow their individual paths, the whole flock keeps moving in the same general direction and, with the peculiar uniformity of action that characterizes flocking birds, they will rise as a unit when something alarms them. They seem to find their food on the surface of the sand or earth and I have never seen them scratch.

Their "horns" are not elevated while they feed but may be erected at any moment in excitement or as they pause to look about. Sometimes they will rush madly at one another one bird apparently driving a trespasser from its own feeding territory. The tail is carried horizontally and the tips of the wings are on a level with the base of the tail or droop a trifle below it, the wings are noticeably long and pointed, as we see them when the birds occasionally leap into the air or when they take flight.

The flight is undulatory and, as each bird moves individually within the flock, some are always rising and others falling but they keep close together, now skimming the ground now rising a little, coming into clear view against the sky or when over the white sand, and vanishing from sight when they cross some old brown field or patch of winter marsh. Samuel Rhoads and I found a large flock in the bitter cold days of January, 1892, on old fields behind the town, where the waterworks now stand, and when we flushed them they would fly round in long ovals coming back eventually to almost the exact spot from which they had arisen, and this happened again and again so loath were they to leave the neighborhood. When they come to rest they usually stand directly on the sandy beach or on the ground but I once saw a number resting on old logs washed up by the tide and buried in the sand above high water mark.

While Shore Larks are excellent examples of protective coloration when feeding in open winter fields they are quite conspicuous against the white sand of the beaches and appear almost black when seen against the sun. In a good light, however, and in proper position, the pink tints of the back are brought out strongly and they seem to acquire a ruddy flush, while the yellow throat and black chest band stand out conspicuously as they face you. There is a perceptible difference in size between the sexes the female being the smaller, and usually more streaked.

Shore Larks are most frequently seen from November to February and are most abundant when a storm brings them from regions where the snow has recently covered the ground and they are seeking new feeding stations; though they are not averse to traveling with the snow. Our earliest Cape May date is October 9, 1921, and we have seen them October 17–18, 1931; October 25, 1929; October 27, 1928. We have only two March records March 6, 1927, and March 19, 1932.

While Snow Buntings are stated to associate with Horned Larks the only bird that I have seen in company with them is the Lapland Longspur, a number of which were feeding with them on the strand at the southern extremity of Seven Mile Beach on November 28, 1926.

*PRAIRIE HORNED LARK

Otocoris alpestris praticola Henshaw

While the Common Horned Lake is strictly migratory and does not breed south of Labrador and Quebec the paler Prairie Horned Lark, a bird of the Mississippi Valley, breeds there and to a lesser extent on the Atlantic coast south at least to Long Island. It occurs in winter flocks in New Jersey either alone or mingled with the more northern race and seems to be extending its nesting range southward.

Nelson Pumyea has recorded many of them present on the golf links at Mount Holly, N. J., during the winter of 1932–33, which remained through the spring with some still present on May 14, six on June 23 and two on July 1. On July 4 he saw four one of which sought food and fed another, evidently a well grown young of the year. This seems pretty good evidence of their breeding at this locality. The birds were seen to soar high up in the air, in the "Skylark" habit of the species, and return to almost the exact spot from which they arose. Julian Potter found two singing at Bridgeport, Gloucester County, on May 31, 1936, one on June 6, 1937; and another at Auburn, Salem County. They were doubtless all breeding.

At Delaware City across the river John Gillespie and others found these birds singing on May 14, 1933, which suggests that they nested there as well. At Cape May, where they in all probability occur, at least in winter, no one has yet detected them.

Beside its paler coloration above, the Prairie Horned Lark can easily be recognized by the white throat and superciliary stripes in contrast to the distinctly yellow coloration of these parts in the other race. About Philadelphia and elsewhere the two races often associate in the same flock during their winter sojourn.

TREE SWALLOW

Iridoprocne bicolor (Vieillot)

The Tree Swallow is the most characteristic swallow of the seacoast and has been observed about Cape May in every month of the year. It is scarcest, of course, in winter and reaches its maximum abundance with the enormous flocks of September and October, while as a breeding bird it is local and not very common. One or two pairs usually nest at Cape May Point and not infrequently we find a pair on the edge of the meadows just back of the golf links. Old Flicker holes in telegraph poles are their favorite nesting sites here, while one pair used a cavity in a piling on the edge of the Harbor, and another bred in a hole in a post far out on the meadow back of Two Mile Beach. This hole was not more than three feet from the ground and the female could be seen perched in the entrance while the male rested on a stake nearby. They paired on top of the nest post. On Seven Mile Beach there used to be an abundance of old dead cedar trees with woodpecker nests or knot holes in their trunks and these harbored quite a number of Tree Swallows. Some still are to be found there. Richard Miller has visited this colony on Memorial Day (May 30) from 1911 to 1920 and has usually found eggs partly incubated although in 1933 and 1934 all nests contained young from one third to one half grown. As young were still in the nests on June 16, 17, and 19, in different years, it is quite probable that two broods are raised in a season, while a nest with eggs found here by Turner McMullen on July 1, 1923, and another on June 30, 1923, in Salem County, lend color to this theory. I have found young still being fed in the nest at Cape May as late as the first week of July while families of young have first appeared on the telegraph wires on July 9, 1923; July 16, 1924; July 4, 1927; June 30, 1932. The nests that I have examined are composed of grass and lined with feathers, the birds of Seven Mile Beach using Laughing Gull feathers, according to Richard Miller; the eggs number from four to six.

(689)

By the middle of July all of the young from local nests are on the wing and begin the formation of the great autumn flocks constantly augmented by arrivals from farther north. Throughout the Pine Barrens, especially along the Egg Harbor River and on flooded bogs and ponds, there are many dead trees of various sorts that have been killed by damming up the water and these furnish abundant nesting places for the Tree Swallows which in South Jersey always seem partial to the vicinity of water. While such areas produce a fair crop of swallows every year they constitute but a small part of the enormous autumn flocks, the bulk of which evidently come from much farther north. Instead of streaming steadily southward as do the Barn Swallows, in late August and September, the Tree Swallows form in large flocks and seem to remain more or less stationary for weeks at a time moving definitely southward only toward the end of the autumn.

A northwest wind in early August generally brings the first flock of migrants to Cape May Point, but even before this a considerable gathering may be seen on the Tuckahoe River from the passing train as it turns southward into the Cape May Peninsula, while an extensive sand pit at the western end of the Stone Harbor Road, where it enters upon the meadows, is another early gathering point. On August 1, 1921, several hundred Tree Swallows swarmed over a thicket of bayberry bushes, where none were to be seen the day before, and hovered close to the branches like monarch butterflies which often perform similar migrations later in the autumn. Dates for the first of these migrant flocks are:

1920.	August 5.	1927.	August 3.	1932.	August 13.
1922.	August 3.	1928.	August 13.	1933.	August 5.
1923.	August 4.	1929.	August 12.	1934.	August 11.
1924.	August 10.	1930.	August 12.	1935.	August 8.
1925.	July 30.	1931.	August 18.	1937.	August 6.

My notes of 1920 give a good idea of the character of the annual assemblages of Tree Swallows at the Point. On August 5, the flock that had gathered there was augmented by large numbers from the north, both old and young. They lined the wires and gathered in masses on a nearby cluster of bayberry bushes. Some clung to the telegraph poles like woodpeckers, looking for an opening on the wires where they might manage to squeeze in, and there was a constant sidling along the wires and wooden cross-trees as the birds packed ever closer together. Some adults were seen to feed full-fledged young and all were busy preening their plumage, holding the wings straight out from the shoulder, one at a time, in the stiff manner of the swallow tribe and reaching over the shoulder to preen the feathers of the lower back. They showed no alarm at my presence and I could walk past on the opposite side of the road without causing them to take wing. They swarmed over certain bayberry bushes which seemed to bear the most fruit but I do not think that they fed upon the berries at this time. If they did they must have simply scraped off the waxy coating from the surface as I found that the berries were difficult to detach from the branches. Furthermore on other occasions I have noticed Tree Swallows swarming in the same way over dead bushes which bore neither leaves nor fruit. Suddenly the whole assemblage of birds would take wing for no apparent reason, or several detachments would leave in succession, flying off a few yards and swarming back again. As I stood perfectly still, close to the bushes, I could feel the whir of the birds' wings as they passed within a foot of my face.

Flocks of Tree Swallows seem to pick some definite object upon which to swarm without any apparent reason, returning to it again and again only to desert it entirely the next day. On August 19, a small bayberry bush on the Fill, slightly taller than the surrounding vegetation, was selected and about thirty birds gathered upon it sitting so close together that it did not seem possible for them to move. On August 3, 1927, a spot at the base of the sand dunes near South Cape May struck their fancy and they crowded in great numbers onto sticks that were projecting from the sand and upon low bayberry bushes.

The 1920 flock at the Point had increased by August 17 to approximately 1200 individuals which thronged the wires and bushes the latter being simply loaded down with them. The long wires stretching across the meadows back of South Cape May were a favorite roost and the birds would line them so closely that at a distance they looked like heavy cables sagging down in the middle with the weight of birds. And so all through September they fluctuated in numbers and in location, some always passing southward along the beach, others gathering in larger or smaller flocks. Whether these were changing in their make-up from day to day or whether the same individuals remained throughout the month it was impossible to say.

On October 3, as I walked down the beach from Cape May, I found them as usual scudding back and forth over dunes and meadow, but arriving at the Point the air was suddenly charged with Tree Swallows. There were several thousand in a driving, swirling mass, quite definite in shape, like a great cloud. As quickly and mysteriously as they came they had gone again, alternately flapping and sailing, far off to the east. Then they were back again immediately overhead, the individual birds performing all sorts of evolutions within the prescribed limits of the flock. Then the cloud-like mass would rise steadily until nearly out of sight when it appeared like a swarm of minute gnats. At other times, as on September 13, 1921, the birds would suddenly scatter from the dense flock and pour over the country in all directions, as if belching forth from some enclosure of which the door had been suddenly opened, and then upon some common impulse they were all back in close formation. While never flying so habitually low as do the Barn Swallows, the Tree Swallows rarely ascend more than thirty to fifty feet until these great autumn swarms are formed when they may mount aloft clear to the limit of vision.

While Cape May Point seems to be the most favored place for these congregations of Tree Swallows they sometimes prefer to gather on the Fill.

On October 21, 1923, Walker Hand estimated that there were from ten to twelve thousand birds assembled there and again on September 26, 1926, I watched a flock of at least five thousand, a drifting maze of birds, now settling on the bayberry bushes, now shifting to the telegraph wires, and then spreading in a long procession far across the sky line like a gigantic swarm of bees. When at rest they covered all the wires on Madison Avenue as well as on some of the cross streets. On October 10 they occupied seventeen spans between the poles and with four wires to a span and two hundred birds to a wire there must have been close to 15,000 present.

They also like to gather over bodies of water and I saw one flock of about a thousand come down from very high in the air to the shallow pond at the Lighthouse skimming the surface in regular order. Then the van turned suddenly and flew through the ranks of their followers until all was confusion, then mounting in air again they circled several times and in three minutes had disappeared utterly. On October 23, 1927, at the same spot a flock was hovering or sailing over a small pool in the meadows, moving back and forth with great regularity, and every now and then a bird would touch the tip of its bill to the surface. Now and then one would advance by a short sail but usually all were flapping and hovering so that, at a little distance, they resembled a flock of butterflies. On November 8, 1930, a flock of one thousand clustered over this same pond in a dense mass, constantly dropping down, possibly to dip the bill, though it was usually the tail and belly that touched the water, and I wondered if this was a form of bathing. The birds that touched the water immediately arose again so that with some dropping and others rising there was a maze of motion, while every few minutes, with that remarkable uniformity of action, the entire mass would rise some ten feet in the air and then settle back to resume their performance. Finally they scattered widely over the meadow everyone for himself. While hovering over the water a Marsh Hawk several times passed close to them but made no attempt to molest them nor did they seem disturbed at his presence.

On some of these swarmings the birds appear to be catching minute gnats or other flying insects but there was no evidence of this on the above occasions. On July 23, 1934, however, a number of Barn and Tree Swallows were observed coursing back and forth over an old briery field catching strawberry flies which were there by the thousand and again on August 4, 1932, I watched both species of swallows as well as Chimney Swifts flying back and forth along a road catching flies of some sort.

Every year I have watched the Tree Swallow gathering on the Stone Harbor Road, and on August 14, 1923, I made a careful estimate of the birds on the wires which showed about ten thousand present, the crowded lines seemed to stretch across the meadows almost to Seven Mile Beach. There was a heavy wind blowing and many of the birds, as on several other occasions, sought shelter on the sides of the old sand pit nearby. On August 10, 1927, the sides of the pit were covered with them and they spread out over the level sandy floor, which was also out of the wind, and some fifteen hundred were ranged there in regular rows. On September 25 and 26, 1921, they assembled on the golf links, sitting close together and forming a great hour-glass some three hundred feet long by one hundred across the end. There is another spot where these swallows roost to their sorrow, on the shore road of Ludlam's Beach above Sea Isle City. Year after year they gather here on the concrete roadbed until they completely cover an area of some twenty-five feet square which appears as if covered by a glistening blue pavement. Why they should select this particular spot I have been quite unable to determine but it shows poor judgment for automobiles are constantly passing and serious slaughter of the birds ensues. John Emlen picked up ninety-six dead swallows on September 26, 1926, and on similar spots in the county Conrad Roland gathered up over one hundred from September 4 to 12, 1934. The killing is usually accidental as it is very difficult to avoid the birds but a few drivers deliberately increase their speed and plow through the flock with apparently no thought but to see how many they can hit.

On September 21, 1929, there was a veritible whirlwind of Tree Swallows on the Fill. With a strong north wind blowing they were beating their way against it, pyramiding high up into the sky, and then flattening down and spreading out into a sheet of birds a hundred yards wide. They flew low over the bushes and poured by close above my head until it seemed as if the whole sky was in motion and I felt dizzy. Then they would turn and drift slowly back again tacking and flapping against the gale and piling up gradually into another dense mass formation, shoulder to shoulder and layer upon layer, each bird holding his position by vigorous wing action. Once again they turn and going with the wind are fairly swept from the skies and we see them far to leeward like a swarm of gnats, gathering once more to face the gale.

Mass action such as this seems to be due primarily to the force of the wind. The swallows depend upon their wings to take them to water and to where insect food may be obtained in the air and a strong gale prevents them from reaching either objective. Their only hope is to keep beating to windward, as they would be blown to destruction if they travelled with the wind for any distance, and the strain would seem eventually to force them to the ground where without food or water they would perish. Walker Hand has described to me such an occurrence which forms a fitting climax to the account that I have just presented. "On October 25, 1930," he writes, "we had more Tree Swallows than I have ever seen before; the north wind was so strong that they were forced to the ground and were sitting on the open ground of the Fill by the acre and, the wind continuing, they were unable to obtain either food or water. On the 26th and 27th probably several thousand died. It was possible to stand in one spot and count over a hundred on the beach, on the Fill, or at the Point east of the Light. There are no ponds on the Fill and the gale prevented them from going to windward in search of water and there is no dew on the trees and bushes. The three days' struggle seemed to exhaust them. Myrtle Warblers and Robins which were in a similar plight flocked to a pan of water in our yard and I counted thirty birds there at one time. Being more or less terrestrial they were better able to cope with the situation than the aërial swallows." On September 16, 1903, a still more unusual incident occurred on the Physick property on Washington St., where the Martins were accustomed to roost, which was reported to me at the time by Walker Hand. The Martins had left and a host of flocking Tree Swallows took possession of the trees during a driving rain storm with very high wind, something that they never did before or since, except perhaps in the case of a few straggling individuals. In the morning the ground was literally covered with drenched, helpless swallows. Large numbers were gathered up in baskets and dried out and eventually all but about seventy-five recovered and flew off. The number on the ground as counted and estimated at the time was between six and seven thousand.

While the great October flocks decrease rapidly through November, some birds usually remain during the winter. In 1923 there were still 1500 present on November 17, 250 by November 24, and the same number on December 11, while from six to twenty were to be seen at intervals in January, February and March following. In 1926, another great Tree Swallow year, there were thirty-five present on November 13 and fifty on the 21st while twenty-three were seen on December 26. We have records of wintering Tree Swallows in 1900–1901; 1902–1903; 1923–1924; 1924–1925; 1925–1926; 1926–1927; 1927–1928; while in other winters no careful observations were made so that it is more than probable that some were present then also. The Christmas census of the Delaware Valley Club follows:

December 26, 1926, twenty-three.   December 24, 1933, eight.
December 26, 1927, two.   December 23, 1934, nine.
December 28, 1930, seven.   December 22, 1935, eleven.
December 26, 1932, fourteen.   December 27, 1936, eleven.

There is something inspiring about these supposedly delicate birds sailing over the frozen ponds in midwinter or rising from the brown bayberry bushes in the patches of yellow dune grass where they had sought shelter against the cold winds. Their food in winter is supposed to consist entirely of the bayberries which they glean from the bushes along the coast but an examination of the stomach of a midwinter specimen disclosed the remains of several large dipterous insects.

While the great autumn flocks of Tree Swallows contain no other species the earliest assemblages associate with Barn Swallows, stray Cliff and Bank Swallows and Rough-wings, and both Barn and Tree Swallows may join the Purple Martins in their roosting grove. Sometimes about the first of September one may be fortunate enough to find all six species resting on the wires at Cape May Point.

It is difficult to ascertain the migration dates of a resident species but there seems no doubt but that there is a definite arrival of north-bound Tree Swallows at Cape May about April 24, and I have recorded such arrivals as follows:

1921. April 24.	1923. April 22.	1931. April 25.
1922. April 23.	1926. April 24.	1932. April 24.
	1927. April 24.	

No great northward flight of the birds has been seen in spring, however, and what becomes of the thousands that pass south in autumn is another of the mysteries of migration.

In the northern counties of the state, and especially along the coast, the Tree Swallow is an abundant autumn transient, less plentiful in spring, and a decidedly local summer resident. (Griscom, Birds of the New York City Region.) It would therefore appear that the autumn hosts that reach Cape May must come from much more northern regions.

On the first appearance of Tree Swallow families on the telegraph wires one notes the great contrast in color between the old and young, the former burnished greenish blue, the latter dull brown, but both alike in the silky whiteness of their breasts. By August 15, the adults are in all stages of molt and soon assume their new plumage with prominent white tips to the wing coverts which are lacking in the worn post-breeding individuals. Later on the brown young assume the plumage of the adults and by late autumn they are indistinguishable. The general resemblance of the young Tree Swallow to the young Rough-wing is rather striking but the whiter under surface of the former and the white, instead of rusty, edgings to the wing coverts will serve to distinguish them. Walker Hand observed a pure albino in one of the large autumn flights on October 10, 1926, and another was seen at Barnegat farther up the coast by William Loetscher on September 6, 1935.

The autumnal swarms of Tree Swallows—and swarm is the only word that will adequately describe their abundance—form one of the most striking features of Cape May ornithology. Only the roosting Martins, the breeding terns, or the feeding shore birds of the sounds, present such a concentrated mass of bird life, while the intricacy of the Tree Swallows' movements, the fluctuations in their numbers and the sudden appearance and disappearance of the great flocks, seemingly at a moment's notice, arouse our admiration and wonder and furnish a constant field for speculation.

BANK SWALLOW

Riparia riparia riparia (Linnaeus)

So far as my observations go the Bank Swallow occurs about Cape May only as a south-bound migrant from late July to early September. While there are apparently no breeding spots to its liking in Cape May County there is no reason why it should not occur in spring on its northward flight, but we have seen very few at this season—four on May 13, 1928, with other swallows near the Lighthouse and one on Pond Creek Meadows. Bank Swallows usually occur in small numbers in flocks of Barn and Tree Swallows which rest at the Point on their southward flight, but sometimes there is a rather extensive migration of this species alone and for several days in August, 1920, numbers of them could be seen over the meadows and cattail swamps all the way from Broadway to Cape May Point. Their small size—the smallest of our Swallows—and the blackish breast band which in flight seems to be broken in the middle, are their most striking characteristics. When lined up on the wires with other species these features are very conspicuous and the breast band is seen to be complete although narrowed in the center. This character will distinguish them from the occasional Rough-wings and the young Tree Swallows, the other two swallows with brown backs and wings.

Bank Swallows are frequently seen, during August, sailing over the surface of Lake Lily, as many as fifty having been counted there on August 3, 1921, and sometimes they are intimately associated with Chimney Swifts which also resort to the lake at this time. While we see them mainly back over the meadows some of them travel down the beach with the Barn Swallows. They have one trait which seems peculiar to the species; when they settle to rest, as they often do, in the shelter of the sand dunes, several of them will be observed picking up small pieces of grass and sedge stems etc. in their bills, and then laying them down again as they waddle a few steps this way and that, over the sand. This habit does not seem to be a search for food as no attempt is made to swallow any of the pieces nor can it at this season (August 3) be a nest building impulse. A pure white swallow seen over the meadows on September 2, 1920, was identified as a Bank Swallow on account of its small size and association with many individuals of this species. Arrival dates of Bank Swallows from the north are as follows:

1920.	August 3.	1925.	July 30.	1930.	August 12.
1921.	August 3.	1926.	July 23.	1931.	August 1.
1922.	July 27.	1927.	August 3.	1933.	August 5.
1923.	July 22.	1928.	July 18.	1935.	August 7.
		1929.	August 2.		

Last dates of observation are:

1920.	September 2.	1923.	August 27.	1927.	August 25.
1921.	September 3.	1925.	August 22.	1932.	August 29.
1922.	August 28.	1926.	August 24.	1935.	September 6.

The Bank Swallow nests at various localities in the state but owing to the necessity for vertical sand banks, in which to drill its holes, it is decidedly local. There are several breeding colonies along the Delaware River above

Camden in one of which, at Delair, full sets of eggs have been found from May 22 to June 6. The nearest colony to Cape May seems to be in a gravel pit about one mile south of Bridgeton in Cumberland County. In June 1936, Julian Potter found about one hundred birds breeding there while in 1937, there were but seventy-five. He tells me that when the sand holes are not worked the tops of the banks cave in and the sides become sloping and the birds desert. Twenty years ago he found Bank Swallows abundant in glass sand pits near Vineland about ten miles to the east.

Like many purely transient species the Bank Swallow is by no means a common or characteristic bird of the Cape. It would seem to vary considerably in abundance and in three of the last sixteen years it was missed entirely. This however may be due to lack of constant observation immediately along the coast, for such continuous observation is necessary if we are to make an accurate record of its activities; most of the individuals that occur here are on the wing travelling rapidly down the shore line to winterquarters in the far south.

ROUGH-WINGED SWALLOW

Stelgidopteryx ruficollis serripennis (Audubon)

A few pairs of Rough-winged Swallows have been found nesting in holes in the bluff along the Bay shore below Town Bank during May of several years and doubtless occur there every spring. Rotten roots of trees, that have been exposed by the action of high tides during winter, fall out from the face of the sandy cliff and leave satisfactory nesting holes for the birds. Doubtless the first July Rough-wings that we see about Lake Lily, at the Point, come from there and also the scattered individuals that are seen in the swallow flights of August. On June 15, 1937, Julian Potter saw a pair of Rough-wings entering the bilge holes of the wrecked concrete boat at the end of the Cape May Point turnpike and it seemed likely that they were nesting there. The bird is here very close to the northern limit of its range and there cannot be many to migrate from the country to the north.

Most of our observations of this little swallow have been of family groups perched on the wires or dead tree tops in July or skimming over the waters of Lake Lily. One flock (July 22, 1923) consisted of ten individuals, perhaps two families, which were foraging over the upper end of the lake, keeping within an area of sixty by sixty feet and skimming the water slowly, directly in the face of the wind, and then drifting back to begin the advance all over again. They rested, three or four at a time, on dead bushes on the bank, all the while calling: *psud, psud, psud.* Another group of seven (July 16, 1927) consisted of two adults and five young. The latter rested on the wires and were fed by the parents which skimmed the lake for food. A third family of five was on the wires at the Point on July 5, 1932, while the earliest appearance was on June 30, 1935, an adult and five young.

All through the summer young birds can readily be distinguished from adults by the rich rust-colored edgings to the wing feathers and the slight brownish suffusion on the breast.

We have spring records at Cape May Point as follows:

1924.	May 10.	1927.	April 30.	1933.	May 6.
1926.	May 1.	1928.	May 5.	1934.	May 12.
		1932.	April 24.		

Our latest occurrences have been:

1924.	August 16.	1927.	August 26.	1929.	August 17.
1925.	August 18.			1931.	August 18.

William Rusling has recorded one on September 6, 1935.

Nests of the Rough-wing with sets of five and six eggs have been found at Palmyra, near Camden, on May 23 and 27, and the bird has been reported nesting as far north as Plainfield, Morristown, Summit, Paterson and High Knob.

1926
·CR

BARN SWALLOW
Hirundo erythrogaster Boddaert

The Barn Swallow is a common summer bird in the Cape May region and a conspicuous and abundant migrant during August and September, especially close along the ocean front where a few are sometimes passing south as late as November. They nest about all the farms in the upland and locally along the coast where suitable situations are to be found. At Schellenger's Landing, where numerous boathouses are clustered, many pairs of Barn Swallows nest every year and during May they may be seen flying in and out through the doors which open on the water, carrying building materials. They are early risers and I recall one mid-May morning as Walker Hand and I were leaving the Landing for a trip up the sounds at about three o'clock when we could hear the swallows calling all about us although it was still too dark to distinguish them. Wherever there are houses close to the shore, with sheds or other outbuildings that furnish acceptable shelter, a pair or two of Barn Swallows are usually to be found, as well as about isolated Coast Guard Stations or deserted summer cottages. Some also nest under the platforms leading to the boathouses at Wildwood Crest and at Stone Harbor, while a pair apparently had a nest for several years under a low stone culvert arch where Cape Island Creek formerly entered the town at Grant Street.

While living in a gunners' cabin out on the meadows back of Two Mile Beach in May, 1922, a few Barn Swallows flew past every day but they did not seem to be nesting in any of the similar shacks which dot the meadows, and which were perhaps too much exposed for their liking; they were evidently wide ranging individuals from some mainland colony like many others that I have seen foraging far out on the meadows.

Turner McMullen gives me the following list of nests examined by him on Seven Mile Beach where the birds nest in boathouses and under elevated boardwalks:

May 30, 1911, four eggs.
May 30, 1912, nest just begun.
May 26, 1929, five eggs.
June 15, 1930, three eggs.
June 18, 1922, four eggs.
June 24, 1928, five fledglings.
July 4, 1929, four half-grown young.
July 4, 1931, five eggs.
July 4, 1932, six eggs.

On June 15, 1937, Julian Potter saw Barn Swallows entering the cabin of the wrecked concrete boat in the Bay opposite Cape May Point and doubtless they nested there.

I saw a pair preparing to nest in a boathouse at Sea Isle City as early as May 9 and Richard Miller has found full sets of eggs on May 30, June 17, and June 20, in different years. On July 2, 1921, I found a nest with young on top of a beam in a small shed north of Schellenger's Landing, a favorite position for coastwise Barn Swallows, possibly because of the danger of disintegration, due to the humid atmosphere, of nests built in the usual style against the side of a wall or rafter. The parents of this family were coming in constantly with food and at intervals could be seen removing the white excrement sacs which they dropped about a hundred feet away. Another pair at Frank Dickinson's barn at Erma fed several large young in the nest on June 21, 1931, and apparently the same pair had another brood in the same place on August 9. Other pairs still had young in the boathouses as late as July 22, 1921, and Barn Swallows are to be seen there all through August.

Fledglings, conspicuous for their swollen yellow gapes and shorter less forked tails, appear on telegraph wires during early July—July 3, 1920; July 1, 1921; July 5, 1922; July 5, 1923; July 4, 1927; July 29, 1932; June 20, 1933; June 30, 1935. One family of four, perched on a wire near an unoccupied cottage at South Cape May, was being fed at intervals by the parents which were gathering food along the ditches of the adjacent meadows As a parent approached with food all of the young fluttered their wings anxiously and one of them usually launched into the air but it seemed always to be one of the others that got the food. They all opened their mouths wide and uttered cries of a single syllable. When the excitement was at its height the other parent often flew up and sometimes two other adults which had been feeding in the vicinity. The commotion over for the time being, the young plumed themselves and stretched their wings to their fullest extent, first one and then the other, until they reached beyond the tip of the tail. Sometimes the young birds will perch on bayberry bushes and once I saw one fed while on the wing.

On July 2, 1923, a family of five young rested on a bare sandy spot on the Fill, one on a low weed, the others directly on the ground, where they were fed by the parents with much fluttering and elevation of wings. On

July 8, 1932, there were several young on the wing near the same spot and one adult repeatedly flew close to my head calling *sheep* as she fairly brushed my cheek with her wing. I could find no young on the ground or bushes and her concern seemed to be for the birds in the air. On June 24, 1924, the entire boathouse colony rallied to attack a Fish Crow which had unwittingly alighted on a nearby post and drove him from the neighborhood.

The flight of the Barn Swallow is marvellously accurate, swift and graceful, and apparently is effected with a minimum of effort. As a rule they fly lower than other swallows, frequently just skimming the ground or the top of the meadow grass, and with just enough wing action to keep them going and to steady them as they tilt now to the right, now to the left, on their drifting, gliding way, often reminding one of a lot of wind blown leaves rather than birds which are consciously directing their flight. Their favorite position is not over two to four feet from the ground but occasionally one will throw its body up into the wind and after a rise and momentary pause will shear off again on another tack and glide away as before. Often they will select some piece of roadway as their feeding ground, skimming back and forth only a few inches from the surface and their habit of flying immediately in front of an automobile proves fatal to many a Barn Swallow, in spite of the efforts of the driver to avoid them. I have sometimes seen single birds follow a path across a field for a hundred yards without deviating from the straight line.

Fields of cut hay furnish attractive feeding grounds for Barn Swallows and here, with Martins and Chimney Swifts for companions, they will gather in numbers, whirling in intricate evolutions. Sometimes where there are swarms of gnats in the air many Barn Swallows will gather in a surprisingly small area to feed upon them and I once saw fifty so engaged in a space not over fifteen feet square. The gauzy wings of the flies flashed in the sunlight and made them very conspicuous, so that it was possible to see the birds snap them up as they passed and repassed in a maze of flight.

On August 13, 1928, I came upon a mass of Barn Swallows concentrated over a patch of short grass on the meadows above the Point which varied their usual method of feeding by darting down continually to snap up some sort of insect, and on July 23, 1934, there were a great many scouring some fields near the Court House where swarms of voracious strawberry flies made progress by humans almost impossible. The swallows were snapping up the flies right and left. Again on August 2, 1934, an old wheat field near Cold Spring, the crop of which had proved a failure, was plowed up and a host of birds assembled to feed on the scattered grain and the insects turned up by the plow. Killdeers, Red-wings, Robins and even Kingbirds were present at the feast and a number of Barn Swallows found some small flying insects to their liking and flew around over the backs of the other birds or rested temporarily on the upturned clods of earth.

At rest on the wires Barn Swallows, like other swallows, have a habit of stretching the wing straight out from the shoulder and so short is the upper arm that it appears as if the long flight feathers arose from close to the shoulder of the bird. They also stretch the wing to full length out past the tip of the tail, spreading the tail feathers on that side simultaneously. They sometimes open the mouth in a wide gape or yawn and will scratch the head with the foot leaning the head back over the shoulder in the accomplishment.

I have seen Barn Swallows during May resting on the muddy shores of Cape Island Creek and in plowed fields gathering mud for their nests and have noticed them in early July, when still feeding young in the nest, alight on the ground to pick up food washed up on the edge of the meadows and even fly out on the sea beach and alight along the high water mark to catch the flies that gather about the ocean refuse left by the waves. They alight on the ground, too, at times of strong wind apparently exhausted by their efforts to keep on the wing under such conditions. I have seen spring migrants resting thus in plowed fields and waddling on their short legs to get to the shelter of clods of earth or tufts of grass, as the small sandpipers do on the meadows during a gale. On July 10, 1921, many Barn Swallows gathered on the side of a railroad embankment for protection from a strong wind, while on July 1, 1922, they settled on a gravel road on the edge of the town in the lee of a thick hedge. On August 9, 1920, during a heavy southwest wind numbers of Barn Swallows collected on a sand flat below South Cape May and skimming the surface of the ground came to rest continually, one at a time, until they covered a considerable area. They acted individually in taking wing and settling again and never all flushed at once. Again in April, 1919, they assembled in the same way in a field by the Cape May Point turnpike where once again they were protected from the wind by a hedge. They were so loath to leave or so exhausted that one could approach to within a few feet of them before they would take wing. On September 2, 1920, many migrant Barn Swallows settled on the sloping roof of a cottage below South Cape May and lay close to the shingles on the side away from the wind; later they flew down to the nearby sand flat and settled there, all heads to the wind, looking exactly like the scattered black and gray oyster shells that were strewn over the ground.

When most of the young are on the wing the Barn Swallows of the upland begin to collect in flocks, feeding together and resting on the roofs of the larger barns. On one barn at Fishing Creek, on July 24, 1923, I counted seventy-five birds, mostly young, and on a nearby house on August 19, about 150. Such assemblages prepare the way for the great southward movement along the coast, which is so characteristic of August. Indeed some of the later gatherings are doubtless migrating birds from farther north pausing to rest, such as a flock of fifty seen on a clump of poplar sprouts on the Fill on August 7, 1923, fluttering as they clustered there until they seemed to be only part of the quivering foliage; and the long lines, sometimes 150 together, which throng the wires at Cape May Point late in the summer.

Barn Swallows migrate by day and the flight at Cape May seems to be mainly confined to the immediate vicinity of the sea beach while most of the birds pass within a strip one hundred yards wide. I have recorded the movement in progress by:

1920. August 5.	1925. August 9.	1929. August 17.
1922. August 3.	1926. July 23.	1932. July 28.
1923. August 4.	1927. August 3.	1933. July 26.
1924. July 18.	1928. August 13.	1935. August 8.

The major part of the flight always occurs before noon. The birds often fly only a few inches above the strand and rarely over six feet. Their course is somewhat erratic, drifting right and left and sometimes tacking back again for a few yards but the general progress southward is steady and rapid. During one of the August flights I stationed myself back on the meadows below South Cape May, some fifty yards from the dunes, where I had a clear view all the way to the sea. Apparently all the Barn Swallows were passing in front of me and selecting a definite line of bushes as a base I counted the birds as they passed it and the average was seventy per minute. Again on August 27, 1926, I counted the swallows that passed along the dunes and beach and the average was forty-six per minute and, so far as I could see, the flight continued at this rate for the better part of the morning and part of the afternoon, although there may have been breaks or pauses in the stream.

On August 24, 1927, I followed the flight all the way to the Bay shore and in spite of a strong northwest wind the birds kept right on, dropping from the higher elevation that they had assumed in crossing the pine covered area of the Point and flying close to the waves, so close in fact that they seemed in imminent danger of being washed under. They seemed to head directly for Cape Henlopen, but on August 1, 1932, they left the coast nearly opposite the Lighthouse.

On several occasions the birds paused on arriving at the Point and lined the telegraph wires, but never in the numbers of the Tree Swallows. On August 26, 1921, I counted eighty-five near the Lighthouse; there were 160 on August 26, 1920, and 150 on August 31, 1925. In the first of these assemblages there were two Tree Swallows and frequently a few Barn Swallows are found in large gatherings of the latter species but for the most part the two remain separate. On August 17, 1920, for instance, an immense host of Tree Swallows had gathered at the Point but there were no Barn Swallows with them although out over the beach the latter were passing steadily southward.

I have referred in my notes to the southward flight as "streaming" or "pouring" down the beach and that is the impression one gets of it. On one day I estimated that one hundred birds were in sight at once and again that upon looking ahead at any minute twenty or more would be in the immediate range of vision. On one or two occasions there was a definite flight across the Fill following the line taken by certain shore birds which coincides with the edge of old Cape Island Sound, which was filled in thirty years ago, and this may be the regular flight line here although on Seven Mile Beach it is along the ocean strand as it is between Cape May and the Point.

On some evenings, during the southward migration, a certain number of Barn Swallows went into the Purple Martin roost which existed at the time on Washington Street, sometimes only two or three and at other times a considerable number. On August 8, 1924, at least 250 gathered and swirled about over the trees, just as the Martins do, and again on August 20 and 21, 1926, there were several hundred present which kept more or less to themselves both in their aërial evolutions and when roosting on the wires. They were later than the Martins in arriving and were the last birds to go into the roost. The northward flight in spring is by no means so conspicuous as the southward migration and it seems possible that the birds do not follow the coast at this season, with the exception of those that summer here. On May 10, 1924, however, there was an exceptional flight. It was a raw cloudy day and the Barn Swallows were as abundant over the beach as in August and curiously enough they were flying south with a strong northeast wind! They would gather on dead snags sticking up on the edge of the meadows while some took refuge on the ground. Others were attracted by a flock of Semipalmated Sandpipers which were endeavoring to walk on the soft mud of a shallow pond and flew close down over them.

The average date of arrival of the Barn Swallow in spring, according to the records kept by Walker Hand from 1901 to 1920, is April 20, while in the eight succeding years my average is April 23.

In the autumn individuals continue to drift down the beach long after they have disappeared from the upland and we have a number of records for November, usually of single birds—November 11, 1921; November 12, 1922; November 16, 1924; November 25, 1928; November 7, 1931; November 14 and 15, 1935. William Rusling recorded 1500 passing the Point on August 26, 1935; 2520 on August 30; 1500 on September 9; with smaller numbers for later and intermediate dates. My latest date is a single bird seen on December 1, 1923. They are widely distributed summer residents throughout New Jersey.

The color of the breast in the adult Barn Swallow varies greatly and I have noticed, in the lines of birds on the wires during the August flight, some in which the breast was almost white and others in which it was deep rusty. There is no molt so far as I have been able to detect before the southward migration and all adults retain the long outer tail feathers until they depart, although the tips are often broken off.

While the Barn Swallow does not exhibit the spectacular massing that we see in the Martin and the Tree Swallow, to one who appreciates what is going on, its steady streaming down the coast, without pause or evident destination, is no less impressive, while the home life of the swallows about our farmhouses is a delightful feature of the bird life of the Cape.

CLIFF SWALLOW

Petrochelidon albifrons albifrons (Rafinesque)

A few Cliff Swallows apparently pass through Cape May every year on both the northern and southern migration, occurring in late April and May and again in August, the south-bound birds being usually young of the year. In August they are usually stray individuals in flights of Tree Swallows or Barn Swallows. I saw one on the western edge of the town on August 7 and another on September 3, 1920, the latter our only September record. An adult bird was seen on the telegraph wires at Cape May Point on August 15 and 16, 1921, in a flock of some three hundred Tree Swallows; while on August 26 one was found sailing back and forth over Lake Lily with other swallows. On August 24 and 25, 1927, I saw a young bird resting on a wire fence at the Point which allowed me to approach to within six feet of it. The top of its head was dull glossy green, back of head and neck dull gray in strong contrast, and there was a bluish black spot under the chin. The forehead and cheeks were chestnut and the rump reddish buff.

In spring the Cliff Swallows seem to occur in small groups by themselves and do not mix with the other species. Julian Potter saw six on April 23, 1922, and Richard Miller ten on May 13, 1923, while the Delaware Valley Club has recorded a pair near the Point on May 30, 1924. I saw two pairs at the Lighthouse Pond on May 13, 1928, and several pairs flying over the fields just back of the town on May 24 to 27, 1892. The lateness of these records might indicate nesting in the immediate vicinity, at least in the past, but we have no positive evidence.

The present day rarity of the bird at Cape May may be indicated by the fact that we have but eighteen records in the past eighteen years. In recent years the Cliff Swallow has very definitely lost ground in New Jersey, for some reason or other, and as demonstrated by B. S. Bowdish (Auk, 1930, pp. 189–193) its breeding range in the state is now restricted to a few barns in Passaic County, while it has become much scarcer as a migrant elsewhere. It would seem probable that the breeding birds pass south along the Delaware Valley rather than down the coast. It is today a rare bird anywhere in the East except in some of the mountainous districts.

The buff rump patch, paler in the adult, is the best identification mark but the bird has a peculiar habit of turning the head upward and mounting with vigorous beats of the wing for some feet and then shearing off on a long downward sail. They also skim the surface of Lake Lily and the Lighthouse Pond and on all of these performances the light rump is conspicuously displayed.

PURPLE MARTIN

Progne subis subis (Linnaeus)

With the possible exception of the Laughing Gull, the Purple Martin is the most characteristic bird of Cape May. Though always local in its distribution it is much more abundant in the coastal plain region of New Jersey than in the upland country farther inland where Martin colonies are few and far between. To many visitors to the Cape, therefore, Martins are quite as much of a novelty as are the birds of the sea.

As usual the Cape May Martins occupy nesting boxes erected especially for their accommodation on poles in yards and gardens. Some of these are quite pretentious affairs with many compartments, low projecting roofs and broad porches; others are more modest structures housing only two or three pairs, while the house most popular with the birds, to judge by its many tenants, was an aggregation of old wooden boxes partitioned off into sections with suitable openings cut in them. Unfortunately this commodious residence was taken down by its owner some years ago and the nesting Martin population was thereby sadly decreased. From careful estimates in recent years there have not been more than fifty pairs of Martins breeding in Cape May while colonies farther up the peninsula are mainly at Dennisville, Court House and other towns and villages, doubtless because only in these places have nesting boxes been erected. The dates for first arrival of

Martins at Cape May, mainly from the records of Walker Hand, are as follows:

1902.	May 2.	1912.	April 3.	1923.	April 22.
1903.	April 8.	1913.	April 12.	1924.	April 3.
1904.	April 1.	1914.	April 17.	1925.	April 8.
1905.	March 27.	1915.	April 12.	1926.	April 11.
1906.	April 4.	1916.	April 14.	1927.	April 24.
1907.	April 25.	1918.	April 16.	1928.	April 1.
1908.	April 18.	1919.	April 6.	1931.	April 19.
1909.	April 11.	1920.	April 14.	1932.	April 10.
1910.	April 10.	1921.	April 24.	1933.	April 20.
1911.	April 6.	1922.	April 23.	1934.	April 15.

In 1924 I saw the first birds for that year, one perched on the railing of the deserted board walk and a pair on one of the nesting boxes on Lafayette Street. The early comers not infrequently make an inspection of the box and then disappear for a few days, or perhaps are replaced by others, and it may be well on in the month before they become permanently established. Even then they may be loath to begin nesting operations until assured that the weather is favorable and as late as April 30, in 1923, I have seen a number of the birds huddled together on the porch of a nesting box, trying to keep warm in the face of a strong, cold wind. May and June are occupied with the nesting and during the period of incubation the females stay pretty closely indoors while the males perch, during part of the day, on nearby wires or tree tops, or even on the porches of the boxes, though they are mainly occupied in sailing about overhead or skimming the marshes in search of food. They do not, however, seem to wander very far afield at this time. The females are also seen occasionally, especially toward evening, feeding over the meadows, leaving the nesting box precipitately and entering immediately upon their return. Males have been seen to enter the compartments at this time but whether to relieve the females I could not be sure. There may be a harsh note or two of greeting as the birds arrive or depart but the general quiet that prevails is in strong contrast to the bedlam of a few weeks later when the young begin to appear.

The Martin on the wing is deserving of careful study: a glorified swallow, he is a perfect master of the air and his flight is at all times a wonderfully graceful performance. I stood on the edge of the meadows, one day in June, while twelve males were skimming the tops of the salt grass, tilting now to this side now to that to maintain their balance in the air. Now one of them turns in his course and passes close to my head, swift as an arrow and uncanny in his blackness, which has no relieving spot of white, not even on the belly where most birds are lighter in color. His call uttered either in greeting or protest as he passes is a harsh *zhupe, zhupe*. Again he will mount upward with rapid strokes of his narrow pointed wings only to return again to the lower level on a long sloping sail. Sometimes at the very summit of the ascent he will come about into the wind and remain stationary, on rapidly beating wings, before sliding away on the long downward sail. When flying high over the town, late in the summer, the Martins' mastery of the air is particularly noticeable. They come in against the wind on set wings like small three-cornered kites, steadying themselves now and then with two or three short wingbeats, and then, apparently tiring of the sport, they will drop through considerable distances and, flapping rapidly, regain their former altitude. While Martins flying low over the meadows are undoubtedly engaged in seeking food many of their aërial evolutions, like those of other expert flyers, seem to be for the shear joy of flight. Certainly they can find no food far out over the sea where we not infrequently see them as we walk along the beach, and the wonderful mass flight at roosting time has nothing to do with food. On July 29, 1916, too, I watched a single Martin associated with a band of Swifts maintaining a position directly over Congress Hall hotel, in the face of a strong south wind, for at least half an hour. The birds would often remain absolutely stationary in the air for several minutes at a time, evidently supported by the upward currents of air deflected by the walls of the building. Here was no search for food but some sort of enjoyment or play.

On June 29, 1920, and on several occasions before the young took wing, I noticed Martins, usually old males, coming down to the sea beach. They came regularly to the high water mark at the edge of the wet sand and picked up food of some sort, probably insects washed up by the waves. They waddled about in a clumsy fashion, as they often do on the porches of their houses, taking a few steps but remaining stationary most of the time, picking up whatever happened to be within reach. They would fly down in the face of a strong wind off the water and sail on set wings until their feet would touch the sand, sometimes poising in the air a foot above the beach for a moment or two. Again on July 6, 1921, I saw them sailing low down over the dunes just before sunset, catching insects of some sort but not alighting on the strand. On August 10, 1922, a large number of Martins settled on the golf links and apparently fed on some sort of small insect but this was a very unusual occurrence.

When the young are hatched excitement in the Martin colony increases daily. The parents are now continually scouring the meadows for food and are constantly arriving at the boxes or taking their departure, all of their movements accompanied by harsh challenges which are later mingled with the shriller calls of the young, as they become strong enough to stick their heads out of the entrances, or waddle boldly out onto the porches. Some young, at this time, are often crowded off onto the ground which adds to the excitement, as the parents swoop over them calling loudly. On July 17, 1922, the porches were full of young and I could see that both males and females fed them but only one young bird was fed on each trip. Young birds are seen poking their heads out of the entrance holes during the first week of July and by about the 20th of the month the first broods leave the boxes, the actual dates for the years 1920–1928 when I first saw the young away from the boxes being July 23, 21, 25, 12, 24, 20, 28, 27, and 25 respectively. By the end of the month all have departed and a strange stillness settles down over the once noisy nest boxes.

The families of young, accompanied by their parents, now take up positions on telegraph wires or on dead tree tops close to the feeding grounds on the meadows. The young seem to be able to fly fairly well as soon as they leave the boxes, as on July 26, 1920, I noticed a late brood being fed by their parents near the box, which upon my approach took wing and followed the old birds with only a suggestion of uncertainty in their flight.

On the wires the young birds sit very erect at first and are evidently not sure of their balance, their legs are spread wide apart and their tarsi are clearly seen from the ground below, as they hold their bodies clear from the perch. As the parents bring in food to the family on the wires there is much excitement, fluttering of wings and loud conversation. The young call *choop, choop,* and the females have a similar note which is sometimes doubled into *choop-choop,* while the old black males constantly cut in with their harsh rasping *zheee, zhudge, zhudge-zhudge-zhudge;* or again, *zhupe, zhupe, zhee-zhee-zhee,* a peculiar quality for a bird voice; almost guttural. When, as was usually the case, all called at once there was quite an uproar and when an intruder approached the whole family took wing in great and voluble excitement and passed on a little farther, alighting on the wires as before.

The increase of families of young on the wires and tree tops is noticeable during the last week of July, and the birds are more widely spread over the country than at any time earlier in the year. The young have a frosted appearance due to the grayish edgings to the feathers and to the presence of down, while their throats are paler, which serves to distinguish them from the old females.

I witnessed the feeding of the young in the air on several occasions and apparently they are induced to take wing in this way either intentionally or through the desire for food. As the young one would approach its parent in the air they would flutter up to one another, make the transfer of food, and then drift away again, the young one often returning to the wire. Several times I have seen an adult bring a large dragonfly in to its young as it perched expectantly on a wire and ram it down the wide open mouth. The insect with a body and wings three inches in length is forced head first down the throat while the young bird gulps spasmodically again and again, in an effort to swallow the meal, and often one will be found standing particularly erect with the tips of the dragonfly's harsh, rough wings protruding from either side of its broad mouth. The old Martins frequently follow closely after the flying dragonflies and once one of them in headlong pursuit of a flying cicada almost struck me in the face.

The dead tree tops are now the favorite resting places for the young and they return to them after their first foraging trips, and there the parents continute to feed them for some time longer. One young bird, roosting in such a situation, was seen to sidle along a branch toward a female that had just alighted, evidently in the hope of securing food. It moved its feet sideways, one at a time, and I have seen them do the same thing when shifting their position on a telegraph wire.

As soon as the young are able to secure their own food the Martins begin to disappear during the daytime and there will be many a day in August when not a single individual can be found between Cape May and the Point. Young and old sometimes return to the nesting boxes after they have left and roost there for a short time. On July 25, 1920, about 6:30 p. m., I noticed a number of Martins flying about one of the boxes and on August 6 following, early in the morning, some of the boxes were covered with roosting birds but this was probably due to heavy rains which had prevailed the night before and prevented the usual flight to distant feeding grounds.

I have seen Martins at sunset skimming back and forth over the sand dunes along the coast, apparently after insects while at Lake Lily on August 1, 1932, five hundred of them were assembled on the wires and every few minutes detachments sailed down over the water and dipped their bills to the surface apparently drinking.

The activities of nesting time and the care of the young, interesting as they are, fall far short of the spectacular performances incident to the roosting of the Martins in August and early September. Cape May has been fortunate in having in its immediate vicinity a roost of notable proportions and of long years standing. For many years it was located on the Physick property on the principal street of the town. Here there is a grove of silver maples about thirty feet in height and covering an area of some two acres, growing so close together that their tops join one another, making a dense canopy with constant shade. Extended grounds, screened from the public and supporting other trees and shrubbery, afford protection to the roosting birds. Between the grove and the sea beach, half a mile away, was formerly a stretch of salt meadow which has for thirty years past been converted into a slightly elevated tract of sandy brush-covered ground, known as the "Fill," and across its level surface the roost stands out conspicuously.

This roosting place was later abandoned in favor of other nearby locations as explained below and in 1936 the birds apparently deserted Cape May entirely as a roosting spot as no trace of a roost could be found anywhere in the county. The following account was prepared when the roost on the Physic property was at the height of its activity and it seems hardly worth while to alter it from the present to the past tense especially as the erratic birds may at any time return to their former haunts.

Here for some twenty years on every night during late July, August and early September, there gathered great flocks of Martins, Grackles, Robins, Cowbirds, English Sparrows and a few Red-wings, and in later seasons an increasing horde of Starlings. The Sparrows come in first, then the Grackles in compact flocks take up their position in the trees, next come the mixed flocks of Cowbirds and Red-wings, and close-packed bands of the Starlings glad of company and ever ready to mix into any assemblage of birds without inquiring into its character. The Robins come next in small detachments or singly, although many young Robins may be found feeding on the ground below the trees early in the evening. Finally come the Martins which outnumber all the rest together.

Were it not for this roost, the only one in South Jersey so far as I know, Martin history at Cape May would come to a close early in August when the last of the fledglings become self dependent and sail away with their parents. But as it is, though there may be many days in August when practically no Martins are to be found for miles around Cape May from sunrise to sunset, they will gather in ever increasing numbers to pass the night in this small grove which, so far as our eyes can detect, offers no advantages over hundreds of similar groves past which the birds must have flown. It would seem that most of these Martins must have come from areas far to the north of New Jersey, as the local breeding Martins could not have yielded such a crop of young. I estimate that there are not more than fifty pairs of the birds in Cape May and perhaps twice that number elsewhere in the peninsula and these hundred and fifty pairs could not produce more than six hundred offspring, making some nine hundred Martins in all, and yet at least 15,000 of the birds come to Cape May every night to roost. In the New York area, including northern New Jersey, Ludlow Griscom states that the Martin colonies are very locally distributed and that the birds are rare as transients, which further complicates the question of where our Martins come from! Another fact of interest is that on July 23, 1926, before any of the young had left the Cape May nesting boxes one thousand Martins had already assembled at the roost.

We have no means of ascertaining whether the same birds gather in the roost night after night, with constant additions to their ranks, and then pass southward when weather conditions seem favorable, or whether individuals are arriving and departing every day. Possibly the roost gathers all the Martins that pass south along the New Jersey coast and is maintained with the same personnel throughout the summer, until the entire assemblage leaves together early in September. There is undoubtedly some fluctuation in the number of birds present on different nights, and this might indicate that individual Martins remained in the roost only a few days, but on the other hand the evident familiarity with the roosting place would suggest that the same birds were present for the greater part of the summer. A curious feature is that, during the height of the roosting, Martins are exceedingly scarce about Cape May during the day and although I have traveled some miles to the north and west I have been unable to find any of the birds or locate any possible feeding grounds. Nevertheless at the approach of evening, at almost the same hour, night after night, they come to the roost by thousands from we know not where and almost always against the wind. It has been suggested that they fly very high, beyond the range of our vision, but it is hardly likely that any insect food could be found at such an altitude. In support of such a theory I must admit that on two occasions, August 28, 1928, and August 10, 1921, they did come to the roost at such an altitude that they looked like mere specks in the sky, while on August 14, 1932, a large number of Martins that had remained on the telegraph wires at Cape May Point until noon, took wing and at once ascended until nearly out of sight. These birds however may have been completing or beginning a long migratory movement.

As early as June 25 Martins have been observed going into the roost but these must have been either barren or non-breeding birds. They included both males and females and gathered on the wires opposite the roost and resorted to the upper branches of the maple trees at dusk. There were six on July 2, 1921; thirty on July 5, 1922; one hundred on July 9, 1922; a like number on July 4, 1923; twenty on July 10, 1924; fifty on July 5, 1925. The numbers present each night increased and fluctuated during the remainder of the month. Since the young Martins of the local colonies do not leave their boxes until about July 20 it is obvious that these roosting birds were not local breeders unless some of the males desert their families during the night. As the season advances the number of roosting birds increases steadily although it is usually August 10 before a really characteristic roosting occurs. The years 1925 and 1926 were exceptions, however, and we had an assemblage of at least 1000 birds on July 26 and July 23 respectively. Counts of the early gatherings indicate that there are about four females (or possibly un-molted young) to one black adult male.

On clear nights the birds come in a little before sunset heading straight for the roost, with their long powerful wing strokes carrying them directly into the wind, and so long as the twilight lasts they perform their remarkable evolutions over the tree tops or the streets of the town, preparatory to resting for the night.

My records show that the first arrivals may settle directly on the roosting trees, on isolated trees in the neighborhood, preferably those with dead tops, or on telegraph wires or the roofs of houses. Then, on some impulse or other, they all rise in the air and circle about overhead gradually dropping back again to their perches. As their numbers increase the repeated risings are like regular explosions of birds and as the whole mass goes circling—or "milling" as it is called, round and round high above the roost it resembles an immense swarm of bees. Finally the birds narrow their orbit with successive circles until they pour down into the tree tops in a great cyclonic cloud and as darkness creeps on they settle for the night. As one enters the grounds the chatter of the thousands of birds makes a great buzzing like the rush of water and if, on some impulse, they again arise in flight the rush of wings, to one standing immediately below, sounds like the crash of some great tree falling in the forest.

I have further details of the roosting during several seasons. On August 11, 1921, the Martins began to gather about 6:10 p. m. on the wires at various points near the roost, and by 7:00 the entire assemblage, numbering probably five hundred birds, had retired to the trees. A bunch of one hundred or more would circle about, drop in, and then take wing again, and spread out widely making a great circle 250 yards in diameter. Sometimes they would return to the wires for a few minutes and then swing around again to the roost. Part of the gathering would be milling regularly and the other detachments passing over the trees in a much larger circuit. On August 20, of the same year, there was an immense gathering, the largest of the summer. Up to 6:15 p. m. there were but a few individuals flying about overhead, but from sunset at 6:45 to dark at 7:15, the great flight arrived from the northeast. They came out of the gloom and haze, far away above Schellenger's Landing, and stretched out over both the meadows and the Fill. None came from the west or south, the wind being from the southwest. There was a continuous scattered flight augmented every now and then by a dense stream of birds, as if a great number had been suddenly liberated from some enclosure. Toward the end of the flight there would be a lull and then definite flocks would come in flying low. These late comers did not mill around like the early flight but plunged directly into the roost. At Schellenger's Landing the birds would pause and line the telegraph wires, doubtless tired by their long flight and in need of rest. Then they would rise all together and pass on to the roost. A few Barn Swallows were with them. On the 25th the wind was from the east and the birds all came in from the west. On the 26th there were absolutely no Martins in sight all day and I covered the country all the way to the Bay shore, yet at 6:00 p. m. they gathered by thousands at the Physick place, from where I could not determine, as I failed to see them arrive. The ground below the trees was by this time white with their droppings.

"August 30. Scattered birds present this morning but only a few. The wind was from the south and between 6:15 and 6:40 p. m. a great steady flight came in from the north. The birds alighted on the wires at the Landing for a few minutes and then passed on making a sudden congestion in the flight. Over the roost they appeared like a swarm of bees, or gnats, whirling round and round. Part of them had settled on the branches by 6:00 p. m. but the bulk did not go in until about 6:30. It became dark before 7 o'clock.

"August 29, 1922. There was a light southwest wind and the Martins began to gather very high over the roost by 5:30 p. m., descending to a lower level as their numbers increased. By 6:30 there were about six thousand over the Physick grove and probably two thousand farther out over the meadows to the west. They formed a huge maze like an immense flock of mosquitos and now and then about one third of the flock would veer off in a sudden surge to the right or left and back again, until finally the flying mass resolved itself into an irregular oval of moving birds. By 6:45 the flock had come down closer to the tree tops and great detachments began dropping into the roost like birds that had been shot and were falling to earth. By 7:00 p. m., the last individual had retired and there was a continuous hissing sound pervading the whole grove. In the morning there was scarcely a Martin to be seen.

"August 30. There was again a large gathering, but the birds came in suddenly and pitched down in large flocks, all within a few minutes, so that none was to be seen on the wing after 6:30.

"August 14, 1923. Great masses of Martins gathered on the wires on Madison Street below the roost and on Lafayette Street opposite. They seemed to occupy first the cross bars of the telegraph poles, and then to line the wires, all facing the wind and sitting as close together as possible. Some of them plumed themselves and stretched their wings one at a time down behind their bodies, while all of them raised and lowered their tails to balance themselves against the wind. By 6:00 p. m. they had all left the wires and the entire mass of birds was over the roost whirling about in a great irregular circle, large flocks pitch down into the trees and hundreds of birds on set wings sail slowly into the wind; suddenly something disturbs those already in the roost and the grove seems fairly to boil over with birds. Then they settle once more and all is quiet.

"August 15. Not a Martin was seen all day and by 5:45 p. m. none had arrived and I began to think that they had gone south when, far away over the golf links to the southwest, a few scattered birds came into view winging their way against the northeast wind, and immediately across the whole southwestern sky they appeared in a great scattered swarm, those in the rear mere specks on the horizon. They swerved about when over the roost and pitched straight down into the trees with no resting on the wires and no preliminary evolutions. The suddenness of their arrival and the short time required to settle in the trees were remarkable. Doubtless their progress had been delayed by the force of the wind." On August 26, occurred the largest roosting of the year 1923 and I estimated that there were at least 15,000 Martins present. On Madison Street they filled nine sections of telegraph wires, six strands wide, until the wires sagged perceptibly between the poles; the birds, with a very few exceptions, faced the wind which came from the northeast and they sat as closely as possible to one another. Suddenly, and with no apparent cause, the entire assemblage poured off of the wires close down over our heads with a great whir of wings, beating their way about fifty yards to windward and then drifting back again, they once more lined the wires. Some of them perched on the top of a nearby smokestack and its guy wires, as well as on the roofs of the houses. Once more they take wing and whirl close over our heads until we become dizzy as we gazed at the maze of moving wings, which like a black snow storm swept over the roost. When directly above us the impression was that the sky itself was in motion.

By counting the birds on one section of wires, and multiplying by the number of sections, one could form a pretty accurate estimate of the number present and tonight this showed four thousand on Madison Street, six thousand on Lafayette Street and at least five thousand in or over the roost. They seem to enjoy close quarters, like most birds of flocking tendencies, and would crowd onto a wire already apparently full in preference to making use of an adjoining wire which was still empty. Needless to say the sidewalks below the wires were all splashed with excrement and walking there was dangerous!

On September 4 it was cloudy and the Martins came to the roost early and by 4:30 they were all whirling high above the tree tops. Then they gathered in several tall silver poplars at the golf club, settling on the twigs and branches until the whole foliage appeared black, while their calls produced a constant twittering like escaping steam, now swelling loudly and then dying away again. When they crossed to the roost they went out in a great wave and the trees seemed to fairly belch forth Martins in a band forty feet wide, which continued long after one would have thought that all had flown. They swirled round and round on a great oval course mounting into the air until they seemed no larger than flies. None used the wires tonight.

In 1924, the flight of August 28 was notable. By 6:00 p. m., there were not over twenty Martins in sight. The Grackles, Red-wings, English Sparrows and Robins went into the roost and finally at 6:30 came the Martins all from the southwest like minute specks high up in the sky. The swarm of birds poured into the grove until 6:50 and by 7:00 all was quiet. The sun had set as a great red ball over the Bay at 6:30 and by the time the birds had retired darkness had settled down over the landscape. There was no preliminary gathering tonight and the birds milled around in a great surging swarm which constantly poured down into the trees like a huge funnel or the tail of a cyclone.

On September 1 the cyclone effect, as the birds entered the roost, was again marked. They seemed very nervous and again and again they rose from the trees to circle once more, so that it was quite dark before the last ones were settled for the night.

When thunder storms occurred at roosting time the birds acted differently and frequently came in earlier than usual, while on days of continous rain they lined the wires early in the morning and remained there the greater part of the day. Whether they spent the night on the wires in preference to the tree tops I was not able to determine. On August 8, 1922, there was a heavy downpour of rain at roosting time and the birds gathered on the wires to the number of about seven hundred and later the whole assemblage coursed low over the golf links apparently feeding on some sort of insects. On August 12, during a northeaster there were 975 Martins lined up on the wires all day; and on August 27, under similar conditions there were 1900 on the wires at 6:00 a. m., but as the rain moderated they left and returned as usual at 5:00 p. m. to roost in the grove. On August

17, 1923, there was rain all afternoon and the birds began to come in to the wires by 2:00 p. m., until they numbered about five thousand. They rested there until time to repair to the roost. On September 2, 1924, just at the time that they usually assemble, a thunderstorm broke over the town and a south wind, which had prevailed all day, changed to a gale from the northeast, the direction of the storm center. The birds were much bewildered and poised on set wings over the houses in broad columns, facing the wind. They did not seem able to circle in the usual fashion but gradually drifted into the roost.

When the time for their departure approaches there will be an evening when the Martins will appear in very much reduced numbers, many of them having apparently gone south after they left the roost in the morning. After that each evening brought a smaller number of the birds until the last stragglers were recorded. My dates for these are as follows; sometimes only a single bird:

1902.	September 6.	1924. September 15.	1927. September 11.
1921.	September 8.	1925. September 3.	1928. September 8.
1923.	September 15.	1926. September 11.	

In 1928, while some Martins made use of the roost, there were not more than two thousand present on any night during July or August and it was evident that there had been some change in their roosting habits. Next year the conditions were similar but the birds coming into the Physick grove had still further decreased. On August 22 I discovered large numbers of Martins circling over the bayberry thickets on the Fill, half a mile east of the roost, which had grown up in the past few years until they formed a dense "forest" eight to ten feet high. In 1930 the birds were going through the same evolutions over the bayberry thicket. Many of the Starlings and Grackles gathered first in the old roost on the Physick property and then crossed over to the bayberry roost all together. It was discovered late in the summer that there was still another roosting place, a tract of woodland in West Cape May. Thousands of Martins and Starlings and quite a number of Grackles repaired to this spot every night, and the bayberry roost on the Fill, as well as the original roost, were deserted. The West Cape May woods being off the main road, with no convenient wires upon which to assemble, the birds came directly to the trees from gathering points at more remote spots but they went through the usual circular flights and other aërial activities, just as they had done over the Physick grove. They used this roost during the summers of 1931, 1932 and 1933.

During the maintenance of the roost in the Physick grove the Martins used to assemble at Schellenger's Landing, at the old farm at the end of Mill Lane, and elsewhere, from which they would repair to the roost just before dark. Similar gatherings now occurred nightly on the wires along the Bay Shore Road at its crossing with the Higbee's Beach Road, and at Cape May Point, and several thousand birds would assemble there nightly preparatory to the flight to West Cape May. As these detachments arrived at the roost the birds that had already settled for the night would take wing and join the milling throng, and great excitement would prevail until all had settled again.

During the severe winter of 1933–1934 the woods were cut down for firewood and when the Martins began to arrive in July, 1934, they returned to the bayberry thicket on the Fill, using the same places of assembly as in the previous year—at Cold Spring and Cape May Point. By locating on an open spot on the Fill one could obtain a wonderful close-up view of the roosting activities of the birds. The Grackles and Starlings came first, individually or in small bunches, followed by Red-wings and Robins and then the Martins, a few scattered bunches at first and finally the great flights from the gathering places already mentioned. They came in force at 7:00 p. m., and all were in by 7:15. There were the same restless flights as already described at the original roost. The birds swarmed high up in the air and circled over a course, one hundred yards of more in diameter, in several distinct flocks. These eventually united and the circuit contracted until it formed a great funnel-like mass of birds which poured steadily down into the thicket. Later the birds burst forth again and circled close over our heads in the increasing darkness, until a final settlement was effected and all was quiet. From previous experience we estimated that 10,000 Martins used the roost this year. In 1935 they again gathered in the Fill but in 1936 and 1937 no trace of a roost could be found. Curiously enough in these two summers no flocks of Grackles, Robins or Starlings have been observed anywhere in the southern part of the peninsula and it would seem that they have all followed the Martins to some new rendezvous wherever it may be!

The Martins left the roost quietly at a very early hour in the morning and I never saw them leave. On rainy mornings, however, a large number of them remained all day on the nearby wires.

The Purple Martins have been the most interesting land birds of the Cape and the study of their activities, especially during the great August assemblages, has been fascinating. In spite of years of close observation we are still in doubt about many of their movements. We should like to know whether the same individual birds return night after night to the roost or whether its personnel changes every day. We should like to know where the birds spend the day, if the same ones that leave the roost in the morning come back at night, and we should like to know just when they start on their southward flight.

BLUE JAY
Cyanocitta cristata cristata (Linnaeus)

While the Jay is a resident of the woodland to the north of Cape May it is always most abundant in late summer and autumn and south of Cold Spring and Higbee's Beach it seems to be a transient only, occurring during September and October and occasionally, though rarely, at other times of year. Prior to 1927 I had never seen a Blue Jay at the Point but on September 25 of that year, to my great surprise, I found a flock of fifteen flying restlessly from one piece of woodland to another. They were repeatedly observed here during October and had increased to forty by the 23d while on October 10 two were seen about a deserted cottage at South Cape May close to the beach. This was our most notable Jay year and although they decreased in numbers as the autumn progressed a few established themselves in the Physick grounds in the heart of the town and remained throughout the winter until about May 1 when they disappeared. At the Point some of them lingered until May 10.

On October 5, 1929, there were at least one hundred Jays at the Point in a great drifting flock screaming as they flew about the pine woods. Smaller flocks of twenty-eight, twelve and eighteen were found on the 13th; nine on December 22 and seven on December 7. These winter birds were very quiet taking refuge in low thickets and feeding on the brown lawns of the summer cottages.

The record of the Delaware Valley Club's Christmas census is as follows:

December 27, 1931, thirteen.   December 23, 1934, four.
December 26, 1932, four.   December 22, 1935, fifty.
December 24, 1933, five.   December 27, 1936, four.

On May 2 and 7, 1932, I found two Jays apparently left over from a small number that had wintered at the Point and which seemed from their actions as if they might be nesting. They were very quiet sneaking through the woods and thickets until they encountered a red squirrel, a rare animal in these parts, which started them to screaming with a vengeance. They were not seen again.

Turner McMullen found a Blue Jay's nest at Rio Grande on May 29, 1926, which contained six young, and two nests near Camden on May 2, 1915, which held five and six eggs. These records will give some idea of their nesting time in this vicinity.

I was at first inclined to regard the scattered occurrences of Jays at the Point as the drifting down of resident birds from farther north in the peninsula but I later decided that they represented a regular southward migration and the observations of William Rusling, who was present during the autumn of 1935 and able to watch their fluctuations, abundantly confirm this view. He found Jays present every day in varying numbers from two to fifty and all apparently passed down the coast.

North of the Higbee's Beach Road Blue Jays are to be found in the deep woods every summer, especially in the coastwise patches of tall timber opposite Bennett, where one may walk through the shaded wood roads that lead from the inland fields to the salt marshes and hear the familiar cry of the unsuspecting Jay from close at hand. The cry is redoubled in its vehemence as the bird detects your presence. Blue Jays occur throughout the state in similar situations and it is probable that the migrants that appear at the Point may come from much farther north. Just where they come from and whither they are bound are problems for the bird banders.

MAGPIE
Pica pica hudsonia (Sabine)

A Magpie was shot on the meadows back of Atlantic City on November 16, 1933, by Mark Reed and reported to me by R. Dale Benson, Jr., and F. W. Laughlin. Unfortunately it was not definitely identified as the American race but it showed no evidence of captivity and in view of other occurrences in the East may well have been a wild bird, a straggler from the West. One had been seen on Edisto Island, S. C., early in May of 1934 (Sass, Auk, 1934, p. 524) and another was taken near Milwaukee, Wis., on November 5, 1934 (Mueller, Auk, 1935, p. 90), while one had been seen at Point Lookout, Md., on June 28, 1931 (Ball and Court, Auk, 1931, p. 604).

In September, 1935, Joseph J. Hickey tells me that two Magpies appeared in the Hudson Valley fifty miles north of New York City, one of which was eventually caught and identified as the American race, and in late November one was found in Vancortland Park, in the Bronx, and was observed by many persons daily until the time of the Christmas census after which it disappeared. On February 12, 1936, William A. Weber writes me that he and others saw doubtless the same bird on the Palisades below Alpine, N. J., directly across the Hudson, which constitutes a definite New Jersey record. There seems to have been a distinct movement of Magpies eastward during the above years as they were seen also in Ontario.

RAVEN

Corvus corax principalis Ridgway

Walker Hand always spoke of Ravens coming out to the sounds above Cape May from somewhere in the interior of the peninsula and there is no possibility of his having been in error as he knew the birds well. This however, was about the time of my first visit to the Cape, in 1890. His latest observations of Ravens on the meadows were on November 8, 1908, and December 31, 1912, a single bird in each instance. On October 25, 1936, James Tanner, Roger Peterson and Robert Allen saw a single Raven at Cape May Point which soared overhead for some time and is the only recent Cape May record so far as I know.

Farther north in the state they probably still exist in very small numbers. One was seen as recently as October 30, 1932, by Fletcher Street, Arthur Emlen and myself, flying from the outer shore of Barnegat Bay to the mainland, about a mile south of Manahawkin. It kept low down over the water and interspersed its flapping with rather long sails in the characteristic Raven manner. Its large size was notable. Charles Urner has seen Ravens in this same general region on several occasions—a single bird on October 9, 1927; two on January 23, 1927; August 11 and November 10, 1928; three on September 7, 1929; while Ernest Choate saw one on February 12, 1935, and there are other records.

In December, 1892, and February, 1893, Samuel Rhoads and I visited May's Landing for the purpose of collecting small mammals and almost every day during our stay we saw a pair of Ravens winging their way from a dense cedar swamp, lying west of the Egg Harbor River, out to the river bank or to the coast, presumably to feed. Natives who were well acquainted with them told us that they had a nest in the swamp. We not only studied their flight and compared them with Crows but on several occasions heard their guttural croaks.

In May, 1889, George Benners secured two young Ravens from a nest in a gum tree, between West Creek and Tuckerton, which he reared in captivity and named them, appropriately, "Never" and "More," unfortunately they lost any love that they may have had for one another and, as they grew older, engaged in several conflicts and eventually "Never" killed "More" and partly devoured him. "Never" is now in the collection of the Academy of Natural Sciences of Philadelphia.

On a subsequent visit to Tuckerton, about 1905, we were informed by Jillson Bros., who procured Benners' birds, that the Ravens still occupied the old nest but we were unable to verify this statement although the later records of Ravens in the Barnegat Bay region tend to confirm it.

When visiting the meadows about Atlantic City with Norris DeHaven in 1892 and 1893, we several times saw a pair of Ravens come from the mainland and settle on the edge of the thoroughfares and had good views of them from our catboat which was anchored for the night nearby.

EASTERN CROW

Corvus brachyrhynchos brachyrhynchos Brehm

Crows are present about Cape May throughout the year but during the summer months they keep for the most part back in the woods and open farming country to the north, or out on the edges of the salt meadows, and shun the vicinity of the town and the beaches. They are at all times alert and wary birds and give man a wide berth, so that we see them mostly at a distance or crossing high overhead.

In May and early June they are scattered in pairs, nesting in the woodland and foraging about the farms for food, approaching quite close to the buildings where there is promise of scraps of garbage or, perhaps, a young chicken which they are not averse to carrying off. I have also seen a Crow making off with a young Robin which was struggling and squeaking while its parents and another Robin which had come to their help pursued the robber viciously.

By early July the Crows have already begun to flock and course over the country in ever increasing numbers, playing havoc with watermelon patches in remote spots where they are screened from view, though they seem to find most of their food at this season along the Bay shore or on the Pond Creek Meadows and other similar pieces of marshy land surrounded by woods. The summer flocks of Crows are almost always seen at evening flying south along the Bay shore and apparently have a temporary roost in some of the adjacent woodland. On July 26, 1920, a long line could be seen flying southward just after sunset when the western sky was still aglow.

They were out over the Bay shore and proceeded as far as Pond Creek Meadows where they suddenly wheeled and flew back north disappearing behind the intervening woods; a bunch of twelve came first, followed by the main flock of forty-two. On August 18 a similar flock at sunset numbered four hundred and on July 9, 1922, and on the same day in 1924, a flock of seventy-five flew by in the same way. Another flock of twenty-five came all the way to the Point flying high overhead but when they sighted the ocean they turned back in bewildered confusion. The same thing occurred on November 16, 1929. Early in August, 1930, if not earlier in the summer, the Crows of Cape May had established a summer roost in an extensive patch of woodland near Bennett where they assembled every evening about 6:00 p. m. On August 19, I saw them coming in from a distance. As they arose far off on the ocean side of the tall woods north of Bennett they looked like a swarm of swallows and I could not realize that they really were Crows until I had focused my glass upon them. They came in two main detachments, great oval masses of birds which later spread out into long streams, about five hundred in each lot; other smaller detachments followed. On October 28, 1928, I found a flock of one hundred in a nearby field of neglected corn which later repaired to the same woods so that the spot may have been a temporary roost for a number of years. In 1930, however, it was abandoned by August 29, after a heavy gale and rain storm. When the Egrets and white-plumaged Little Blue Herons roosted in the coastal woodland opposite Bennett the Crows gathered with them and although in small numbers they made quite a contrast to the masses of snowy white birds that assembled there.

During the day these summer flocks seem to feed on the ocean side of the peninsula and on September 7, 1924, about eight hundred of them were gathered in an old field of stunted corn which the farmer had left to its fate. They were having quite a feast and finally betook themselves to a nearby woods where they settled and engaged in a curious low-voiced conference. Walker Hand and I, who had been watching them, managed to work our way through the underbrush until we were almost among them but they detected us and instantly became silent, retreating precipitately in great confusion. On October 2, following, this same flock was seen from a distance arising from the ground in a great square mass, the birds milling about like Martins at their roost. At the same time a mass of Starlings arose and apparently flew back and forth right through the Crows, looking in the distance like a swarm of gnats. Then the whole assemblage settled down again. On September 4, the flock was divided in three, each in a square mass not drawn out in a long stream as usual, and in this formation numbering 512 individuals it drifted across to the Bay just at sunset.

On September 20, 1924, and on several days preceding, Walker Hand saw this flock resting on the rails of the branch railroad that carries the fishermen's train to Schellenger's Landing, and on the adjacent marsh and fences. I later examined the spot and it seemed likely that the birds had spent one or more nights there. They had evidently been sitting very close together on the rails and all facing the wind, for on that side of the track was a long row of ejected pellets consisting mainly of poke berry seeds with some sumac seeds and fragments of corncob fibre. On the opposite side the ties and roadbed were splashed all over with white excrement evidently voided by the long line of Crows as they sat there.

As winter approaches the Crows seem to resort to the open marsh back of Hereford Inlet between Five Mile and Seven Mile Beaches where I have been informed that they have long roosted in winter. On January 2 and February 21, 1926, there were evening flights to the northeast from the vicinity of the Cape and on November 28, 1926, from the mainland opposite Stone Harbor I saw a long scattered line of Crows beating their way across from the northwest heading for the inlet and too far south to reach the wooded part of the island which one would think was more suitable for a roost.

On December 27, 1931, about one thousand Crows were counted by members of the Delaware Valley Club on their Christmas census, all of which were flying steadily south to the roost and on February 6, 1932, large numbers were gathered on the mainland edge of the meadows, as early as four o'clock, from which detachments left every few minutes flying southeast, and on December 26 of the same year Julian Potter found some five hundred on the meadows at Jenkins Sound which had evidently assembled *en route* to the same spot. This seems to be the only winter roost in the vicinity of Cape May although on November 8, 1930, I saw three Crows crossing over the Bay above the Point flying low over the water in the face of a light southwest wind, which may have been bound for one of the roosts in Delaware. Farther north, the vicinity of Salem, and, formerly, Pea Patch Island in the Delaware River, were notable Crow roosts.

Alexander Wilson, writing in 1810, says of the latter: "The most noted Crow roost with which I am acquainted is near Newcastle, on an island in the Delaware. It is there known as the Pea Patch, and it is a low flat aluvial spot, of a few acres, elevated but a little above high water mark, and covered with a thick growth of reeds. This appears to be the grand rendezvous, or headquarters, of the greater part of the Crows within forty or fifty miles of the spot. It is entirely destitute of trees, the Crows alighting and nestling among the reeds, which by these means are broken down and matted together. The noise created by those multitudes, both in their evening assembly and reascension in the morning is almost incredible." He also comments on the tenacity of the birds when, on the occasion of a severe storm and consequent rising of the water, the roost was entirely submerged and most of the Crows drowned. "Thousands of them were next day seen floating in the river; and the wind shifting to the northwest drove their dead bodies to the Jersey side, where for miles they blackened the whole shore. The disaster, however, seems long ago to have been repaired, for they now congregate on the Pea Patch in as immense multitudes as ever." (American Ornithology, Vol. IV, p. 82.)

In January, 1899, I attempted to locate a roost that was supposed to exist in the woods some five miles south of Salem, and which we thought might be the successor of the Pea Patch which had by this time been long abandoned. "By following the flight line taken by Crows passing Salem we finally located near a woods that we were assured was the one used by the birds, but although Crows were abundant flying about the adjacent fields there were none in the woods. We began to think that we had missed the roost after all when, upon emerging from the far side of the woods, we found an immense flight just beginning to pass overhead from the westward; evidently the river Crows had concluded that bedtime had come. They did not, however, alight on the trees, but passed over and dropped noiselessly into the low fields just before us, seeming to select a black burnt area on the far side. To our amazement this 'burnt' patch proved to be a solid mass of Crows sitting close together, and in the gathering gloom it was difficult to see how far it extended. Four immense flights of the birds were now pouring into the fields, in one of which we estimated that five hundred Crows passed overhead per minute, during the height of the flight, and the rate of the others seemed to be about the same.

"It was now quite dark, and we began to think that the birds had no intention of retiring to the woods, so we determined to vary the monotony of the scene and at the same time to warm our chilled bodies. We, therefore, ran rapidly toward the nearest birds and shouted together just as they first took wing. The effect was marvelous; with a roar of wings the whole surface of the ground seemed to rise. The birds hovered about a minute, and then entered the woods; we soon saw that but a small portion of the assemblage had taken wing. Those farther off had not seen us in the darkness and doubtless thought that the action of their companions was merely the beginning of the regular nightly retirement into the trees. The movement, once started, became contagious and the Crows arose steadily section by section. The bare branches of the trees, which stood out clearly against the western sky but a moment before, seemed to be clothed with thick foliage as the multitude of birds settled down. After all had apparently entered the roost we shouted again, and the roar of wings was simply deafening; another shout brought the same result in undiminished force and even then probably not half the birds took wing. They soon settled again and we were glad to leave them in peace." (Stone, Bird-Lore, 1899, p. 177.)

These winter Crow roosts and the long lines of Crows on their way to roost constitute one of the most impressive phenomena of our bird life. Unfortunately the popular prejudice against Crows, aided and fostered by manufacturers of fire arms and ammunition, has resulted in organized Crow shoots at many of the roosts which have brought about the slaughter of hundreds of the birds and the abandonment of several of the roosts. As long ago as 1810 Alexander Wilson describes the organization of farmers in Delaware to effect if possible the extermination of the birds and, while he claims that Crows are in many ways beneficial, he says "to say to the man who has lost his crop of corn by these birds, that Crows are exceedingly useful for destroying vermin, would be as consolatory as to tell him who had just lost his house by the flames, that fires are excellent for destroying bugs." While neither Crows nor corn fields are so abundant now as in his day the prejudice remains the same, even in spite of the careful analysis of the Crow's food by the U. S. Biological Survey which proves that the bird is just as beneficial to the farmer as it is injurious. We fear that it is the lust for killing, just as in the case of the hawks of Cape May Point, that is at the bottom of the Crow shoots, not the desire to protect crops on the one hand and song birds on the other! One Cape May farmer having provided himself with a stuffed owl as a decoy to bring Crows within shooting range found that the birds were diligently hunting and devouring cut worms on his fields and promptly abandoned all thought of killing them. Would that there were more like him! Protection against Crows and efforts to control their numbers are necessary in specific cases but may be accomplished by scarecrow methods and limited sl. oting, where the birds are actually committing depredations, but the who esale slaughter at the roosts at a time of year when the birds are wholly beneficial is an outrage.

Julian Potter, in January 1924, described the historic roost at Merchantville farther north in the state: "Twenty years ago" he says, "it harbored from twenty to thirty thousand Crows. At that time one line of flight which passed over Collingswood was so long that it stretched in either direction as far as the eye could see. This roost was investigated on March 20, 1924. As near as could be determined only about three thousand Crows came in. Instead of the long flight lines of former years, the birds came in in flocks of two to three hundred. This roost does not contain one-tenth of the number of Crows that it did twenty years ago which means that the Crow population for a certain number of miles around has decreased that much." (Bird-Lore, 1924, p. 188.) And now, twelve years later, he tells me it has utterly disappeared.

One of the most curious things about roosting Crows is their habit of roosting in trees in one place and on grass-covered marshes or islands in others. Do the same birds resort to different styles of roosts or are the components of one roost always the same birds and their offspring? These are some of the many questions about roosting Crows that we should like to solve. Also why it is that in spite of rain, snow, sleet or gales the devoted birds will persist in seeking the regular roosting place and pass over many a spot apparently far better sheltered and far more suitable for their night's lodging. But no, they would even perish in the effort, as they often do, before they will abandon the ancestral roost. This trait makes the wanton assaults upon the roosts all the more despicable.

At all times of year, especially in winter, Crows about Cape May feed about the hogpens that flourish here and there along the edge of the salt meadows, where refuse from the hotels is carted in great quantity, while they join the Herring Gulls and Turkey Vultures to feed upon fish cast up on the shores of the Harbor or the sounds, and do not disdain to share in the Vultures' feast of carrion.

On the salt meadows their range meets that of the Fish Crow, easily distinguished when the two are side by side, by its smaller size and, when in voice, by its harsh choking call as if suffering from a severe cold, as compared to the clear cut *caw, caw, caw* of the larger bird. On April 29, 1923, I found both species feeding on the Fill, perched side by side on the top of a signboard. The difference in size was clearly recognized but there seemed to be no difference in actions. Investigation disclosed that the Fish Crow was feeding on a recently killed hop toad while the other had the carcass of a bird apparently killed sometime before by a hawk. Solitary silent Crows from an area inhabited by both species are not always easily identified and I fear that on some "daily lists," where both are recorded, the desire to add another species gets the better of strict accuracy.

The slaughter of hop toads on the concrete roads, that have come into being with the automobile, has been enormous and the toad once so numerous about the Cape is actually becoming rare. At the height of destruction, between Cape May and the Point, there was scarcely a square foot of roadway in over two miles that had not its flattened and disembowled toad and I then saw many Crows each day frequenting this turnpike to feed upon the luckless batrachians. Crows often pursue one another in their capture of choice bits of food and in January I saw two relentlessly following another which was possessed of a choice tidbit, wheeling and twisting with wonderful agility, and finally forcing the victim to drop his plunder, whereupon all three dived to the ground after it.

The deserted winter beaches are favorite resorts for roving bands of Crows and on March 20, 1924, I found a flock of some fifty of the birds accompanied by Herring Gulls, gathered on the beach at South Cape May, feeding upon the trash which the waves had washed up, while a similar gathering of forty was seen at the same place on March 17, 1927. As they fed they walked about in all directions each one for himself with no uniformity of action. When standing erect on the beach the Crow's body is roughly triangular in outline, the breast coming straight down to make an angle with the abdomen which is parallel to the ground. The wings are folded tight and their tips show above the base of the tail only when the bird is bent over forward in feeding. The upper line of the body is then a perfect curve from rump to bill. There are always some members of the flock which are temporarily alert with head erect so that at any given moment there is always at least one on the lookout for danger; he promptly gives the alarm and all respond instantly. In this feeding flock some individuals were always flying a few feet in advance and settling again but there was little or no quarreling, although occasionally one bird would rush at another which had gotten a choice morsel. Crows coming in from the marsh, where many more were scattered about, proceded with the usual slow flapping until near the beach when they set their wings and sailed gradually down to earth. When about two feet from the ground they would give several rapid flaps and, with legs extended, alight on the strand. Several Crows gathered around a gull which was endeavoring to break open a mussel evidently hoping to share in its disposal.

It is difficult to approach Crows without being seen and my ability to study this group was due to the fact that I was partly concealed by a shed when they arrived and they did not see me. The moment I moved, however, they were off. I came upon two Crows on August 26, 1920, which were walking about in a pasture field looking for such food as might be there and, in the shelter of a hedge, I was able to approach within thirty feet of them undetected. They were excedingly wary holding their heads high and constantly on the lookout for danger. Presently they caught sight of me and, although I remained perfectly still, they immediately took wing. Their feet were at first dangling but after a few wingbeats were gathered up against the under side of the tail with the toes tightly clenched.

Single Crows passing the nesting places of Red-wings and Kingbirds are at once attacked and on one occasion fourteen of the former birds pursued a luckless Crow that had ventured to cross their marsh. On the other hand Crows will attack a roosting hawk or owl and pursue him relentlessly, gathering in a vociferous ring about his roosting place, and will apparently taunt him into flight and then follow to the next stand. Such attacks are usually in the autumn or winter and I can see no apparent reason for them. Wintering Red-tailed and Red-shouldered Hawks are frequent victims. On September 27, 1926, thirty Crows were busy mobbing a Barred Owl in some cedars on the Bay shore while on July 26, 1920, another band was harrassing a Great Horned Owl in the tall woods east of Bennett, flying at him with loud cries every time he took wing. Even the Bald Eagle does not escape their persecution; William Baily saw thirty Crows pursuing an Eagle on Seven Mile Beach in November and a large flock at Cold Spring followed another Eagle from perch to perch in early July.

There are few birds that present such opportunities for the study of behavior as does the Crow and it is unfortunate that his occasional lapses from grace have brought upon him so many enemies bent upon his extermination. As a matter of fact in the East all economic studies of the food of the Crow by qualified investigators show that he is as beneficial as he is injurious and he is economically, as we so often see him in actual life, on the fence!

Crows are resident birds throughout New Jersey and while most of their winter roosts have been located in the southern part of the state one was established as far north as Union County. Samuel N. Rhoads has published an excellent account of New Jersey Crow roosts in "The American Naturalist," 1886, pp. 691 and 777, while Herbert L. Coggins contributed a later account to "Cassinia," 1903, pp. 29–42.

FISH CROW

Corvus ossifragus Wilson

The difficulty in deciding whether a lone and silent crow is a Fish Crow or a Common Crow may as well be admitted for it is hard to judge of relative size unless two or more birds are present for comparison. If the bird is in voice the harsh, nasal, more or less subdued, *kouh, kouh,* of the Fish Crow is in strong contrast to the clear cut *caw, caw,* of the larger species and if in exceptionally good light and very close at hand the greenish instead of purplish reflections of the plumage may be distinguished.

The true resort of the Fish Crows is the vast expanse of salt meadows which stretches from the mainland to the coast islands and from the Harbor northward along the entire seaboard of the state, while they also follow what is left of the meadows south along Cape Island Creek to Cape May itself. At almost any time of year they may be seen flying low over the grass and, now and then, dropping to the ground in search of food which in spring and early summer consists to some extent of birds' eggs and young birds. Back of the golf links I have seen them on July 7, 1931, and June 27, 1932, evidently searching for nests of Red-winged Blackbirds and marsh sparrows with several pairs of the former in hot pursuit. In days gone by when the Clapper Rails were abundant Fish Crows were responsible for the destruction of large numbers of their eggs which they carried to a stake on the meadows and devoured, leaving the shells scattered all about. Fish Crows also associate with the Herring Gulls and Turkey Vultures to feed on refuse fish thrown out on the meadows by the fishing boats and here they sometimes meet the Common Crows which sally forth from their home on the mainland. While living in a little shack on the meadows in May, 1922, I had abundant opportunity to become acquainted with this small crow and not infrequently one would pause and alight on our roof, beating a hasty retreat when, to his astonishment, he found the building occupied.

Fish Crows resort to the woods of the mainland or the coast islands to breed. Many still make use of the scattered trees on Seven Mile and Two

Mile Beaches as well as a grove on a point projecting out into the meadows south of Mill Lane, known locally as Brier Island. The best known nesting place, however, is the pine woods at Cape May Point east of Lake Lily where for the past thirty years I have found from one to three pairs located. As soon as one approaches this woods the Fish Crows fly out to meet him and circle round and round overhead uttering their guttural cries as long as he remains.

On June 22, 1924, I found in the top of one of the trees a brood of four young about which the parents were greatly concerned. They were dull black, in strong contrast to the glossy plumage of the adults while the bills were short with swollen yellow bases. In other years the old birds have kept up their outcries as late as August 7, 1922; July 19, 1924; July 16, 1927; August 8, 1929; July 22, 1931. On two occasions a young bird was detected in one of the tree tops but on others no trace of young could be found although I am of the opinion that they were somewhere about. By the middle of August the wood is deserted. At the other breeding places the behavior is the same and on July 4, 1927, I saw two young at Brier Island which although just out of the nest managed to fluff up their plumage until they actually looked larger then their parents. Turkey Vultures or other birds which appear near the nesting trees are viciously pursued.

The following list of nests which they have examined on Seven Mile Beach has been submitted by Turner McMullen and Richard Miller:

May 13, 1927, seven nests with four to five eggs.
May 14, 1916, four eggs.
May 15, 1926, five and six eggs.
May 18, 1924, four, five and six eggs (seventeen nests).
May 21, 1922, three and five eggs.
May 31, 1920, two with five eggs each.
June 11, 1922, four eggs.
June 20, 1926, three, four, four, and five eggs; these were second layings from nests robbed on May 15.
June 21, 1925, young birds.
June 24, 1925, young birds, one third grown.

Most of these nests were built in red cedar trees, a few in hollies.

Miller tells me that Fish Crows never use earth in nest construction as does the Common Crow.

In winter the Fish Crows collect in small flocks on the meadows and on the island beaches and Julian Potter found no less than thirty roosting on the roof of a closed cottage on the edge of the Fill on November 14, 1920. The Delaware Valley Ornithological Club's count of Fish Crows on their Christmas census from 1928 to 1936 are as follows:

December 23, 1928, twenty-four.	December 24, 1933, sixty.
December 22, 1929, four.	December 28, 1934, fifteen.
December 28, 1930, ten.	December 22, 1935, seven.
December 27, 1931, thirty.	December 27, 1936, eleven.
December 26, 1932, ten.	

While the Fish Crow is chiefly maritime and is resident all along the New Jersey coast it also occurs on the Delaware River and large numbers of the birds have been caught in winter for trapshooting, just north of Philadelphia. As to the extent of their nesting along the river I do not have satisfactory information. (cf. Stone, Auk, 1903, pp. 267–271.)

Nests have been found as far north as Hudson and Union Counties (Eaton and Urner).

* BLACK-CAPPED CHICKADEE

Penthestes atricapillus atricapillus (Linnaeus)

The common Black-capped Chickadee of northern New Jersey occurs as a resident as far south as the Raritan River according to Waldron Miller, and reaches Haddonfield and Princeton during the periodic autumn migrations which bring it as a winter resident to the vicinity of Philadelphia and elsewhere in southeastern Pennsylvania. Farther south in New Jersey I know of but one positive record, a specimen obtained at Vineland by Dr. Max Peet on February 1, 1914; the Carolina is the common Chickadee throughout southern New Jersey.

The Black-cap has been recorded once or twice in winter at Cape May Point by visitors to the Bird Sanctuary but the members of the Delaware Valley Club have been unable to detect it there or elsewhere in the vicinity and no Cape May specimen has ever been taken. The two species can readily be distinguished when in song as the Black-cap has but two notes while the Carolina has four. Relative length of tail, amount of white on the edges of the tertials and secondaries, and amount of frosting on the black throat patch—the most striking differences, are not very satisfactory field marks and I am inclined to think that in winter errors might easily be made, unless one had given a great deal of study to the birds.

CAROLINA CHICKADEE

Penthestes carolinensis carolinensis (Audubon)

The Carolina Chickadee is a characteristic bird of the pine woods and swamps of the Cape May peninsula. It is a permanent resident and nests in old woodpecker holes or natural cavities in dead tree trunks, while in autumn and winter it ranges more widely, often in small parties, and may be found in orchards and gardens or wherever there is promise of hibernating insects. We have found Chickadees in every month of the year at Cape May Point and in late summer, after the nesting season is over or during autumn and winter, they will come into Cape May foraging in the shrubbery of the yards or in the shade trees along the streets. Philip Laurent found them throughout the year on Five Mile Beach before the woods were removed.

During a large part of the year we more frequently see Carolina Chickadees in pairs or as single individuals but in the colder months we may come across bands of six or eight, doubtless family groups, or they may associate with Myrtle Warblers, Tufted Tits, Red-bellied Nuthatches or Brown Creepers, all drifting through the woods in a concerted search for food, and now and then a Downy Woodpecker may join the group.

We may discover several such groups in the course of a winter's walk and on January 25, 1925, Henry Gaede counted no less than twenty-five Carolina Chickadees, while on December 27, 1927, Julian Potter saw twenty. On the Christmas census, when members of the Delaware Valley Ornithological Club cover all the country south of the Court House in a combined count, much larger totals are recorded:

December 23, 1928, sixty-eight.

December 22, 1929, one hundred and five.

December 28, 1930, fourteen.

December 27, 1931, forty-two.

December 26, 1932, eleven.

December 24, 1933, thirty-five.

December 23, 1934, fifty.

December 22, 1935, one hundred and forty-seven.

December 27, 1936, eighty-six.

While there is some fluctuation in numbers this is apparently due to local shifting of Chickadee population and to food or weather, and there is no evidence of regular southward migration which would cause a definite scarcity. The bird is here close to the northern limit of its distribution and there could be no influx from farther north to take the place of possible emigrants to the south. William Rusling, present daily at the Point from October 10 to November 14, 1935, saw the species on only nine days, from two to seven per day, which simply represents the normal population there at this season.

As we see the Carolina Chickadees at the Point they go drifting leisurely through the pine woods, stopping here and there to explore a branch from above or swinging upside down below an attractive bunch of pine needles, while they pick off minute insects that may be lurking there. Then we see one dangling from the end of a slender branch while he probes between the opening scales of a cone, and now he will drop into the undergrowth of bay-berry bushes or even onto the ground where he hops about for a few minutes among the tussocks of grass ever on the alert for food. Their actions are sudden, almost instantaneous, and they will often turn completely around in a single jump with a rapid flirt of the tail. After swallowing an insect a Chickadee will wipe the bill on a convenient twig, first on one side then on the other, and is on his way again with the usual rapid low *ts'dee, dee, dee, dee*,—as it sounds to us close at hand. The song of nesting time is a double call "*tsee-dee, tsee-dee*" the second couplet being pitched a trifle lower than the first. I have heard it as late as August 7 but it is unusual after May or June.

Carolina Chickadees, like the Red-breasted Nuthatches, often feed on the seeds of the pine cones and carefully trim off the wings before swallowing the kernel so that when several are feeding at once there is more or less of a shower of discarded seed-wings floating down from the treetops. In the winter of 1905–1906 Walker Hand found a flock of fifteen Carolina Chickadees which took up their quarters in a privet hedge in the heart of the town and were seen daily for several weeks. Others have wintered in the shrubbery of the Physick place on Washington St. which offers adequate shelter and protection to many winter birds.

On February 7, 1932, I came upon two of these Chickadees in a swampy thicket at Cold Spring feeding on the berries of the poison sumac. They would seize a berry in the bill and, hopping out on a horizontal limb, would hold it against a twig with both feet while they picked at it and apparently ate it piece by piece as I saw no fragments drop. On March 5, following, I saw two feeding in the same way on berries of poison ivy which grew on the trunks of pine trees. They never carried berries away but clinging to the vines leisurely picked them to pieces and swallowed them bit by bit. The great abundance of poison sumac and poison ivy about Cape May should ensure these little birds a regular food supply when the store of hibernating insects grows less as the winter advances.

I have a record of a nest at Swain with five eggs on May 15, 1924; another at Rio Grande with six young on May 24, 1925; and one at Whitesboro, May 10, 1925, with six eggs (McMullen and Miller).

There is something very attractive about these little Carolina Chickadees in winter when we find them most abundant. They are very trim and sleek in their gray, black and white plumage which seems particularly fresh at this season., We note the duller gray instead of pure white edgings to the wing and tail feathers and the clear cut line of demarkation between the black throat and the white breast which distinguish this little southerner from his slightly larger cousin of the north—the Black-capped Chickadee. The two are, however, not so easily distinguished in life as might be supposed.

Waldron Miller, who was better acquainted than anyone else with the bird life of Plainfield, told me that the Raritan River apparently marked the division between the breeding ranges of these two Chickadees. The Carolina is resident at Princeton but apparently not abundant in the breeding season although Babson (Birds of Princeton) records a nest and six eggs found May 29, 1901.

*ACADIAN CHICKADEE

Penthestes hudsonicus littoralis (Bryant)

The unusual flights of this brown-headed Chickadee to New England and New York during the winters of 1913–1914 and 1916–1917, brought several to northern New Jersey. Waldron Miller recorded one shot November 1, 1913, at Ramsay (Auk, 1920, p. 593) and two were seen by him on December 17, 1916, in the vicinity of Plainfield and single birds on January 7 and 28, 1917, while another was seen by Charles Rogers on February 4. In the vicinity of Englewood one was found on December 23, 1916, by Lester Walsh and on January 1, 1917, by Charles Rogers (Auk, 1917, p. 218). At Princeton four were observed by Henry L. Eno from November 18, 1916, to March 31, 1917 (Auk, 1918, p. 231).

No others have been recorded, so far as I know, and it has, of course, not been seen at Cape May but, inasmuch as Evening Grosbeaks and Red Crossbills occurred with the Hudsonian Chickadees at Plainfield, both of which have been recorded from Cape May, its occurrence there, in some future wave of northern bird life, is not beyond the range of possibility.

TUFTED TITMOUSE

Baeolophus bicolor (Linnaeus)

While the Tufted Tit occurs regularly throughout most of the peninsula of Cape May it is by no means universally abundant and becomes rather rare at the southern extremity, about Cape May City and the Point, and there may be some years when none at all are seen there. It seems more at home on the Delaware Valley side of South Jersey from Salem north but family parties drift down to the Point at intervals and some remain through the winter while a pair now and then nests there. There seems to be no true migratory movement and even at the northern limit of its range the Tit is resident. Its movements are apparently merely in search of food when parents and young unite in a foraging party.

On January 19, 1922, I came upon a flock of twelve, busily feeding in the pine woods beyond Lake Lily. They hopped from branch to branch with a rapid, nervous flip of the wings, carefully inspecting the trunks and limbs of the trees. Sometimes one of them would hold himself erect and peck at the bark like a woodpecker or hold a seed in his bill and hammer it against the branch. Now and then all would resort to the ground, hopping about with neck held high and crest erect, pecking right and left and rustling the dead leaves as they went. Their flight from tree to tree was somewhat undulating.

On April 3, 1924, there were two Tits associated with a Carolina Chickadee and some Myrtle Warblers. One of them worked for some time at a large scale of bark which he finally dislodged and then joined the warblers on the ground hopping about and feeding among the scattered pine needles which covered the woods floor.

The call of the "Tom Tit" is a loud clear whistle which sounds at a distance like the cry of a young Turkey—*weet, weet, weet, weet*, but which upon closer approach resolves itself into *peeto, peeto, peeto, peeto*, the "o" being unaccented which accounts for the difficulty in detecting it at a distance. In the midst of repeating this call over and over, the bird may suddenly alter its notes entirely substituting the more rapid, though less strongly accented, *toolée, toolée, toolée, toolée*. There is also a low *dee-dee* note like the latter part of the Chickadee's call.

In April I have records of single birds in six of the past seventeen years and sometimes a pair. In May, June and July all of my records are from the country north of Cold Spring and Higbee's Beach, where the Tits are always more plentiful. For August and September I have no records at all, doubtless due to the silence of the birds at this season, for we always hear more Tits than we see. In the late autumn more records have been obtained; two in October and two in November and the Christmas lists of the Delaware Valley Club give us December records as follows:

December 26, 1927, two.	December 26, 1932, five.
December 23, 1928, two.	December 24, 1933, nineteen.
December 22, 1929, thirty-three.	December 23, 1934, five.
December 28, 1930, one.	December 22, 1935, forty-one.
December 27, 1931, thirty-four.	December 27, 1936, seventeen.

The far carrying whistle of the Tit is its most striking characteristic and is a delight in cold winter days when so many birds are virtually silent. Like the Cardinal and the Carolina Wren, fellow Southerners, the Tit is here close to the northern limit of its range, being of only irregular occurrence in the northern counties, yet it shows no disposition to retreat southward in severe seasons, and doubtless some individuals pay the penalty and years of absence of any of these species may often be correlated with severe winters immediately preceding. It has bred as far north as Union and Essex Counties.

WHITE-BREASTED NUTHATCH

Sitta carolinensis carolinensis Latham

While one race or other of the White-breasted Nuthatch is resident from northern North America all the way south to Florida the species is absent, except as a rare transient, and still rarer winter resident, in southern New Jersey. In the Middle States the high hard wood forests seem to appeal more strongly to it than do the coastal pine lands. While a few probably pass through Cape May every autumn we have records for only seven of the twenty odd years that careful attention has been given to the birds of the Cape and only twice has an individual been seen in spring, one by William Yoder on April 11, 1926 and another on April 23, 1932. While this scarcity of records may be due in part to lack of continuous observation,

William Rusling's experience during the autumn of 1935 bears out our designation of the White-breasted Nuthatch as a rare species in this region. He saw it on but four days. Three on September 24 and two each on October 5, 17 and 19. Philip Laurent saw but two on Five Mile Beach before it was built over, both in autumn. It is a resident species in the northern counties.

Members of the Delaware Valley Club have seen White-breasted Nuthatches at Cape May as follows:

1923. September 16, 30.	1925. October 18.	1928. October 7.
1923. October 1.	1927. October 23.	1929. November 16.

All single birds evidently on migration.

Wintering birds as recorded on the Christmas census trips follow:

December 23, 1928, one.	December 26, 1932, one.
December 22, 1929, one.	December 22, 1935, three.
December 27, 1931, three.	December 27, 1936, two.

*BROWN-HEADED NUTHATCH

Sitta pusilla pusilla Latham

While this southern nuthatch occurs in Delaware across the Bay we have never been able to find it at Cape May. Beesley gives it in his list of Cape May County birds (1857) and Turnbull (1869) states that it is a rare visitant in the southern counties, but does not specify whether he refers to Pennsylvania or New Jersey. Neither writer is explicit as to the source or nature of his information.

The only definite record for the state is a single individual observed at Haddonfield by Samuel Rhoads in winter about 1876. The bird came to feed on suet fastened to a tree near the window and was carefully studied.

RED-BREASTED NUTHATCH

Sitta canadensis Linnaeus

The Red-breasted Nuthatch is an irregular or erratic autumnal transient about Cape May, abundant in some years and rare or absent in others; a varying number remain through the winter but there is no definite return flight in spring although occasionally an individual bird may be seen at this time. What becomes of the hordes of Red-breasted Nuthatches that move on to the south in some years and apparently never come back is one of the standing puzzles of bird migration. (cf. Nichols, Science, August 16, 1918.) The pine woods at the Point constitute one of the favorite resorts of these little birds but they are common elsewhere in the peninsula and at the height of their migration occur in numbers in the town, wherever trees are to be found. It was plentiful on Five Mile Beach in September and October when the island was wooded.

On September 9, 1921, a year when these birds were here in great abundance, a great many could be seen in the pine woods, where they crept about the branches exploring the clusters of pine needles and probing between the scales of the cones. Sometimes one would pause on the under side of a cone cluster and stretch his neck upward and backward, until he was looking out over his shoulders, while another would cling upright to a cone and probe into it after the manner of a woodpecker. A third bird followed out an old broken branch clear to the end, examining every inch of it with the greatest care, and then was off on his somewhat undulatory flight in which the shortness of his tail was conspicuous. Alighting on the trunk of the next tree he went round and round head down creeping like a mouse; now he pauses to pry off a loose scale of bark, seeking some lurking insect hiding beneath, now he stops and daintily picks off a number of gray aphids from a bunch of pine needles and then is off to other feeding grounds. One of the few spring birds that I have noticed was hovering in the air as he pursued a small moth in the manner of a flycatcher.

On October 23, 1927, and on several previous occasions I have seen Red-breasted Nuthatches feeding on seeds from freshly opened pine cones. Picking the seed out with the bill the bird would carry it to a definitely selected spot on a nearby limb where he placed it in a notch or under the edge of a bark scale and proceeded to hammer it with the point of the bill, deftly cutting off the wing, and perhaps breaking the seed covering, before swallowing it. Each bird visited again and again the spot which he had selected to prepare his seed for eating, and when several were working in the same tree the loose seed-wings were constantly fluttering to the ground.

Several times I have seen a Red-breasted Nuthatch flying wildly round and round above the tops of the pines in an ellipse probably one hundred feet in length, covering exactly the same course several times before coming to rest. I could not guess at the meaning of this unless it is the recurrence of a flight song performance incidental to the mating time.

Our records for Red-breasted Nuthatches at Cape May are as follows:

1918.	July 18.	1927.	August 26–October 23.
1921.	August 13–October 9.	1928.	September 30–October 21.
1923.	August 26–October 21.	1929.	September 20–October 25.
1923.	November 11.	1930.	August 11–October 5.
1924.	October 2 and 3.	1931.	August 27–October 18.
1925.	August 20–October 18.	1931.	November 28.
1926.	July 18 and August 8.	1932.	September 25–October 8.
1926.	November 13.	1933.	July 8–October 29.
		1935.	September 7–November 14.

In some years it will be noticed that we had only one or two records during the entire season while in others there were days when the nuthatches fairly swarmed, such as September 16, 1923; September 22, 1925; etc.

While the usual time of arrival, on years when a pronounced migration occurs, is about the middle of August on years there were three years when a much earlier arrival was noted. On July 18, 1918, I found a single bird on a pine tree near the head of Lake Lily and on the same date in 1926, William Yoder saw one in the same vicinity, while on July 8, 1933, I found at least three in the pine woods at the Point.

Our winter records are as follows: On December 1, 1921, Fletcher Street saw a single bird while I saw perhaps the same individual on January 9 associated with a party of Tufted Tits. On January 25, 1925, Henry Gaede saw ten in the woods at the Point and I found one there on February 14. Records of the Christmas census are as follows:

December 26, 1927, four	December 22, 1935, twenty
December 23, 1928, eight	December 27, 1936, one
December 27, 1931, two	

In spring I found one in a spruce tree on the Physick grounds in Cape May on May 10, 1924, associated with a Tennessee and a Black-poll Warbler; and on May 2, 1932, there were three or more in the pine woods at Cape May Point and another in the Physick place, with several more there on May 5, 6 and 7 and several at the Point on the last date. On May 12, 1934, I found one at the latter spot and this record further illustrates the irregularity in the movements of the bird.

We naturally associate the Red-breasted Nuthatch with the crisp days of September and October when it appears in greatest abundance and with the pond pines of the Point which seem to be its favorite haunt.

As the Red-breasted Nuthatch does not breed anywhere in New Jersey the host of transients that arrive periodically in autumn must come from the north woods of Canada and the northernmost United States.

BROWN CREEPER

Certhia familiaris americana Bonaparte

The Brown Creeper is a common and regular autumnal transient throughout the Cape May Peninsula, and formerly on Five Mile Beach, while occasional individuals remain as winter residents. There is no noticeable return flight in the spring and our five April records, all single birds, may just as well have been wintering individuals as migrants from farther south; I should certainly so regard the few March occurrences.

Our autumn records for the past fifteen years are as follows, the November dates being after the main flight may possibly indicate wintering birds:

1921.	September 14–October 16	1931.	October 18. November 7.
	November 11.	1932.	September 25–October 14.
1923.	September 30–October 21.		November 13.
	November 4.	1933.	September 30–October 29.
1925.	September 22–October 18.	1934.	September 19. November 3.
1927.	October 30. November 21.	1935.	September 29–October 29.
1928.	September 27–October 27.		November 3.
1929.	October 25.		

The abundance of the Creeper varies considerably in different years. In 1935 William Rusling, with the exception of three days, when his count reached seven, saw only one to four per day, but Julian Potter saw at least one hundred on September 30, 1923. On the next day I found Brown Creepers all over the Point as well as in the shade trees of the town, sometimes three or four were to be seen hitching up the trunk of a single tree, while out on the Fill they were climbing up the stems of bayberry and marsh elder bushes. On October 18, 1925; October 23, 1927; October 25, 1928; they were also abundant.

On the Christmas census the Brown Creeper has been reported as follows:

December 26, 1927, one
December 24, 1933, three
December 23, 1934, one

December 22, 1935, eleven
December 27, 1936, twelve

Those that attempt to winter seem to meet with disaster as we have only a few records later than December—i. e. January 25, February 14, March 21–22, 1925, and March 19, 1932. Our April records are April 23, 1922; April 11, 1926; April 24, 1927; April 2, 1933.

While it breeds mainly north of New Jersey, P. B. Philipp has found nests in a tamarack swamp at Newton, Sussex County, in the summers of 1906, 1907 and 1908. Elsewhere it is a regular transient and less frequent in winter.

The Brown Creeper shows little variation in habit as we see him here on migration. He sticks to the tree trunks in a never ending search for hibernating insects or insect eggs concealed in the crevices of the bark, occasionally trying his luck on the stems of saplings or low bushes if he lands among them as he rests from his strenuous night flight. He not infrequently tries telegraph poles and once I saw one laboriously ascending an iron standpipe by the Lighthouse, following the perpendicular line of rivet heads that reaches from the ground to the top. But there was method in his madness as a search disclosed the fact that many spider eggs, with their thick white covering, and some insect pupae were to be found in the cracks between the rivets and the sheet iron.

Hitching ever upward like a diminutive woodpecker or a climbing mouse the Brown Creeper furnishes a wonderful example of protective coloration, disappearing from sight the moment he comes to rest and never conspicuous even when in motion. He is also the personification of patience and no sooner does he reach the limit of profitable foraging on one tree trunk than he drops to the base of the next one and begins all over again.

HOUSE WREN

Troglodytes aëdon aëdon Vieillot

The House Wren is a common summer resident at Cape May Point and about farmhouses throughout the peninsula while there are usually one or two pairs in Cape May itself; Laurent found a pair on Five Mile Beach in 1891. In the northern parts of the state it is common and generally distributed.

It nests in bird boxes, when available, or in old woodpeckers' holes in trees and telegraph poles, as well as in holes in stumps or in old buildings. The most usual nesting materials are twigs with a lining of chicken feathers but a pair that occupied the box in Walker Hand's garden built a nest which was composed almost entirely of bits of wire from a discarded piece of chicken wire which had rusted to pieces on the ground nearby. A nest with six eggs was found by Richard Miller at Swain on June 6, 1920, and another with five eggs at Cape May Point on June 4, 1922, while nests with young were examined at Cold Spring on June 17, 1933; July 5, 1932; August 5, 1932; the last obviously a second brood. Young of the 1932 nest were on the wing by July 9 and another family on the 5th.

When the young are able to shift for themselves House Wren families accompanied by one or both parents often resort to the woods and thickets and we usually find one or two groups during the latter part of the summer at the Point, or along hedgerows back in the farmland. I have found young and old at the former locality on July 8, 1921, and July 24, 1920. The parents were much excited hopping nervously about in the thickets and constantly uttering their alarm note, a complaining, rasping *chesh-esh-esh-esh-esh-esh* etc. or *shee-shee-shee-shee-shee-shee* etc. as I have written it on another occasion. They constantly bob or duck the body up and down as if their slender legs were wire springs and give an instantaneous flutter to the wings, so rapid that the eye catches it only as an indistinct blur. While these birds were tending full-fledged young another pair was caring for a nest, with the male in full song.

Birds have been heard in song as late as: July 24, 1920; July 27, 1922; July 22, 1923; July 16, 1927; July 19, 1931.

Our latest dates for the occurrence of House Wrens at Cape May are:

1913. October 13.	1927. October 10.	1932. October 8.
1920. October 31.	1928. October 21.	1935. October 13.
1925. October 18.	1929. October 25.	

The majority have left by early October but we have had definite migratory flights on October 1, 1923; September 30, 1928; September 28, 1930. In other years there seemed to be no marked concentration.

Walker Hand has recorded the arrival of the House Wren at Cape May up to 1920 and I have recorded its appearance in several subsequent years. It is probable that his May records are late or that the birds arrive in the surrounding country a few days before they put in an appearance in the town itself. The list follows:

1902. May 5.	1914. April 26.	1922. April 23.
1904. May 7.	1915. May 7.	1923. April 30.
1906. May 5.	1916. May 2.	1926. April 24.
1907. April 30.	1917. April 24.	1927. April 24.
1908. April 26.	1918. May 4.	1928. April 30.
1910. May 2.	1919. April 23.	1933. April 30.
1912. April 8.	1920. May 2.	1934. April 23.
1913. May 3.	1921. May 8.	

An exceptional date was one seen by Edward Weyl at Heislerville on November 29, 1931.

The House Wren is a common summer resident throughout New Jersey.

WINTER WREN
Nannus hiemalis hiemalis (Vieillot)

The Winter Wren is not a common bird at the Cape and our records are almost entirely of single individuals seen during the autumnal migration or during December.

Our autumn records are:

1923. September 30, October 7, 21.	1930. October 5, November 8.
1925. November 14.	1932. November 24.
1928. October 12, 21, 27.	1935. October 5, 8, 10, 12, 19, 25.
1929. October 5, November 16.	

The 1935 observations are those of William Rusling who was present throughout the autumn. He saw quite a number of the birds on October 5, and there may have been similar concentrations in other years, had someone been constantly present at the Point to record them.

The observations on the Christmas census follow:

December 22, 1929, one.   December 23, 1934, five.
December 28, 1930, one.   December 22, 1935, two.
December 25–26, 1932, one.   December 27, 1936, four.
December 24, 1933, three.

In spring we have but few records: February 19, March 12, April 1, 1933. Winter Wrens are usually found in dense thickets especially along stream banks and I once found one in the very heart of the tall woodland at Cold Spring. Sometimes they frequent old buildings and one in search of spiders or insects was led to enter a shed through a crack in the wall and remained a prisoner there for some time, failing to find a way out. Another was found in a hedge in the Physick garden, in the town, where he went like a mouse in and out among the roots, so rapidly that it was difficult to decide whether he used his wings or legs in his progress.

The Winter Wren is a common transient and less common winter resident elsewhere in the state.

*BEWICK'S WREN

Thryomanes bewicki bewicki (Audubon)

The records of Bewick's Wren in New Jersey published by Dr. C. C. Abbott are so involved that I feel that an error in identification has been made (cf. Stone, Birds of New Jersey, 1909, p. 299). The only other record of the occurrence of the bird in the state is one seen by Samuel Rhoads at Haddonfield in 1890. William Baily secured a specimen at Wynnewood, Pa. not far from Philadelphia on April 12 of the same year. It is mainly a bird of the Middle West and southern Alleghanies and breeds regularly as far east as Fulton County, Pa., but seems to become scarce as the House Wren increases in abundance.

CAROLINA WREN

Thryothorus ludovicianus ludovicianus (Latham)

There are nearly always a pair or two of the big rusty Carolina Wrens about the old buildings of Cape May and especially in the negro quarters of West Cape May where they seem to find an environment more like that of the South. In the thickets of the Point, too, we come upon a pair or two and their clear whistle comes from the dense tangles of green brier and honeysuckle which flank the old farm buildings at Higbee's Beach. Everywhere in fact, throughout the peninsula, where old tumble-down sheds and dense thickets of vines or shrubbery are in proximity, we may look for the Carolina Wren. Philip Laurent found two on Five Mile Beach in autumn and I saw one on Seven Mile Beach in winter.

After very cold winters there may be a diminution in the number of Carolina Wrens and as they do not migrate I fear that most of them fall victims to the weather and it may be several years before they regain their normal numbers. They are here close to the northern limit of their range and with the Cardinal they hold to it regardless of all obstacles. Like the opossum, the persimmon tree and the country negro these birds are outposts of the South and have their boundary line beyond which they seldom pass—the northern limit of the Carolinian Faunal Zone. They are rare as far north as Princeton and Plainfield and absent in the more northern counties except in the lower Hudson Valley.

The Carolina Wren is a resident species and we have records of its presence in every month of the year while under normal conditions there is very little seasonal variation in its numbers except what may be due to the appearance of the young birds during the summer. It is an active nervous bird delighting in brush piles and hedges and thick shelter of any sort, moving in and out like a mouse, or we may find it, when singing, perched on the top of some tree or bush in full view. I have several times seen them fluff up the plumage and once in May this was apparently the action of a male before a female and was probably part of the sexual display.

Their feeding is usually done in the depths of the thickets, but one individual mounted the trunk of a pine tree and fed from the crevices of the bark like a creeper.

Holes in old buildings seem to be the favorite nesting places of this wren. One pair had built on the flat sill above the door of an old shack at the Point and the nest was found with a full set of eggs by Charles Page when he came to occupy the place for the summer. The opening and shutting of the door was too much for the nerves of the birds and they deserted. A roll of roofing paper was immediately put up on the trunk of a nearby pine with a tomato can at the end and within two hours the wrens had take possession and were soon at work on a new nest in which a brood was raised. A nest seen by Turner McMullen at North Cape May on May 14, 1927, contained five eggs.

The song of the Carolina Wren, which is perhaps his outstanding character, is exceedingly variable but always reducible to a series of clear penetrating whistles, except for the long drawn rolling trill—*chirrrrrrr, chirrr* which he utters on occasion.

There is some resemblance to the song of the White-eyed Vireo, a frequent companion of the Carolina Wren in low damp woodland. The resemblance lies in the clear quality of the notes and in the ability to make an instant change in the character of the utterance, but the make-up of the two is entirely different. It is the call and song of the Cardinal, however, which the Carolina's music most closely resembles. The most frequent song is the *twéedle, twéedle, twéedle, twéedle* with which all fellow residents of wren territory are familiar. This I have written, on other occasions, *teedle*, and *cheedle*, as there is a good deal of variation in the syllables of different individuals. The other familiar song of the bird, which seems to be fundamentally different since it is accented on the first of three syllables instead of the first of two, is—*túleeah, túleeah, túleeah, tòo*. This has also been rendered effectively as *téa-kettle, téa-kettle, téa-kettle*. One bird (August 16, 1921) was endlessly reiterating the *twéedle, twéedle* note when he suddenly stopped and sang *sweét-tel-ee, sweét-tel-ee, sweét-tel-ee* (a slight variant of the second song) and then as suddenly reverted to the first phrase to which he finally added two syllables: *twéedle-dee-dle, twéedle-dee-dle*, etc. Other phrases that I have recorded are *whéet-a, whéet-a, whéet-a; séetilo, séetilo, séetilo; tsíleo, tsíleo, tsíleo;* and *pée-to, pée-to, pée-to*.

Several times I have heard these wrens shorten up the long trill into *chéer-up, chéer-up, chéer-up* and once (March 22, 1925) a bird executed a peculiar rolling song something like the gurgling of a Long-billed Marsh Wren or like the beginning of a House Wren's song: *tlit'l, tlit'l, lit, lit, lit, lit, lit.* I was quite at a loss to guess the author until I caught sight of him in the act.

A Carolina Wren at the Point (July 18, 1923) suddenly started a series of clear penetrating notes: *quip, quip, quip, quip,* etc. separated by several seconds. This was one of those bird calls, which, perhaps because of its peculiar character, attracts the attention of other birds whose curiosity is aroused. At any rate in the course of two or three minutes, while this performance was going on, there were gathered near the singer, four Towhees, four Thrashers, a Flicker, a House Wren, a Yellow Warbler and a White-eyed Vireo. Then as if satisfied the Carolina abandoned this call and launched into his familiar *túleeah, túleeah, túleeah,* and the other birds lost interest and immediately scattered. It would be interesting if we could interpret bird songs and calls or at least know just how the song of one species affects others.

While Carolina Wrens sing most frequently in the spring and summer it is not unusual to hear them burst into song at other seasons and even in midwinter we hear them on fine mornings, despite a snow covered landscape, vie with the wintering White-throated Sparrows in voicing their cheerful phrases. We have recorded their song in every month except January and February and doubtless constant observation would find them whistling then as well.

H. B.

LONG-BILLED MARSH WREN

Telmatodytes palustris palustris (Wilson)

None of our summer resident land birds has suffered such depletion in numbers within my recollection as has this little denizen of the marshes and cattail swamps. In 1890, when I first visited Cape May, Marsh Wrens were abundant all the way from Broadway to the Lighthouse nesting in the marsh elder bushes and in the taller sedges all along Cape Island Creek and in similar situations through the marshes behind the town nearly to Schellenger's Landing. On both sides of the turnpike as it traversed West Cape May their bubbling song could be heard continually while the great cattail swamps to the south of the road were alive with them.

By 1920 the draining of these marsh lands in the interests of mosquito extermination had sadly depleted the numbers of the Marsh Wrens and destroyed most of the fields of cattails so that the birds were found only in a limited area just east of South Cape May and around some of the ponds and sluices nearer to the town. Then came the unfortunate exploitation of this area of marsh which resulted in its being almost entirely filled in so that by 1926 Marsh Wrens were driven completely away from the Cape May district with the exception of a small colony immediately behind the golf links, in the marshes which border the railroad. So tenaceous of their nesting sites are these little birds, however, that there may be a few scattered pairs elsewhere which from their secretive habits and limited range have escaped notice.

On the upper part of the South Dennis marsh and doubtless at other spots on the peninsula, where "improvements" and ill-advised drainage operations have not ruined their native habitats, the Long-billed Marsh Wrens still abound and small colonies existed a few years ago on the upper part of the Pond Creek Meadows and along a creek that flows from the mainland opposite the middle of Seven Mile Beach and they probably still survive. Another colony formerly existed on this beach near Eighty-fourth St., not far from the ocean (McMullen).

While the songs of the Marsh Wrens are heard on every side as we approach one of their colonies it is not so easy to see the birds as they keep well down in the thick sedge but presently curiosity will bring one of them into view. He grasps an upright reed with both feet and leans out almost horizontally with tail cocked over his back until it nearly touches his head, the head is bent upward with the throat extended and the whole body sways slightly up and down. He has a short quick alarm note: *chip; chip; chip; chip;* which is uttered continually when the young are on the wing and is accompanied by an instantaneous flip of the wings with each note.

Every now and then one of the wrens will launch forth in flight, just skimming the top of the sedge like a great bumble bee, with wings beating so rapidly that they almost make a blur; the slender bill held out in front and the closely appressed tail behind make the bird appear distinctly pointed at each end. The flight is not long sustained and the bird drops back into the depths of the sedge. In the resting season a male will be seen to mount six or eight feet in the air, ascending in a curve over the marsh with head and tail held high and throat distended, pouring out the continued bubbling medley of the flight song and then sinking slowly down on fluttering wings into the grass.

The song has a peculiar spluttering, bubbling quality, like water gurgling out of a pipe or the rippling of a brook over a pebbly bottom. I have written it on several occasions: *possa-wat'l-wat'l-wat'l-wat'l-wéegal.* The first two short notes are unaccented and low, the four gurgle notes are higher, with the last couplet suddenly descending. Sometimes the preliminary notes sound more like *tick, tick.*

Again I have recorded the song: *witla-witla; watt'l—watt'l—watt'l—watt'l-was-it* and again: *possa-psilla—psilla—psilla—psilla—wisset.* It is difficult to detect how many low preliminary notes there are and the impression is that the bubbling might have been going on for some time in an undertone and has suddenly become audible. So, too, there is often a gurgle of incompleted song from the depth of the sedge. I have heard the song frequently to the middle of July and fragments of it as late as August 3, 1920; August 5, 1921; August 11, 1925.

I have seen young birds perched in the marsh elder bushes near their nest holding on to two twigs, one with each foot, with the result that their legs were spread far apart with the tail twitching nervously, and once a parent perched on a telegraph wire reiterating the alarm note as it turned

to the right and left with a quick flip of the wings. The general appearance
of the Long-bill in flight is dull brown but when perched head-on the silky
whiteness of the throat and breast is very conspicuous and they fairly shine
in the sunlight. The full-fledged young look very black with a rusty patch
on the back and are much darker than the worn and faded adults of late
summer.

I found many nests with fresh eggs in the cattail swamps on May 25,
1892, situated about three feet from the ground or water and Richard Miller
has found them still with eggs on June 8, 1919. The regular nests are
rather oblong with more or less cattail down for lining and this often pro-
trudes from the entrance hole on the side. The sham nests are usually
nearly spherical, unlined and often built in the low sedge. The usual number
of eggs in a full set is six. One nest evidently deserted was found with two
eggs on July 12, 1923, and another, possibly a sham nest, was started on July
22 and completed on July 31, but no eggs were deposited in it. I have also
seen the Marsh Wrens carrying nesting material as late as August 7.

Young just able to fly were found on July 14 and in full juvenal plumage
by August 28, 1917. From the last days of August through September
there seems to be an increase in the number of Long-billed Marsh Wrens
and they occur in the marshes nearer to the Lighthouse and even in the
coarse grass and sedge immediately back of the sand dunes. These are
obviously migrants from farther north and it is quite likely that the individuals that have been found here in winter are of this sort rather than birds
that have nested at Cape May. On October 1, 1923, they were more abundant than I had ever seen them, taking wing constantly as I tramped
through the short sedge, or broke my way through the marsh elder bushes.
Sometimes from three to six flushed at once and in a short walk I counted
more than one hundred yet they would not have been seen at all had I not
happened to enter the area in which they had sought shelter after their
night flight. They flew heavily and dropped back to the ground like miniature rails. Judging from their labored flight on such occasions one wonders
how they can accomplish such long migrations. Similar flights occurred on
September 21, 1929; September 28, 1930; September 19, 1934; while in
several years a number of dead Marsh Wrens were picked up along the roads
which had struck overhead wires or had been struck by automobiles in the
night.

Our dates of arrival are not very satisfactory as few observers have
been in the haunts of the birds in the early spring but they are obviously
late migrants and our earliest dates are May 11, 1908; May 17, 1914; May
10, 1924. The breeding birds seem to leave by late September or early
October but we have records for November 6, 1921; November 21, 1926;
November 9, 1930; and on several Christmas censuses of the Delaware
Valley Club there are December records of single birds—December 24, 1923;
December 26, 1932; December 22, 1935; December 27, 1936.

The occurrence of the Long-bills at Cape May in winter has always interested me. They seem so out of place at this season. We look upon them as among the latest spring migrants, birds that seem to revel in the hottest and sunniest of open marshes and yet here they are thriving in the coldest days of the year. Samuel Rhoads and I first established them as winter residents in January, 1892, during a bitter gale from the northwest when the marshes were all encrusted with ice and one could scarcely stand against the wind. Making our first trip to Cape May at this season, we plowed through the acres of tall cattails back of South Cape May to see what might be lurking there. As we parted the close ranks of dry yellow blades we sent clouds of soft down from the ripened fruit drifting through the air and suddenly one of these little birds appeared before us and, pausing for a moment, plunged again into the thick vegetation. Then we found them every few steps swaying up and down on the dead stalks with tail cocked over back in the characteristic manner. Evidently there was insect food here sufficient for their support—hibernating flies, perhaps, on warmer days, and pupae of various sorts waiting to be sought out. We thought the occurrence of the Marsh Wrens at this time entirely unusual but subsequent experience has shown them to be regularly present in winter in small numbers. Thus was a summer resident species shown to be a resident but I still doubt that the birds that breed at the Cape are the same individuals as those that are to be found in the cold days of January.

The occurrence of Long-billed Marsh Wrens in winter is not peculiar to Cape May as Charles Urner found one at Manasquan on the northern coast on February 26, 1931, and John Gillespie saw another on the Tinicum marshes below Philadelphia on January 26, of the same year.

While they are summer residents in suitable localities throughout New Jersey their distribution is dependent upon the presence of cattail swamps. They are to be found in marshes all along the coast and up the Delaware as far at least as Trenton and along the larger rivers, sometimes above tide-water; above Mays Landing, near Princeton (Phillips), Plainfield (Miller), and even at Newton, Sussex County (Philipp). Their abundance in former times may be judged from the statement of B. B. Haines (O. and O., 1883, p. 204) that he knew a collector to obtain from 400 to 500 eggs in a day in the vicinity of Elizabeth!

SHORT-BILLED MARSH WREN

Cistothorus stellaris (Naumann)

This very different Marsh Wren is a local summer resident of open semi-brackish marshes about Cape May, inhabiting areas of fine low grass— *Distichlys spicata* and *Spartina juncea*—usually back toward the upland where stray bushes occur scattered about and never out on the main salt meadows; nor is it found in the cattail thickets where the Long-bill abounds. Like the latter it also occurs in winter usually in smaller numbers but frequenting the same haunts as in summer and never mixing with the other species.

Curiously enough it was in midwinter that I first discovered this little Marsh Wren at Cape May. Samuel Rhoads and I had been trapping shrews and mice on the marshes then lying between Cape May and the Lighthouse on January 28, 1892, when we flushed one of these diminutive birds from an open grassy spot back of South Cape May and, as we did not know it, it was promptly shot. The wind was blowing a gale from the northwest and it was bitter cold, hardly the weather to expect such a bird as this. But subsequent experience has shown that it is of regular occurrence here in small numbers every winter.

While Richard Harlow had found the Short-billed Marsh Wren nesting on the edge of the salt meadows in Burlington County in 1913 and Dallas Lore Sharpe saw two on the marshes of Salem County at the head of Delaware Bay, at a much earlier date, it was not until 1917 that we found them at the Cape in summer. On August 4, of that year Fletcher Street and I heard one singing near the site of my winter record. There was a large patch of fine grass flanked by an impassible muddy bog on one side and a cattail thicket on the other, where a few Long-bills were established. Search as we would we were unable to find a nest nor did we see more than a single bird and up to date no one has been able to locate a nest in this region. I visited this bird many times during the summer and studied it carefully, but failed to find a mate or young. Sometimes it would not be in evidence when I arrived, then suddenly I would hear the song and there was the bird on the top of a low bush or weed right in front of me, but when I looked again it was gone, dropped back into the dense mat of grass whence it had come, and where it remained silent, while no amount of tramping about would cause it to take wing. The habits of this and many other individuals observed on subsequent occasions was always the same, ever elusive and always singing from one or more definite perches, flying low over the grass when changing its position and eventually plunging into it to seek refuge and concealment.

A bird watched on August 28, 1929, gave a flip to the wings and tail that was so rapid that it seemed like a blur while the head was jerked to right and left so quickly that the action almost escaped the eye. Another individual was seen to climb up the stalk of a sedge by rapid jumps of an inch or two facing first to one side, then to the other, while the feet changed position with each jump and the head and body were "bobbed" up and down between them.

The notes of the rapid song are fine and insect-like but sharp and emphatic and give the impression of being chopped off short. The most frequent song as I have transcribed it is: *tchip; tchap; tchip, tchu-tchu-tchu* and at a distance when only a part of it is audible it recalls the *chís-eck* of the Henslow's Sparrow which inhabits the same region. On other occasions the song terminates with more of a trill: *tchip; tchap; tchip; churrrrrrrr* or *chip; chap; chip; zizl-zizl-zizl.* One bird varied its song considerably, sometimes shortening it to *chap; chap; chappy-chappy,* or *chip; chap; zee-zee-zee-zee-zee.* Next day he sang *chap; chap; zeeda-zeeda-zee* and *chip; che-che-che-che,* the last notes of this song being much more deliberate than in any of the other renderings.

Another individual watched on August 15, 1921, began with *tchip; tcha; chu-chu-chu-chu-chu,* then changed to *tchip; tcha; tchip; churrrrrrrr* and again to: *tchip; che-tzee-tzee-tzee-tzee,* with a noticeable pause after the first note. The alarm note *tchip* is often interpolated and repeated between the renderings of the songs. A short song was heard as late as September 5, 1926, by John Gillespie and a mid-winter bird was heard to render the alarm cry *chick, tick.*

A bird singing on August 13, 1921, had a mouth full of dry grass strands three or four inches in length and managed to utter his notes without dropping them; another carrying a longer strand was also singing. These birds may have been building sham nests although Richard Harlow's nests farther up the coast were found on August 4; one with four eggs, one with five and a third with five young. He considered them as probably second breedings which may have been the case since a nest found in Salem County by H. M. Harrison contained eggs on June 4, 1922, and W. B. Crispin found another in the same vicinity on June 5, 1909. Turner McMullen has examined nests near Cape May Court House on May 30, 1930 (seven eggs), June 7, 1931 (six eggs), and at Fortesque on Delaware Bay on May 28, 1932.

While Short-billed Marsh Wrens have been found in breeding colonies in Morris and Sussex Counties in the upper part of the state and scattered individuals have been recorded from various localities the bird seems to be rare and we have no definite migration records. It may be that the summer birds at the Cape are the same individuals that winter here, certainly there is no such influx of transients as we see in autumn in the case of the Long-bill and birds summering to the north may pass to the southwest or down the Delaware Valley as seems to be the case with other species that are rare at the Cape. (cf. L. K. Holmes, Cassinia, 1904, pp. 17–25.)

Our records follow, but absence of continual observations during the spring and autumn probably has much to do with the paucity of data at these seasons.

1917.	August 4–30.	1928.	May 30–July 14. September 1; October 7.
1920.	August 15.		
1921.	July 30–August 19.	1929.	July 27–August 28. October 25.
1922.	July 8.	1930.	July 25–August 20, September 28.
1923.	May 30–July 7.		November 9.
1924.	June 22–August 9.	1931.	April 19, May 30, August 7.
1925.	May 30–July 6.	1932.	May 8, June 22.
1926.	August 24–September 5.	1933.	July 8.
1927.	May 22–August 26.	1934.	December 23.

As a rule our records are of single birds but on May 30, 1925, a colony of six was found back of South Cape May as well as several individuals in other localities.

The counts on the annual census by members of the Delaware Valley Club are:

December 23, 1923, one.	December 24, 1933, ten.
December 26, 1927, one.	December 28, 1934, twelve.
December 23, 1928, one.	December 22, 1935, four.
December 28, 1930, one.	December 27, 1936, ten.
December 27, 1931, one.	

The Short-bill may be more abundant at all seasons than we suspect since its secretive habits, and ability to resort to shelter from which it is difficult to dislodge it, are characteristic.

MOCKINGBIRD

Mimus polyglottos polyglottos (Linnaeus)

We like to think of the Mockingbird as one of those austral birds, like the Gnatcatcher, Yellow-throated Warbler and the occasional Purple Gallinule, whose presence seems to make Cape May almost a bit of the South, but, as a matter of fact, it is hardly more plentiful or of more regular occurrence than it is at several points much farther north and it is only in unusual years that we can go looking for Mockingbirds at the Cape with any certainty of finding them.

With the exception of several nestings and the visits of family parties in late summer or early autumn our observations have been of single birds only, and so many of these have been in the colder months of the year, that we are forced to look upon the Mockingbird as a winter visitant quite as much as a breeding species. We have records at the Cape in all but two of the past twenty years—1919 and 1924, but the total number of our observations is not large. Arranged by months they are as follows: January, three; February, two; March, two; April, two; May, three; June, two; July, two; August, seven; September, four; October, two; November, one; December, five.

There are many tales of nesting Mockingbirds in the country about Cape May in the past and of men who, a generation or so ago, took the young from the nest and reared them for cage birds or for sale, but I find very few of the present day residents who know the bird at all and Otway Brown, who tells me of trapping Cardinals in the late eighties, says that no Mockingbirds were obtained at that time. Dr. Samuel W. Woodhouse some years before his death, when asked about the abundance of Mockingbirds in past years, told me that from about 1840–1850 when he was constantly in the field he and his associates, John K. Townsend and George Leib, never found one north of Maryland or Delaware. In William Bartram's manuscript diary, 1802–1820, he mentions occurrences of single birds in his garden at Philadelphia but all in winter with the exception of one April and one October record. All of this testimony makes one wonder whether Mockingbirds were ever really plentiful anywhere in Pennsylvania or New Jersey. Peter Kalm, it is true, mentions them as cage birds in this vicinity as early as 1748 and Alexander Wilson, writing in 1810, says: "the eagerness with which the nest of the Mockingbird is sought after in the neighborhood of Philadelphia has rendered the bird extremely scarce for an extent of several miles around the city. In the country around Wilmington and New Castle, they are very numerous, from whence they are frequently brought here for sale. The usual price for a singing bird is from seven to fifteen and even twenty dollars." He also speaks of a man whom he had met on his rambles "with twenty-nine of these birds old and young, which he had carried about the fields with him, for several days, for the convenience of feeding them while engaged in trapping others," but he does not state whether this was in Pennsylvania, or in Maryland.

I have had several excellent opportunities for studying Mockingbirds at the Cape. One usually sees them perched on some telegraph wire or on the top of a low pine or shrub or perhaps you may catch a glimpse of one on the wing. If it is your first view of the bird you think at first of a Robin—but too light a gray, you say, and there is the longer tail, and as seen from the back the conspicuous white on the wings, and finally the upward swoop as the bird reaches his objective perch and comes to rest—all very unlike the Robin. You then think of the Thrasher—yes the pose and the proportions are more like his but the color is of course very different and then there are characters not shown by any other species which after all make the Mockingbird what he is. The more you know him the better you realize his individuality and see that your first comparisons were merely the attempt to describe the unknown in terms of the known.

Now he is up on the top of a pole standing high on his slender legs with tail straight out behind, vibrating slightly as if moved by the breeze, a veritable balancer; and now he is pouring out his song: *chuley, chuley-chuley-chuley shee, shee, chu-aleé chu-a-leé chu-a-leé chee, chee, chee, tzit-tzit; tzit-tzit* etc. It is not the careful and consistent duplication of notes as given by the Brown Thrasher but a repetition apparently dependent entirely upon the pleasure of the performer. Some notes he abandons at once, the next perhaps will strike his fancy and he repeats it over and over, far beyond the usual limit of repetitions, closing perhaps with some curious attenuated *sotto voce* duplications of the note which he rolls out with great apparent satisfaction. Now he is hopping on tiptoe, as it were, along the fence, singing all the while and then launches forth across the road, the notes still bubbling from his throat, and with wings flapping in an exaggerated effort as if it were all that he could do to transport such a load of song. With the usual upward swoop he comes to rest again on the telephone wire and begins all over again: *chuley, uley, uley, uley, cheer, cheer, cheer, whe-oo, whe-oo.*

The call of the Cardinal will be recognized in this effort but master mimics as are some Mockingbirds I have not detected the songs of many other birds on the few occasions on which I have heard Mockers sing at Cape May. The best singer that I have heard here was one of a pair that nested near the fire house at the Point in June, 1931, and was always in evidence singing at any hour of the day. I heard him on June 28, at 7 p. m. and on June 29 at 9 a. m. while William Shryock heard him at 9 p. m. on July 7, apparently the last day that he was in song. Sometimes he would leap into the air from his perch and turn an almost complete somersault making a great display of the white of both the wings and tail.

Single Mockingbirds that I have studied at other than nesting time usually perch on the top of some low bush or small pine tree or on telegraph wires or even on the ridge pole or chimney of a house, and from such a stand will drop suddenly to the ground, or to some fence post, and flap back again.

Sometimes I have been fortunate enough, in late August or early September, to have Migrant Shrikes and Mockingbirds under observation on the wires at Cape May Point at the same time. While the similarity in coloration causes those previously unfamiliar with these birds to mistake one for the other, their perching position and manner of flight are so strikingly different that they can easily be distinguished at long distances. To the alert, slender-legged Mockingbird, the perching Shrike, short-legged and heavy-headed, with tail drooping and head usually bowed over, is in striking contrast. The flight of the Mocker is steady and low, the beats reminding one of the leisurely strokes of the oars in an old row boat, and as the objective perch is reached the bird will set its wings and sail with a slightly undulatory motion terminating in the sharp upward rise that is so characteristic. The Shrike's flight, on the other hand, consists of a series of rapid wing beats and then a short pause in which the body falls slightly before the next series of pulsations. In both birds there is a conspicuous show of white in the wings but the slower flight of the Mocker produces a very different effect from the rapid flicker of the Shrike.

In one Mocker family that I studied an adult and young spent much of their time in the heart of a thicket, busy pluming themselves and the situation permitted close approach. Their quietness, and the length of time they remained under cover, demonstrated how easy it would be for them to escape the observation of one making but occasional visits to their haunts, and convinced me that they had thus escaped my attention on several previous visits to the spot.

Some dense thicket or cedar tree will furnish a wintering Mockingbird with satisfactory shelter and into these he will plunge when alarmed and refuse to emerge and, in spite of all our efforts to dislodge him, will remain safely out of sight in the depths of his retreat. I have observed this on not a few occasions.

Reviewing the several actual or probable nestings of Mockingbirds at Cape May; Reynold Spaeth, whose family lived at Cape May Point during his boyhood, told me that he found a Mockingbird's nest in the shrubbery of their garden in June, 1899, which contained four eggs. He took one of them and the birds promptly deserted. In other years about that time the birds were present and probably nested and that they did so in earlier years is suggested by the fact that I found a very worn bird there on August 27, 1891.

On August 25, 1917, Julian Potter found two old birds and a speckled-breasted young one at the Point. Two days later I found that there were five Mockers in this party, one a molting bird minus a tail, but with an exceptional amount of white on the wings; one in nearly full fresh plumage, possibly a bird of the year; two adults in worn plumage and the young one still being fed by the adults.

In May, 1918, a male was reported in full song near David Baird's residence on the western edge of Cape May not far from the beach, possibly the same bird that Walker Hand had reported as present during January and February. It disappeared, however, before summer set in.

On March 25, 1928, a Mockingbird was found in a thicket back of Second Avenue where it perched on the topmost spray of a bush and frequently indulged in those peculiar aërial somersaults for which these birds are famous. It was seen again on May 6 and by June 28 a pair had taken up residence amongst the shrubbery on the Mercur (Dougherty) place on Washington Street in the heart of Cape May and apparently nested, although I was unable to definitely locate a nest. Walker Hand told me that he had heard the male singing there since the end of April on nearly every time that he passed the premises and I heard it daily from June 28 to July 4 when both birds suddenly disappeared. Possibly their nest had been destroyed by one of the numerous marauding cats that are left every year by departing summer visitors to enrich the already too numerous cat population of the town. On June 26 I came upon this pair at 6 a. m. on Corgie Street immediately back of the yard in which the nest was supposed to be. They hopped about the low shade trees and down on the sandy ground where they progressed by a sort of running hop, with head and tail up, wings drooping slightly with the tips held just below the base of the tail. They exhibited the characteristic swooping, sailing flight of the species and the slow, measured beats of the wings. On the following day they were on the ground moving rapidly with heads stretched out in front seeking food of some sort.

On July 2, 1929, one was heard singing on Lafayette Street near Schellenger's Landing but was not seen later. On September 7, however, a family of four put in an appearance on the wires along Broadway near the beach, two adults one of which was molting, and two young with speckled breasts although in one the marks seemed more like streaks, perhaps due to the beginning of its molt. While perching they would now and then turn the head and point the bill diagonally upward. They also flirted the tail when excited and cocked it up over the back as if to aid in retaining their balance in the breeze. When one of them descended to the ground it literally dropped like a plummet with legs dangling. On October 3, there were five Mockingbirds in a yard on Lafayette Street, possibly the same family.

In June, 1930, William Schwebel told me that one had been singing at night at his cottage at the Point and kept his family awake and he thought that it must have had a nest nearby. Another sang from March 23 to April 25 in Walker Hand's yard in Cape May but was not seen later in the year.

In 1931 a pair nested at the Point near the fire house and the young were out of the nest by May 30 and were seen by William Shryock during June, the male being still in song and the young feeding in a garden close by. I noticed that they did not hop but ran rapidly over the ground pausing occasionally and spreading their wings, arching them slightly away from the body so that the white areas were displayed. On July 7 I saw six of this family at once. A single bird was there on July 17 and again on August 7 but none afterwards.

In 1932 we looked in vain for the pair of the previous year but no Mockingbirds were seen about the Point. On August 9, however, Charles Page found a pair with several full grown young among the dune thickets northwest of the Pond Creek Meadows and one or two were seen there later in the summer by James Bond and myself on several occasions.

On August 25, 1933, I found an adult bird with two spotted young, again in the same vicinity, and a single adult on August 20, 1934. My opportunities for following up the history of these birds were limited and they may have been present both earlier and later than the dates given.

June 1935, found a male Mockingbird in full song in a dense thicket in a yard on North Street which continued to sing until July 7 and was last seen on July 15 but no mate or young were detected. This may easily have been due, however, to lack of continued observation and the ability of the birds to conceal themselves. In the summer of 1936 the male was there again and was heard singing at various times, once as early as 1 a. m. by Ernest Choate.

During September and October one or two Mockingbirds have frequently visited the yard of David Baird on the western edge of Cape May where dense shrubbery close to the house furnished congenial shelter. One individual roosted there nightly from September 20 to October 2, 1925, and voiced his complaint when a light was turned on in the window near to his retreat. Two adults and several young appeared there on September 17, 1927.

In winter the Mockingbirds seen at the Cape are solitary individuals which select a suitable spot and are resident until spring time when they apparently depart to seek mates. One remained in the vicinity of Walker Hand's home all through the winter of 1929–1930, coming regularly to a table on the back porch, where a bunch of Christmas holly had been placed, to feed on the berries. Another individual located in a yard on Lafayette Street in January 1922, and came regularly to fight his image in a window of the house next door.

As many of these wintering Mockingbirds and the others that are seen singly or in family parties, in late summer, are not observed earlier in the year, when many persons are making daily studies of the bird life of the region, it would seem that they must have drifted in from more remote localities back in the country. This could hardly be termed migration, especially as the young birds in the late summer groups are still in the juvenal plumage and are being fed by their parents. Furthermore they could not have come from greater distances to the north because there are no breeding Mockingbirds there, while a northward movement, say from Delaware or Maryland, such as we see in the case of various herons after the breeding season is over, is hardly likely.

On the other hand they may have come from quite near at hand since Mockingbirds are so local, when nesting, that they might easily escape notice except from those immediately on the spot, who as a rule would pay no attention to them. The pair found on the Dougherty property, for instance, scarcely wandered away from the yard in which they had established themselves during the whole time that they were under observation, and not more than half a dozen persons in the town recognized them. Again, a single bird seen by Alexander Wetmore, at Pierce's Point on the Bay shore on August 11, 1926, might easily have reared a brood there without anyone being the wiser.

While departure from the Cape, after the summer is over, has prevented my continous observation of the activities of the late summer family groups, week end trips by myself and various members of the Delaware Valley Club during the autumn and winter have furnished fairly accurate records of the occurrences of Mockingbirds throughout the year. It would seem from these that the bird is in no way gregarious and after the young have been reared to the point that they are self reliant, they scatter and become solitary hermits during the winter months.

That Mockingbirds have some preference for the immediate vicinity of the sea is suggested by the records of a pair on Long Beach in the summer of 1906, seen by Norris DeHaven; and on Sandy Hook at least until 1892, according to Rev. Samuel Lockwood; at Point Pleasant, through the winter of 1902–1903, reported by Miss Caroline Murphy; at Barnegat through the summer of 1900 according to John Lewis Childs; at Stone Harbor, September 4, 1903 seen by David McCadden and on Five Mile Beach on December 27, 1903, seen by William Baily. The latter also found a young spotted-breasted bird just able to fly on Five Mile Beach on September 14, 1895, and I saw one in full juvenal plumage with wings and tail fully developed on

Seven Mile Beach on September 10, 1927. These last records would seem to indicate nesting on the two islands.

There are scattered records of Mockingbirds in winter and more rarely in the breeding season in northern New Jersey as well as north to New England, but as the bird is a resident wherever found, there is no definite migratory movement.

In spite of the rather numerous records of the Mockingbird at the Cape it is still a red letter day when the bird watcher is lucky enough to find one, whether a solitary winter sojourner or a family in the warm days of late summer and early fall, and the puzzling irregularity of the occurrence of the species always arouses our curiosity and interest.

CATBIRD

Dumetella carolinensis (Linnaeus)

Catbirds are as universally distributed throughout the Cape May Peninsula as are the Thrashers but are more plentiful. In the town they nest in most of the gardens where there is thick shrubbery, while at the Point and back in the country nearly every thicket has its pair of Catbirds. The Catbird is more sociable than the Thrasher and delights in the immediate vicinity of old houses if suitable cover is left for its shelter.

It does not seem to occur close to the beaches except in migration, as bayberry bushes are not adapted to its needs, and Laurent saw only a few on Five Mile Beach in the old days when the island was wooded.

While mainly a summer resident and abundant transient, a few Catbirds usually remain through the winter in the dense tangles of greenbrier and holly about Cape May Point and along the Bay shore, and to a lesser extent in similar situations elsewhere. The familiar song of the Catbird is in evidence from late April to mid-July—July 16, 1927; July 14, 1923; July 19, 1924; July 23, 1926; July 16, 1927; etc, while fragments of song have been heard as late as August 9. The long drawn mewing alarm call—*tweeee* or *shreeee* is heard at all times through the summer when the bird is disturbed by an intruder. Catbirds sing frequently from the interior of a thicket but also from the top of a bush, with head up, wings slightly drooping and held a little away from the body. Perhaps no bird is more ready to respond to a cry of distress and a call of this sort from any bird will bring all the Catbirds in the vicinity "mewing" in sympathy, or more likely in curiosity.

The abundance of Catbirds at the Cape varies greatly in late August and during the autumn. Definite increases have been noted on August 20,

1927; August 21, 1922; etc., and then with the apparent departure of the summer population they may be almost absent until a great migratory wave brings others from the north and we marvel at the number of Catbirds that must summer in eastern North America! Such a flight occurred on October 5, 1927, when Walker Hand reported the town full of them and counted eleven in one small yard. Again on October 8, 1932, when I found them swarming in the thickets of Cape May Point, the wood edges were fairly vibrant with them as they continually darted up into the sour gum trees and dropped back again into the bushes. I estimated that there were at least one hundred in two gum trees which bore a generous crop of berries. There were many also at Cold Spring and a large number in the Physick yard while I picked up three on the road, victims of the wires. William Rusling observed a similar flight on October 5, 1935, and I saw a congested migration on September 28, 1930, and September 27, 1926.

I have seen fully fledged young being fed by the parents as late as July 17, 1920, and have seen an adult feeding with a Water-Thrush on the dry bottom of a woodland pool near Bennett, tossing the leaves aside vigorously. Another alighted on our porch roof in search of caterpillars which dropped there from an adjoining tree and he also tossed away the leaves that often covered his prey.

Five nests with fresh eggs, three or four in number, were found by Richard Miller at Rio Grande on May 27, 1923, and others on June 3 with eggs partly incubated. At the same locality Turner McMullen has found four nests with eggs as late as June 10 and one on May 23; two on June 8 contained young.

Walker Hand's arrival dates for the Catbird are as follows, with several additional records for recent years:

1902.	April 24.	1908.	April 26.	1914.	April 23.
1903.	April 20.	1909.	April 6.	1917.	April 23.
1904.	April 27.	1910.	April 28.	1920.	May 1.
1905.	April 23.	1911.	April 24.	1923.	April 30.
1906.	April 21.	1912.	April 30.	1926.	April 24.
1907.	April 25.	1913.	April 30.	1932.	April 30.

Concentrated spring migrations of Catbirds have been noted on May 1, 1914; May 8, 1915; May 6, 1930; May 7, 1932; but these did not compare with the autumnal flights. From their abundance as transients I assume that the coast line is one of their chief avenues of migration.

The usual date of departure is about the middle of October:

1920.	October 31.	1925.	October 18.	1931.	October 19.
1921.	October 16.	1927.	October 23.	1932.	October 14.
1922.	October 1.	1928.	October 28.	1933.	October 29.
1923.	October 21.	1929.	October 25.	1935.	October 29.
		1930.	October 11.		

We have later records which may probably all be regarded as wintering birds:

1903.	March 9.	1926.	December 26, two.	1930.	December 28.
1914.	December 9.	1927.	January 23.	1932.	March 5.
1922.	December 9, 30.	1927.	February 3.	1932.	December 26.
1923.	December 2.	1927.	December 26, three.	1934.	January 24.
1925.	February 14.	1928.	December 6, 9.	1934.	December 23.
1926.	November 14, 21.	1928.	December 23.	1935.	November 4.
		1929.	November 7.		

The February 1925 bird was in company with a small flock of Cedar Waxwings and some Robins. Winter records elsewhere are: Atlantic City, December 26, 1892 (Rhoads); Seven Mile Beach, February 11, 1894 (McCadden) and December 31, 1905 (Hughes); Five Mile Beach, several in the winter of 1897–8 (Baily). Also Plainfield, December 30, 1897 (Miller), and Moorestown, December 25, 1903 (Evans).

The Catbird is a common summer resident throughout New Jersey, including the coastal islands.

BROWN THRASHER

Toxostoma rufum (Linnaeus)

Thrashers are common summer residents throughout the Cape May Peninsula all the way down to the pine woods and thickets of the Point. They also occur here and there in the gardens of the town and in clumps of bayberry and brambles just back of the sand dunes, from which they occasionally repair to the sea beach itself in their search for food. I have found them breeding on Two Mile Beach and on the other island beaches to the north where they find satisfactory shelter. Occasional Thrashers also remain throughout the winter in dense thickets at the Point and along the Bay shore.

They are in full song from their arrival in April until the middle of June. It is their habit here, as elsewhere, to mount through a tree from limb to limb until they attain the topmost twig and there, with head up and tail straight down, they pour out the familiar medley of couplets that constitutes their song. They stop singing, I think, sooner than any of our other breeding birds and my latest dates are June 22, 1924; June 30, 1920; July 2, 1921; July 4, 1928.

As there is usually more or less of a breeze near the shore it is sometimes amusing to see the efforts of the Thrasher to maintain his exposed singing perch. One observed on May 9, 1925, swung his long tail up and down like a veritable balancer, now almost vertical above his back and then straight down beneath his body. Another on May 24, 1918, was pouring out his song in the face of half a gale and had great difficulty in keeping his balance. At one moment his tail would be blown forward over his back until it nearly touched his head and then he would bring it around below his belly while he fairly leaned back against the wind.

The gait of the Thrasher on the ground varies. I have seen them, when feeding deliberately on the lawns on Washington Street, walk slowly and occasionally take a half-running step, while others on the sandy roads of Cape May Point progressed by a series of vigorous hops for a few yards and then seemed to combine a hop and running step until it was difficult to define their progress as either. Thrashers delight in sandy and dusty roads and before these were so generally replaced by hard concrete they were much more in evidence than they are today. They come to such places to dust themselves and rid their bodies of the lice that infest the nest. A full plum-aged young bird of the year was found on July 29, 1920, squatting full on his belly in a hollow in the sandy road east of Lake Lily. He fluffed out his feathers in the dust and every now and then would preen his plumage, prob-ing the bill well under the wings, after which he would poke it down into the sand as if to cleanse it. Then he arose and hopped vigorously for a few yards returning to the same spot where he rolled partly over on one side and remained so for several minutes holding his bill open the while. I saw an adult bird going through the same performance in a sandy wood road on July 7, 1928, and was in some doubt whether he was cleansing his bill or swallowing some of the sand. I have several times seen Thrashers bathing in rain water pools by the roadside.

Nests with four fresh eggs were found by Richard Miller at Rio Grande on June 6, 1920, and June 3, 1923, while in another nest found on May 27, 1923, the eggs were just hatching. Two nests found there by Turner McMullen on May 24, 1925, contained eggs and two on May 29, 1932, held young. I have seen full-fledged young birds in gardens on Washington Street on July 11, 1922, and July 5, 1925; in the former case they were feeding themselves. The duller coloration of the young Thrashers and the looser structure of their feathers are quite noticeable but not so striking as the different color of the iris which is gray in the young and bright yellow in the adult.

On July 3, 1920, I came upon two Thrashers in a thicket which were very much concerned at my presence and constantly uttered their alarm note—*chut, chut*, a sound almost exactly like that produced by sucking the tongue against the roof of the mouth and suddenly removing it. Doubtless they had young in the vicinity as on July 15, I found them there again, both carrying food in their bills although one of them soon swallowed his mouth-ful and wiped his bill on a twig. They again uttered the monotonous *chut, chut*, but this time there was also a peculiar mellow note *chuuurl* which had a curious ventriloquial character. I heard precisely the same note on July 8, 1916, and July 21, 1917, and on each occasion the birds were feeding young. Again I detected it on July 18, 1923, and recorded it as *churrly*, interposed among the numerous sucking alarm notes.

One individual came to our bird bath in mid-July 1937, and fed on scraps of bread thrown out to Grackles, Red-wings, etc. The former of these were cock of the walk so far as Red-wings, Starlings and Sparrows were concerned but they always deferred to the Thrasher when he appeared on the scene.

A family of three young Thrashers visited our garden on July 5, 1925, showing that they soon begin to wander from the nest site as we had had none present earlier in the summer. Another family was full-grown, but still in the juvenal plumage, on September 1, 1928. Thrashers are most abundant after the middle of July when the broods of young are on the wing. Later, in August, they become quite scarce, perhaps due to the retiring habits of the adults while molting, but it is my opinion that Thrashers like a number of the breeding birds begin to move south before the migrants from farther north arrive along the coast; with the great flights of late September and October they again become plentiful.

Walker Hand's records of first arrivals at Cape May up to 1920 with a few other records for more recent years are:

1902.	April 24.	1911.	April 27.	1920.	April 25.
1903.	April 20.	1912.	April 5.	1921.	April 24.
1904.	April 24.	1913.	April 18.	1922.	April 23.
1906.	April 21.	1914.	April 13.	1923.	April 22.
1907.	April 25.	1916.	April 23.	1926.	April 24.
1909.	April 23.	1917.	April 19.	1932.	April 30.
1910.	April 28.	1918.	April 16.	1934.	April 23.
		1919.	April 23.		

Our latest dates are:

1913.	October 13.	1925.	October 18.	1930.	November 9.
1920.	October 17.	1927.	October 30.	1931.	October 19.
1921.	October 16.	1928.	October 13.	1932.	October 12.
1923.	October 21.	1929.	October 25.		

We have several November and December records and a few for January and February, all apparently of wintering birds; and usually single individuals; Christmas census data are included.

1915.	February 3.	1926.	November 21.	1932.	March 5.
1919.	January 19.	1926.	December 26.	1932.	December 26.
1923.	February 22.	1927.	December 26.	1933.	December 24.
1923.	November 4.	1928.	December 23.	1934.	December 23.
1924.	December 28.	1929.	December 22.	1935.	December 22.
1925.	January 25.	1931.	December 27	1936.	December 27.
1925.	December 27.		(eight).		

On Five Mile Beach William Baily saw several during the winter of 1897–98, one on February 22, 1894, and two on December 27, 1903.

The abundance of Thrashers at the Point during the autumn migration is evidence of the great numbers of these birds that must nest in northern New Jersey and the states to the north, while it also indicates that the shore line is a regular migration route for south-bound Thrashers.

ROBIN

Turdus migratorius migratorius Linnaeus

While varying in numbers from year to year, the Robin seems to have increased very noticeably in Cape May within my recollection, and is now the most plentiful and generally distributed breeding bird within the town limits. Originally there were probably few if any summer Robins on "Cape Island," judging by their status on those parts of the other coast islands where original conditions still prevail, but the wonderful development of shade trees in Cape May and the steady increase in dwellings with gardens and shrubbery and well-kept lawns have caused a corresponding increase in Robins. We find them feeding on the lawns and building their nests in the trees along the streets; we hear their cries of alarm when the young are starting to shift for themselves; their evening song in the days of early spring and, during May and June, their early morning chorus from 4:00 to 4:30 a. m., notable for its volume and drowning out all competition. In fact an occasional Red-wing, Catbird or House Wren voice is about all that can be detected in the prevailing wealth of Robin song. Laurent found them only in spring and autumn on Five Mile Beach in the eighties.

Robins in the East have become so closely identified with man's habitations that it is in close proximity to dwellings that we usually look for them, and so we find them about the houses at Cape May or at the Point, or in the orchards and gardens about the farms of the upland, nor do they stray far except for food until the flocking begins and the drifting bands of autumn and winter are forming.

Robins are incubating eggs by the last week in April at Cape May and young are hatched by May 8. They, however, have two and often three broods in a season and freshly hatched young have been found as late as July 10, while young were still being fed in the nest on August 13 and August 23, in different years, and parent birds have been seen on the lawns caring for full-fledged, active young as late as September 1, 1925. The young in a nest in a sycamore tree in front of our cottage took their first flight early in July, 1932, and the old birds were soon busy building another nest in the next tree pulling material from the first one to aid in its construction. The second brood had hatched by July 28 and were still being fed by August 1. A nest in the same tree held nearly fledged young on July 29, 1934, as did another on July 10, 1935.

Robins have been heard in full song as early as March 13, 1921; March 19, 1924; March 21, 1925; March 11, 1929; March 23, 1935; while our latest dates for singing are July 26, 1922; July 26, 1925; August 7, 1926. While there is more variation in bird song than we realize there are usually certain dominant phrases that are always the same and this gives us the impression that all songs of a species are identical. Occasionally, however, one individual will utter a song so peculiar that we at once recognize it as different. So it was with a Robin at the Point on July 14, 1923, and on several days following which sang *chíl-o-wit*, *chíl-o-wit*, *chíl-o-wit*, each triplet exactly like the others, with none of the customary change of pitch. In autumn, winter and early spring, while not singing, the Robins have a series of subdued calls, *twit*, *twit*, *twit*, three or more together and accompanied by a flip of the wings as they sit on a perch or as they take flight. The loud screaming alarm cry seems to be limited to the breeding season and uttered mainly when nest or young are threatened.

The first full-fledged young have been seen on the lawns as early as May 23, 1921, and May 28, 1923, while others evidently but recently out of the nest, seen on June 21, 1932, and June 23, 1929, were probably second broods. By early July—July 8, 1921; July 11, 1922; July 11, 1924; July 18, 1925; they assemble in considerable numbers. On the shaded grounds of the Physick place I have frequently counted upwards of fifty or sixty on a single evening with very few adults, the latter doubtless still engaged with later broods. Some of the full-fledged young are in beautiful plumage, the dark spots below and the white ones on the back arranged in perfect rows. I have seen one of these speckled-breasted young bathing with an adult in Lake Lily on July 8, 1921. He stood in shallow water facing the shore, ducked his head completely under and fluttered his wings, fluffing up his plumage meanwhile. Another young Robin sat on a lawn with wings partly spread in bright sunshine while still another rested flat on his belly while he preened his plumage, both probably trying to rid themselves of lice.

In June Robins make free with the garden cherries, carrying them away one at a time in their bills and they also feed on the black mulberries which are frequently found along the wood edges, especially those that face the salt meadows. Later on wild cherries form an important part of their diet while in autumn poke berries are a favorite food and in winter the berries of the holly. They are to be seen too in July feeding on fields of new mown hay where insects are evidently the attraction. Their main staple of food however, throughout the spring and summer, is the earthworm and Robins may be seen constantly in search of them on the lawns and on the greens of the golf links. They gather in numbers on the close-cropped sod, running here and there, especially where a water sprinkler has brought the worms to the surface. They stop suddenly with head on one side as they detect their prey; and soon they are busy dragging the luckless worms from their burrows. With the advent of the Japanese beetle the Robins vie with the Starlings and Grackles in digging out the larvae from the ground. Unlike these birds, however, the Robin is not interested in scraps of bread thrown out on the lawn and while they are enjoying a feast of this sort nearby Robins persist in their search for worms or beetle larvae.

The usual gait of the Robin is a run, the feet moving with great rapidity, but they also hop, and a bird about to take wing from the ground usually changes from the run to the hop seeming to get a better start from both feet at once.

The roosting habit of Robins in late summer is well known and at Cape May, at this time, they associate with the Martins, Grackles and Starlings. At the time my observations began, they made use of the grove on the Physick property on Washington Street. Later they followed the other birds to a patch of woodland in West Cape May and finally back to the tall bayberry thickets on the Fill just east of the original roost. The young birds that assembled in July on the lawns of the Physick property increased in numbers as summer advanced and each evening flew up into the trees which shaded their feeding grounds. Hosts of other Robins both old and young came to the roost from farther off, some of them resting for a short time on the golf links where they stood listlessly and fed but little. I have counted thirty on a single green on July 29, 1921, eighty-six on August 21 and sixty on August 29—approximately one for every two foot square. Another gathering place was a lawn immediately opposite the roost where I have often seen as many as 250 crowded into a comparatively small area. All of these birds eventually repaired to the roosting trees while the latest comers went directly into them pitching down precipitately onto the branches. They seem to be the last birds to go to roost, with the possible exception of the Martins, and often I have seen flocks of Robins come in after the last Martin had settled for the night, while in the roost on the Fill I have seen Robins flying about from bush to bush in the gathering gloom when it was almost impossible to distinguish them ten feet away. They leave the roost with the other birds shortly before sunrise and scatter over the country, sometimes perching for a short time on chimneys, house tops and isolated trees or out once more on the golf links.

Young Robins have been found molting by August 12 and 17 in some years and adults were in full molt during the first two weeks of September, 1927. Old birds have been seen entering the roost on August 21, 1926, and

August 27, 1924, which exhibited forked and wedge-shaped tails indicating successive stages in the progress of the annual molt. As the young birds do not molt the flight feathers in their first year they showed no such peculiarity.

In late August there seems to be a diminution in the number of Robins about Cape May as the breeding birds begin to flock and scatter farther afield for the day's feeding. We then find them more common on the edges of the woodlands, in the pine groves of the Point, and even out on the mud flats of Pond Creek Meadows with the Killdeers. Later the breeding Robins of Cape May probably leave the neighborhood entirely and their places are taken by migrants from farther north.

These great flights of northern Robins begin to arrive in late October and continue to pass throughout the first two weeks of November. We have records of great flights on October 31–November 7, 1917; November 7, 1922; November 11, 1923; November 6, 1927; October 25, 1929; November 8, 1930; November 7, 1931; October 24–25; November 3 and 14, 1935. They are often accompanied by Bluebirds and, both being day migrants, their time of occurrence does not coincide very accurately with the flights of the smaller night migrants. On October 2, 1924, for instance, when the woods of the Point were thronged with small birds that had arrived during the night, there were only a couple of Robins to be seen anywhere in the vicinity, and on October 8, 1932, I could find only one. Indeed October, with the exception of the last few days, seems to mark the lowest point in Robin population at the Cape; and the contrast of great flights of November is striking.

William Rusling who was present at the Point continually during the autumn of 1935, watching the hawk shooting as the representative of the Audubon Association, has given me an excellent account of the Robin flight of that year, he states that "on October 24 there was a drop of twelve degrees in the temperature accompanied by a fairly strong north-northwest wind. A great flight of Robins came in flying north across the turnpike and into the wind, the birds were everywhere in the air from the ground up as far as the naked eye could see, while the glass showed more Robins at still greater altitudes. The flight lasted until about 10 o'clock in the morning." He estimated that, between 6 and 10 o'clock, fifteen thousand Robins passed; his actual count being 9815 with three thousand more between 5 and 5:30 p. m. The next day there was a similar flight between 6 and 6:30 a. m. and he counted 9150 all flying northwest, his total count of south-bound Robins for the autumn was 30,334; some were still passing November 14–16 but the bulk had gone before that. His daily counts from the first big flight until the end of his stay are as follows:

October 24, 9815.	November 2, 480.	November 6, 1100.
October 25, 9150.	November 3, 2385.	November 14, 2150.
October 30, 650.	November 4, 800.	November 15, 2000.
November 1, 665.	November 5, 75.	November 16, 100.

On the other days the number present was negligible.

The Robin flights are exactly comparable to the flights of Flickers, Kingbirds, Woodcock etc., and their direction is always northward along the Bay shore. The theories advanced to explain this apparent reverse migration are discussed elsewhere (cf. pp. 38–46). Otway Brown records a similar concentrated flight which took place on November 7, 1931, when very large flocks passed the Point all morning flying, as usual, into a northwest wind. The birds were also plentiful at Cold Spring, and many rested there on the ground fighting and feeding on fallen apples. At the time of the visit of the American Ornithologists' Union to Cape May Point on October 25, 1929, a similar flight, but of lesser proportions, was in progress with many flocks of Bluebirds mixed with the Robins. Sometimes these flights, just as in the case of the night flying warblers, get blown out to sea by the northwest winds and struggle valiantly to beat their way back to shore. I was fortunate enough to witness one of these occurrences on November 6, 1927, when from 10 a. m., and probably earlier, until 2 p. m. Robins came in great flocks across the boardwalk from far out at sea heading directly into the gale. They appeared at first as mere specks high up in the air, fifty, one hundred, and two hundred, in scattered flocks with gaps and pauses between them. As they came nearer one could see that they were birds and buffeted this way and that, rising and falling, they struggled on, coming nearer and nearer and finally when approaching the shore they dropped low over the beach, probably to get in the lee of the buildings and hedges. Then they coursed over the marshes and sand flats in a generally northwest direction crossing Lake Lily and the pine woods at the Point and on northward along the Bay shore. Apparently they all came in where the beach faces south, between the pier and South Cape May. Now and then some of them were turned about or blown back by the wind when they had all but reached their goal and, drifting along with the wind, losing ground every moment, they would once more face about and make another supreme effort.

In spring Walker Hand's records show a distinct northward movement of Robins on March 9, 1904; March 10, 1905; February 25, 1906; March 14, 1907; March 9, 1908; March 3, 1909; March 16, 1916; and the migration continues after the earlier arrivals are established in the town for the summer and are in full song. On April 2, 1922, for instance, a flock of these silent transients passed north over the Fill and on April 2, 1924, I came upon a similar flock resting in the pine groves of the Point near Lake Lily. With some Robins present all winter, however, it is difficult to give actual dates of first arrival or of latest departure. In some years, when very severe weather has prevailed during January and February, the usual appearance of migrating Robins, which we recognize about the middle of the latter month, is delayed until March, as was the case in 1936. There are also curious influxes of Robins in the late winter months, at times of severe cold, which are certainly not composed of birds from farther south, and which appear to be birds that had wintered to the northward and had been forced back by severe weather, snow and lack of food.

Wintering Robins are usually to be found about Cape May in small flocks in the dense thickets and wooded dune hollows between Higbee's Beach and Cape May Point on the Bay shore, where red cedars, hollies and greenbrier make an impenetrable shelter with food in abundance. On calm, sunny winter days, too, they may be seen in the same general neighborhood, flying overhead, pausing in the tree tops as they pass from one piece of woodland to another, or chasing one another over the bare fields, and stopping to feed on the low berry bearing bushes. Then in exceptional years we may have the winter influxes referred to above as in mid-January, 1922; January 14, 1923; January 25, 1925; January 3, 1926; when flocks of one to three hundred coursed over the open country sometimes to remain for weeks in the same neighborhood, sometimes to disappear as mysteriously as they came.

The numbers of Robins counted in the Cape May region on the Christmas census of the Delaware Valley Ornithological Club are as follows although with such active birds there is probably some duplication:

December 22, 1929, 130.
December 28, 1930, sixty.
December 26, 1927, thirty-four.
December 23, 1928, 279.
December 27, 1931, twenty-four.

December 26, 1932, 450.
December 24, 1933, eighteen.
December 23, 1934, 154.
December 22, 1935, 597.

It is always one of the pleasures of winter bird study at the Cape to come upon these bunches of the familiar Robins in their winter quarters, familiar and yet somehow different from the sociable Robins of summertime; no longer dependent upon human environment but shifting for themselves in the wild Bay side dunes and marsh-lands; their breasts dulled by the brown suffusion of the winter plumage; their voices stilled but for the occasional flight call.

*VARIED THRUSH

Ixoreus naevius meruloides (Swainson)

It is a remarkable thing, one of the mysteries of bird migration, that three individuals of this bird of the far Northwest should have found their way to New Jersey. As early as March, 1848, Dr. Samuel Cabot procured a specimen in Boston, Mass., which had been shot in New Jersey (Proc. Boston Soc. Nat. Hist., 1848, p. 17) and in December 1851, another was taken at Hoboken, and recorded by George N. Lawrence (Ann, Lyc. Nat. Hist., N. Y., 1851, p. 221).

While three specimens were recorded from Long Island no others were seen in New Jersey until November 26, 1936, when one appeared at Mr. M. L. Parrish's cottage at Pine Valley, near Clementon. It was identified from plates in standard bird books and from an examination of specimens at the Academy of Natural Sciences and was later seen by Mr. and Mrs. E. S. Griscom. It remained until March 20, 1937 (Auk, 1937, p. 395). Curiously enough on November 24, 1936, another Varied Thrush appeared in the garden of Mrs. John H. Boesch in Richmond, Staten Island, N. Y., where it remained until December 6, and was observed by many persons. (Bull. Staten Isl. Inst. Arts and Sci., January, 1937.) While none of these individuals was identified subspecifically it seems probable that they all belonged to the more eastern race which ranges to northwestern Montana.

WOOD THRUSH

Hylocichla mustelina (Gmelin)

The Wood Thrush occurs as a summer resident in moist woodlands from Higbee's Beach, Cold Spring and Bennett, northward through the peninsula, but never about Cape May or the Point except as a very rare transient in autumn. It is not adapted to the dry pinelands of south Jersey and does not find congenial surroundings close to the farmhouses, while in the Pine Barrens to the north it occurs only occasionally in deep cedar swamps. It is a bird of West and North Jersey and occasional in the narrow coastal strip where deep deciduous woods are to be found, as opposite to Rio Grande, and Bennett and in the Manahawkin Swamp in Ocean County, farther north. We have heard the song of the Wood Thrush as late as July 12, 1925, at Goshen; July 14, 1927 at Bennett; July 28, 1922 at Erma; and as early as May 2, 1926, at Cold Spring.

In the autumn, at Cape May Point, William Yoder saw a Wood Thrush on October 25, 1929, and David Baird one on October 8, 1932, while one was picked up at the Lighthouse on September 18, 1932. William Rusling through the autumn of 1925 saw it but four times—September 8, one; September 10, three; September 17, one; October 23, one. Our only spring record for the town is a single bird that I saw in the Physick garden on May 10, 1929.

We have records for three nests all at Rio Grande and all containing four eggs, one on May 27, 1923, seen by Richard Miller, and two on May 30, 1925, and May 29, 1926, by Turner McMullen.

When we realize the numbers of Wood Thrushes that breed in the northern counties and in the states to the north of New Jersey we are at a loss to explain the scarcity of the bird in migrations at Cape May and are inclined to think that it must pass in a southwesterly direction north of the Pine Barrens and down the Delaware Valley or along Chesapeake Bay, and back in spring by the same route.

HERMIT THRUSH

Hylocichla guttata faxoni Bangs and Penard

The Hermit Thrush is a common transient and a less abundant winter resident. It occurs in suitable localities all through the peninsula and even in the gardens of Cape May, but is most plentiful in the dense thickets of the Point and it is there that wintering individuals are most likely to be found. It was also found in migration on Five Mile Beach (Laurent). We may see many Hermits during the autumn migration but as we traverse the wood road through the sanctuary at the Point there are so many other birds present at this time that the silent south-bound Hermits do not impress us as do those that we come upon in midwinter. Then each individual seems to have selected a territory of his own where others shall not intrude, and where he may live up to the name that Alexander Wilson bestowed upon him. The presence of such an apparently delicate bird at such a season also impresses us and makes us associate the Hermit more with the stillness of his winter retreat than with the bustle of the autumn migration.

On one of those rare days of December or January when no wind blows and the sun warms up the pine woods and holly thickets until we think that spring has come, we may be fortunate enough to startle a Hermit Thrush from one of the dark tangles of vines and briers which are to be found along the Bay shore and we stop instantly to watch. With a few strokes of his wings he has alighted again upon some low limb or fallen tree trunk and so suddenly does he come to rest that the effect is almost startling. There seems to be no movement after his feet first touch the perch and with wings slightly drooping he remains motionless in his statue-like pose, the brown russet of his back blending with the dead leaves of the woods floor until we all but lose sight of him. Presently there is a slow lifting of the tail, an interrogatory *put* and reassured he slips lightly to the ground. Now he pecks rapidly at some choice morsel among the leaves and is suddenly still again as if carved out of stone, or occasionally he will raise his head high up and stand as if on tiptoe to gain a better view when some sudden noise startles him. At this season the term Hermit is peculiarly appropriate. Even during the migrations he does not mingle intimately with his kind. There may be several Hermits in the woods but each one seems engrossed with his own affairs and there is no flocking in the true sense of the word. The Hermit, moreover, does not seek the vicinity of man's abode and when, through some accident of migration, he finds himself in the garden he shows none of the confidence of the familiar home birds but holds himself aloof full of an air of mystery, as if aware that the spell of the "wanderlust" is upon him and that his place is not with us.

Silence is one of his characteristics, when with us in the autumn and winter, but on a bright spring morning when a warmth as of midsummer makes all nature leap forth into early leaf and flower, there has come to me a vision of hemlock woods, of dark forest floors with sunlight filtering through the canopy of boughs high over head, and the tinkling of a mountain brook—and as I try to account for the mental picture I realize that a Hermit Thrush is singing—not the ethereal melody that will mark his homecoming in the north woods but a delicate imitation of it—his whisper song. There he stands on a bough not ten feet away, his bill closed but his throat swelling out with his song half-formed and ventriloquial but with sufficient of the true quality to bring back at once the vision of his mountain home. Just as the breath of summer has drawn forth the buds and blossoms a bit before their time so it seems to have forced our silent transient to try his voice before he reaches those surroundings amidst which he is accustomed to pour forth his music. In autumn Hermits, like many other transients, linger longer than in spring. They mingle more with other birds, exhibit more activity and are less secretive than on their northward flight. We find them pulling at the pokeberries and belated sassafras fruits or flitting about the dogwood trees looking for the red berries, but there is always the same grace, the same dignity, and the same statue-like pose in moments of inactivity that are characteristic of this and the other small thrushes that grace our woods in the flights of spring and autumn.

The December Hermit Thrush records as reported on the Christmas census of the Delaware Valley Club are as follows:

December 24, 1922, one.	December 27, 1931, thirteen.
December 26, 1926, two.	December 26, 1932, three.
December 26, 1927, one.	December 24, 1933, six.
December 25, 1928, one.	December 23, 1934, two.
December 22, 1929, three.	December 22, 1935, thirty.
December 28, 1930, four.	December 27, 1936, seven.

We have scattered records of Hermit Thrushes for January, February and early March and on January 25, 1925, Henry Gaede saw as many as twelve in the course of a morning's walk at the Point. As in the case of several other wintering species there is little doubt but that they would be found to be quite as common in January and February as in December, had we been present continuously or had made more visits to the Cape at this time. It is not very likely that there is a southward migration of birds of this sort after December but some may fall victims to predatory enemies or perish from unusual weather conditions as winter progresses.

William Rusling's observations throughout the autumn of 1935 show the height of the autumn flight to be in late October. His daily counts are:

October 12, three	October 24, three	October 29, four.
October 14, fourteen.	October 25, thirty-two.	November 3, three
October 19, seven.	October 26, fifteen.	November 9, two.
October 21, three.	October 28, eight.	

The spring migration seems to take place in late March and April and Hermits were reported as very numerous on April 24, 1926; April 1, 1928; March 19, 1932. On March 3, 1935, twenty-five were seen in the Physick garden by Otway Brown which may have been a migrating group but more likely wintering birds driven in for shelter or food by the excessive cold and snows of the preceding February.

While the Hermit Thrush is a summer resident in the mountains at New-foundland, Passaic County, near the northern boundary of New Jersey (Miller, Auk, 1922, p. 116), and perhaps elsewhere in the same vicinity, it is only a transient or casual winter resident in other parts of the state, and the host of Hermits that spread over the state during migrations must come from much farther north.

OLIVE-BACKED THRUSH

Hylocichla ustulata swainsoni (Tschudi)

A regular transient especially in the autumn at times of strong north-west winds and less abundant on the return flight in May. They come with the Veeries in late August and early September although the peak of their migration seems to be a little later, and like them they frequent the gardens of Cape May, although most common at the Point. They nest entirely north of New Jersey and migrate both down the coast and through the interior.

Our autumn dates are as follows:

1913. October 13.	1928. September 5–October 21.	
1920. September 2.	1932. September 7–October 14.	
1921. August 31–September 14.	1934. August 26.	
1924. September 4–October 2.	1929. September 6–October 5.	
1925. September 20–October 18.	1930. September 28.	
1926. September 11–27.	1931. September 7–October 19.	
1927. August 26–October 16.	1935. September 7–October 5.	

Spring records are:

1915. May 15.	1916. May 16.	1930. May 15.
	1927. May 20.	

Philip Laurent found them on Five Mile Beach in spring and autumn in the eighties.

GRAY-CHEEKED THRUSH

Hylocichla minima aliciae (Baird)

Occurs with the Olive-back but apparently not so abundant. I have seen them at Cape May Point on September 25, 1927; September 27, 1928; September 19 and 21, 1934; while William Rusling as a result of his continued observations during the autumn of 1935 reports them on September 7, 29 (seven), 30, October 4 (nine) and 5 (six) with one on the unusual date of November 2. These were all single birds except as indicated. I also saw several on May 20, 1927. The Gray-cheek breeds far to the north and northwest and probably only a small proportion of them traverse our coast line in migration.

BICKNELL'S THRUSH

Hylocichla minima minima (Lafresnaye)

William Rusling found five thrushes that had been killed by flying against wires or the Lighthouse, and that agreed in measurements with this small edition of the Gray-cheeked Thrush which breeds in the mountains of New England and New York. It would be impossible to satisfactorily identify them in life unless one had the two forms close together for comparison. His birds were found on September 23 (two), 24, 30 (two), 1935. At Barnegat light Charles Urner found a single Bicknell's Thrush among some five hundred birds killed there during a storm on October 28, 1925. Babson has recorded it at Princeton on September 10 and October 5 (Birds of Princeton).

VEERY

Hylocichla fuscescens fuscescens (Stephens)

An autumnal transient occurring in numbers at times of northwest winds in late August and September especially at Cape May Point but also in the gardens of the town and on the Bay shore north to Higbee's Beach. Walker

Hand and William Rusling have both commented upon the frequency of the call of the Veery on still nights when the fall migration is in progress and it would seem that far more of these birds pass overhead than are brought down by adverse winds. Veeries exhibit an unexpected spirit of hostility on occasion, and I have seen two of them, that settled on a sassafras tree to feed on the berries in late August, drive off Kingbirds that came to the tree for a similar purpose. Both species feed largely upon these berries during the autumn flight.

Our inclusive autumn dates for Veeries are as follows:

1891. August 26.	1927. September 12.
1920. September 1–2.	1928. September 5–8.
1921. August 19–September 4.	1929. September 2.
1922. August 29–September 17.	1930. August 11–September 28.
1923. August 27.	1931. September 1–7.
1924. September 4–7.	1932. August 29–September 17.
1925. August 29–September 7.	1934. September 6–13.
1926. September 1–27.	1935. August 30–October 4.

The earliest date, August 11, was of three in the Physick garden and my only spring record was of two that I saw in the same place on May 11, 1929. William Rusling's continued observations during the autumn of 1935 showed a single Veery on August 30, thirty-five on September 2, ninety-two on September 7 and from two to eight on five dates up to October 4.

While the Veery nests in the mountainous parts of northern New Jersey and as far south as Plainfield, Paterson, and possibly Princeton, the number of the transients that pass the Cape indicates that many of them come from still farther north. The Atlantic coast seems to be a regular highway of migration.

*WILLOW THRUSH

Hylocichla fuscescens salicicola Ridgway

While all of the specimens of Veeries that I have seen from Cape May County are the familiar eastern race one found on the campus at Princeton by Charles Rogers, on September 10, 1934, proved to be the darker western form known as the Willow Thrush (Auk, 1935, p. 191).

So far as I know this is the only record for the state.

BLUEBIRD

Sialia sialis sialis (Linnaeus)

Bluebirds are quite local about Cape May in summer. There are usually a pair or two nesting along the railroad between the town and Cold Spring and sometimes one or two at the Point and during July and August family parties may be seen on the telegraph wires but of late years none has been seen there. Their favorite nesting places seem to be old woodpecker holes in telegraph poles or cavities in apple trees in the orchards. Philip Laurent found them in spring and autumn on Five Mile Beach in the eighties but none bred there.

At the time of their arrival in March they sometimes occur in flocks of a dozen or more, most of which pass on farther north, and in autumn when the late October and November flocks of Robins are flying south we always find flocks of Bluebirds mingling with them. It is at this season that they reach their maximum abundance at the Cape although they are then only passing transients and escape general notice. A few remain through the winter along the hedges and fence rows of the uplands usually associating with winter sparrows, Juncos, Goldfinches, and other hardy species. As the winter wanes these birds break into fragments of song but the "melting music of the Bluebird" is never at Cape May the herald of springtime as it is, or used to be, farther north. Indeed our harbinger of spring would seem to be the male Red-winged Blackbird returning to his nesting ground close to the town and announcing the fact to the world.

It is difficult to distinguish spring arrivals from winter resident Bluebirds but I have listed all of our February and March dates below:

1921.	March 31.	1930.	March 22.
1923.	February 22.	1931.	February 1, twenty.
1924.	March 20.	1932.	March 5.
1925.	February 8.	1933.	February 19 (with a flock March
1927.	March 18.		12).
1929.	March 11.	1934.	March 8, flock.
		1935.	March 23.

The state of the weather affects the time or arrival of the Bluebird and in some years continued snow and ice throughout February undoubtedly set them back, while in other years, when upwards of twenty species of wild flowers bloomed in this month, the Bluebirds pushed northward early.

Our winter Bluebird records are as follows:

1921.	January 9, three; January 10, four.	1929.	December 22, six; December 26, four.
1922.	December 24.	1930.	December 28, twenty-one.
1923.	December 2, three; December 23, two.	1931.	December 27, thirty-one.
		1932.	December 26, twenty-nine.
1924.	January 19, three.	1933.	December 24, forty-two.
1926.	December 26, four.	1934.	December 23, sixty-six.
1927.	January 23, two.	1935.	December 22, eighty-four.
1928.	December 23, sixteen.	1936.	December 27, 165.

The late December counts were made on the Christmas census of the Delaware Valley Club and represent the combined observations of an increasing number of individuals. Most of the individual returns were of a few birds only but flocks of as many as twenty were seen in several instances. The possibility of duplications by several observers must also be considered in comparing the totals.

Winter Bluebirds either associate with other species or are sought out by them, and on January 19, 1924, three Bluebirds seen along a fence row, back of the town, were associated with two Myrtle Warblers and a Palm Warbler. While steadily decreasing as a breeding bird about Cape May the heavy autumn flights and the number of winter residents indicate that they must still breed in abundance in the country lying to the north of us, more especially north of New Jersey, and that the Atlantic coast is one of their regular migration routes.

I have seen Bluebirds with nests in telegraph poles on May 10, 1931, and May 25, 1922, while Turner McMullen has seen a nest with a full set of five eggs at Wildwood Junction on May 9, 1926, and one in the same place on May 8, 1927, containing five young. In Salem County and farther north, in West Jersey, Bluebirds have eggs as early as April 22 but those nesting near to the ocean seem to be a little later. I have seen the speckled-breasted young on the wires being fed by the parents from June 30 to July 27 and once on September 2, 1920, although they are most frequently observed during the first week of July.

The south-bound flocks have been recorded, usually with Robins, on:

1923.	October 21 and 26.	1931.	October 31.
1927.	October 22.	1932.	November 12.
1928.	October 27 and November	1933.	October 28.
	1 and 10.	1934.	November 25.
1929.	October 25.	1935.	October 24–25, November 1, 2,
1930.	October 26.		6, and 14–15.

The 1935 record is that of William Rusling who was present at the Point throughout the autumn. His counts were as follows September 19 (seventy-five); October 24 (352); 25 (440); November 1 (710), 2 (220), 3 (forty-five), 6 (245), 14 (125), 15 (135), with scattering individuals on other days. The big flights occurred with Robins on days of north winds.

BLUE-GRAY GNATCATCHER

Polioptila caerulea caerulea (Linnaeus)

The Gnatcatcher is one of those southern species whose presence at Cape May adds to the interest of its bird life and gives to the southern counties of New Jersey a character all their own. It is most frequently and regularly seen in the pine woods of Cape May Point surrounding Lake Lily though it occasionally visits the gardens of the town and may be seen as a rare transient farther north in the peninsula. On the Bay shore it ranges north at least to Bridgton where it has nested for several years and doubtless has always been a summer resident. Bennett Matlack has found four nests in this vicinity; one in Bridgeton Park on May 6, 1910, on the horizontal limb of an oak, twelve feet up and another back of Fortesque on a similar limb fifteen feet from the ground. In the latter instance "both birds carried nesting material and through the glass the shiny cobwebs and other material could be plainly seen." On May 27, 1933, a third nest was found in the park on an oak limb eighteen feet up and still another May 30, 1933, on a small limb of a tree which held four eggs. We have another Bridgeton record, May 12, 1935. The first nest found in the state was discovered by William Baily in the same neighborhood on May 19, 1885. For the vicinity of Cape May we have but two breeding records, a nest with four eggs found at the Point by Samuel Rhoads on May 17, 1903, and a brood of young just out of the nest June 13, 1937, seen by Frederick Schmidt and others. Julian Potter and Fletcher Street saw a family being fed near Court House on June 16 of the same year. It probably nests also at Dias Creek and elsewhere in the intervening country as Walker Hand and Henry Fowler found six birds there on May 28, 1911. Julian Potter found a Gnatcatcher singing at the Point on June 9, 1929, and another at Medford Lakes on June 16, 1935, while Fletcher Street showed me a nest at Beverly on May 2, 1936, which contained eggs. On the Atlantic side of the state it has been recorded on Five Mile Beach on April 12, 14 and 15, 1879 (W. L. Abbott), and April 7, 1901, while Norris DeHaven and I found a number in the strip of woodland that formerly stood back of Chelsea, just below Atlantic City, on April 8 and 16, 1893. While these birds may have been migrants the fact that the Gnatcatcher does not breed north of New Jersey, in the east, makes it likely that they spent the summer where they were found. Straggling migrants have been seen farther north, as one at Manasquan on September 5, 1927 (Urner); Princeton, April 28, 1875 (Scott); Woodbury, May 1, 1880 (Abbott); Haddonfield, April 10, 1882) Rhoads); and Mt. Holly, April 25, 1920 (Pumyea); also four records for Essex Co. (Eaton) and two for Union (Urner).

The Gnatcatcher is an exceedingly active little bird, hopping from twig to twig as it moves through the woods, peering from right to left with its prominent black eyes and swinging its long tail forcibly from one side to the other, usually keeping the feathers tightly closed although they are sometimes slightly spread. The tail seems entirely out of proportion to the body and its constant action is most characteristic; sometimes the bird is swung completely around by the force of the effort. The head is usually bent over with the bill pointed downward, and the tail, between flirts, is cocked up at an angle while the wings droop slightly.

Once I saw two Gnatcatchers fly up fighting in the air with a great show of white tail feathers (September 1, 1926). And again one was seen hovering over a stalk of golden-rod catching some small insects in the air. Seen from below the Gnatcatcher displays only the silky whiteness of its breast and belly but from the side, especially in spring, the blue of the upper parts is quite intense. Young of the year are grayish blue not nearly so brilliant as old birds in spring. A young bird in full juvenal plumage was seen on August 7, 1931, and adults in molt, with new tail feathers not fully grown, on August 7 and August 4, in different years.

The wheezy squeek which constitutes the Gnatcatcher's song has been heard in the spring and early summer but in August and September when the little bird is most frequently seen at the Cape it is silent.

One that I studied intimately on May 2, 1932, remained for some time in a small bush, and on the lower limb of an adjoining pine tree, feeding diligently. It peered intently with its beady black eyes at every twig and as it hopped about, the tail was flipped slightly up and down, and every now and then there was a much more violent rotary swing to one side or the other. The bird then came to rest and plumed the feathers of the breast and back, down to the base of the tail; it also scratched the head with its left foot. After this it became more active than before with much more flirting of the tail. The colors were very bright and contrasting, the blue of the back, white breast and black tail. The white feathers of the latter were seen only when the bird flew from one twig to another.

With the exception of the records already referred to our earliest dates for summer Gnatcatchers at the Point have been:

1922.	July 15.	1927.	July 29.	1934.	July 29.
1923.	July 18.	1928.	July 18.	1935.	August 4.
		1932.	July 17.		

They become more plentiful in August, though never abundant, and our latest dates are:

1922.	September 17.	1928.	September 1.	1934.	September 20.
1924.	September 1.	1929.	September 29.	1935.	September 6.

While most of our observations of Gnatcatchers have been of single birds or parties of two or three, John Gillespie saw five on September 5, 1926, Julian Potter eight on April 23, 1922, and Ernest Choate five on August 3, 1935.

Our earliest spring arrival dates for Gnatcatchers at the Point are:

1903.	April 11.	1926.	May 1.	1932.	May 2.
1921.	April 24.	1927.	April 24.	1934.	April 14.
1922.	April 23.	1928.	April 6.	1935.	April 15.
1923.	April 22.	1929.	April 14.		

With no competent observer in the field continuously through the spring some of these dates are undoubtedly late.

GOLDEN-CROWNED KINGLET
Regulus satrapa satrapa Lichtenstein

The Golden-crowned Kinglet is a regular autumn transient, especially about Cape May Point, but occurs also in Cape May and in the wooded country to the north as well as, formerly, on Five Mile Beach. A small number remain through the winter in cedar groves and orchards but it is much less abundant then, as well as on the northward migration in spring.

Our earliest arrival dates are:

1921.	October 9.	1925.	October 1.	1930.	October 5.
1923.	September 30.	1927.	October 10.	1932.	September 24.
1924.	October 2.	1928.	September 30.	1935.	October 5.
		1929.	October 5.		

We have many records for November and December but our Christmas census lists usuually show only a small number present:

December 23, 1928, five.	December 24, 1933, one.
December 22, 1929, two.	December 23, 1934, eight.
December 27, 1931, one.	December 22, 1935, eighty.
December 26, 1932, five.	December 27, 1936, eleven.

I seldom see over six on a day's walk in autumn and on only three days did the number ever exceed ten.

In spring we have only two records—April 1, 1928, one seen by David Baird, and March 18, 1933, when I saw several. This is in part due to lack of field work in early spring and partly, no doubt, to the fact that the bird passes through rapidly at this time or travels farther inland, but it is unquestionably scarce.

The Golden-crown is at home in the north woods and the higher mountain forests of the eastern United States but does not nest in New Jersey. For some reason or other it does not seem to be as abundant in either migration as it is farther inland.

RUBY-CROWNED KINGLET

Corthylio calendula calendula (Linnaeus)

The Ruby-crowned Kinglet is a regular transient, more abundant in autumn; occasionally an individual remains through the winter, or at least through December, but this is very unusual. While here its habits and haunts are like those of the Golden-crown and they often associate.

Arrival dates:

1923.	September 30.	1928.	September 27.
1924.	October 2—numerous with Chickadees.	1929.	October 5.
		1930.	October 5.
1925.	September 22.	1934.	September 28.
1926.	September 27—abundant.	1935.	September 27.

Our latest dates for migrants are:

1923.	November 11.	1934.	November 7.
1931.	November 7.	1935.	November 3.
1932.	November 13.		

Single individuals were seen December 26, 1927 by Julian Potter and December 23, 1934, on the Christmas census.

In spring we have the following records:

1922.	April 23.	1932.	May 8—three.
1927.	April 30.	1933.	May 7—several.

The Ruby-crown breeds still farther north than the Golden-crown while its route of migration appears to be the same.

PIPIT

Anthus spinoletta rubescens (Tunstall)

Flocks of Pipits are of regular occurrence about Cape May from October to January but curiously enough none have been recorded during the northward flight of spring. Most of our records are of two or three birds flying overhead, or flocks of a dozen, thirty, or forty, but in December and January there may be great flocks of several hundred apparently spending the winter in the Cape region. Whether we have simply failed to come upon them in February and March, or whether the flocks have moved off to the south and then gone north rapidly or by a more inland route, I cannot say.

To me the Pipit seems to be the embodiment of three impulses of bird life: gregariousness, the restless spirit of migration, and protective coloration. On some day of midwinter when there has been no blanket of snow such as sometimes covers the landscape, even at such a supposed "semi-tropic" region as Cape May, we gaze over the broad monotonous expanses of plowed fields and conclude that here at least bird life is absent. We contrast these silent brown stretches with the swamp edges and their bursts of sparrow conversation or with old pasture fields where Meadowlarks are sputtering. But let us start to cross these apparently deserted fields and immediately with a weak *dee-dee, dee-dee*, a small brown bird flushes from almost beneath our feet, then another and another, displaying a flash of white feathers in the tail as they rise. In a moment they have settled again farther on and are lost to sight against the brown background as suddenly as they appeared. We advance again and now the ground before us seems fairly to belch forth birds, as with one accord, the whole flock takes wing, and with light, airy, undulating and irregular flight, courses away over the fields, now clearly defined against the sky, now swallowed up in the all pervading brown of the landscape.

On January 3, 1893, I came upon an enormous flock of Pipits back of South Cape May feeding on black grass stubble. After following them about for a few minutes they all took wing and mounting high in air, until they looked no larger than flies, they disappeared to the southward. On January 7, 1922, I encountered a similar flock on the Fill feeding in a burnt over area several acres in extent. In places the grass had been burned down to the ground, which was black with the ashes and traversed everywhere by the winding runways of meadow mice, now mere channels in the earth. There were scattered bayberry bushes, clumps of low willows and occasional tufts of ruddy buff Indian grass which had escaped the fire. I had deliberately approached the spot with the expectation of finding Pipits but they had all taken flight before I was able to distinguish a single bird on the ground. A large part of the flock arose first and the others joined them in succession, rank after rank. They formed a long rather dense flock and did not fly high above the ground, as they often do, the individual birds rising and falling slightly which produced a somewhat undulating effect, while all were chirping in their thin wheezy manner.

After circling in a large arc they came drifting back and settled down near where they were before. Several times later they flushed but always returned to the burnt area. By watching exactly where they alighted I was able to detect them scattered all over the ground, about one bird to each square foot, where thickest. Their backs had a distinct olive cast in the strong light but the streaks on the under parts were only seen clearly when the birds were breast on. They all walked deliberately or sometimes took half a dozen steps in rapid succession, almost a run, though less regular. They all moved in the same general direction and as I moved parallel with them I could see them pressing straight ahead through the grassy spots and between the grass tufts and the stems of the bushes that had escaped the fire. They kept their heads pretty well down on the shoulders and leaned forward, dabbing at the ground with the bill, to one side or the other, apparently picking up scattered seeds of grasses and sedges. The tail was carried parallel with the ground or tilted up a trifle while the tips of the wings hung just below its base. The tail moved a little as the bird advanced but there was no distinct tilting as in the Palm Warbler or the Water-Thrush. Only just as the bird took wing was it possible to see the white on the outer tail feathers. Sometimes the neck was outstretched and the head held erect while the bird made a hasty survey all around. Just how the Pipits took wing I could not ascertain as the individuals that broke into flight were never the ones upon which I had my glass focussed, and once the start was made it became contagious and all were on the wing in a moment.

Once or twice a few Pipits took after a nearby Meadowlark that had flushed at the same time but usually they made no mistake and hung closely together. Eventually, three days later, the flock disappeared, rising high in air as did the other flock and made off for the shores of Delaware. They have a certain sparrow-like appearance on the ground but their walking and their systematic and concerted activities are characteristic.

The Pipit is a typically gregarious species and exhibits that peculiar simultaneous response to some signal that we cannot detect but which causes the entire flock to act as a unit with such beautiful precision.   The action, however, is not developed to quite the perfection that we see in some of the smaller sandpipers.

The Horned Lark is the bird with which the Pipit would naturally be confused as they have much the same habits.   Julian Potter writes me that in his opinion the flight of the Pipit is more uncertain and erratic than that of the Lark while the call note of the latter is higher pitched and clearer, the Pipit's call being thinner and slightly harsh.

It seems to me that the Pipits look much darker, almost black in certain lights, when on the ground, while they stand higher and do not appear to "creep" as the Larks seem to do.   /

Our arrival records are as follows:

1921.	November 11.	1926.	November 21.	1933.	October 29.
1922.	November 7.	1927.	October 23.	1934.	November 25.
1923.	October 21.	1928.	September 30.	1935.	October 8.
1924.	September 22.	1929.	October 25.		

They have remained every year until the last of December and our Christmas census returns are:

December 26, 1926, forty-four.	December 28, 1930, nineteen.
December 26, 1927, one hundred.	December 26, 1932, one hundred
December 23, 1928, fifty.	and ten.
December 22, 1929, one.	December 23, 1934, flock.
	December 22, 1935, four.

While I have no spring records for Cape May and only two for January, Julian Potter saw a flock of thirty near Salem, on April 28, 1918, which indicates the time of their northward flight.

Our Pipits come to us from breeding grounds north of the United States and while they occur inland, as well as on the coast, in migration, the latter is evidently one of their principal highways.

*BOHEMIAN WAXWING

Bombycilla garrula pallidiceps Reichenow

There have been several records of the Bohemian Waxwing in New Jersey but none of them is satisfactory.   Dr. C. C. Abbott records one obtained in Cape May County and another in Morris County (Birds of New Jersey, 1868) and adds two more from Mercer County in his list of 1884, but there are no details regarding any of them.   T. M. Trippe states that a pair was "observed in the vicinity of Orange, April 28, 1867" (Amer. Nat., II, p. 380), whether by himself or someone else is not stated.

These Waxwings often accompany the Evening Grosbeaks in their southward winter flights from their home in the far northwest but we have no record of them among the recent flights of the Grosbeaks.

CEDAR WAXWING

Bombycilla cedrorum Vieillot

Cedarbirds about Cape May are found mainly in the pine woods of the Point although they occur irregularly and sporadically northward along the Bay shore and in the farming district farther inland. In Cape May itself I have seen them but once or twice; a flock came to rest in a wild cherry tree in the Physick garden on August 30, 1932; and Walker Hand told me that a flock of seven spent the winter of 1904–1905 there; while during March 19–24, 1906, they were common all over the town feeding on the berries of the Japanese honeysuckle and he counted fifty-two in one flock. They were also common during the winter of 1921–1922.

At the Point I have records of Cedarbirds for every month of the year but they are irregular there, as elsewhere, and vary greatly in numbers. They seem to be most abundant from May 20 to June 1 and from late July or early August to October. During some years there may be periods of several months when no Cedarbirds are seen and in some summers they have been recorded but once or twice. Usually, in summer, they occur in flocks of six or eight or as many as twelve or fifteen may be seen together; our largest flocks, however, have been in spring or fall: seventy on May 22, 1927; sixty on March 20, 1924; forty-five on May 10, 1924; thirty on September 11, 1924; while the largest winter flock was on December 26, 1926, when twenty-seven were counted. The largest flocks are obviously on migration and William Rusling who watched the Cedarbirds continuously during the autumn of 1935, found that during that year, at least, the height of the flight was in September.

His counts per day were as follows:

September 6, 125.	September 16, 160.	September 29, 120.
September 7, 830.	September 17, 225.	September 30, 180.
September 10, 531.	September 23, 191.	

and in smaller numbers to November 15.

While the spring Cedarbirds are in beautifully fresh plumage, many of them with conspicuous red "wax" tips to the secondaries, the flocks that arrive in August are composed of both old and young.

On only one occasion have I any record of the Cedarbird's nesting in the Cape May district. On August 19, 1921, I found a nest on the horizontal branch of a pine tree about ten feet from the ground upon which the female was sitting. It contained no eggs and by the 25th had been deserted. There were migrant flocks of the birds in the woods at the time.

Cedarbirds are usually seen flying rapidly overhead in compact flocks, passing from place to place, and uttering their faint wheezy notes. With the glass it is usually possible to distinguish the yellow tips to the tail feathers of the flying birds, especially in strong sunlight. When at rest the birds occupy some dead pine top or in winter any leafless tree, where they perch upright, usually with crest erect and slightly recurved forward. When preening the wings and body plumage hang loosely, at other times they are closely appressed and the neck is drawn out so that with the erect crest the bird looks somewhat top heavy. A single Cedarbird when it reaches a perch will often remain rigid for long periods. On September 7, 1931, one was detected on the very top of a dead pine at 10:30 in the morning which failed to move for over ten minutes. It was observed at intervals until 12:40 and although it turned once, so as to face the other way, it apparently never left its perch.

When feeding on flying insects, as Cedarbirds often do, especially over ponds, they will fly in circles like swallows remaining on the wing for some time but returning at short intervals to their perches. In flight the Cedarbird rises and falls, there is a rapid flicker of wings and then they are closed against the body for an instant, but there is not the marked undulation of the Goldfinch. They will also set the wings and sail as they whirl into a tree to rest, much in the manner of a flock of Starlings, and present somewhat the same triangular wing outline.

I have seen Cedarbirds in early August feeding on wild cherries and on August 27, 1934, they were associated with migrating Kingbirds devouring the berries of the sassafras which are abundant at the Point. On March 20, 1924, there was a large flock busy picking off the holly berries that still persisted on the bushes. They held the berries between the mandibles for some little time, whether to break them or simply preparatory to swallowing I could not determine. On May 10, 1924, a flock of forty-five was feeding in a patch of shrubbery eating the soft leaf buds; one bird was noticed moving sideways along a limb to another which he proceeded to feed apparently ejecting a partly masticated bud from the throat into the bill. Other birds in a wild cherry tree in early August craned their necks high up and twisted the head half way round as they did so. Several times they were noticed to break off little pieces of twig or pull off fine shreds of bark with their bills for what object I was unable to determine. I have seen them in November persistently devouring chicken grapes at spots where the vines were growing abundantly and also feeding on the berries of Ampelopsis on the walls of a house, remaining until the crop had been entirely consumed.

Here as elsewhere Cedarbirds seem to be erratic both in habits and occurrence and might well be rated as vagrants rather than as migrants.

The first dates for south-bound migrants for a number of years are:

1921.	August 5.	1925.	July 18.	1930.	July 24.
1922.	July 27.	1926.	August 9.	1931.	July 22.
1923.	August 4.	1927.	July 19.	1932.	August 9.
1924.	August 9.	1929.	July 20.	1935.	August 12.

The results of the Christmas census of the Delaware Valley Club follow:

December 27, 1931, one. December 23, 1934, nine.
December 26, 1932, two. December 27, 1936, twenty-six.
December 24, 1933, twelve.

NORTHERN SHRIKE

Lanius borealis borealis Vieillot

An occasional winter visitant on the Fill and about Cape May Point, perching on low bushes or telegraph wires.

All that I have seen in southern New Jersey are birds in the plumage of the first year with dusky vermiculations on breast and belly.

Walker Hand reported single individuals on November 6, 1917; December 27, 1919; December 13, 1928; while I saw one daily on the Fill January 8–10, 1922, which was eventually collected and another December 24 to 27, 1929. Julian Potter saw one on December 26, 1926, and David Baird another on January 22, 1921.

At Beach Haven, farther north along the coast, where Charles Urner had counted sixteen on January 22, Richard Erskine and I saw five perched at intervals on the telephone wires on March 13, 1927. These were probably north-bound migrants as were single birds seen at Cape May on March 18, 1927, and March 23, 1930.

Northern Shrikes breed far to the north and while they occur more frequently in the northern part of the state New Jersey is near the southern limit of their winter wanderings.

MIGRANT SHRIKE
Lanius ludovicianus migrans Palmer

This representative of the Loggerhead Shrike of the South is a regular transient at Cape May from mid-August to late September. We have five records for October and I saw one in an orchard at Cold Spring on November 13 and 14, 1932, which may have been established there for the winter, as probably the same individual was seen in the same place on February 5, 19 and March 12, following. Julian Potter found two at Pennsville on December 7, 1930, and we have a few midwinter records of shrikes which were not positively identified as to species but were probably referable to the Northern Shrike which seems to be the more likely bird at this season. In spring I identified a Migrant Shrike on March 25, 1923, and another on April 2, 1933, and William Baily saw one at Ocean View on April 30, 1901. Farther north in the state Charles Urner has recorded one as late as May 11, 1929. They are, however, much rarer at this season than in late summer and early autumn.

Elsewhere in the state they are transients, just as they are at Cape May.

Migrant Shrikes are solitary birds and we usually see but one at a time but in the course of a day's walk we may come upon two, three or five while Conrad Roland saw six in the Cape May Point region on September 16, 1934. Their favorite perch is a telegraph wire or the top of a bush or small tree, always in a location commanding a clear view of the surrounding country. The bird's position is erect usually with the head bent over forward in the posture of a Bluebird, as if in deep meditation, while the tail hangs straight down with feathers tightly closed so that it appears very slender.

When the Shrike first alights it seems to stand higher while the tail is held horizontally and flirted sharply once or twice before the usual attitude of rest is assumed. Occasionally the bird will turn its head when danger threatens and may also drop diagonally to the ground, apparently after a grasshopper or other prey, but returns at once to its perch.

I have not often observed Migrant Shrikes caching their prey on thorns nor can I testify as to their returning to feed upon it later. One bird watched for some time on August 13, 1921, had a large grasshopper in its beak which it was trying to impale upon some of the twigs of a bush. Failing in this it wedged the insect into a fork of a branch and left it there. It held its prey in its beak during the attempts to impale it but between these efforts it grasped it with its feet holding it against the side of the branch. Another bird dropped onto a haycock in pursuit of a grasshopper and remained there for several minutes apparently searching for it. Another Migrant Shrike, that was in the usual attitude of repose on a wire just west of the town, was struck at by a passing Sparrow Hawk and it was amusing to see how quickly he dodged and how amazed he seemed to be as he again took up his position on the wire. The hawk did not renew the attack.

The flight of the Migrant Shrike is low and distinctly undulating with several rapid wing beats between the dips, producing a conspicuous flicker of white and recalling that produced by the Mockingbird although the flight of the latter is not undulatory, the wing action more deliberate, and the flashes of white continuous.

Adult Migrant Shrikes have very white breasts while immature birds are much darker due to the narrow gray vermiculations on the lower parts. The black eye stripe in this species runs clear to the nostril a character which distinguishes it from the larger Northern Shrike in which the lores are pale. This difference in the adults, at least, can usually be detected with the glass.

Our dates of arrival at Cape May for a number of years run as follows:

1917.	August 25.	1922.	August 21.	1930.	August 22.
1918.	August 23.	1923.	August 23.	1932.	August 20.
1920.	August 15.	1926.	August 27.	1934.	August 27.
1921.	August 13.	1927.	August 19.	1935.	August 29.
		1929.	August 20.		

Our latest records for migrants are:

1917.	September 29.	1925.	October 2.	1932.	September 25.
1920.	October 3.	1928.	October 21.	1935.	October 14.
		1929.	October 5.		

STARLING

Sturnus vulgaris vulgaris Linnaeus

The Starling, which man so unwisely introduced from the Old World in 1890, is now one of the most abundant birds of South Jersey and bids fair to be the most destructive. Our other introduced species, the English Sparrow, seems to be distinctly on the wane, particularly in the towns and cities. The Starling, however, has been increasing ever since its first appearance and with its gluttonous appetite and generally hardy nature is in danger of crowding out many of our native species by sheer force of numbers, while by consuming the winter food supply it makes it difficult for many of our former winter residents to survive the cold months.

Introduced in Central Park, New York City, on March 6, 1890, they had reached Red Bank and Princeton by 1894, and Tuckerton and Vineland by 1907, if not before.

Walker Hand recorded the first Starlings at the Cape on April 4, 1909, when a pair nested in the woodwork of the old Stockton Hotel. By 1911 they had increased rapidly and in the summer of that year flocks of at least one hundred were to be seen. Since then they have been familiar birds at all times of year and are as regular in occurrence as the English Sparrows about every farm in the country. Single birds or pairs may be seen flying continually back and forth over the town from early spring to the end of July. They nest about buildings, in all sorts of holes and cavities, and take advantage of broken weather boarding which opens a way to inner spaces in walls. Flicker holes in telegraph poles along the railroads, or in old trees, are also acceptable nesting places, and Richard Miller tells me that they occupied one telegraph pole nest for six consecutive years and had young there on May 18, 1924. Old birds are busy all through the spring feeding on the lawns and in the gardens, and I have seen them carrying food as early as May 9, while others were feeding well-fledged young in the nest as late as June 20.

When the young first leave the nest they follow a parent, presumably the female, on the lawns of the town, running rapidly after her and jostling one another in their greed to get the food that she is finding for them. This often continues until the birds are full-grown and seem perfectly able to shift for themselves. By July 1 we may see flocks of young numbering one hundred or more, all in their plain gray dress, arising from the fields of cut hay where they find an abundance of insect food; other flocks all composed of young were seen on July 2 and 7 in different years about the hogpens which are to be found on the edges of marsh and woodland, where the garbage from the town is hauled. Similar flocks have been recorded as late as July 27 which had not yet begun to molt and young about half molted have been seen by August 5, while others on August 16, 1925, had conspicuous patches of glossy black feathers appearing in the gray plumage.

Adults and young seem to flock separately from the time the latter are on the wing, at least, until the late summer molt when they assume the black plumage. Flocks of both young and old birds are increasingly common throughout the summer and autumn and on the Fill, before the bushes attained such a growth, Starlings occurred in immense flocks during the autumn, many of them doubtless moving south as winter approached. On September 1, 1920, I saw at least a thousand rolling along over the ground, the rear ranks constantly rising and settling down in front, and curiously enough two Meadowlarks kept always in the van separated from the black horde by a small interval as if acting as leaders. Starlings also gather in flocks on the golf links, and about the aviation station, where they swarm over the roofs of the buildings and perch in rows on the fences, doubtless attracted by some refuse food.

As soon as the Grackles and Martins began to gather at their summer roost, formerly on the Physick place and later in the bushes on the Fill, the Starlings joined them every evening and spent the night with them. I have seen them coming in as early as July 7, 1923, and by August there were good sized flocks. On August 19, 1921, from 7 to 7:20 p. m., I counted 382 in detachments of from ten to one hundred and by September upwards of 1500 came in every night. Probably double that number came to the roost later. They usually come in before the Martins and did not rise and settle again as the latter do for some time before they rest for the night. When the Starlings did arise they took wing in a solid mass. When the Martins, after many years occupancy of the Physick grove, left for a piece of woods in West Cape May and later adopted the bushes of the Fill for a roosting place, the Starlings always went with them. It would seem possible, in the absence of any other explanation for this shifting of roostings, that the Martins may have resented the intrusion of the Starlings since their constantly increasing numbers must have crowded the Martins severely. They may also have forced the Martins to finally abandon their local roost. When occupying the Physick property the Starlings were seen to gather in considerable numbers on the roof of the greenhouse and if left to themselves they may prefer buildings to trees as night quarters, as they evidently do when they fly from the open country for miles around to roost on tall store and office buildings in the heart of Philadelphia.

A Starling flock is a unique sight. The birds seem to be the only land birds that can equal the lesser shore birds in unity of action and the sight of a flock of Starlings rising and falling, turning and pitching in perfect harmony cannot fail to arouse enthusiasm. Every bird moves instantaneously in exactly the same evolution that all the others are performing, yet we detect no possible guiding signal. There are no stragglers, no ragged edges to the flock, and at a distance it maintains its square, round or oblong shape, as the case may be, and drifts about like a swarm of bees. A great flock arose over the mainland, back of the Lighthouse on August 31, 1925, which looked from a distance exactly like a cloud of smoke. It came across the meadows and settled near the pond and later arose and covered the branches of some low dead trees, nearby, until they seemed to be clothed with foliage.

On July 27, 1920, a flock of some two hundred Starlings mingled with a large number of Red-wings on a field of freshly cut hay. When the whole mass of birds arose together they seemed inextricably mixed but in a moment the Starlings were in their own compact flock, still associated with the Red-wings but in a unit of their own. And so it is when they came into the roost with the Grackles, the bunches of Starlings formed compact units which cut through and around the slower moving Grackles in an astonishing manner, traveling, apparently, at least twice as fast.

The great flocks of Starlings drift about the country until the approach of winter when they apparently move off as they are much less frequently seen in cold weather although small flocks are to be found about villages and farm houses all winter.

As soon as the nesting period is over the Starlings become social birds willing and anxious to associate with any flocking species that will tolerate them and apparently forcing themselves upon some. There is rarely a flock of Red-wings, Grackles or Meadowlarks that does not contain Starlings and sometimes I have seen a stray Starling take up with a flock of English Sparrows. Starlings, too, have, to some extent, replaced the Cowbirds in the cattle yards and fields where cows are grazing, and during June and July most of the birds attendant upon the cattle will be found to be Starlings. On July 3, 1935; July 6, 1928; and July 5, 1931; such was the case and on the last occasion, only, were there any Cowbirds present. The Starlings, moreover, were all adults.

Starlings are omnivorous and seem to be adapted for feeding in a variety of ways; insects of various sorts and garbage they pick up from the ground; where Japanese beetle larvae are plentiful they will dig them out of the soil; while when other fare is scarce they will strip berry-bearing bushes of their fruit. So, too, when the flying ants take to the air on their nuptial flight the Starlings rise to the occasion and catch them on the wing with the agility of a flycatcher. On August 12, 1929, there were at least one hundred of the birds out over the Fill, circling continually as they snapped up the insects which were swarming everywhere. Many of the old birds were molting and their tails presented a peculiar wedge-shaped outline while the wings of both old and young, especially the latter, appeared curiously semitranslucent with the strong sunlight behind them. One young Laughing Gull joined them in their evolutions as these birds, too, seem to be not averse to catching insects in the air or to engaging in aërial evolutions.

Starlings are easily recognized in flight. Their wing action is more rapid and more sustained than that of the other "blackbirds" and there is no dropping of a beat as in the Grackle, no undulatory motion as in the Red-wing, while the long tail of the Grackle is missing; indeed the Starling's tail is so short that the body appears to be shaped the same at each end and to taper symmetrically to bill and tail with the rather narrow wings in the middle. The wings are pointed and when the bird is at rest they seem to be joined closely at the shoulder, with a short upper arm, like that of a swallow; when spread they are strikingly triangular.

While the Starling has gained appreciation on account of its sociable nature and its fondness for Japanese beetle larvae, those who can see further into its life history realize that it is a serious menace to our native birds by devouring the food that formerly served as their winter support, while Robins and Grackles, relieved from the pressure of Starlings, would, I believe, be just as efficient in destroying the beetles. In addition there is the danger that the Starling will become a destroyer of garden products, as it has in its native country. All in all the addition of the Starling to our avifauna seems to me an excellent example of the folly of introducing the birds of one country into another and thereby upsetting the balance of life that nature has so carefully developed.

WHITE-EYED VIREO

Vireo griseus griseus (Boddaert)

The White-eye is the characteristic vireo of the Cape district and every piece of low swampy woodland has a pair or more. It is common at Cape May Point and northward along the Bay as well as in the woodland strip that borders the meadows above Cape May on the ocean side, and in moist spots inland—a summer resident from mid-April to late September and occasionally until early October.

Julian Potter counted thirty White-eyes in thickets bordering the Pond Creek Meadows on June 21, 1923, and nests, each with four eggs, have been examined by Turner McMullen and Richard Miller on May 20, 1916; May 21, 1926; and one with three eggs and one of the Cowbird on May 29, 1932. They were swung from forks of low bushes, rarely more than three feet from the ground. I have seen old birds feeding young on July 15, 1922, and a female carrying food on July 18, 1923, while young were shifting for themselves on the same date in 1935.

The White-eye is a favorite fosterer of the Cowbird and I found one feeding a well-fledged young parasite on August 3, 1927, while another on August 5, 1929, was being closely followed by a husky young Cowbird begging for food. On August 5, 1933, F. C. Lincoln and I found a pair feeding two Cowbirds! The actions of the White-eye are slow and deliberate. He hops leisurely from twig to twig of the low bushes with an occasional flirt of the tail, always peering this way and that, turning his head to one side or the other. A pair studied on July 13, 1920, apparently had young nearby as they were very nervous and excited. Their plumage was constantly ruffled so that their bodies seemed larger than normal; the tail was carried partly erect. Another individual, with plumage smooth and sleek, gave a nervous flip of both wings and tail as it hopped heavily from one branch to another.

The white iris, the blue gray feet, and the light wing bars are satisfactory identification marks but the vocal accomplishments of the bird are still more characteristic and one is at once attracted by the peculiar, emphatic, penetrating song—if song it may be called; it is surely not soft enough to be termed a warble.

After repeating one phrase a dozen times or more the bird will suddenly stop short or, after a pause, may change to an entirely different phrase so that one would feel sure that he had been listening to another individual had he not had the performer constantly in view.

The peculiar character of the songs renders them easy for syllabic representation and I have recorded quite a number. All have the same loud penetrating quality which makes the White-eye, as it were, the voice of the swamp. The simplest songs run: *tick, che-weéo, sick; chuck, che-weé-ju; ill, chee-whéo, chíp; seé it, see, weé-ah;* and a shorter: *chíck, wíssa.* Other more complicated phrases are: *chíck, che-weary-o-wísset; chuck, see, chur-a-lury, stíck; che-wírra, chu, weé; che swítz, ah, weé; pechítcha wirra, wée; sweé-a, stpititulia, stp.* The last was shortened in various ways sometimes the first note only was uttered, *sweé-a* or *sweé,* while the middle phrase was often cut in a variety of ways.

As an illustration of the changing of phrases, a bird heard singing on May 8, 1925, began with: *chéri-lo-wísset* and then added a couple of preliminary notes: *chíp, sweé, chéri-lo-wísset* and then after several repetitions changed entirely to *weét, se-cheéa, stk.* In all of the "songs" the sharp preliminary note *tick, chick,* etc., is always followed by a pause while the similar final note when present is preceded by a pause.

Other notes are an interrogatory *cheee* which is often uttered as the bird catches sight of an intruder on his domain, and a curious guttural *zchurrrrrr,* to which is sometimes added one of the common final phrases, *wée-ah.* A bird with young kept up a constant chatter *shi-shi-shi-shi-shi* resembling one of the notes of the Chat when heard far off, while a young bird had a somewhat different chatter note: *chip-chip-chip-chip.*

Conrad Roland heard one still in song on September 6, 1934, which is unusually late.

Young birds were seen on the wing on July 18, 1935, and in the molt by August 13. An adult in the molt on August 14, 1932, had faded to a very pale gray and the new yellow feathers on the sides shown out in very brilliant patches. Another adult with a bob tail was seen on August 27 and a fully molted individual on August 30.

We have spring arrival records for the White-eye as follows:

1908. April 26.	1921. April 24.	1923. April 30.
1919. April 23.	1922. April 22.	1927. April 30.

Late autumn records are:

1928. September 27.	1930. September 28.	1932. October 12.

The height of the migration seems to be during the first week of September.

While the White-eye breeds as far north as southern New England, it is more plentiful in southern New Jersey than in the northern counties.

YELLOW-THROATED VIREO

Vireo flavifrons Vieillot

I saw one Yellow-throated Vireo on October 5, 1929, at Cold Spring during a remarkable flight of migrants, while Fletcher Street found one singing in the woodland west of Cape May Court House on June 7, 1936, which would indicate that it was nesting there. It is apparently much rarer in the Cape district than the preceding.

It would seem that this species and the Warbling Vireo, of which we have no records for Cape May County, must migrate along the Delaware Valley rather than along the coast, although they probably cross the state north of the Pine Barrens. While it is a summer resident in most parts of the state, it seems to be rather local in its distribution.

BLUE-HEADED VIREO

Vireo solitarius solitarius (Wilson)

The Blue-headed or Solitary Vireo is an uncommon transient at Cape May and is more frequently seen in autumn. It is a bird of the north woods and does not breed in New Jersey, being everywhere a transient, although Dr. F. M. Chapman did find it at High Knob, Sussex Co., in June, 1890 (Abst. Proc. Linn. Soc., 1890–91, p. 4). From its abundance in the north and its scarcity along the New Jersey coast I infer that its migrations follow the Delaware Valley or regions farther west and that it pushes north in spring up the Susquehanna and Delaware Rivers to reach its summer home.

Our Cape May records, all single birds, are as follows:

1920. October 31 (Stone).	1934. May 5 (Stone).
1930. October 5 (Potter).	1935. October 10 and 27
1932. September 25 (Roland).	(Rusling).

RED-EYED VIREO

Vireo olivaceus (Linnaeus)

While not an abundant summer bird about Cape May there are usually one or two pairs of Red-eyed Vireos breeding in the shade trees of the town and a like number at the Point. In the woodlands farther north in the peninsula they are more frequent.

Nests with incomplete sets of eggs have been found on June 3, 1923, and June 8, 1924, while parents were seen feeding young in the heart of the town on August 11, 1924, and again August 7, 1934. On the latter occasion they made a vicious attack on an English Sparrow which had alighted on a wire close to where the young Vireo was perched and drove it away. Later one of them caught a white moth and after clipping off the wings fed it to the young bird. Red-eyed Vireos have been heard singing in the shade trees along Washington Street as late as August 21, 1924, and August 10, 1926, and another at the Point on August 7, 1922, but they are usually through singing by the middle of July.

The northwest winds of late August and September bring many Red-eyes along with the host of migrant warblers, flycatchers etc., and it is then that the bird is most conspicuous in the avifauna of the Cape. This influx was especially noticeable on September 1–2, 1920; September 4, 1921; August 29, 1922; August 31, 1923; August 27, 1924; September 1, 1925; September 27, 1926; September 27, 1928, September 28, 1930; September 17, 1932; on all of these dates the birds were abundant in the woods at the Point and in some seasons there are several "waves" in which Red-eyes are conspicuous. Their deliberate movements and slow progress through the foliage, with no nervous action of wings and tail, is in sharp contrast to the warblers which are migrating with them. During these congested flights they crowd together in remarkable numbers, where food is obtainable, and are not particular as to its character. On one occasion as many as twenty-five were found in one small Sassafras tree, some of them hanging upside down while they pecked at the fruit, and I have seen them fluttering in the air close to berry bearing branches in a manner very unlike the usual habit of a vireo.

We have records of their arrival in spring on:

1924.	May 10.	1932.	May 7.
1928.	May 5.	1934.	May 5.

The latest records for autumn are:

1924.	October 2.	1929.	October 5.
1927.	October 10.	1935.	October 13.

The Red-eyed Vireo is nowhere in Cape May County the characteristic woodland bird that it is in the northern counties or in Pennsylvania and the states to the north. It is from the latter region that come the great waves of Red-eyes that throng the Point and the Bay side woods in autumn— silent berry eating birds, very different from the vociferous "preacher" of the more northern forests who keeps up his vocal efforts long after other singers have subsided in the heat of midday.

*WARBLING VIREO

Vireo gilvus gilvus Vieillot

While the Warbling Vireo is a common summer resident of Warren and Sussex Counties according to Griscom and doubtless nests in other sections of northern New Jersey, the village shade trees which constitute its favorite habitat in most sections where it breeds, do not seem to attract it in South Jersey, and I have been unable to obtain a single nesting record. Julian Potter has recorded transient individuals at Camden on May 10, 1909; May 25, 1913; May 15, 1914; and Fletcher Street has observed it in spring at Beverly but none has been yet recorded from Cape May County although autumn stragglers should certainly occur here.

PHILADELPHIA VIREO

Vireo philadelphicus (Cassin)

The Philadelphia Vireo does not breed in New Jersey but occurs as a rare transient throughout the state. We have but two records for Cape May County, a specimen shot on Five Mile Beach by H. W. Wenzel on September 21, 1889, and contained in the Laurent collection in the Reading (Pa.) Museum; and one shot on Seven Mile Beach by Dr. William E. Hughes, on September 11, 1898, now in the collection of the Academy of Natural Sciences of Philadelphia.

Specimens were taken at Princeton by W. E. D. Scott on September 21 and 28, 1876 (Babson, Birds of Princeton, 1901) and others, I believe, have been seen and collected in the northern counties.

WARBLERS OF CAPE MAY

All of the warblers recorded for the state of New Jersey have occurred at Cape May with the exception of the Mourning and the Cerulean. There is every probability that a few of the former occur occasionally in the migrations but the latter, which is an accidental straggler from the south and west, is of unlikely occurrence.

Six warblers occur as summer residents and nest at the Cape while several others nest in small numbers a little farther north in the peninsula. These in the order of their abundance as breeders are:

Maryland Yellow-throat.	Yellow-breasted Chat.
Yellow Warbler.	Black and White Warbler.
Prairie Warbler.	Parula Warbler.
Pine Warbler.	Blue-winged Warbler.
Hooded Warbler.	Louisiana Water-Thrush.
Ovenbird.	

Others which may nest in the Cape May region are:

Yellow-throated Warbler.	Redstart.
Prothonotary Warbler.	

The most abundant transients are:

Black-and-White Warbler.	Northern Water-Thrush.
Redstart.	Palm Warbler.
Myrtle Warbler.	Yellow Palm Warbler.
Black-throated Blue Warbler.	Ovenbird.
Maryland Yellow-throat.	

We do not have at Cape May, in either spring or autumn, the great warbler waves that are so characteristic of the migrations in the vicinity of Philadelphia and elsewhere, except so far as the above species are concerned, and the many other species are usually represented by but a few individuals. It must be admitted that we have not had competent observers constantly on the ground during the periods of migration but we have gathered enough data, I am confident, to justify the above conclusions.

As stated elsewhere, I am of the opinion that in the spring flight the great majority of the northern-breeding warblers turn north through the Susquehanna Valley or along the western border of the Delaware Valley, leaving only a few to reach the coast, while in the autumn the presence of these species on the coast of Cape May is largely due to strong northwest winds during the progress of their flight which either blow them from a more westerly course or force them to concentrate at the Point.

As W. W. Cooke has shown (Distribution and Migration of N. A. Warblers, Bull. 18, Biol. Survey, 1904), there are two main highways of warbler migration in the East; one up the coast and the other crossing the Gulf of Mexico and extending northward along the Alleghanies to cross somewhere in Virginia or West Virginia and thence, by the great river valleys mentioned, and across central New Jersey, to the Hudson and Connecticut Valleys.

The warblers that are supposed to follow the coast line and winter mainly in the West Indies (although some individuals take the other route and winter in Central America) are:

Black-and-White Warbler.

Redstart.

Parula Warbler.

Myrtle Warbler.

Black-throated Blue Warbler.

Palm Warbler.

Northern Water-Thrush.

Maryland Yellow-Throat.

Ovenbird.

This group it will be noticed contains all of the species most common during the migrations at Cape May and all of them occur on the coast of South Carolina (*cf.* Wayne, Birds of South Carolina) both of which facts confirm the theory of their coastal migration route.

The species that are supposed to cross the Gulf and proceed north along the mountains are:

Golden-winged Warbler.

Blue-winged Warbler.

Magnolia Warbler.

Chestnut-sided Warbler.

Bay-breasted Warbler.

Blackburnian Warbler.

Black-throated Green Warbler.

Canada Warbler.

None of these is a common transient at the Cape and none I believe has yet been recorded from the South Carolina coast. The former fact would seem to show that in coming from their summer home in Canada and the mountainous portions of our northern states, where all but the first two breed, they must cross central New Jersey in a southwesterly direction and proceed down the western edge of the Delaware Valley, or even farther west, and then down the Alleghanies, thus for the most part keeping clear of Cape May, while their absence from the South Carolina coast shows that such individuals as do reach the Cape must, like the rest of their kind, cross the mountains in Virginia on their way to the Gulf of Mexico.

In the southward flight, as already suggested, the greater abundance of warblers at the Cape and along the coast to the north is probably due to the action of northwest winds in diverting them from their normal course.

BLACK AND WHITE WARBLER
Mniotilta varia (Linnaeus)

The Black and White Warbler and the Redstart are the most regular and abundant transient warblers at Cape May on the southward flight and from the first week of August until the end of September they may be found in the pine woods at Cape May Point almost continuously, although they vary in numbers from day to day, in accordance with changes in the weather and direction of the wind. During the latter part of this period they may be found also in the shade trees and gardens of the town as well as throughout the wooded parts of the peninsula to the northward, though their center of abundance is always at the Point.

Upon the first northwest wind of August we shall find the Black and Whites circling the trunks of the pine trees and exploring the limbs and cone clusters, sometimes hanging over them like Chickadees. Their strongly contrasted black and white striping makes them very conspicuous against the dark trunks and branches of the pines. Their progress through the woods is more deliberate than that of the Redstarts with which they are so often associated and they spend much time climbing out and exploring the limbs of one tree and then moving to another in a long undulating flight, showing much white when on the wing. They are abundant on one day and rare or absent on the next as successive waves of migrants pass through. The dates of first arrival of south-bound birds at the Point for sixteen years are as follows:

1920.	August 1.	1925.	July 30.	1931.	August 8.
1921.	August 4.	1926.	August 14.	1932.	August 9.
1922.	August 9.	1927.	July 30.	1933.	July 26.
1923.	August 4.	1928.	August 13.	1934.	July 28.
1924.	July 28.	1929.	July 27.	1935.	August 5.
		1930.	July 24.		

Our latest autumn records are:

1917.	September 29.	1924.	October 2.	1929.	September 20.
1921.	September 14.	1925.	September 29.	1932.	September 25.
1922.	September 17.	1926.	September 27.	1934.	September 19.
1923.	October 1.	1928.	September 30.		

In spring Walker Hand and I have recorded the Black and White Warbler on:

1911.	April 26.	1916.	April 30.	1926.	April 24.
1912.	April 28.	1917.	April 29.	1927.	April 24.
1914.	April 24.	1923.	April 30.		

The Black and White Warbler breeds occasionally just north of Cape May but cannot be regarded as a common summer resident anywhere in the peninsula. Richard Miller found a nest with four fledglings at Rio Grande on June 3, 1923, which was tucked away among dead leaves at the base of a huckleberry bush in thick dry woods, and also saw a young bird fed by a parent at the same place on June 10. Turner McMullen found two nests with three eggs each in the same locality on June 8, 1924, and May 27, 1928, and one with four young on May 30, 1925.

Mark Wilde saw a number of the birds at South Dennis on May 18, 1894, and found one nest. I saw an adult and young at Erma on July 7, 1929, and other individuals at Goshen and South Dennis as early as July 12, 1925, while Fletcher Street and others, on an all day survey of the central part of the peninsula on June 7, 1936, saw no less than five of these warblers in the woodland from one to two miles west of Court House, one near Bidwell's Ditch and another near Seaville. Julian Potter informs me that he found a single bird in the woods on the eastern side of Pond Creek Meadows on June 11, 1923, and another at Ocean View on June 10, 1928. It is probable that all of these birds were nesting.

In northern New Jersey the Black and White Warblers are common summer residents in most tracts of woodland but judging by the numbers that pass through Cape May in the autumn the migrant flocks must include birds from wide areas to the north of the state as well.

PROTHONOTARY WARBLER

Protonotaria citrea (Boddaert)

The Prothonotary Warbler is well known to nest along the Choptank River near the Delaware-Maryland line, at Marydel, where it was first detected by Gordon Smith on July 18, 1898, and at Seaford, Delaware, where Samuel Rhoads found it on June 18, 1903, but that would seem to be the northern limit of its normal breeding range in the East. The bird however has the habit of occurring, and even breeding, sporadically farther north, as witness the nest recorded by Howland and Carter (Auk, 1925, p. 138) in Morris County, in the northern portion of New Jersey, on June 30, 1924, and records of single birds by Samuel Rhoads at Haddonfield in the eighties; by A. H. Phillips, at Princeton, on May 8, 1894; by H. C. Oberholser (Auk, 1918, p. 227) at Morristown, June 14, 1888; at Mt. Holly, by Nelson Pumyea, April 25, 1920; by Reimann, at Marshalltown, April 29, 1934; by R. K. Haines, at Pemberton, May 11, 1930; and by Roger Peterson, at Pennsville, on June 9, 1935.

Arthur Emlen and I also saw a brilliant male on the Egg Harbor River near the Fourway Lodge above Mays Landing, on May 16, 1931, which dodged in and out of the shrubbery that overhung the banks of the river, keeping just abreast of our canoe as we paddled slowly up the stream. We have two instances of the possible breeding of this southern warbler in the lower counties of New Jersey although no nest has ever been found. On June 19, 1914, while Julian Potter was exploring a wooded swamp close to the Maurice River, two miles west of Vineland, a Prothonotary Warbler appeared, attracted by the chirping of a pair of Redstarts with young (a noteworthy record in itself!). The bird not only gave every opportunity for careful observation, but uttered its characteristic alarm note which Potter likened to that of a Water-Thrush. The swamp was overflowed by a few inches of water, a most likely place for a Prothonotary to breed.

On June 15, 1924, Richard Miller, in the heart of the Timber and Beaver Swamp at South Dennis, Cape May County, came upon a pair of Prothonotaries the female carrying a bill full of insects. They approached to within a few feet of him and gave every indication of having young in the immediate neighborhood. This swamp, with which I am well acquainted, is an ideal habitat for these warblers, with deep sluggish streams and many overflowed areas.

The Prothonotary is not a bird of the pinelands and judging by the character of the country I should think that the southwestern section of the state in Salem and Cumberland counties, or northwestern Cape May County, would be the most likely area in which to expect its nesting.

WORM-EATING WARBLER

Helmitheros vermivorus (Gmelin)

I have only twice detected this bird in Cape May County, both times during the early autumn migration. One was seen in a moist thicket at the Point on August 27, 1926, and the other in a tangle of greenbrier in the tall moist woodland east of Bennett, on September 2, 1930. It is a regular summer resident in parts of southern Pennsylvania and in some of the northern New Jersey counties but is not a bird of the pineland and is very rare even in West Jersey, across the river from Philadelphia. I am inclined to think that its normal route of migration is down the Delaware and that my Cape May birds had been diverted to the coast by northwest winds. As but few birds breed north of New Jersey, however, there must be very few migrants passing through the state.

GOLDEN-WINGED WARBLER

Vermivora chrysoptera (Linnaeus)

A rare but probably regular transient at Cape May Point in autumn and twice observed in spring. Our records are as follows:

1921. August 15–16, two.	1929. August 17, one.
1922. August 21, one.	1934. August 27, one.
1927. May (early), one.	1935. August 23, three (W. Rusling).
1927. May 20, one.	1935. September 7, one (W. Rusling).

It nests in Essex, Passaic and Sussex Counties in northern New Jersey.

BLUE-WINGED WARBLER

Vermivora pinus (Linnaeus)

It is probable that a few Blue-wings occur at Cape May Point every autumn and a lesser number in spring, although we have observed none at that season. Our records are as follows:

1920. September 1, two.	1927. August 3, 6 and 17, one each.
1921. August 15–16, several.	1929. August 17, six.
1922. August 9 and 28, one each.	1931. August 7, one.
1923. August 31, one.	1932. August 29, one.
1925. August 18, one.	1934. August 7, one.
1926. August 27, one.	1935. August 30, two (W. Rusling).

A few may nest in moist thickets farther north in the peninsula as I found several singing at Goshen on July 12, 1925, and Fletcher Street found one in the same vicinity and three in swamps west of Court House on June 7, 1936. While a common breeder in southeastern Pennsylvania and in the central and northern counties of New Jersey the Blue-wing does not range north of the Carolinian Fauna and would naturally not be a common transient at Cape May.

*LAWRENCE'S WARBLER
Vermivora lawrencei (Herrick)

*BREWSTER'S WARBLER
Vermivora leucobronchialis (Brewster)

While no individuals of Brewster's or Lawrence's Warblers, hybrids between the Blue-wing and Golden-wing, have been recorded in Cape May County, or elsewhere in southern New Jersey a number have occurred in the northern part of the state where both of the parent species nest. The type specimen of *Vermivora lawrencei* (Herrick), essentially a yellow-breasted Golden-wing, was taken by Aug. Blanchet at Chatham, May 1874 (Proc. Phila. Academy, 1874, p. 320), another by D. B. Dickinson, at Hoboken, September, 1876 (Bull. N. O. C. 1877, p. 19) and a supposed cross between *lawrencei* and *pinus* by Frank Blanchet, at Morristown, May 15, 1884 (Auk, 1886, p. 411).

Specimens of *V. leucobronchialis* Brewster, a white-breasted Blue-wing with yellow wing bars, have been taken or seen at Morristown, May 1859, (A. Blanchet) and May 15, 1887 (E. C. Thurber) and at Englewood, May 15, 1886, June 26, and July 31, 1887, May 11, 1890 (F. M. Chapman) and May 13, 1905 (G. E. Hix) (cf. Auk, 1886, p. 411; 1887, p 348 and 349; 1890, p. 291 and 1905, p. 417). C. B. Riker has recorded a cross between *leucobronchialis* and *pinus* at Maplewood, May 11, 1883 (Auk, 1885, p. 378) and Dr. Chapman a pair representing these two forms breeding at Englewood in 1892 (Auk, 1892, p. 302). There have, I believe, been other later records.

TENNESSEE WARBLER
Vermivora peregrina (Wilson)

A rare transient during late August and September with a single record in spring. Observed mainly at the Point but found also in trees in the town.

1920.	September 1–2, one each.	1931.	September 1, one.
1922.	August 21, two.	1932.	August 29, one.
1922.	August 29, six.	1932.	September 17–18, an unusual
1923.	September 16 and 30, one each.		number.
1924.	September 7, one.	1932.	October 8, three.
		1924.	May 9–10, one.

The spring individual was feeding continually with some Black-polls in an ornamental spruce tree in the Physick garden.

It occurs only as a transient in New Jersey.

ORANGE-CROWNED WARBLER

Vermivora celata celata (Say)

I have always expected to come upon this warbler in the flocks of Myrtles feeding in the bayberry bushes in midwinter but have been unsuccessful. Philip Laurent however collected a specimen on Five Mile Beach (now in the Reading, Pa., Museum) on October 6, 1889, while it has been seen in winter at other points along the sea coast both to the north and south of New Jersey and Frank E. Watson has recorded one at Point Pleasant on the northern part of the coast on December 5, 1926. There is also a specimen in the American Museum of Natural History secured at Hoboken, May 1865, by C. S. Galbraith (Howell, Auk, 1893, p. 90); two May records for Union Co. (Urner) and three records for Essex Co. (Eaton). It is a very rare transient in the Delaware Valley—at Philadelphia about March, 1876 (McIlvaine, cf. Bull. Nutt. Orn. Club, 1879); at Haddonfield, New Jersey, March 22, 1883 (Samuel Rhoads, Bull. Nutt. Orn. Club, 1883) and February 25, 1909 (R. T. Moore, Cassinia, 1909, p. 53); at Rancocas Creek, New Jersey, February, 1860 (Turnbull).

NASHVILLE WARBLER

Vermivora ruficapilla ruficapilla (Wilson)

One seen at Cape May Point by Julian Potter on September 30, 1928. It is a regular transient north of the Pine Barrens and in the Delaware Valley while Dr. F. M. Chapman took a breeding female at Englewood, on June 16, 1887, the only summer record for the state (cf. Auk, 1889, p. 304).

PARULA WARBLER

Compsothlypis americana pusilla (Wilson)

The Parula is a spring and autumn transient at Cape May Point and in the trees of the town, occurring usually from September 8 to 18, with scattering individuals as late as October 14; and in spring from May 1 to 20, with a few throughout the month.

They vary in abundance and while our records are for the most part of a few individuals they have occurred in abundance on October 2, 1924; September 20 to October 2, 1925; September 27, 1926; September 17 to 18, 1932; and in spring on May 20, 1927, and May 5, 1934. We have but one August record—August 23, 1924.

Immediately about Cape May I know of only one nest which I found on May 26, 1892, at the Point. It was about ten feet up in a tree and contained three eggs and, curiously enough, one of the Cowbird, although how the latter bird managed to deposit its egg in such a frail structure without damaging it I am at a loss to explain.

Parulas have been seen at the Point on May 29, 1921, and at Higbee's Beach on May 30, in 1923, 1924, and 1925, while Julian Potter heard one in full song at the latter place on June 21, 1923, and I heard several at Goshen on July 12, 1925, all of which indicate probable nesting.

The swampy woods of the Great Cedar Swamp, a little farther north, and the borders of the extensive lake at South Dennis were favorite nesting places of the bird in years past but lately they do not seem to be so plentiful there, while similar localities at Mays Landing, Browns Mills and elsewhere in the Pine Barrens were, and doubtless still are, breeding grounds of the bird. Many also nested in the wooded swamps that existed on Five Mile Beach in the eighties, according to Philip Laurent.

The Parula's nesting seems to be governed by the presence of the pendant lichen or "beard moss" (*Usnea*) which formerly festooned all of the bushes that grew along the edge of the lake, and in the water, at South Dennis. In May, 1892, I spent some days exploring this attractive spot in company with Charles Voelker and Harris Reed and I never saw the Parula so abundant as it was at that time. We pushed our flat-bottomed boat in among the maze of moss-draped bushes, shoving it through billowy, water-soaked beds of sphagnum, with cranberry vines and clumps of pitcher plants all about, and through great patches or wiry Cassandra tangles until we seemed to have penetrated to the innermost recesses of the Parulas' retreat and to be able to share the secrets of their home life. They flit about on every side; here we see one entering its dainty moss basket nest while in another the female is sitting deep down in the delicate pocket engaged in incubation, her bill just protruding from the narrow opening. Besides the constant *buz-z-z-z-z-z----zic* of the little warblers there comes to our ears the harsh cry of a Kingbird balancing himself on the topmost twig of a partly submerged gum tree; the emphatic song of the White-eyed Vireo, rises with startling intensity from the bushes only a few feet from us and from farther away, close to the black shade of the cedar swamp, come the clear phrases of the Hooded Warbler. Hummingbirds dart back and forth and Martins, black against the sky, pass on rapid wings, while far overhead an Osprey with labored flight carries a fish from the coast to his nest farther inland.

Mark Wilde who made a thorough study of the bird at this spot has written as follows of the nesting of the Parula: "The nest is invariably placed in a hanging position usually in a bush on which the beard moss grows quite thickly, and here, within a dense tuft, the birds loop and weave the strands to form the nest. They are careful to leave the moss on the bush hanging about the nest so that it may be well concealed and it can only be found by diligent search. Building material is carried from other bushes and not taken from the one in which the nest is located, and while many nests are lined exclusively with strands of the moss others contain horsehair and yellow down from the stems of swamp ferns. The entrance, always level with the top of the bowl, is through the moss on the side, very often directly under the limb from which the tuft is hung. Very rarely the entrance is from the top and occasionally a nest is hung from a limb with very little moss, all of the material being carried from elsewhere; such a nest is much more conspicuous" (Auk, 1897, pp. 289–294). Of the thirty-three nests that Wilde examined sixteen were from one to four and a half feet from the ground or water; eleven from six to fifteen feet up and one twenty feet high. The eggs are usually four in number sometimes three and the average date for full sets was found to be May 20 while young were hatched in one nest by June 4.

The vegetation on the South Dennis lake has grown much taller in the past few years and the moss has to a great extent disappeared which doubtless accounts for the scarcity of the birds here, and as migrants at Cape May Point. They still nest, however, where suitable environment can be found and Turner McMullen has found nests with full sets of eggs at Swain on June 6, 1920, and June 15, 1924, and Richard Miller at South Dennis on June 13, 1926.

The Parula is a common transient throughout the state and nests locally in the northern counties, at Newton, Highknob, etc.

YELLOW WARBLER

Dendroica aestiva aestiva (Gmelin)

The Yellow Warbler is the most generally distributed summer warbler about Cape May unless it be the Maryland Yellow-throat. Ever a lover of the vicinity of water, it is found in the thickets of marsh elder and bayberry which border Cape Island Creek, as well as in the shrubbery surrounding the ponds at the Lighthouse and along the banks of Lake Lily at the Point. In the gardens of Cape May, especially those which back on the meadows or on the Fill, it is a familiar bird and its cheerful *sweet, sweet, sweeter, sweeter,* is constantly to be heard in the spring and early summer. It has always been common on Two Mile Beach and the other island beaches to the north, and in a small area at the southern extremity of the former island I counted ten pairs on May 22, 1922. In the orchards and gardens of the farmhouses of the upland a pair or two are almost always to be found. To those who know the bird only in such surroundings, among a wealth of green foliage or by clear ponds and little brooks, it seems strange to find it equally at home in marsh elder thickets overhanging the black mud of the tidewater creeks.

While usually a bird of low bushes the Yellow Warbler is often seen exploring the branches of the pine trees of the Point, after the nesting season is over, and I have also seen it feeding on the ground in the plum thickets of the Bay shore dunes.

In early May when the bushes of the coastal region, always backward as compared with more inland localities, are largely bare of foliage, the golden yellow of the bird's plumage stands out most conspicuously but later in the season it is not so noticeable. It is in full song during May and June and was once heard as late as July 24, doubtless a belated breeder.

Turner McMullen found a nest with four eggs at Cape May on June 4, 1922, another with a like number of young on June 8, 1919, and two with eggs, at Rio Grande, on May 20, 1923, and June 8, 1924.

I have seen full-plumaged young as early as July 6, 1925, and July 26, 1923, and an adult in freshly molted winter plumage on August 7, 1920. I have come upon very few birds in this plumage, however, and am inclined to think that the old birds move southward as soon as their annual molt is completed, or perhaps before, as almost all Yellow Warblers seen in late summer and autumn are birds of the year. Perhaps the breeding birds from farther north travel down the Delaware or Susquehanna Valleys.

Walker Hand's dates of spring arrival with supplementary dates by others are as follows:

1905.	May 8.	1917.	April 24.	1926.	April 24.
1908.	April 26.	1918.	April 30.	1927.	April 24.
1909.	April 29.	1919.	April 23.	1932.	April 30.
1910.	May 4.	1920.	May 1.	1933.	April 30.
1911.	April 30.	1922.	April 23.	1934.	May 1.
1914.	April 27.	1923.	April 29.		

Our latest autumn records are:

1922.	September 17.	1925.	September 30.	1930.	September 28.
1926.	September 25.	1927.	October 10.		

The birds are, however, scarce after September 5.

MAGNOLIA WARBLER

Dendroica magnolia (Wilson)

Apparently a regular transient passing through on the southward flight, mainly in September, at times of concentrated waves of small passerine birds, and seen on the return migration in May.

Our records are as follows, single birds unless otherwise stated:

1921.	September 1, 9, 14.	1930.	September 27.
1922.	September 17.	1931.	September 7.
1923.	August 26.	1932.	August 29, several.
1924.	September 6.		September 17, 18, many.
1925.	October 1.		October 12.
1927.	September 1, 5, 8.	1934.	August 27, several.
	October 7, several.		September 6, 19, several.
1929.	October 5, four.		

In spring they seem to be much rarer and we have but four records:

1927.	May 20 and 22.	1932.	May 8.	1934.	May 15.

The Magnolia Warbler is a common transient throughout northern and western Jersey and nests sparingly in the mountains of Sussex and Passaic counties.

CAPE MAY WARBLER
Dendroica tigrina (Gmelin)

A specimen of this little bird was shot in a maple swamp in Cape May county by George Ord, later president of the Academy of Natural Sciences of Philadelphia, in May 1809, and presented to his friend Alexander Wilson, who was at the time engaged in publishing his classic "American Ornithology." The bird was quite unknown to him and he prepared a painting and description of it which appeared in his great work. He named it in honor of the place at which it had been found, and the specimen was preserved in the famous Peale's Museum in Philadelphia. The bird thus became known to scientific men throughout the world and served to advertise the name of Cape May probably more widely than has been done in any other way.

It later developed that a specimen of this same bird had flown onto a vessel sailing off the coast of Jamaica and had been painted and described by an English naturalist George Edwards some years before Ord secured his specimen. This resulted in the publication of an earlier technical name but Wilson's English name, Cape May Warbler, has been retained for it ever since. We now know that the Cape May Warbler is a native of New Brunswick and other parts of Canada and that it occurs in New Jersey and other Eastern States only on its migrations to and from its winter home in the West Indies.

Curiously enough it seems never to have been recorded again at Cape May until September 4, 1920, when I recognized one in a shade tree on Perry Street in company with some Chestnut-sided Warblers. Since then we have seen a few nearly every year in spring and fall both at Cape May and at the Point. The birds had doubtless always been passing the Cape and the lack of any records has been due to the lack of ornithologists.

(815)

Our autumn records follow:

1920.	September 4, one.	1926.	November 21, one.
1921.	September 16, one.	1927.	September 25, one.
1922.	August 28, one.	1929.	October 25, one.
1924.	October 1 and 2, one each.	1930.	October 5, one.
1925.	September 30, two.	1932.	September 17–19, six.
1925.	October 1, six.	1934.	September 28, one.

The November record was a bird found by John Gillespie which was studied at leisure for some time as it searched for insects in a sycamore tree near the depot. It had a rather loud squeaking note which attracted attention to it.

From September 17 to 19, and probably later, in 1932, there were several young birds of the year feeding in the flower beds of the Physick garden on Washington Street, plain gray birds with very faint dusky streaks below and the telltale lighter crescent-shaped mark on the ear-coverts. They had a constant slight tilting motion of the tail which was also noted by Julian Potter in birds that he observed.

In spring I saw an exceptionally brilliant male in the pines at Cape May Point on May 1, 1926, in company with Myrtle Warblers, and on May 15, 1928, two beautiful males in an old deserted nursery near Schellenger's Landing. In these birds the tail tilting was very pronounced. They were so tame that they approached within arm's length of me as they fed in the privet bushes.

BLACK-THROATED BLUE WARBLER

Dendroica caerulescens caerulescens (Gmelin)

The Black-throated Blue Warbler is a regular transient through the wooded parts of the Cape May peninsula and is seen most frequently at the Point and along the Bay shore. Its line of migration follows the coast to its winter quarters in the West Indies.

Our autumn records are as follows:

1921.	August 27.	1925.	October 1.
	September 4.	1926.	September 11, 27, two.
1922.	September 17.	1928.	September 30.
1923.	September 30.	1929.	October 4, 5.
	October 1.	1931.	September 1, 2.
1924.	August 30.	1932.	August 29, nine.
	October 2.		September 17, 18, 25, 30.

Spring occurrences:

1925.	May 8.	1929.	May 10.
1927.	May 8, 20, 22.	1932.	May 7.
1928.	May 6, 20.	1934.	May 5.

It is a common transient throughout the state and breeds in the mountains of Sussex and Passaic Counties.

MYRTLE WARBLER

Dendroica coronata (Linnaeus)

The Myrtle Warbler is an abundant winter resident and autumn transient throughout the Cape May area and may justly be termed the most conspicuous small land bird of the coastal district in late fall and winter. As one follows the shore line from Cape May to the Point they arise from the low bushes back of the sand hills at every step, while the dense growth of bayberry near the Lighthouse simply swarms with them. Out on the fields, east of the pine woods at the Point, they collect in large numbers in the shelter that is there offered against the wind, and as we advance they get up in a continuous succession and drift onward through the woods and into the bushes bordering Lake Lily.

Following the old sand road leading north from the Lighthouse Pond it seems at first as if birds of various kinds were taking wing ahead. Some of them dart out of the thickets like scattered flocks of sparrows; others cling to the pine cones like Chickadees; some cross the sky line like drifting swallows; and still others pause to flutter in the air in the manner of flycatchers; but we realize ere long that they are all Myrtlebirds and we marvel at their abundance and at their varied activities. In color, too, they vary not a little; the birds of the year in dull brown dress; old males showing slate-blue below the brown fringes of the winter plumage; females in a plumage between the two; while there is not a little individual variation in the intensity of shades. In some the black blotches and yellow breast spots of the spring livery show up strongly while in others they are scarcely evident, but when flying directly away from us, so that a good view of the back is afforded, the yellow rump always catches the eye and it, with the white spots at the extremity of the outer tail feathers, make a triangle of identification marks that stamp them all as Myrtle Warblers.

We soon learn to identify their rather jerky flight as they rise from the bushes, and with a series of short wing flips turn now to the right now to the left, in their zigzag progress, rising somewhat with the beats, and falling in the intervals. Sometimes a bird will go but a short distance, flitting from bush to bush, while others will climb higher and higher in the air, drifting on their jerky way cross the sky like wind-blown leaves.

I have seen the Myrtle Warblers down among the bases of the bayberry bushes gleaning from the ground, as well as up among the clusters of berries which are supposed to form the greater part of their winter food, and I have also seen them dropping down onto the muddy flats along the banks of Lake Lily where some sort of choice food was evidently available. As soon as a Myrtlebird alights on a bush there is a short, sharp flip of the tail, not a seesaw action, but one involving the body as well, and as it comes to rest the head is drawn in and the plumage ruffled up making the outline more nearly globular, while the wings are dropped slightly so that their tips are a little below the base of the tail.

October 31, 1920, was a characteristic Myrtle Warbler day. All day long they were present in abundance. The air seemed full of them wherever one went. Thousands were flitting here and there in the dense growth of rusty Indian grass (*Andropogon*), in the bayberry thickets, in pine woods and in dune thickets. As the dusk of evening began to settle down I found them flitting on their zigzag way to the bushes about the Race Track Pond where they were apparently settling for the night.

On October 13, 1913, Julian Potter encountered a great flight of Myrtle Warblers which he estimated at 3000. Upon reaching the coast he found numbers of them right in town, and, walking up the boardwalk, he encountered an almost continuous flight going down the coast in loose flocks. Some appeared to be coming in from the ocean as if they had been blown out to sea by the strong northwest wind of the night before. Many of them were very much exhausted and alighted on the beach, on the boardwalk, and on any conceivable perch, permitting one to approach them very closely. The flight continued until 11 a. m. when it had practically ceased, although there were straggling flocks to be seen all day. Some Juncos and White-throated Sparrows were pursuing the same course and along the boardwalk he found a number of dead birds evidently killed by striking the telephone wires. They consisted of three White-throats, three Juncos, four Savannah Sparrows, two Myrtle Warblers, one each of the Golden-crowned and Ruby-crowned Kinglets, and a Field Sparrow. At Cape May Point the Myrtle Warblers fairly swarmed.

On October 23, 1927, they were again in full flight, a steady scattered stream crossing the town and continually alighting on houses, hedges, fences etc., and winging their way on down the coast. Passing through the woods and thickets they made a rustling like falling leaves. Many were on the ground and clustered about rain water pools on the roads.

While the Myrtlebirds are most abundant on the southward migration an immense number remain all winter and shift here and there so that one may not realize how numerous they are until he comes upon one of the main flocks. On January 9, 1922, the majority of the Myrtlebirds of the Point were feeding on the ground along the road above the Lighthouse hopping rapidly about and picking up food like sparrows. Then all flew up on the wires and perched for a moment, and then off again over the meadows, and later they were seen crossing the turnpike into thickets east of the pine woods. While we speak of flocks of Myrtle Warblers they do not move in the simultaneous manner of true flocks but rather each bird for himself so that the movement of a flock is a continuous straggling performance, some birds alighting and others taking wing at every moment.

At times, often in midwinter, they swarm on the Fill east of the town where bayberry bushes have formed many dense thickets and from here they may flood the town in stormy or windy weather, flocking into gardens and shade trees and close about the houses, even thronging the porches of unoccupied buildings. They also, at times, cover the old weedy fields back of the meadows flying up onto fences and hedgerows when disturbed. These wild winter flights often seem to be without rhyme or reason but they are so extensive and so spectacular that they challenge our ability to explain them.

Entirely aside from the usual feeding habits of the Myrtlebirds was the case of an individual observed on April 2, 1924, which was systematically visiting the holes just drilled in some hickory trees by a Yellow-bellied Sapsucker, feeding either upon the sap that exuded from them, or possibly upon small insects that might have been attracted by it. The bird made regular rounds and could be seen repeatedly poking his bill into the holes, but he retreated whenever the Sapsucker returned to make his own rounds.

The Myrtles arrive from the north in late September or early October and our records of first arrival run as follows:

1920. October 3.	1925. September 22.	1930. September 28.
1921. October 9.	1927. October 10.	1932. September 17.
1923. September 30.	1928. October 7.	1934. September 28.
1924. October 2.	1929. October 5.	1935. September 29.

The great flights, however, take place toward the end of the latter month. In the spring we do not notice any marked increase in the winter population and they begin to thin out and then pass on to the north almost before we realize that they are going. In late April and May, however, there is evidence of a flight from farther south, birds that are associated with migrant warblers of various other species and obviously not left-overs from the local winter population. Our latest records for the spring are:

1921. April 24.	1926. May 1.	1931. May 9.
1922. April 23.	1927. May 20.	1932. May 7.
1923. April 30.	1928. May 14.	1933. May 7.
1924. April 20.	1929. May 11.	1934. May 5.
1925. May 10.	1930. May 6.	

In the summer of 1921, Earl Poole discovered some Myrtle Warblers in the pines at the Point, associated with Red-breasted Nuthatches, on the extraordinarily early dates of August 10 and 11, and they were seen there frequently until the time of my departure on September 5, but with no increase in numbers, never more than six present at a time. They had evidently come with the Red-breasted Nuthatches and these birds, notoriously irregular in the time of their migration, may in some way have influenced the warblers to come south far in advance of their usual time of flight. They had evidently left their breeding grounds before the beginning of the annual molt as they were all changing plumage and completed the molt while here. In 1924, three were found in the same wood on August 29, busily exploring the pine cones and bunches of needles and on August 20, 1925, two were present associated again with the Nuthatches; a few were present on August 26, 1927, and August 30, 1932, so that this early movement may be more frequent than was at first supposed.

In the spring the birds of the year shed their dull brown plumage and assume the more brilliant dress of the adults. I have collected specimens in all stages of this molt in the woods that formerly stood on the edge of the meadows between Atlantic City and Longport on April 9, 1892, and April 8, 1893, and birds at Cape May at this season were evidently in the same condition although the late migrants of early May are all in full nuptial dress.

While the Myrtlebirds are always found in abundance on the Christmas census of the Delaware Valley Ornithological Club it is interesting to note the estimates of numbers seen on these occasions although the relative number of observers and the thoroughness of the search accounts for much of the variation, the census of 1935 being by far the most thorough. In the case of such free moving birds, however, there is bound to be duplication.

1931. December 27, 7,500.	1934. December 23, 820.
1932. December 26, 6,000.	1935. December 22, 7,257.
1933. December 24, 996.	1936. December 27, 1,556.

The Myrtle Warbler does not breed in the state but is everywhere an abundant transient and winters all along the coast and less commonly inland.

BLACK-THROATED GREEN WARBLER
Dendroica virens virens (Gmelin)

The Black-throated Green Warbler is a rather rare transient. Spring records: May 20, 1927; May 14, 1928; May 8, 1932; May 5, 1934 (three).

Autumn records: September 8 and 25, October 8 and 9, 1932; September 19, 1934.

While outside the limits of Cape May County, the record of a Black-throated Green Warbler in the Pine Barrens of Ocean County on July 4, 1935, by Julian Potter, and another in the same region on June 6 of the same year by Charles Urner, are of particular interest since the former, especially, is too late to be regarded as a belated transient. If the bird was located there for the summer it may possibly have been nesting and that raises the question of its relation to Wayne's Warbler (*Dendroica waynei* Bangs) of the South Carolina lowlands and the Dismal Swamp of Virginia, which is at best scarcely more than a Black-throated Green summering far south of its normal breeding area.

Another was seen near Camden, N. J., however, as late as June 2, 1929 (Potter).

The Black-throated Green Warbler is a common transient through most of the state and breeds in the mountains of Sussex and Passaic Counties.

BLACKBURNIAN WARBLER

Dendroica fusca (Müller)

This beautiful warbler is a rare transient in the Cape May region. Autumn records: September 30, 1928; October 6, 1929; September 4, 1930; August 25, 1931; August 29, September 17, and October 8, 1932.

Our only spring record is a bird found in song in the pine woods at the Point by Julian Potter on the unusually late date of June 9, 1929.

In most parts of the state the Blackburnian Warbler is a common transient and it breeds locally in the mountains of the northern counties.

* CERULEAN WARBLER

Dendroica cerulea (Wilson)

The Cerulean Warbler is an exceedingly rare bird in New Jersey. The only published records of specimens taken in the state are one at Trenton, C. C. Abbott (Birds of New Jersey, 1868) and one taken by S. D. Judd, at Boonton, Morris County on September 1, 1887 (Auk, 1897, p. 326). The record of five seen at Cape May Point on May 23, 1936, which is our only Cape May record, need hardly be taken seriously (Bird-Lore, 1926, p. 299).

The Cerulean Warbler has been found nesting on the Hudson River in Dutchess County, N. Y. by M. S. Crosby (Auk, 1923, p. 104) and G. W. Gray (Auk, 1924, p. 161) and the birds of this, the most northern breeding place, may travel through New Jersey, but most likely down the Delaware or Susquehanna Rivers. The bird also nests on the Choptank River near the Delaware-Maryland line and at Seaford, Delaware (Rhoads, Auk, 1905, p. 203), also on the lower Susquehanna River near the Maryland-Pennsylvania line. Stragglers might be expected to reach Cape May from Delaware but we have as yet no satisfactory evidence.

YELLOW-THROATED WARBLER

Dendroica dominica dominica (Linnaeus)

This southern warbler is a rare summer resident at Cape May Point, where it apparently reaches its most northern limit. While no nest has yet been found, nor young seen, it seems hardly possible that the numerous individuals that have been observed can all be solitary birds that have overshot their normal limit in the northern migration, like the one found by Col. Theodore Roosevelt on Long Island, on May 8, 1907 (Scribner's Magazine, 1907, p. 387). There are several records for North Jersey.

Our first record for the Cape May district was made on July 13, 1920, while I was making a census of the birds at the Point. I had been studying some Pine Warblers in the tops of the trees when I noticed a bird agitating a bunch of pine needles at the extremity of one of the limbs and focussing the glass on it I was amazed to have a Yellow-throated Warbler emerge into the field. The exceptionally long bill, blue gray back, yellow throat and black stripes on the sides of the breast, were all clearly observed. Just previous to seeing it I heard a fragment of song coming from the same tree and resembling that of the Indigo Bunting but I did not hear the bird sing again. I watched it for nearly ten minutes and during this time it moved deliberately about the branches, especially among the cone clusters, now and then fluttering in the air for a moment or two opposite the bunch which it was examining. Finally it caught a green caterpillar about an inch in length and hopped with it back of the main trunk of the pine amongst some ampelopsis vines. From there it apparently flew away in the opposite direction as I lost it completely.

While I watched it it came down to the lowest branches of the tree and followed them out to their tips until it was within ten feet of my face and showed no trace of fear. On July 15 I returned to the spot and searched for about half an hour when I again found my bird. It was moving deliberately as before and seemed to prefer the larger limbs and the clusters of needles and cones which grew close to them, in its search for food. Regarding this bird as a purely accidental straggler I collected it and found that it was a male in worn plumage but with no signs of molting. I saw no other individual during the summer although the region was pretty thoroughly studied. Since then we have had twelve observations of the bird, all at Cape May Point.

On May 30, 1923, one was seen near Lake Lily by members of the Delaware Valley Ornithological Club. The song was heard at the time but the bird was not found again although looked for all summer. On June 8, 1924, another individual, attracted by the cries of a pair of Ovenbirds with a nest of young, flew close to me in the woods to the west of the lake and I watched it for some time. The long bill and heavy black stripes on the sides, recalling those of the Black-and-white Warbler, were striking field marks.

On July 5, 1925, Julian Potter saw one west of Lake Lily which was in full song, probably the bird seen there on May 30, previous, by Henry Collins, while I watched one, possibly still the same individual, in the pine woods east of the lake on August 27, following, which fed among the pine needles like a Pine Warbler and sometimes hung to them like a Chickadee. Later observations are: April 7, 1929, by Julian Potter; May 24, 1931, by Mrs. John Gillespie; May 27–30, 1931, by Julian Potter; August 27, 1931, by Julian Potter (these three possibly the same bird); April 24, 1932, by Conrad Roland; March 29, 1936, by Harvey Moore. All of these were single birds except in Roland's case when two were seen together. Moore's record, the earliest that we have, probably indicates the normal time of first arrival as the bird is known to be an early migrant.

The occurrence of this bird at the Point is of particular interest as, like the Mockingbird, it gives a distinctly southern tinge to the bird life of the region.

CHESTNUT-SIDED WARBLER

Dendroica pensylvanica (Linnaeus)

Apparently a rather rare transient, occurring both at Cape May and at the Point. Our records follow:

In autumn:

1920.	September 4, two.		1928.	September 27, one.
1921.	August 15, several.		1932.	August 29–30, several.
1922.	August 21, one.		1932.	September 17, several.
1924.	September 10, one.		1934.	August 27, one.
1927.	August 25, one.			

In spring:

1927. May 20.   1928. May 10.   1934. May 5.

In the northern half of the state the Chestnut-sided Warbler is a summer resident and has been found breeding as far south as Plainfield (Miller) and probably Princeton (Babson) while Turner McMullen found a nest and four eggs at Haddonfield, on May 30, 1914, with the female incubating.

BAY-BREASTED WARBLER

Dendroica castanea (Wilson)

A rare transient at Cape May and the Point as in other parts of the state. Our autumn records are:

1920.	September 2, several.		1928.	September 30, one.
1921.	September 14, one.		1932.	September 17, several.
1925.	September 22 and 29, one each.			

In spring: May 10, 1924, one; May 26, 1929, one; and two seen by Richard Miller at Cape May Court House, May 21, 1916.

BLACK-POLL WARBLER

Dendroica striata (Forster)

A regular transient; never so numerous as in the Delaware Valley; apparently more common in spring. Spring records:

1892. May 24, several.	1925. May 30, two.
1915. May 16, two.	1927. May 20, abundant.
1916. May 21.	1927. May 22, two.
1918. May 24, one.	1928. May 13, two.
1919. May 25.	1928. May 30, several.
1921. May 29, one.	1930. May 13.
1922. May 30.	1932. May 5.
1923. May 20.	1933. May 12.
1923. May 25–27, one.	1935. May 19.
1924. May 10 and 30, one each.	

Autumn records:

1924. October 2, several.	1931. October 17.
1925. October 11, one.	1931. September 17 and 25.
1928. September 27.	1933. September 28.
1929. October 5 and 25.	

The Black-poll breeds entirely north of New Jersey.

PINE WARBLER

Dendroica pinus pinus (Wilson)

Few birds are as restricted in habitat as the Pine Warbler. True to its name it is found almost exclusively in pines; and about Cape May its range is limited to the pond pine woods at the Point. Here throughout the summer some of these birds are always present. Farther north in the Pine Barrens it is, of course, everywhere an abundant summer resident. In the northern half of the state it is a very local breeder and a colony was found by Dr. F. M. Chapman at High Knob on June 10, 1890.

The action of the Pine Warblers in the trees is deliberate, and they hop from one limb to another and creep about the cone clusters like mice. The wings usually droop slightly so that the tips are a trifle below the base of the tail. One bird (July 13, 1921) searched the larger limbs and apparently fed exclusively from the clumps of needles; more usually however they are out on the ends of the limbs among the clusters of small twigs and bunches of cones. One was seen to pull an insect of some sort from the scales of a cone and smash it against a limb. Sometimes they will fly clear of the tree in pursuit of prey and flutter in the air for a moment or two, or will descend to the ground after it.

On their first arrival in spring they often feed on the ground, and on April 2, 1922, half a dozen males were found in an old field near the pine woods feeding diligently. They hopped about pecking at dead leaves and scattered pieces of bark, deliberately putting the bill under them and turning them over, so as to expose such insects as were lurking beneath. When disturbed they flew into low shrubs, three or four feet in height, and rested, hopping down again in a few minutes. Again on April 1, 1933, they were feeding in a field with Juncos. A female was seen feeding on the ground in the woods August 22, 1925, and another August 6, 1927, but they usually do so only in early spring.

The song, a long, nearly monotone trill, is most strongly emphasized on the first two notes while the trill seems to rise during the first three pulsations and then gradually descend—*zit, zit, ziz-ziz-ziz-ziz-ziz-ziz-ziz-ziz-ziz-ziz*, at a distance it sounds like a clear monotone, somewhat like that of a Chipping Sparrow. When the bird sings his whole body seems to vibrate in unison. One studied (April 4, 1924) would sit on one twig and trill four to six times and then fly to another perch. He raised his head as he sang and the entire body trembled.

Several times I have found them feeding full-fledged young,—July 6, 1925; July 22, 1923; August 17, 1920. In the last instance the female dropped like a plumet to the ground at my feet as I paused beneath the tree; this behavior is seen when they are disturbed at the nest—an action akin to the "broken wing" ruse of ground nesting birds.

A young bird, July 13, was in full juvenal plumage, while a young male, August 5, another July 28, and a female, August 17, were molting to the adult dress, which was completely acquired by other individuals by August 18 and 27. An adult female on the latter date had not begun to molt and was dull sooty brown and very much worn, as was still another seen on August 17. An adult male September 1, was in full winter plumage, of unusually bright yellow. The yellow of the male Pine Warbler is not the usual yellow of our warblers but a sort of wax yellow and often rather dull but, even so, the plain dull brown female with only a shade of light yellow on the breast, if any, is quite a different looking bird. The gray, not white, wing bands however form an excellent identification mark for both.

Our earliest arrival dates for the Pine Warbler at Cape May Point are:

1925. March 21.	1921. April 3.	1930. April 6.
1923. March 25.	1924. April 4.	1932. March 25.
1922. April 2.	1928. April 1.	1935. March 23.
	1929. March 18.	

Latest autumn dates recorded:

1924. October 2.	1929. October 5.	1935. October 5.
1925. October 10.		

In the Pine Barrens occasional individuals remain throughout the winter and one was seen at Haddonfield by Samuel Rhoads on January 30, 1898.

Julian Potter found the only nest at Cape May Point of which we have record on May 27, 1931. A number of pairs of the birds must breed there every summer but the nests are exceedingly hard to find as they are usually situated on limbs near the tops of the pine trees and well screened by clusters of cones or needles. In the Pine Barrens, at Chatsworth, Turner McMullen has found nests with sets of three and four eggs on May 12 and 13 of different years, and one at Mays Landing on May 9, while George Stuart found a nest with four eggs at the former locality as late as May 24, 1918. McMullen also found young in nests there on May 13, 1934.

Robert T. Moore, describing a nesting of the Pine Warbler in Griscom's Swamp in the Pine Barrens, May 22, 1908, writes: "With my glass I surveyed the supposed nest for several minutes. No sign of a bird was visible, but a knock on the tree brought her down, tumbling almost into my arms. Like a dropping plumet she fell straight to the earth, fluttering to a log a few yards distant from my feet. On it she crouched acting the broken wing in motion, a pathetic picture of trembling love boldly acting deception. During several hours spent about this nest I only once saw the male, and never heard a note from either bird.**** The cones on the tree were bunched near the top, and the nest was placed directly in their midst, not dissimilar from them in breadth at the bottom. A thick network of pine spills, five to six inches long, screened it below and above, forming a mass of umbrage impossible to pierce. This considered with the secretive movements and surprising artifices of the bird make it clear why the nest is so seldom discovered." (Cassinia, 1908, p. 39.)

Philip Laurent records Pine Warblers as common in the woods of Five Mile Beach in the eighties.

PRAIRIE WARBLER

Dendroica discolor discolor (Vieillot)

Prairie Warblers are common summer residents in the pine woods and oak scrub of Cape May Point and in similar situations in the peninsula to the northward, occurring also throughout the Pine Barrens where they are one of the most characteristic species. In northern New Jersey they seem to be rare and local. They increase in numbers in the latter part of August when the young have molted and migrants are drifting down from farther north. At this time, too, they appear on the Fill and even in the trees and shrubbery of Cape May gardens.

They are normally birds of the scrub and undergrowth and in the pine groves of the Point we often find them seeking minute insects in the clumps of wild indigo, while some of them actually feed on the ground. The low branches of huckleberry and bayberry bushes are favorite resorts and when disturbed they may fly up into the trees, and occasionally one will be seen feeding there twenty feet from the ground. They are active birds and their movements from twig to twig are accompanied by a constant up and down tilt of the tail, such as we see in the Palm Warbler but not so energetic. Sometimes, too, there is a nervous flutter of the wings and tail, both being spread to some extent so that the white marks on the rectrices come prominently into view. Sometimes they will pursue one another vigorously through the woods long after the excitement of the breeding season is over.

Data on twelve nests found by Richard Miller about Rio Grande, a few miles north of Cape May, show that the usual number of eggs is four, rarely five. Full sets of fresh eggs were noted on May 27, June 3 and 10, of various years while some on June 3 were partly incubated and three nests on June 10, contained recently hatched young. Other nests found on May 20 and 31 had incomplete sets. The nests were located from eighteen inches to four feet from the ground in holly, azalea, huckleberry and oak bushes.

Turner McMullen reports twelve nests in the same neighborhood found from May 20 to 27, most of which contained four eggs, one five, while three held also an egg of the Cowbird. Another found on May 30, contained young birds.

I have found adult Prairie Warblers feeding fledglings in the bushes on July 15 and August 5, and young were molting into the winter plumage on July 15, while the molt had been completed by August 5, 1921, and August 7, 1925. Our earliest spring records of the Prairie Warbler are April 30, 1923; May 8, 1925; May 1, 1926; April 30, 1927; May 2, 1927. In the autumn we have recorded it as late as October 16, 1921; October 2, 1924; October 1, 1925; September 30, 1928; October 5, 1929; October 13, 1935.

This active little warbler with its wheezy insect-like notes which Dr. Coues has likened to those of "a mouse with a tooth-ache," and its ever moving tail, is one of the most characteristic summer birds of the Point.

WESTERN PALM WARBLER

Dendroica palmarum palmarum (Gmelin)

Palm Warblers, recognized at once by their exaggerated tail dipping, arrive at Cape May about the middle of September and are present through October in varying numbers while a few individuals remain during the winter.

Unfortunately it is impossible to refer all of our sight records of these birds subspecifically. It is generally conceded that the vast majority of the duller colored western race travel north along the Mississippi Valley and as all of our spring birds have the bright yellow breasts of the eastern form— the Yellow Palm Warbler, it is certain that they represent that race. In autumn, however, the case is not so simple as both forms apparently occur, but it is my belief that the western form is the more abundant, immediately along the coast, especially in the earlier flights, and Julian Potter while making a careful study of the birds of the Cape region, September 20 to October 2, 1925, came to the same conclusion. All wintering individuals seem also to be referable to this race, as also the one seen on March 13, 1921, which was probably a winter resident.

Both races breed far north of New Jersey.

These little birds are distinctly ground birds found in old bushy fields and in low shrubbery or along the roadsides, always in open country, and alight in low trees or bushes only when disturbed at their feeding. Winter individuals have been found in company with Bluebirds along old fence rows back of the town, or seeking shelter under the deserted boardwalk on the beach front.

Our arrival dates in autumn are:

1893. September 11.	1927. September 25.	1930. September 21.
1922. September 16.	1928. September 23.	1932. September 7.
1925. September 20.	1929. September 21.	1934. September 28.
1926. September 26.		

While most of them have passed on by the middle of October we have five records for November—November 14 and 21, 1920; November 26, 1922; November 22, 1929; November 9, 1930.

In winter it has been recorded as follows:

| 1920. December 12 and 26. | 1922. December 31. | 1924. January 19. |
| 1921. December 31. | 1923. December 2. | 1926. December 26. |

Usually these have been single birds but on December 2, two were seen and on the Christmas census of the Delaware Valley Club the combined observations of a number of men yielded four individuals on December 23, 1934; ten on December 22, 1935; ten on December 27, 1936.

YELLOW PALM WARBLER

Dendroica palmarum hypochrysea Ridgway

This race of the Palm Warbler is the prevalent form in the East and the only one that we have seen in the spring. It occurs throughout the latter half of April and in the southward migration occurs a little later than the western race, our records being mostly in October. While all wintering birds at Cape May seem referable to the latter, one collected by Samuel Rhoads at Mays Landing in the Pine Barrens on December 2, 1892, is unquestionably the yellow-bellied eastern form.

OVENBIRD

Seiurus aurocapillus (Linnaeus)

A typical bird of moist deciduous woodland, the Ovenbird is a common summer resident farther north in the peninsula but is not abundant immediately about Cape May. A pair or two nest at the Point between Lake Lily and the Bay and others in the woods bordering Pond Creek Meadows, while they occur also in the wooded strip back of the marshes on the ocean side of the peninsula from opposite Bennett northward and the bird is a common summer resident throughout the state.

A nest found at the Point by William Baily on May 31, 1924, contained five young and when I examined them on June 8 they were arranged in two tiers completely filling the mouth of the domed nest, the birds in the rear looking out over the backs of those in front. Their bills with yellow swollen

(832)

gapes were very conspicuous, while their brown plumage, so different from the olive green of the adults, blended beautifully with the surrounding dead leaves. The parents were greatly excited and flew about, one with food in its beak, calling *tíck; tíck; tíck;* and with each utterance there was a sudden elevation of the tail.

An hour or two after studying them I found the nest empty, the young being apparently concealed in the adjoining thicket as the old birds were still very anxious. They are always extremely solicitous and the cries of any bird in distress will bring Ovenbirds if there are any in the vicinity.

Nests were examined by Turner McMullen as follows:

Mayville, May 16, 1915, three eggs.	Rio Grande, May 30, 1925, four eggs.
Rio Grande, May 17, 1925, five eggs.	Rio Grande, June 3, 1923, five young.
Rio Grande, May 25, 1925, four eggs.	

A bird that I found in the pines on July 12, 1929, was walking and jumping along the limbs and seemed much disturbed. It gave utterance to alarm notes similar to those of the other pair but which at a little distance sounded more like *chip; chip; chip;* with a distinct pause after each note. This bird stretched its neck to the fullest extent and erected the dull orange feather of the crown, and its pale pink tarsi were conspicuous against the dark limb of the tree. It apparently had young somewhere in the vicinity in spite o the lateness of the date.

Another individual on August 16, 1921, acted in much the same way, calling and walking along the limbs of a pine tree, and the nesting season being long past, its actions could hardly have had anything to do with the care of young. It had a peculiar habit of squatting down on the limb every few minutes.

The Ovenbird is one of those species that is more frequently heard than seen and the loud ringing crescendo of its song: *cher, téa-cher, téa-cher, téa-cher, téa-cher,* announces its presence during the nesting season wherever it may be. With the wane of the singing period the bird seems to disappear, so secretive are its habits, and we realize that we had been in the habit of noting its presence more by ear than by eye. I have heard the song as late as July 6, 1925; July 6, 1928; July 9, 1921.

In late August there seems to be an influx of Ovenbirds from farther north—August 21, 1922; August 27, 1923; August 27, 1924; after which they fluctuate in numbers until the middle of September. Our latest records are:

1921. September 13.	1924. September 11.	1932. September 25.
1925. September 29.	1930. September 28.	

Dates of first arrival in spring are:

1923. April 30.	1926. April 24.	1932. May 2.
	1927. April 30.	

NORTHERN WATER-THRUSH

Seiurus noveboracensis noveboracensis (Gmelin)

The Water-Thrush is the first transient land bird to be seen in the Cape May district on the southward migration. Any time after the first of August, or even a few days earlier, we may expect to hear its metallic *pink, pink,* from the black muddy banks of Cape Island Creek where it winds through the meadows between Cape May and the Point and to see the bird perched on some projecting snag or overhanging branch with tail wagging up and down like a veritable wagtail.

Coming from its breeding grounds along the clear mountain brooks and cascades of the north it seems strangely out of place on these foul smelling tidewater creeks, which, where they flow through coastwise towns, are little better than open sewers. Yet here the Water-Thrushes remain through August and September and seem to find an abundance of food.

They have several times been flushed from the edge of the open salt meadows, near some shallow pool, flying at once to the nearest bushes for shelter, and a shady thicket at the Point, where a sewage pipe empties into the Lighthouse Pond, is a favorite resort. On September 1, 1920, two Water-Thrushes and two Solitary Sandpipers took possession of a small rain water pool well surrounded by weeds, on a vacant lot in the town, and remained until the water had entirely dried up. At best the pool was only three feet in diameter. Another Water-Thrush was seen feeding with a Least Sandpiper on the dry mud of a shallow pond, running this way and that and darting its bill right and left to the ground. Every summer a few of them take up their residence in the gardens in the heart of Cape May and we see their characteristic tail-tilting as they walk about among the flower beds. On September 7, 1924, I watched one feeding on the bottom of a dried up pool in the woods east of Bennett where a long arched wood road leads from a farm out to the salt meadows. The spot was covered with a carpet of dried decaying leaves left as the water evaporated, and the bird was diligently turning these over with its bill and tossing them to one side as it sought for insects beneath.

In a thicket where other birds are present in August, the alarm cry of a Catbird, Vireo or Towhee will at once bring all the Water-Thrushes in the vicinity to the scene of trouble and they fly nervously from branch to branch adding their cries to the miniature babel. Frequently, too, they will pursue one another vigorously through the thicket, turning and twisting with remarkable agility. When very close at hand the sharp emphatic note of the bird seems to lose the metallic quality that we note when farther off and I have recorded it as *pist!, pist!*

The Water-Thrush can be readily identified when flying in the open. Not only is it darker and apparently blacker than any other small bird seen against the sky or the meadows, but its flight is characteristic. The body is long and slender and the long swoops between the series of short wingbeats produce a diving, somewhat undulatory movement, but more irregular and less pronounced than that of the Goldfinch. I once saw a Water-Thrush hover for at least a minute in and over a bunch of cattails growing in water where there was no opportunity for the bird to alight.

Usually one sees not more than one or two Water-Thrushes together, but in the thicket about the sewer opening at the Point I have had six in view at once, and in a morning's walk on August 20, 1922, I listed twenty-five.

Our dates of arrival on the southward flight are:

1920.	August 2.	1925.	July 30.	1931.	August 8.
1921.	August 5.	1926.	August 4.	1932.	July 26.
1922.	August 7.	1927.	July 29.	1933.	August 5.
1923.	August 9.	1928.	July 29.	1934.	July 23.
1924.	August 4.	1929.	July 27.	1935.	August 1.
		1930.	July 30.	1937.	July 15.

Latest observations in autumn are:

1914.	September 13.	1925.	September 30.	1930.	September 28.
1921.	September 14.	1926.	September 11.	1931.	September 7.
1923.	September 16.	1927.	September 11.	1932.	September 25.
1924.	September 14.	1928.	September 8.	1935.	October 5.
		1928.	September 29.		

In spring we have but two records: May 22, 1927, two seen by Julian Potter and May 6, 1928, one.

While a common transient in most parts of the state the Northern Water-Thrush nests in the mountains of Sussex and Passaic Counties.

*GRINNELL'S WATER-THRUSH

Seiurus noveboracensis notabilis Ridgway

Two specimens of this western race of the Water-Thrush have been recorded from New Jersey—Raritan, May 30, 1889 (Southwick, Auk, 1892, p. 303), and Princeton, September 10, 1879 (W. E. D. Scott, Babson, Birds of Princeton). Its slightly larger size, larger bill and darker coloration are not easily recognized in life.

LOUISIANA WATER-THRUSH

Seiurus motacilla (Vieillot)

A very rare transient with similar distribution to that of the Worm-eating and Kentucky Warblers and rare anywhere in southern New Jersey. I saw one on the shore of Lake Lily on August 9, 1932, and William Rusling reported one in the same place on September 7, 1935.

One was found in a swamp near Cape May Court House by Richard
Miller on June 6, 1920, and David Harrower once found a pair apparently
nesting in Timber and Beaver Swamp at South Dennis.

This Water-Thrush is a summer resident in some of the northern counties
—along the upper Delaware and lower Hudson and at Lake Hopatcong
(Dwight), also in Essex and Union Counties, and possibly Princeton (Babson).

KENTUCKY WARBLER

Oporornis formosus (Wilson)

A very rare transient in the Cape May district and my only records are
one seen by Julian Potter in the woods near Lake Lily on May 26, 1918, and a
single individual that I saw in the undergrowth of the pine woods at Cape
May Point on May 13, 1928, as it hopped about the roots and lower branches
of the bayberry bushes.

The distribution of the Kentucky Warbler corresponds closely with that
of the Worm-eating Warbler, a common summer resident of southeastern
Pennsylvania but absent from southern New Jersey. Richard Harlow did,
however, find a pair in the extensive low woodland at Manahawkin in the
coastal strip which seemed to be located for the nesting season, and Chreswell Hunt found a few breeding on Pensauken Creek nine miles from Camden. Dr. F. M. Chapman found it nesting at Englewood on the lower
Hudson and there are records for Essex and Union Counties (Eaton and
Urner).

CONNECTICUT WARBLER

Oporornis agilis (Wilson)

A rare transient, in autumn. Our only records are birds found dead at
the Lighthouse; three on September 24, 1932, reported by Conrad Roland;
one found the next day by Julian Potter, one on September 20, 1936, by
John Hess and Hampton Carson and another by Hess on September 23, 1937.

The Connecticut Warbler is a common autumn transient locally throughout the state to the north and west of the Pine Barrens but very rare in
spring. Samuel Rhoads secured a male at Haddonfield on May 20, 1882.

*MOURNING WARBLER

Oporornis philadelphia (Wilson)

The Mourning Warbler has never been seen in Cape May County so far
as I am aware and the nearest occurrence was one shot by George Morris
when with me and others on the Pensauken Creek, near its mouth above
Camden, on May 30, 1897. Specimens have also been taken in the more
northern counties at Englewood, Morristown, Summit etc. It does not
breed in the state and would appear to migrate along the Delaware Valley.

MARYLAND YELLOW-THROAT

Geothlypis trichas brachidactyla (Linnaeus)

The Maryland Yellow-throat is a generally distributed summer resident throughout the Cape May Peninsula. It is common in the bramble patches and moist thickets between Cape May and the Point; in the denser thickets of the latter locality; around the borders of the various ponds; and even in bayberry thickets on the edge of the meadows and in plum and sumac thickets behind the dunes of the Bay shore. They also come into the gardens of the town, especially where they back upon the meadows or on the Fill. On Two Mile Beach and the other coast islands they are common. In days past when there were almost continuous thickets along the turnpike to the Point one could hear six or eight Yellow-throats singing along the two mile stretch and a like number if one followed the coast line behind the dunes.

The Maryland Yellow-throat is not a bird that can be approached undisturbed. Ever alert he usually sees us before we can catch a glimpse of him and our first knowledge of his presence is his sharp complaining *shick, shick,* as he hops about in the thicket near the ground watching our every move. It is only when the male is singing that one has an opportunity to study him at rest. I saw one perched on a telegraph wire along the turnpike singing with vigor and quite oblivious to the traffic passing almost below him. Another was sitting on a dead twig at the base of a bush at the edge of a thicket not more than a foot from the ground and well surrounded by tall grass. He would glance first to one side and then to the other, occasionally picking at something on a leaf or twig within reach. Then he would raise his head and pour forth his familiar warble. The flight song is more elaborate and is uttered as the bird mounts six or eight feet into the air from his perch and flutters gradually down again. I have heard it as late as July 7.

If one thinks that all Yellow-throats sing alike it will pay him to jot down some careful records of their songs. I have gathered several at Cape May and no two are alike, although each carries a similar phrase which is characteristic, and gives to all the songs the impression of identity.

It was moreover quite easy to identify individual birds after their songs had once been memorized. Here are some of them:

Group I

1. *Tzu-za-wítsky, tzu-za-wítsky, tzu-za-wítsky.*
2. *Tsivit-swéeah, tsivit-swéeah, tsivit-swéeah, tsivít.*
3. *Chit-wissa-whít, chit-wissa-whít.*
3a. *Chissa-wissa-whít, chissa-wissa-whít.*
4. *Chissa-wítso, tsu-wítso, tsu-wítso.* [*tsu-wítso*]

Group II

5. *Wisset, seé-wisset, seé-wisset, seé-wisset.*
6. *Sit-seea, weét-see-a, weét-see-a, weét-see-a.*

Group III

7. *Tse-wítsa-seéa, tse-wítsa-seéa, tse-wít.*
7a. *Tse-wítsa-séea, tse-wít.*
8. *Tsi-vít-sa-vía, tse-vít-sa-vía, tse-vít.*
9. *Tsi-vít-sa-réah, tsi-vít-sa-réah, tsi-vít-sa-réah.*
10. *Tsi-vít-sa-wée, tsi-vít-sa-wée, tsi-vit-sa-wée, tsi-vít-sa-wée.*
11. *Tsu-wítsua-weéah, tsu-wítsua-weéah. tsu-wít-su* [or *tsu-wit*].

I last heard the song of the Yellow-throat on July 27, 1920; August 4, 1923; July 28, 1924. These probably marked the last nestings of the season though early broods were of course on the wing long before this. I noticed bobtailed young with their parents on July 23, while birds, probably young of the year, were full-fledged and in close association with Prairie Warblers and Yellow Warblers, in a mixed flock, on August 26, 1920, while on August 13, 1922, and August 18, 1924, they were in all stages of molt.

The actions of the adults with the brood of bobtailed young were studied for some time. The male was very active, jumping from twig to twig, and flitting from one bush to another. He held the body well up with the tibio-tarsal joint clearly exposed, head down and tail cocked up while the bill was slightly elevated. There was a constant flirt of the tail to one side or the other accompanied by a nervous flap of the wings, so rapid as almost to escape detection, and a chirp of alarm uttered at regular intervals but not so frequent as the flirt of the tail.

Nests with fresh eggs have been found on May 13, partly incubated sets on May 22 and June 15, and young on June 8. The usual number is four.

After the singing ceases the Yellow-throat figures much less frequently in the bird life of the Cape district, but it can readily be called forth from its favorite thicket by the sucking sound used to attract birds—along with Catbirds, and certain other species which respond to this sound with equal alacrity. The rich olive green of the upper parts of August Yellow-throats is conspicuous in contrast with the dull brownish gray of breeding individuals, and marks the full-plumaged young of the year or early molted adults.

About the middle of August (August 16, 1921; August 17, 1927; August 18, 1924; etc.) there seems to be a great influx of Yellow-throats and from then on they vary greatly in abundance, as each succeeding migratory wave brings additional hosts until one is astounded that so many of the birds exist. They throng the countryside and the town gardens as well, and yet between flights it is sometimes difficult to find a single Yellow-throat!

During the great migratory flight of September 1 and 2, 1920, Yellow-throats fairly swarmed, and up on the Fill to the northeast of the town, as well as in the fields about the Race Track Pond, they arose like grasshoppers from the ground and low brambles as one advanced. On September 25, 1926, I noted but one in a morning's walk about the Point, while the next day when the wind shifted to the north they were present by hundreds.

Our latest autumn dates run:

1921. October 9.	1927. October 23.	1931. October 18.
1923. October 21.	1928. October 21.	1932. October 14.
1924. October 2.	1929. October 25.	1934. October 19.
1925. October 18.	1930. October 5.	1935. October 26.

The bulk has left by October 5 but on December 27, 1931; December 22, 1935; December 27, 1936; single individuals were seen which indicate that a few may winter.

Dates of spring arrival, kept up to 1920 by Walker Hand, with subsequent records by others are:

1908. April 26.	1916. April 28.	1923. April 22.
1909. April 29.	1917. April 23.	1926. April 24.
1910. April 16.	1918. April 30.	1927. April 24.
1911. April 27.	1919. April 23.	1932. April 23.
1912. April 25.	1920. April 25.	1933. April 30.
1913. April 30.	1921. April 24.	1934. April 20.
1915. April 20.	1922. April 23.	

The Maryland Yellow-throat is a common summer resident throughout the state. Whether both races G. t. trichas and brachidactyla occur in the state has not been definitely determined the differences are very slight and not recognizable in the field while New Jersey covers the area of intergradation between the two. Dr. H. C. Oberholser informs me that it seems probable that all New Jersey Yellow-throats are referable to brachidactyla the more northern race.

YELLOW-BREASTED CHAT

Icteria virens virens (Linnaeus)

The Chat is a familiar summer resident of the Cape May Peninsula as well as all other parts of southern New Jersey, arriving a little later than most of our summer birds and leaving somewhat earlier. It is an inhabitant of low scrub growth both in dry areas and on the borders of swamps, but is never found in the woodland nor is it seen in protracted flight, indeed we are lucky to see it on the wing at any time.

A pair or two are to be found every year established in some of the more remote thickets of the Point and there will be others on the borders of Pond Creek Meadow, as well as along the edges of the wooded strip on the ocean side of the peninsula from Brier Island and Cold Spring northward, while in the vicinity of Rio Grande they seem always to be particularly plentiful. Nesting pairs are usually scattered at intervals and do not colonize nor do the birds associate in groups after the breeding season is over, so that we do not see more than one or two at a time, although when in full song I have listed twenty Chats in a morning's walk back through the country north of Cape May.

The Chat is probably the most eccentric and erratic of our passerine birds both in postures and in vocal expression. He is far more often heard than seen, yet few birds are more certainly identified by voice alone. We hear his characteristic medley of notes coming from some dense thicket but we strive in vain to get a glimpse of him; restless and ever alert he manages to keep the thick foliage between us and, while watching our every move, he keeps himself unseen. He is likely to be calling at any time of day, or even at night, but in the heat of midday when most other voices are stilled his modest efforts dominate the scene. As one approaches his haunts amid

(840)

the greenbriers he utters a deliberate *kuk-kuk-kuk-kuk* in a high key and then on a much lower note, and still more deliberately, *caw-caw-caw* followed by several whistles and a high pitched *kek-kek-kek-kek*. There is a long pause and we stealthily approach to gain a view of the performer when suddenly, from another clump of briers farther on, comes a derisive *tscheet, tscheet, tscheet, tscheet*, in a low almost guttural tone, if such is possible in a bird. He has been keeping a watchful eye upon us and has slipped deftly out on the far side of the thicket to a new place of shelter farther on. If we persist in following him it becomes largely a game of hide and seek among the brier patches at which the Chat is well able to hold his own, dodging about with much agility and occasionally uttering a *chuck* of satisfaction. Should we remain perfectly quiet we may excite his curiosity and as he approaches closer and closer we get a glimpse through the foliage of the brilliant yellow of his breast, the bright green of his back and the strong black and white markings about the eye, and realize more clearly what a beautiful bird he is.

After the young are hatched the vocal efforts of the Chat begin to wane and we hear but scraps of his spring repertoire while some of the sounds that he does produce seem different from those heard earlier in the year. On July 13, 1920, I found a Chat at the Point which was much excited and probably had young in the immediate vicinity as, contrary to custom, he kept in plain view for some time. As he shifted his position, hopping from one perch to another, he would elevate his tail with each jump and allow it to drop slowly back again, while he kept up a continuous series of calls such as I had never heard a Chat utter before—*scape, scape, scape*, etc., about two seconds intervening between the calls. The note more closely resembled the cry of a Nighthawk that any other bird call that I could think of. With each cry he would turn his head sharply to one side or the other, ducking the body slightly at the same time. This individual, or possibly another, seen a few days later in the same vicinity, acted in exactly the same manner but had an altogether different note—*tuck, tuck, tuck*, at intervals, and then much more rapidly *tuck-tuck-tuck-tuck*, and finally the call of the previous day but with a break in the middle of each note *skee-uk, skee-uk, skee-uk*, etc.

Another peculiar cry was uttered by a bird in the pine woods at the Point (July 24, 1920)—*whit, whit*, he called and then a quick *keu, keu, keu, keu*, like the rapid cry of the Osprey, finally relapsing into the familiar Chat call *tscheeet, tscheeeet, tscheeeet, tscheeet*.

Another bird still singing the medley of the nesting season had in addition a rapid call like that of the Kingfisher *ki-ki-ki-ki-ki-ki-ki* followed by several flute-like notes and a single *tlut* and finally the characteristic *tscheeet, tscheeet, tscheeet*. On June 5, 1923, one bird had a single cry possibly in the nature of an alarm note—*whee-to-lit*.

In spring (May 10, 1924), perhaps before the birds were mated, I found three Chats together in a thick hedgerow. One singing from inside a wild cherry bush had a trill like that of a tree toad, a *pheu, pheu*, call like a Greater

Yellow-legs, and a strange note resembling a distant automobile horn. One of the other birds sat on the top of a dead bush in full view, all hunched up as if its back were broken and with tail hanging straight down. Every now and then it would stretch up its neck, which appeared very thick and out of proportion, with feathers all ruffled up and on end, and utter a triple note *hoo-hoo-hoo*.

In the spring more than at other times I have come upon Chats up in trees, especially in the open pine grove of the Point, and on May 25 one was seen diligently exploring the ends of the pine branches apparently for food, jumping clumsily from one twig to another. On July 3, 1921, another bird launched forth from the top of a rather tall tree with his characteristic flight song, wings flapping lazily and feet hanging straight down and wide spread. I have seen this performance as late as July 8, 1923, and the regular song until July 14, 1927, and July 11, 1930.

Turner McMullen has found nests of the Chat at Rio Grande on June 3, 1923, and May 31, 1924, with three and four eggs respectively and one at Court House on June 6, 1920, with two eggs. The usual number of eggs seems to be four. The nests have been placed in oak scrub, bayberry bushes and greenbrier thickets. Recently hatched young have been found on June 16 and 19 in different years.

I have found young birds out of the nest on July 5, 1932, and July 15, 1922, dull gray birds with a lemon yellow suffusion on the center of the breast. On the former occasion a parent was feeding them. Other young, molting to the winter plumage, were seen on July 20, 1929, and August 4, 1923, and showed new bright yellow feathers coming in as an inverted V on the breast.

Our dates for spring arrival of the Chat are:

1924. May 10.	1927. May 8.	1931. May 10.
1925. May 8.	1928. May 6.	1933. May 7.

I have never seen an adult Chat in the molt and as we rarely see them after the middle of August, and not often after the first week of the month, I sometimes think that they must move south before they molt, if not they certainly become very secretive. Our latest records are August 23, 1923, and August 26, 1927, but we have found several dead birds at the Lighthouse and in Cape May, at later dates, which were probably all migrants from somewhat farther north. They were picked up on September 21, 1934; September 29, 1935; all in full winter plumage. As New Jersey is about the northern limit of the Chat's range they could not have come from far.

While rare in the northern half of the state the Chat has been found breeding north to the state line on the lower Hudson as well as at Summit (Holmes), Morristown (Thurber), Lake Hopatcong (Dwight), Greenwood and Beaver Lakes (Rhoads) and High Knob (Chapman).

HOODED WARBLER

Wilsonia citrina (Boddaert)

This beautiful warbler is a summer resident of swampy woodlands all over the Cape May Peninsula as far south as the borders of the Pond Creek Meadows, and on the edge of the cedar swamps of the Pine Barrens farther north. In the Pond Creek area Julian Potter counted twenty on June 21, 1923, but I have never found it so abundant there in recent years. Immediately about the Point I have never found it until mid-August when birds from a little farther north move down the peninsula and, with the first Black and White Warblers and Redstarts, mark the beginning of the autumn flight. Occasionally in migrations they have been seen in the gardens of Cape May.

I first made the acquaintance of the Hooded Warbler on its breeding ground in May, 1891, when with Harris Reed and Charles Voelker I spent some days in the heart of the Timber and Beaver Swamp at South Dennis where the birds abounded. Here among thickets of sweet pepper bush and swamp magnolia, bordered on the one hand by brown cedar water of the lake and on the other by the sandy upland covered with oak scrub and holly, I came to know it intimately. The swamps resound with the song of the bird throughout May and June—a full toned emphatic warble, quite different from the lisping buzzing song of the Parula or the still more wiry strain of the Prairie, which are here its most intimate associates. *Swee, swee, tsip, tsip, se-wit-su*, I interpreted it; now on this side, now on that, it breaks the solitude of the swamp and suddenly right in front of us appears the bright yellow face and frontlet of the bird framed in its velvety black hood. With tail partly spread he gazes intently at us for a moment and then with a sharp *tschip* he has plunged back again into the shelter of the bushes and presently he once more voices his song, over on the far side of the swamp; or perhaps he has changed it to a shorter rendering, lacking the two preliminary notes, with which he is wont to vary his performance. In the fork of a holly four feet from the ground is the nest and four eggs and now the female appears on the scene, usually duller of plumage and with less of the black hood developed. The male abandons his song and both birds resort to the sharp *tschip* of alarm. While singing the male is very active moving constantly from bush to bush and covering a wide area so that with the song coming now from one angle, now from another, the performer is difficult to locate. Every moment or two there is a sudden nervous flirt of the tail as the bird hops from one twig to another and a great show of white when the feathers are widely spread. Sometimes I have seen a Hooded Warbler cling for a moment to the trunk of a pine tree, as the Pine Warblers often do, in pursuit of insects.

The nest in the Beaver Swamp contained four eggs on May 27, 1891, and two others near the same spot on June 1, 1907, held three and four eggs respectively. At Rio Grande, a few miles north of Cape May, Turner McMullen found nests with three and four eggs on May 27, 31, and June 8, in different years, and other nests containing young on May 30 and 31, 1924. Richard Miller found one at Rio Grande with two eggs as early as May 23, 1923, while I found a female in a swamp on the Rutherford farm northeast of the Point which gave every indication of having young in the vicinity as late as July 8, 1925.

The dates of the appearance of Hooded Warblers in the pine groves at the Point, which may usually be interpreted as the first indication of the southward migration, are:

1920.	August 5.	1923.	July 26.	1931.	August 6.
1921.	August 11.	1925.	July 30.	1932.	August 18.
1922.	August 9.	1927.	August 3.	1934.	July 28.
		1929.	August 17.		

Our latest records are: September 27, 1928; September 7, 1931; October 5, 1935 (William Rusling).

First arrival dates for the spring migration are:

1912.	April 30.	1915.	May 4.	1926.	April 25.
1913.	April 30.	1919.	May 5.	1932.	May 7.
1914.	May 1.	1920.	April 25.	1933.	May 7.

Over much of New Jersey outside of Cape May and the Pine Barrens the Hooded Warbler is a casual transient as at Princeton and Plainfield but at Englewood on the lower Hudson Dr. Chapman has found it a regular summer resident and it breeds at Demarest (Bowdish) and apparently at Alpine (Rhoads).

WILSON'S WARBLER

Wilsonia pusilla pusilla (Wilson)

A rather rare transient at Cape May Point during September and the last few days of August, and in late May. Once seen in shade trees of Cape May. Autumn records are:

1917.	September 29.	1928.	September 5.
1920.	September 2.	1929.	August 31 and October 5.
1922.	August 21.	1930.	August 28 and 30.
1924.	September 11.	1932.	August 29 and 30;
1925.	September 29.		September 17 and 18.

In spring we have only a single record, a bird in full song observed by Julian Potter on May 22, 1927, while quite extraordinary is the observation of another on June 4, 1933, on Seven Mile Beach, by John Gillespie, also in song but evidently a belated migrant.

Many of the birds seen in autumn at the Point show only a slight trace of the black cap and some lack it altogether. They are easily identified by their small size and nearly uniform greenish yellow coloration above and below which covers both breast and belly. The Wilson's Warblers are active little birds when feeding in the bushes, with a nervous flip of the wings and an irregular wag to the tail.

They breed entirely north of New Jersey.

CANADA WARBLER

Wilsonia canadensis (Linnaeus)

A regular transient in the Cape May district, more common on the southward flight. Practically all of our records are at the Point:

1920.	September 2.	1927.	August 17 and 25.
1921.	August 15.	1928.	August 13.
1922.	August 21, 28 and 29.	1929.	August 17 and September 6.
1923.	August 26.	1931.	September 1.
1924.	August 27, 29.	1932.	August 29 and 30.
1924.	September 3 and 7.	1934.	August 27.
1926.	August 27 and September 1.	1935.	August 31 and September 3.

On August 17, 1927, five were seen and on September 3, 1935, two, all other records are of single birds. The dates of August 13, 15 and 17, indicate how early some of the transient warblers from far northward reach localities as far south as Cape May although as some individuals nest in the mountains of northern New Jersey they may have come from there. Over most of the state the Canada Warbler is a common transient.

We have but three spring records—May 20, 1917; May 24, 1892; May 20, 1927.

REDSTART

Setophaga ruticilla (Linnaeus)

The Redstart is a regular and abundant transient in the Cape May area and throughout the peninsula while it appears to breed sporadically in Cumberland and probably also in northwestern Cape May Counties. It is, however, unusual in the nesting season anywhere in southern New Jersey.

About the first of August Redstarts appear in varying numbers in the pine woods at Cape May Point and are present, off and on, until the end of September. With the Black and White and Hooded Warblers and the Northern Water-Thrush they usher in the southward flight of land birds at Cape May and are the most regular and consistant of our migrant warblers. During most of August and September they occur regularly in the shade trees of Cape May as well as throughout the woodlands of the interior upland.

The days of marked flights are always days with wind out of the northwest and one has but to glance at the pine groves of the Point to learn whether a migration is on, so conspicuous are the fluttering yellow and salmon colors of the Redstarts against the somber blue green of the trees. Throughout the day the birds seem to be everywhere and as they pass on only a few stray individuals will be seen until the next great wave arrives. As we see him in our late summer woodlands the Redstart is the very embodiment of nervous activity, flirting the tail to the right and left as he darts this way and that and then spreads wide the feathers like a fan. Some birds are always fluttering at the ends of the branches where the rapid movements of wing and tail make a constant flicker.

On August 21, 1922, a typical Redstart day, the birds were abundant both in the trees and in the shrubbery while along an old wood road through the pines dozens of them fed on the ground, darting a few inches up into the air, now and then, to catch a passing insect and dropping back again, all the while flaring their fan-tails and jerking them from side to side. Others were constantly coming down from the trees in a sort of "tail spin" and as the sun struck them the light areas of the wings and tail seemed almost transparent. Occasionally a bird would cling for a moment to the tree trunks like a creeper.

Again on August 29 of the same year Redstarts were swarming at the Point. They hopped from limb to limb as they passed through the pine trees and, after two or three hops, would flutter out into the air in a spiral, keeping at about the same level, or continuing in a prolonged flutter almost to the ground, in pursuit of some escaping insect. I have seen them whirl down in this manner for a distance of ten or twelve feet, but in any case they return to the tree and resume their feeding. When working their way through the foliage, with wings and tail closed, they seem so sleek and slender that we take them for a different kind of bird until the tell-tale yellow fan is flashed out and their identity proclaimed. In deciduous trees the actions of the Redstart are somewhat concealed by the foliage and they are not always so conspicuous, but wherever they are, in the tree tops, on the ground, or in the air, they are graceful in every movement and the personification of nervous energy.

Our evidence of nearby nesting of the Redstart is not extensive. On June 19, 1914, Julian Potter found a pair feeding young in a wooded swamp two miles west of South Vineland, Cumberland County, and Walker Hand saw a female feeding a full-fledged young at Eldora, Cape May County, on July 24, 1932. The occurrence of a single bird in the Physick garden on July 15, 1933, seems too early for a northern migrant and probably was a bird that had bred or been raised somewhere in the southern part of the state, and the same may have been the case with those seen at the Point in four years, on August 1 and 2.

While the bulk of the Redstarts seen in the southward flight are females or young, with pale yellow on the wings, tail and breast, the bright salmon and black males are by no means uncommon and a number of them may be seen on a single day. In spring while old males and yellow females make up the bulk of the flight there are some deep yellow males in their first breeding plumage with scattered black spots on the breast and back.

Arrival dates on the southward migration run as follows:

1920. August 5.	1925. July 30.	1930. August 11.
1921. August 11.	1926. August 14.	1931. August 1 (next 18)
1922. August 9.	1927. July 30.	1932. August 1 (next 9).
1923. July 26.	1928. August 13.	1934. August 2.
1924. August 8.	1929. August 2.	1935. August 12.

Latest autumn dates of occurrence:

1921. September 16.	1926. September 27.	1931. September 27.
1922. September 17.	1927. September 25.	1932. October 8.
1923. October 1.	1928. September 30.	1934. September 19.
1924. October 2.	1929. October 5.	1935. September 21.
1925. October 2.	1930. September 28.	

In spring Walker Hand has recorded their arrival on May 5, 1914, and May 8, 1908, while I have records of May 8, 1925; May 7, 1932; May 8 and 9, 1925; May 20, 1927; May 14, 1928; May 27, 1931; while they have been seen as late as May 30 in 1912, 1921 and 1928. They are not nearly so abundant on the northward flight as in the autumn, and I have seen them common on only one occasion—May 20, 1927.

While mainly a transient in New Jersey the Redstart breeds in the northern half of the state—at Plainfield (Miller), Patterson (Clark), Summit (Holmes) etc., and has occurred in the nesting season sporadically at a number of localities farther south.

ENGLISH SPARROW

Passer domesticus domesticus (Linnaeus)

The English Sparrow had become abundant at Cape May long before I first visited the town and I have never been able to ascertain just when it first put in an appearance. The paving of the streets has probably had some effect in reducing their numbers, but not to the extent that we have seen in the large cities, and the advent of the Starling has also been a factor in the same direction. Sparrows are most evenly distributed in the town at nesting time in early spring, and by winter have gathered into several rather well defined flocks with definite roosting places, especially on ivy clad walls of buildings. I doubt if there is any actual change in the numbers of the bird at different seasons of the year except for the advent of young in early summer.

After the harvest they go out on the grain and hay fields and mingle with Red-wings, Starlings and Grackles as they feed among the stubble, or gather immediately on the new mown hay and alfalfa where insects abound. They are also to be seen about Schellenger's Landing where the shores of the Harbor furnish a variety of food and, after the young are on the wing, great flocks visit the various piggeries on the edge of Pond Creek Meadows and elsewhere. Not missing any possible source of food they come frequently to the sea beach where they catch blowflies on the lines of trash washed up by the waves and search the latter for bits of refuse. They alight on the wet portion of the beach just above the water and, when a wave rolls up a little farther than usual, they flutter back to safety. I never have seen one hop away from an oncoming wave, they always take wing. One day a ghost crab (*Ocypoda*) came out of his burrow for food and passed within an inch or two of a sparrow without any interest being shown on either side.

When on the ground the English Sparrow hops and never walks or runs, while it assumes an attitude quite different from that of any of our native species. The body is held erect with the head high and the tail just touching the ground and the bird seems, at times, almost to lean over backward in its exaggerated erectness. Sometimes when certain insects are abundant in the shade trees I have seen the sparrows clinging upright to the trunks like woodpeckers in their efforts to catch their prey. This habit and their attitude on the ground emphasize the fact that they have no near kinship with the true sparrows and that, as is shown by their anatomy, they are really related to the Weaver Finches of Africa.

In spite of their usual unsavory reputation I have seen the English Sparrow devouring mature Japanese beetles, clinging meanwhile to the flower stems and shrubbery in our yard. Surely they deserve some of the credit that is so lavishly accorded to that greater nuisance, the Starling, for devouring the larvae of this imported pest; had neither of the *three* pests been brought to America we should have been far better off!

By mid-summer the English Sparrows gathered in the evening at the Martin and Grackle roost. Some of them went in as early as July 2 while later they flocked there by thousands taking up their positions in the trees before any of the other species had arrived and making a prodigious noise by their continued chirping.

They nest about houses, under the eves or in any cavity or shelter that may be offered, and not infrequently build a globular nest in the top of an evergreen tree—another Weaver Finch character. One nest at Cold Spring as large as a peck measure was well out on top of a horizontal limb of a tulip poplar tree and was constructed in a few days after an old nest in a shed had been destroyed. I have seen them busy carrying nesting material as early as March 18 and by June 30 flocks composed wholly of young birds are on the wing. On the other hand I have seen parents feeding young as late as August 15, 1935, and September 6, 1927, illustrating the fecundity of the species, as they must build several nests in a season. While they are notorious for occupying the nest boxes erected for other birds I have seen a pair or two sharing a large Martin house with the rightful owners without any apparent conflict, each pair, of course, using its own apartment, while several times I have seen sparrows nesting in the lower parts of the Fish Hawks' great nests, going in and out between the larger branches which composed the structure while the Fish Hawks paid them no attention whatever.

The English Sparrows are pugnacious and resent the intrusion of other birds on their feeding grounds and I have seen them attack Goldfinches and native sparrows of several sorts. Those that come to feed on scraps of bread thrown out in our garden for other bird visitors defer to Starlings, Grackles and Red-wings but they do not leave the field and by watching their opportunity will snatch up a crust from under the very bill of one of the larger birds and fly off with it.

Sparrows are by no means confined to the town and may be seen even in winter flying far afield in search of food and I have seen them about closed cottages at South Cape May in mid-winter associated with Juncos, and out on the Fill at all seasons.

Back in the country there will be a good sized colony about every house or barnyard, smaller nesting groups being formed about dwellings where winter forage and shelter are not so plentiful. About the home of Otway Brown at Cold Spring and the adjacent farm buildings of David McPherson there is a flock which is doubtless characteristic of many others. About two hundred birds winter here and probably the summer residents number about the same. They nest about the various buildings and constantly seek shelter in the privet hedges, often escaping the pursuit of a Sharp-shinned Hawk in this manner, in autumn or winter. They also frequent loose piles of brush in the orchard and as I approached one of these I have counted upwards of one hundred leaving in small bunches. Sometimes the entire winter flock will fly in a dense mass from one shelter to another and once, on February 1, 1931, a bitter cold day, the birds arose *en masse* and circled overhead for nearly half an hour following exactly the same oval course and maintaining their dense flock formation. This performance seemed so unusual that I was in some doubt at first whether they really were English Sparrows. There seemed to be no cause for their action unless it was the cold. They finally settled in the trees by the house and on the barn roof.

Even this familiar bird would seem to warrant further study!

BOBOLINK

Dolichonyx oryzivorus (Linnaeus)

In the Reedbird plumage the Bobolink is an abundant transient from mid-August to late September traveling mainly by day and crossing at the Point to the coast of Delaware, while little groups stop for a few days on the marshes and in old fields west of the town. The Delaware River marshes, with their abundant growth of wild rice, have always been notable as hunting grounds for Reedbirds, when this sport was permissible, and the birds are as abundant there as ever, down to the head of the Bay. Whether the Cape May migrants come from the Delaware or along the coast from their breeding grounds to the north I am unable to say.

In spring the Bobolink is not nearly so abundant and occurs mainly farther back in the farming country, in fields of clover or alfalfa, occurring, as always, in flocks. Otway Brown tells me that some twenty years ago a flock of Bobolinks established themselves in an old brier field close to his home at Cold Spring and evidently bred there, the males singing all season from his apples trees, and on May 9, 1935, a flock of some twenty-five settled in an alfalfa field in almost the same spot and remained for two weeks, but when the crop was cut the birds immediately departed. There was every indication that they would have bred there. As additional evidence of the probable breeding of the birds in the Cape May district may be mentioned the arrival of eight individuals, still in breeding dress, on the marshes back of South Cape May on July 18, 1924, and fifteen on July 24, 1925. They remained for a week or more. In the latter group, when I first discovered them, were three males in full black and white plumage, one in an interesting molting condition with broad buff bands on the breast forming a great inverted V, while the rest were females and young of the year. It seems hardly likely that these birds would have traveled far before the molt was completed and I am inclined to think that they had nested in some out of the way field on one of the farms immediately to the north. Julian Potter also saw six on July 25, 1930.

The late August birds that we come across immediately about the town are often found perching on the tops of bayberry or marsh elder bushes, or among patches of reeds and tall grass, and swing back and forth in the wind bracing themselves to maintain their position. When they obtain a secure perch, they crane the neck up clear of the foliage to get a better view of the surroundings, and eventually either drop to the ground to feed or take wing to some more distant point. The striped buff and brown pattern of plumage makes the Reedbird really very conspicuous among the sparrows with which it may be associated and when the head is stretched upward the stripes seem to run continuously from bill to tail. The perching birds that we see are resting rather than feeding and, although they do feed on seeds of weeds and ranker grasses, we never see them stripping seeds from the tall stalks of wild rice which is a familiar sight on the Delaware marshes, as this plant does not grow about Cape May, being a strictly fresh water species. The *spink, spink,* call of the Reedbird is heard both from birds at rest and from flocks passing overhead.

The flight of the Reedbird somewhat resembles that of the Goldfinch in that there is a flicker of several short wingbeats followed by a drop of the body and then another series of beats causing a rising and falling, undulating motion in the flock. The Reedbird, however, swerves from side to side making its flight far more erratic and quite different from the straightaway course of the Goldfinch. Two that I saw on August 20, 1927, pitched down from a considerable height in zigzag drops to the tops of some bayberry bushes where they rested for a long time. They were apparently migrants which hesitated to cross the Bay, although on August 30, 1920, I saw several flocks of from one hundred to one hundred and fifty individuals put out across the water without a moment's hesitation.

On some late August or early September days I have seen a number of flocks passing between Cape May and the Point but have never kept count of them for any considerable period. In 1935, however, William Rusling made almost daily counts of their numbers and their flights and he tells me that they fly most frequently during north or northeast winds and upon reaching the Bay shore often circle about several times before crossing. Those that come on a northwest wind, however, fly north along the Bay as do all other birds under similar conditions, and do not cross until later. His counts are as follows:

August 23, 800.   September 6, 2,615.   September 16, 66.
August 24, 1,000.   September 7, 1,575.   September 17, 391.
August 28, 75.   September 8, 150.   September 21, 350.
August 29, 175.   September 10, 330.   September 23, 121.
August 30, 4,000.   September 12, 35.   September 29, 27.
September 2, 350.   September 13, 250.   October 4, 4.

My dates of arrival of Reedbirds on the southward flight are:

1920.	August 16.	1925.	August 22.	1931.	August 25.
1921.	August 15.	1926.	August 25.	1932.	August 20.
1922.	August 14.	1927.	August 20.	1933.	August —.
1923.	August 24.	1928.	September 1.	1934.	August —.
1924.	August 27.	1929.	August 25.	1935.	August 23.
		1930.	August 17.		

In spring we have Walker Hand's dates of observation for seven years:

1907.	May 11.	1909.	May 13.	1916.	May 7.
1908.	May 8.	1910.	May 11.	1917.	May 2.
		1914.	May 6.		

I have but few spring records for Cape May and am inclined to think that the birds travel farther inland on their northward journey.

Seven were present at Cold Spring on May 29, 1933, and I heard one in full song on the Fill on May 20, 1927, while another individual was seen on May 30, 1923, but appeared sick and emaciated, and was probably unable to travel further.

While most conspicuous as transients Bobolinks nest locally in several of the northern counties.

MEADOWLARK

Sturnella magna magna (Linnaeus)

The Meadowlark is a common resident bird of the open grasslands about Cape May, from the farms of the upland down to the brackish marshes, and on the grass-covered dunes of the coast islands, while it strays occasionally onto the salt meadows themselves. For some years after the formation of the Fill it was common among the grass and clover which covered the original dredgings but as this has been succeeded by a forest of bayberry bushes it has to a great extent disappeared. Not infrequently we see Meadowlarks flying back and forth over the housetops of Cape May in their passage from the Fill to the golf links and in autumn and winter we may flush them from vacant lots and garden patches in the town.

All through the spring and early summer Meadowlark song is in the air as we sit on our porch, or it may come floating in at the open window, and we come to regard the bird as a familiar everyday species rather than an inhabitant of the more remote open fields with which, in most localities, it is associated.

While habitually a ground bird the Meadowlark frequently alights on the top of a small isolated tree or bush or on fence posts, or even on telegraph wires along the roadside. I have also seen them on several occasions perched on the peak of a cottage roof in full song, and on December 12, 1920, a flock of fifteen that flushed from the Fill lined up on the ridge pole of a closed cottage near the beach. When perched on a wire the Meadowlark settles its head well down on its shoulders and at intervals flirts its short tail as if to steady itself against the wind, for it is a heavy-bodied bird and it must be no easy matter for it to hold its balance on a slender wire. They sing from their perches at regular intervals and between songs I have frequently seen a bird bend its head over until the bill almost touched the breast and apparently gaze at something below it or at its feet, then the head will be raised again, the mouth will open, and the clear call floats out over the meadows.

On the ground the Meadowlark is by no means easy to study as its back is a fine example of protective coloration and effectively conceals it so long as it remains quiet, and it is usually careful to keep its brilliant yellow breast turned the other way. A mixed flock of Meadowlarks and Starlings feeding in an old field back of the town, on January 19, 1924, was an interesting study. The Starlings in their more or less black plumage, stood out clearly while the Larks with their mottled brown and buff backs were almost invisible against the dead grass that formed the turf. Even after I thought I had counted all that were present half as many more were disclosed as they took wing.

When moving about in a field of grass stubble the Meadowlark humps its back, draws its head down close on its shoulders and runs, almost like a rail, in and out among the tufts and roots, but its gait is not so rapid or direct as that of a rail and is rather more of a waddle. Now and then it will pause and holding the body nearly vertical will crane its neck upward giving it a slight twist and pointing the bill skyward as if straining every muscle to see over the top of the surrounding grass; but curiously enough I have seen them do exactly the same thing when perched on a wire. In winter too I have seen them acting in much the same way, walking rapidly with head and neck extended and back humped, when suddenly they pause and the vertical craning of the neck follows with an instantaneous flicking of the tail feathers, open and shut, displaying for a moment the white edgings. One bird walked about with the tail about half spread and nervously jerked it farther open and then shut again, showing the white very conspicuously.

In early July I have on several occasions observed Meadowlarks carrying insects about in the bill and later flying off apparently to feed young in the nest. On July 7, 1931, I watched two pairs of Meadowlarks with six young in a low grassy meadow back of the golf links. The adults were walking about like quail their heads moving back and forth with each step and were catching young grasshoppers on the ground sometimes flying up a foot or two in pursuit of prey. The young, while they could fly perfectly well, remained stationary and were apparently still being fed by their parents.

The contrast between young and old was striking. The latter had white or pale gray head stripes, quite different from the rich buffy stripes of the young, while the short almost conical bill of the young birds and their generally suffused buff plumage were characteristic. A curious melanistic specimen in the collection of the Academy of Natural Sciences of Philadelphia, which was secured at Haddonfield, on October 6, 1857, has all of the yellow breast plumage replaced by black.

A bird noticed pluming itself on July 2, 1921, poked the bill down among the feathers and spreading its shoulders ruffled up all the plumage. It also spread its tail feathers and while holding them so, shook the tail forcibly from side to side.

Except for perching birds our experience with Meadowlarks is usually limited to flushing them from the grass as we walk through the old fields, and watching them sail away to alight again on the ground farther on, and it requires some care and keenness of sight to see them first and to study them before they take wing. When taking flight from the ground the Meadowlark vibrates its wings rapidly three times and takes a short sail followed by another series of vibrations and another sail, all the time ascending at a slight angle with the head held out in front and turned from side to side in constant observation. When hurried, the bird sometimes flaps its wings continuously for a moment or two, like a Starling, and the rapidity of its wingstrokes is about the same as in that species. Having attained a moderate altitude there is a long descending sail and sometimes a resumption of the short periods of beats and sails as if the bird had changed its mind about alighting at that particular spot and was extending its flight a little farther.

When a strong wind is blowing I have noticed that a Meadowlark will take off against it and, gaining a little altitude, will veer off and, on set wings, sail a long distance with neck stretched out and up, giving one the impression that the heavy body is drawing the bird down and that it is trying desperately to keep afloat. Other individuals going straight away, on flights of a quarter of a mile or more, kept up a pretty continuous flapping with an occasional short sail and one watched on August 20, 1920, flapped its wings continuously for nearly seventy-five yards, gaining altitude all the time and then began the alternate sailing and flapping with a gradual descent.

The song of the Meadowlark is usually rendered by the words *can't see me* but in reality there are four notes, not three, and it seems to me that the syllables *see—aar see—eee*, the second descending and the third and fourth on the same pitch, better represent it. The song is rendered either from the ground or from a perch and I have heard it as late as August 13, while its revival in late February is one of the first signs of spring. The Meadowlark has another call or song, which it utters from a perch. There is a single *tzud* or *zhud* followed by a sibilant trill *dzzzzzzzzzzzzz* then a pause and the notes are repeated, sometimes with the opening *tzud* duplicated. At other times this note is repeated at appreciable intervals without the trill.

The Meadowlark nests in grassland especially in old fields or in open places on the Fill and nests have been found as early as May 21, 1921, with five heavily incubated eggs and May 22, 1920, with four fresh eggs, and as late as June 17, 1922, and June 20, 1909, with three and five fresh eggs respectively. Another nest found on May 31, 1926, by members of the Delaware Valley Club, contained three normal eggs, one runt and four smaller eggs apparently those of a Sharp-tailed Sparrow! Young birds have been seen on the wing with their parents as early as June 18.

On the evening of July 6, 1923, and on two evenings previously, about dusk, a Meadowlark was seen to fly several times in a circle over a field where its mate apparently had a nest and finally to alight on a nearby telegraph wire; doubtless a courtship display. On another occasion in early spring I came upon two Meadowlarks, presumably rival males, tumbling about on the ground on their backs with feet firmly locked together as they reared up and struck at each other with their bills.

The best identification mark of the Meadowlark, aside from its song and manner of flight, is the flash of white in the tail as the bird takes wing. The bright yellow breast with its black crescent is only seen when the bird is perched on a bush or wire, and even then it does not seem nearly so bright as we might expect unless the sunlight strikes it at the proper angle. In winter the yellow is largely obscured by brownish edgings to the feathers, the back is browner, and the whole bird gives one the impression of being larger and fluffier, as if in a shaggy winter coat.

From April to July or early August the Meadowlarks are scattered in pairs but by August 15 they begin to collect in flocks and parties of twenty to twenty-five are not uncommon. Often at this season they gather on fields of second-crop hay, recently cut, or on patches of black grass mowed and left lying on the marsh. Later in August or in early September they may become very scarce so that only two or three are seen on a day's walk or perhaps none at all. The disappearance being possibly coincident with the gathering of the late hay crop or to retirement of the birds during the molt, both old and young having a complete change of plumage at this time.

From October on through the winter, however, Meadowlarks occur in large flocks. On October 19, 1924, Walker Hand and I flushed no less than three hundred from an old weed-covered field near the old toll gate. They arose, as they normally do, in small bunches one after another, not all together. On October 21, 1923, a similar flock of one hundred was seen; on October 23, 1927, sixty-five; one hundred on October 19, 1931; and on October 27 and 28, 1928, at Cold Spring, flocks of twenty, thirty and fifty. One flock flew directly over Cape May travelling from one feeding ground to another. The birds when on the ground, wherever they could be approached closely, seemed to be catching grasshoppers.

While many of these large October flocks pass on farther south some remain throughout the winter, associating with Cowbirds, Grackles and Red-wings and forming a single mixed flock which drifts here and there over the uplands and may be missed entirely by those who make single day trips to the shore at this season. We have records of a flock of one hundred on November 14, 1926; seventy-five on January 25, 1925; sixty on February 8, 1925. The counts of the Christmas census of the Delaware Valley Club are as follows:

1922.	December 24, ten	1930.	December 28, seventy-two
1923.	December 23, 200	1931.	December 27, 246
1926.	December 26, 135	1932.	December 26, 100
1927.	December 26, 115	1933.	December 24, 300
1928.	December 23, 275	1934.	December 23, 123
1929.	December 22, 158	1935.	December 22, 294
		1936.	December 27, sixty-seven

Sometimes the October flocks fly in a regular stream in passing from one spot to another and on October 31, 1920, fifty passed over the marsh back of South Cape May in a long line. Later, upon being disturbed where they were feeding, they flew back again picking up other individuals which had not been seen before. The flock then divided and each section settled to feed. Sometimes in midwinter I have seen flocks acting in the same way.

The most frequent associate of the Meadowlark is the Starling but I am inclined to think that the companionship is always sought by the latter which is a notoriously social bird, while the Meadowlark seems perfectly satisfied to associate with its own kind. On September 1, 1920, I found on an open space on the Fill two Meadowlarks and about a thousand Starlings. The former kept some distance in advance as if trying to escape from the black throng which constantly followed them up, rolling over one another like a flock of blackbirds in spring. Cowbirds and Flickers are sometimes associated with Meadowlarks when feeding on stubble fields or on freshly cut grass. A curious association was that of a couple of Kingbirds which persistently attacked a pair of Meadowlarks that had flushed from the grass near where their nest tree was located. All four birds alighted on a telegraph wire each Kingbird close to his Lark and as soon as the latter took wing the attack was renewed.

In winter Meadowlarks become much tamer than at other times, especially when food is scarce and it is then that they most frequently enter the town. On one occasion I found six feeding with English Sparrows on a vacant lot close to the street and at another time I approached within fifteen feet of some that were feeding in an old cabbage patch. During the severe blizzard of January, 1918, Walker Hand reported a single Meadowlark living in a narrow alleyway alongside of the post office on Ocean Street and feeding on some old bones that had been left there. During a storm on March 11, 1934, when the whole countryside was covered with snow, two Meadowlarks rested in an orchard at Cold Spring with a flock of Grackles and I have seen them doing the same thing during ice storms although in that case they had much trouble perching on the ice-covered twigs.

While I presume that Meadowlarks roost on the ground at night we really know little about the matter. On January 8, 1922, I flushed fifteen of the birds on the Fill and at the spot where one of them took wing I found a form in the grass much like those used by cotton-tail rabbits but it contained many droppings of the bird which seemed to demonstrate that it was used as a shelter or roosting place. On August 21, 1920, too, I flushed several young Larks from the edge of the marsh which took wing just as I was about to step upon them and I found that each one was resting in a grass form similar to the one just mentioned. As evidence of a possible roosting in another situation, a single Meadowlark was seen to enter the Martin and Grackle roost on the Physick place in the heart of the town on the evening of August 18, 1921, doubtless attracted by the tremendous flight of various birds that was pouring into the trees but I have no evidence that it remained through the night. Throughout New Jersey the Meadowlark is a summer resident or resident retiring to the river valleys or coast districts in winter, with possibly some migrants from states to the north.

I regret to see a definite decrease in Meadowlarks not only about Cape May but elsewhere in southern New Jersey. The uncalled for draining operations have ruined many of their haunts as have the recent "developments" and I cannot but wonder whether the Starling, a bird of much the same size and feeding habits, and which is fitted to fill the same ecologic niche, may not have been a factor in the matter. One species may effect the decrease and even extermination of another without active antagonism.

YELLOW-HEADED BLACKBIRD

Xanthocephalus xanthocephalus (Bonaparte)

Several individuals of this bird of the prairie lands of the North Central States have wandered as far east as New Jersey having doubtless become associated with flocks of Red-wings which were bound for the Atlantic Coast, or blown by storms from their normal migration route. One was shot by a gunner at Tuckerton about 1890, and later secured for the Delaware Valley Club Collection at the Academy of Natural Sciences. Another was shot on Newton Creek near Audubon by J. Kelton on September 1, 1917, and was shown to me by Joseph Tatum.

A third individual was seen in mid-August, about 1923, by Fletcher Street on the meadows between Cape May and the Point. All three were in immature plumage with duller colors than those shown by the adult males.

RED-WINGED BLACKBIRD

Agelaius phoeniceus phoeniceus (Linnaeus)

Red-winged Blackbirds are to be found about Cape May in every month of the year but from the time that the last young are on the wing, in mid-July, until the arrival of the males on their breeding grounds, in February, they occur in roving flocks of varying size and there may be many days in autumn and winter when none can be found in the immediate vicinity of the Cape. The appearance of the first migrant male swinging on the top of some dead cattail on the marsh, or perched on a bayberry bush on the Fill, and the sound of his gurgling, wheezing song give us the first intimation of the passing of winter; for it is the Red-wing rather than the Bluebird that is the harbinger of spring at Old Cape May.

There is often a large flock of "blackbirds" composed of Red-wings, Grackles, Cowbirds, and latterly Starlings, wintering on the open fields north of the town, especially about Cold Spring, but most of these roving winter assemblages seem to drift farther south, although they always linger longer on the shores of Delaware Bay and may be found on the extensive river marshes in the vicinity of Salem long after they have left the ocean side of the peninsula. I saw one of these flocks near Elmer on November 28, 1926, which was typical of such gatherings. It consisted of probably five thousand birds which fairly blackened the ground as they alighted in close ranks and advanced over the fields in a rolling progress, the rear lines constantly rising and settling in front of the flock, the others following in regular sequence. When large detachments would occasionally fly up into the bare trees of some orchard or grove of swamp maples it appeared at a distance as if the branches were once more clothed in dense foliage. As the mixed flock went streaming past against the background of winter woods the bright epaulets of the male Red-wings would flash in the sunlight giving a touch of color to the black horde while a solitary albino looked strangely out of place. On March 8, 1919, I saw a similar flock at Bombay Hook on the Delaware shore which was rolling over the marsh in the same manner but when the birds settled in the trees they burst into the great blackbird medley of song so characteristic of early spring.

When the Red-wings start to move northward in February, and apparently during most of the winter, the males and females are, for the most part, in separate flocks. Of winter gatherings I remember one in late October at Mauricetown which contained one hundred females and a single male while another smaller group at Cape May in late January, 1891, was wholly composed of females. The males are the first to reach the nesting grounds in spring and it may be several weeks before the females arrive, so there may be a long interval between the appearance of the first migrants and the actual beginning of nesting operations. As early as February 21, 1926, solitary males were singing on the Fill as well as on the marshes at South Cape May and back of Schellenger's Landing, while on March 14, 1931, they were singing in the same way on the cattail marshes of the Lighthouse Pond, spreading both wings and tail as they gave vent to their characteristic vocal efforts. The extent to which the wings were spread, and the consequent amount of red shoulders to be exposed, varied greatly in different individuals but in every case the display was best seen from the front.

Walker Hand's dates for the arrival of singing males with some later records are:

1903.	February 26	1921.	February 10
1904.	March 9	1923.	February 22
1905.	March 1	1925.	February 14
1908.	February 16	1926.	February 21
1911.	February 27	1928.	February 14
1912.	February 21	1929.	February 15
1913.	February 23	1931.	February 9
1914.	February 27	1932.	February 22
1915.	February 15	1933.	February 19
1916.	February 18	1934.	February 10
1917.	February 22	1935.	February 23
1918.	February 12		

In the very cold Februaries of several recent years the Red-wings were much later than usual, and even if an early arrival came before the freezing temperature or the ice and snow storms, he disappeared again.

All through March additional Red-wings take up their positions about the nesting grounds but as late as March 20, 1924, and March 21, 1925, no females were in evidence although on the latter date two flew rapidly over the marsh at South Cape May and were vigorously pursued by a male which, however, soon gave up the chase and returned to his stand. During this same period some males will be found still associated with other blackbirds feeding out in the fields. On March 19, 1924, six were feeding with Cowbirds and Starlings just back of the town, and on February 14, 1925, there were ten associated with thirty Starlings and fifty Meadowlarks in the same place, while at Cape May Point on March 13, 1921, there was a typical flock of spring migrants such as we often see moving through to points farther north. There were twenty-eight Red-wings, five Cowbirds and two Starlings, all feeding on a damp piece of ground near a swampy thicket, a favorite resort at this time of year. Presently they swirled up into the low trees and broke into a medley of blackbird music. The spring song of the Red-wings—the *o—car—ée* and their liquid gurgle and harsh *chuck* note mingled with the coarser efforts of the Cowbirds, and some whistles from the Starlings.

Transient flocks of Red-wings seen as early as February 6, 1932, settled in the wooded borders of the New England Creek marshes and broke into song in the same way, while as late as April 2, 1922, and on the same date in 1924, I saw feeding flocks composed of males only, which seemed evidently to be migrants passing through. By the end of the month, however, both sexes are established on the breeding grounds and there is a scene of wild excitement as the males leave their perches to pursue females or rival males, while the air is full of their songs and cries of alarm.

The Red-wings have many notes and there is in addition great variation in their expression. Most characteristic is the song of the male so frequently interpreted as *o—car—ée* but which to me, especially when heard near at hand sounds more like *kill—géeeze;* then there is a delightful liquid, flute-like gurgle difficult to represent syllabically which is usually uttered as the bird settles down on its perch—*zhu-lu-lu-lu-lu-lu-lu.* Another "song" consists of two similar notes followed by a series of short notes uttered rapidly on a descending scale *zhurr, zhurr, chip-chip-chip-chip-chip-chip-chip-chip,* the short notes repeated sometimes as many as twenty times. When one approaches a breeding colony early in the season, so as to arouse the birds' suspicion, the males will lapse into an occasional interrogatory *tuk, tuk, tuk.*

On April 29, 1923, there were some thirty Red-wings about a cattail pool on the Fill and some of them were evidently mating. One male sitting near a female on the branch of a low tree was uttering the *kill—géeze* song. At each utterance he would begin by bending the head over forwards and spreading the tail, at the same time bringing it forward under the perch.

Then, just as the song began he would swell up his shoulders and the brilliant red epaulets would flash out in a wonderful display of color, which was conspicuous for quite a distance, but which immediately disappeared as the bird resumed his normal attitude, with only the buffy edges of the shoulder patches in evidence. Several times the performer would drop to the ground and was immediately followed by the female who had been perched close by. Another male, with no female near, sang the *kill—géeeze* song continually from his perch, with wings drooping slightly and tail spread and with a little of the scarlet shoulder patch in evidence all the time, even between his songs, but he never attempted the spectacular display of the other individual.

When feeding on the ground the male Red-wing walks slowly about with tail elevated at an angle of 45° and every now and then will flirt it up and down uttering a harsh *chuck*. Sometimes when walking on soft ooze he will flap the wings suddenly and perhaps rise a few inches into the air to keep his feet from sinking in the mud. When perched among the cattails both males and females will grasp the stems with both feet one above the other while they hold the body close against the stalk and when they wish to descend nearer to the ground or water they will allow themselves to slide gradually down. At other times a bird will support itself on two stems holding one in each foot and spreading its legs widely, and I have seen a male perched thus on two stalks of wild carrot.

The pugnacity of the Red-wing is always in evidence during the nesting season, and not only does he resent human intrusion but woe betide other birds that venture near. I have seen five males vigorously pursue a Marsh Hawk which had inadvertently approached their territory and as he beat a hasty retreat and passed other colonies of Red-wings they too took up the chase. A passing Crow also invites attack and even a Grackle, with which most friendly association is maintained at other times of year, may be viciously set upon during the nesting season. On July 6, 1931, a pair of Red-wings even pursued a Bald Eagle which ventured to cross the meadows back of Five Mile Beach, where they had established a colony.

Red-wings formerly occurred in great abundance throughout the extensive cattail swamps that stretched away from Cape May to the Lighthouse and around the borders of the Race Track Pond and Pond Creek Meadows, but with the ruin of most of this territory by draining operations and real estate speculation, birds have sadly decreased in numbers. The Red-wings have not been entirely exterminated, as have some of the other marsh loving species, but persist in much smaller numbers on the edges of the old marshes where some available spots remain, and back of the town where some small ponds and marshy ground still are to be found. The Fill which has in the course of time developed some small cattail pools of its own, has been a favorite breeding place, but more recently the oiling of these waters has driven the birds away. The tall privet hedges and thickets of Japanese cane (*Arundinaria*) on the Physick place, which backs on the Fill, continue to be a favorite nesting place for the Red-wings as they are not so particular about building over water or wet ground, providing that such environment is not far away. Therefore we find both old and young frequenting the wet spots on the Fill after the latter are on the wing and also the lawns of the Physick garden and sheltered yards of adjoining or nearby properties, and coming up to the kitchen door in search of food. In Walker Hand's garden one male Red-wing came regularly to the chickenyard and entered a large coop through a hole in the top of the wirework to share the food furnished to the young chickens and he later brought his family to the spot. During July Red-wings, Grackles and Starlings come regularly to our garden for scraps of bread that are thrown out for them and to use the bird bath. A female Red-wing repeatedly carried bread from the ground to young birds on nearby telegraph wires which later came down and fed themselves. The Red-wings usually peck at the bread on the ground tossing it about this way and that until a piece is broken off, but occasionally they will hold it down with their feet as the Grackles always do.

A deserted nursery north of Schellenger's Landing, which flanks the salt meadows, is another favorite nesting place for Red-wings and here, too, they make use of the privet bushes while rows of English Holly are entirely neglected. The earliest date for eggs that I have is May 22, but by the 30th most nests have full sets. Richard Miller has found eggs as late as June 20 and I saw two nests with eggs in a cattail swamp on the Race Track Pond as late as July 3, 1920; whether there are two broods in a season I am not sure but am inclined to think that these late nestings are by birds whose first nests were destroyed. I have found young just hatched on June 9 and also as late as July 6, while full-fledged birds were on the wing and being fed by their parents by June 18, and other families, still in care of adults in thickets along the marsh edges, on July 7, 9 and 10 in different years.

Richard Miller has submitted a record of nests examined on Seven Mile Beach where the birds build mainly in bayberry bushes and generally in colonies:

> May 14, 1916, one nest with three eggs.
> May 18, 1924, three nests with four eggs, one with one.
> May 26, 1929, four with two eggs, one with three.
> May 30, 1911, one with one egg and one with two, others just begun.
> May 30, 1913, two with four eggs, one with two.
> May 30, 1915, one nest with three eggs just pipped.
> May 30, 1919, one with three young.
> June 2, 1929, two with three eggs.
> June 7, 1925, one with four eggs.
> June 22, 1924, one with three young.
> July 3, 1927, two with three.

Other Cape May County nests examined by Turner McMullen are:

May 23, five nests with eggs (Corson's Inlet).
May 25, 1924, four nests with eggs (Court House).
May 30, 1921, eleven nests with young (Cape May Point).
May 30, 1926, one nest with three eggs (Ocean City).
June 2, 1935, three nests with eggs (Ludlam's Beach).
June 7, 1934, one nest with four young (Ludlam's Beach).
June 8, 1919, one nest with four eggs and one with young (Cape May Point).
June 14, 1925, one nest with three eggs (Corson's Inlet).

Nests have been found in various locations besides the privet hedges. Bayberry bushes, marsh elders (*Iva frutescens*), rose bushes and tussocks of sedge are all used and I saw one in a clump of goldenrod six inches from the ground and another in a small cedar bush one foot up. The most attractive nests are those swung between cattail stalks over the water and I have seen similar ones in marsh elder bushes six feet from the ground swung from the branches like nests of the Orchard Oriole.

After the nests have been completed and contain either eggs or young the parent birds become exceedingly agitated upon the approach of an intruder and their vocal efforts are redoubled. As one approaches the nesting site the male launches into the air and begins to call *sheep; sheep, sheep; sheep;* each call separated from the next by an interval. Then as the excitement increases there is a long drawn *zeeet* interpolated irregularly thus: *sheep; sheep; sheep; sheep; zeeet; sheep; sheep; sheep; zeeet; sheep; sheep; zeeet* etc. the bird all the while poised on rapidly beating wings directly overhead, and now and then swooping down still closer. The female, arising from her perch on a cattail, has a similar note but less harsh than the *sheep* of the male, and she also utters a much more rapid and differently pitched series of notes; *chip-chip-chip-chip; chip-chip-chip-chip-chip;* etc., then both birds alight on a bayberry bush and call together, the female seeming to relieve the male entirely from the first part of his cry and to her repeated *chip-chip-chip-chip* etc. he contributes only the long drawn *zeeet* at regular intervals so that the combination is almost like his opening effort. Taking wing again the male varies his cry. It is now *tuk; tuk; tuk;* etc. at intervals of several seconds or perhaps better rendered *put; put; put;* with the same interpolated *zeeet* as before. Another bird had both of these notes and used them together, thus: *sheep; sheep; tuk; tuk; sheep; tuk; tuk;* etc., while his long wheezy note was so very thin as to be hardly audible a few yards away and sounded like a whisper. In the distance, where the birds were not disturbed I could hear the *kill—géeeze* and the liquid *lul-lul-lul-lul-lul-lul-lul-* with a slight pause after the first note. On another occasion I have recorded the male's alarm note as *tseeet; tseeet; tseeet;* and the wailing cry as *tzeeee* while the cry of the female as she takes wing was written *sh-sh-sh-sh-sh-sh* etc. I have heard the *kill—géeez* song as late as July 3; the gurgle to July 16 and 26 and the long drawn *zeeeet* to July 17 and once on August 3. As they sit on the bushes uttering their alarm cries both male and female give a spasmodic flirt to the tail with every call.

By early July many young Red-wings are collected in bushes along the edge of the marshes where they sit rather stupidly, with their legs wide spread, quietly pluming themselves and paying no heed to the frantic cries of their parents, occasioned by our approach. When they finally realize the presence of danger they utter a low *chuck; chuck;* and fly straightaway for some distance, generally flapping their wings rapidly and ascending at a rather steep angle. They seem to be anxious to gain considerable altitude and then make for some thicket or group of low trees to which they gradually descend. The launching forth of the young always caused renewed clamor on the part of their parents and sometimes they accompany them. Usually, however, they remain at the nest site, possibly to protect other young birds still hiding in the vicinity.

Bobtailed young were being fed by parents in bayberry bushes on the Fill on June 24, 1928, and full-fledged young birds on July 7, 1921, were also apparently under the care of the adults, as a female was present with a green grasshopper in her bill nervously flirting her tail. These young were very sluggish and remained in the bushes, only flying when approached quite closely. They were very red buff below with broad buff edgings to the plumage of the upper parts which easily distinguished them from the duller, more worn adult females. In a group of five young the males were distinguished by their larger size, being larger indeed than the adult female. A female on August 19, 1927, was still carrying food and showed much concern at my presence but I did not locate any young. If they were present, as seems likely, I think that they must have represented a second brood. In spite of these late observations of parental feeding I found young Redwings shifting for themselves and feeding about a stable on the edge of the town with no adults about on July 3, 1920.

From about the first of August, or earlier in some years, many Red-wings repaired to the roost at the Physick place at dusk, along with the Martins, Grackles and Robins, but they also gathered in the cattail swamps just behind the town and later in the season most of the Cape May Red-wings apparently went there to spend the night and were seen rising in large flocks early in the morning. During late August and early autumn similar flocks pass high overhead bound apparently for the shores of the Delaware or marshes near the head of the Bay. Julian Potter observed them coming into a thicket of wild rice and reed grass on Oldman's Creek near Pedricktown on August 10, 1935, at 6:40 p. m. and estimated that for three quarters of an hour one thousand birds came in every minute.

There may be days at this time when it is difficult to find any Red-wings immediately about Cape May and my notes show that only a single individual was seen during a morning's walk on September 6, 1924, while on October 1 and 2 of that year and September 25, 1927, there were none at all.

When the molt is completed the Red-wings have practically all left to form the characteristic winter flocks, but during its progress they have certain haunts in which they may be found regularly. One of these is a thicket at the head of the Lighthouse Pond where they flush from the muddy flats surrounded by cattails and resort to the low trees, peering down at the intruder with little thought of danger. On August 23 the old males in this gathering were in all stages of molt and on August 21, 1922, a young bird was seen with a newly acquired red shoulder patch although his wings and tail were still in the brown streaked juvenal plumage.

On September 9 and 10, 1921, a flock still resorting to the Physick roost used to stop every evening to feed in a patch of neglected corn close to my window. There were in addition to the Red-wings many Grackles, Cowbirds and Starlings. The Red-wings were for the most part tail-less, adult males in full molt. Every now and then they would jump at one another or at one of the other birds like little game chickens. This corn patch was finally cut down and after visiting the site the next evening, and eating the grain on the ground, the flock came no more. We must give them credit where due, however, for not only do they destroy many a grasshopper but I have seen a Red-wing busy on an ear of corn pull out and devour a large fat caterpillar that was feeding on the grain.

Red-wings do much damage to patches of corn that are planted in out of the way places or close to the edge of the meadows. The husks are ripped down and the exposed grain devoured while still in the milk. Earlier in the year they feed on open mud flats often in company with Killdeers and Spotted Sandpipers, retiring to the surrounding cattails upon the approach of danger. In the same way those that nest near the salt meadows will drop down into the grass to feed and fly up to the thickets where their nests are located. These birds are, I think, searching for insects or minute crustacea and those that I see hovering over the surface of the Lighthouse Pond, with the swallows, must be seeking a similar diet. When they have gathered in flocks the Red-wings frequent the grain and hay fields both before and after harvest and haymaking.

I have never noticed Red-wings visiting the sea beach for food as the Grackles do but they come to the dunes and hover like Sparrow Hawks catching grasshoppers in the dune grass and flying off with them to feed their young. Red-wings have always been shot in the autumn when the great flocks are formed, and by erecting poles and branches on the marshes near to his blind, the gunner was able to kill three or four dozen at a shot according to "Homo" writing in "Forest and Stream" in August 1881. He adds that dead birds were stuck on the branches as decoys.

The Red-winged Blackbird is an abundant summer resident throughout the state and becomes resident at many places along the shore or in the river valleys, while many of the birds that make up the great flocks of autumn must, I think, come from much farther north, finding their goal, or winter quarters, in southern New Jersey, Delaware and Maryland.

ORCHARD ORIOLE

Icterus spurius (Linnaeus)

Two or more pairs of Orchard Orioles nest regularly about Cape May Point and in some years there is a pair established in the heart of Cape May while back in the farming district to the north they are of regular occurrence in orchards. Walker Hand has recorded their arrival in spring as follows:

1907. May 4.	1912. May 1.	1915. May 4.
1908. April 21.	1913. May 3.	1917. May 6.
1910. May 8.	1914. May 5.	1920. May 1.

On the southward migration they seem to leave early and my latest dates are August 28, 1922; August 21, 1923; August 24, 1924; August 26, 1926; August 25, 1927.

During May and June the Orioles are in full voice, a rollicking song somewhat reminiscent of the Bobolink. A fine old chestnut and black male used to sing continually from the topmost branch of a giant silver poplar on Washington Street near the Martin roost, during June of 1925, and once at 8 p. m. I heard the song coming from the roost itself. At the Point I have heard a bird singing as late as July 11. On May 27, 1929, I listened for a long time to a singing bird in an orchard in West Cape May and after recording and checking it a number of times I regarded the following as an exact syllabic representation: *teétle—to—wheéter-tit-tíllo-wheétee, chip, chip, cheer.* The song varies with different individuals, however, and one that I heard at the Point, on July 6, 1927, called *choop, choop, choolik* as if trying to start a song and failing in the effort.

By July 1 in some years the young are on the wing and we soon find family parties traveling along the fence rows and hedges ahead of us, the olive yellow young very quiet but the parents much excited, flirting the tail up and down but never spreading it. One female on July 18, 1923, was particularly disturbed and kept up an almost continuous *chi-chi-chi-chi-chi*, etc. accompanied by exaggerated tail action. Sometimes these parties are headed by a chestnut and black male and sometimes by an olive green one with a black throat, a bird in its first breeding season. Such individuals have been seen quite as frequently as the older chestnut and black ones. The young birds are fed by the parents for some time after they are perfectly capable of finding their own food. Even as late as July 12 and August 7, in different years, I have seen a female feeding young and in the latter case the young bird rested quietly in a bush and awaited the visits of the parent. During August I have seen young Orchard Orioles perched on the large red flowers of the trumpet creeper feeding on something that they picked from the outside of the corolla near the base, probably minute insects. I have also seen them feeding among the flowers in the garden.

A nest with five eggs was examined by Turner McMullen at Cape May Court House on June 18, 1922.

While a summer resident in most parts of the state the Orchard Oriole is much more plentiful in the southern counties.

BALTIMORE ORIOLE

Icterus galbula (Linnaeus)

The Baltimore Oriole is a regular transient during late August and early September and much less numerous in spring. The south-bound birds are partly gray-coated young of the year and partly old birds in burnt orange plumage. They occur in some numbers in the orchards and shade trees of the town and country and in moist woods at the Point, where I have seen them feeding on the trumpet creeper flowers like the Orchard Oriole, apparently seeking small insects. In May, 1928, there was an unusual number present, four being seen at a time in the trees on Washington Street.

There are reports of the Baltimore Orioles nesting in tall trees in the town in former years, but I have been unable to verify this and they certainly have not done so in the past twenty years.

Dates of spring occurrence are:

1924.	May 10.	1932.	May 1.
1925.	May 10.	1934.	May 12.
1930.	May 6 and 15.	1928.	May 12 and 15.
1931.	May 9.		

In autumn:

1922.	August 21.	1930.	September 4.
1924.	September 4, 5 and 12.	1931.	September 1, 2, 3, 5.
1927.	August 25, 26.	1932.	August 29, September 17.
1928.	September 1, 5, 8.	1934.	August 27.
1929.	August 30, 31, September 2, 3.	1935.	August 29, September 7, 8.

Exceptional dates are single birds seen on September 21, 1925, at the Point by William Rusling and on October 5, 1929, at Cold Spring, by myself.

I have seen twenty-five feeding on apple boughs in an orchard back of our cottage on August 29, 1932, while William Rusling saw 125 passing at Cape May Point at the height of migration on September 7, 1935.

It is a transient in the Pine Barrens and southward, and a summer resident only in the northern counties and the upper Delaware Valley.

RUSTY BLACKBIRD

Euphagus carolinus (Muller)

Curiously enough I have been unable to find a Rusty Blackbird at the Cape although there are several sight records by others. On the Delaware Meadows below Philadelphia, they occur regularly in both spring and autumn, especially at the latter season when they are present in flocks, often mingling with the abundant assemblages of Red-winged Blackbirds which at that time are feeding in the rank growth of weeds and coarse grasses. Single individuals have also been seen in winter with flocks of sparrows.

I have carefully examined many mixed flocks of blackbirds about Cape May in autumn and winter in the hope of finding a Rusty but without success. Julian Potter, however, saw two near Cape May Courthouse on December 24, 1935, which is our only definite record for the county.

BOAT-TAILED GRACKLE

Cassidix mexicanus major (Vieillot)

Philip Laurent states (Ornithologist and Oologist, 1892, p. 88) that "Two birds of this species made their appearance on Five Mile Beach in company with a number of Purple Grackles; one was shot by Samuel Ludlam who had it mounted." This, so far as I know, is the only occurrence of the Boat-tailed Grackle in New Jersey. It is a southern species and until quite recently was not known to breed north of Virginia. It is everywhere a coastal bird and I have seen large flocks on the marshes at Chincoteague, Virginia, composed of the big black males and the smaller brown females.

In 1930, Herbert Buckalew saw one of these Grackles near Milford, Delaware, and in 1933, found a pair near Cedar Beach on the shore of Delaware Bay, opposite Cape May, while on May 5, he found four pairs present and located a nest with three eggs. Young birds were on the wing by June 18. On May 30, 1934, I visited the spot with other members of the Delaware Valley Club and saw the birds as well as a nest containing young. This colony was doubtless the source of the birds that visited Five Mile Beach.

PURPLE GRACKLE

Quiscalus quiscula quiscula (Linnaeus)

Purple Grackles may be seen at Cape May in every month of the year, though they occur less frequently, or with less regularity, during November, December and January, and it is by no means certain, or even probable, that the same individuals are permanently resident.

There may be a few Grackles about the town in winter, in spots where food and shelter are to be found, or there may be scattered individuals in the great roving bands of Red-wings, Cowbirds and Starlings that course over the bare fields, while occasional flocks composed entirely of Grackles visit the vicinity of the Cape in the heart of winter often following a snow storm that has driven them from some favorite feeding ground. A flock of several hundred birds arrived on the meadows back of South Cape May on the day following such a storm in mid-January, 1922, and drifted about in search of bare ground and then disappeared, probably moving farther south. On December 26, 1927, Julian Potter saw a flock of some five hundred birds and on December 27, 1929, Otway Brown tells me that upwards of one thousand flew over Cold Spring, traveling northward in detached flocks and long lines, taking some fifteen minutes to pass. During a snow storm a few days previously, probably this same flock devoured practically all of the corn left on the cob in a nearby field. On December 26–27, 1932, a flock of seventy-five remained about Cold Spring roosting in a maple swamp but later disappeared.

The variability in the number of Grackles present in the Cape May district in winter is shown by the counts on the Christmas census of the Delaware Valley Ornithological Club which are as follows:

1927.	December 26, 500.	1932.	December 26, 450.
1928.	December 23, 121.	1933.	December 24, eleven.
1929.	December 27, 1,000.	1934.	December 23, five.
1930.	December 30, three.	1935.	December 22, seventeen.
1931.	December 27, fifty.	1936.	December 27, sixteen

The possibility of many of these winter Grackles being Bronzed Grackles is discussed elsewhere. Dr. Frank M. Chapman, who has made an exhaustive study of the Grackles, considers that there is a recognizable race intermediate between the breeding birds of New Jersey and the Bronzed and that the name proposed for the former belongs to this. He has therefore named the New Jersey bird Stone's Grackle (*Quiscalus q. stonei*) (cf. Auk, 1935, pp. 21–29).

It is mid-February before the regular northward migration of Grackles at Cape May begins and we find them present continually either at their usual haunts or passing overhead in small or large detachments. Even then weather conditions have much to do with the time of their arrival.

Walker Hand's records of arrival of migrants at Cape May to 1917 and those of others in later years are as follows:

1904.	February 29.	1910.	February 23.	1918.	February 27.
1905.	March 5.	1911.	February 27.	1921.	February 10.
1906.	February 22.	1912.	February 27.	1923.	February 22.
1907.	February 2.	1913.	February 26.	1926.	February 21.
1908.	February 21.	1915.	February 15.	1932.	February 28.
1909.	February 13.	1916.	February 18.	1933.	February 19.
		1927.	February 22.		

In 1934 and 1935 when February was marked by sub-zero weather with much snow the Grackles did not put in an appearance until March. Even after the flight is well underway a change in weather may halt them or drive them back again. So on March 10, 1924, a storm covered most of the peninsula with a deep coating of snow and a flock of one thousand Grackles settled down in the orchards at Cold Spring and remained for an hour or more as if uncertain where to go. From March on, however, Purple Grackles become a regular and conspicuous feature of the bird life of the Cape. Like the drifting assemblages of winter these early spring flocks are often associated with Red-wings, Cowbirds and Starlings and one gathering, resting in a field near Schellenger's Landing, on March 25, 1923, contained in addition a Meadowlark, a Flicker and some fifty English Sparrows.

While the transient flocks of March are passing to the northward smaller groups of Grackles establish themselves about the nesting sites of former years—in the shade trees of the Physick place on Washington St., in a grove of pond pines at Cape May Point, and about old farm houses a little farther north where they seem particularly fond of evergreens. The actions of these nesting communities are interesting. The flock at the

Point is constantly changing its position; now the birds are streaming out in a long line bound for plowed ground near the Lighthouse to feed, or out to the edge of the salt meadow, where the last year's crop of black grass was cut, and then suddenly back they go to the pine grove. As they approach the trees they sail down on set wings which form a triangular, kite-like outline, with the long tails of the males deeply depressed into the characteristic boat or keel. As early as March 13, many of the Grackles are flying in pairs, the male just behind the female and at a slightly lower level. They are noisy, too, about the nest trees and there is a constant chorus of harsh alarm calls; *chuck; chuck; chuck;* like the sound produced by drawing the side of the tongue away from the teeth, interspersed with an occasional long-drawn, *seeek*, these calls being uttered by birds on the wing as well as those that are perching. Then at intervals from a perching male comes the explosive rasping "song" *chu-séeeek* accompanied by the characteristic lifting of the shoulders, spreading of the wings and tail, and swelling up of the entire plumage.

As early as March 5 I have seen evidence of mating and sometimes two males have been in pursuit of a single female, resting near her in the tree tops, where they adopted a curious posture with neck stretched up and bill held vertically. On April 1 the birds in the Physick yard were in pairs, the males going through their courtship "song" but every now and then they all flew off together in true gregarious fashion although they always separated in pairs on their return.

By April 21 (1921) there were flimsy skeleton nests in the pine tops at the Point but even though nest building was under way the flock would often stream away to the feeding grounds. Grackles are walkers and as they stalk about the ground they carry the tail, at this season of the year at least, cocked up at an angle of 45° and the wings of the males slightly drooped. The head is bent over toward the ground and they turn it continually as they walk deliberately about. Sometimes one of them will turn his head completely around so that he looks back over his shoulders or, pointing the bill up nearly vertically with neck still twisted, he will gaze intently at some other bird passing overhead. Those feeding on the marsh were observed to jump a few inches clear of the ground every now and then, as if something below had alarmed them.

While I have no data on time of laying at Cape May, three nests seen by Richard Miller on Seven Mile Beach, contained six, four and six eggs on May 12, 1934, and two in Salem County, May 4 and 9, held four and five respectively. Farther north in West Jersey dates for complete sets range from April 20 to May 12 and the sets seem to be normally six or five eggs.

By May 20 the young in the colony at the Point had apparently hatched as the adults were flying back and forth from the feeding grounds to the nests, individually or in small bunches, and there seemed to be always Grackles in the air. Those nesting in Cape May evidently had young by June 21 in 1932, in the trees or still in the nests, as they were constantly coming in with green cherries, grasshoppers etc.

The birds of this colony may seek food on the Fill, in the gardens, or on the streets, and especially about the refuse dumps, while some individuals regularly visit the sea beach and the shores of the Harbor, and even as late as July, I have seen birds coming from inland colonies to the beach for food and returning northward with their bills full. They usually fed on the beach early in the morning and on July 19, 1925, there was a flock of thirty so engaged. They walked down onto the wet beach, and when shallow water flowed over their feet, they merely fluttered their wings or hopped up into the air for a moment. Larger waves caused them to run clumsily or fly a few feet up the beach but they came back after the retreating wave in true beachcomber fashion. On June 26, 1932, there were six feeding on the beach; they walked into the shallow water with their tails elevated and allowed them to wag from side to side as they progressed. One of them caught a small crab and retreated to the hard sand to eat it. In June I have seen them taking cherries from trees in the town and carrying them to the house tops to eat. On July 16, they were catching May beetles on the golf links, hopping vigorously with a peculiar rushing gait, in an attempt to overtake the flying insects. And once a Grackle was seen pursuing a flying beetle on the street, an unusual performance; the bird was exceedingly clumsy in turning on the wing and after following its erratic prey for several minutes without result it gave up the chase. On August 31, several Grackles were observed darting up in the air from the tree tops in pursuit of flying ants in which activity they also proved very clumsy.

Favorite feeding places visited by both old and young Grackles, until well into the summer, are the piggeries on the edge of the meadows where scraps of refuse food are always to be had. Many Grackles in the town also learn to come to gardens where scraps of bread are thrown out for them and for some years I have seen the birds in July and August, pick up bits of bread and deliberately fly up onto our bird bath and after soaking them for a few moments fly off with them to their young or eat them on the spot. Later the young came to the garden and were fed on the wires and fence tops, fluttering their wings and begging for food, although perfectly able to pick it up for themselves. I have seen this food solicitation on the part of full-fledged young on the Fill as late as June 25 and July 12, in other years.

Young are out of the nest by the early part of June and one family was seen as early as May 22, 1927. On June 30, 1920, thirty young, with perhaps a few adult females, were seen in a cattail swamp on the edge of the town holding their bills wide open as they often do on very hot days. There were also scattered bunches of old males, three to six together, feeding by themselves or passing overhead. Later in the day the entire colony from the Point, old and young to the number of sixty, joined forces and located in a nearby cornfield from which came their familiar *chuck, chuck*, though the "song" of springtime had long since been silenced. On July 3, the same flock was in a grassfield a little to the west where they were joined by four Domestic Pigeons, all of them feeding on the ground. On July 26, what was presumably the same gathering was feeding with Red-wings and Starlings in a field of new cut hay along the railroad, where grasshoppers and other insects abounded. When disturbed the Grackles always took to the thickets and hedgerows while the Red-wings resorted to a nearby cattail swamp. On another occasion the Cape May Point flock was feeding on the shores of a pond dug in the dunes of the Bay side where sand had been removed.

The flocks of old and young drift farther afield as the summer advances in search of new feeding grounds, although some few may remain about the refuse dumps during August, and some old males may be seen on shady lawns along Washington Street. Here they walk about looking particularly sleek in their glossy plumage. As we view them from the rear their legs stand wide apart and they seem almost bowlegged. They are busy investigating the dead leaves, turning them deftly with the bill and tossing them to this side or that to see what prey may be lurking beneath, their yellow eyes gleaming all the while.

By August 10, foraging flocks have scattered so widely that there may be days when no Grackles are to be seen from dawn to dark but at night they always come in to roost with the Martins, Robins, Cowbirds etc., which form the great summer roost of Cape May. For years this was located on the Physick place, then in West Cape May, and lately in the bayberry thickets of the Fill. The Grackles come in as early as July 1, if not earlier, apparently as soon as the young are on the wing, and probably the first to gather at the roost are birds that have bred in the vicinity, but as the season advances the roosting movement becomes more pronounced and definite. Early in the season the birds generally arrive from 7:00 to 7:30 p. m., while by mid-August they come in as early as 6:00 p. m. the actual time however varies considerably according to weather conditions and they always gather earlier on dark rainy days. Many Grackles, especially local birds, gather on the golf links or in nearby cornfields or orchards before entering the roost to feed on pears, or other fruit, or corn on the stalk. A neglected corn patch close to our cottage on Queen Street was a favorite rendezvous for flocks bound for the roost in early September 1921. Here Red-wings and Cowbirds, old and young in all sorts of molting plumages, gathered every evening for a hasty attack on the stunted ears of corn. With them were molting Grackles, some brown with patches of glossy black feathers coming in on the breast, others with scarcely a tail feather left and others, still, with necks almost denuded. Now and then they sprang at one another like miniature gamecocks when one encroached on the food of another. Flocks arriving at the roost from more distant points seemed to come at a higher altitude and when over the trees plunged precipitately down and at once took up their position for the night. On August 18, 1921, and succeeding evenings, I watched them from a location on the Fill which commanded a wide view from north of Schellenger's Landing to the roost. The birds could be detected coming in from far away to the north as tiny specks, then the wing action could be distinguished and finally they streamed past, one long flock after another. From 6:15 to 6:42, 854 Grackles entered the roost, there were 326 in the first five minutes, 150 in the next, and 297 in the next, making 773 in fifteen minutes. On August 20 they began to arrive a little earlier and I missed a part of the flight but from 6 to 6:40, 810 birds were counted. They flew much lower than on the previous evening and paused in the trees about the Landing. The next night the wind changed from southwest to northeast and not a bird came in from the north, like the other roosting birds they seemed always to come in against the wind.

In the morning they leave the roost at daybreak, or before, and Walker Hand has seen Grackles feeding under the electric lights on Washington Street before they were turned off at dawn. Often, especially on rainy mornings, they linger about town and on September 15, 1921, at 6 a. m., I saw about a thousand gathered on the golf links which rolled up in a great sheet of birds onto the wires and trees on Lafayette Street. Grackles continued to come to the roost until the middle of September but by October 1 all had left. Probably the presence of foliage on the trees has something to do with the length of their occupancy. The roost was entirely abandoned in 1936.

The progress of the molt in Grackles can easily be noted by the appearance of the wings and tail as the birds fly overhead, although the new and old body plumage of the adults are the same. They show gaps in the flight feathers as early as July 18 and some are still molting as late as September 8, 11 and 16 in different years. When the tail molt begins the long central feathers drop out first so that the tail appears split or forked, this gap becomes wider as successive pairs of feathers are lost, but by the time the outer pair is dropped the new central feathers have grown out and the outline of the tail is pointed or wedge-shaped. Young birds show a similar molt of the flight feathers and in addition experience a distinct change in the color of the body plumage, from dull brown to glossy black, and birds were seen on August 10, 24 and 17, in different years, with large patches of the new plumage supplanting the worn sooty juvenal dress. Whole flocks of birds in this transition plumage have been seen walking about on the lawns of the Physick place early in the evening before taking up their position in the trees for the night. There is considerable variation in the color of the glossy feathers of the head and neck and I have seen birds showing blue, green and purplish reflections all in the same flock. One albinistic male that visited our garden in July 1927 had the secondaries grayish white.

Grackles are of regular occurrence in flocks of varying dimensions about Cape May until the middle of October, at least, but soon after that they pass on to the southward and we have left only the shifting winter representatives. These seem to be more abundant in West Jersey toward the Delaware River and northwest of the Cape May Peninsula, and the great autumn and winter flocks seem to linger there longer, possibly finding more abundant food along the river swamps and their tributaries than in the pinelands farther east, and the occasional Cape May flocks may drift over from this area. On November 28, 1926, I saw a great flock near Elmer which contained many thousand birds. They covered the ground in great black sheets, the rear ranks constantly arising and flying over to take their place in the van which gave the impression of rolling over the ground. When they took wing in force the long procession streamed past shutting off from view all that lay beyond and when they alighted in the trees the bare branches appeared to be clothed with a dense black foliage.

With the departure of the October and November flocks from the Cape May region the Grackle becomes no longer one of the characteristic birds of the town until on some fine morning in February we awake to hear the familiar harsh call and find them once more in the tops of the ornamental spruce trees in the gardens of Washington Street or in the pines of the Point and when we once more listen to their guttural attempts at song and witness the grotesque swelling up of the body plumage we are apt to forget their ravages of the cornfield as we welcome them back and delight in the familiar voices which usher in another spring.

BRONZED GRACKLE

Quiscalus quiscula aeneus Ridgway

This northern and western race of the Grackle has been detected several times in the Cape May region in autumn and winter. On March 11, when the ground was covered with snow, a dozen or more Grackles which had been perching in the orchard at Cold Spring came down into the driveway to feed on the bread that had been thrown out for the birds. They were carefully studied from a window of the house at a distance of twenty feet or less and every one proved to be a Bronzed Grackle. It therefore seems likely that a fair proportion of the wintering Grackles of southern New Jersey, possibly all of them, belong to this race.

When we studied the birds referred to above we could distinguish at a glance the brassy bronze color of the back and rump sharply separated from the greenish or purplish blue of the head. The Purple Grackle on the other hand has the back bottle green and the rump purplish, both with iridescent bars or spots, making a varied or broken pattern in sharp distinction to the plain, uniform bronze of the other bird.

The Bronzed Grackle, the bird of the Mississippi Valley, has extended its range eastward to the coast of New England and far to the north, and in autumn a number of these birds come southeast of the Alleghanies on migration and evidently winter with us.

COWBIRD

Molothrus ater ater (Boddaert)

Throughout the year Cowbirds may be found about Cape May. In April and May we come upon single birds or bunches of two or three, sometimes several males in pursuit of a single female. They usually are seen flying overhead or perched on the tops of trees but sometimes they are on the ground and on April 24, 1921, I came upon two pairs feeding in a plowed field. On May 7, 1933, I saw a solitary female skulking about the shrubbery in the Physick garden evidently seeking nests in which to deposit her eggs and on May 9, 1931, another female similarly engaged was closely attended by a male. Cowbirds' eggs have been found in the nests of several species of birds about Cape May. A deserted Field Sparrow's nest found on May 27 contained one as did a Parula Warbler's nest, on May 26, 1892, and I have been puzzled to know how the Cowbird managed to deposit an egg in such a delicate structure with only a small side entrance (pl. 117, p. 855).

The Prairie Warbler is another favorite victim of the Cowbird and Turner McMullen has examined nests of this little bird each of which contained an egg of the parasite—on May 31, 1934, four eggs of the warbler; May 24 and 27, 1925, with five eggs. He also found a nest of the Hooded Warbler with a Cowbird's egg on May 27, 1928.

Julian Potter came upon a young Cowbird in the woods at the Point which was calling lustily for food on July 4, 1928, and on August 3, 1927, and again on August 5, 1929, in the same woods, I found a full-fledged young Cowbird following a pair of White-eyed Vireos, which were busy supplying him with green caterpillars. On August 5, 1933, I was surprised to find a pair of these Vireos, also at the Point, busy feeding two young Cowbirds which were apparently full-grown and perfectly able to fly. As early as August 2 in 1934, four young Cowbirds were found feeding in a plowed field which showed scattered black feathers beginning to appear in the gray brown juvenal plumage and feeding with them were three adult males. On August 22, 1929, a flock of forty young were found in a rag-weed field back of the town which were feeding together on the ground and which flushed in a compact flock, flying for three or four yards and settling again. They were in true flock formation and arose and wheeled to right or left like Starlings or sandpipers. Their wings appeared curiously translucent as the sunlight struck them and about half of them were males showing black feathers here and there in their plumage; the rest were either females or males that had not yet begun to molt. They were pecking at the ground and apparently feeding entirely upon grass or weed seed. It is remarkable how these young Cowbirds, which are widely scattered when they leave the nest, and which have been reared by fosterers of different species and habits, are able so quickly to find and recognize their fellows and form into definite flocks. Other flocks of molting birds both young and adult have been found on August 18, 1921; August 28, 1924; August 17, 1925; on lawns on Washington Street in the early evening where they were waiting to enter the Grackle and Martin roost for the night. Another large flock of very much mottled individuals was seen in a barn yard near Erma on August 31, 1928, and a number of birds about half-molted used to come every night with Red-wings in a similar condition to a neglected corn patch close to a window in our cottage during the first week of September 1921.

Flocks of Cowbirds were seen entering the roost as early as July 5, 1923, and July 11, 1927, and they continued to do so until September 29, sometimes numbering one hundred or more to a flock. After the first of October most of the other roosting birds had departed and the foliage of the trees had to a great extent fallen, but a flock of twenty-five Cowbirds continued to come in at night as late as October 18 and joined some Starlings in a row of Norway maples on the sidewalk which still retained their leaves. Some of the birds also roosted on the tops of nearby houses. In the summer flocks we find both males and females and sometimes young mingled indiscriminately, although there is a tendency for the young to flock separately at first, and for the sexes to separate later in the season.

During the daytime these Cowbird flocks are usually found in attendance on cattle in the open fields or cowyards close to the barns. I have observed them so associated from early July to the end of October, and on two occasions, in April, when other Cowbirds were still flocking in the fields with Starlings and Red-wings. I think that the reason for not seeking the companionship of the cattle from November to March is due to the fact that the stock is not turned out in the fields during the cold weather, and not to any disinclination on the part of the birds. The association is peculiar and has generally been explained on the ground that the cattle, as they walk along and browse, stir up insects upon which the birds may feed. As a matter of fact, however, after many careful studies of such flocks with the binoculars at close quarters, I have been unable to detect the birds in the act of catching any insects nor have I seen the cattle stir up any form of insect life from the short grass of the pastures. On one occasion (October 8, 1932) when I was but a few yards away, I could see that the feet of the cows were covered with flies but the Cowbirds continued to peck at the ground and paid not the slightest attention to them nor were the flies disturbed by the progress of the cows. Again, on April 27, 1931, when a strong cold wind was blowing, there was no trace of insects on the ground of the pasture yet the Cowbirds were there feeding apparently on grass or weed seeds as usual.

While the birds sometimes walk quite close to the nose of the browsing cow, or to its feet, I could not see that they derived any benefit from the association. Most of the birds, moreover fed some feet away and seemed to pay but little attention to the activities of the cattle, except that, when the latter were slowly walking along as they browsed, the birds would walk in the same direction and if outdistanced they would fly on a few yards and settle again in front of the cattle. Sometimes I have seen Cowbirds in attendance on cattle that were lying down in the field, in which cases they wandered off from them and fed several yards away. Once I saw a small flock in attendance on a horse that had been turned out to pasture.

When on the ground the Cowbirds always walk and both males and females hold their tails more or less erect, sometimes almost vertical with wings slightly drooped or with the tips below the base of the tail. The appearance of one hundred or more of these rather small birds walking about with tails in the air is striking. I have never seen a Cowbird alight on the back of a cow as they are reported to do in some localities in the Middle West. Starlings often associate with Cowbirds in the fields or cowyards where cattle are gathered, having apparently learned the habit from them.

After the cattle have been retired to their winter quarters the Cowbirds continue to feed in the open fields throughout the cold weather. There is usually one large flock numbering perhaps, five hundred individuals which shifts from place to place between Cold Spring and Cape May and with which numerous Starlings, Red-wings and Meadowlarks associate, while smaller bands of Cowbirds may be found in large flocks of Red-wings or Grackles that now and then drift across the country.

On January 19, 1924, a flock of one hundred and fifty was found in fields just behind the town which contained a single pure white albino. This flock would fly up into the trees that grew along the fence row, or onto the telegraph wires, and later drop back to feed on the ground. The birds were still there on March 19 having moved but little during the winter. Snow, when heavy enough to cover the ground, disturbs the Cowbird flocks not a little and on March 11, 1934, a number of them were forced to repair to the orchard on Otway Brown's place at Cold Spring and flew down close to the house to feed on scraps that we threw out to them. A heavy snow on February 5, 1933, drove a flock of five hundred Cowbirds out on the beach at Cape May Point, about the only spot in the entire countryside where bare ground could be found. These winter flocks seem to be made up of both males and females although one sex or the other will sometimes predominate. While Cowbirds are common in Cape May County in winter their occurrence farther north in the state at this season is more or less sporadic but they have been recorded at Princeton, Plainfield, Yardville, etc.

The young birds in juvenal plumage are much brighter and browner than the drab gray females and have all of the feathers bordered by rich buff, while they are somewhat streaked below. They have acquired the adult plumage, however, by early September.

SCARLET TANAGER

Piranga erythromelas Vieillot

A regular but not abundant transient, apparently more plentiful in the autumn flight. In the pine woods of the Point the spring males, in their gorgeous scarlet, stand out most conspicuously against the dark foliage though I have never seen more than one or two at a time. Philip Laurent saw a few in the spring migration on Five Mile Beach in the eighties and Fletcher Street saw one in spring on Seven Mile Beach in 1916. We have records for spring arrival as follows:

1928. May 5.	1929. May 9.	1934. May 5.
1924. May 10.	1932. May 5.	

and earlier records by Walker Hand.

In the autumn they have been seen on:

1913. October 13.	1925. September 26.	1933. October 28.
1924. October 2.	1929. October 5.	1934. September 19.
	1932. September 17–18.	1937. August 22.

We do not often see old males in autumn and one that I observed on October 8, 1932, is therefore of special interest. It was very bright yellow olive with jet black wings in contrast to the young of the year with darker green back and wings and only the shoulders black.

The Scarlet Tanager occurs as an occasional summer resident farther north in the peninsula, since it has been seen on May 30, near the Court House by Richard Miller, while two were detected in the woods between there and the Bay Shore Road and one near Seaville by Fletcher Street on June 7, 1936. In the northern half of the state it is a more common summer resident.

SUMMER TANAGER

Piranga rubra rubra (Linnaeus)

The Summer Tanager seems to have been very much more plentiful in southern New Jersey one hundred years ago than it is today and we have but meager information on its decrease. Alexander Wilson writing in 1807 says: "In Pennsylvania they are a rare species, having myself sometimes passed a whole summer without seeing one of them; while in New Jersey, even within half a mile of the shore opposite the city of Philadelphia, they may generally be found during the season (May to August)." He adds: "it delights in a flat sandy country covered with wood, and interspersed with pine trees * * * I have frequently observed both male and female, a little before sunset, in parts of the forest clear of underwood, darting after winged insects, and continuing thus engaged till it was almost dusk." He also alludes to its fondness for "whortle-berries" and for "humble bees."

In 1857, Beesley gives it as a rare breeder in Cape May County and in 1869 Turnbull lists it as rather rare. George N. Lawrence (Ann. N. Y. Lyceum of Nat. Hist. VIII, p. 286) states that he found it as late as 1866 in magnolia swamps near Atlantic City, but no farther north. Charles C. Abbott writing in 1868, says that up to 1850 it was as abundant as the Scarlet Tanager but that he had seen no nest since 1855 and no bird since 1862. His statements are rendered unsatisfactory by his later mention in 1870, that it was abundant until 1857 (Amer. Nat., IV, p. 536) but he still later records a pair nesting near Trenton in June, 1884! John Krider states that he had found the nest in New Jersey in former years. Thurber (1887) mentions it as an accidental visitant at Morristown in the northern part of the state and Babson (Birds of Princeton) mentions one taken by W. E. D. Scott at Princeton on August 5, 1880.

Mrs. Cordelia H. Arnold saw a Summer Tanager at Somers Point during a great flight of spring migrants on May 8, 1935. She writes: "Our greatest find was a young male Summer Tanager. He had much bricky red but his wings were yellowish green. We had near views for at least five minutes. On May 9, he was near the same spot. We studied him a long time and finally came away and left him. He was singing a low throaty song. In the twenty years that I have been hunting for birds at Somers Point never before have I seen a Summer Tanager." This is the only recent record for southern New Jersey.

Whether it was as plentiful as Wilson infers may be open to question as it is difficult to explain how such a bird should have been practically exterminated. We have a few records from the Pennsylvania side of the Delaware in more recent years and the bird still occurs in Delaware and along the Choptank River though even there it does not seem plentiful.

CARDINAL

Richmondena cardinalis cardinalis (Linnaeus)

Cardinals are common residents all over the Cape May Peninsula where swampy thickets overgrown with wild grape, or congenial wood edgings, afford them satisfactory shelter. They will also come close about the farmhouses to pick up food thrown out for the chickens and we hear the clear whistle of the male from the trees in the orchard. There are usually several pairs at the Point and others in thickets bordering Pond Creek Meadows and in the dense tangle surrounding the buildings at Higbee's Beach. There is usually a pair nesting in one of the gardens along Washington Street in Cape May and Miss Mary Doak informs me that in 1935 she was sure that three broods were raised by the pair that nested back of her house and whose activities she had watched carefully. Stray cats left behind by departing summer visitors to the resort have on one or two occasions been responsible for the death of one or more of the Cardinals of the town and for the rest of that year none will be in evidence. In the summer of 1936, a small boy with a gun destroyed the male Cardinal of the local pair but the female continued to come to our yard and eat the scraps of bread that were thrown out, and eventually a male appeared from somewhere.

The Cardinals of the town become quite tame but the birds of the countryside are very secretive and hold pretty well to their cover and when one is fortunate enough to come upon them unawares they leave precipitately the moment they detect our presence. Sometimes we may come upon a pair feeding out in the middle of a road, usually in the early morning, but this habit is becoming rarer as the delightful old shady sand and gravel roads are being replaced by concrete "boulevards" upon which the birds find little but spots of gasoline or oil!

I came upon one female Cardinal in the heart of a thicket busily picking at some sumac berries. She was very active and with each movement would spread her tail and flirt from side to side. One fine day in March, 1932, Otway Brown and I entered a sheltered cornfield near Cold Spring in which the crop had proved a failure and had never been gathered, while many of the stalks had broken or fallen down. There were no less than twelve Cardinals busily engaged in eating the corn from the ears thereby demonstrating their right to the sobriquet of "corn bird" bestowed upon them in parts of the South. They made a wonderful display of color as we saw them against the dark swamp which bordered the field. Usually however when we see a male Cardinal flying from one thicket to another or venturing a flight in the open it is surprising, especially when the light is behind him, how inconspicuous his brilliant color appears. Indeed he seems to be only red when there is a suitable background to show him off.

Otway Brown tells me that, before the law was enacted forbidding the caging and sale of Cardinals, there was a woman at Cold Spring who raised Canaries and who made a standing offer to the boys of her neighborhood of two dollars for every male Cardinal that they could catch. What she received for them from dealers in Philadelphia was not revealed. The boys made little criscross cages of light cedar splints which were used as deadfalls, the trigger being baited with the end of an ear of ripe corn and grains of corn scattered about. They also had a pair of buckskin gloves handy with which to take the bird from the trap as the Cardinal's powerful beak can inflict a bite that will be remembered. Many of the birds that they caught were females which had no market value whatever.

The Cardinal has quite a repertoire but all of his vocal efforts come under the head of whistles rather than songs. There is the loud emphatic call—which I have recorded as *whoit, whoit, whoit*, often followed directly by the longer drawn out *cheer, cheer, cheer*, and sometimes a bird utters quite a different call *cheedle, cheedle, cheedle, cheedle*. On one occasion a bird called rapidly and continuously *whit, whit, whit, whit, whit*, etc., like the Flicker's rapid call, while another had a very low modification of the *cheer* call—*pheu, pheu, pheu*.

Nesting records furnished by Richard Miller and Turner McMullen are as follows: full sets of three eggs on May 10, 16, 25 and 27 in different years with one of two on May 25 and another of three on June 10. They were all about six feet from the ground in greenbrier thickets or holly bushes.

The Cardinals formerly nested in the forest on Seven Mile Beach and were present also on Five Mile Beach but recent "developments" have apparently driven them from these islands.

I have seen females feeding young on June 19, 1933, and August 14, 1935, and on September 2, 1920, I found a young male in a most interesting phase of molt; most of the dull juvenal plumage remained but the red crest and irregular patches of red on the body had already been assumed. Otway Brown saw two young in a similar condition in the Physick garden on September 1, 1921.

There is no migration of Cardinals since the bird is resident wherever found and southern New Jersey is near to the northern limit of its distribution as a regular and common species. In most of the northern counties it is absent or rare but nests have been found as far north as Plainfield and South Orange.

The counts of Cardinals in mid-winter are shown in the records of the Christmas census at Cape May.

December 24, 1922, two.
December 23, 1923, twenty.
December 26, 1926, four.
December 23, 1928, forty-five.
December 22, 1929, fifty-two.
December 28, 1930, nine.

December 27, 1931, forty-two.
December 26, 1932, twenty-five.
December 24, 1933, fifty-four.
December 23, 1934, thirty-five.
December 22, 1935, fifty-nine.
December 27, 1936, twenty-one.

The "Redbird" is one of the birds that give a distinctly southern touch to the avifauna of Cape May and, although it does not occur in quite the abundance that it attains in the lowlands of Virginia and the Carolinas, it is sufficiently plentiful to constitute one of the most characteristic resident species, while its outstanding color and voice are ever a delight, whether in the green of summer or in the snowy days of winter.

ROSE-BREASTED GROSBEAK

Hedymeles ludovicianus (Linnaeus)

The Rose-breasted Grosbeak is a transient at Cape May but by no means common. As a rule only one is observed at a time and in many years none is recorded. Philip Laurent saw two on Five Mile Beach on May 11, 1890, but it was rare there even then.

Dates of occurrence at Cape May are as follows:

1927. May 20.
1931. May 10.
1925. September 22.

1932. September 17–18.
1934. September 19.
1935. September 7, 29, 30, and October 1.

It nests in the more northern counties, south to Haddonfield and Beverly while a single bird was seen and photographed at Rutherford, by Clarence Brown from January 26 to February 13, 1908, which had doubtless been injured and was unable to migrate (Bird Lore, 1908, p. 82).

BLUE GROSBEAK

Guiraca caerulea caerulea (Linnaeus)

We have but a single record of the Blue Grosbeak at Cape May, a female seen at Cape May Point by Julian Potter on the unseasonable date of November 11, 1923. He says: "While looking for birds at Cape May Point, N. J., I heard a loud metallic call which I soon found came from a bird in a nearby thicket which I had no trouble in identifying as a female Blue Grosbeak *(Guiraca caerulea caerulea)* I approached to within fifteen feet of it and studied it for fifteen or twenty minutes as it showed no disposition to fly away. The large heavy bill, reddish brown of the wings and all the details of brown markings on the plumage were clearly made out. Previous acquaintance with the western race in Arizona precluded any possibility of mistake and skins of the bird were examined the next day which agreed in the minutest detail with my observations." (Auk, 1924, p. 159).

Audubon describes in detail a nest with young which he found near Camden, N. J., in the summer of 1829, and which appears in the plate of this species in his "Birds of America." Thurber (1887) records a specimen seen by Mr. Fairchild at Morristown and there are records for the vicinity of New York and Philadelphia, mainly without details, but nothing further as to the occurrence of this distinctly southern bird in New Jersey.

It is of casual occurrence in Delaware and doubtless breeds in that state. Charles Pennock and Samuel Rhoads recorded a specimen taken near Dover in 1882 by Walter D. Bush (Auk, 1905, p. 202) and John Emlen and Benjamin Hiatt saw one at Rehobeth across the Bay from Cape May, on May 12 and 13, 1928. On the Choptank River, near the Delaware-Maryland line it is known to breed. John Carter found a male, female and young there on July 3, 1904, and Clifford Marburger saw one in full song on June 26, 1932.

INDIGO BUNTING

Passerina cyanea (Linnaeus)

The Indigo Bunting is a decidedly rare summer resident about Cape May and but slightly more plentiful in the migrations, nor is it common anywhere in the peninsula or in the pinelands of southern New Jersey. Usually we have a pair nesting in the thickets along the edge of the meadows at Brier Island below Mill Lane where I have heard the male in song in early July in several years. One found singing on the wires along the railroad at Rutherford's Branch, a little to the west, on July 3, 1935, doubtless was one of these. Another pair is usually to be found in thickets along the southern edge of the Pond Creek Meadows and possibly two pairs bred there in the summer of 1928. A brilliant male found bathing in a rain water pool in the pine grove at the Point on July 2, 1930, probably came from there as I never saw it at the Point again. Still another pair was located in an orchard on the Town Bank Road on May 29, 1921. Others were seen in the same vicinity or at the Point on May 26, 1929, May 30, 1925, and 1928 all of which may well have been migrants. Undoubted migrants were males recorded at Cold Spring on May 8, 1927; May 10, 1929; May 8, 1932; May 6, 1933, May 7, 1934. One of these was feeding with Goldfinches on dandelion seed on a lawn.

In the autumn a few are seen nearly every year especially at Cape May Point and our records follow:

1921.	October 9.	1928.	September 30.
1923.	October 1.	1929.	September 29.
1924.	October 2, three.	1930.	September 28.
1925.	September 22, two.	1932.	September 25, several.
1926.	September 27, several.	1933.	October 9, two.
1927.	October 10, three.		

In those parts of the state north and west of the Pine Barrens the Indigo Bunting is a common summer resident.

DICKCISSEL

Spiza americana (Gmelin)

So far as I am aware the Dickcissel has only once been found in Cape May, a male bird trapped on the edge of waste ground in the town by James Tanner, Audubon Association representative during the autumn of 1936, on September 28, when trapping English Sparrows. In adjoining counties it has occurred as an erratic transient. The history of the species in the East, its former local abundance and later disappearance, has been treated by Samuel Rhoads in "Cassinia" for 1903. It has always been primarily a bird of the Middle West.

Recent nearby records are as follows: Maurice River, September 18, 1890, William L. Baily; Bridgeton, January 18, 1925, B. K. Matlack; Sharpstown, Salem Co., June 10–11, 1928, one seen in an alfalfa field by Leland I. Warner. (Auk, 1928, p. 509).

Curiously enough on May 26, 1928, Miss Mary Wood Daly found a pair of Dickcissels breeding in a clover field in Delaware County, Pennsylvania (Auk, 1928, pp. 507 and 509) and on June 23, 1934, Victor Debes found several Dickcissels in song near Mt. Pleasant, Delaware, and again on May 19, 1935, a single bird was singing at the same spot.

The last nesting in New Jersey was July 3, 1904, when Waldron Miller found a pair with young at Plainfield (Auk, 1904, p. 487).

EVENING GROSBEAK

Hesperiphona vespertina vespertina (Cooper)

The first record of the Evening Grosbeak at Cape May was a female picked up dead on the grounds of the Physick place on Washington Street by Otway Brown on January 14, 1930, which is now in the collection of the Academy of Natural Sciences of Philadelphia. This so far as I am aware constitutes the farthest south record for the range of this far north species, in the East.

On March 30, 1932, Raymond Otter saw a pair of Evening Grossbeaks at his home in West Cape May and identified them at once by comparing them with the plate in Eaton's "Birds of New York." They were seen again in the same vicinity by two other residents of the neighborhood on April 16, and 25, and finally on May 1. There are a number of box elder trees at the spot which undoubtedly attracted the birds as the seeds of this tree are their favorite food. Our third local record is of four birds found by Conrad Roland on December 24, 1933, during the Christmas census of the Delaware Valley Club. They were perched in a tree on the edge of the dunes along the Bay shore, north of Cape May Point, and were studied by several other members of the Club. They were so tame that Roland was able to sketch one of them. Strangely enough in one of the years in which these grosbeaks occurred at Cape May they were not recorded anywhere else in the state. There have however been many other New Jersey records and a resumé of them seems of interest in connection with the Cape May occurrences.

1890. A flock of eight was found at Summit on March 6, 1890, by W. O. Raymond (Ornithologist and Oölogist, 1890, p. 46).

1910–1911. A flock of eight seen by Blanche Hill at Andover, Sussex Co. on December 13, 17 and 18. Mary F. Knouse saw them at Newton in the same county on January 6, 1911, and Stephen D. Inslee reported a flock of twenty-five at the same place on February 5. Waldron DeWitt Miller found a flock of thirteen near Plainfield on January 29 and on February 12 and 19 at least twenty were present. They lived in a grove of red cedars and flowering dogwood and fed on the berries of the latter rejecting the soft part and cracking open the stones with their powerful beaks to get at the kernels. On April 26, Belle C. Cooke found three feeding on cherry buds at Fair Haven on the Shrewsbury River four miles from the coast. They sat in stolid quiet for fifteen minutes at a time.

1916–1917. The flight this winter was more extended. On December 3 Charles Evans saw one at Cinnaminson, Burlington Co. They were at Smithville, Burlington Co., on December 24; at Mt. Holly on the same day (Pumyea); Westville December 25, six (Potter); Browns Mills, January 10, 1917, sixty-five (Darlington); New Lisbon, January 29, seventy-four (Scoville). They were seen again at the last locality February 11 (forty), 17 (thirty), February 22, March 2 (fifty), March 12 (thirty), March 20 (six) (Scoville and Huber). They were at Hammonton February 22 (Bassett) to March 26 and April 18. Birmingham March 2 (five) and 6 (four) (Huber). Lumberton, March 21 (Clayberger). Lakewood March 21 (twelve) (N. C. Brown). The latest records were, April 2, Wenona (Erskine); Rancocas, April 11–12 (Miss E. Haines); Mt. Holly, April 8 (six), 26 (eight) (Pumyea); Yardville, May 7, (Rogers). Also recorded from Blairstown, Morristown and Hackettstown in the northern counties from January to March.

1918–1919. New Lisbon, a flock of twenty-seven, February 22, 1919. Lakewood, February 20, 1919 (N. C. Brown).

1919–1920. Princeton, February 16, 1920 (four) to March 13 (eight) (H. L. Eno). Pt. Pleasant, early February, a pair came in a snow storm and remained for some days (A. P. Richardson). Vineland, February 21 to May 7, a flock of from six to forty on different days (Price). Morestown, April 16-23, a flock of twenty-six. Mt. Holly April 21, two pairs (Pumyea).

1921–1922. Mt. Holly, November 20, a male (Pumyea). Elizabeth, November a flock of 121 (Urner). Vineland, January 5 to March 18, from twelve to forty (Prince). Moorestown, April 21, 25, 27 (Linton).

1922–1923. Mt. Holly, November 20, 1921, a flock of 110 (Pumyea).

1926–1927. Elizabeth, January 16 and February 6 (Urner).

1929–1930. Tuckerton, February 2, a small flock (J. Emlen).

1933–1934. Mount Holly, December 15—April 18, forty (Pumyea). Moorestown, February 22, twelve (Street). Medford, April 13–30, fifty-two (Quay).

No doubt there are other records but probably all the winters of Grosbeak invasion are mentioned.

PURPLE FINCH

Carpodacus purpureus purpureus (Gmelin)

The Purple Finch is an autumn transient and somewhat irregular winter resident. In the spring we have but one record, three birds one in full song observed by William Yoder on March 6, 1927.

In 1935, when William Rusling was present at the Point in the interests of the Audubon Association, he recorded Purple Finches on nineteen days between September 16 and November 14, usually less than a dozen although on October 14, twenty-five were seen. Whether his observations represented a flock constantly present or small bunches on migration it is difficult to say.

Our records are as follows:

1922–1923. November 12, two; January 14.
1923. November 11, six; December 23.
1925. February 8.
1925–1926. September 22 and 26, two; October 18, three; February 8.
1926–1927. November 21, six; December 26, two; March 6.

1927. October 23, seven; November 6, fourteen; December 26, two.
1928. October 27, December 23, two.
1929. October 25, December 22, 26, five.
1931. October 18; November 7, two; December 27.
1933. October 28, two.
1935. September 16; 17, sixteen; 20; 21, ten; 23, eleven; 29, twelve; October 2; 4; 7; 10, 12, twenty-five; 14; 15; 24; 25; November 1; 3; 6; 14.
1936. December 22, December 27, seven.

While mainly a transient or winter visitant the Purple Finch has bred at several localities in the northern counties.

*PINE GROSBEAK

Pinicola enucleator leucura (Muller)

The Pine Grosbeak is a rare visitant from the Far North in severe winters. It has been recorded on several occasions in the northern parts of the state but has never, so far as I am aware, occurred in Cape May County; indeed the most southern record seems to be a flock seen at Princeton by Prof. A. H. Phillips in 1886 (Babson, Birds of Princeton). There were notable southward flights of these birds in the winters of 1836–37; 1884–85; 1899–1900; 1903–04; but they only reached the northern edge of New Jersey.

Charles Urner records them present in Union County in the winters of 1916–1917 and 1929–1930, a flock of forty on January 5, 1930 and they were present in Essex and Hudson Counties in December, 1913.

*BRITISH GOLDFINCH

Carduelis carduelis britannica (Hartert)

This bird was introduced at Hoboken according to Warren Eaton in 1878 and a few were still to be seen as late as 1913. John T. Nichols found them at Englewood, a little farther north, from 1912 to 1915 and thought that they probably bred there. He says "a flock of about eight on January 28, 1912 (at Englewood); about six at Leonia on February 16, 1913; one on February 21, 1915, seven, one in full song, in a heavy wet snow storm on March 6; a flock of about five at Coytesville on March 13, unusually common in the Englewood region this year, singing on March 23, 1915." (Auk, 1936, p. 429.)

They were also introduced on Long Island but do not seem to have spread beyond the immediate vicinity of New York City.

COMMON REDPOLL

Acanthis linaria linaria (Linnaeus)

The Redpoll is a rather rare and irregular winter visitant over most of the state but seldom ranges south of the Pine Barrens. Notable flights have been recorded in the winters of 1836–1837; 1878–1879, March 1888, 1899–1900, 1906–1907 and 1908–1909. Prior to 1930 Charles Urner had found them in Union County in seven out of twelve years from October 18 to March 21 (extreme dates).

Redpolls have occurred as far south as Haddonfield in 1888 (Rhoads) Swedesboro in 1909 (Lippincott) and down the coast to Brigantine where Brooke Worth found four with Siskins on February 22, 1936, during a spell of exceptionally cold weather and after a six inch fall of snow on the 21st. They were in bayberry bushes at the southern point of the island with Snow Buntings also present.

On the same day, a flock of fifty was seen at Beverly by members of the Delaware Valley Club. They appeared very dark-colored against the snow differing from the Goldfinches in this respect just as do the Siskins. Indeed the Redpolls are essentially Siskins with crimson crowns and a pink flush on the breast and rump, and even these marks are absent in the female, but they always lack the yellow wing flashes of the latter bird.

So far as I am aware we have but a single record of the Redpoll in Cape May County—two seen by John Carter and Edward Marshall on Seven Mile Beach on March 26, 1937.

*GREATER REDPOLL

Acanthis linaria rostrata (Coues)

Babson, in his "Birds of Princeton" has recorded two specimens of this slightly larger race of the Redpoll which were shot at Princeton on February 6, 1872. Prof. A. H. Phillips and Charles Rogers have assured me that they were correctly identified and thanks to Mr. Rogers I have been able to examine the birds and quite agree with them.

PINE SISKIN

Spinus pinus pinus (Wilson)

For the most part the occurrence of the Siskin in the Cape May region is irregular. Some individuals may occur every autumn but in only a few years has there been a large flock present throughout the winter. Several times there has been a notable migration of these birds down the coast but whether this is a regular occurrence I have been unable to determine. On November 12, 1922, Julian Potter saw them passing down the coast at intervals, all day long, in flocks of from ten to two hundred. They did not alight and scarcely any were seen except on the wing. On November 16, 1929, unaware of this record, I saw exactly the same thing. The flocks, consisting of from ten to one hundred birds, kept passing me all morning between South Cape May and the Point traveling in rapid flight along the dunes or the edge of the meadows just behind. They looked very dark but were difficult to study as they never paused or rested either on the ground or on bushes. I was unable to ascertain whether they came in off the ocean, as the Robin flocks do at the same spot, or whether they crossed the Bay immediately or cruised up the western side of the peninsula. On November 7, 1931, a similar flight occurred and I estimated that at least five hundred passed in an hour or two. Some of these birds alighted on the bayberry bushes by the Lighthouse and then continued on their way. On October 18, 1925, we have record of one hundred Siskins along the beach and 1500 observed at Barnegat by Charles Urner, on November 29 of that year, which may have been part of another of these flights.

In only the winters of 1922–1923 and 1925–1926, did they remain at Cape May as winter visitants and on the former occasion they were present until April 30. They associated with Goldfinches on the dunes and meadows back of South Cape May where Spencer Trotter and I found them feeding on cockleburs on December 30, 1922. We had a few present on November 6 and 11, 1921; October 7, 1922; November and December 27, 1931 (two); December 22, 1935 (two).

The most striking features of the Siskins are their general dark coloration and very heavy streaking below and, when they fly, their long pointed wings and flash of yellow as they are successively spread and closed again.

Throughout New Jersey the Siskin is a transient and irregular winter visitant varying in numbers in different years. While their occurrence in West Jersey along the Delaware Valley is usually between October 15 and April 25 I have seen considerable flocks as late as May 17.

GOLDFINCH

Spinus tristis tristis (Linnaeus)

The Goldfinch is a resident about Cape May but of somewhat irregular occurrence in the town and about the Point and not a bird that will be seen on every day afield. It is often seen or heard as it passes overhead in its undulating flight or pauses for a moment on a telegraph wire or tree top. In spring or early summer pairs or small flocks may come into the gardens or to the shade trees along the streets, and a pair or two are usually established on the Fill where they doubtless nest, as also in the fruit trees of the Physick place, where a few usually summer.

Occasionally we see a male Goldfinch flying high in the air more or less in circles, and after covering this imaginary track several times he will relapse into the usual undulating flight and drop back to his perch. This performance is apparently a display, incident to the mating season. I saw it once on July 9, 1923, and again on August 1, 1926, when the bird was tossing himself through the air in perfect abandon. Out on the Fill I watched other Goldfinches going through this display on June 26, 1928; August 6, 1925; June 30, 1935; circling high over the dense growth of bayberry bushes where the female was concealed.

I have watched Goldfinches in May feeding on dandelion seeds on the lawns of dwellings in the open country while in July and August they may be found feeding on thistles and chicory in waste ground near the town, or along the roadsides, and in late August and September they frequent sunflowers, picking the seeds from the great round heads and often feeding them to young birds already full-fledged and on the wing. These youngsters, apparently perfectly able to shift for themselves, have been seen begging for food and twittering their wings as they followed their parents, as late as September 12 and 29 in different years. In wintertime the Goldfinches feed in flocks on cockleburs along the dunes or edges of the marshes while I have watched some individuals exploring pine cones in the groves at the Point, and others in the center of a tall woods at Cold Spring picking seeds from the burs of the Sweet Gum trees. Yellow males resting against the sunflowers and primroses in the garden are almost invisible and would easily be passed by.

By the middle or end of August, but more especially from October to March, Goldfinches gather in larger flocks and drift about the country, often leaving the Cape May district entirely in midwinter when snow and ice render food difficult to obtain. Dr. William L. Abbott and I found an immense flock largely birds of the year along the roadsides near Higbee's Beach on August 27, 1918. On November 21, 1926, John Gillespie saw a flock of one hundred or more and I encountered a similar large gathering on March 15, 1931, while on December 30, 1922, there was a great flock associated with Siskins on the marshes below the town where they remained all winter and were present as late as April 30 though in much reduced numbers. Walker Hand records a flock of over two hundred on February 9, 1919, and refers to "thousands" present, November 21–25, 1918. While usually forming their own flock scattered winter individuals will join a miscellaneous assemblage of sparrows along the fence rows and thickets and feed with them on the ground.

I have seen males in the yellow and black breeding plumage as late as September 7, which showed no sign of molting but others observed on October 8 had undergone the change to the gray-brown winter dress. Spring birds were molting on March 25, 1933, while others by April 30 were essentially in full nuptial plumage.

The variation in the number of Goldfinches present in midwinter is indicated by the counts on the Christmas census of the Delaware Valley Club.

December 30, 1922, 100.
December 26, 1926, thirty-one.
December 26, 1927, forty-one.
December 23, 1928, fifty-seven.
December 22, 1929, 120.
December 28, 1930, seventy-six.

December 27, 1931, 145.
December 26, 1932, 300.
December 24, 1933, 144.
December 23, 1934, 210.
December 22, 1935, 154.
December 27, 1936, ninety.

Elsewhere in the state the occurrence of the Goldfinch is similar to that at Cape May—a resident species most abundant during the migrations.

RED CROSSBILL

Loxia curvirostra pusilla Gloger

We have but one record of the Red Crossbill at Cape May, one found dead at the Point by Turner McMullen on March 4, 1923. Julian Potter found some at Barnegat on December 21, 1919, and it has been recorded at various times and places in the northern parts of the state although always irregular and erratic in its occurrence.

Dr. William E. Hughes found a flock of about two dozen of these birds at Forked River on June 6, 1900, which were feeding on the pine cones while on May 10, 1894, he saw three at Lewes, Delaware. These very late records might indicate the breeding of the birds in the pine lands were it not for their well-known erratic habits. William Evans saw some at Hanover on May 6, 1900; George E. Hix in northern Somerset County, July 6, 1903; and Samuel Rhoads at Wawayanda Lake, June 5, 1909.

Nelson Pumyea has found them wintering at Mt. Holly, where they were observed from January 21 to March 18, 1923.

WHITE-WINGED CROSSBILL

Loxia leucoptera Gmelin

A female White-winged Crossbill was brought in to Walker Hand by a pet cat on February 5, 1909, the specimen is now in the collection of the Academy of Natural Sciences of Philadelphia. No others have been seen at any time in Cape May County.

Elsewhere in the state they have occurred as irregular winter visitants especially in the northern counties. Audubon writing at Camden, in the first week of November, 1827, says: "they are so abundant that I am able to shoot, every day, great numbers out of the flocks that are continually alighting in a copse of Jersey scrub pine, opposite my window." John Cassin states that they were present in the winter of 1836–1837, and were not seen again until the winter of 1854–1855 when they were unusually plentiful among the pines about Camden, and so tame that they could be killed with stones. Samuel Rhoads found a small flock at Haddonfield in the winter of 1896–1897 and in that of 1899–1900 they were abundant at Plainfield, Princeton, Englewood, etc. On December 25, 1906, Charles Rogers records four at Leonia, while Nelson Pumyea saw some at Mt. Holly on January 21, 1923, which, as is often the case, were associated with the Red Crossbills.

RED-EYED TOWHEE

Pipilo erythrophthalmus erythrophthalmus (Linnaeus)

The Towhee or Chewink is a common summer resident of scrubby woods and clearings throughout Cape May County and northward. It is plentiful at Cape May Point and along the wood edges on the eastern side of the peninsula as far south as Cold Spring and Brier Island. It spends much of its life in the heart of the thickets from which comes its loud metallic alarm note "*chewink*" whenever one approaches its domain, for it is ever alert and we are lucky if we are able to steal a march on it. Occasionally we may manage to creep up unawares and get a view of the bird busily scratching a living from the dead leaves that cover the woods floor. His relatively large feet seem to be well adapted for the work as he jumps up a few inches and kicks out with both at once.

In early spring, and throughout the nesting season, the male Chewink will mount to the top of some small tree or occasionally a larger one, such as a Thrasher might select, and with head erect give voice to his familiar song. His usual habitat, however, is close to the ground and as we pass along some wooded road, where the present deplorable craze for clearing has left a fringe of underbrush, we may catch a glimpse of him crossing ahead of us on the wing and get a momentary flash of his striking black, white and chestnut livery, or he may cross on foot with head and neck extended and making prodigious hops. In the short flights of the Chewink the tail is always flailing up and down as if hinged at the base, producing a sort of broken-backed flight such as we are familiar with in the Song Sparrow. When perched for singing, however, the tail is pendent and the feathers of the crown are slightly elevated, while one bird hopping along the horizontal limb of a tree held the tail cocked up at an angle and kept flapping the

(895)

wings as if to balance itself. In many of its actions, especially in its infrequent sustained flight, the Chewink recalls the Cardinal.

The normal song of the bird may be represented as *chúck, burr, chéeeeeee* the second note lower than the first and the following trill higher than either. This, however, is subject to much variation in different individuals. One seemed, to my ear, to say, *seé, tschu, chéeeee* while the song of a third I have recorded *whíp, o, shéeeee*. Still another added a second trill *chíck, o, chéeee, chéeee*, and in one song that I heard the unaccented note came first *che, sék, chée* while in another there was only a single note before the trill, *chúck, chéeee*. Then there was a bird with a peculiarly harsh preliminary note somewhat like the single call note of the Flicker, *skéuk chéeeee* and one which sang the first two notes on exactly the same pitch. There is practically no variation in the alarm note although I once heard two birds, perhaps a pair, calling alternately with a slightly different pitch and in a way supplementary to one another, one call rising and the other falling, *ka-wéek; che-wínk; ka-wéek; che-wínk*. Young birds, still in the dull gray brown juvenal plumage, have an alarm note like the latter part of the parent's call but weaker, *wink, wink;* while an adult female, heard on August 24, 1927, uttered an entirely different alarm note, a rapidly repeated *shwee, shwee, shwee, shwee,* etc. which immediately brought a dozen birds to the spot—Carolina Wrens, White-eyed Vireos, Thrashers and Catbirds.

I have heard the Towhee's song as late as July 16, 1927; July 23, 1926; July 27, 1922; July 29, 1924; while the *che-wink* call may be uttered at any time, even by the occasional winter resident birds.

Richard Miller has examined a number of nests of the Towhee in the Cape May district and states that they are usually on the ground, often at the base of a bush or small tree, but one that I found at Erma, on July 29, 1924, evidently a delayed nesting, was in a huckleberry bush three feet from the ground and contained only two eggs. All full sets consisted of four eggs; and of Miller's nests three were found on May 27 and one each on May 25, 30, June 6 and 10; two nests found on May 27 and one on May 31 contained four young each.

Young birds in full juvenal plumage were seen on the wing on July 15, 1920 and 1922; July 31, 1921; July 28, 1934; looking, except for the tail, like plump gray brown sparrows. Some in the same condition were seen on August 9 and August 27, while on August 17, 1891; August 29, 1922; September 1, 1931; September 2, 1920; birds were found that were in all stages of molt. Some had black collars on a gray brown plumage and others irregular patches of black and chestnut producing a remarkable piebald pattern A little later old and young become indistinguishable.

Walker Hand's arrival dates are as follows up to 1920 with other records for more recent years:

1902. March 10.	1911. March 11.	1916. April 18.
1903. March 9.	1912. March 27.	1928. April 1.
1904. April 4.	1920. April 20.	1929. April 7.
1905. April 9.	1921. April 24.	1930. April 6.
1906. April 3.	1922. April 23.	1932. March 25.
1907. March 18.	1923. April 22.	1933. April 1.
1908. March 14.	1924. April 20.	1934. April 20.
1909. March 8.	1925. March 22.	

Some, at least, of the March dates probably refer to wintering individuals. Our latest autumn dates are:

1921. October 16.	1927. October 27.	1931. October 19.
1922. October 1.	1928. October 23.	1932. October 12.
1923. October 21.	1929. October 25.	1934. October 19.
1924. October 2.	1930. October 5.	1935. October 29.
1925. October 18.		

We have several records of single birds in November, and for February 22, 1923; February 8 and 14, 1925; February 6, 1932; while the counts on the Christmas census show the probable number of birds that wintered:

1922. December 24, six.	1932. December 26, five.
1923. December 23, one.	1933. December 24, two.
1928. December 23, one.	1934. December 23, three.
1929. December 22, two.	1935. December 22, four.
1930. December 28, eleven.	1936. December 27, two.
1931. December 27, ten.	

Additional winter records are one seen on Five Mile Beach December 27, 1903 (Baily) and one at Moorestown, December 25, 1907 (Evans).

Like many other summer resident birds Towhees seem to start southward rather early and often about the middle or end of August there are days when none are to be found—August 27, 1925; August 28, 1926; August 17, 1929; August 29, 1932; while later they again become common or even abundant as hordes of migrants from farther north throng the woods, usually to stay a few days and then pass on, while their places are taken by later visitors.

IPSWICH SPARROW

Passerculus princeps Maynard

This striking pale sparrow of the dunes has an interesting history. It was first obtained by Alexander Wilson on the New Jersey coast, probably on Peck's Beach where Ocean City now stands, the region known to the older ornithologists as "Great Egg Harbor." Wilson considered his bird as merely a fully adult male of the Savannah Sparrow and figured it under that name in his classic "Ornithology." In 1892 my friend Norris DeHaven called my attention to the similarity of Wilson's plate to the Ipswich Sparrow and I saw at once that he was right and that what we had always regarded as an exaggerated representation of a Savannah was in reality an excellent portrait of this interesting bird (cf. Osprey, 1898, p. 117). It had meanwhile been "rediscovered" on Ipswich Beach in Massachusetts by Charles J. Maynard who described and named it and his appellation still stands. The next New Jersey specimen is one found in the collection of Bernard Hoopes, without data but probably also from Great Egg Harbor, doubtless secured in the "sixties." No others were recorded from this state until Dr. W. L. Abbott shot one on Five Mile Beach, December 30, 1879, and Samuel Rhoads got another at Atlantic City on March 15, 1888. On April 3, 1889, John Sterner obtained one on Five Mile Beach for Philip Laurent and I secured two on January 29, 1892, at Cape May and two more on January 5, 1893. Since that time many more specimens have been obtained and many others seen, as ornithologists increased in numbers and the bird became better known.

It was later discovered that the summer home of the Ipswich Sparrow was Sable Island on the coast of Nova Scotia.

The Ipswich Sparrow is a winter resident of the sand dunes close to the beach and probably no other bird has such a restricted range. It runs like a mouse through the yellow dune grass or occasionally rests on logs or trash washed up by the waves. When standing still the light colors of the bird exactly match the tints of the sand and it is hard to distinguish it from its surroundings. Some individuals, probably young of the year, have brighter bay tints on the wings and the cheeks are deeper buff, while spring birds have the yellow eyebrow much more conspicuous. Ipswich Sparrows occur at Cape May singly, as a rule, and not more than four have been seen during a day's tramp but on the wilder stretches of Seven Mile Beach Dr. William E. Hughes found them present in greater numbers and secured as many as nine on January 1, 1899. He has recorded them as present from November 14, 1897, to March 27, 1898 and from November 13, 1898, to January 2, 1899. His interesting series of specimens is now in the collection of the Philadelphia Academy.

The earliest dates of autumn arrival of the Ipswich Sparrow for the Cape May region are: November 21, 1926; October 23, 1927; October 27, 1928; November 16, 1929; November 7, 1931.

The latest dates of observation in spring: March 31, 1921; March 12, 1922; March 20, 1924; March 18, 1927; April 12, 1930.

To show the variation in the midwinter population we may quote the figures of the Delaware Valley Club's Christmas census:

December 31, 1921, one.	December 28, 1930, eight.
December 30, 1922, three.	December 27, 1931, two.
December 23, 1923, four.	December 26, 1932, four.
December 26, 1926, one.	December 23, 1934, two.
December 26, 1927, four.	December 22, 1935, three.
December 23, 1928, four.	December 27, 1936, four.
December 25, 1929, one.	

SAVANNAH SPARROW

Passerculus sandwichensis savanna (Wilson)

The Savannah Sparrow is a common winter visitant to the brushy marshes and old fields of Cape May County as well as on the dunes and open sandy spots of the coast from about September 7 to May 10.

A specimen secured by F. D. Stone, Jr., on the borders of Cape Island Sound on July 6, 1891, raised the question of its possible occurrence as a breeder at Cape May and in 1921 I found two pairs established on the Fill, which had replaced the old Sound, and for several summers thereafter one or more pairs evidently bred on the edge of the swampy spots that had developed there. Finally Julian Potter found a nest on the Fill on June 19, 1923, containing four fully fledged young which promptly took wing. The nest was in a slight depression in the ground. A male was still in full song on July 4, 1927, in the same locality. Meanwhile Waldron De Witt Miller had found a nest and three young on Seven Mile Beach, on July 8, 1903, and W. B. Crispin one near Salem and on June 3, 1934, Edward Reiman found one with four eggs at Brigantine farther up the coast. While these are the only nests so far recorded for southern New Jersey, Charles Pennock obtained an adult and a young bird in the juvenal plumage at Delaware City, Del., on June 24, 1911.

Savannah Sparrows are most plentiful in the spring and autumn migrations but a few remain through every winter and in some seasons they occur in large numbers, as in the week of January 29, 1891, when Samuel Rhoads and I found them fairly swarming on the meadows between Cape May and the Point during a spell of extremely cold weather with a gale from the northwest. Like many other sparrows they always take wing as individuals, not in concentrated flocks, and fly from the ground in a low somewhat undulating but direct flight, soon dropping down to the ground again. They usually crouch perfectly still like a Snipe and flush again before one can get near to them. Their flight is totally unlike that of the Song Sparrow lacking the tail action while the tail itself appears distinctly shorter. Compared with the Grasshopper Sparrow the Savannah lacks the twisting character of flight as it takes wing, and the tail is longer. The migrant and winter birds run like mice through the grass and it is almost impossible to study them on the ground, but summer breeding birds, resenting one's intrusion on nest or young, are much more in evidence. They run rapidly with tail slightly elevated and after taking wing will alight on the top of a bayberry bush or weed stalk from which perch I have heard them sing as late as July 21, 1921. On January 17, 1932, I followed a single bird which I found feeding in an old bean field, for fifty yards, walking at a medium gait some fifteen feet behind it. The bird would pause when I did and then run on, sometimes giving a slight flip to the wings or spreading them a little to steady itself and maintain its balance. It never attempted flight although it would, of course, have done so had I quickened my pace. Other birds that were flushed later alighted on bushes giving an instantaneous flirt to the tail, as they reached their perch, so quick that the eye could scarcely catch it.

The breeding Savannahs appear very pure white below and the black streaks stand out boldly while the yellow eyebrow is conspicuous. In winter there is more of a buff tint on the breast and whitish edgings to the feathers of the back emphasize the streaked appearance, while there is an olive yellow cast to the sides of the head. The young are darker than the adults with a strong buff suffusion on the breast and heavier streaking below. The feet in all are pinkish flesh-color.

Spring migrants that I saw feeding on an open sandy flat on Ludlam's Beach on May 9, 1925, seemed to spread the legs very far apart as if to bring the head nearer to the ground as they picked up food, flirting the tail nervously all the time. They ran with a jerky gait.

Autumn dates for arrival of migrants are:

1921. September 9.	1925. September 22.	1929. September 21.
1922. September 17.	1926. September 25.	1930. September 21.
1923. August 21.	1927. September 25.	1932. September 17.
1924. September 7.	1928. September 23.	1935. September 19.

The variation in numbers recorded on the Christmas census are shown below:

December 28, 1921, three.
December 24, 1922, six.
December 23, 1923, twenty.
December 26, 1926, sixteen.
December 26, 1927, ninety-two.
December 23, 1928, nine.
December 27, 1929, two.

December 28, 1930, twenty.
December 27, 1931, nine.
December 26, 1932, seven.
December 24, 1933, forty-two.
December 23, 1934, forty-nine.
December 22, 1935, twelve.
December 27, 1936, nineteen.

While mainly a transient in New Jersey the Savannah Sparrow breeds at various localities in the northern counties—Morristown, Patterson etc.

GRASSHOPPER SPARROW

Ammodramus savannarum australis Maynard

The Grasshopper Sparrow is a common summer resident in old fields of the uplands and ranges down almost to the edge of the salt meadows. It also occurs on sandy grassy areas back of the dunes and is abundant on the Fill, although becoming less common there as the bayberry bushes overgrow the open grassy spots.

It is essentially a ground bird and progresses by quick hops intermingled with running steps and occasionally breaking into a regular run. When it stops it often ducks the body nervously like a wren. The head and neck seem out of proportion to the rest of the body and as the bird leans far forward, in running, the legs seem set too far behind and it looks as if it were about to fall over. The tail is very short and slender and is elevated at an angle of 45°.

When feeding or resting in the short grass the bird will hold its position until almost trodden upon and then dart away in its characteristic erratic, twisting flight, apparently turning on one side or the other like a Snipe. After a short flight it drops back to the ground. During the nesting season both parents and young will perch on the bushes or weed tops and the former will utter their insect-like buzzing song from such positions. While singing the body is scarcely raised but the head is thrown back a little as the song is uttered. One individual, singing from a telegraph wire, where it could be seen more clearly, threw back its head and simultaneously spread its tail, the feathers standing out like a semicircle of bristles which closed as soon as the song ended.

The most characteristic markings of the Grasshopper Sparrow are the "strings" of round black spots on the wing-coverts, the yellow on the bend of the wing, the plain unspotted buff breast and the pale pinkish feet. The young with their spotted breasts look quite different but they have the same top-heavy build, and the black spots on the coverts.

The nest is placed in a shallow depression in the ground alongside a tussock of grass and is often so arched over that it is impossible to see into it from above. From observations of Richard Miller, Turner McMullen and myself I have prepared the following summary of nests:

> May 20, one nest of four eggs.
> May 22, one nest of four eggs.
> May 30, four of four eggs and two of five.
> June 17, 20, 21, one set of four each.
> June 29 and July 10, one set of five each.

Nests with young were found on June 8 (five young) and July 19 (four young). Young birds are on the wing by June 18 and 30 in different years while the juvenal plumage is retained until August 5.

First observations in spring are April 29, 1923; May 9, 1924; May 10, 1929; but judging from the nest data most of these must be late.

Last dates for autumn are: September 16, 1921; October 21, 1923; October 2, 1924; October 5, 1929; September 28, 1930.

In build and habits the Grasshopper Sparrow can be compared only with Henslow's but the chestnut tints on the back and the streaked breast of the adults of the latter will distinguish it.

Through most of the state it is a regular summer resident in open country.

HENSLOW'S SPARROW

Passerherbulus henslowi susurrans Brewster

Henslow's Sparrow is a common summer resident of the Cape May Peninsula but distinctly local in its distribution and sometimes associates in little colonies. Old fields overgrown with patches of bayberry, brambles and wild rose, interspersed with little moist or boggy spots, and reaching down to the edge of the salt marshes are the home of Henslow's Sparrow—that region where upland and meadow meet. The Fill also offered attractive quarters before the bushes grew so tall. The range of the bird, however, is not entirely coastwise as it is found in similar spots far inland.

If we stop for a moment on the edge of one of these boggy spots that the Henslow's frequents we shall hear immediately the explosive ventriloquial note of the bird and scanning the field we shall see him perched on the top of some dead weed stalk standing perhaps a foot above the surrounding grass and sedge or on some briar twig. If approached he will fly to another similar perch, for he has several regular singing stands, and as long as we are in the vicinity he will continue to move from one to another. The female seems rarely to flush unless we almost tread upon her and this trait helps to make the nest so difficult to find. The young, when first on the wing, may roost in the lower branches of some nearby bush but once the nesting season is past Henslow's Sparrow becomes exclusively a ground bird but does not move far from its breeding spot. As we cross the field we may flush them repeatedly but they fly only a few yards and plunge back into the grass. They apparently run a few steps after alighting, as they never arise from the spot where we expect them to be, and after a while they may fail to flush at all and run like mice through the tufts of grass until a safe shelter is found where they crouch until danger is past, and our best efforts fail to dislodge them.

Henslow's Sparrow presents a peculiar appearance but is in many respects an exaggerated Grasshopper Sparrow. The bill is very heavy and the head large in proportion to the body which tapers rapidly to a slender wisp of a tail. As the bird perches, hunched up, perhaps, on the top of a wild carrot plant, he appears to be all head and neck but when excited the tail is cocked up at an angle and there is a nervous twitch of both body and wings. As he flies with his short rapidly pulsating wings the tail is all but lost sight of and he looks decidedly bobtailed. When seen in good light, the rich mahogany brown tints of the back contrasted with the olive of the head and the minute streaking of the sides of the breast, distinguish the Henslow's from any other of our sparrows; but color aside, its peculiar build, its perch on the weed stalks and its never to be forgotten note will always identify it.

The note of the Henslow's Sparrow has been variously transcribed but after long and careful study of the Cape May birds the syllables *chis-eck* seem to my ear best to describe it. The accent is surely on the first syllable although there is not really much question of accent about it as the two syllables often almost run together into a single—*chisk*. Only once have I seen a female perch out in the open. She took up a position near a singing male and uttered a weak almost inaudible *tick, tick*, evidently an alarm note. Her life seems to be spent almost entirely on the ground. Conrad Roland has frequently heard the note of the Henslow's Sparrow at night.

The young are strikingly different in color from the adults having plain yellow buff breasts with no trace of streaks, but there is the same top heavy appearance as if the bird had run entirely to neck and head. I have found them barely able to fly, in fields by the old Race Track Pond, on June 8, 1919, while they were still in the juvenal plumage as late as July 20. Two nests found by Turner McMullen on May 30, contained newly hatched young.

Nests with three or four eggs have been found at Cape May on May 22, May 24, May 30 and one on July 7, and at Ocean View on June 7 and 17.

Our earliest spring records for the bird at Cape May have been:

April 24, 1921.
April 29, 1923.

April 24, 1926.
April 24, 1927.

and Richard Miller found it at Pennsville on April 17. Our latest Cape May dates are September 16, 1921; October 1, 1923; September 28, 1924; October 7, 1928.

This interesting little bird entirely escaped the notice of Alexander Wilson nor did Audubon know it from New Jersey except for the information given him by James Trudeau and Edward Harris who obtained a specimen as early as 1838. The next definite record of the occurrence of the bird in New Jersey was of specimens obtained on Seven Mile Beach in June 1875 by John McIlvain, a notable Philadelphia collector of early days (Atlantic Slope Naturalist, I, p. 79). Harry G. Parker records a nest on this same beach on May 27, 1885 (O. and O., XI, p. 140) while Sylvester D. Judd secured a young bird at Boonton, Morris County (Auk, 1897, p. 326). Frank L. Burns found a nest on Peck's Beach on May 30, 1895 (Auk, 1895, p. 189) and W. E. D. Scott obtained several specimens near Princeton prior to 1900, while Thurber (1887) recorded it as a local summer resident at Morristown. Waldron Miller found Henslow's Sparrow breeding in the mountains north of Plainfield and located colonies at various locations in the Passaic Valley, Great Swamp etc. in the same vicinity. A specimen had been obtained by Dr. Amos P. Brown at Point Pleasant on August 16, 1886. I visited the spot with Dr. William E. Hughes on May 30, 1895, and we found quite a colony although we failed to locate a nest. Subsequently Samuel Rhoads and I found a colony near Bayside; Stewardson Brown at Forked River; and William Baily at Ocean View, while I detected a few in a bog at Lindenwold, Camden County. Of course they may have increased very much in later years but I am inclined to think that they have always been in these haunts as their secretive nature would make it very easy to overlook them and ornithological knowledge of southern New Jersey land birds was never very extensive until comparatively recent times.

Samuel Rhoads has published an excellent history of the Henslow's Sparrow in New Jersey, in "Cassinia" for 1902, pp. 6–14.

Later nests have been found at Marlton (Carter); Millville (Hunt), and New Lisbon (Stuart) etc.

SHARP-TAILED SPARROW

Ammospiza caudacuta caudacuta (Gmelin)

This bird like the Seaside Sparrow is entirely limited to the salt mead-ows and has suffered the same restriction in range immediately about Cape May through the influence of man. I found it quite as abundant as the other species in 1891, nesting in the same areas south of the town and about Cape Island Sound and on the meadows above and below the Harbor. I have been able to detect little if any difference in the habits of the two species except that the Sharp-tail seems to prefer the rather dryer spots, where the grass is shorter, while the Seaside frequents the taller vegetation along the tide-water creeks, although both nest in areas of short grass.

The Sharp-tail perches on the grass stems with its head protruding just as does the other species and runs rapidly through the grass cover with neck extended and head held low while it has the same heavy flight for short distances. Its song is much less powerful than that of the Seaside Sparrow and does not carry well nor does the bird seem to sing so frequently and is rarely heard after the first of July. A late singer heard early in the morning just back of the town on July 7, 1923, had four notes in its song: *chuck, char-éee-zik.* The first separated by a short interval and the accent on the third. Julian Potter who has given the species much study describes the note as a faint wheezy ditty scarcely audible at fifteen yards, suggestive of the song of the Grasshopper Sparrow but of much less volume.

When in good light the Sharp-tail is a handsome bird and very different from the somber gray Seaside. It has the usual white underparts of most sparrows with dusky streaks while the upper parts have a yellow brown tone with a bright orange yellow patch on the side of the head. The young of the year are entirely different being suffused with ruddy buff above and below, the back striped with brown, and the breast plain.

There seems to be an influx of migrants in the autumn and on September 3, 1925, I counted upwards of one hundred on the edge of the meadows opposite Bennett. On the meadows back of Atlantic City in 1894 and 1895 I found the growth of tall grass and sedges along the salt creeks simply swarming with Sharp-tailed and Seaside Sparrows which the boatmen of the sounds used to refer to indiscriminately as "meadow wrens" and were disgusted that anyone should regard them as game when they found us collecting some specimens! The Seasides of these assemblages were doubtless mainly of local origin as the New Jersey coast is close to the northern range of the species but many of the Sharp-tails probably came from farther north as they range up along the coast of New England. Turner McMullen and Richard Miller have furnished me with many records of nests of the Sharp-tailed Sparrow and I have found a number myself. Four eggs seem to form the regular set and nests with full complements have been examined on:

May 22, one.	June 18, one.	June 29, one.
May 26, two.	June 20, two.	July 3, one.
May 30, one.	June 22, two.	July 4, one.
June 2, two.	June 25, one.	July 15, one.
June 4, one.	June 26, one.	July 19, one.
June 12, three.	June 28, one.	

Three sets of five were found on May 30, June 10 and June 24.

Sets of three were found from June 9 to July 22, most of them doubtless incomplete.

Young birds were found in nests on June 2, 4, 10, 15, 17, 22, July 1 and 11 while they have been seen on the wing by July 1.

I have few records of the spring arrival of the Sharp-tails at Cape May but have found them present in small numbers on March 31, 1921, and April 11, 1926. In the autumn, last dates of observation are September 5, 1921; October 21, 1923; September 28, 1924; October 18, 1925; October 25, 1929; October 12, 1932.

Single birds have been seen on January 13, 1924; January 25, 1925; January 23, 1927; while the Christmas census shows:

December 28, 1921, one.	December 26, 1932, two.
December 24, 1922, three.	December 24, 1933, seven.
December 23, 1923, one.	December 23, 1934, fifteen.
December 22, 1929, one.	December 22, 1935, eight.
December 28, 1930, two.	December 27, 1936, fourteen.
December 27, 1931, two.	

ACADIAN SPARROW

Ammospiza caudacuta subvirgata (Dwight)

In the large flocks of Sharp-tailed Sparrows that occur late in summer or early autumn, specimens of the smaller Acadian Sharp-tail which breeds in Nova Scotia are to be found every year. They are grayer and duller in color than the native race and lack the white streaks on the back.

Specimens have been taken as follows:

Atlantic City meadows, October 1 and 2, 1892, four (I. N. DeHaven and W. Stone).

Atlantic City meadows, October 21, 1894, I. N. DeHaven.

Point Pleasant, May 30, 1895, W. L. Baily.

They have been recorded in Essex and Hudson Counties (Eaton) and in Union County (Urner).

I have seen them at South Cape May on several occasions in October.

*NELSON'S SPARROW

Ammospiza caudacuta nelsoni (Allen)

Nelson's Sharp-tail is a fresh water race which breeds in the upper Mississippi Valley and on rare occasions individuals are to be found in the mixed Sharp-tail flocks on the Atlantic coast. While I have never detected this bird at Cape May it must occur there as Norris DeHaven and I have taken specimens at Atlantic City as follows:

Atlantic City meadows, October 2, 1892, three, I. N. DeHaven and W. Stone.

Atlantic City meadows, October 21, 1894, I. N. DeHaven.

Atlantic City meadows, September 27 1896, W. Stone.

Atlantic City meadows, May 9, 1892, I. N. DeHaven.

Charles Urner records one on the marshes of Union County.

SEASIDE SPARROW

Ammospiza maritima maritima (Wilson)

The Seaside and Sharp-tailed Sparrows are the only breeding land birds peculiar to the maritime environment and both are absolutely restricted to the salt meadows where they are regular summer residents, and where a few individuals usually manage to remain throughout the winter. Up to 1890, when I first visited Cape May, and for some years later, the Seaside Sparrow bred abundantly on the meadows between Cape May and the Lighthouse as well as on the more extensive meadows surrounding Cape Island Sound, the body of water that later disappeared when the great Fill was created. This operation drove practically all of the sparrows from the latter area although for a few years several persisted in marshy spots that developed temporarily on the filled-in land. Then began the draining and "development" of the marshes to the west until today not a single pair of Seaside Sparrows is left of the thousands that once nested about Cape May, except a few pairs immediately behind the golf links along Cape Island Creek, and with draining activities started there, during 1936, they too will probably be doomed. The same thing has been going on all along the coast until this species, the Sharp-tail and the Marsh Wrens are actually threatened with extinction so far at least, as most of the New Jersey coast, is concerned.

The Seaside Sparrows perch on the tops of marsh elder bushes, or more frequently, on the stems of the grass and rushes, grasping them low down so that most of the bird is concealed by the surrounding vegetation, and all that we see is the head stretched upward to get a view of the surroundings. When perched among the grass a bird will often grasp a different stem with each foot and as they sway apart in the wind its legs are drawn out almost at right angles to the body. When resting on a more substantial perch the slender tail is often held parallel to the ground, otherwise it hangs vertically.

When the male sings the head is thrown well back and the bill pointed upward and the harsh explosive note *che-zheéeege* bursts forth—after which the bird drops back into its normal position. This song coming often from a dozen different throats at once is one of the most characteristic sounds of the marsh. There is also, of course, the clapper of the Mud Hens, heard mainly at dusk or at night, the whistle of the Fish Hawk, and the cry of the Common Tern, but the two latter belong more properly to the mainland and to the strand respectively, but the humble efforts of the Seaside Sparrow we hear on all sides and all day long at the proper season, and more than all the others it is to me the voice of the marsh. I have recorded it at different times as: *che-zhée; che-wéege; chur-zhée;* and *too-szhéee,* the first syllable is short and the other long drawn and accented. One bird that I studied for twenty minutes, very close at hand, seemed to have a very short preliminary note, which may be characteristic of all the songs, and a curious sound— *let, let, let,* like water breaking through a pipe as the air escapes, which was uttered simultaneously with the last long note and gave the impression that it came from some entirely different source. The bird really seemed to be producing two sounds at once. In its entirety this performance might be written *e-dulllt-tzéeee, let, let, let.* Sometimes a male will mount a few feet into the air and give utterance to a slightly more complicated flight song dropping down again as the spluttering vocal effort is completed. As we gaze over the meadows we can see the dark heads of the birds stretched up from the grass on every side, while later in the season, when the southward migration is about to start, they concentrate in the taller growth of sedge along the tidewater creeks. The ordinary flight of the Seaside Sparrow strikes one as laborious, the impression being of a heavy body that is being dragged along, while the rapid wing beating seems necessary to keep it afloat. The body really is heavy and the short and rounded wings seem not to be built for a long flight so that the bird usually makes only short efforts, low over the tops of the grass, and drops back quickly into shelter. It seems strange that a bird that makes no further attempts at flight during the entire summer will, when the proper time arrives, launch forth on the long journey to its winter quarters many miles to the south.

While making but little show in the air the Seaside Sparrow is very much at home on the muddy bottom of the marsh and its large feet are well adapted for running over the soft ooze while they, as well as the short tail, shape of the body, and somewhat elongated bill, all recall the structure of the rails, which are co-tenants of the meadows. The Seaside Sparrow can run very swiftly, threading its way in and out among the coarse stalks of the Spartina grass that grows along the edges of the creeks. As it runs the legs seem rather long and the body is held well up from the ground while the tail is always pointed downward. Occasionally we see it slow down and pick at food on the edge of the water.

As we look at the perching Seaside it appears very dark, almost black, indeed, and stands forth very clearly. This is due largely to counter shading produced by the uniform darkness of the under surface; for in white-bellied birds the light on the dark upper surface and the shadow on the white under parts equalize each other and the bird appears unicolor and is not conspicuous. When we are reasonably close to one of the birds in full nuptial plumage we can see the light loral spot and the gray streaks on the lighter gray breast, while the white throat is often quite conspicuous. The wearing effect of the constant contact with the harsh grass stems is very pronounced and by August most of the birds are almost uniform sooty gray. The new autumn plumage shows richer tones and more olive on the back.

The totally different young birds, with brown backs and streaked breasts, somewhat resembling young Song Sparrows appear on the wing by early July and parents were seen feeding them on July 14, 1920, and August 18, 1924, although the latter date was very unusual and was probably due to a delayed nesting.

Richard Miller and Turner McMullen have given me data on a number of nests of this species which were all built in the grass either on the ground or over shallow water at an elevation of from three to eight inches. The normal number of eggs is four but one nest found on May 26, contained five and some of those holding three were apparently complete sets. Sets of four were found as follows:

May 18, one.	June 17, one.	June 28, two.
May 30, two.	June 18, two.	June 30, one.
May 31, one.	June 19, one.	July 1, one.
June 2, three.	June 20, two.	July 2, two.
June 7, three.	June 21, five.	July 3, one.
June 9, two.	June 22, one.	July 4, one.
June 11, two.	June 23, two.	July 5, one.
June 12, ten.	June 24, six.	July 15, one.
June 16, five.	June 26, two.	July 19, one.
	June 27, one.	

Sets of three were found from May 30 to July 22, many doubtless incomplete but no less than six were observed on the last date. A second nest found on the early date of May 18 held two eggs. Young birds were found in nests on May 26, June 7, 8, 11, 12, 14, 17, 24, 26, July 2 and 4.

I have no satisfactory data on the arrival dates of the Seaside Sparrow at Cape May in spring but they are established there by early May and doubtless by late April. In the autumn we have recorded them as late as October 7, 1923; October 25, 1929; October 19, 1931 and 1934; while birds seen on November 9, 1930, and November 21, 1926, may have been wintering individuals; as was one observed on January 23, 1927. William Baily who found the bird on Five Mile Beach on February 22, 1892, was the first to establish the species as a winter resident. The counts on the Christmas census follow:

December 26, 1927, one.
December 23, 1928, three.
December 22, 1929, two.
December 27, 1931, three.

December 26, 1932, six.
December 24, 1933, eleven.
December 22, 1935, two.
December 27, 1936, three.

The Seaside Sparrow was discovered and named by Alexander Wilson who, however, termed it a finch and not a sparrow, and probably got his specimens near where Ocean City now stands about 1810. He writes of it as follows:

"It inhabits the low, rush covered sea islands along our Atlantic coast, where I first found it; keeping almost continually within the boundaries of tide water, except when long and violent east or northeasterly storms, with high tides, compel it to seek the shore. On these occasions it courses along the margin, and among the holes and interstices of the weeds and sea wrack, with a rapidity equalled only by the nimblest of our Sandpipers, and very much in their manner. At these times also it roosts on the ground, and runs about after dusk.

"Amidst the recesses of these wet sea marshes it seeks the rankest growth of grass, and sea weed, and climbs along the stalks of the rushes with as much dexterity as it runs along the ground, which is rather a singular circumstance, most of our climbers being rather awkward at running." (American Ornithology, Vol. IV, p. 68.)

VESPER SPARROW

Pooecetes gramineus gramineus (Gmelin)

The Vesper Sparrow is a regular but not very abundant transient, more plentiful in autumn than in spring, while a few remain throughout the winter. We have no records during the breeding season and no evidence of its nesting in the county but that it does nest sporadically not very far to the north of Cape May is suggested by the following observations. One was heard on May 30, 1925, near Cold Spring by Julian Potter; another was seen by William Yoder near Cape May on July 18, 1926; and several on the meadow below Higbee's Beach on August 12, 1932, by James Bond and Conrad Roland. At Ocean View one was seen by Eliot Underdown on May 20, 1928, and another by myself on July 27, 1932, while Fletcher Street saw one near Goshen station on June 7, 1936. It is a common summer resident of west Jersey as far south as Vineland, Bridgeton and Franklinville (Potter).

Vesper Sparrows occur during migration in open areas about the Lighthouse and elsewhere at the Point; in old fields along the railroad back of the town and at Cold Spring and Higbee's Beach. On April 4, 1924, moreover, I found a spring flock feeding among the dune grass in front of Congress Hall Hotel.

When flushed these sparrows will usually alight in a tree or bush where they are easily studied. The most striking characteristics are the prominent line of black spots on the wing coverts and the clear cut line separating the streaked breast from the white abdomen, all the streaks stopping at the same point and not running farther back on the sides as in the Song Sparrow. Then there is the chestnut on the bend of the wing and white on the outer tail feathers; although this last, is considered the best field mark, it is not always so easily seen as might be supposed unless the bird is below our line of vision and flying away from us.

Our records of autumn arrival are:

1921. October 30.	1924. October 19.	1929. October 25.
1922. October 22.	1925. October 15.	1930. October 12.
1923. October 21.	1928. October 13.	1932. October 12.

Last dates in spring are: April 4, 1924; March 21, 1925; March 18, 1933. Midwinter records are as follows, mainly from the Christmas census:

February 22, 1923, five.	December 26, 1932, three.
December 23, 1923, two.	December 24, 1933, three.
December 26, 1927, one.	December 23, 1934, four.
December 28, 1930, seven.	December 22, 1935, twenty-one.
February 6, 1932, twelve.	

I first determined the Vesper Sparrow to be a winter resident at the Cape on January 27, 1892, when I found a flock of twenty in a field close to the town.

North of the Pine Barrens the Vesper Sparrow is a regular summer resident and has been found occasionally in winter as far north as Princeton (January 21, 1879, Scott), Haddonfield (December 29, 1880, Rhoads), etc.

LARK SPARROW

Chondestes grammacus grammacus (Say)

This western bird was first detected in Cape May County by Brooke Worth who found one in the hotel yard at Avalon on Seven Mile Beach, on August 21, 1927, and studied it with a glass for some time at a distance of fifteen feet. On September 10, following, I flushed an unfamiliar sparrow from the roadside near the Lighthouse at the Point and as it alighted on a wire and later on a fence post I had ample opportunity to study it at short range. I recognized it at once as a Lark Sparrow. The head markings were rather indistinct but distinguishable with the glass as was the rounded shape of the tail and the white edgings to all the rectrices but the middle pair. The back reminded one of that of a female English Sparrow the rump and lower back were plain brownish ash, breast pale gray with a central black spot. The feathers of the crown appeared to be slightly raised. After a short time the bird plunged again into the thicket and could not be flushed. Julian Potter and I saw another at the same place on August 15, 1937. On May 12, 1935, Victor Debes saw a Lark Sparrow on the Bay side of the peninsula near Price's Pond.

Daniel Lehrman reported one at Brigantine farther north on August 15, 1935. The first New Jersey specimen was taken by Dr. Frank M. Chapman at Schraalenburg, in the extreme northeastern corner of the state on November 26, 1885. There is a record for Essex County, September 19, 1934 (Eaton) and one for Union County, October 28, 1928 (Urner).

SLATE-COLORED JUNCO

Junco hyemalis hyemalis (Linnaeus)

A common winter visitant throughout the peninsula occurring somewhat irregularly in the town itself and frequent at the Point and along the coast and marshes from there to Cape May. "Snowbirds" as they are usually termed, here, always occur in flocks, sometimes by themselves, sometimes mingled with the various winter sparrows and often drifting about with flights of Myrtle Warblers in windy weather and in snow flurries. Like the Tree Sparrows they are essentially ground birds and only fly up into the trees when disturbed. Their slaty blue plumage and white outer tail feathers make them conspicuous and easily recognized, and they are therefore better known to the residents of the Cape than are any of the other sparrows.

On December 12, 1920, I watched a large flock of Juncos feeding in the open pine woods of the Point. While I remained perfectly still they hopped close around me, picking up food from the ground among the grass and dead plant stems, causing a rustling among the fallen leaves and a constant movement of the tops of the dead vegetation all about. Upon the first alarm they swarmed up into the trees and I presently saw several of them hopping out along the pine branches and pecking at the cone clusters. There are days during the southward migration when they are particularly abundant and on November 7, 1931, I counted two hundred at the Point while all along the stretch of Seven Mile Beach they fairly swarmed.

The earliest autumn records that we have are:

1921.	October 30.	1927.	October 10.	1932.	September 30.
1922.	November 12.	1928.	September 30.	1933.	October 28.
1923.	October 1.	1929.	October 5.	1934.	September 28.
1924.	October 19.	1930.	September 25.	1935.	October 12.
1925.	September 29.	1931.	October 19.		

In spring we have recorded them as late as:

1922.	April 2.	1928.	April 1.	1933.	April 1.
1923.	March 25.	1929.	April 7.	1934.	March 25.
1924.	April 20.	1930.	April 12.	1935.	March 23.
1926.	April 24.	1932.	March 26.		

Mid-winter counts on the Christmas census are as follows:

December 24, 1922, twenty.

December 23, 1923, twenty-three.

December 26, 1926, eleven.

December 26, 1927, forty.

December 23, 1928, 412.

December 22, 1929, fifty.

December 28, 1930, 170.

December 27, 1931, 112.

December 26, 1932, 120.

December 24, 1933, 366.

December 23, 1934, 256.

December 22, 1935, 463.

December 27, 1936, fifty-six.

The Junco does not nest in New Jersey but is everywhere a winter visitant.

TREE SPARROW

Spizella arborea arborea (Wilson)

The Tree Sparrow is a common winter visitant along wood edges, fence rows and swampy thickets in the Cape region and to the north, but not in the town itself, as it is distinctly a bird of the open and not of the gardens or door yards. It is frequent on the Fill and in thickets along the edge of the meadows back of Cape May, and north of Schellenger's Landing. The Tree Sparrow is found in flocks all winter long and sometimes as many as two hundred may be found together. It forms the bulk of most of the mixed sparrow flocks of winter but is as frequent in flocks of its own. I have found Field, Song, Swamp, Savannah and Chipping Sparrows mingling with the Tree Sparrows in the winter assemblages along with White-throats, Juncos, Myrtle Warblers and an occasional Cardinal or Bluebird.

We usually come upon flocks of Tree Sparrows feeding on the ground and they fly up into the bushes or trees as we approach dropping back again when assured that danger has passed. This habit was the basis for the name of Tree Sparrow given to them by Alexander Wilson. As they perch they give a nervous flirt to the tail and the neck is craned up with the feathers of the crown slightly elevated. Other individuals settle themselves in a hunched attitude with all the plumage fluffed up. The prominent wing bars, chestnut crown and black breast spot, and the yellow under mandible, are their most striking field marks.

We have records of Tree Sparrows in autumn as early as:

1921.	November 11.	1928.	October 28.	1931.	November 28.
1926.	November 21.	1929.	November 16.	1934.	November 25.

Our latest spring dates are:

1924.	April 2.	1934.	March 11.
1932.	March 26.	1931.	March 15.

Some doubtless remain every year until early April.

The Christmas census gives some idea of the numbers in the Cape May area in midwinter:

December 29, 1921, six.

December 24, 1922, twenty.

December 23, 1923, fifty-four.

December 26, 1927, 110.

December 23, 1928, 165.

December 22, 1929, 300.

December 28, 1930, 135.

December 27, 1931, eighty-five.

December 26, 1932, 120.

December 24, 1933, 360.

December 23, 1934, 256.

December 22, 1935, 463.

December 27, 1936, 153.

Elsewhere in the state the Tree Sparrow, as in Cape May, is a regular winter visitant.

CHIPPING SPARROW

Spizella passerina passerina (Bechstein)

The "Chippy" is a common summer resident in the open farming country north of the town; every year there is a pair or two about the home of Otway Brown at Cold Spring, feeding on the lawn and nesting in the shrubbery or in the apple trees in the orchard, and the same is true at every farm house or country residence in the peninsula. The Chippies range south as far as the turnpike to the Point, and a few pairs nest about the houses at the Point itself, but strangely enough on only three occasions has a pair bred in Cape May although one would suppose that the shady lawns with gardens and abundant shrubbery would furnish ideal quarters for them.

In the summer of 1921 a pair bred on Lafayette Street near Queen and on July 9 an adult was feeding a full-fledged young in our garden nearby, on July 2 the birds were seen pairing so that they probably raised two broods. No others were seen in the town in summer until 1932 when a pair nested in a yard on Washington and were seen feeding young on July 1. Another pair was seen in July 1937. Occasional transients have been noted in the Physick garden and elsewhere in spring and fall but they passed on at once.

When Five Mile was a "wild beach," back in the eighties, Philip Laurent found Chipping Sparrows nesting in the numerous cedar trees.

Richard Miller has examined a nest at Mayville which contained three eggs on May 25, and another at the Court House with four eggs on May 21 while one at Cape May Point held half grown young on June 8. The last nest was on the branch of a pine tree twenty-one feet up, the others in cedars five and six feet from the ground.

Walker Hand has recorded the arrival of Chipping Sparrows in the country north of the town from March 9 to April 8, with an average of March 21, and I have seen them at Cape May or the Point on:

1921. April 3.

1922. April 2.

1924. April 2.

1925. March 22.

1928. April 1.

1932. March 26.

1933. April 1.

Our latest dates for southbound migrants have been:

1921.	October 30.	1928.	October 27.	1933.	October 28.
1924.	October 2.	1929.	October 25.	1930.	November 9.
1923.	October 18.	1932.	October 12.	1931.	November 7.

The striped-breasted young are on the wing as early as June 8 and are seen frequently during July and early August. By this time the Chippies have gathered in flocks and mingled with the Field Sparrows along the brushy fence rows, but while most of them pass on to the south some remain every winter in the mixed sparrow flocks about the Cape as well as in gardens and door yards at Cold Spring.

We have the following Christmas census and other winter records:

December 24, 1922, one.
December 28, 1924, one.
February 21, 1926, one.
December 26, 1926, one.
December 26, 1927, six.

December 22, 1929, two.
December 28, 1930, seven.
February 1, 1931, one.
February 6, 1932, one.
December 22, 1935, seven.
December 27, 1936, one.

The Chipping Sparrow is a common summer resident throughout the state.

*CLAY-COLORED SPARROW

Spizella pallida (Swainson)

While there is no record of the occurrence of this western bird at Cape May, Charles Urner and James Edwards saw one below Beach Haven on May 8, 1932, and, as it has occurred in New England and in Florida, there is always a possibility of its presence here.

The Beach Haven bird was recognized by both observers as something with which they were unfamiliar. They immediately noted that it was unstreaked below but with a definite dark wash on the sides of the breast, a medial line through the darker crown and a white line over the eye. The upper parts showed little if any rufous and were finely striated with dusky and creamy buff, the general effect was a rather light-colored bird.

A Field Sparrow, to which species it is most closely related, was near where this bird was feeding and offered an opportunity for comparison (cf. Auk, 1932, p. 491).

FIELD SPARROW
Spizella pusilla pusilla (Wilson)

A common and universally distributed summer resident species through-out the dry sandy fields and thickets of the interior down to the very edge of the salt meadows, the numerous bramble-covered stretches on neglected farms being ideal cover for them. Immediately about Cape May one finds them along the railroads and highways while they approach the outskirts of the town at Schellenger's Landing and along the south side of Broadway. When we walk down the beach Field Sparrows are encountered as soon as one reaches the Lighthouse and they are common all over the Point in the pine and oak scrub. For some reason they never adapted themselves to the Fill as did the Song and Grasshopper Sparrows but, taking all things into consideration, I am inclined to think that the Field Sparrow out-numbers the Song Sparrow in the Cape May Peninsula and is probably our most adundant sparrow.

The Field Sparrow's song is heard on every side in spring and early sum-mer at the Point and from the roadsides and fence rows of the open country. As it sounds to me it may be expressed in three double and one single notes and a final trill *sée-o, sée-o, sée-o, see-seseseseseses* although when very close there seems to be a slight preliminary unaccented note so that the first note above might be written *se-sée-o* on other occasions I have recorded it as *fée-o, fée-o, fée-o, trrrrrrr*, the fourth note being absorbed in the trill, and again, there have been four notes all on the same pitch followed by the trill *chíp, chíp, chíp chíp, trrrrrrrrrr* or the four have seemed very slightly broken, approaching the first form, *féu, féu, féu, féu, feffffffffffffff*. The trill is always ascending and gradually wears out as it were. I have heard the song as late as August 7, 9, 11, 18, 20 and 28 in different years but the birds are usually silent after August 1; and they have been heard singing in spring by March 13.

From the records of Richard Miller and Turner McMullen I have compiled the following data on nesting. Four eggs are the usual complement and nests with that number have been found on:

May 16.	May 24.	May 27
May 22, three.	May 25.	May 30.
May 23, two.	May 26.	

One nest with five eggs was found on May 21 and sets of three, often doubtless incomplete, on:

May 17	May 22, three.	June 3, three.
May 20, two.	May 30.	

Nests with young were examined on May 20, 22, 23 and 29.

I have recorded what appeared to be arrivals of Field Sparrows from farther south on:

1921. April 3.	1924. April 2.	1929. March 31.
1922. April 2.	1925. March 22.	1930. April 6.
1923. March 25.	1928. April 1.	1933. April 1.

The Field Sparrow is regularly present in winter in varying numbers. Sometimes it is in little flocks of its own but is more often found joining forces with the composite sparrow flocks which one finds feeding on the ground in the winter sunshine, in the lee of some wood edge or fence row, and which may contain Juncos and Cardinals and a few Bluebirds as well as the somber colored sparrows. It is therefore difficult to determine when the migrant Field Sparrows leave but the following dates seem to mark the last of the transient flocks:

1921. November 21.	1929. November 17.	1933. October 28.
1922. November 12.	1931. November 7.	
1927. October 30.	1932. November 12.	

The numbers of wintering individuals is well shown by the Christmas census of the Delaware Valley Club.

December 30, 1922, twenty.	December 28, 1930, ten.
December 22, 1923, thirty-five.	December 27, 1931, twenty.
December 26, 1926, twenty-six.	December 24, 1933, twenty-three.
December 26, 1927, six.	December 23, 1934, twenty-four.
December 23, 1928, fourteen.	December 22, 1935, fifty-four.
December 26, 1929, twelve.	December 27, 1936, twenty-eight.

Elsewhere in New Jersey the Field Sparrow is a common summer resident and individuals have been seen in winter in many localities.

WHITE-CROWNED SPARROW
Zonotrichia leucophrys leucophrys (Forster)

A rather rare transient seen more frequently in autumn. While adult males have been seen in spring most of our autumnal records are birds of the year in their olive gray plumage strongly contrasting with the rusty tints of the Song Sparrows with which they have been found feeding. The sides of the neck appear quite gray and there is narrow streaking on the saddle of the back; the tail appears long and the bill quite pink and small in proportion to the size of the bird. I have watched them feeding on the ground at the Point, where most of our observations have been made. They scratch vigorously and can run rapidly.

Our records are as follows:

Autumn:

1923.	September 30.	1930.	October 11, two.
1924.	October 2, several, nineteen.	1931.	October 18.
1928.	October 13, two; 15, two;	1932.	October 8; 12, four.
	21, two; 28.	1935.	October 27.
1929.	October 13, two.		

Spring:

1924.	May 10, two.	1933.	May 7, 14.

In other parts of the state the White-crowned Sparrow is a rather uncommon transient.

WHITE-THROATED SPARROW
Zonotrichia albicollis (Gmelin)

This large plump sparrow is a regular winter visitant to the Cape region and a more abundant transient in autumn. Small flocks take up their residence in dense greenbrier thickets along the wood edges, old brush heaps in orchards and gardens, or other similar shelters, even in the town itself; and while they may venture out into open ground to feed they do not stray far from their favorite retreats into which they fly at the first approach of danger. Usually there are not more than six to twenty in these winter resident flocks but in the migrations they throng the thickets and woods, although their numbers vary from day to day. On October 10–11, 1927, they were to be seen everywhere on the Bay shore but on October 16, following, none could be found.

The brilliant coloration of the old males with their black heads, and conspicuous white crown stripes is in strong contrast to the duller females and young males and has caused more than one novice to take them for the

White-crowned Sparrow, but the white throat in distinct contrast to the gray breast will always identify them as will the yellow spot before the eye. Our earliest autumn dates are:

1923. October 1.	1926. September 27.	1931. September 26.
1924. October 2.	1929. October 4.	1932. September 25.
1925. September 22.	1930. September 28.	1935. September 29.

Latest spring dates of observation are:

1917. May 20.	1927. April 30.	1932. May 7.
1924. May 10.	1928. May 14.	1933. May 6.
1926. May 2.	1931. May 10.	1934. May 5.

Numbers counted on the Christmas census are:

December 22, 1923, ten.	December 27, 1931, fifty-seven.
December 26, 1926, twenty-four.	December 26, 1932, seventy-five.
December 26, 1927, 100.	December 24, 1933, 174.
December 23, 1928, 255.	December 23, 1934, 105.
December 22, 1929, fifty.	December 22, 1935, 870.
December 28, 1930, eighty-four.	December 27, 1936, ninety-three.

The White-throat is a common transient throughout the state wintering more or less frequently especially in the southern half. One was heard in song on July 7, 1935, in the Secancus Swamp, Hudson Co. (Eaton).

FOX SPARROW

Passerella iliaca iliaca (Merrem)

This, the largest of our sparrows, seems to be a regular transient in March and November while a varying number remain through the winter. The migratory flight would seem to pass quickly and in some years, at least, is very heavy and concentrated immediately along the shore. In the vicinity of West Creek, Ocean County, farther up the coast I have found them fairly swarming in early March, 1906, when every thicket seemed full of them and they alighted all over fences, chicken houses and elsewhere along the roads. I have never had the opportunity to see such a flight at Cape May but others have found them abundant on November 16–17, 1929, and at other times.

Our records for Cape May are as follows:

Autumn:

1921. November 6, eleven.	1929. November 16–17, abundant.
1922. November 12.	1931. November 7, several.
1923. December 2, five.	1932. November 13, ten.
1926. November 21, twenty.	1933. October 21, six.

Spring:

1924. March 19, six; April 3, three.	1931. March 15.
1928. March 25, two.	1932. March 5, two; 25,
1930. March 22, six.	1934. March 11.

The counts of the Christmas census show how many of these birds remain through the winter:

December 11, 1921, two.
December 30, 1922, six.
December 26, 1927, twelve.
December 23, 1928, twenty-one.
December 22, 1929, nine.
December 28, 1930, thirty-three.

December 27, 1931, thirty-nine.
December 26, 1932, nineteen.
December 24, 1933, fifteen.
December 23, 1934, ten.
December 22, 1935, forty.

We have also records for January 14, 1923 (one) and January 25, 1925 (four).

I have seen Fox Sparrows several times in gardens along Washington Street where they seem to enjoy digging in the flower beds, jumping several inches in the air and scratching with both feet at once among the dead leaves like the Towhee. Like the White-throat the Fox Sparrow seems to spend most of the time of his winter sojourn in some favorite thicket and does not wander far afield.

Elsewhere in the state it is a common transient wintering in small numbers at least as far north as Princeton.

LINCOLN'S SPARROW

Melospiza lincolni lincolni (Audubon)

We have but one record of this rare sparrow in the Cape May district, a single bird found at the Point by Julian Potter and Conrad Roland, on October 12, 1932, on the lawn at the Lighthouse where it was studied for some time.

Lester Walsh found another at Seaside Park on the upper part of the coast on September 12, 1926, which, so far as I am aware, is the only other record for the coastal region. It would seem probable that this and many other "inland" species cross the state north of the Pine Barrens in a south-westerly direction and proceed south along the Delaware Valley.

Lincoln's Sparrow is a rare transient in New Jersey. Specimens have been taken at Princeton on October 25, 1875; September 21, 1878; and October 7, 1879, (Scott) and May 8, 1894 (Phillips), and it has been seen several times in the northern counties.

SWAMP SPARROW

Melospiza georgiana (Latham)

A regular transient, most plentiful in autumn, while some remain through the winter. It is an inhabitant of swampy spots where there is a shelter of sedges or low bushes, but during large flights it may be found in almost any situation. Such flights were observed on October 1, 1923; October 18, 1925;

September 27, 1926; October 4, 1929; October 12, 1932; and doubtless in other years as well had I been present at the time. On January 28–29, 1892, Samuel Rhoads and I found the meadows back of South Cape May simply swarming with Swamp Sparrows during a spell of very severe weather but I have not seen such an assemblage in winter again. Sometimes, however, a few individuals will be found in the mixed sparrow flocks of midwinter when they gather on moist ground.

We often see Swamp Sparrows in growths of tall reeds along some pond where they hold to the upright stalks and often flirt their wings and tail like a wren. At nesting time they are as characteristic of fresh water marshes or swamps as are the Seaside and Sharp-tailed Sparrows of the salt meadows and consequently are not found summering at Cape May. On the marshes of the Delaware, however, for some distance south of Philadelphia they breed commonly and Turner McMullen has examined nests at Marshalltown in Salem County; one on May 15, 1921, with two eggs; one on May 29 of the same year with four; and two on June 3, 1922, containing five and four eggs.

Our dates of observation in the Cape May area are as follows.

In autumn:

1921.	October 16–November 11.	1928.	September 30–October 21.
1922.	November 26.	1929.	September 21–October 25.
1923.	September 30–November 11.	1930.	October 5–November 9.
1924.	October 2–19.	1931.	October 18–November 7.
1925.	September 23–October 18.	1932.	September 25–November 13.
1926.	September 27–November 21.	1933.	October 28.
1927.	September 25–October 23.	1935.	October 5–November 3.

In spring:

1921.	April 3.	1929.	May 11–12, March 31.
1924.	March 20–May 10.	1930.	April 6.
1926.	April 24–May 1.	1931.	March 14.
1927.	May 8.	1932.	April 30–May 7.
1928.	May 13–14, April 8.	1933.	May 6.

The counts of the Christmas census are:

December 23, 1923, four.	December 26, 1932, seven.
December 26, 1927, four.	December 24, 1933, fourteen.
December 23, 1928, three.	December 23, 1934, twenty-four.
December 28, 1930, ten.	December 22, 1935, forty-four.
December 27, 1931, six.	December 27, 1936, three.

Elsewhere in the state the Swamp Sparrow is a summer resident in suitable localities.

R.

EASTERN SONG SPARROW

Melospiza melodia melodia (Wilson)

The Song Sparrow may be regarded as a regular resident about Cape May but whether the birds that we find in winter are the same as our summer breeding birds is open to question. There is a pronounced migration in autumn and in early spring when more Song Sparrows are in evidence than at any other time, and the summer population is greater than that of winter. But while the Song Sparrow may be the most generally distributed sparrow of summer it is doubtful it if exceeds the Field Sparrow in actual number of individuals.

It is an inhabitant of thickets all along the coast where it may be heard in song from mid-March to August; also in the gardens of the town and all about Cape May Point where moist thickets are to be found. Farther back in the country it is found everywhere in similar situations and, while not a woodland bird, may be found in shrubbery along the wood edges. In winter it collects in flocks, often associating with other sparrows, and seems to desert its ocean front habitats for less exposed shelters farther inland.

The relatively long tail of the Song Sparrow and its jerky "broken-backed" flight, as if the tail were hinged at the base, serve to distinguish it from our other sparrows, while the song is, of course, always characteristic and renders it familiar to many who do not trouble to study its markings.

In late March and April the air seems simply filled with Song Sparrow song and at this time we see male birds flying from bush to bush with neck stretched out, head and tail held high, and wings vibrating rapidly. This seems to be a part of the courtship display and as soon as the bird alights it bursts into song. On March 21, 1925, and April 2, 1914, I have noted this performance and the birds were evidently paired although at the same time there were flocks of migrant Song Sparrows present nearby. I have heard the song as late as August 10 and 21 in different years. Richard Miller and Turner McMullen have furnished me with data on the nesting of the Song Sparrow in the Cape May area upon which the following summary is based.

Nests with five eggs have been found on May 15, one; 23, one; 30, one; June 17, one. Four eggs: May 20, two; 21, one; 23, one; 26, three; 30, two; June 25, one. Three eggs: May 14, one; 18, two; July 5, one; 15, one. Young birds were found in the nest on May 10, 25, 30, June 6 and 7. Full grown young in juvenal plumage have been recorded on June 28, July 7 and 10 in different years.

The curious variation in the day by day numbers of the Song Sparrow is noticeable in autumn as in other species, due to the irregularity of the migratory flights or "waves" at the Point. On October 5, 1929, they were estimated to be present by the hundred while on the 13th, only two could be found. On October 13, 1928, I found but six while on the 15th and 21st they were abundant, but on the 27th only two could be found.

While we see Song Sparrows most frequently in their passage from one bush to another or perched conspicuously as they sing, or later when they protest our approach to nest or young, they spend much time when undisturbed feeding on the ground and it is there that we see them in winter and realize how well adapted they are for a terrestrial life and how rapidly they can run, mouse-like, through the grass.

With Song Sparrows present in varying numbers in every month of the year it is useless to try to designate dates of arrival and departure. The counts of the Christmas census will show the numbers present in midwinter:

December 30, 1922, two.
December 23, 1923, four.
December 26, 1927, 100.
December 23, 1928, sixty-two.
December 22, 1929, forty-two.
December 28, 1930, forty.

December 27, 1931, forty-five.
December 26, 1932, thirty-two.
December 24, 1933, 172.
December 23, 1934, 108.
December 22, 1935, 446.
December 27, 1936, seventy-three.

The Song Sparrow is a common resident throughout the state.

ATLANTIC SONG SPARROW

Melospiza melodia atlantica Todd

The study of the Song Sparrow at the Cape is further complicated by the fact that we have two races of the bird present as breeders, the only case of the sort in the bird life of New Jersey. In the interior is found the Eastern Song Sparrow, the same form that occurs all over the Eastern States but on the coast islands, and possibly the inner edge of the salt meadows as well, lives the larger, grayer Atlantic Song Sparrow, recognizably different when series of specimens are compared but not distinguishable in life, nor can the eggs be differentiated. This form would seem to be permanently resident while the bird of the interior is migratory and to the latter belong the migrant host that comes through in the autumn. Winter Song Sparrows naturally include both races as well as intermediates between the two, from areas where upland and meadow meet. To illustrate how environment affects the distribution of these birds it may be mentioned that a series of breeding Song Sparrows collected in 1891 on the edge of the old Cape Island Sound and on the salt meadows that formerly existed southwest of Cape May are all typical of the Atlantic Song Sparrow with its grayer coloration and much heavier bill. Since the Fill was formed and the meadows replaced by dry ground with thickets of bayberry, etc., the common Eastern Song Sparrows of the interior have spread out and occupied the area. Earl Poole's drawings of the two races show very clearly the difference in the size of the bill.

I have secured specimens from as far north as Atlantic City.

Nests supposed, from locality, to belong to this race are reported by Richard Miller and Turner McMullen: from Seven Mile Beach:

May 14, 1827, three eggs.	June 6, 1926, one with four young.
May 18, 1924, two with three eggs.	June 7, 1925, one with four eggs.
May 20, 1922, two with four eggs.	May 22, 1932, two with four eggs.
May 21, 1922, one with four eggs.	June 16, 1935, one with young.
May 23, 1926, one with four eggs.	

LAPLAND LONGSPUR

Calcarius lapponicus lapponicus (Linnaeus)

An irregular winter visitant to the coasts of southern New Jersey, where a few might be seen every winter were competent observers present continuously. I first saw them on the strand of Seven Mile Beach on November 28, 1926, when in company with Richard Erskine. There were three of them in a flock of Horned Larks which seem to be their usual associates. They ran or walked like the Larks and were similar in all of their actions but easily recognized by their darker coloration and chunkier build. Against the light they looked quite black as compared with the grayer hue of the other birds, while in good view the chestnut tints of the back and the black of the breast could easily be seen with the glass. Once recognized, the Longspurs could readily be picked out whenever the flock took wing and settled again. On December 26, 1932, Joseph Tatum saw four Longspurs on the open sand flat on the southern part of Five Mile Beach.

On Brigantine Beach farther up the coast they seem to be of more frequent occurrence or else the bird watchers have given more attention to the spot. On December 26, 1930, Stuart Cramer saw twelve with twenty-five Horned Larks and Donald Carter found four present on February 15, following. On February 22, 1935, Joseph Tatum found a flock of thirty and on March 24 of the same year Julian Potter found the same flock some of them in song, "a medley of tinkling, sputtering notes recalling the Bobolink's song." It seems probable that in both these seasons the birds were winter residents there.

Elsewhere in the state it seems to be a rare winter visitant except on the salt meadows where it is more regular and more plentiful. Specimens have been taken at Princeton, February 13, 1895 (Phillips), Washington Park on the Delaware, February 14, 1895 (Reiff), Salem, December 28, 1898 (Warrington) and in Union County where Charles Urner has found it present on the meadows in five out of the seven years, 1921–1927, extreme dates being November 14 and March 22.

Across the river in Delaware Charles Pennock and I have found Longspurs on several occasions at Delaware City associated with Horned Larks on the marshes in midwinter.

SNOW BUNTING

Plectrophenax nivalis nivalis (Linnaeus)

An irregular winter visitant along the coast occurring on the sand dunes and upper beach or on the sand flats and marshes behind them.

Spencer Trotter and I came upon a pair on the dunes below South Cape May on December 30, 1922. They ran over the sand with head erect and jumped up at the seeds of the dune grass. When flushed they flew in long undulating sweeps showing to perfection the contrasting black and white of the plumage, the two colors being in approximately equal proportions.

The female was browner but the amount of white in flight was about the same. There were chains of delicate footprints in the sand where the birds had been feeding and the marks of the long hind claw were very evident.

The next day William Yoder and Henry Gaede saw a flock of fifty near the same spot while I saw a single female on December 26, 1929. Prior to these observations Julian Potter saw four at Cape May on December 18, 1921; and a flock of fifty on January 19, 1919, while he saw two flocks totalling twenty-eight birds on Seven Mile Beach on January 23, 1927 and Conrad Roland found them abundant there on February 16. He also saw them with Potter on January 19, 1919, and wrote me that their note was a trilled whistle and that on the ground they resembled rats as they crept along but "in flight they seemed to forget their terrestrial affinities and to impersonate the spirit of the wind their movements resembling wave crests although they seemed to ricochet rather than proceed in undulations. They had an erratic, lark-like habit of taking flight, not from alarm, but apparently to regale themselves by flying in the face of the unspeakably cold winds and frequently returned to the very spot from which they started."

The counts on the Christmas census of the Delaware Valley Club are as follow:

December 23, 1928, one.   December 23, 1934, eighteen.
December 22, 1929, three.   December 22, 1935, one.
December 26, 1932, fifty-one.   December 27, 1936, six.
December 24, 1933, one.

Charles Page saw two at Cape May Point on November 27, 1922, and William Rusling one on November 6, 1935, two on the 10th and one on the 11th while Dr. Henry Wharton found them on the Salem marshes on the Bay side as early as November 7, 1919 and Richard Bender found a flock of thirty at Fortesque on December 7, 1930.

A favorite resort of late years has been the broad sand flats of Wildwood Crest (formerly Two Mile Beach) where 450 were seen on February 22, 1930, and seventy-five on December 2, 1933, by members of the Delaware Valley Club.

Farther north on the coast Julian Potter found a few at Corson's Inlet as late as March 12, 1922, and also on November 18, 1923. On Brigantine Clifford Marburger found a flock of seventy-five on November 15 and December 20, 1925, while Potter found fifty still there on February 22, following, which would indicate that the birds were winter residents there. Brooke Worth found a flock of twenty-nine at Brigantine in a blinding snow storm on February 21, 1929. They seem to be more abundant and regular on Barnegat Bay and John Emlen has found them there as early as November 2, 1923. On January 5, 1935, Arthur H. Howell, who had visited Ocean City, Maryland, on the day before, wrote me that he had seen great flocks of Snow Buntings flying north along the coast and wondered if the flight had been noticed at Cape May. Unfortunately none of our observers had been at the

Cape at the time and I could get no information from residents. It is of course possible that the Snow Buntings travel over the sea from Brigantine to the Delaware or Maryland coast, passing by the retreating shore line of the Cape May Peninsula which has been suggested in the case of several of the water birds.

Elsewhere in the state the Snow Bunting occurs as an irregular winter visitant. Large flocks were observed about Princeton in the winter of 1895–1896 and it has been recorded at various other localities, but it seems to be much more plentiful and regular in its occurrence on the shore and in Union County Charles Urner states that it occurs every year, his extreme dates being November 6 and March 18.

H. B.

BIBLIOGRAPHY

A reasonably complete bibliography of books, magazine articles and short notes dealing with the ornithology of New Jersey from 1753 to 1909 was published in the author's "The Birds of New Jersey" (Ann. Rept. N. J. State Museum for 1908), while lists of later publications on the subject will be found in the several issues of "Cassinia" (Proc. Delaware Valley Ornithological Club) for subsequent years.

It seems unnecessary to republish this matter in the present volume and it is the intention to present below only such leading works on the subject as have been frequently quoted in the preceding pages and the journals upon which we have depended for local records.

ALEXANDER WILSON. American Ornithology, Vols. I–IX. Philadelphia, 1808–1814.
  This is the original edition of this classic work, the first publication to present an account of the bird life of the New Jersey coast. The author died before the work was completed and his young friend, George Ord, later president of the Academy of Natural Sciences of Philadelphia, superintended the publication of the last two volumes from Wilson's notes. (Cf. p. 29.)

GEORGE ORD. American Ornithology, Vols. I–IX. Philadelphia, 1808–1825 (= 1824–1825).
  This is the Ord reprint of Wilson's work. Vols. I–VI follow the original very closely and retain the original dates of publication. Vols. VII–IX, which contain most of the matter relating to the birds of the coast, were carefully revised by Ord and much original material incorporated. Vols. I to VIII were issued in 1824 and Vol. IX in 1825.
  Curiously enough the great works of Bonaparte, Audubon and Nuttall added very little to our knowledge of the birds of the New Jersey coast nor have any of the numerous sportsmen who later frequented the region left any record of the bird life. The next work dealing with coastal birds is the following:

J. P. GIRAUD. The Birds of Long Island. New York, 1844.
  Contains casual mention of New Jersey birds.

THOMAS BEESLEY. Catalogue of the Birds of the County of Cape May. (Appendix to "The Geology of the County of Cape May, N. J.," by William Kitchell.)
  A briefly annotated list of 196 species on pp. 138–145. The author was evidently personally acquainted only with the commoner species and records many birds as breeders in the county which do not even breed in the state. In the chapter of the work dealing with the history of Cape May by Dr. Maurice Beesley, there are several interesting references to birds observed by the early explorers.

CHARLES C. ABBOTT. Catalogue of the Vertebrate Animals of New Jersey. (In Cook's "Geology of New Jersey," App. E., pp. 751–830.) Trenton, 1868.
  Like the preceding author, Dr. Abbott has recorded many species as breeders in the state which occur only as migrants. In later publications (Amer. Nat., IV, 1870, pp. 536–550; and "A Naturalist's Rambles about Home," 1884), he

endeavors to substantiate some of these statements but presents no satisfactory data while he corrects or contradicts other statements.

WILLIAM P. TURNBULL. The Birds of East Pennsylvania and New Jersey. Philadelphia, 1869, pp. i–viii + 1–50.

An exceedingly accurate and painstaking list, the first reliable work to cover the region. We can only regret that he did not amplify his comments and present more individual records.

JOHN KRIDER. Forty Years Notes of a Field Ornithologist. Philadelphia, 1879, pp. [4] + xl + 84.

Had the author possessed more literary ability this could have been made a work of the greatest importance as he had wide experience, but it was compiled by another, apparently from the recollections by the famous gunsmith in his later years when his memory was fading, and it contains many obvious errors.

W. E. D. SCOTT. Notes on Birds Observed at Long Beach, New Jersey. *Bull. Nuttall Ornith. Club*, 1879, pp. 222–228.

This is a most important paper dealing exclusively with coastal New Jersey but at this date the abundant bird life of earlier times had suffered a great decrease from over-shooting and millinery collecting was at its height.

CHARLES S. SCHICK. Birds Found Breeding on Seven Mile Beach, New Jersey. *The Auk*, 1890, pp. 326–329.

An important picture of the bird life of this famous island before it had been ruined as a bird haven by draining and "developments." Unfortunately the paper was never finished. The author was primarily an egg collector and a few of his identifications may be open to question.

PHILIP LAURENT. Birds of Five Mile Beach, N. J. *Ornithologist and Oölogist*, 1892, pp. 43, 53, 88.

An excellent and thoroughly reliable account of the birds of this island at a time when most of it was a "wild beach." Mr. Laurent's collection is now in the Reading, Pa., Museum.

WITMER STONE. The Birds of Eastern Pennsylvania and New Jersey. Philadelphia, 1894. Pp. i–vi + 1–176.

A publication of the Delaware Valley Ornithological Club in which the attempt was made to bring our knowledge of the birds of the region up to date. Unfortunately the members of the Club had as a rule very little personal acquaintance with the birds of the coast at this time.

WITMER STONE. The Birds of New Jersey. (Annual Report of the New Jersey State Museum for 1908.) Trenton, 1909. Pp. 11–347.

Published, like the preceding, at a time when bird students were very few in New Jersey, especially along the coast, while modern methods of observation and transportation had not developed and information was lacking or difficult to obtain. Nevertheless it presents a fairly accurate summary of conditions as they were at the time.

GRISCOM, LUDLOW. Birds of the New York Region.

This work, like its predecessors by Dr. F. M. Chapman, 1894 and 1906, includes many records from New Jersey, mainly from the northern part of the coastal strip.

THE AUK. Organ of the American Ornithologists' Union. 1884–1937. A Quarterly Journal of Ornithology.

Many articles and short notes from this journal are quoted in the text almost always with direct reference.

BIRD-LORE. Organ of the National Association of Audubon Societies. New York. 1899–1937.

Most of our quotations from "Bird-Lore" are from the Philadelphia section of "The Season" compiled by Julian K. Potter and including records from the New Jersey coast; and from the Christmas Censuses. Most of these are observations of members of the Delaware Valley Ornithological Club and the Linnaean Society of New York.

CASSINIA. Proceedings of the Delaware Valley Ornithological Club. Philadelphia, 1891–1934.

Most of the records from "Cassinia" are from the annual migration records kept by members of the Club and others, covering southeastern Pennsylvania and southern New Jersey.

PROCEEDINGS OF THE LINNAEAN SOCIETY OF NEW YORK. 1889–1936.

Records quoted from this journal are mainly from the annual summary of observations for the New York region.

In No. 39–40 will be found Charles A. Urner's "Birds of Union County" and in No. 47 the late Warren F. Eaton's "Birds of Essex and Hudson Counties," excellent accounts of the bird-life of the extreme northern part of the coastal region of New Jersey.

Migration and occurrence records taken from these three journals are simply credited to the observer; in many cases they have been published in two journals; the place of publication if desired may readily be found by looking up the month in "Bird-Lore's" "The Season" or the yearly list in the other journals. Many records of this sort, however, are first published in the present volumes.

For information on the birds of the mountainous portion of the state one should consult:

WILLIAM L. BAILY. Breeding Birds of Passaic and Sussex Counties, N. J. *Cassinia*, 1909, pp. 29–36.

WALDRON DeW. MILLER. The Summer Birds of Northern New Jersey. *Natural History*, October, 1925, pp. 450–458.

INDEX

[This comprehensive index covers both volumes of the work. Volume I contains pages 1 through 484 and Volume II contains pages 485 through 932.]